T0186684

COMPUTER APPLICATIONS IN THE MINERAL INDUSTRY

PROCEEDINGS OF THE FIRST CANADIAN CONFERENCE ON COMPUTER
APPLICATIONS IN THE MINERAL INDUSTRY / QUEBEC / 7-9 MARCH 1988

Computer Applications in the Mineral Industry

Edited by
KOSTAS FYTAS & JEAN-LUC COLLINS
Université Laval, Québec

RAJ K. SINGHAL
CANMET, Coal Research Laboratories, Devon, Alberta

A.A.BALKEMA / ROTTERDAM / BROOKFIELD / 1988

The texts of the various papers in this volume were set individually
by typists under the supervision of each of the authors concerned.

Published by

A.A.Balkema, P.O.Box 1675, 3000 BR Rotterdam, Netherlands

A.A.Balkema Publishers, Old Post Road, Brookfield, VT 05036, USA

ISBN 90 6191 760 3

© 1988 A.A.Balkema, Rotterdam

Printed in the Netherlands

Computer Applications in the Mineral Industry, Fytas, Collins & Singhal (eds)
© 1988 Balkema, Rotterdam. ISBN 90 6191 760 3

Table of contents

3 Mine evaluation

4 Mineral processing I

5 Orebody modelling – Geostatistics

Computer Applications in the Mineral Industry, Fytas, Collins & Singhal (eds)
© *1988 Balkema, Rotterdam. ISBN 90 6191 760 3*

Foreword

Canada is a country rich in mineral and personnel resources. However the survival and prosperity of its mineral industry in our highly competitive world will depend to a large extent on the degree of the development and adoption of new innovative technology, computerization being a major part of this technology.

In an effort to enhance the development, dissemination and application of new ideas in this area the Department of Mining and Metallurgy at Laval University and the Coal Research Laboratories of Canada Centre for Mineral and Energy Technology (CANMET) in Edmonton have organized the First Canadian Conference on Computer Applications in the Mineral Industry (1st CAMI) in collaboration with the Mining Engineering Departments of University of Alberta, Ecole Polytechnique, Laurentian University, McGill University, Technical University of Nova Scotia, Queen's University, University of British Columbia and the Canadian Institute of Mining and Metallurgy (CIMM).

Laval University appreciates the opportunity to present the first CAMI. The timing of this conference coincides with the celebrations of the 50th anniversary of the Department of Mining and Metallurgy at Laval University. Subsequent conferences in this series will be held every two/three years.

Over 80 papers selected for presentation by authors from Canada, Australia, Brazil, China, France, Nigeria, Portugal, Spain, Sweden, UK, USA and West Germany offer a broad perspective of the best efforts of industry, university, consultants, private and government research in applying computer technology in the mineral industry.

In addition to the main technical part of 1st CAMI, an exhibition has been organized during which software products have been presented. Most of the conference topics have been covered: Exploration; mine planning and scheduling; mine design; mineral processing; expert systems; mine monitoring; information management systems and geotechnical applications.

The organization and success of such a conference is due mainly to the tireless efforts of many individuals, authors included. All members of the organizing committee and symposium chairpersons have contributed greatly. Words of encouragement and guidance received at all time from Prof. Gilles Barbery are gratefully acknowledged.

Kostas Fytas
Jean-Luc Collins
Raj K. Singhal
Editors

The following companies (registered as of 87/12/22) have presented their software products during the three day exhibition:

Centre de Recherches Minerales (CRM), 2700 rue Einstein, Ste Foy, Quebec, Canada, G1P 3W8

Gogema, Groupe CEA, 2 rue Paul-Dautier, BP 4.78148, Velizy-Villacoublay Cedex, France

Derry, Michener, Booth & Wahl, Suite 410-20 Richmond St.E., Toronto, Ontario, Canada, M5C 2R9

Energie, Mines & Ressources Canada, Recherche et Technologie, 555 rue Booth, Ottawa, Ontario, Canada, K1A OG1

Gemcom Services Inc., 801-1030 West Georgia Street, Vancouver, BC, Canada, V6E 2Y3

Geomines, 300 rue Léo Pariseau, suite 305, Montreal, Quebec, Canada, H2W 2P4

Mincom USA Inc., 15 Technology Pkwy South, Suite 225, Norcross, Ga., USA, 30092

Minemap (North America) Ltd, 1111 West Hastings Street, Suite 700, Vancouver, BC, Canada, V6E 2J3

Mineral Industries Computing Ltd. (MICL), 1 Langham Place, Hamblin House, London, UK, WIN 7DD

Mintec Inc., P.O. Box 31420, Tucson, Ariz., USA, 85751

Queen's University-Mining Engineering Department, Kingston, Ontario, Canada, K7L 3N6

Sacda Inc., 605-7575 Trans Canada Ouest, St-Laurent, Quebec, Canada, H4T 1V6

Knowledge Systems Products, Unisys, 300-1661 Portage Avenue, Winnipeg, Manitoba, Canada, R3J 3T7

Université Laval-Groupe GRAIIM, Pavillon Pouliot, Ste Foy, Quebec, Canada, G1K 7P4

Wenco International, 600-1130 West Pender Street, Vancouver, BC, Canada, V6E 2R6

Whittle Programming Pty. Ltd, 42 Yenada Street, North Balwyn, Victoria 3104, Australia

Computer Applications in the Mineral Industry, Fytas, Collins & Singhal (eds)
© 1988 Balkema, Rotterdam. ISBN 90 6191 760 3

Organization of the conference

Honorary Chairman

Gilles Barbery, Professor, Département de mines et métallurgie, Université Laval, Ste Foy, P.Q., G1K 7P4, Canada

Technical Program

Kostas Fytas, Chairman, Assistant Professor, Département de mines et métallurgie, Université Laval, Ste Foy, P.Q., G1K 7P4, Canada

Raj K.Singhal, Co-chairman, Group Leader, Coal Research Laboratories, CANMET, P.O.Bag 1280, Devon, Alberta T0C 1E0, Canada

Exhibition Coordinator

Jean-Luc Collins, Assistant Professor, Département de mines et métallurgie, Université Laval, Ste Foy, P.Q., G1K 7P4, Canada

General Committee

Peter N.Calder, Professor, Department of Mining Engineering, Queen's University, Kingston, Ontario, K7L 3N6, Canada

Pierre Choquet, Associate Professor, Département de mines et métallurgie, Université Laval, Ste Foy, P.Q., G1K 7P4, Canada

Jorgen Elbrond, Professor, Département de Génie minéral, Ecole Polytechnique, Case Postale 6079, Succursale 'A', Montreal (Quebec), H3C 3A7, Canada

André Galibois, Professor, Département de mines et métallurgie, Université Laval, Ste Foy, P.Q., G1K 7P4, Canada

Daniel Hodouin, Professor, Département de mines et métallurige, Université Laval, Ste Foy, P.Q., G1K 7P4, Canada

Frank Jerabek, Associate Professor, Department of Mining and Metallurgical Engineering, Technical University of Nova Scotia, P.O.Box 1000, Halifax, N.S., B3J 2X4, Canada

Yves Lizotte, Assistant Professor, Department of Mining & Metallurgical Engineering, McGill University, McConnell Eng. Building, 3480 University Street West, Montreal, Quebec, H3A 2A7, Canada

Mary-Ann Roberts, Lynx Geosystems Inc., 357 Bay Street, Suite 801, Toronto, Ontario, M5H 2T7, Canada

S.Paul Singh, Associate Professor, School of Engineering, Laurentian University, Ramsey Lake Road, Sudbury, Ontario, P3E 2C6, Canada

C.O.Brawner, Professor, Department of Mining and Mineral Process Engineering, The University of British Columbia, 6350 Stores Road, Vancouver, B.C., V6T 1W5, Canada

Noel Billette, Mining Research Laboratories, CANMET, EMR, 555 Booth Street, Ottawa, Ontario, K1A 0G1, Canada

Tad Golosinski, Professor, Department of Mineral, Metallurgical and Petroleum Engineering, University of Alberta, Edmonton, Alberta T6G 2G6, Canada

Computer Applications in the Mineral Industry, Fytas, Collins & Singhal (eds)
© 1988 Balkema, Rotterdam. ISBN 90 6191 760 3

Session chairpersons

Mary-Ann Roberts, Lynx Geosystems Inc., 357
Bay Street, Suite 801, Toronto, Ontario,
M5H 2T7, Canada

Yeong Su, Senior Operations Research
Engineer, Falconbridge Limited, Sudbury
Operations, Falconbridge, Ontario, P0M 1S0,
Canada

Jim H.Gray, Manager, Fording Coal Limited,
#200, 205 Ninth Avenue S.E., Calgary,
Alberta, T2G 0R4, Canada

Charles W.Pelley, Assistant Professor,
Department of Mining Engineering, Queen's
University, Kingston, Ontario, K7L 3N6,
Canada

Andrew L.Mular, Professor, Department of
Mining and Mineral Process Engineering,
University of British Columbia, 6350
Stores Road, Vancouver, B.C., V6T 1WS,
Canada

H.A.Hamza, Manager, Fuel Processing
Laboratory, Coal Research Laboratories,
CANMET, P.O.Bag 1280, Devon, Alberta,
T0C 1E0, Canada

Michel David, Professeur, Ecole Polytechnique,
Département de génie minéral, Case Postale
6079, Succursale 'A', Montreal, P.Q., H3C
3A7, Canada

Ray Mason, Senior Mining Engineer, Iron Ore
Company, P.O.Box 1000, Labrador City,
Newfoundland, A2B 2L8, Canada

Daniel Laguitton, Senior Research Scientist,
Energy, Mines and Resources Canada,
CANMET, 555 Booth Street, Ottawa,
Ontario K1A 0G1

Michel Garon, Directeur de Division Opemiska,
Minnova Inc., Division Opemiska, C.P. 10,
Chapais, Quebec, G0W 1H0, Canada

Peter N.Calder, Professor, Department of
Mining Engineering, Queen's University,
Kingston, Ontario, K7L 3N6, Canada

Malcolm Scoble, Professor, Department of
Mining & Metallurgical Engineering, McGill
University, McConnell Engineering Building,
3480 University Street West, Montreal, H3A
2A7, Canada

Jean-Marc Robert, Adjoint au Chef, Service de
Technologie minière, Centre de Recherches
Minérales, Complexe Scientifique du
Quebec, 2700, rue Einstein, Ste Foy, PQ.,
G1P 3W8, Canada

Julian Partyka, Professor, School of
Engineering, Laurentian University, Ramsey
Lake Road, Sudbury, Ontario, P3E 2C6,
Canada

Leo Nenonen, Senior Research Officer,
Systems Laboratory, Division of Mechanical
Eng., Montreal Road, Bldg. M-3, Ottawa,
Ontario, K1A 0R6, Canada

Rick Kolada, Research Scientist, Coal Research
Laboratories, CANMET, P.O. Bag 1280,
Devon, Alberta, T0C 1E0, Canada

Tad Golosinski, Professor, Department of
Mineral, Metallurgical and Petroleum Eng.,
University of Alberta, Edmonton, Alberta,
T6G 2G6, Canada

T.Smales, Senior Mining Engineer, Federal
Government, RR1, Box 14, Site 4, Sydney
Forks, N.S., B0A 1W0, Canada

Bernard Coulombe, Directeur, J.M. Asbestos
Inc., C.P. 1500, Asbestos, P.Q., J1T 3N2,
Canada

D.Bruce Stewart, Manager, Coal Research
Laboratories, CANMET, P.O. Bag 1280,
Devon, Alberta, T0C 1E0, Canada

Denis Côte, Directeur, Service de la
Technologie Minière, Centre de Recherches
Minérales, Complexe Scientifique du
Quebec, 2700 rue Einstein, Ste Foy, P.Q.,
G1P 3W8, Canada

G.Blackwell, Associate Professor, Department
of Mining Engineering, Queen's University,
Kingston, Ontario, K7L 3N6, Canada

René Dufour, Professeur, Département de génie
minéral, Ecole Polytechnique, Case Postale
6079, Succursale 'A', Montreal, Quebec,
H3C 3A7, Canada

Richard Jauron, Service d'environnement
minier, L'Association des mines de métaux
du Quebec Inc., 2 Place Quebec, Suite 704,
Quebec, P.Q., G1R 2B5, Canada

Claude Bourgoin, Professeur, Département de
mines et métallurgie, Université Laval, Ste
Foy, P.Q., G1K 7P4, Canada

Yves Lizotte, Assistant Professor, Department
of Mining & Metallurgical Engineering,
McGill University, McConnell Engineering
Building, 3480 University Street West,
Montreal, Quebec H3A 2A7, Canada

J.K.Szymanski, Research Scientist, Coal
Research Laboratories, CANMET, P.O.Bag
1280, Devon, Alberta, T0C 1E0, Canada

Jean-Luc Collins, Assistant Professor,
Département de mines et métallurgie,
Université Laval, Ste Foy, P.Q., G1K 7P4,
Canada

Raj K.Singhal, Group Leader, Coal Research
Laboratories, CANMET, P.O.Bag 1280,
Devon, Alberta T0C 1E0, Canada

Gilles Barbery, Professor, Département de
mines et métallurgie, Université Laval, Ste
Foy, P.Q., G1K 7P4, Canada

1 Plenary session

Computer Applications in the Mineral Industry, Fytas, Collins & Singhal (eds)
© 1988 Balkema, Rotterdam. ISBN 90 6191 760 3

Evolution of operations research techniques in the mining industry

J.Elbrond
Department of Mineral Engineering, Ecole Polytechnique, Montreal, Quebec, Canada

ABSTRACT: Operations Research was mainly developed during World War II at the allied side by civilian scientists. At the demobilization, OR rapidly found its way to peace time business and is now a generally spread set of quantitative management tools. OR also found its way to the mining industry, where the growth of its applications has been somewhat slower than in most other industries because of obvious problems of quantifications in the mining industry. OR constitutes now an important part of evaluation and management tools in larger mines with inroads being made into smaller mines.

RÉSUMÉ: La recherche opérationnelle était développée en grande partie durant la 2e guerre mondiale au côté des alliés par des scientifiques civils. À la démobilisation, RO a rapidement trouvé son chemin aux entreprises de paix et est présentement un ensemble généralement distribué d'instruments de gérance. RO a aussi trouvé son chemin à l'industrie minière, où sa croissance a été moins rapide que dans la plupart des autres industries causée par les problèmes de quantification dans l'industrie minière. Aujourd'hui, RO constitut une importante part des instruments d'évaluation et de gérance dans les grandes mines et les mines de grandeur moyenne avec une certaine pénétration dans les petites mines.

1. INTRODUCTION. THE WARTIME EXPERIENCE

It is generally accepted that Operations Research grew out of the allied services, initially of the British, during the second world war. Scientists of various specialities were called upon by fore-sighted military men to evaluate military operations at the staffs. The emphasis was on the analysis and evaluation of operational data by classical statistics, which had a fertile ground at the English universities. The scientists also brought with them the critical scientific method and they participated with that even in the development of tactics. Some typical subjects were the size of convoys, the maintenance of fighter planes, the measures and countermeasures in the submarine warfare, and artillery and aerial bombing patterns. Without any doubt, operations research made a tremendous difference in warfare by making use of the weaponry more efficiently. It is interesting to note that the military of the German side did not employ civilian scientists in the same capacities as the allied side. The more rigid German military traditions did not allow such extravagances. There were also many more reserve officers called in on the allied side than on the German side allowing for fresh air into the services.

The strategic planning, however, was out of reach of the civilian scientists, remaining the exclusive domain of the professionals: the generals and the politicians.

The autobiographies by Solly Zuckerman and R.V. Jones, both civilian scientists in the British wartime Services, contain in anecdotal form many valuable observations about the beginning of OR.

2. FROM WAR TO PEACE

At the end of hostilities, when the scientists and some of their military counterparts returned to civilian life, these men brought with them their wartime experiences of operations research to their civilian jobs and we see OR being introduced in all sorts of civilian activities: industry, economy and administration.

Quite early, already during the war, it became obvious that OR techniques often were in need of the processing of large amounts of data. Some of the military OR groups had a large number of clerical personnel attached to them in order to make the sums (as it was called). The electronic computer saw the light of day during the war years in England. It was, however, another even more crucial task than OR, namely the breaking of the German codes, which started this development. Scientists from the same English Universities that had produced the OR men played a vital role in it. But it was not until some years later that the computer became an industrial tool. The accountants had already since decades used the punch card machines but with this programmable new tool, the OR people got what was needed to advance the new techniques economically, without having to employ large numbers of clerks for calculation.

Special OR groups were formed and a vivid debate was conducted about how best to introduce the new thinking to management. As it was the case during the war, emphasis was at the beginning on the statistical analysis of operations such as materials flow and quality of production which to a certain extent conducted to systems analysis. The book by P.M. Morse and K. Kimball is the classical account of this period. See also O.L. Davis about the use of statistics in industry. But quite early more active optimization techniques such as linear programming made their entrance and very soon after the war OR comprises techniques which had seen the day decades before, the wartime developments, and new techniques.

As one would expect it is in England, the pioneering country in the field, that OR first spread in its civilian form. There the steel industry and the coal mining industry were early examples of industries using OR techniques.

3. THE START OF OR IN THE MINING INDUSTRY

It is difficult if not impossible to document all phases of the evolution of OR in industry because of confidentiality and the mining industry is no exeption. However, there are some early publications which are of interest in order to fix some dates related to the mining industry. At the first international mining congress after the war in 1955 in Paris, an American, Mr. Th. M. Ware of International Minerals and Chemical Corporation, gave a conference: "OR and the Mine of Tomorrow", which in fact traces the possibilities and the prospects of OR in all phases of mining from exploration, over ore reserve estimation through to planning and operations.

At the first international conference on OR in 1957 in Oxford, a Canadian, Mr. R.D. Hypher of the Canadian Defence Research Board, gave a paper with the title "Operational Research in Mining" where he, after having suggested various strategic (planning) and tactical (operational) uses of OR, in some detail describes the use of the queuing theory and simulation, at that time called the Monte Carlo method, in the analysis of the production system of a smaller mine with the objective of increasing the capacity. He did not mention the name of the mine where the study was made.

At the subsequent international OR conferences, papers on OR in mining were presented. But the best forum became the series of symposia on OR and computer applications in the mineral industry, which began at the University in Tucson in 1961 and which since then has been held in the U.S.A. during the first 10 years and which since have alternated between countries overseas and the U.S.A. with the first outside U.S.A. being held in Montreal in 1970. Until now, 20 symposia of this series, which are now called APCOM (Application of Computers and OR in the Mineral industry), have been held and their planning stretches through to the end of the next decade with new countries lining up to bid for its organization. At these twenty APCOM symposia alone, about 1200 papers have been presented and the attendance has been in excess of 8000 (a list of APCOM Symposia transactions is given under references).

The mining industry had early used tabulators for ore reserve calculations (as well as for the accounting tasks). The computers brought new possibilities particularly for OR. The first uses were most probably in simulation by the Monte Carlo method of mine transportation systems in the mid fifties and since then OR and computers have been inseparable. Later the optimizing techniques of mathematical programming made the need for large computer memory capacity even more felt.

4

Besides the APCOM symposia, there are more regional symposia reoccuring, of which the German have been particular active and providing a useful connection with mining OR in the eastern world. The various national mining societies often include OR sessions in their annual meetings.

Through the publications from these symposia and meetings, it is possible to trace the evolution of OR in the mining industry. It should, however, be born in mind that many applications (the large majority) are not made public, many being confidential.

From the begining, the probabilistic techniques gained interest for two reasons. Probabilistic methods became well accepted during the military period in determining how to get a bigger bang for the £ (or $). Mining operation bear some similarity with military operations. It is particularly the element of uncertainty (although for different reasons) which fall in mind.

The heterogeneity of grade and physical properties of orebodies render the geological as well as the economical (price and costs) evaluation difficult. Therefore, probalistic analysis plays an important role in mine operations research and mine valuation.

Geological evaluation which for many decades struggled with the problem of uncertainty became amenable to scientific numerical analysis by the advent of geostatistics, which is an optimizing procedure (minimizing the error of estimate) and as such can be called an OR technique. It was introduced by Matheron at the Paris Mining Conference in 1955 and has now gained wide acceptance and a large field of activity. Geostatistics has been tryed on all types of deposits and has apported great benefits in many cases. Since its first appearance, it has not stopped to develop. At each new OR Symposium, the geostatisticians are revealing new tricks.

Another aspect of the uncertainty of mining data is that of operations. It soon became clear that the mere use of averages in the estimate of capacities conducted to overestimation and very soon simulation and queuing theory were used in the analysis of mine production systems. It is difficult to point at a specific first lieu of the use of these techniques. They were so obvious that their use probably were initiated at several places simulta-

neously. This took place in larger organisations such as the National Coal Board in the UK and in the French and Swedish iron ore mines during the 50's. Later, the coal mining industry of the U.S.A. and other countries became users of OR for systems' engineering and operational evaluation.

This conducted to very elaborated simulation studies for example of room and pillar mining where all operations from the mining front through to the washery are connected. In the coal mining industry, global operational optimization has been introduced by monitoring of the state of machines and installations and using the gathered data for optimizing control in real time.

The use of the optimizing technique of OR in the mining industry began with modest attempts of linear programming to solve some production composition problems.

In 1964, H. Lersch and I.F. Grossman published in Canada their paper about Optimum Design of Open-Pit Mines, which signalled the real start of the strategic OR studies in mining. The Lerch and Grossman algorithm finds the true optimal final contours of open pit mines using the graph theory. This has developed into a major application and has undoubtedly had an important influence on management's understanding and acceptance of OR techniques. The pratical use of the algorithm required highly rationalized computer programming to go from the 600 hours computer time for the Lac Janine mine of Quebec Cartier, which was the first open pit optimization in the world, to our days' much shorter computer times for pit optimization. Much research has gone into the production planning problem, which apart from the very short range (next day) planning, where mixed integer programming can be used, has escaped strict mathematical solution until now. Optimizing techniques are now being introduced for the truck dispatching, which in its final development stage comprises several merging OR techniques such as now-linear programming, assignment procedures, blending, and queuing theory. The grey area of long, medium, and short range planning is now a field for artificial intelligence.

A very particular mining problem is that of the optimal cut-off grade, which was recognized several decades ago but which has joined with other problems to be

5

partially solved by OR techniques. Since 1964, when K. Lane presented his first paper about optimal cutoff grade, there has been a growing awareness about the possibility of solving the problem. It is, however, a hush-hush subject, the domain of top management and as such it is difficult to get a clear picture of the use of OR techniques in solving the problem in practice.

4. CONCLUSION

OR techniques supported by computers have now wide applications in larger mines. It is impossible to carry out a feasibility study of an open pit mine, without making extensive use of OR techniques.

These techniques are also penetrating the medium size mines. However, the small mine, is the new frontier for mining OR people. There is now a upshoot of integrated computer programs using OR techniques designed and available for small mines.

With the increasing automation and robotization in mining, OR techniques are being used in the modelling and evaluation of the changes as is the case in other industries and as has been the case in the modelling and evaluation of mechanization and rationalization in the mineral industry.

The further integration of operation and maintenance is an important new/old field of OR in all industries.

In the mining industry, a field which merits more attention is the simulation of mining methods, integrating the rock mechanical with the operational aspects of mining.

REFERENCES

DAVIES, O.L. 1947. Statistical Methods in Research and Production. London. Oliver & Boyd.

HYPHER, R.P. 1957. Operational Research in Mining. Proceedings of the First International Conference on Operational Research (Oxford). London. The English Universities Press Ltd.

JONES, R.V. 1978. Most Secret War. London: Hamish Hamilton.

KRIGE, D.G. 1977. The Human Element in APCOM's Development. Proceedings of APCOM 77, Brisbane. The Australian Institute of Mining and Metallurgy.

LANE, K.F. 1964. Choosing the Optimum Cut-Off Grade. Quarterly of the Colorado School of Mines. Vol. 59, No. 4 (Transactions of the 4th APCOM Symposium).

LERCHS, H. and GROSSMAN, I.F. 1965. Optimum Design of Open-Pit Mines. CIM Bulletin (January 1965). 4th APCOM Symposium, Colorado School of Mines.

MATHERON, G. 1955. L'utilité des méthodes statistiques dans la recherche minière. Transaction du Congrès du Centenaire de la Société de l'industrie Minérale. La Recherche Minière. Paris, R.I.M.

MORSE, P.M. and KIMBALL, K. 1951. Methods of Operations Research. London: Chapman & Hall, Ltd.

RAMANI, R.V. and PRASAD, K.V.K. 1987. Worldwide Trends in Computer and Operations Research Applications in the Mineral Industry. Proceedings of the 13th World Mining Congress in Stockholm. Rotterdam. A.A. Balkema.

WARE, Th.M. 1955. Les recherches opérationnelles et la mine de demain. Transactions du congrès du centenaire de la Société de l'Industrie Minérale. La Mine Future. Paris, R.I.M.

ZUCKERMAN, Solly 1978. From Apes to Warlords. London: Hamish Hamilton.

THE LIST OF APCOM SYMPOSIA
(Some transactions are out of print)

1st Symposium 1961. The University of Arizona, Dept. of Mining & Geological Engineering, Tucson, Arizona.

2nd Symposium 1962. The University of Arizona, Dept. of Mining & Geological Engineering, Tucson, Arizona.

3rd Symposium 1963. Stanford University, Department of Applied Earth Sciences, Stanford, California.

4th Symposium 1964. Colorado School of Mines, Mining Engineering Department, Golden, Colorado.

5th Symposium 1965. The University of Arizona, Dept. of Mining & Geological Engineering, Tucson, Arizona.

6th Symposium 1966. Pennsylvania State University, Minerals Engineering Dept. University Park, Pennsylvania.

7th Symposium 1968. Colorado School of Mines, Mining Engineering Deptartment, Golden, Colorado.

8th Symposium 1969. Society of Mining Engineers of AIME, Salt Lake City, Utah.

9th Symposium 1970. Canadian Institute of Mining and Metallurgy, Montreal, Canada. Decision making in the mineral industry.

10th Symposium 1972. APCOM, South African Institute of Mining and Metallurgy, Johannesburg.

11th Symposium 1973. The University of Arizona, Dept. of Mining & Geological Engineering, Tucson, Arizona.

12th Symposium 1974. Colorado School of Mines, Mining Engineering Department, Golden, Colorado.

13th Symposium 1975. Clausthal Techmical University, Verlag Gluckauf, Essen.

14th Symposium 1976. Pennsylvania State University.

15th Symposium 1977. Brisbane. The Australasian Insitute of Mining and Metallurgy.

16th Symposium 1979. University of Arizona, Tucson.

17th Symposium 1982. Colorado School of Mines.

18th Symposium 1984. Institute of Mining and Metallurgy, London.

19th Symposium 1986. Pennsylvania State University.

20th Symposium 1987. The South African Insitute of Mining and Metallurgy. Johannesburg.

Computer Applications in the Mineral Industry, Fytas, Collins & Singhal (eds)
© *1988 Balkema, Rotterdam. ISBN 90 6191 760 3*

Mechanization, automation and computerization – Mining will never be the same

J.H.Nantel

Centre de Recherche Noranda, Pointe Claire, Quebec, Canada

ABSTRACT: Relentless pressures placed on the Canadian minerals industry force mine operators to become more productive and to produce at lower unit costs than anyone else in the world. This has generated an intense drive to adopt techniques that demand a greater degree of mechanization, automation and computerization. The author reviews the developments in these three areas over the last decade and demonstrates that there is no turning back; computers, computer controls and monitoring, information management systems, expert systems are here to stay and will continue to develop at an unprecedented rate over the next decade.

1 INTRODUCTION

Canadian mineral producers will prosper only if they can operate mines which are squarely located in the bottom part of the comparative costs curves. It is my strong belief that the mines better equipped to succeed, will be the mines using the best technology. Clearly, mine operators need to keep their productivity at a higher level than their competitors. In the mineral sector, a competitor is anyone that produces the same commodities anywhere in the world. Increased use of new and better technology is generally accepted as a prime factor in increasing productivity. In the Canadian mining sector, this translates into using and possibly developing better mining machines and improving the mining methods.

2 TECHNOLOGY EVOLUTION CURVES

At the Sixteenth Centre for Resource Studies Policy Discussion Seminar held in Kingston, Ontario, April 7-9, 1986, Margot Wojciechowski (1986) explained how new types of technologies may be required to bring an industry into new levels of performance. Figure 1 illustrates the concept.

We see three types of technology evolution curves. T1 represents incremental changes, up to now the dominant sort of technology changes in the mining industry; for example:

1. Better drills (i.e., electric-hydraulic drills versus pneumatic drill, the drilling rate is twice as fast).
2. Better steel and better bits, to accomodate the higher performance drills.
3. Improved chemical explosives.
4. Bigger trucks in open pits.
5. Etc., etc., the list could be made much longer.

We could call this the evolutionary type of innovations.

T1 changes generally arise directly from the improvement to existing operations. Low risks are associated with them.

T2 curves represent major changes in existing technologies. They include new combinations of existing technologies. They will bring about new plants (or mines), involve a higher degree of risk and are the results of more intensive research within the industry.

An example of this type of technology in the mining industry could include bigger blast holes in underground mines, this one was responsible for the development of a new mining method called Vertical Crater Retreat mining.

Another one is the introduction of the flexible rubber wheeled load haul dump

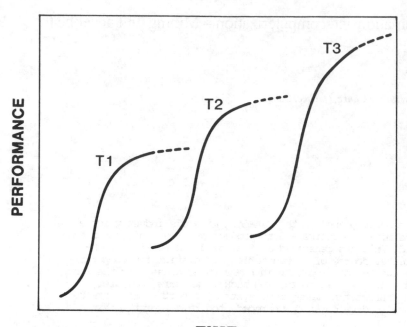

TYPE OF			
TECHNOLOGY :	Incremental advances	Product & process development	Radical innovations
R&D RESULTS :	Supports existing businesses	Extends the business	Revolutionizes existing and/or creates new businesses

Fig. 1

and trucks underground, which has vir-tually changed the way underground mines are designed.

T3 curves result from major break-throughs, and most often will require interdisciplinary work effort. This type of new technology tends to revolutionize the industry, it may make existing plants and processes obsolete. Such examples can be found in several indus-tries, they are rare in the mining sec-tor: the advent of the jet engine, changed aviation. The transistor, the computer, etc., have had a major impact in several sectors.

I believe that a more intensive use of computers and automation in the mining industry will launch the mining industry on a new S-Curve that will take the industry to heights which would have been impossible to reach with mechaniza-tion alone.

3 MECHANIZATION, AUTOMATION AND COMPU-TERIZATION

The object of my paper is to look at these three headings and see how the mining industry is evolving by first, having made great progress in the area of mechanization; secondly, by having reached a critical period where some of these highly mechanized mining opera-tions can now be integrated and made more efficient through automation; thirdly, without the development of com-puter technology to push and support mine automation, further progress will be very slow. We may be able to con-clude that, in some mining sectors, automation has indeed taken some techno-logies on new curves, and that higher performances have resulted.

3.1 Hard Rock Drilling

Looking back a few years, a really sig-nificant change to effect the drilling function has been the introduction of the hydraulic percussion hammer in the early 1970s. This technique replaced the conventional pneumatic drill. Several improvements became readily evident, namely: less noise, less dust, absence of oil/air mist, less energy per metre drilled, and most important for the mine operators, significantly higher penetration rates, up to twice the pneumatic drills' rates. Undoubtedly, hydraulic drilling brought the drilling process to a much higher level of mechanization.

Mr. Patrick C. MacCulloch (1987), in his keynote address to the CIM delegates at the Annual General Meeting in Toronto in May 1987, calculated that pneumatic drilling with a jack leg would produce 13 tonnes per man-hour, while the use of a two-drill jumbo would produce 107 tonnes per man-hour; that is indeed productivity.

Some mine operators believe that the performance of such drilling equipment could be increased further with the use of electronics and computer technology. The following parameters could be optimized: drilling pattern including hole contour in tunnels, drifts and stopes, smooth and speedy movement from one hole to the next, boom utilization, penetration rates, which depends on percussive energy, feed pressure, rotation speed, rock types, etc.

Micro-processor controlled hydraulic drilling jumbos have been successfully developed and used by civil and mining contractors. It is only a matter of time for the mine operators to start to use them on a larger scale. Equipment cost is still high, but this difference in costs can be fully absorbed by productivity improvements in several instances.

Operators using automated drill jumbos have experienced the following: i) they can use inexperienced operators, ii) training periods are shorter, iii) wear and tear on the boom is negligible compared with experience on other rigs, iv) maintenance problems for the electronics and sensors is small, as well as for hydraulic and mechanical parts of the rig, v) productivity is higher, vi) costs per metre of hole drilled is lower.

Performance for the drilling functions can be taken to new summits with the addition of electronic components and computers; after an initial period of debugging, the industry will never go back to earlier techniques.

3.2 Hard Rock Loading and Haulage

Draw point loading with scooptrams is still the preferred method in most Canadian mines. Some operators are experimenting with continuous machines, pulling ore from bulk mining stopes. This operation sometimes proves difficult, especially when large boulders fill the drawpoints. As with scoop mucking, this problem can be eliminated with better blasting practices.

There is a strong movement in the mining industry to fully integrate all functions of materials handling to permit continuous operations. Elimination of discontinuities and establishment of a constant flow of material means that such systems lend themselves to remote supervision and automated control. (Wyllie 1987).

Development work, which is an important feature of hard rock mining, will continue to be done with LHD equipment, loaders, trucks, rail haulage, etc. The LHD and truck combination is still unmatched for its flexibility. Improvements in technology are steadily enhancing their capacities, speeds, cycle times and overall control and operations. A recent development is the machine health monitoring systems. Fault prognosis and diagnosis produces significant reductions in unplanned downtime and thus maximizes the degree of utilization of the machine. The early display of defects that might arise makes it possible to plan the appropriate maintenance and repair measures and carry them out at suitable times. If faults arise suddenly, immediate diagnosis of the cause minimizes the time needed for trouble shooting. (Wyllie 1987)

Some Canadian operators, working in collaboration with universities and high-technology companies would like to take the electronics monitoring on scooptrams one step further; they envisage an LHD operating totally on its own from dump point to mucking area, back to the dump area, unaided and entirely automatically. The business rationale can be characterized by lower maintenance costs, higher availability and longer operating time.

Existing electronic and mechanical control devices associated with remote operation, coupled with extra servo-drivers and limit switches, on-board microprocessors and television cameras, can be used to develop automated LHD operations.

Continuous ore handling is gaining

11

momentum, as indicated above. In soft rock, the cutting machine is the front end of the continuous system. In hard rock formations, until continuous mining machines are developed, the front end of the system is the continuous loader. Haggloaders were used for mucking in Canada with some success in development headings. Their use in stope mucking is not well documented. Recent developments in continuous loaders are encouraging. Continuous Mining Systems have developed a continuous loader with an oscillating lip which digs into the bottom of the muck pile. It loads onto a chain conveyor which discharges into a truck or other appropriate equipment such as a conveyor belt. In Europe, Voest Alpine has also developed a continuous loader. These are exiting developments which point the way towards a smoother flow of ore from stopes to surface.

An ideal continuous system would incorporate the following elements:

1. A continuous loader,
2. a low profile crusher, to make the material more amenable to uncomplicated transport,
3. an ore sorter, to eliminate the undesirable fraction of the material, thereby reducing transport and processing costs,
4. a conveyor belt and/or train or trucks, delivering the material all the way to a surge bin ready for automatic conventional hoisting, or
5. a hydraulic hoisting system, to deliver the material to the processing plant from its underground location.

All the elements of such a system are available at the moment. Some components require further development. It is, in my opinion only a matter of time before all the parts are assembled into an integrated continuous mine ore handling system.

The author is aware that work is on going on the weak parts of the system, namely: ore sorting and hydraulic hoisting; but the other components are available today. Microprocessors and remote controls will undoubtedly play a major role in making such systems work reliably and economically.

3.3 Continuous Hard Rock Mining

Continuous mining is a reality in soft rock formations such as salt, potash and phosphate mining. Canadian hard rock mine operators are envious of the productivities achieved by the potash operators. The following numbers will illustrate:

MINE NAME	MINING METHOD	TONS HOISTED PER DAY	NO. OF EMPLOYEES		PRODUCTIVITY (TONS/MANSHIFT)	
			OTHERS	U/G	OTHERS	U/G
CENTRAL CANADA POTASH	ROOM & PILLAR	11,110	274	151	73.6	26.1
POTASH CORP.OF SASK.	ROOMS	10,000	199	249	40.0	22.3

As the cost of drilling and blasting continues to increase, there has been some rapid and significant advances in the technology of continuous mining machines for hard rock. It is not therefore too foolhardy to think that one day mining machines can be designed to operate successfully and economically in the harder rock formations and under adverse geologic conditions.

For example, San Manuel reports very interesting results with roadheaders in development headings with rock formations varying from 125 to 200 MPa; they claim a reduction in costs of 30% and an increase in the rate of advance of 37% compared to conventional drill and blast methods.

Several research organizations have been working with auxiliary processes that could increase the performance of mining machines. High pressure water jets have been retrofitted on machines in the USA, Germany, France and England. Preliminary results indicate an increase in cutting rates and reductions in the specific energy requirements. Reductions in vibrations and dust, are also noted.

Testing of the Robbins' Mobile Miner began in March 1982. In 1984, Mt. Isa Mines Limited in Australia started to use the Mobile Miner to drive a 1.2 km decline in very abrasive quartzites, with an uniaxial compressive strength varying from 110 and 260 MPa with over 60% silica. The machine underwent major modifications and is again at work at Mt. Isa, on some development work.

HDRK Mining Research Limited, is a joint research and development corporation formed by three major Canadian mining companies: Falconbridge, Inco and Noranda, to conduct major mining research and development. HDRK has embarked on an ambitious program to develop within ten years, a continuous mining machine which will eliminate the present cyclic nature of the hardrock mining process. Excellent progress has been made to date.

Eimco, England Division, has tested a

prototype of a continuous miner for hard rock formations in Sardinia, Italy. Plans are to introduce such a machine in a Canadian mine soon.

The author is aware of several organizations which are at present addressing the challenges arising from the development of a continuous machine for hard rock formations.

Continuous mining machines will, at the beginning be bulky and expensive. It will be imperative to extract the most productivity from each machine. Manufacturers are making plans to have an extensive monitoring system on each machine.

3.4 Mine Engineering

Under this heading, the author will attempt to regroup several mining activities where the use of computers have seen an impressive ascent over the last decade. It will not be possible to cover all areas of interest included in mine engineering functions, only some will be highlighted.

3.4.1 Mine Ventilation

The International Mine Computing Magazine, in its July/August 1986 Issue, (Petruska 1986) lists at least 8 computer mine ventilation programs that can be used for underground mining operations. With the addition of real time monitoring for ventilation functions such as: air volumes, speed and direction, contaminants, status on mine fans, motors, doors and other regulators, etc., it is now possible to fully understand and control all mine ventilation functions. Timely detection of underground fires is now possible with such systems; warnings and help can be dispatched quickly to miners.

3.4.2 Geological Models and Ore Reserve Estimation

Several computer-assisted geological interpretations and orebody modelling exist on the market. These programs facilitate the handling of diamond drilling information with the creation of geological information data base. The programs can then manipulate the data into a flexible and efficient system. Interactive graphics software offer a powerful tool in making geological interpretations. Ore reserve calculations,

can also be part of the computer package. Having this information computerized is of obvious benefits when it comes to making development and production plans.

3.4.3 Mine Design

The design of open pits with computers is surely the area of mine design where pioneering efforts have been made. Computerized open pit design has demonstrated in the early days, that computers could be of tangible value to mining along with data processing for accounting and payroll functions. Remarkable progress has been made over the years; current programs can simulate just about any feature of open pit mining including: ore reserves, ore values, benches, ramps, wall slopes, etc. Their graphics features are not the least as impressive: plans, sections, three dimensional views, superpositions, etc.

Computer-aided underground mine design by contrast is a relative newcomer to the mining scene. Kidd Creek is obviously a pioneer in the field. (Harris 1976). Recently a few programs have been produced, notably the Geomin and the Noranda Research Mine Layout System. Much remains to be done and with the increasing power of microcomputers/minicomputers and their relatively low costs, it is anticipated that this vacuum will be filled shortly.

3.4.4 Mine Operations: Maintenance, Others

A number of such computer programs have been developed for the broad industry; programs more specifically addressing the requirements of the various types of mines is still wanting. MAG International, a consulting subsidiary of the M.A. Hanna Company, has developed a comprehensive maintenance program which contains 10 sub-programs. These sub-programs provide all the information required by maintenance management and supervisors to effectively run their departments, control costs and maximize productivity (Rushton 1986). Another software package for maintenance management, MAPCON is described in the May/June 1987 Issue of International Computing (Proctor 1987).

There appears to be a need for getting the various pieces of the puzzle into a more coherent program. Lynx Geosystems

Inc. of Vancouver has taken a step in that direction by combining three well known programs, i.e., GEOLOG, GEOMIN and CADMIN.

All of the above is a sample of programs now available to the mine operator. Well used, these computer programs can make a marked difference on the efficiency of a mining operation.

3.5 Mine-wide Communications

Mine-wide vocal communications has been identified by mine operators and mining authorities as one of the most promising technologies to improve worker safety and increase over-all mine productivity. (Falconbridge-CANMET-Sherritt Gordon 1985). The author would add that a first rate data communication system is essential for the development of tele-chirics or centralized remote control of mining machinery. Without belittling the recent developments in the area on mine communications over the last 10 years, more reliable systems need to be developed.

4   A CANADIAN PROGRAM IN MINE AUTOMATION

The participants at the two consultative workshops organized by CANMET in March and October 1986 were in agreement on the desirability of increasing the degree of automation in mining operations. It was also stressed that to achieve this goal, two important ingredients were needed:
1. high availability of equipment, and
2. reliable, multi-function underground communication systems, capable of linking all features of the mine to surface through voice, data and image transmission.

The participants strongly suggested that a fast track approach was required for the development of a Canadian program in mine automation. An organization should be created to accomplish the following:
1. transfer the existing know-how,
2. bring the specialists from various disciplines together, and
3. generate new knowledge through research for future applications.

I am pleased to report the following actions that have taken place over a relatively short period of time on the general topic of mine automation.

4.1 Formation of the National Advisory Committee on Mining Automation (NACMA)

This committee comprises members from the major mining companies, equipment manufacturers, hi-tech companies, universities and governments. The goals of the Committee are:
1. to promote the development and use of automation in Canadian Mining,
2. to provide a forum for communications between users,
3. to coordinate efforts on mining automation between other national and international organizations,
4. to organize workshops on mining automation,
5. to establish a Canadian Centre in Mining Automation, and
6. to establish priorities for research and innovation reflecting the needs of the Canadian mining industry.

4.2 Symposia on Mining Automation

Two symposia on mining automation were held in Sudbury; in October 1986 and 1987. These symposia attracted over 200 participants each from mining companies, equipment manufacturers, hi-tech firms, governments and universities. A real sense of movement could be felt amongst the participants and speakers.

4.3 Formation of a Canadian Centre for Mining Automation and Robotics

At the time of writing, the creation of a Canadian Centre for Mining Automation and Robotics in Montreal, with the combined resources of Ecole Polytechnique and McGill University, was in the final stages of negotiation. Support for the project has been obtained from MAC, QMMA and the initiative is fully endorsed by the National Advisory Committee on Mining Automation. NACMA will become part of the technical advisory committee for the new Centre.

The prime objectives of the new Centre are to conduct research directed towards the application of automation to the mining process and its constituent component activities and to develop teaching in mining system dynamics and control at undergraduate and Masters levels with a view to producing the professional mining automation engineers needed to design and implement comprehensive automation schemes in the field.

4.4 Collaboration Between CATA and MAC

An initiative started in 1985, to bring together the members of these two organizations (Canadian Advanced Technology Association and Mining Association of Canada) has produced several valuable contacts and contracts for their members. Mining automation will not be accomplished by mine workers alone; a multidisciplinary approach is required. Members of CATA and MAC are now working on several joint projects.

4.5 Mining Association of Canada Research Initiatives

In May 1987, MAC announced the formation of the Mining Industry Technology Council of Canada (MITEC). Some of the objectives of the organization are to:
1. increase the amount and effectiveness of cooperative mining research and development in Canada by improving cooperation within the industry and between industry, governments and universities while reducing duplication of efforts,
2. to focus research on Canadian mining industry needs and priorities in the medium and long term, and to promote and identify funding available from mining companies and governments for such research,
3. to increase the awareness at all levels in Canadian mining companies of the need for the availability of advances in technology and methods in order to regain the Canadian industry's international competitiveness, and to encourage the rapid implementation of advances in technology and methods, and
4. to suggest national priorities for major technology thrusts and to promote public and corporate policies that favour the development and implementation of appropriate technologies and methods in the mining industry in Canada.

NACMA has been asked to advise MITEC on matters dealing with mine automation. This request was strongly endorsed by the members of NACMA.

4.6 Centre for Advanced Resource Technology, Prince George, B.C.

In December 1983, an advanced technology training centre oriented towards the needs of the forest-related and mining resource industries was formed in Prince George, British Columbia. The mission of the Centre (CART) is to assist in the improvement of productivity in the Canadian forestry, mining and related industries, through an effective process of technology transfer and to stimulate the development of software and manufacturing of equipment for resource industries in Canada.

CART has currently several million dollars invested in computer aided design and manufacturing hardware and software, plus an additional $1.2 Million of software donated by a company involved in the CAD/CAM industry. The Centre will apply CAD/CAM, automation, laser, fiber optics, process control and instrumentation to the forestry and mining industries.

4.7 National Research Council and CANMET

NRC and CANMET have been strongly encouraged to participate fully in the Canadian mine automation efforts. Their responses have been very positive; several initiatives have taken place in their respective thrusts.

All of the above actions indicated strongly that Canada is ready for an all out effort in mine automation development effort.

CLOSING REMARKS

The author has attempted to paint a picture of modern mining; with the emphasis being placed on the increasing mechanization heavily supported by microprocessors and computers. Computers are extensively used in mining functions of all sorts. Their use, however is not as generalized as in some other spheres of activities. This situation will change over the following years. As the benefits of automation start to take effect, computers will become an integral part of modern mining. The author predicts an explosion of computer applications in the mining industry over the next decade.

The changes described in the paper are only the beginning of a new area for mining. In the near future, developments will take place on the following fronts:
i) rapid mine development, through the commissioning of continuous mining machines for hard rock formations. These machines will be capable of remote control operations and will incorporate limit switches and sophisticated monitoring devices for rapid self-diagnosis.

ii) new mining methods applicable to
the new generation of non-chemical
explosives or continuous mining equipment. These new mining methods will aim
at: a) reducing the ratio of development
waste to ore, b) reducing waste rock
dilution in work areas, c) eliminating
the cyclicality of mining functions
iii) computer applications will become
conventional practice in all aspects of
mining activities.
There is room to innovate in all
sectors of the mining industry. The
battle to improve productivity will
always exists; only our continued
efforts will ensure relative growth.
There is no turning back, Canadian
mining is here to stay.

REFERENCES

Falconbridge Limited, 1985. Communica-
tion system for isolated areas mined -
Final Report - Phase 1, prepared by
Falconbridge Limited, Sudbury Opera-
tions, July 1985.
Harris, A. 1976. A brief description
of the Kidd Creek mine engineering
system. 14th APCOM Symposium Pro-
ceedings, p.781-791.
MacCulloch, P.C. 1987. Productivity.
Address given at the CIM Annual
General Meeting Plenary Session,
May, 1987.
Petruska, T. 1986. Ventilation soft-
ware: mine health and safety
applications. International Mine
Computing, July/Aug. 1986.
Proctor, D.H. 1987. MAPCON Software
for maintenance management. Inter-
national Mine Computing, May/June
1987.
Rushton, J.W. 1986. Computers and
mine maintenance, the MAG story.
International Mine Computing,
Jan/Feb. 1986.
Wojciechowski, M.J. 1986. Technology
policies in the Canadian mineral in-
dustry. Proceedings of the Sixteenth
CRS Policy Discussion Seminar, April
7-9, 1986, Kingston, Ontario, p. 15-45.
Wyllie, J.M. 1987. Underground load-
ing and haulage. Engineering and
Mining Journal, May 1987, p. 22-33.

2 Mine planning and scheduling

Computer Applications in the Mineral Industry, Fytas, Collins & Singhal (eds)
© 1988 Balkema, Rotterdam. ISBN 90 6191 760 3

Mine scheduling using interactive graphics

Gordon H.Jardine & Bryan J.Evans
Mincom USA Inc.

ABSTRACT: Considerable progress has been made in utilising interactive graphics in geological modelling and mine layout, but mine scheduling and production scheduling has typically been treated as a "spreadsheet" exercise, rather than a graphical exercise. Mincom's MINESTAR system utilises the interactive graphics capabilities of workstations to provide the mining engineer with an extra dimension in addressing mine scheduling challenges. MINESTAR has been used on a range of mines - coal and metalliferous; open pit and underground.

1 INTRODUCTION

Mine scheduling is the process of simulating the extraction of a deposit over time. The process comprises:

a. Defining the deposit as a group of mining blocks, and establishing attributes for these blocks (volumes, tons, qualities, grades, etc). This is typically done in a block modelling program, and this information is the usual input to a scheduling system like MINESTAR.

b. Establishing rates of removal for the materials in the mining blocks, and the sequence in which the blocks are to be extracted.

c. Simulating this extraction sequence.

d. Reporting the results of the schedule.

Steps b through d are commonly iterative as refinements are made to obtain the "best" schedule.

Because the spatial relationships between mining blocks usually play an important role in the feasibility of an extraction sequence, the mining engineer can benefit from a graphical representation of the schedule, and MINESTAR provides this capability.

Another drawback with conventional scheduling programs is that the "logic" is inbuilt and cannot be modified by the engineer to suit a specific set of scheduling challenges. By contrast, MINESTAR enables the engineer to fully control the scheduling logic so that MINESTAR can be used to schedule ANY mining operation. This means that all types of schedules can be addressed - long term, medium term, short term.

2 MAIN FEATURES OF MINESTAR

The major features of MINESTAR are:

a. a flexible mining block DATABASE
b. DYNAMIC removal rate simulations
c. comprehensive INTERACTIVE GRAPHICS
d. USER TAILORABLE reports and graphs
e. powerful extraction SEQUENCE
capabilities.

3 BLOCK DATABASE

The amount and type of data which can be stored for each mining block is entirely under the control of the user. Each mining block has a name and as many attributes as desired. Attributes might include:

. tons
. volume

. grades
. qualities
. rock types
. position (co-ordinates)
and so on.

The mining engineer can enter and store those attributes which are critical to his scheduling challenge.

Another advantage of this flexible database is that MINESTAR can import its data from an ASCII file produced by the user's block modelling system.

Prior to undertaking any scheduling, the mining engineer can access this database to:

a. produce statistics for the deposit
b. produce plans of the mining blocks highlighting areas of interest (high or low grades, high ratios, rock types, etc)
c. combine existing database items into new items (eg calculate mining ratios, compute dollar values, etc).

4   DYNAMIC REMOVAL RATES

In order to simulate the extraction of the mining blocks over time, the mining engineer needs to apply removal rates to the various types of material in the mine, or to the various groups of equipment used in the mine.

In real life, the removal rates vary according to a variety of factors such as the height of the mining face, the nature of the material being mined, and so on.

MINESTAR enables the engineer to simulate these real-life conditions by allowing the removal rates to be defined as conditional formulae, tables, graphs and so on. When the scheduling encounters a mining block of a certain material, the removal rate will auto-matically be matched to that appropriate to the material.

This dynamic selection of rates provides for realistic simulation of complex mining operations.

5   INTERACTIVE GRAPHICS

Within MINESTAR, interactive graphics takes a number of forms.

A graphics terminal can be used to display the progress of the schedule as it happens. Blocks which have been extracted can be shaded according to user-selected patterns, so that the engineer can track the scheduling progress visually.

The graphics terminal can also be used to select mining blocks or sequences to be extracted.

A common arrangement is for mining blocks to be selected for scheduling from plans mounted on a digitizer. This method is more friendly for the engineer than having to type the names of blocks at a terminal. It enables the engineer to work with the mining blocks in plan form - an approach which is more comfortable than dealing with lists of block names.

The best arrangement is a combination of digitizer (for selecting blocks to be mined), a graphics screen (for displaying the progress of the schedule) and a regular terminal (for displaying the computed values of the schedule in tabular form). This "workstation" provides the engineer with maximum graphical and non-graphical information.

6   REPORTS

Each mining operation has its own set of scheduling challenges. For this reason, an engineer is interested in reporting the parameters that are key to his particular operation.

MINESTAR accommodates that by means of user-defined reports and graphics. The engineer can decide which items he wants reported and the format in which that report is presented. The capability includes business graphics output such as graphs and bar charts.

7   MINING SEQUENCES

The real power of computer-based mine scheduling systems lies in the ability of the engineer to rapidly assess variations of a mining strategy. MINESTAR enables the engineer to compile, store and modify sequences of mining blocks to be extracted. These sequences can be "run" to produce reports and plans of the schedule, modified, then re-run to assess what differences in output would result from a change in the basic inputs, eg a change in production rate, or the number of operating shifts.

These mining sequences are also used for TARGETING - for establishing an extraction sequence which, when scheduled, will meet certain pre-defined targets for tons, grade and so on.

8 OTHER CAPABILITES

8.1 Breakpoints

MINESTAR uses "breakpoints" to interrupt a schedule if a pre-defined condition arises. These pre-defined conditions are set by the engineer to incorporate key parameters upon which the acceptability (or otherwise) of the schedule depends.

Most mines have certain product specifications which must be met. When the scheduled production does not meet these specifications, the engineer will want to modify his mining sequence. Within MINESTAR, breakpoints are used to advise the engineer whether his key parameters are being met or not.

8.2 Backtracking

MINESTAR enables the engineer to "backtrack" or "undo the schedule" back to a point where the schedule was acceptable so he can try some alternative approaches. In this way, multiple options can be rapidly evaluated.

8.3 Equipment levels

MINESTAR can be used to ascertain the equipment requirements to meet the production schedule.

8.4 Costs

MINESTAR can also be used to determine mining costs for each production schedule generated.

9 CONCLUSION

By virtue of its integral use of interactive graphics and the degree of control over the scheduling logic given to the engineer, MINESTAR represents a major advance in mine scheduling systems. Its use worldwide in mines of all types affirms the need for this approach to mine scheduling.

Computer Applications in the Mineral Industry, Fytas, Collins & Singhal (eds)
© *1988 Balkema, Rotterdam. ISBN 90 6191 760 3*

Two approaches to the computerized planning of dragline operations

N.J.Stuart
CANMET, Energy Mines & Resources Canada, Devon, Alberta

Q.Cobb
Manalta Coal Ltd, Calgary, Alberta, Canada

ABSTRACT: There are several approaches to the development of computer programs to aid in the planning of mining operations. This paper examines two such approaches, the use of interactive graphics and expert systems, by reference to a common surface mining problem, planning of dragline operations.

The basic rules defining a dragline operation are clearly set out in commercial operators manuals and are predicted by the mine geology and production requirements. The problem tends to require a repetitive arithmetic solution and is ideally suited to the application of computers.

In 1983, Manalta Coal Ltd. developed an interactive, graphical program in HP BASIC to run on workstations to solve this problem. The inputs to the program are factors associated with dragline geometry (stored for reference in a dragline database), pit geometry factors such as slope angles and pit widths and the geological structure. The program produces an animated dragline which is manipulated on the screen until the operator has produced a satisfactory solution. The output is a range diagram plotted to screen or to paper and the calculation of factors such as percentage rehandle, swing angles and times. Thus the program stores and retrieves dragline data, automates certain routine calculations and greatly speeds the production of range diagrams. It was designed specifically not to apply any decision making or optimization.

The second program uses the features of the Prolog programming language that lends itself to finding multiple solutions and to the application of heuristic rules. The operator selects sets of input values (such as sets of draglines that may be considered) and criteria for eliminating unacceptable solutions. The program lists all possible mine designs that meet the criteria. This program extends the rule base of the operators manual to estimate productivity in terms of bucket size, swing angles, etc.

In summary, the first program assists the operator to examine an option in detail and decide on a satisfactory configuration. The second program is intended to find and describe all feasible configurations.

The paper discusses the advantages and drawbacks to the two design philosophies and the methodologies involved.

1. INTRODUCTION

The selection of a dragline is a major decision the consequences of which a mine will live with for up to twenty or thirty years. This paper candidly discusses two approaches by which computer software may aid in the decision process. Two appendices attempt to give the feel of sample sessions using each program.

2. THE ROLE OF THE COMPUTER IN THE SELECTION OF DRAGLINES FOR PRAIRIE COAL MINES

Coal mines in the Canadian Prairies gener-ally produce thermal coal for utility power generation. The geological setting is typified by level seams with little tectonic disturbance overlain by glacial deposits. The topography is typically flat. Coal mines on the Canadian Prairies are therefore planned with good knowledge of both geology and production requirements. This environment favours the use of a dragline as the prime excavator. Such machines are large and require a large initial capital investment to be recouped over a number of years.

This paper examines two approaches to this problem, the first uses the techniques of interactive graphics to assist the engineer, the second uses an expert system approach.

The starting point of the process is the generation of a production schedule based on the predicted burn requirements of the utility company. In the past this would have been generated by hand, but now it is usually produced by geological modelling and mine planning software. The schedule will detail the coal required by period and the associated overburden broken down by mining unit together with any rehandle required by the mining method (ramps, sidewall slopes etc.). The aim is to select a dragline of optimum size to remove overburden fast enough so that peaks in demand can be satisfied. This is a slight over statement since under extreme conditions remedial measures can be taken, such as the use of mobile equipment for prestripping. However, as these methods are considerably more expensive than dragline stripping, it is generally preferable to select a dragline sufficiently large to meet the needs of the power utility.

From the production requirements it is possible to determine the required bucket size by selecting appropriate values for equipment availability, available operating hours, walking time and swing time. Various rules of thumb based on operating experience are used to assist in the selection of these parameters. As an example, swing time is usually assumed to be independent of bucket size. Though a judicious selection of values is important, the process is straight forward mathematically.

Having established the required bucket capacity one must then select a dragline configuration. This is done by considering the geometrical constraints by examining range diagrams and calculating reach, stacking height and digging depth based on overburden and coal thicknesses, highwall and spoil pile angles and the swell factor for the broken material. The mining sequence and method would be determined at this stage, if it proved necessary to use an extended bench, for example, the additional rehandle required would have to be included in the production schedule and the process repeated. Other factors which may need consideration are any geotechnical conditions which may require a larger tub size or increased set backs from the slope crest.

3. THE ROLE OF THE COMPUTER

The main contribution of the two programs presented here is the calculation and display of the geometric aspects of the decision. All of the input values to the process require careful consideration, it is presumed that such values as slope angle and swell factors have been determined elsewhere although the programs do suggest initial default values.

3.1 The first program

The first program presented was developed by Manalta Coal Ltd. as part of a larger geological modelling and mine planning system. The program is written using HP BASIC, a typical procedural language, using the techniques of interactive graphics. The philosophy was to automate the preparation of range diagrams and volumetric calculations. It was a strict design criteria that the computer not make engineering decisions and is aimed at assisting the engineer reach a design satisfactory to him.

The program maintains a data base of draglines and can immediately retrieve a selected dragline for display or manipulation. New draglines may be stored by entering the vital statistics from the keyboard or by graphically modifying an existing dragline (eg. by altering the boom angle or stretching the boom length). The pit geology is entered either from the keyboard or by entering the coordinates of the section in which case the program will look up the section from the geological model. Pit geometry is entered from the keyboard.

When the information about the pit has been entered the dragline may be "walked" to any position on the screen using the graphics input device (arrow keys or mouse) and various calculations performed. Values such as excess reach, swing angle, rehandle quantity or percent (if extended bench is used) or if the dragline has insufficient reach the material volume to be moved by other methods may all be determined by calling the required function. In addition any distance or volume may be scaled from the screen using the screen pointer. The final design may be sent to the plotter or printed on a dot matrix printer.

The challenge in making such a program easy to use is in the user interface. The user wishes to be able to move between the multitude of functions in any sequence without the necessity to memorize complex command structures. This is achieved by assigning the commands to function keys and displaying descriptions at the bottom of the screen. The results of any actions are displayed immediately on the screen.

3.2 The second program

The second program adopts an experimental approach to the problem. It too principally addresses the problem of the geometrical constraints of extraction but does so in a completely different way. Using a dialect of the Prolog computer language, the program consists of a set of rules and facts. Running the program consists of giving the system a logical goal that it attempts to prove and find all the solutions for using these facts and rules.

Let us examine how this works in practice. The central rule to the program is onw which says "A dragline will work in a given geology if it has a wide enough operating radius, it can lift the spoil above the top of the spoil pile and it can work to sufficient depth to dig the overburden from the top of the coal." These three conditions are three other rules in the rule base which further divide the problem down. At some stage the "unification" (rule solving) process consults facts (which have no "if" component to them) which state the dragline dimensions slope angles etc. to be assumed. This rule can be "run" in four different ways. One may ask:

1) Will dragline "A" work with geology "G"?
2) Which geologies can dragline "A" cope with?
3) Which draglines can work at geology "G"?
4) Which draglines work with which geologies?

In (1) both the dragline and geology variables are said to be fixed. In (4) they are said to be free and in (2) and (3) one is free and one fixed.

There is another similar rule that contemplates the case where extended bench working can be contemplated with up to a given percentage of rehandled material.

It is clearly very simple to implement rules that take productivity requirements and bucket size into account. The program can thus grow as required as an ad hoc collection of rules relevant to the problem which may or may not cross reference each other. The prolog allows goals that combine rules -- for example one might ask "Which draglines can work in geology "G" by extended bench that can't work without it?

One of the problems with the program as it now stands is that it requires a knowledge of prolog syntax and the rule base to use it. The tendency is to using "natural language" interfaces with prolog rather than menus. This means that you type in the goal in english. It is not hard to develop some software to handle restricted versions of the english language but the cost is often a restriction on the complexity of the question asked.

We can divide the rules of this program into three classes:
1) The fundamental rules that determine which draglines work where.
2) Service rules that allow the decomposition of (1).
3) Additional rules. The program contains additional features that allow graphics (a simple range diagram), sorting of lists etc.

4. CONCLUSIONS

The most obvious difference between the two approaches is the role of the user. The first system is "non-expert" and does not impose logical constraints on the user, who is expected to provide the expertise. Different engineers might well produce different solutions to a design problem. It is also quite possible that a proposed solution might not be feasible as although the program will inform the user that for example "the dragline has insufficient reach (-2.34m)" it does not prevent the user from positioning the dragline there. The main advantages are the inherent flexibility of the approach and the immediate visual feedback, which gives the engineer a good feel for the problem. The penalty is one of speed, the process of examining a number of draglines in a range of pit configurations is tedious.

The second program may be regarded as an expert system as the engineering logic is built into the system and will always offer the same advice for a given circumstance. It will always apply all of the rules and checks specified by the rule base and the results will therefore be independent of the user. This is a more rigorous technique.

Furthermore, once the data base of draglines and pit configurations has been set up, it is extremely rapid to query the system for workable solutions satisfying various goals, including quite complex compound goals. Therefore while one cannot expect the solution to be innovative, one can be sure that all possibilities have been analyzed.

The dangers of this approach are that the rule base is unlikely to be complete for all circumstances. For example, it would be misleading if a dragline/pit combination "failed" because it was 1cm too short (as an example). While new facts and rules may be added to the knowledge base at any time and this particular deficieicy could be corrected, that is always the concern that others may be present. As the system appears as a "black box" to the user, there is the danger that a solution may be accepted without a good understanding of the limitations of the techniques and processes which produced it. It is not envisioned that this system would replace the expert engineer and the user should have a general knowledge of both the technical problems and the techniques used by the program.

The strengths of the two approaches compliment each other well. Ideally the techniques could be combined to use the expert system from within the interactive process. It could be used as pre-processor for example, to generate all of the possible solutions to be examined in greater detail. It would also be desirable if a solution generated interactively could be immediately checked against the rule base. This would require programming in mixed languages, an option which is currently available in theory, but quite difficult to implement.

5. ACKNOWLEDGEMENTS

The authors wish to thank Paul Scott of Manalta Coal for advice on the selection procedure, Carol Neumiller of the University of Alberta for programming assistance and the Manalta Coal Company for the use of resources and information.

6. REFERENCES

1. Bucyrus-Erie Company, 1977. "Surface Mining Supervisory Training Program".
2. Seymour, C.A., "Dragline Stripping: Extended Bench Method". World Coal. pp. 23-26, April, 1979, Vol. 5, No. 4.

7. APPENDIX "A"
   SAMPLE SESSION WITH GRAPHIC PROGRAM

Fig. A1 shows the list of draglines in the data base. This is essentially duplicated for program 2 except that multiple occurrences of a given model were removed. Fig. A2 show the data stored on a sample dragline.

One starts a session by entering a geology (1a) either interactively or from a digital model. One then enters the pit geometry (2a) in terms of angles. Thirdly, one consults the index file (3a) of known draglines and selects a model or create a new entry. If the dragline is too small and the extended bench method is to be considered, it is moved to a position from which it can reach (5a) and the extended bench constructed beneath it (6a). The program permits zooming in on details (7a), calculating bench volumes and percentage rehandle and will generate reports summarizing the selected plan (8a).

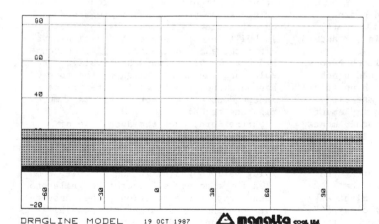

DRAGLINE MODEL   19 OCT 1987   manalta coal ltd   Fig. 1a Input Geology

```
                      DRAGLINE INDEX
------------------------------------------------------------------------
MODEL :BE1260-W   35 Deg    30.5 m3 Bucket(40 CuYd)    FILE NAME : B1260W
MODEL :BE1570-W   35 Deg    44.0 m3 Bucket 105 m Boom  FILE NAME : B1570V
MODEL :BE 800-W 195' Boom @35 Deg                      FILE NAME : B800Wm
MODEL :PAGE 740    68m(223') Boom @35 Deg              FILE NAME : P740_m
MODEL :MARION 7620-W Spec -8R 83.8m(275') Boom, 35 Deg FILE NAME : M7620m
MODEL :BE 2570-W   110m(360')Boom @35 Deg  COSTELLO MINE FILE NAME : B2570c
MODEL :BE1570-W 34 Deg 99m Boom COSTELLO MINE(WIDE TUB)  FILE NAME : B1570c
MODEL :BE 2570-W 122m (400') Boom @30 Deg Willow Bunch  FILE NAME : B2570w
MODEL :PAGE 736-I                                      FILE NAME : P736I
MODEL :PAGE 736-II                                     FILE NAME : P736II
MODEL :BE 680-W                                        FILE NAME : BE680
MODEL :BE 880-W                                        FILE NAME : BE880
MODEL :BE1570-W 34 Deg 99m Boom Wide Tub               FILE NAME : B1570T
MODEL :BE 380-W 52m(170') Boom @ 30 Deg                FILE NAME : BE380w
MODEL :BE 480-W 59m(195ft)Boom @ 30 Deg                FILE NAME : BE480W
MODEL :M8200 W                                         FILE NAME : M8200W
MODEL :BE 680-W                                        FILE NAME : BE680w
MODEL :BE 380-W 43m(140') Boom @ 30 Deg                FILE NAME : BE380W
MODEL :MARION 8750 H                                   FILE NAME : M8750H
MODEL :MARION 8050 (325')Boom 37 Deg                   FILE NAME : M8050m
MODEL :BE 1360-W   99m(325')Boom 38.25Deg              FILE NAME : B1360h
MODEL :BE-2570-W LOT15 POPLAR RIVER MINE               FILE NAME : LOT15
MODEL :BE-2570-W LOT 38 POPLAR RIVER MINE              FILE NAME : LOT38
MODEL :M8200T                                          FILE NAME : M8200T
MODEL :BE 2570-W   110m(360')Boom @31.5 Deg  UTILITY MINE FILE NAME : B2570u
MODEL :M7800       73.m(240')Boom @30.0 Deg  UTILITY MINE FILE NAME : M7800f
MODEL :BE 2570-W   110m(360')Boom @31.5 Deg  UTILITY MINE FILE NAME : B2570f
MODEL :BE1570-W 34 Deg 99m Boom Wide Tub  COSTELLO MINE FILE NAME : B1570C
MODEL :M7800       73.m(240')Boom @34.0 Deg  UTILITY MINE FILE NAME : M7800u
MODEL :M8200                                           FILE NAME : M8200L
MODEL :STACKER                                         FILE NAME : STACK
```

Fig. A1   List of Draglines in the data base

```
                    =========================
                    DRAGLINE SPECIFICATIONS
                    =========================
MODEL :BE 380-W 43m(140') Boom @ 30 Deg            FILE NAME : BE380W
------------------------------------------------------------------------
Boom Length         :       53.00  m   Operating Radius :      47.50  m
Boom Angle          :       30.0   Deg Boom Point Height:      29.76  m
 -Foot Radius       :        1.60  m
 -Foot Height       :        3.26  m
Dumping Clearance   :        7.62  m   Stacking Height  :      22.14  m
Digging Depth       :       22.86  m
Bucket Size         :       20.00  m3  Bucket Mass      :   1000      Kg
Max Suspended Load  :    36300     Kg
Net Working Mass    :   378200     Kg
Tub Diameter        :        8.53  m   Bearing Pressure :       .0649 MPa
Shoe Area           :       25.00  m2  Walking Pressure :       .1484 MPa
Cab Clearance Rad.  :        9.45  m   Dia over shoes   :      10.00  m
          Height    :        1.25  m
Notes :
```

Fig. A2 Sample Dragline data file

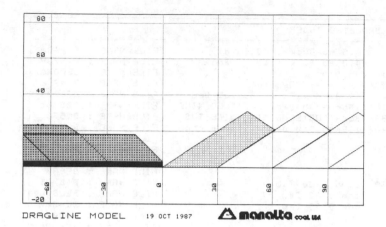

Fig. 2a Input Pit Geometry

```
                      DRAGLINE INDEX
--------------------------------------------------------------------
MODEL :BE1260-W  35 Deg    30.5 m3 Bucket(40 CuYd)    FILE NAME : B1260W
MODEL :BE1570-W  35 Deg    44.0 m3 Bucket 105 m Boom  FILE NAME : B1570V
MODEL :BE 800-W  195' Boom @35 Deg                    FILE NAME : B800Wm
MODEL :PAGE 740    68m(223') Boom @35 Deg             FILE NAME : P740_m
MODEL :MARION 7620-W Spec -8R 83.8m(275') Boom,35 Deg FILE NAME : M7620m
MODEL :BE 2570-W  110m(360')Boom @35 Deg COSTELLO MINE FILE NAME : B2570c
MODEL :BE1570-W 34 Deg 99m Boom COSTELLO MINE(WIDE TUB) FILE NAME : B1570c
MODEL :BE 2570-W 122m (400') Boom @30 Deg Willow Bunch FILE NAME : B2570w
MODEL :PAGE 736-I                                     FILE NAME : P736I
MODEL :PAGE 736-II                                    FILE NAME : P736II
MODEL :BE 680-W                                       FILE NAME : BE680
MODEL :BE 880-W                                       FILE NAME : BE880
MODEL :BE1570-W 34 Deg 99m Boom Wide Tub              FILE NAME : B1570T
MODEL :BE 380-W 52m(170') Boom @ 30 Deg               FILE NAME : BE380w
MODEL :BE 480-W 59m(195ft)Boom @ 30 Deg               FILE NAME : BE480W
MODEL :M8200 W                                        FILE NAME : M8200W
MODEL :BE 680-W                                       FILE NAME : BE680w
MODEL :BE 380-W 43m(140') Boom @ 30 Deg               FILE NAME : BE380W
MODEL :MARION 8750 H                                  FILE NAME : M8750H
MODEL :MARION 8050 (325')Boom 37 Deg                  FILE NAME : M8050h
MODEL :BE 1360-W  99m(325')Boom 38.25Deg              FILE NAME : B1360h
MODEL :BE-2570-W LOT15 POPLAR RIVER MINE              FILE NAME : LOT15
MODEL :BE-2570-W LOT 38 POPLAR RIVER MINE             FILE NAME : LOT38
MODEL :M8200T                                         FILE NAME : M8200T
MODEL :BE 2570-W  110m(360')Boom @31.5 Deg  UTILITY MINE FILE NAME : B2570u
MODEL :M7800     73.m(240')Boom @30.0 Deg  UTILITY MINE FILE NAME : M7800f
MODEL :BE 2570-W  110m(360')Boom @31.5 Deg  UTILITY MINE FILE NAME : B2570f
MODEL :BE1570-W 34 Deg 99m Boom Wide Tub  COSTELLO MINE FILE NAME : B1570C
MODEL :M7800     73.m(240')Boom @34.0 Deg  UTILITY MINE FILE NAME : M7800u
MODEL :M8200                                          FILE NAME : M8200L
MODEL :STACKER                                        FILE NAME : STACK
```

Fig. 3a Dragline Index File

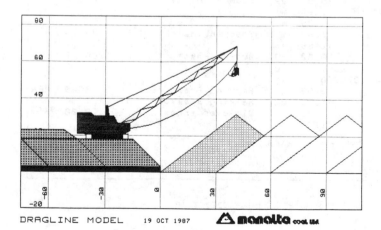

Fig. 4a Selected Dragline

28

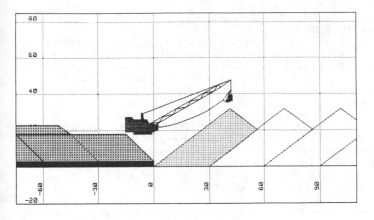

Fig. 5a Dragline In a
Position to Reach
Spoilpile.

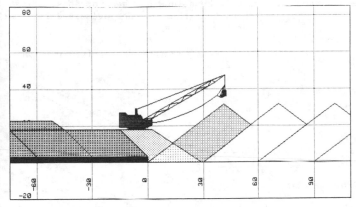

Fig. 6a Extended Bench
Constructed

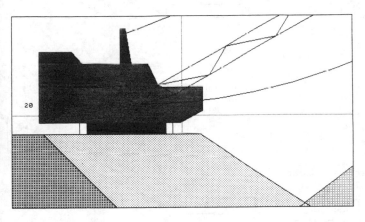

Fig. 7a Zooming Function

Rehandle: 240 m2 (31%)   Bench Total: 389 m2 Dragline Rehandle: 31%

Highwall Geometry

BENCH	WIDTH	HEIGHT	ANGLE	OVERALL SLOPE ANGLE
4	7.00	5.00	35.54	23.86
3	30.00	0.00	0.00	21.80
2	15.00	15.00	45.00	50.19
1	0.00	3.00	90.00	90.00

Spoilpile Geometry

SWELL FACTOR	ANGLE OF REPOSE	OVERALL ANGLE	SLOPE HEIGHT	BENCH HEIGHT
1.30	38.00	38.00	31.86	0.00

Fig. 8a Report

29

8. APPENDIX "B"
   SAMPLE SESSION WITH PROLOG PROGRAM

To use the program one sets its goals.
for example, the clause "works" determines
whether a dragline - pit combination will
work without rehandle. Variables starting
with capital letters are unbound i.e.
Prolog must find values for them that make
the goal true.

In the following examples, we have a
knowledge base of 15 dragline types and 4
sample geologies (test 1 - test 4). The
comments enclosed in slashes and start /*
like this */ are supposed to indicate the
thought process of the user. The goal,
e.g. GOAL Works(A,B) is the entry made by
the user, the remainder is the programs
reply:-

/* which draglines work in the area with
geology "test2"?*/

GOAL works (Dragline, test2)
Dragline=b1570v
Dragline=b2570c
Dragline=m8200w
Dragline=m8750h
Dragline=m8050h
Dragline=b1360h
6 solutions

/* what pits will dragline b1770v work at
?*/

GOAL' worksBb1570v,Geology)
Geology=test1
Geology=test2
2 solutions

/* which draglines will work in geologies
test1, test2 and test3?&/

GOAL works(D,test1),works(D,test2),works
D(test3)
D=m8750h
1 solution

/*That too big a dragline - lets consider
using extended bench working in case of
test3 geology -test2 correspond to the
normal worst case, test3 is a channel
deposit where the coal is unusually thin
and deep*/

GOAL works(A,test1),works(A,test2),works_
with_extended_bench_(dragline(A,),test3,_,
_)
A=b1570v
A=b2570c
A=m8750h
A=m8050h
A=n1360h

5 solutions

/* Well lets look at only the draglines
where the rehandle is 0.25%. The goal now
becomes: "choose a dragline that works in
geology test1 and test2 without rehandle
and in test3 with rehandle but no more
than 25%." */

GOAL works(A,test1),works(A,test2),works_
with_extended_bench(dragline(A,),test3,
Rehandle,_),Rehandle>0,Rehandle<0.25
A=b1570v, Rehandle=0.246436
A=b2570c, Rehandle=0.173867
2 solutions

/* well these sound like sensible options.
Lets take a look at these options. First
check that they are not oversized for test3
stratigraphies*/

GOAL range_diagram(dragline(b1570c),_),
test2)

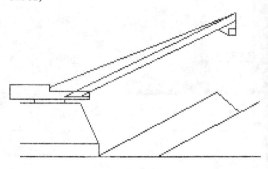

Fig. 1b

GOAL range_diagram(dragline(b2570c),_),
test2)

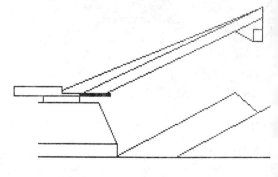

Fig. 2b

/* now look at the extended bench situations under test 3 conditions *?

30

GOAL extended_bench_range_diagram(dragline
(B1570v,_)test3)

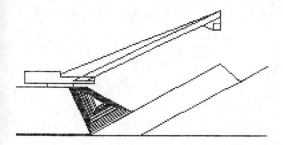

Fig. 3b

GOAL extended_bench_range_diagram(dragline
(b2570v,_)test3)

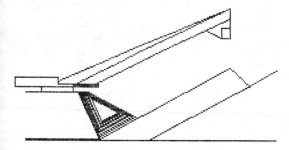

Fig. 4b

/* the b2570v requires rehandle only
because of the distance that the tub is
set back from the highwall - this option
might be chosen with the hope that rehandle could be avoided or else the smaller b1570v selected. The program can also
give bench widths etc*/

Computer Applications in the Mineral Industry, Fytas, Collins & Singhal (eds)
© *1988 Balkema, Rotterdam. ISBN 90 6191 760 3*

Computerized underground mine planning (Abstract)*

John C.Davies & A.Fredrick Banfield
Mintec Inc., Tucson, Ariz., USA

ABSTRACT: Two major problems exist when applying computers to underground mine planning. The first is the design, display, and accurate calculation of tons and grade for three dimensional volumes such as stopes and production headings. The second problem is in calculating a realistic production schedule given the reserves within multiple operating areas with different constraints. Underground mines each tend to be unique in their layout and design, and so operating constraints can cover a wide range of possibilities. An enhancement of existing software has allowed Mintec Inc. to address both of the problems above, without introducing unnecessary complexity. Stopes are interactively designed on a graphics screen on a slice by slice basis. Only slices that represent major changes in geometry of the volume need be designed and intermediate slices are generated at a desired resolution. The volume is then superimposed over a computer model of reserves and information required for production scheduling is generated. The production scheduling allows for multiple operating areas, providing a preliminary sequence of mining has been manually imposed. Targets can be production rate, blend, or economics.

*Paper not received on time for publication.

Computer Applications in the Mineral Industry, Fytas, Collins & Singhal (eds)
© 1988 Balkema, Rotterdam. ISBN 90 6191 760 3

An open-pit planning and scheduling system

Mark Gershon
Temple University, Philadelphia, Pa., USA

ABSTRACT: This paper describes the recent implementation, through the use of the microcomputer, of a new approach to open-pit planning and scheduling. This approach, although not a true mathematical optimization, attempts to approximate the complex optimizations while producing more practical results with much less effort. The approach is easily understood by mining engineers with little or no training in operations research and the menu-driven multi-window screen requires litle program documentation. A companion program for ultimate pit limit analysis runs on the same data bases.

1 INTRODUCTION

Most recent work in the field of mine planning and production scheduling has focused on either computerization of traditional methods or development of sophisticated mathematical optimization models. Yet it is clear to all concerned that the optimizers need to be made more practical, and the traditional approach needs to be more optimal. The program described in this paper opens a middle ground between these two approaches.

The traditional and optimization approaches are so far apart that advocates of each have difficulty identifying with the other. This is especially true concerning the difficulty of understanding the optimization models for those not trained in operations research. This paper takes the position that the proper role of the operations researcher is to provide not only useful and advantageous approaches, but ones that are practical, understandable, and implementable as well. This is the goal of this paper in describing the computer implementation of a new approach, consistent with the traditional methods and aimed at approximating the results of the optimizers.

The program described herein relies on the best features of both current approaches. The traditional approaches available all tend to computerize the "by hand" methods, putting the engineer in a position of working faster but not necessarily better. This so-called "scheduling software" does no scheduling. This new program does the scheduling automatically, or provides guidance to the engineer working in the traditional way. In either mode, or in combination, it provides the capability to conduct planning and scheduling activities both faster and better.

The next section provides the background discussion of the production scheduling problem and the approach that the program takes to solve it. Then, the program is described in detail. The remainder of the paper describes briefly the ultimate pit limit analysis program and provides conclusions.

2 DESCRIPTION OF PROBLEM AND APPROACH

In this section, the problem addressed is described and put into the context of a more general discussion of mine planning and production scheduling. The second half of the section summarizes the basis for solving this problem and sequencing the removal of the mining blocks (Gershon 1987).

2.1 The production scheduling problem

The production scheduling problem, in mining, concerns the sequence of removing the mining blocks within the limits of the mine plan. In other words, the problem is one of block sequencing, taking into account operational criteria such as minimum pushback widths and haul road access. The sequence of mining that yields the largest net present value return is the one that should be chosen.

This simplistic problem statement is not sufficient once the production scheduler considers the many limitations on the actual schedule. For a variety of reasons, only a small subset of all possible sequences can be implemented. The rest are not practical. These reasons include working and final slope considerations, restrictions on the mobility of equipment, the need to balance strip ratios, and blending considerations on the qualities. Thus, the problem becomes one of finding a practical schedule - one that meets all of these requirements.

This is the level at which most production scheduling projects are conducted. The difficulty in achieving a practical schedule overrides the problem of finding the best schedule. With these considerations in mind the problem definition used in this paper is:

Find the schedule (sequence of mining blocks) that maximizes the net present value return within the physical, logistical, operational, and contractural limitations by which the scheduler is constrained.

This problem definition does not necessarily fit into either the category of mine planning or production scheduling as practised at many mines, but falls in between. When referring to mine planning, ultimate pit limit analysis is usually used. This provides a broad picture, but no scheduling. While some may attempt to schedule in this way by placing an expanding time limit on each pit, there is no basis for this and there are many reasons why it is erroneous to do so. Production scheduling usually involves daily or hourly equipment assignments, a level of detail not considered here.

The problem just described, too often ignored, is the aggregate production scheduling problem. An aggregate plan, describing the sequence of block removal, is the mechanism by which it is possible to link the mine plan and the production schedule.

2.2 The scheduling approach

There are many advantages to using a heuristic approach for the mine production scheduling problem. First, it uses a common sense approach that is easily understood by the mine engineer who may have been using a manual or computer-assisted approach or a simulator. It can even be used to guide a manually derived schedule. Finally, since the optimization methods are so difficult to apply to this production scheduling problem (Gershon 1983), a method that can quickly and efficiently approximate the results of the optimizer is needed. The heuristic approach described here accomplishes this task. Therefore, the heuristic provides most of the power of optimization within the framework of the more traditional scheduling approaches.

This heuristic is based on the concept of a ranked positional weight. First developed for use in designing industrial production lines (Helgeson and Birnie 1961), the concept is used here to accomplish automatically many of the common sense ideas that every mine engineer takes for granted.

In pit design, cones are generated upward from each block to approximate the shape of a pit and to determine whether or not the block in question should be part of the ultimate pit. This same concept of generating cones is applied here, with one twist. The cones are inverted.

From the production scheduler's view, there are many blocks that cannot be mined because one particular block has not yet been mined. Until this key block is mined, all those behind it (or below it, really) cannot be mined. Following a desired pit slope, the aggregate of all the blocks that a single block prohibits mining forms a cone with that single block at its apex. Thus, for each block this heuristic generates cones downward, not upward.

Figure 1 shows these downward cones for blocks within the ultimate pit. No block in this cone can be mined until the single block at its apex is mined. For each block in the ultimate pit, a similar cone can be generated downward.

Now comes a key question facing the mine scheduler. What makes it more desirable or less desirable to mine an exposed block at the present time? The answer within this heuristic is the same as it is when using a more traditional approach. That is, if the block is an ore block, valuable in itself, it is more desirable than a block of lesser quality ore. This is a short-term reasoning. A longer-term reasoning is that a block must be mined if it enables high-grade ore beneath it (within its cone) to be reached. The mine scheduler is always trying to balance these short- and long-term needs.

It would seem then that the total quality

of ore within a cone, in a simplistic
sense, is a good measure of the
desirability of mining the block at the
apex of that cone. Calculating this
desirability index must be refined to be
more meaningful and practical. But,
clearly, it now provides a means to achieve
the desired sequencing of blocks.

This index of desirability is called the
positional weight or the positional grade.
As developed thus far, it is the sum of the
ore qualities within the cone generated
downward from a block within the ultimate
pit. That is, for any given block, the
positional grade is its quality, plus the
qualities of the blocks below that would
require its stripping. Once these
positional weights are determined for each
exposed block, they are ranked. The block
with the highest positional weight should
be mined first. Then the positional
weights are recalculated, based on the
revised set of exposed blocks, and the
sequencing continues in this manner.

Figure 1 shows a cross section of an
ultimate pit to demonstrate the most basic
calculation. The numbers in the blocks
here represent ore grades, but these
numbers could easily be generalized to some
priority values based on a combination of
multiple qualities. The shaded area (////)
shows the cone used for calculating the
positional weight for the ninth block on
the first level. The other shaded area
(\\\\) shows the cone used for the same
calculation for the 15th block on that same
level. The crosshatched area is used in
both calculations.

Block Quality	= .3	Block Quality	= .2
Cone Below Block	= 7.7	Cone Below Block	= 4.2
Total in Cone	= 8.0	Total in Cone	= 4.4

Fig. 1 — Downward cone calculations

While the entire calculation is provided in
Fig. 1, it suffices here to say that the
ninth block has a positional weight of 8.0
and the 15th block has a weight of 4.4.
Given this particular cross section, common
sense tells the engineer that an initial
pit should be developed in a somewhat
central position within the ultimate pit,
and then this pit should be expanded to the
right to reach the best ore. This is
exactly the sequence that results from the
use of this heuristic.

In a general way, an approach to the mine
scheduling problem has been developed. The
steps of this approach are:

1. Determine the set of blocks currently
available for mining.

2. Calculate the positional weights for
each of these blocks.

3. Decide which block to mine and return
to step 1.

3 THE HOPS SOFTWARE

HOPS is an interactive scheduler developed
specifically for open-pit mining, but
applicable to other kinds of surface mining
as well. The program incorporates two
major areas of capabilities required for
production scheduling and mine planning.
These are the ability to simulate the
traditional practices of mine scheduling in
an interactive environment and the ability
to use advanced scheduling techniques to
guide the scheduler toward the best
possible schedule. It is this second
capability that differentiates HOPS from
the "trial and error" schedulers that are
available today.

In the simulation mode, the scheduler can
develop a pit block by block or in large
sections such as pushbacks. A major
addition over most mine schedulers is that
haul roads can be designed within the
program. Advanced graphics features allow
for display of the pit at any time and
summary data concerning blends, strip
ratios, etc. are constantly available. Pit
slope and cutoff grades are incorporated
into the program, taking these burdens off
of the scheduler.

The advanced scheduling techniques use
the heuristic analyis described in the
previous section to add intelligence to the
program. Priorities are assigned to each
block indicating how important it is to
mine that block at any given time. These
priorities are dynamic in that they change
as the pit is developed. The scheduler can
list the top priority blocks to use the
program as guidance or may let the program
generate schedules automatically.

Preferably, the automatic and manual scheduling modes are combined in practice.

The program operates from an initial screen that gives the user a plan view of part of the mine surface. The blocks are represented by integers ranging from 0 to 15 that correspond with the grade of that block. The third dimension (depth) is indicated by textual color which corresponds to a legend directly below the display area, providing a 3-D view at all times. The user can scan the mine surface by moving the cursor from block to block using the numeric keypad. When the cursor is positioned on a block the status of that block is indicated in a window at the lower right corner of the screen. This window indicates the position of the block in the mine, the weight given to the block relative to blocks in the cone below it and the status of the block. A status of NO indicates that the block cannot be mined due to slope requirements or the position of the block. A primary command menu on the right edge of the screen lists all of the available commands from this display screen. These commands allow the user to enter a graphics display mode, list top surface blocks, list summary data on the mine, change the viewing area, load or save a mine set-up, reserve or unreserve a block, display a cross-section or plan view of mine block grades, and perform mining operations. A sample screen is provided in Figure 2.

3.1 Mining operations

Mining operations are conducted in one of three ways:

1. Mine one block at a time
2. Mine all available blocks
3. Mine according to positional weight priority.

To mine one block at a time, it is only necessary to hit the "M" while the cursor is positioned at the appropriate block. All summary statistics and priority weights are then upgraded automatically. Using the program in this way is just a computerization of the traditional approaches of trial and error. A convenient feature in this context is the "CTRL M" key, which is used to unmine a previously mined block.

The "T" command displays a pop-up window listing the top priority blocks that are currently available for mining. In this way, the heuristic does not do the scheduling, but provides guidance to the scheduler.

The sweep mine feature ("S") is used to mine all available blocks. By combining the "R" feature (to reserve, or prohibit, a block from being mined) and the "C" feature (to limit the range where the available blocks can be located), the sweep mine can be a useful and powerful tool. Pushbacks can be quickly evaluated, as one example of its use.

The auto-mine ("A") is used to implement the positional weight heuristic, automatically choosing to mine the block having the highest priority at each step. This works very well in more of a mine planning mode. As the plan is converted to a schedule, there will often be a small number of blocks that would be impractical to mine by the generated sequence in the

```
 2   7  6   7   0   8   6   1  15 8   3   5   7  15 4  12 0   4   2  14      1   COMMAND MENU:
 7  10 11 15 13 13   5  12 9   3   0   8   2   3   4  11 15 6   9  12      2   <V>iewing area
 8   4  1   4   3   7   4   3  10 0  12 5   1   8   1   1   1   7  10 0      3   <T>op ten
 1  10  4   0  12 14 12.14 11 9  15 4  15 7   2   8  15 15 14 3      4   <W>eights
 0  10  5   2  15 10 13 2  10 11 12 5  11 6   2   9   3   8  14 13      5   <M>ine block
 1   1  1   3   5   2  12 10 13 1  14 10 1   9   2  12 5  10 4   8      6   <^M> unmine
 3  11  5   0  11 4  14 1   4  12 4  11 3  12 6   1  14 0   8  12      7   <R>eserve block
 1   7 14  5  14 12 1   3   9  R  5   1  15 13 12 13 14 0   2   3      8   <^R> unreserve
 9   0  8   1  11 9   1  12 14 13 6  11 2   9   2  11 5   1  13 6     R9   <C>urrent range
 9   5  2   1   0   6  13 9   0  13 7   5   8  13 3   9   1  10 3  14    O10  <A>uto-mine
13  2   9   0   8   2   0  14 13 5  11 7   0   6  13 5  13 7   2  14    W11  <S>weep mine
 2   5  8  13 2   9  12 11 14 12 10 1   6   1  13 10 8   7   7   8    S12  <D>ata display
 7  10 15  5  11 1  14 11 13 11 13 10 7  13 1   7  14 6  15 9     13   <G>raphics menu
 6  15  9   4  12 0   5   1  13 14 11 12 8   4   8  13 2   8  15     14   <I>nput/Output
15  1   4   8  13 3  10 2   5   9   4  15 15 11 11 5   1   7  15 7     15   <ESC> EXIT
14 13 13 14  0   3   7  13 1   8  10 12 7   0  15 11 10 3   0  13     16   <^D>ata summary
15  0   1   3   7  11 0   6  14 4   8  14 15 0  10 2   0  11 9   0     17   <^K> free bytes
 1   7 13  5  13 7   1  13 9   6   7  12 15 4   4  13 4   1  11 18     18
 1  15 15  0   1   4   7  12 2   8   0   0   4   5  10 8   8   0   6     19   CURRENT BLOCK
11  6   3   1   0  15 0   2   5   8  13 3  13 1   7   5   3   7   8  12     20   level 2
                              columns                                         col 1  row 1
 1   2  3   4   5   6   7   8   9  10 11 12 13 14 15 16 17 18 19 20            cone wt. 50
Level legend = 1234567891011121314151617181920          RANGE                status NO
Movement NUMLOCK right<6> left<4> up<8> down<2>  C 1-30,R 1-30                   MINE DATA
COMMAND ?                                                                    # mined   4204
```

Fig.2 Sample screen

short term. To counter this effect, experience has shown that the best practice is not to develop an entire schedule by this method. Instead, the auto-mine should be used for a smaller number of blocks in each pass, allowing the engineer to review and refine the schedule before proceeding.

At this point, it may seem that selecting the best block to mine is quite straightforward. Unfortunately, this is not the case. While this heuristic approach is successful in considering both long- and short-term factors in mine production scheduling, there are still many site specific and time specific considerations that it fails to take into account. For this reason, the best way to use this approach is to use it interactively, with the engineer always deciding which block to mine next. While the rankings provide useful guidance, the ultimate rule is: **heuristic recommends – engineer decides.** The ability of the engineer to accept or override the recommendation is a major feature of this approach.

3.2 Other concerns

The HOPS program without the heuristic algorithm would be just like other scheduling programs on the market. For this reason, more discussion is required in the calculation of the positional weights, the key priority values assigned to each block. The goal should be to calculate these priorities in such a manner so that they match most closely with the profitability concerns of the mine. Until this point, the priorities have been calculated as the sum of the ore grades in the downward cone. This is too simplistic for practical use. A high grade on the 15th level, for example, is not as valuable, in present value terms, as a similar grade ore on the third level. To accurately model this situation, a discount factor must be applied at each level so that the value of blocks decreases with depth. On many properties, distance across the property is also a factor that must be considered. If the mine schedule is to be developed within a long-term mine plan, as is usually the case, this mine plan will also yield information concerning approximately how long it will take to reach a given area of the mine. Obviously, this is the best possible basis for assignment of the discount factors.

Thus, the positional weight of each block is a function of: quality of the block, position of the block, quality of ore behind the block, and discount factors. In practice, the same set of discount values is not used at each different mine. Their appropriate assignment must be part of every implementation of this approach.

The program uses a cone discount factor to handle different discount schemes. A discount factor of zero gives each block grade the same weight (no discounting). Increasing the cone factor increases the rate of discounting, in effect limiting the depth of ore considered in the priorities.

Another factor of concern is the selection of the blocks available for mining. Obviously, due to pit wall stability concerns, equipment availability, and other operational concerns, not all exposed blocks can be mined. If the only consideration is pit slopes, then it is possible to automate determining the set of blocks available.

This is what the program does. Despite this, the engineer can gain more control over the set of blocks available, either by reserving blocks or by limiting the mining area.

The cutoff grade can also be specified in the program. This is especially useful when providing graphical displays with the cutoff grade masking the display.

Finally, the summary and graphics displays are important to any type of program. The data summary window displays valuable information about all mining operations that have taken place and the values of critical mining variables. Summary data consists of the total blocks mined, the cutoff grade, the blocks mined above and below cutoff, the average grade of blocks mined above and below cutoff, and the discount factor used to determine block weight. Graphics displays include both plan and cross-sections, showing the pit as developed, the maximum possible pit, and having the capability to show qualities. A 3-D view of the pit can also be obtained at any stage of development.

4 THE ULTIMATE PIT PROGRAM

Most mine planners will want to work with their mine scheduler in conjunction with their ultimate pit program for more general planning purposes. For this reason, an ultimate pit program has been developed that runs from the same data base as the HOPS program and provides the same graphics displays. The program can handle up to 250,000 blocks on the IBM PC AT and is, in most respects, a standard kind of program for this application.

It does differ, however, in one major respect, avoiding the use of the floating

cone approach. The flaws involved in the
use of floating cones are well known, but it
is still the most widely used technique
because of the ease of programming it and
the ease of understanding how it works. The
approach used here, a true optimization, is
the network maxflow approach. This approach
was developed elsewhere and is also known to
be equivalent to the 3-D Lerchs-Grossmann,
which has been gaining in popularity.

5 SUMMARY AND CONCLUSIONS

The currently available scheduling software
does not do any scheduling, only acting as
a tool to aid the engineer by taking some
of the tediousness out of this task. The
software described in this paper includes
these same capabilities, but also adds a
scheduling method on top of this foundation
to enable it to really accomplish
scheduling aimed at long term
profitability. The scheduling method is
based on the heuristic concept of the
ranked positional weight, this weight being
used to assign a priority to each block.

REFERENCES

Gershon, M.E. 1983. Optimal Mine Production
   Scheduling: Evaluation of Large Scale
   Mathematical Programming Approaches.
   International Journal of Mining
   Engineering. 1:December:315-329.
Gershon, M.E. 1987. An Open-Pit Production
   Scheduler: Algorithm and Implementation.
   Mining Engineering. 39:8:August:793-796.
Helgeson, W.B. and D.P. Birnie. 1961.
   Assembly Line Balancing Using the Ranked
   Positional Weight Technique. The Journal
   of Industrial Engineering. 7:No.4.

Computer Applications in the Mineral Industry, Fytas, Collins & Singhal (eds)
© 1988 Balkema, Rotterdam. ISBN 90 6191 760 3

Advanced computer simulations – The key to effective blast designs in open pit and underground mines

B. Mohanty
CIL Inc., Explosives Technical Centre, McMasterville, Quebec, Canada

J.P. Tidman
ICI Explosives Group Technical Centre, Beloeil Site, McMasterville, Quebec, Canada

G.K. Jorgenson
CIL Inc., Technical Services, North York, Ontario, Canada

ABSTRACT: The evolution of computer applications in the design of blasts in underground and open pit mines is reviewed. This is shown to be a direct outcome of the scale and complexity of modern blasting practice. The use of analytical techniques in modelling the blasting process has largely supplanted the empirical and semi-empirical approaches of past in the design of blasting rounds and prediction of blast results. The advent of powerful microcomputers has now made it possible for every mine, employing these techniques, to design cost-effective blasting operations on a routine basis. Accuracy of prediction from computer models, however, depends critically on the extent of characterization of relevant properties of the explosive as well as the target rock. The relative strengths of various computer models in current use such as SABREX are examined, and the key areas requiring additional research efforts highlighted. Examples of blast designs and optimization of results achieved through such computer simulations are also described.

1 INTRODUCTION

The advent of very large scale mining during the last two decades has brought about a significant change in the approach to blast designs. Computer simulations of blasting process and prediction of blast results in advance are now standard practice at all major mining operations. In a large open-pit mine, a single blast may involve up to a million tonnes of rock. Such scales preclude the use of trial-and-error method of blast design, as even an occasionally poor blast can significantly affect the economy of the operation. The introduction of computer-aided blast designs has also been greatly facilitated by the recent advances in explosives technology. It is now possible, through novel initiation systems and wider spectrum of explosives, to closely match the explosive system, the blast geometry, and the rock type to achieve optimum results.

Modelling of the blasting process is not a new technology. A blasting model is any relation between blast design and blast results, and therefore, must have been inherent to all practical blasting operations since their advent. What has changed recently to this simple approach is our increased understanding of the blasting process, the properties of the explosives and rock, and the easy access to powerful computers. Consequently, several blast prediction models of varying sophistication have been proposed or are in current use (Langefors and Kihlström, 1979; Favreau, 1980; McHugh, 1983; Tidman, 1984; Chung and Jorgenson, 1985; Demuth et al, 1985; Kirby et al, 1987). This paper addresses some of the key issues in the modelling process, and describes the essential features of our most advanced blasting model SABREX, along with several case studies.

2 ROCK FRAGMENTATION DUE TO BLASTING

There are four elements in a blasting system, which interact to produce the desired results – the explosive, the rock, the blast geometry, and finally the cost (Fig. 1). The interactions among the first three elements determine the blast results, which must always be weighed against the cost.

The relevant properties of the explosive in this case are its density and detonation properties (velocity of detonation, pressure and energy). The corresponding properties of rock are its strength, density and geological characteristics such as joints, faults, dip and inhomogeneity. The

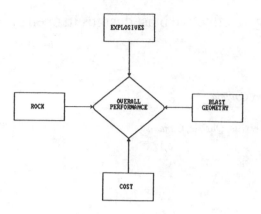

Fig. 1: Blasting System

blast geometry includes borehole diameter, hole depth, spacing, burden, stemming, coupling, and mode of and timing of initiation of explosive in borehole. Some of these parameters as shown in a later section can also profoundly affect the detonation properties of explosives. The blast results themselves also require further qualification as these consist of several inter-related parameters such as, degree of fragmentation, extent of backbreak, rock movement and fly rock. The blasting system is thus a very complex one in which several interacting parameters determine the final blast results and the cost of the mining operation.

3 MODELLING OF BLASTING PROCESS

In its simplified form the general concensus on the blasting process is the following. Detonation of explosive column in the borehole gives rise to a strong shock wave transmitted to the borehole wall. The amplitude of this shock is several orders of magnitude greater than the characteristic strength of the surrounding rock. Consequently, the rock in the immediate vicinity of the borehole is intensely fractured. The shock wave decays very rapidly however as it propagates away from the borehole. Upon reflection at free face or any other discontinuity, its compressive components transform into tensile ones, which can lead to spalling at the free face and can also create additional cracks and extend existing ones in the intervening rock mass. In the typical scale employed in large diameter ($\sim$30 cm) open-pit blasts (8 - 10 m burden), this phase of fragmentation is completed within 2 $\sim$ 3

msec. Immediately behind the shock front the high pressure explosion gases continue to expand into these cracks (Fig. 2). In the final stages, some of the internal energy in the gases is transferred on to the fragmented rock mass - resulting in mass motion, fly rock and the formation of the muck pile. The effect of explosive action in blasting can thus be seen to consist of two distinct phases - one due to explosion shock energy and the other due to explosion gas energy.

The blasting process can thus be modelled on the basis of total explosive energy or the energy partitioning and the relative contributions of the two energy fractions to rock fragmentation and heave. Of the three basic modelling approaches - empirical, phenomenological, and analytical (Mohanty and Chung, 1986), the first two employ the total energy concept (or the total weight of the explosives). The empirical approach relies heavily on experience factor and employs a simple criterion based on the ratio of weight of the explosive and the weight or volume of rock to be broken (e.g. Powder Factor).

In the phenomenological approach, a correlation is sought between certain blast parameters (usually the degree of fragmentation or the volume of rock broken) and the energy or weight of explosive. No explicit information on failure mechanism for rock, explosive energy utilization in the various facets of blasting process, or

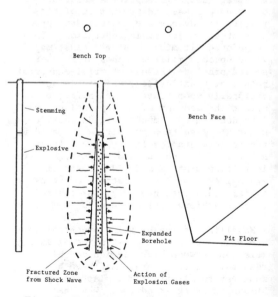

Fig. 2: Initial Fragmentation Process in Blasting

the geology of rock mass is required in this approach. Crater tests and small scale fragmentation studies fall into this category. The results of tests are extrapolated on the basis of dimensional analysis to predict the outcome of actual blasts. The following relation represents this approach in its most general form for a given explosive,

$$Q = k_1 X + k_2 X^2 + k_3 X^3 + k_4 X^4 \qquad (1)$$

where Q is the explosive charge weight, X is any linear dimension of the blasting pattern (usually the burden), and the constants k_1, ----- k_4 are fitting parameters related to blast geometry and the rock. The main drawbacks of both these approaches stem from their neglect of the dynamic processes underlying fragmentation and the nature of explosive energy partition.

The analytical approaches, on the other hand, do take into account these factors, by explicit representation of a failure mechanism for rock in response to explosive action, the pressure-time relationships specific to each explosive, and finally, the interaction between exploding holes in underground or open-pit blasts. The extent to which these factors can be realistically incorporated in the blasting model and blast results predicted depends on the precision of the model. Because of these reasons, such models have the best potential for predicting blast results by combining known detonation properties of the explosive, the strength and elastic properties of target rock and the blast geometry. The computer model SABREX is the latest among commercially available blasting models, which incorporate these features, and predict blast results and provide cost-effective blast designs.

4 SABREX BLASTING MODEL

The SABREX model (Scientific Approach to Breaking Rock with Explosives) consists of three essential elements; detonation properties of explosives, strength and elastic properties of rock, and blast geometry. the treatment of detonation properties is unique to this model, as it can accommodate the non-ideal behaviour of commercial explosives. Unlike the molecular and other military explosives, the commercial explosives are multi-component systems, dictated by the needs of convenient packaging and economy. As a result, these explosives can sustain 'ideal' chemical reaction

only in very large diameters. In practice, the same explosive composition is packaged or used in a range of diameters. In addition to the diameter of the borehole, non-ideal detonation is strongly controlled by the extent the explosive fills the hole (decoupling), the mode of initiation and the confinement provided by the surrounding rock. The extent of this non-ideality is also reflected in the energy partitioning (between shock and gas). In the reduced scale of practical charges, the reaction rates of commercial explosives are influenced by the surroundings. The loss of energy through lateral expansion results in lower detonation velocities and concurrent reduction in pressure and energy.

The flow chart for calculating the non-ideal detonation parameters and the energy partitioning is shown in Fig. 3. The division between gas and shock energy in a specified explosive for a given charge diameter and rock type is shown schematically in Fig. 4.

The complete SABREX architecture, as shown in Fig. 5, consists of several interacting modules. The external supporting modules such as information from high-speed films, photogrammetric studies of muck pile and vibration analysis used to confirm and refine SABREX predictions are also shown. In the absence of experimental data to run the non-ideal detonation code (CPEX), there is provision for utilizing the output from an ideal-reaction code, as an initial step to prediction.

A simplified flow diagram for SABREX is shown in Fig. 6.

The properties of rock normally used as inputs to SABREX are, i) density, ii) Poisson's ratio, iii) Young's modulus, iv) unconfined static compressive strength, v) unconfined static dynamic strength, vi) dynamic tensile strength, and vii) the shock attenuation coefficient. The latter two are obtained experimentally under shock loading conditions (Mohanty, 1987). The input on blast geometry are usually the following, i) borehole diameter, ii) blasting pattern, iii) borehole inclination, iv) bench height, v) depth of subgrade drilling, vi) collar distance, vii) initiation pattern and delay times. Information on unit costs of explosives, accessories and drilling complete the required input file.

The SABREX predictions consist of the following,

i) fragment size distribution
ii) muck pile profile

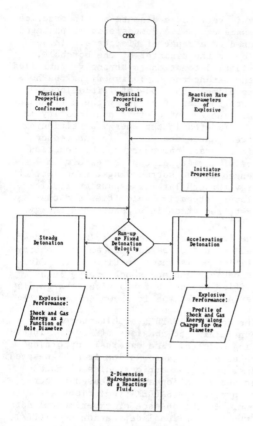

Fig. 3: Flow Chart for Calculation of Non-ideal Detonation Properties

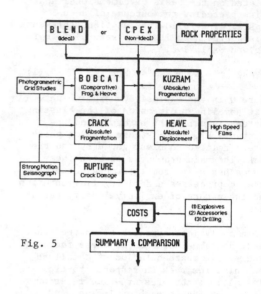

Fig. 5

iii) grade level fragmentation
iv) collar block fragmentation
v) flyrock control
vi) backcracking, including damage below subgrade
vii) cost per unit volume of rock broken

The size distribution of blasted rock can be predicted with either the 'Crack' or the KUZ-RAM module (Fig. 5). The 'Crack' module (Harries, 1977) is based on postulated crack patterns around each borehole as a function of the equilibrium explosion pressure in the expanded borehole during the shock phase, and the strength and attenuation properties of target rock. The failure mode employed in this model is that due to tension. The size distribution analysis is carried out in two dimensions on any specified plane perpendicular to borehole axis by the Monte Carlo sampling method. The space bounded by the cracks and sampled in orthogonal directions by this technique essentially represents the fragment size in that space (Fig. 7). The KUZ-RAM (Kuznetsov-Rammler) model provides an alternate method of estimating fragment size distribution. It is based on comminution principles and assumes the available explosive energy to have a functional relationship with fragment size distribution (Cunningham, 1983). This follows the well established Rosin-Rammler distribution,

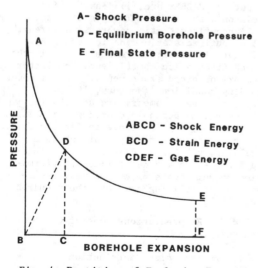

Fig. 4: Partition of Explosive Energy

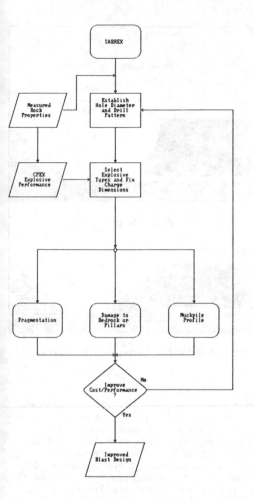

Fig. 6: Simplified Flow Chart for SABREX

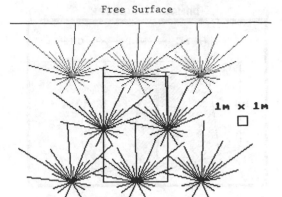

Free Surface

1m × 1m

Fig. 7: Crack Pattern

$$y = 1 - \exp[- (x/x_o)^n] \qquad (2)$$

or rewritten as,

$$\log\log[1/(1-y)]=n\log x-n\log x_o+\text{constant} \qquad (3)$$

where, y is the fraction of fragments with dimension smaller than x, x_o is the dimension representing 37% of the material, and n a constant obtained by plotting Eqn. (3). The validity of the above distribution function in treating fragmented rock has been established by several workers (Hjelmberg, 1983; Kuznetsov, 1973).

The 'Heave' module can yield either a relative measure of rock movement with respect to some reference blast or the complete profile of the muck. The choice is dictated by the desired precision in the blast simulation exercise. The detailed profile is generated by calculating the velocity and time of initial movement of blocks within the burden and tracking their path until they come to rest. The initial velocity is calculated from a knowledge of the explosion gas expansion curve. Because of the enormous complexity of multi-block motion accompanied by collision, the simulation is based on the simplification that the motion of the first row limits the motion of subsequently fired rows, and on the assumption of a swell condition and angle of repose for broken rock.

The 'Rupture' module yields the extent of cracking of rock (i.e. the rupture envelope) behind the blast hole as well as below subgrade. This is obtained by calculating the limiting distance from the borehole for failure of rock by tension, compression or shear from the hole pressurized at the equilibrium pressure. The rupture envelope as well as the grade break-out angle are generated from the postulated crater dimensions in this region. As mentioned previously, there are a variety of blast monitoring techniques which can be used to verify SABREX predictions and in the process, increase the reliability of its predictions.

A typical rupture envelope simulation for an open pit operation is shown in Fig. 8. In this case BT and BC represent the burden at toe and collar respectively. The back-break profile is represented by CB, BB and SB at three key regions of the bench. The case shown consisted of only one continuous charge; different explosives for the toe

45

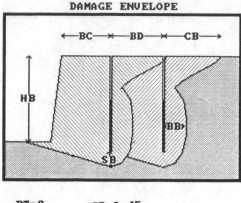

DAMAGE ENVELOPE

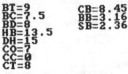

BT=9
BC=7.5
BD=8
HB=13.5
DH=15
CO=7
CC=0
CT=8

CB=8.45
BB=3.16
SB=2.36

Fig. 8: Predicted Damage Envelope

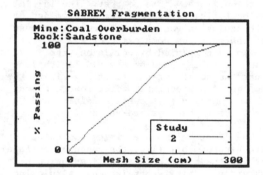

SABREX Fragmentation

Mine:Coal Overburden
Rock:Sandstone

Fig. 9: Fragment Size Distribution

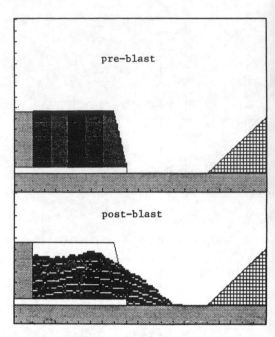

Fig 10: Muckpile from Cast Blasting

and collar regions, as well as deck charges, can also be accommodated by the model. These results have important applications in berm and slope designs.

A fragment size prediction by the model is shown in Fig. 9. Graphical display of size distribution helps in rapid selection of the optimum explosive-blast geometry configurations. Photogrammetric survey of the muck pile can be used to refine these predictions and make them more sensitive to changes in some key blasting parameters.

Fig. 10 shows the resulting muckpile from a coal mine cast blasting simulation. Each shaded region represents an individual delay row. The coal seam is represented by the blank strip at the grade level. Verification of the model predictions against surveyed muck profiles, although limited in number, has confirmed the essential soundness of the approach. This is particularly significant in view of the complex nature of the problem and the simplifying assumptions made in treating motion of broken rock mass.

A typical example of cost analysis obtained with SABREX is shown in Table 1. The cost figures are weighed against specific performance figures, i.e. fragmentation, heave, etc. to arrive at the most desirable explosive-blast geometry combination.

5 CONCLUSION

The computer blast model SABREX represents a powerful new tool in blast and cost optimization for the mining industry. It combines the analytic approach in describing explosive-rock interactions with computer graphics to highlight the role of all relevant blasting parameters on blast results. The micro-computer version of the model allows rapid simulation of blasts at almost any mine site by a qualified blasting en-

Table 1

SABREX
COST ANALYSIS

Location: TEST DEMONSTRATION   Rock: VERY HARD

```
DRILLING DATA- Diameter      (mm):    310.0
               Depth          (m):     14.0
               Bench Height   (m):     12.0
               Burden         (m):      6.0
               Spacing        (m):      7.0

BLEND EXPLOSIVES
         CHOSEN AT:         TOE         COLUMN
                             1
      Type:              MFRAC 500       ANFO
      Density (g/cc):       1.00         0.84
      VoD (m/s):         5445.0        4937.0
      Length (m):           2.0           6.0
      Mass (kg):          151.0         380.4

      Overall Power Factor = 1.054 kg/m3

SELECT INITIATION SYSTEM:

                                        # PER HOLE
      NONEL XT SHOT DELAY                    1
      PROCORE BOOSTERS                       1
      SURFACE DELAYS - 2m NONEL              1
      SURFACE HOOKUP - TRUNKLINE

TOTAL COST OF INITIATION SYSTEM :   12.7895 $/HOLE
                                     0.0254 $/m^3
                                     0.0094 $/TONNE

COSTS SUMMARY : Study no. = 1

      Explosives:   696.1873     1.3813     0.5116
      Accessories:   12.7895     0.0254     0.0094
      Drilling:     280.0000     0.5556     0.2058
      TOTAL:        988.9768     1.9623     0.7268

      Powder Factor :            1.054   kg/m3
      Drilling Utilization :    36.000   m3/m drilled
```

gineer. No blasting model can however claim to simulate all blasting situations accurately and yield unequivocal results every time. Considerable additional research would be required to characterize rock mass in terms of its response to shock loading, especially those with pronounced inhomogeneities such as bedding planes, faults and joints. The same applies to describing a more comprehensive rock failure criteria under combined dynamic loads of compression, shear and tension. The exact role of stemming and its quantification in the modelling process would require further investigation. Our knowledge of crack propagation and branching in inhomogeneous rock mass remains largely qualitative. There is also an urgent need to develop a more accurate method of assessing fragment size distribution in very large muck piles than the photogrammetric techniques currently employed. Despite these gaps in our knowledge, we can already foresee interactive systems operating on the bench or underground consisting of drill monitors, borehole logging equipment, and explosives mix and pump trucks. The integrated system, with the help of real time blast simulations, can

deliver customized explosives product not just for each region of the pit, but even for each borehole.

6 ACKNOWLEDGEMENTS

The authors would like to thank S.H. Chung for many helpful discussions, and to their colleagues at other ICI Group companies for their contributions to this co-operative research effort.

7 REFERENCES

Chung, S.H. & Jorgenson, G.K., 'Computer Design and Field Application of Sub-seam and Multi-seam Blasts in Steeply Dipping Coal Seams'; Proc. 11th Conf. on Explosives and Blasting Tech.; San Diego, pp. 42-48; January 1985.

Cunningham, C.V.B., 'The KUZ-RAM Model for Prediction of Fragmentation from Blasting'; Proc. 1st Int. Symp. on Rock Fragmentation by Blasting; Luleå, Aug. 1983.

Demuth, R.B., Margolin, L.G., Nichols, B.D., Adams, T.F., and Smith, B.W., 'SHALE: A Computer Program for Solid Dynamics'; Los Alamos National Lab. Rept. LA-10236; May 1985.

Favreau, R.F., 'BLASPA - a Practical Blasting Optimization System'; Proc. 6th Conf. on Explosives and Blasting Tech.; Tampa, February 1980.

Harries, G., 'Theory of Blasting'; Australian Mineral Foundation Workshop on Drilling and Blasting; Adelaide, 1977.

Hjelmberg, H., 'Some Ideas on how to Improve Calculations of the Fragment Size Distribution in Bench Blasting'; Proc. 1st Int. Symp. on Rock Fragmentation by Blasting; Lueå, August 1983.

Kirby, I.J. and Lieper, G.A., 'A Small Divergent Detonation Theory for Intermolecular Explosives'; Proc. 8th Int. Symp. on Detonation; Albuquerque; July 1985.

Kirby, I.J., Harries, G.H. and Tidman J.P., 'ICI's Computer Blasting Model SABREX - Basic Principles and Capabilities'; Proc. 13th Conf. on Explosives and Blasting Tech.; pp. 184-198; Miami, February 1987.

Kuznetsov, V.M., 'The Mean Diameter of the Fragments formed by Blasting Rock';

Soviet Min. Sci., vol. 9, pp. 144-148, 1973.

Langefors, U. and Kihlström, B., 'The Modern Technique of Rock Blasting'; J. Wiley & Sons, New York, 1979.

McHugh, S., 'Computational Simulations of Dynamically Induced Fracture and Fragmentation'; Proc. 1st Int. Symp. on Rock Fragmentation by Blasting; pp. 407-418, Luleå, 1983.

Mohanty, B. and Chung, S.H., 'Developments in Blasting Physics - the Current Research Focus'; J. Mines, Metals and Fuels; pp. 477-487, November 1986.

Mohanty, B., 'Strength of Rock under high Strain Rate Loading Conditions Applicable to Blasting'; Proc. 2nd Int. Symp. on Rock Fragmentation by Blasting; Keystone, August 1987.

Tidman, J.P., 'ICRAX - A Blasting Model for Open Pit Mines'; C-I-L Tech. Memo, 1984.

Computer Applications in the Mineral Industry, Fytas, Collins & Singhal (eds)
© 1988 Balkema, Rotterdam. ISBN 90 6191 760 3

Multiple resource constrained underground mine scheduling

Donald E.Scheck & Ilango Sankaralingam
Ohio University, Athens, USA

Prashanta K.Chatterjee
Ohio Automation, Athens, USA

ABSTRACT: Network theory and heuristics seem to be the most promising techniques for solving the production scheduling problem in underground mines. A heuristic model has been developed based upon scheduling rules commonly used by mining engineers. It provides the engineer with a powerful tool for evaluating alternate schedules. The only requirement for the program is that the section definitions are available in a machine readable format.

INTRODUCTION

The productivity of coal mining operations has increased due to higher capacity mining machines and more effective utilization of equipment within a section. The benefits from the improved technology and organization may not be fully achieved if production planning and scheduling introduce unnecessary delays and idle time. Computer-aided scheduling can help management achieve the highest possible productivity that is consistant with the available resources and contractual requirements.

Scheduling of coal mining operations can be classified as a multiple resource constrained project scheduling problem in which the objective is to maximize the present worth of the total mining operation. Contract specifications, delivery requirements and the availability of equipment are constraints on the schedule decisions.

The number of possible schedules for any given combination of reserves, contract specification and equipment is so large that finding the best schedule with manual methods is highly unlikely. Evaluating an alternative or scenario is so time consuming that only a few can be considered. On the other hand, mine scheduling is too complex to model the problem by using a classical optimizing technique such as linear programming. The only possible approach is to use some combination of theoretical and heuristic method that will assure a good if not optimal schedule.

The difficulty of finding an optimal schedule is not so much to meet the production requirement in the near future but to insure that production can be sustained at the required level throughout the life of the mine. The variability of the material and the difficulty of estimating the variability in advance of mining adds to the complexity of the problem. The initial schedule may be a near optimal schedule for the conditions that are known at the time of scheduling but as more information becomes available, the schedule may need to be revised. The dynamic nature of the mine scheduling problem requires that the schedule has inherent flexibility. The development work should be timed so that the mine planner has many practical options for deviating from the original schedule and re-optimizing future plans.

BACKGROUND

Many techniques have been used to solve production scheduling prob-

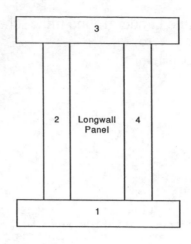

Figure 1 : Precedence constraints for longwall mining.

lems for mining operations. The techniques have included theoretical operations research (OR) methods and heuristic procedures that are based on experience.

Numerous applications of computers and OR techniques to open-pit mining appear in the literature (Weiss, 1979), but the applications in underground mining are few. West-Hansen, Sarin and Topuz (1986) have suggested a sequencing theory approach for the problem of long-term production planning in an underground coal mine. They have reduced the scheduling problem to the general problem of non-preemptive scheduling of precedence constrained jobs on parallel processors. Bandhopadhyay and Sundararajan (1987) have simulated a longwall-development extraction network.

Most of the existing techniques require considerable manual input before a schedule can be generated. They all assume that a precedence network, that is, the technological sequences in which the sections can be mined is available. None of the published techniques have a automatic or computer-aided method for generating this information. But developing the precedence network itself is a major task that is very

time consuming if done by manual methods.

DESCRIPTION OF PROBLEM

Given the mine layout, the problem is to schedule the mining so as to achieve the following objectives.

(1) minimize the deviations from the desired production levels over a period of time.

(2) minimize the time to open up longwall panels which ensures the continuous use of the equipment and consequently reducing the overall production cost.

(3) maximize the total production that can be obtained from the mine and increase opportunities to open up more mining sections in the future.

The objectives are listed in their hierarchical order of preference. Although all the objectives are important, consistent production is essential to meet contractual obligations and thus has the highest priority. The use of stockpiles to meet future demand must be limited due to the time value of money invested in the product, the cost of stockpiling facilities and the risk of spontaneous combustion of some coals.

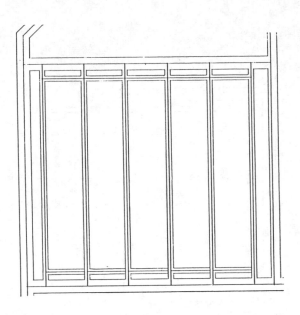

Figure 2 : Sample mine layout

Mining sections are divided into three categories which depend upon the mining methods; longwall, room-and-pillar advance and room-and-pillar retreat mining. Advance room-and-pillar mining creates entries and pillars and all or part of the pillars are removed during the retreat phase. The normal procedure is to advance and retreat mine in opposite directions.

In this analysis the following are assumed :

1) the mining rates and time to mine each section depends on the geologic and environmental conditions and on the mining method.

2) the procedure assumes that all support facilities such as ventilation and power are already in place for the sections scheduled to be mined.

3) mining machines can work at the same time, independently of each other.

4) a continuous miner can mine in only one room and pillar section at a time, and a section can be mined by atmost two such machines at the same time.

The sections are scheduled subject to constraints that are stored in the form of a precedence network that is constructed from the mine layout. Mining of each room-and-pillar section, except the first section, should be preceded by mining another room-and-pillar section. The mining of each longwall section must be preceded by mining the four development sections which surround the longwall panel. Figure 1 is an example in which the development sections numbered from 1 to 4 must be completed before starting to mine the longwall panel. Advance mining of each room-and-pillar area must be done before retreat mining.

Very real constraints such as the availability of ventilation, power and coal quality have been assumed out of the problem. Since some of these restrictions have not been considered in the scheduling process the final schedule may not be practical. Such problems can be overcome by manually adjusting the schedule or by changing the precedence network and rerunning the scheduling problem.

This study follows present practice, that is, the mine layout is an input to the scheduling process. Since the layout must be in a

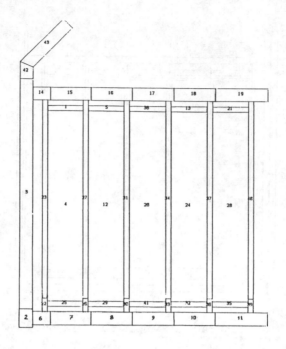

Figure 3 : Layout with section definitions

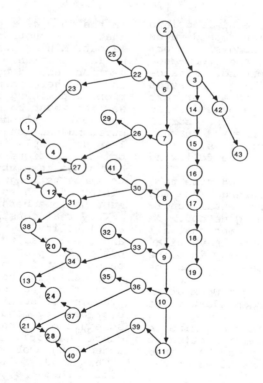

Figure 4 : Network representation of sample layout

52

machine readable form, the layout was generated by ICAMPS (Intergrated Computer Aided Mine Planning System, Chatterjee, Scheck and Sridhar (1986)).

A detailed layout is not essential for long-term scheduling purposes. The mining sections have been assumed to have three or four sides due to ease of representation and computation effort. Only the coordinates of the corners of each main roadway, longwall panel, development section and room-and-pillar panel are necessary for the network determination. The interior details -entry, pillar and crosscut dimensions can be based upon average design parameters. A sample layout and the section definitions are shown in Figures 2 and 3. The layout can be represented by a directed network, in which the nodes represent sections and the directed arcs represent the precedence constraints. Figure 4 is the network for the sample layout in Figure 2.

For the purpose of simplicity only underground mines that use longwall mining were considered. This limitation is valid because longwall mining is extremely capital intensive. Downtime due to scheduling conflicts is much more costly in longwall mining than in room-and-pillar mining. In addition, room-and-pillar equipment is much more flexible and work stoppages due to poor scheduling are less likely to occur.

PROPOSED METHODOLOGY

A heuristic technique is applied to a network that is developed from the mine plan. Unlike a typical project management network, the arcs in a mine network can be bidirectional.

Since the user of the proposed model would be an experienced mine planner, the model is based upon the scheduling rules that mining engineers commonly use for manual scheduling. This approach makes the system more acceptable to the user since the logic is more easily understood than procedures based on abstract operations research techniques.

The cost of producing a schedule is directly related to the computation time and the time to prepare the input. The use of heuristic rules lowers computation time at some potential loss in schedule optimality. ICAMPS overcomes the input difficulties because it provides the design-data,i.e., roadway layouts and section designs in a machine readable format. Since this information must be generated for other planning purposes, the input data is available at no additional cost.

The heuristic approach treats mine scheduling as a resource allocation problem rather than resource balancing. The solution routine consists of scheduling rules which determine when to start mining in each section and which resources to assign to each working section.

The scheduling rules employed in the model can be summarized as follows:

1) A section is scheduled provided that its technological constraints have been met, that is, all predecessor sections are already completed before the section is scheduled to start.

(2) The sequence in which the longwall panels will be mined determines the sequence of mining in the development sections.

(3) If two or more development sections are ready to be mined, the development sections that are predecessors for the next longwall panel to be mined will be given preference over the other sections.

(4) Machine locations do not play a significant role in the assignment of equipment to a section since the move time for the machines are insignificant when compared to the mining time.

5) As many sections are scheduled as available resources permit. The number of sections scheduled is only limited by the constraint that development work is undertaken for only two panels ahead of a longwall panel that is currently being mined. Even if machines are available, the sections are not mined due to the large amount of investment involved in the development work. This condition would justify the objective of maximizing the present worth of the mine.

SCHEDULING PROCEDURE

The algorithm can be summarized in the following steps:

1) Sections are scheduled, starting with the first one in the precedence network.

2) The most critical sections are the predecessors for the next longwall panel and have the highest probability of being scheduled next.

3) In case of ties, the sections are ordered according to the second criteria - minimizing deviations from desired production levels.

4) After all planned longwall panels have been scheduled the remaining sections are scheduled based on the third criteria - maximizing future production from the unscheduled areas.

SCHEDULE EVALUATION

The schedule can be evaluated in terms of its effects upon the related costs. The relevant costs depend upon the type of project, but in general they include :

(1) Resource costs (e.g. cost of machinery and unused resources);

(2) Overhead costs (e.g. auxillary equipment and ventilation);

(3) penalty costs associated with deviations from prescribed production levels.

After the system generates a schedule, the following information is provided :

1. machine utilization data
2. deviation from contract delivering requirements.

Achieving an optimum schedule using this model is an iterative process. Based on the above information the user can change the number of machines and re-run the scheduling procedure. Since the cost of not meeting production requirements may be extremely high, the deviation from planned production summary report will be used to establish the number of machines. Insufficient production would necessitate the use of more machines. Excess production would require a reduction in the number of machines. The user can also alter the machine capacity and the number of working shifts per time period to improve the schedule.

PROGRAM DESCRIPTION

A block diagram of the scheduling program is shown in Figure 5. The section definition module allows the user to define the room-and-pillar and longwall sections. The network generation module converts the mineable section definitions into a network format. This network is used to determine predecessors and followers for each section. The number of shifts required to mine each section is then calculated using the coal seam quality data base and machine characteristics. The user specifies the production required from the mine and the permissible deviation from the production goal. Based on the scheduling rules stated earlier, the mining units are assigned to available sections and the schedule is generated.

The user must have a working knowledge of ICAMPS to execute the scheduling program because ICAMPS generates the layout and defines the sections in a machine readable form. The program generates the schedule in the form of a timing map and also as a data file.

The scheduling program has been designed to be executed sequentially in separate modules. The smaller modules within the program can be executed independently so that the user can run the program in stages. This approach does not tie up the computer for extended periods of time. Considerable attention has been given to user-friendliness and minimization of user input.

There are utility options in the program such as drawing or plotting the layout and timing map and resetting the mine boundary. These options do not appear in the block diagram.

RESULTS

A computer program has been developed and implemented on a MICRO-VAX-II. The coding was done in FORTRAN 77 with graphic and hardcopy interfaces. The summary outputs generated by the program for machine utilizations are shown in Tables 1 and 2. The tables contain the number of machines used, the total number of days worked, the

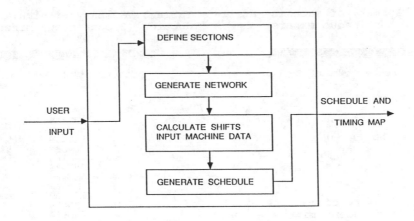

Figure 5 : Block diagram of scheduling program

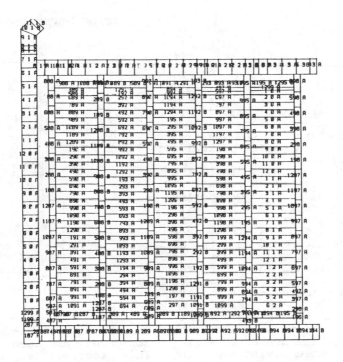

Figure 6 : Timing map (2 continuous miners and one longwall shear)

Table 1. Summary of machine utilization (2 continuous miners)

Machine #	Total days Worked	Setup time	Idle Time
1 (R/P)	2454	162	1328
2 (R/P)	1189	111	2356
1 (L/W)	3038	147	540

SUMMARY OF MACHINE DATA

Table 2. Summary of machine utilization (3 continuous miners)

SUMMARY OF MACHINE DATA

Machine #	Total days Worked	Setup time	Idle Time
1 (R/P)	1787	143	1958
2 (R/P)	1240	74	2271
3 (R/P)	616	50	1108
1 (L/W)	3038	145	484

Table 3. Summary of simulated production (2 continuous miners)

SUMMARY OF PRODUCTION DATA

Period No.	Actual Production	Required Production
1	285620.000	200000.000
2	357908.000	200000.000
3	888103.000	500000.000
4	871229.000	500000.000
5	697608.000	500000.000
6	1016639.000	500000.000
7	804056.000	500000.000
8	815131.000	500000.000
9	915291.000	500000.000
10	755198.000	500000.000
11	781741.000	500000.000
12	818134.000	500000.000
13	649720.000	500000.000
14	927810.000	500000.000

Table 4. Summary of simulated production (3 continuous miners)

SUMMARY OF PRODUCTION DATA

Period No.	Actual Production	Required Production
1	390802.000	200000.000
2	401780.000	200000.000
3	1027365.000	500000.000
4	788476.000	500000.000
5	806469.000	500000.000
6	910142.000	500000.000
7	750806.000	500000.000
8	905128.000	500000.000
9	837310.000	500000.000
10	682315.000	500000.000
11	896946.000	500000.000
12	781977.000	500000.000
13	693356.000	500000.000
14	927810.000	500000.000

setup times and the number of days the unit was idle during the scheduling period. Tables 3 and 4 are summaries of the simulated production from the mine and the desired production for a period of twelve months. These results will help the user to evaluate the schedule and to suggest alternative scenarios. Table 3 contains the production results for a schedule that utilizes two continuous miners and one longwall shear. The required production levels can not be obtained by using one continuous miner and one longwall shear. Increasing the number of continuous miners to three did not increase the production appreciably. As might be expected, the percentage utilization of the machines decrease with an increase in the number of mining machines. A sample timing map using two continuous miners and one longwall shear is shown in Figure 6. The blocks shown in the figure indicate the simulated mining on a monthly basis. The type of mining machine assigned to a section is designated within each block. The letter A represents the first longwall or continuous miner and B represents the second continuous miner.

CONCLUSIONS

The ability to generate long-term schedules has been demonstrated by this program. The computer generated schedule will be free of conflicts and provides the mining engineer with a tool to experiment with different scenarios. This program depends upon the section definitions being available in a machine readable format and thus the mine planner must layout the mine before executing the program. The time taken to generate the schedule, once all necessary input is available, is insignificant compared to the manual method.

The complexity of the scheduling problem suggests the use of an expert system to generate better schedules. Constraints on coal quality, power and ventilation requirements could also be incorporated into the scheduling process.

REFERENCES

Bandopadhyay, S. and Sunderarajan, A., July 1987, "Simulation of a Longwall Development Extraction Network ", Canadian Institute of Mining Bulletin, pp. 62-70.

Chatterjee, P.K., Scheck, D.E., Sridhar, C.K., March, 1986, "Optimum Scheduling of Room and Pillar Operations," Pre-Print 115th AIME Annual Meeting.

Weiss, A., Computer Methods for the 80's - In the Mineral Industry, 1979, New York, Society of Mining Engineers.

West-Hansen, J., Sarin, S.C., and Topuz, E., 1986, "Long-term Production Scheduling in Underground Coal Mines - An Application of Sequencing Theory," Proceedings, 19 th APCOM, The Pennsylvania State University, PA.

Computer Applications in the Mineral Industry, Fytas, Collins & Singhal (eds)
© 1988 Balkema, Rotterdam. ISBN 90 6191 760 3

Use of Geostokos and Planner for daily production

Ginette Reid, Jean Jacques Bouillon & Jean Luc Chouinard
Les Mines Selbaie, Joutel, Quebec, Canada

ABSTRACT: At Les Mines Selbaie, located in northwetern Quebec, the mining of the complex A-1 zone's orebody is done using geostatistics. Two main software packages treating the assay values coming from the samples collected from the blastholes are Geostokos and a local kriging program. On the engineering side, the scheduling aspect of the pit is taken care of by another software package called the Planner. The time saved resulting from the use of these computer programs gives more time to examine other alternatives which will result in a maximization of the revenues.

At Les Mines Selbaie, the daily ore production extracted from the open pit is based on information produced by Geostokos and a local kriging program.

In order to separate the blasted muck in the 2 categories: ore or waste, we sample the cuttings produced by the production holes (200mm in diameter). From the cone of cuttings around a hole, three (3) samples of about 4 kg are taken, as close as possible to the crest of the cone. A sampling tube is used to collect the sample. Three portions taken equidistantly around the cone complete a sample.

Since the mineralization in the open pit is polymetallic, assay results for copper, zinc, lead, silver and gold are incorporated in the data base. The arithmetic averages for the metal grades are calculated for each set of three samples. These averages will be used as grades for each production hole and constitute the data base for the kriging. The production holes are located in space by digitizing or by entering the coordinates obtained by the surveyors.

Two steps are still required prior to the kriging of an individual blasting area. First, with the use of Minitab, a statistical investigation is performed on all the assay results of the production holes that are available at that time. With histograms for each metal, the distribution of available grade values is verified. Generally, the distribution is log normal. At the same time, we obtain the average, the standard deviation and the variance of that group of values. Finally, with plots, the correlation between the logarithmic values and the NSCORE is verified and should be normal.

After, with "Geostokos", the kriging parameters are obtained. With the values still expressed logarithmically, we create a set of semivariograms controled by predetermined variations of the strike and dip. We are presently using the spherical model. We get one sill, with several ranges and nugget effects. By plotting ranges versus strike, we established the anisotropism of grade distributions. The nugget effect is obtained by averaging values. Then, for each type of mineralization, a set of kriging parameters is obtained with the nugget effect and sill put back in real values.

When the kriging parameters are available, an individual blasting area is kriged for each metal except for gold, taking also into account the mineralization type codes. "Geostokos" takes care of special mineralization trends and of georegressions.

The production holes are distributed on a 5.2 metre grid and the local kriging is done on 2.75 X 2.75 metre blocks, bench height is now 8 metres.

It is possible with the "local kriging" program to produce block kriged drawings at 1:250 scale. For each metal, these blocks contain the kriged grade with the variance. From these kriged grades and with the mill recovery curves of the different metals, we calculate a dollar value for the material corresponding to each block. The cut off dollar value is as a rule $12.00.

The dollar value in combination with all other geological information obtained by the bedrock and face mapping as well as the diamond drilling are used to trace the ore/waste boundaries.

When the individual types of muck are outlined, the different polygons are digitized and an option of the local kriging program is used to calculate the tonnage and average grades of the different rock volumes of the blast.

For each blast, the kriging parameters are verified by cross-validation since with the increase of the number of production holes assayed, kriging parameters for each one type could vary slightly. All the semivariograms recreated to obtain new kriging parameters to be used for the next blast, should the cross-validations not meet our standards.

II PRODUCTION SCHEDULES

1. DEFINING THE NEED

Needless to say the optimization of a mining sequence for an orebody constitutes the most important criterion for the economical performance of a mining operation. This means a maximization of the cash flow and of the rate of return on the investment in a project. The generation of a long term production schedule which would be practical and optimized is a complex task in which a series of variables and restraints have to be taken into account. In addition, one must include the objectives and the operational policy set by the management.

To achieve these goals, the planning engineer must have a tool which integrates the latest developments in the computer field to react quickly to any change. Such a software would eliminate repetitive and time-consuming calculations while allowing the examination of a great number of options and their effect on the economy of the project. These programs must be very interactive and flexible in order to take into account all mining conditions.

On top of that, the optimization of the schedule must be left completely in the hands of the user. The programs must be only seen as tools which will help the user to obtain more quickly his objectives, avoiding errors and obtaining more accuracy.

Such a system becomes indispensable in the case of an orebody similar to Selbaie's A-1 zone. In addition to the restraints previously noted, the planning engineer must deal with four metals as well as six different materials. An additional metal recovery of 1% could mean over $6 million more in profits for Selbaie. The data from the mill output should also be included and updated on a regular basis. All this would be difficult to include in a completely manual process.

2. SOFTWARE DESCRIPTION

The description of such a system could be subdivided into three levels, such as:

    i)   Input data
   ii)   evaluation
  iii)   results

2.1 Input data

Five types of information are the basis of the requested short term and long term scheduling programs. First of all, a condensed mining model has to be built from the basic model (MEDS). For each block, it will contain the Cu, Zn, Ag, Au grades, the type of material, its density's average value at the mine, the percentage of the block left to be mined and the percentage of rock in the block.

In order to have the percentage of the block left to be mined continually updated, a procedure was set up. After each monthly survey, an average of the 25 elevations of the survey grid calculated by Z-Map is made. The elevation of the mining block is then calculated and an update of the percentage of the material remaining in the block is done. All the software is menu driven. The only information required are the names of the files, the grid cell size and the coordinates of the selected area. The final result of the procedure is a bench plan file which contains the necessary information required by the Planner.

The data of the pit's contours of the mid-bench also have to be retrieved from the VBM or manually digitized. That information will go in a file and will be transferred with the bench plan file to the working area on our MicroVax.

2.2 Evaluation

Before proceeding to the evaluation of a cut the user must fix the objective for the period under examination. This objective is made up of the ore tonnage or metal content and the stripping ratio from which the quantity of low grade, overburden or waste which could be extracted can be calculated.

The delimitation of a first cut takes place after the user has put the desired plan on the digitizer and has performed the required calibration. The digitizer's cursor will generally be used to put the contour of a cut in memory, but the user could also make use of the cursor of a graphic terminal to do this.

After this verification, the evaluation of materials contained in the cut will take place. The results, on the second screen connected to the graphic terminal, summarize the Cu, Zn, Ag and Au average grade, and the value of the ore, low grade and waste as well as the tonnage of the overburden and of the 3 previous categories. The up-to-date results of the cuts for the current period will also be displayed for each class of materials.

The user will make his decision after having consulted the results. If he decides to reject the cut and consider another alternative, he could simply go back and digitize another cut. In a case where he realizes that it would be impossible to obtain his objective after some tries, he could eliminate the preceeding cuts up to a point that he judges to be satisfactory in which case another avenue would be explored.

The graphic terminal will permanently display, at the desired scale, the area of the bench being examined. A colour code permits the distinction of the different classes of materials. In addition, the outline of the cut appears when it has been declared satisfactory. A colour change of the blocks inside the cut will be made afterwards in order to represent the mining of the blocks. The bench data file will also be automatically updated.

With the intention of making the job easier for the user, the system will offer the possibility of tracing, on top of the blocks, the contours of the topography related to the bench examined. It also allows a display of the limits of the excavation on the upper bench or any desired outline. The possibility of an enlargement (zoom) of a particular area is offered. As many as four selected screens can be kept in memory. This saves a lot of time when the user has to work on different benches for the schedule. The user also has the option of obtaining on the screen the contents of a particular block by placing the cursor on the desired block.

2.3 Results

Once the schedule is finished, the user will obtain automatically the reports and plans relative to the schedule that he has just completed.

The reports present a résumé of the results obtained along the way. For each bench, each period and each cut, the tonnage of overburden, ore, low grade and waste will be given as well as the grades and the mineral value related to each class of materials, the total mined to date and the remaining reserves. This information is found in an ASCII file and could be sent directly to the printer.

In addition to the original data, the output plans show the limit of each cut with its identification. Those limits appear on the bench plan of the mining model.

3. FUTURE ENCHANCEMENTS

This software was developed as per Selbaie's specifications by Mintec, Tucson, Az.

This specialized firm is owner of the pit design package called Meds. The Planner is new for them and they are continuously adding more features.

In the near future, the major addition will be the inclusion of permanent or temporary ramps in the volume calculation. Presently, the user has to modify his results when a cut includes a ramp. A drainage slope will also be the subject of work later on. This possibility is interesting for larger pits. A flat slope may lead to some grade evaluation discrepencies, especially if the body is flat laying.

At Selbaie the next improvement will be a connection with the local blast hole model built for each blast by Geotokos. The krigged values obtained by Geostokos are more precise and should be used when available. A cash flow analysis is another step which will follow. This option will allow the user to quickly analyse and select the best schedules that he has just performed.

CONCLUSION

Pit design packages have been available for a few years but concerning the schedule, no complete software existed before the Planner. It took time to develop such a tool mainly because of the lack of a graphic library like GKS and also

because of the quality of the hardware.

Both the Planner and a procedure using a financial model to update the value of the ore each month in the A-1 zone mining model will allow Selbaie to mine the pit, closely following the metal market and the operating conditions in order to extract the ore with the maximum profitability.

Computer Applications in the Mineral Industry, Fytas, Collins & Singhal (eds)
© 1988 Balkema, Rotterdam. ISBN 90 6191 760 3

Short range scheduling for open pit mining

Phil Edmiston
Minemap Ltd, Vancouver, BC, Canada

ABSTRACT:
An overview of the computer techniques utilized in short range planning at a
number of operating mines where practical and innovative computer solutions must
be developed in close association with operating personnel at the mine site.

1. THE SCOPE OF SHORT RANGE PLANNING

Engineers have typically broken down the planning process into three broad categories:

i) Short Range Planning where the day to day planning process is involved. This time frame can typically range from one day to one or two months depending on the type of operation and the tonnage of ore extracted.

ii) Medium Term Planning which may extend from one month up to two years. It is here that the conceptual pit designs are turned into detailed designs to be given to the short range planners.

iii) Long term Planning which assesses the overall profitability of a proposed mining operation. Here pits are designed with sufficient detail to provide the necessary information as to whether a deposit is of value to consider a more detailed analysis. The time frame is extended to life-of-mine.

In any mining operation there is considerable overlap in the above categories. One merges into the next rather than any clear definition of what is universally acceptable. Each mining department nominates a particular time frame to be given to each planning process and these frames of reference may change depending upon the direction of the company and other economic or political influences. This paper focuses on the short range planning process and the day-to-day needs of the geoligists, surveyors, blasting engineers, production engineers and planning engineers. The problems and solutions discussed herein have been taken from a variety of mining operations with differing deposits. Each mine has its unique problems where it has been necessary to have flexible computer software tools in order to handle the variety of problems. The range of topics that the short range engineer must address includes:

a) Grade Control
b) Surveying
c) Pit Design
d) Computerized Model Update
e) Volume Calculations and Reconciliation
f) Blending
g) Dump Design and Reclamation
h) Equipment Selection and Performance Monitoring
i) Haul Route Optimization
j) Scheduling

This list is by no means exhaustive but it represents the wide range of activity which is encompassed by the short range category.

2. COMPUTER TECHNIQUES

Computerized mine planning systems have been evolving rapidly over the past ten years. Software engineers have been improving the computer/user interface with great advances in the graphical representation of data and highly interactive programs. Many mining engineers know how they want to see a particular mining situation computerized and it is up to the software designers to develop a solution that meets these needs. Each of the program subsystems should be integrated into a total system so as to provide the greatest benefit to the variety of different departments that will want to access the data.

At any gold mining operation grade control is of extreme importance. In order to provide an efficient grade control system it is necessary to incorporate a variety of different approaches that include geology, surveying, pit design, and reconciliation. The current expansion of gold mining operations produces a situation where many deposits are being developed with minimal drilling information. Grade control is based on available drillhole data plus additional channel samples that have been taken from the pit during the previous shift.

In order to handle this lack of drillhole information situation from a computer software orientation, a three-dimensional graphics modeling program was developed that allowed the engineer to combine survey data, drilling data, pit designs and the latest in-pit samples, and thus, produce a graphical model of the geology.

"Pods" of ore are introduced to the model in the plan or section. With the basic unit of a graphical model being a polygon, the geologist can analyze all data and correlate between sections, and digitize the different polygons of grade on each of the benches or each of the sections. At any time the user can flip between section and plan view in order to build up a wire frame model of the deposit. All input data is taken into consideration. The user can then select a particular polygon (ore with a specific grade) or, he may choose to digitize the polygon and let the computer break the polygon into discreet regions of grade based on any intersecting data within the required viewing window. Thus all "pods" of ore are generated for the deposit. The user selects whether he wishes to link between the sections and create a wire frame model or simply specify polygons of ore in plan or section which he will interpolate halfway to the next section or bench.

A block model can also be created from the graphical model with the polygons of grade being inserted into the model on a bench by bench basis. The geologist and engineer have a variety of choices to model the deposit. The graphical polygon model may be maintained for short range planning needs and its data may be used to update the block model for medium to long term planning models.

The grade control process is based on the gathering of drillhole and channel sample data. A channel sample is taken in the pit on the previous shift. The sample is analyzed in the lab and the resulting data entered directly into the lab computer from where it is transmitted to the planning computer.
Sample data (above) is entered into the graphical model and merged with the drillhole data. The geologist and engineers then decide on the "pods" of ore to be mined by examining the data and selecting the required data points and selecting the required data points on the computer graphics screen which outlines the ore pod. The coordinate positions are printed out beside the digitized outline. The plan of the "pods" with their respective grades are given to the surveyors who survey and mark the pod outlines on the pit floor. The "pods" are mined and reconciled against expected results. In this way the grade function has the latest information to provide an accurate survey for the extraction process and the delineation of ore and waste.

The polygons introduced above have grade as one of their values. These graphical elements can have any number of other attributes such as rock type, cost of mining, recovery, etc. It is then possible to optimize the extraction process based on a number of specified parameters.

Three dimensional and four dimensional pit optimization techniques are available which the planning engineer can employ to optimize the pit design. It is possible that the overall pit disign will be changed based upon the actual results of the mining operation versus the predicted values.

The graphical model can be incorporated into both a block model or a gridded vein model. One is able to calculate the tonnages and grades from within the graphical model, or, overlay it on a block or vein model and hence achieve the results from those models. This graphical approach allows a three-dimensional view of the deposit and the planning process. Any pit design, tunnel, slope, or text can be overlayed against the graphical model and the intersection of the two calculated. A particular bench in an open pit can be viewed with the corresponding optimal pit design within a window either side of the bench.

Some mining operations have detailed drilling data and do not find it necessary to go to a graphical pod model of the deposit. A block model may have been developed over a number of years and the engineers and geologists will have confidence in its use.

Survey reconciliation at the end of each mining period provides important feedback to the applicability of the model and its predictive value. Most current EDM surveying equipment have the capacity to reduce the data down to northing, easting and relative level. The data collector is connected to the computer with information being directly transferred into the graphical model. Surveyors manage their data collection so that the data is arranged into strings that define crest, toe, roads, berms, and so forth. This data is then available immediately after the transfer takes place.

The current pit position can be calculated from the new data and a tonnage reconciliation performed against the last survey. It is possible to compare the existing topography against the previous survey as well as comparing planned dumps against the current surface. Stockpiles are frequently surveyed and these are entered into a graphical model to calculate their volume against the known ground surface.

A graphical model has the ability to be incorporated into the scheduling process. The topography has basic haulroad information and a particular path can be digitized for a load/haul cycle. The haul gradients are calculated from the graphical model and together with the rim pull, retarder, and resistance data, a haul profile can be generated for a particular set of equipment. It is possible to incorporate this data into a real time model to assist in the automated scheduling of dumping and equipment movement.

3. CONCLUSION

A graphical model was developed in close cooperation with mine site personnel to integrate the short range grade control and surveying needs into a computerized system. This modeling approach caters to a wide variety of different situations and is very beneficial to the short, medium, and long range planners in their representation of the geology and optimiztion of the entire planning and scheduling process.

(A series of slides will compliment this written text at time of presentation.)

3 Mine evaluation

Computer Applications in the Mineral Industry, Fytas, Collins & Singhal (eds)
© 1988 Balkema, Rotterdam. ISBN 90 6191 760 3

Economic evaluation programs at Falconbridge Ltd – Sudbury operations

Y.L.Su & M.Musson
Falconbridge Ltd, Sudbury, Ontario, Canada

ABSTRACT: At Sudbury Operations major emphasis is being placed on economic analysis and determination of stope, mine, project and Integrated Nickel Operations cashflows. It is essential that programs be available for Mine Engineers to use to produce the necessary information. The programs must be valid in their logic, consistent in their terminology, clearly documented for user application and maintained up-to-date. It is vital that consistent input data, such as metal prices, downstream treatment costs, exchange rates and inflation factors, are used by all to ensure that results are valid. This paper describes three programs, namely PROECO, PEP and SUDSIM that form the base for Sudbury evaluations.

1. INTRODUCTION

In the free market economy, mines are developed and ore is produced with the goal of providing profit to the company. Mine Engineers play a major role in achieving such goals. At Sudbury Operations, where a number of mines contribute to overall production and where various constraints exist, a Mine Engineer must ensure that not only is his mine making its maximum contribution, but also that this contribution is the optimum for the district as a whole.

Major emphasis is being placed on economic analysis and determination of stope, mine, project and Integrated Nickel Operations cashflows. It is essential that programs be available for the Mine Engineers to use to produce the necessary information. The programs must be valid in their logic, consistent in their terminology, clearly documented for user application and maintained up-to-date. It is vital that consistent input data, such as metal prices, downstream treatment costs, exchange rates and inflation factors are used by all to ensure that results are valid.

Economic evaluation programs are available from many outside sources. These programs are designed for straightforward general-ized studies. For more comprehensive applications, which must take into account a variety of detail and consider a company's unique requirements, programs are likely to be developed by in-house computer support groups or by consultants under contract.

At Sudbury Operations, three in-house programs (Farrish, 1972, Su, 1977, Moruzi, 1985) have been in use for several years. These programs (code names: PROECO, PEP and SUDSIM) were designed for use in specific application areas. Recently, the programs have undergone major modification in order to adapt to current economic parameters and to allow user-friendly application with micro-computers. In addition to technical modifications, organizational changes included a review team to ensure that standardized programs satisfied user requirements and assignment of specific individuals to provide input data that is correct and consistent for use at all minesites.

2. SUDBURY OPERATIONS

Falconbridge Ltd. is a major base metal producer. In 1986, 38,200 tonnes of electrolytic nickel were produced from the Integrated Nickel Operations (INO). The INO consists of Sudbury Operations, the refinery in Kristiansand, Norway and the appropriate Head Office personnel which

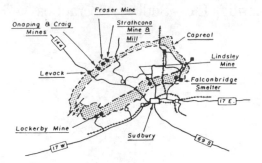

Fig. 1 Sudbury basin

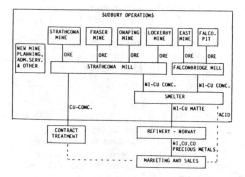

Fig. 2 - The integrated nickel operations

include the marketing and sales group. Sudbury Operations, located in the Sudbury basin (Figure 1), currently operates one smelter, two mills, five underground mines and one open pit.

These operations extract sulphide ores, produce Ni-Cu matte for processing at the refinery and make Cu concentrate for contract treatment. The accountable metals produced from Sudbury Operations include nickel, copper, cobalt, gold, silver and the platinum group of metals, as well as sulphuric acid.

Figure 2 shows the major components within Sudbury Operations and their overall relationship to the INO. The saleable products produced from each component are listed in the figure.

3. THE MINE CASHFLOW MODEL

The pretax cashflow is the mathematical sum of the revenue received from marketing and sales less the costs incurred at unit operations within the INO.

Cashflow derived in this manner is not considered as profit or loss to the corporation but rather as a 'corporate contribution'. The financial evaluation models are formulated within such a framework.

Since revenue is based on actual selling prices, it is necessary to incorporate several sub-models into the cashflow model. These sub-models, dealing with milling, smelting and refining, generate tonnage and recovery values at various stages of mineral processing. Final products from each processing stage are then combined with the corresponding unit costs and ultimately selling prices in order to arrive at cashflows.

The cashflow model is graphically shown in Figure 3.

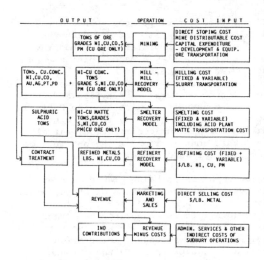

Fig. 3 - Mines cashflow model

3.1 Mill recovery sub-model

The two concentrators produce a total of three products from the mill feed. The useful product is Ni-Cu concentrate; the waste streams are pyrrhotite tailings and final tailings. Grades and tonnages of Ni-Cu concentrate are calculated from mass balance, correlation and control equations, using the following nonlinear relationship:

70

		Three Products		
	Feed	Po	Ni-Cu	Final
	Ore	Tails	Conc.	Tails

Ore: $\quad t_f = t_p + t_c + t_t$

Nickel: $\quad n_f t_f = n_t t_p + n_c t_c + n_t t_t$

Copper: $\quad c_f t_f = c_p t_p + c_c t_c + c_t t_t$

Sulphur: $\quad s_f t_f = s_p t_p + s_c t_c + s_t t_t$

Correlation: $\quad 0 = n_p - b n_c + a$

Control: $\quad 0 = c_c + n_c - d$

't' stands for tonnages and 'n,c and s' stand for grades of nickel, copper and sulphur, respectively. The subscripts 'f,p,c and t' stand for feed, Po tailings, Ni-Cu concentrate and final tailings, respectively.

The mill recovery model involves solving six unknowns (n_p, n_c, c_c, t_c, t_p and t_t) from input variables (n_f, c_f, s_f, and t_f) and constants which are based on historical data.

Recovery of precious metals is less dependent upon ore characteristics, therefore historical recovery values are assigned to these metals.

3.2 Smelter and refinery recovery model

The smelter and the refinery use various models to predict and to control metal recovery. Over the years, the actual recoveries have remained relatively constant. These constants, varying for different metals, are used in the programs at this time.

3.3 Contract treatment of Cu concentrate

A set of complicated contract formulae, for example "concentrate Au grade, oz/ton, minus 10% or 0.02 oz/ton, whichever is greater", are incorporated to calculate the final metal values paid by the contractor who smelts and refines Cu concentrate produced by the mill.

Cost inputs to the model include the following:

3.4 Mining cost

This includes direct stoping costs and mine distributable costs for mining ore (in $/ton). The distributable costs are those common to all stopes such as crushing, conveying, hoisting, pumping and shaft operation. Capital expenditures consist of operating development, permanent development and mine, plant and equipment. The cost of ore transportation from mine to mill is considered a mining cost.

3.5 Treatment and other costs

These include the following: milling($/ton ore), slurry transportation ($/ton conc.), smelting ($/ton conc.), matte transportation ($/ton matte), refining ($/lb. of metals, $/oz. of precious metals), direct selling ($/lb. of metals) and administrative services. Administrative services costs are based on $/$ of expenditure excluding mines plant and equipment capital.

4. GENERAL PROGRAMS CONSIDERATION

The cashflow model is the basis for decisions on investment in mine development and plant and equipment when the time value of money must be considered. The key information being sought includes: total cashflow, net present value, ROI, the cost of producing a pound of nickel and production statistics.

Annual cashflow is the summation of revenue and cost for any given year within the duration of the project. This is then summed up into total project cashflow. Metal prices are specified in US currency while costs and cashflows are in Canadian dollars. Therefore, the exchange rate is introduced as a variable in the programs.

Net present value (NPV) is the discounted value of the future cashflow which involves a simple calculation utilizing a selected discount rate. Guidelines for rates are set by corporate policy but rate is considered a variable in the program.

ROI is the discount rate which yields a zero NPV. This is obtained by iteration, utilizing the Newton-Raphson approximation. In some cases, more than one ROI exists for a given cashflow stream. The first positive value closest to zero is considered the genuine solution.

Cost per pound of nickel is a key parameter during current times of depressed nickel prices. The program obtains these values by deducting by-product values (copper, cobalt, precious metals and acid) from the total operating cost and then dividing this amount by the total pounds of nickel produced.

In addition, equivalent nickel grade, breakeven stoping cost, sensitivity and risk are also considered valuable outputs for users. The programs provide a limited number of options for such determinations.

Three programs have been developed and standardized for application at the minesites. In order to achieve valid results, input data dealing with mining, treatment and overhead costs, metal prices and exchange rates must be accurate and consistent. The approach taken to ensure the consistency and correctness of the input data is as follows:

- the data for each mine is provided by the Mine Engineer who has a complete understanding of cost structure and mining options at his mine.

- the Chief Planning Engineer within the central planning group is assigned the responsibility of assembling the treatment and other costs as well as metal prices and exchange rates. This data is updated and disseminated to mine site engineers on a regular basis.

- unless specifically identified as "incremental", all production from a new mine or a new ore block is considered as replacement ore and assumes full costs of treatment, administrative services and other overhead costs.

- all costs and prices are in constant dollars; current dollar evaluations with defined inflation rates are not commonly used but can be accommodated in the PEP program.

- cashflows are determined on a pretax basis. Assessment of taxes and after tax cashflows is done by the corporate accounting group because of the complex tax position of the Company.

5. PROECO PROGRAM

This is a spreadsheet program in LOTUS-123. It is intended to cover short-term mine economic evaluations of no more than two years. Typical applications are the planning of a new stope or evaluating the performance of a mine over a budget year. Time value of money is ignored.

The program is designed to display all input and output information on an 8 1/2" by 14" sheet. Metal recovery calculations are shown in areas of the same spreadsheet. The results are readily suited for presentation purposes.

Figure 4 shows a typical example of the program layout. This consists of four sections: title, prices and metal production, costs and summary.

```
Title:
Stope Name:
Date:

Exchange Rate:  $1.35 Cdn$/US$
Acid Credit:    $xxxx/ton acid
Version:        Imperial

                         Nickel  Copper  Cobalt  Others  Sulphur
Price (US$/lb.)          $xxxx   $xxxx   $xxxx   $xxxx
Price (Cd$/lb.)          xxxx    xxxx    xxxx    xxxx
Stoping tons.%   10,000  1.28    0.40    0.036           7.87
Develop. tons.%       0  0.00    0.00    0.000           0.00
Total   tons.%   10,000  1.28    0.40    0.036           7.87
Conc.   tons.%    1,394  xxxx    xxxx    xxxx
Matte   tons.%      206  xxxx    xxxx    xxxx

Lbs. in ore              xxxx    xxxx    xxxx
Recovery
Lbs. in conc.            xxxx    xxxx    xxxx
Recovery
Lbs. in matte            xxxx    xxxx    xxxx
Recovery                 xxxx    xxxx    xxxx

Net metal lbs.           xxxx    xxxx    xxxx
  ex refinery

                          Costs

Mining costs                            Total x 1000's
Mining ore              $xxxxx/ton      $xxx
Permanent dev.          xxxx/ton         xxx
Operating dev.          xxxx/ton         xxx
Mine plant & equip.     xxxx/ton         xxx
Misc. mine costs        xxxxx/ton         xx
Sub-total mining        $xxxxx/ton      $xxx
Treatment costs
Transport to mill       $ xxxx/ton      $ xx
Receiving at mill         xxxx/ton        xx
Milling costs             xxxx/ton        xx
Transp. to smelter        xxxx/ton conc.  xx
Smelter costs             xxxxx/ton conc. xx
Sub-total treatment                     $xxx
Admin. services         $xxxxx/$          xx
Sub-total Sudbury       $xxxxx/ton       $xxxx
Other Costs
Matte transportation    $xxxxx/ton matte $ xx
Refining Ni               xxxx/lb.       xxx
         Cu               xxxx/lb.        xx
         Co               xxxx/lb.        xx
Dir.Sell.Exp. Ni          xxxx/lb.        xx
                          xxxx/lb.         x
                          xxxx/lb.         x
Sub-total others          xxxxx/ton      $xxx
Total costs             $xxxxx/ton       $xxx

                         Summary

                 Total      /Ton
Revenues         xxxxx      xxxxx
Costs            xxxxx      xxxxx
I.N.O. Contrib.  xxxxx      xxxxx
Mining ore cost for breakeven   $xxxxx
Ni. equiv. in ore                1.42%
Cost After       $ xxxx/lb. Ni
   Credits       $ xxxx/lb. Ni US$
```

Figure 4 - PROECO example

Information shown in the summary includes INO contribution, cost per pound of nickel, breakeven mining cost and nickel equivalent in ore. Breakeven mining cost informs the Mine Engineer of the sensitivity of his mining cost. Nickel equivalent is useful in establishing cut-off grades because it can be related to an ore value that produces the minimum acceptable level of contribution.

The program is simple and is used by many people.

A variation of the standard PROECO program is the Cu ore version, code name PROECOCU. This program is specifically designed for the Cu ore zone which contains high values of precious metals. A difference between the two programs is that all precious metal grades, prices and treatment costs are displayed in the PROECOCU program. This program also contains a routine for handling custom treatment of concentrate.

6. PEP PROGRAM

This program is used to perform long-term evaluation on projects, ranging from development of new mines to developing new ore zones within an existing mine. The program is written in Fortran 77 for IBM/AT computers.

The program deals with large amounts of input data, various types of output and provides for other special utilities. Therefore, it is carefully structured to provide a user-friendly environment and easy access for future modifications.

The necessary input data to the program is created by DBASE III software. Data entry procedure and default values are displayed on the screen to assist in quick and correct data input. All input information is then re-arranged and presented as part of the overall program output. This allows easy verification of input.

Various outputs are generated by the program and stored in disk files. The user can select any or all of these files for printing. The available output files include the following:

Page 1: project identification, run numbers and general comments.

Page 2: summaries of mine production and mine data input costs (see Fig. 5). Annual mine production includes tons or tonnes, grades of Ni, Cu, Co and S. Because Fraser Mine is a metric mine, a metric version of the PEP program has been created. In the case of Cu ore, all precious metal assays are printed out. Costs are arranged in various categories in order to facilitate data checking and for a future taxation subroutine.

YEAR	ORE TONNES	GRADE % NI	CU	CO	S	DEVELOPMENT OPER	PERM	PLANT & EQUIP CAPITAL	STOPING $/TONNE	ROYALTIES $/TONNE	$/LB.	K$/YR
1988	69516	2.09	0.64	.077	14.21	1722	1216	2878	xxxxx	0.000	0.000	0
1989	262123	1.51	0.67	.051	10.53	929	2498	3218	xxxxx	0.000	0.000	0

Fig. 5 - Mine input data

Page 3: contains treatment and other non-mine input costs together with metal sale prices (Fig. 6). When the program is used to evaluate a copper rich project, the input treatment costs and selling prices for a range of precious metals will also be included.

YEAR	TRANS TO MILL	MILL OPTG.	SLURRY TRANS.	SMELT. CHARGE	INDIRECT COST	MATTE TRANS.	REFINING COSTS ($/LB. METAL) NI	CU	CO
1988	0.00	xxxx	xxxx	xxxxxx	xxxxx	xxxxxx	xxxx	xxxx	xxxx
1989	0.00	xxxx	xxxx	xxxxxx	xxxxx	xxxxxx	xxxx	xxxx	xxxx

YEAR	DIRECT SELLING EXP. (US $/LB. OF METAL) NI	CU	CO	METAL PRICES (US $/LB. OF METAL) NI	CU	CO	ACID CREDIT	PM NI CREDIT	CAN/US EXCHANGE
1988	xxxxx	xxxxx	xxxxx	xxxxx	xxxxx	xxxxx	xxxxx	xxxxx	1.350
1989	xxxxx	xxxxx	xxxxx	xxxxx	xxxxx	xxxxx	xxxxx	xxxxx	1.350

Fig. 6 - Beyond headframe input data

Page 4: provides the summary of cashflow, cost of producing a pound of nickel for each year together with NPV and ROI for the project (see Fig. 7). On this page the 'beyond headframe cashflow' is summarized into one column which is detailed on page 5.

YEAR	OPTG. DEV.	PERM. DEV.	PLANT EQUIP.	STOPING	SUBTOTAL MINING	TOTAL ROYALTY
1988	1722	1216	2878	1514	7330	0
1989	929	2498	3218	5871	12516	0
TOTAL	10608	11560	10249	86359	118776	0
NPV	5697	6688	6968	38594	57947	0

YEAR	BEYOND HEADFRAME	TOTAL	PRETAX REVENUE	PRETAX CASHFLOW	DISCOUNTED CASHFLOW	COST/LB. NI $US
1988	4657	11988	8127	- 3861	- 3448	xxxx
1989	13542	26059	22918	- 3142	- 2505	xxxx
TOTAL	194500	313276	340675	27399	7872	xxxx
NPV	88594	146541	154413	7872	7872	

Fig. 7 - Cashflow report

Page 6: contains production statistics at various stages of treatment and page 7 (Fig. 8) the overall summary of the project evaluation.

	MINING ORE	SURFACE HANDLING	MILL	SMELT	ROYALTY	MATTE TRANSP.	REFINE	DIRECT SELLING	TOTAL OPERATING
NPV ($1000)	38594	2788	10247	20842	0	0	0	6888	79360
TOTAL ($1000)	86359	6187	21933	46244	0	1	0	15190	175913
US$/LB. NI	xxxx	xxxx	xxxx	xxxx	xxx	xxx	xxx	xxxx	xxx

	BI-PROD CREDIT	NET OPERATING	PRE-PROD & DEV.	PLANT & EQUIP	TOTAL MINE CAPITAL	TOTAL PROJECT COST	INDIRECT COST	INO BEFORE CORP
NPV ($1000)	xxxx	xxxx	xxxx	xxxx	xxxx	xxxx	0	xxxx
TOTAL ($1000)	xxxx	xxxx	xxxx	xxxx	xxxx	xxxx	0	xxxx
US$/LB. NI	xxxx	xxxx	xxxx	xxxx	xxxx	xxxx	xxxx	xxxx

```
            VALUE OF PRODUCTION REPORT (1000'S)
            NET NICKEL   xxxxx LBS.  $  xxxxx
            NET COPPER   xxxxx LBS.  $  xxxxx
            NET COBALT   xxxxx LBS.  $  xxxxx
            NET PMS               $  xxxxx

            NPV ($1000) = 55700
            RATE OF RETURN = 30.3%
            PRESENT VALUES DISCOUNTED @ 12.0%

            INO UNIT COST WITH CREDITS (US $/LB. NI)
            * xxx @ 1.35 CDN/US EXCHANGE RATE
```

Fig. 8 - Project summary

At present, the program is installed at all minesites and at central planning computers. Mine Engineers run the program for situations when investment and revenue span more than two years. Training requirement is minimal (less than 1 hour).

6.1 Risk analysis

Mine Engineers will run the program to generate information for various alternatives. The standard approach for sensitivity analysis is to vary input data one at a time. If each variable is tested for three values (most likely, most pessimistic and most optimistic), then a total of 729 computer runs are needed to complete the sensitivity analyses on six variables. The number of computer runs and the presentation of the results is a formidable task. Therefore, a risk analysis routine is available in the PEP program (see O'Hara, 1981, Townsend, 1977 for application of risk analyses).

An example of risk analysis results is shown in Fig. 9.

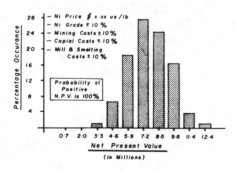

Fig. 9 - Risk analysis example

This analysis consists of varying the five most sensitive variables simultaneously. Each variable is assumed to vary within $\pm10\%$ from the regular input data. It is further assumed that the variation is distributed within a range of ±2 standard deviations. This approximation is statistically sufficient when actual distribution is impossible to obtain (see central limited theorm, Meyer, 1975). The program samples input data from these distributions one thousand times and recalculates cashflow from each input. The final results are output and plotted by the graphic package of LOTUS-123.

7. SUDSIM PROGRAM

It is recognized that when an individual mine is treated as an independent entity, the optimum production alternative for that mine does not necessarily correspond to the optimum solution for the combined results of several mines. This can result from arbitrary allocation of shared overhead costs and shortages or excesses in plant capacity. The main objective of the SUDSIM program is to determine the best alternative within the context of the total Sudbury Operations.

The forerunner of the SUDSIM program (Su, 1977) was formulated as a linear program model. This program worked well in the simulation mode but was not satisfactory in the optimization mode. The present SUDSIM approach is one of simulation and is based on the knowledge that each Mine Engineer knows best the realistic alternatives available at his mine. However, he has little knowledge of options at other mines. Therefore an interactive, consultative procedure model was selected as the basis for the modified SUDSIM program.

This model is currently in the final stages of modification and is not at present in use at the mine sites.

All users will have direct access to the program and data. When evaluating an option, an engineer can alter his own data but that relating to other mines and treatment plants is 'locked out'. Once the user determines the option that best satisfies both his mine and the INO as whole and this receives management acceptance, the complete set of data will be updated. All users will then access this updated set of data.

Figure 10 shows the program structure. The standard dataset contains basic cost and production information usually obtained from the Budget or Long-term Forecast figures.

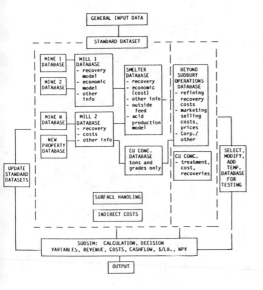

Fig. 10 - SUDSIM program structure

Costs are divided into fixed and variable portions in order to properly reflect the economic aspects of 'incremental' production. Administrative services and other indirect costs are considered separate entities and are not distributed to each individual operation.

Each individual dataset also contains capacity limitations. Any alternative which violates constraints is not automatically rejected as unfeasible. A flagging procedure is used in such cases. Realistic solutions to constraints can often be found when sufficient incentive exists.

The modified program has been developed and tested, and is being scrutinized by the review team. No actual application experience is presented at this time. However, since the proposal for the modification of this program has been favourably reviewed by all potential users, it is anticipated that this financial program will become a key standard evaluation tool at Sudbury Operations.

8. CONCLUSIONS

At Sudbury Operations, financial evaluation programs have been developed in-house and are used by Mine Engineers. The programs are valuable in that they simulate closely the way operations function. They allow on site personnel to relate the performance and options at their mine directly to metal prices and contributions to corporate cashflow. This enables those most knowledgeable about specific mine operations to make appropriate decisions on an ongoing basis.

The use of a review team of users and program developers has resulted in user friendly programs that meet the demands of those that make the evaluations and decisions.

Standardization of terminology, programs and input data are all contributing to the accuracy and the acceptability of results. Not only is this process resulting in better decisions but it has created a much greater awareness of unit costs and focussed attention on the 'bottom line'.

9. ACKNOWLEDGEMENT

The authors wish to thank the management of Falconbridge for permission to publish this paper. Gratitude is also expressed to our many colleagues at Falconbridge who contributed to the development and implementation of these programs.

REFERENCES

Farrish, D.J. 1972. Financial analysis for mining project evaluation, Falconbridge Ltd. internal report.
Meyer, S.L. 1975. Data analysis for scientists and engineers, John Wiley & Sons. Inc.
Moruzi, G.A. 1985. PROECO program, Falconbridge Ltd. internal report.
O'Hara, T.A. 1981. Analysis of risk in mining projects (Hudson Bay Mining & Smelting Co.), 83rd CIM annual meeting.
Park, C. 1977. Computer methods for the 80's - production planning in metal mines, Inco Ltd.
Su, Y.L. 1977. Optimization model for the Sudbury operation, Falconbridge Ltd. internal report.
Townsend, H.W.R. et al, May 1977. We used risk analysis to move our computer, I.E., p.32-39.

Computer Applications in the Mineral Industry, Fytas, Collins & Singhal (eds)
© *1988 Balkema, Rotterdam. ISBN 90 6191 760 3*

The application of computers in mining taxation

Kumara S.Rachamalla
Mining Tax Office, Corporations Tax Branch, Ontario Ministry of Revenue, Toronto, Canada

ABSTRACT: Mining activity in Ontario represents over $4 billion in annual production. This sector is encouraged by the government of Ontario to undertake mineral development through financial incentives offered under mineral resource taxation. Mining companies continue to explore and develop mineral deposits through generous write-offs and allowances under the Mining Tax Act.

As a result of the complex nature of mineral taxation legislation, computers are playing an increasingly important role in the operational, administrative and planning functions of the Mining Tax Office in the Ministry of Revenue. The use of state-of-the-art technology has resulted in significant improvements in productivity in areas such as assessment of mining tax returns, modelling of the Mining Tax Act, sensitivity analysis, management decision-making, and office administration. Implementation of local area networks, micro-computer technology, fourth generation languages and integrated micro-mainframe information processing has resulted in greater efficiency with lower costs.

As a result, the financial services provided to the mineral industry are now delivered in an efficient and cost-effective manner utilizing the latest available computer technology.

1. INTRODUCTION

Mineral production last year accounted for roughly 20% of Ontario's export earnings. Compared to other provinces Ontario holds an impressive leadership position in this sector. Furthermore, the value of metallic mineral production for the province is just under 40% of the United States. This lead in mineral production can be attributed to factors such as:

– the stability of mineral policies,
– the security of land tenure, mineral rights, and
– the attractive incentives under the Mining Tax Act and the Ontario Mineral Exploration Program (OMEP) Act.

The size of this industry and the increasing complexities of the Acts governing mineral taxation and incentives have resulted in the need for improved information management. With the aid of computers, the tasks of assessing mining tax returns and modelling the Mining Tax Act, have been performed with greater efficiency and accuracy. This paper will examine the application of computers and state-of-the-art information management tools in the Mining Taxation area.

2. THE MINING TAX ACT

The Ontario Mining Tax Act deals with the taxation of mining profits of operators in the province. The taxes are levied on mining income, at a fixed rate of 20% for amounts exceeding $500,000. Allowances for straight-line depreciation are provided in the Act, as well as up to 100% deduction of expenditures in Ontario for exploration and development. A processing allowance is also allowed at rates rising to 20% of the cost of processing assets, and is designed to encourage processing in the province beyond the concentration stage. Higher rates are allowed for operations in Northern Ontario to encourage

development of that region.

In addition to encouraging investment in processing assets, the Act provides for up to 100% write-off of new mining assets associated with a new mine or a major expansion of an existing mine.

The Act is quite complex and many of the calculations are inter-related. It is for this reason that the Mining Taxation unit has adopted the extensive use of micro-computers and mainframe computer resources.

3. APPLICATION OF COMPUTERS IN MINING TAX

The application of computers is supported by the Queen's Park Computing Centre as well as by micro-computers utilizing various software packages.

3.1 Hardware

The facilities at the Queen's Park Computing Centre consist of IBM mainframes. These handle all input/output communications between the CPU (Central Processing Unit) and the user terminal connected by modems.

In addition, the Mining Tax unit utilizes IBM PC's, PC/XT's, PC/AT's, compatibles and laptop portables. The portables are used outside of the office at the taxpayer's offices in the course of field audits. Laptop portables have been utilized since their introduction to the personal computing market. Their use began from first generation "transportable" types to present equipment consisting of fully equipped laptops with expanded RAM and hard disks all contained in a truly portable machine. The other micro-computers are grouped to form a Local Area Network (LAN) utilizing Novell Advanced Network 286 software. This arrangement provides multiple access to the central files and results in greater productivity. Local area Networks allow sharing of central disk storage as well as printer and plotter equipment. Not only does this arrangement allow more flexibility, it is very secure. Features such as multi-level password protection and sophisticated file locking secure the highly sensitive information stored on the computers.

The Senior Manager utilizes a Kontact electronic mail terminal to communicate with other managers and senior Ministry officials located in offices as far away as 60 kms. In addition to mail features, these terminals provide user-friendly word-processing, database access and scheduling features.

By utilizing telephone and computer capabilities, messages can be sent by pre-programming the autodialing feature and messages can be received and stored on diskette or forwarded to another Kontact machine. All these features provide more effective communication channels among the senior management of the ministry.

3.2 Software

Various software packages utilized by the Mining Tax Section are:
- Multimate Word Processor,
- Lotus Symphony,
- Focus and PC/Focus,
- IFPS,
- Novell Netware
- Chart-Master,
- Sign-Master, and
- DBase III
- Microsoft Windows
- Microsoft Word
- Other multi-functional, integrated software.

Of the software mentioned above, Symphony is probably the most extensively utilized for tax audit purposes.

Symphony is used by Tax Auditors in two major areas:
1. Analyzing mining tax returns and preparing mining tax assessments;
2. Administratively, to keep an ongoing record of the progress of:
- audits and assessments
- statistical information, and
- revenues derived from mining tax.

3.3 Auditing

Symphony permits the auditors to create related schedules, carrying forward the results of previous calculations for later use. The taxpayer's filing can be modelled and adjusted to agree with the letter and intent of the Act and the resulting tax and interest calculations are executed automatically. It also allows negotiated changes in tax liability to be incorporated in assessments instantaneously.

The printing of schedules is quick and easy and eliminates the typing problems usually associated with the production of large and complex schedules. As a result, assessments are produced on a more timely basis.

An additional benefit of the model has been the standardization of the format used to prepare working papers. This has resulted in greater consistency in the treatment of different taxpayers' assessments. Audit files are better organized, look neater and are more effectively integrated with the taxpayers' returns through the use of computerized work schedules.

3.4 Administration

Symphony is used to monitor the progress of audits and assessments on a regular and frequent basis. The model used to prepare the report is updated monthly, or more frequently if required, and reports can be printed with minimal effort.

Completed tax assessments are analyzed and reported in detail as a control measure.

The following reports are produced:
- Memorandum to the Senior Manager,
- Audit Status Report,
- Completed Assessments Report,
- Revenue Generated Report,
- Analysis of Revenue, and
- Analysis and Reconciliation Report of Completed Mining Tax Assessments to Mining Tax Revenue.

A breakdown of the amount of tax generated by major mineral commodities under the Mining Tax Act is reported to the Federal Government as part of the requirements under transfer payment arrangements between the Federal Government and the provinces. Thus, the reporting system required internally has a useful application for reporting requirements to other levels of government.

3.5 Financial Modelling

The Mining Tax unit has also developed an industry mining tax model using IFPS running on the Queen's Park IBM mainframe computer. This model allows sensitivity analyses to be performed to test the effect of proposed changes to the Mining Tax Act. The effect of changes in the Act can be seen immediately and various alternatives can be assessed. For the purpose of the model, producers are grouped into categories by size, mineral substance produced and degree of processing performed. The model has resulted in a greater understanding of the effect of proposed changes in the mining taxation system on segments of the industry. Management decision tools of this kind were instrumental in recent task force studies to propose revisions to the Mining Tax Act.

3.6 Statistical Database

Currently under development is a statistical database which will provide current status of taxpayers' files on-line to the users of the computer system. This way, up-to-date queries can be addressed avoiding undue delays for either government officials, the auditors, or the public. Maintenance of the database will be accomplished as the changes arise since each auditor will be responsible for several specific files. This will help provide improved communications and reduce research time.

3.7 Computer Training

Training courses are available both through the Ministry of

Government Services, Computer and Telecommunication Services (CTS) Branch, as well as from external sources. Courses provided at CTS include those for typical wordprocessing packages, general computer usage, Disk Operating System (DOS), and spreadsheet packages. More sophisticated training on advanced topics such as database or networking is usually provided by external companies. As well, one-to-one instruction is often provided to the staff by the systems officer employed by the Mining Tax Office.

3.8 Other Developments

The Mining Tax office is currently commencing testing of a remote dial-up system to link communications with head office networks. This feature will permit file transfer, electronic mail and other communication features with a wide user base.

On-line database retrieval from public databases such as Info Globe will eliminate the technical information clippings kept on file relating to the mining industry. Information could be retrieved with greater efficiency and accuracy to keep the staff abreast of industry developments.

All of these state-of-the-art developments will be instrumental in offering more efficient and effective service to taxpayers.

4. CONCLUDING REMARKS

With the use of state-of-the-art computer technologies, the Mining Tax Office has been able to increase productivity while reducing the cost of computer services. The computerization of complex data and mineral statistics has significantly reduced the turnaround time of providing accurate information to government and to the private sector. Furthermore, the use of computers in administering the Mining Tax Act has greatly increased the efficiency of processing returns and has also facilitated policy evaluation and decision-making. With these improvements, better service is being offered to the mining companies, to geologists, and to the industry as a whole.

BIBLIOGRAPHY

Ontario Mineral Score. 1986. Ministry of Northern Development & Mines, Mineral Statistics Section, December.

Rachamalla, Kumara, S., M.B.A., M.Eng., P.Eng. 1987. Interregional United Nations seminar on "Electronic Data Processing Methods in Mineral Exploration & Development", October.

Rachamalla, Kumara, S., M.B.A., M.Eng., P.Eng. 1986. "The Application of Computers in Ontario Mineral Taxation and Incentive Programs", 19th International Symposium on Applications of Computers and Operations Research in the Mineral Industry, April.

Rachamalla, Kumara, S., M.B.A., M.Eng., P.Eng. 1983. "Financial Incentives to Producers and Non-Producers in Ontario", Proceedings, 87th Annual Meeting of the Canadian Institute of Mining and Metallurgy, April.

"1985 Annual Software Review", Winter 1985. PC World, Volume 2, Number 14.

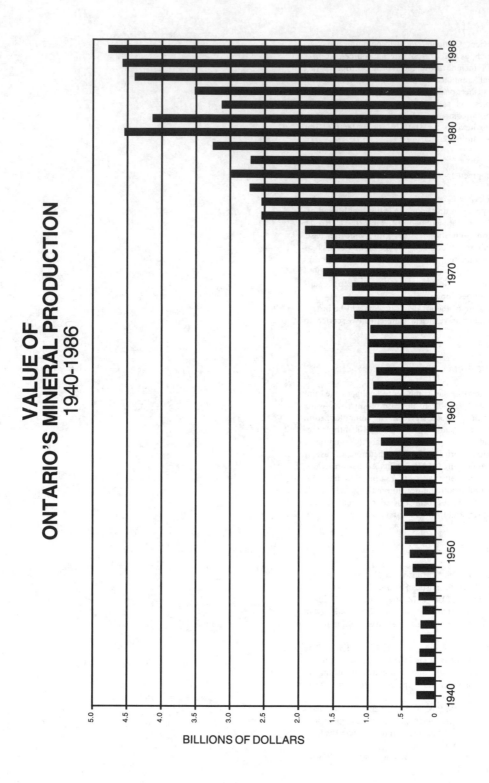

VALUE OF
ONTARIO'S MINERAL PRODUCTION
1940-1986

BILLIONS OF DOLLARS

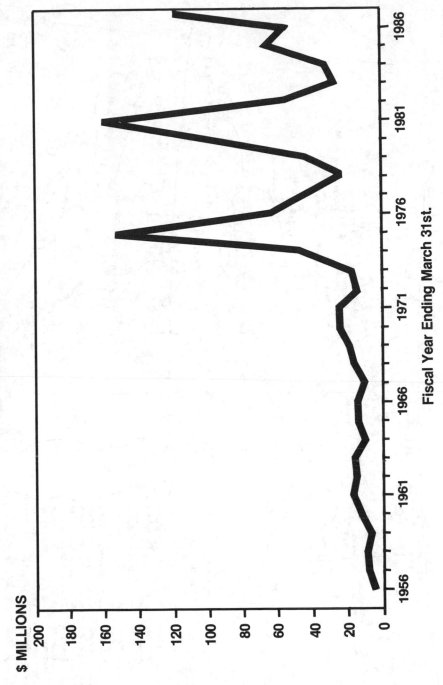

ONTARIO MINING TAX REVENUE

$ MILLIONS

Fiscal Year Ending March 31st.

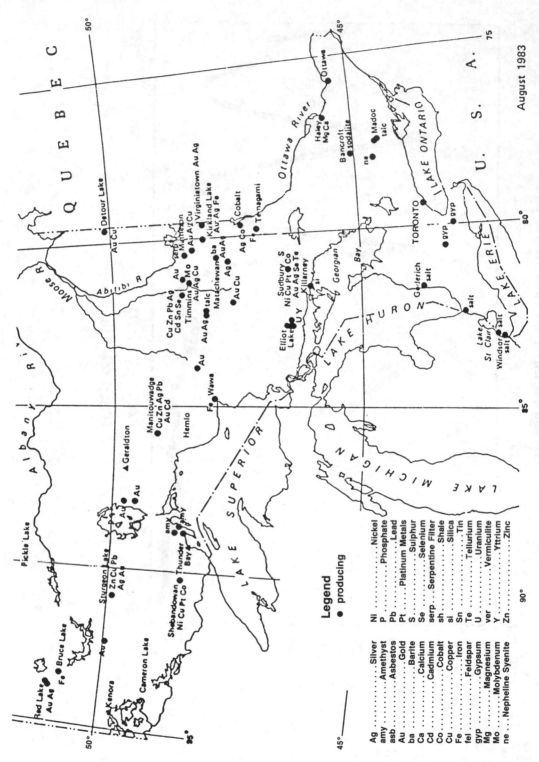

ONTARIO'S PRINCIPAL MINERAL AREAS

August 1983

Legend

● producing

Ag	Silver
amy	Amethyst
asb	Asbestos
Au	Gold
ba	Barite
Ca	Calcium
Cd	Cadmium
Co	Cobalt
Cu	Copper
Fe	Iron
fel	Feldspar
gyp	Gypsum
Mg	Magnesium
Mo	Molybdenum
ne	Nepheline Syenite

Ni	Nickel
P	Phosphate
Pb	Lead
Pt	Platinum Metals
S	Sulphur
Se	Selenium
serp	Serpentine Filter
sh	Shale
si	Silica
Sn	Tin
Te	Tellurium
U	Uranium
ver	Vermiculite
Y	Yttrium
Zn	Zinc

Computer Applications in the Mineral Industry, Fytas, Collins & Singhal (eds)
© 1988 Balkema, Rotterdam. ISBN 90 6191 760 3

Computer-aided mine investment analysis

Michel L.Bilodeau & Yearn H.Park
Department of Mining and Metallurgical Engineering, McGill University, Montreal, Quebec, Canada

ABSTRACT: PCDEP and PCSIM are user-friendly software packages to assess the economics of exploration projects or of mine development proposals at the pre-feasibility stage. The programs are coded in BASIC and operate on IBM or IBM-compatible micro-computers.

PCDEP consists of a data entry/edit module combined with a deterministic cash flow computation module. Available options are controlled with keyboard function keys. A menu displayed at the bottom of the screen indicates the function performed by each key. The geological, cost and market estimates necessary for the evaluation of a mining project are entered interactively, by way of pages displayed on the screen. This information is validated before any computations are performed, and can be saved if desired. Canadian federal and selected provincial mineral taxation systems are currently available in the package. The results of the economic analysis, consisting of standard before or after-tax profitability indicators, are displayed after computations are completed. A detailed year-by-year cash flow spreadsheet can be obtained if a printer is available. The system can alternate between input and result modes, thus allowing the modification of various parameters in order to perform sensitivity analysis.

PCSIM performs risk analysis based on a probabilistic simulation model. From the deposit data created by PCDEP, and with supplemental information quantifying the reliability of key uncertain variables, this module generates distributions of profitability indicators from which expected value and economic risk criteria are assessed.

The results of both programs are designed to provide the exploration geologist and/or mining engineer with a sound quantitative basis for decision-making.

RÉSUMÉ: PCDEP et PCSIM sont des logiciels qui permettent l'analyse économique et financière de projets miniers rendus au stage de pré-faisabilité. Les programmes sont codés en language BASIC et opèrent sur micro-ordinateurs IBM ou compatibles.

PCDEP se compose d'un module d'entrée et d'édition de données combiné à un module de calcul des flux monétaires en mode déterministe. Le contrôle des fontions du logiciel s'effectue avec les touches programmées du clavier. La clé d'utilisation de ces touches est indiquée au bas de l'écran. Les paramètres géologiques et techniques du projet, ainsi que les prévisions des cours du marché pour les éléments présents sont entrés d'une façon interactive, à l'aide de tableaux exposés à l'écran. Les données sont validées avant les calculs et peuvent être sauvegardées par l'utilisateur. Le système canadien d'impôts corporatifs fédéral ainsi que certains systèmes d'impôts corporatifs et de droits miniers provinciaux sont disponibles. Les indices économiques du projet avant ou après impôts apparaissent à l'écran lorsque les calculs sont terminés. Un tableau détaillé des flux monétaires annuels peut être obtenu si une imprimante est disponible. Le système peut alterner entre mode entrée de données et mode résultats, permettant ainsi le changement de certains paramètres dans le but d'effectuer une analyse de sensibilité.

Le logiciel PCSIM effectue l'analyse du risque économique du projet d'après un modèle de simulation probabiliste. A partir des données créées à l'aide de PCDEP, et basé sur des données supplémentaires quantifiant l'incertitude d'estimation des paramètres clés du projet, PCSIM génère des distributions d'indices économiques qui servent ensuite à établir des critères de rentabilité et de risque.

Les résultats de ces deux programmes fourniront au géologue d'exploration et/ou à l'ingénieur minier une base solide pour une prise de décision éventuelle.

1 INTRODUCTION

PCDEP and PCSIM are user-friendly software packages designed to assess the economic feasibility of either exploration projects, or mine development proposals at the pre-feasibility stage. These programs have been developed on IBM hardware and are coded in the BASIC language. The source is compiled with Borland's Turbo Basic product. Thus, the packages can operate on any IBM compatible microcomputer as well.

The system described here is largely based on the economic evaluation model that the senior author developed at the Centre for Resources Studies of Queen's University, with the assistance of Brian W. Mackenzie.

2 OVERVIEW OF MINE INVESTMENT ANALYSIS

2.1 The mining project decision process

The process leading to the development decision of a mining project consists of many stages. These are illustrated in figure 1.

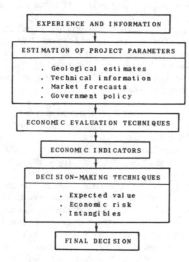

Figure 1. The mining project decision process

The starting point for the economic evaluation of a mineral project to support an investment decision is the gathering together and development of relevant experience and information. These provide the basis for estimating the future conditions anticipated if a decision were to be made

to proceed with the investment. Relevant estimates depend on the type of investment being evaluated but may include, for example, geological parameters such as ore reserve tonnage and grade, engineering plans for mining and processing methods and the capital and operating costs associated with their application, mineral market forecasts of demand and price conditions, and government policies related to such factors as taxation, environmental controls, and the provision of physical and social infrastructure.

Economic evaluation techniques are then applied to reduce these estimates to a few indicators of the economic attractiveness of the project. These indicators portray the quantitative economic dimensions of the investment in terms of expected value and risk criteria. At the same time, attention must be given to important non-quantifiable or "intangible" factors. The investment measures thus obtained support the decision which is required on whether or not to select and implement the project.

2.2 Economic evaluation techniques

Economic evaluation techniques are applied to reduce a complex set of estimates to a relatively few indicators of economic value and investment attractiveness. Thus, they assist in comparing and selecting among mineral project alternatives. These techniques, outlined in figure 2, are based on the concepts of cash flow and time value of money.

Based on projected benefits and costs, cash flows are initially estimated on a before-tax basis. Taxation considerations may be introduced to convert before-tax cash flow estimates to an after-tax basis. Tax credits and payments and, thus, the determination of after-tax cash flows, are usually affected by the presence of inflation.

The payback period criterion, a weak but commonly used economic indicator, depends only on the estimation of future cash flows associated with a project. The basic concepts of cash flow and time value are combined in various ways to evaluate discounted cash flow criteria such as net present value, present value ratio, and rate of return. Single-point estimates of expected future conditions for relevant geological, engineering, market and government policy parameters are combined to assess the expected values of the discounted cash flow indicators. This usually constitutes the first stage of the economic evaluation of an investment alternative.

Then, sensitivity analysis may be used to examine the sensitivity of the economic indicators to possible variations in parameters away from their expected values. Sensitivity analysis may also be applied to determine the breakeven conditions required to justify investment. This is a helpful approach when a high degree of uncertainty is associated with one or more project parameters.

Risk analysis translates perceived uncertainties concerning the project parameters into probability distributions of possible values for the discounted cash flow indicators. In this way, it is possible to assess the risks associated with realizing the expected value outcomes. What is the probability of economic loss if the investment alternative is undertaken? What is the lower confidence limit of the economic outcome? These questions are ultimately answered by risk analysis.

The expected value, sensitivity, and risk results, together with an appreciation of the non-quantifiable intangible factors, are ultimately used to support the investment decision.

The PCDEP package assists the decision-maker in the expected value and sensitivity analyses stages. PCSIM performs the risk analysis required to complete the economic assessment of the project.

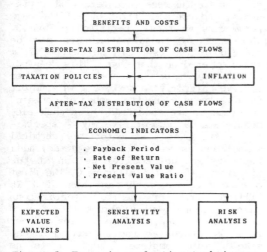

Figure 2. Economic evaluation techniques

3 DESCRIPTION OF THE PCDEP SOFTWARE PACKAGE

The PCDEP package consists of a data input/edit module, a deterministic comput-

ation module, and an output module. In addition, the system contains several built-in utilities. A brief description of these modules and utilities follows. All menus and messages in the current version of PCDEP are displayed in the English language.

3.1 Data Input/Edit Module

The parameters necessary for the economic evaluation of a mining project are entered interactively on fixed-format pages displayed on the micro-computer screen. Data entry and system control is achieved through the use of the keyboard function keys. A menu displayed at the bottom of the screen indicates the function performed by each key. In input/edit mode, the user can enter or modify the value of a parameter, move a cursor from one parameter to another on a given page, or skip to the next or previous page. There are six types of information pages in the current version of PCDEP. Each page contains a series of elements related to a particular aspect of the project.

Page 1 contains information concerning the preproduction period. The layout of this page is shown in figure 3. As indicated, the preproduction period, capital costs and investment schedule are specified here. Capital costs include exploration, development, plant and equipment, mill facility and infrastructure expenses. Working capital and salvage value may also be specified. For the illustrative example used in this paper, the total preproduction capital expenditure requirements, including working capital, are $45 million. The investment schedule indicates the proportion of total capital costs actually spent in each year of the preproduction period. Working capital is excluded from this schedule and is assumed to be required at the very end of the preproduction period. The reference point or "time zero" is the start of production. Thus, points in time during the preproduction period are specified with negative values. As indicated in figure 3, the investment schedule calls for an equal allocation of capital costs in both preproduction years -- 50 percent at time 0 (the first value), and 50 percent at time -1 (the second value). The 0 percent at time -2 (the third value) indicates that no expenditures are required at the very beginning of the preproduction period.

The number of ore reserve categories and information related to production period capital expenditures are entered on Page

2. The ore reserve categories are a means of varying operating parameters during the production period. Production investments are specified with the same breakdown as that used for preproduction capital costs. In addition, the starting time and duration must be given for each production investment specified. A maximum of five ore reserve categories and independent production investments may be specified. The layout of Page 2 is illustrated in figure 4. As indicated, two ore reserve categories and one production period investment are specified. This investment begins one year after the start of production and lasts over a two-year period. A total of $12.5 million are to be spent on development and equipment, and an additional $0.75 million of working capital is required. Production period capital costs are assumed to be evenly allocated over the period of time specified.

```
    MINERAL DEPOSIT EVALUATION PACKAGE        PAGE 1
    ===================================

PREPRODUCTION PERIOD INFORMATION :

    Preproduction Period        2.00 year(s)

    Capital Costs :
      Exploration               0.000 thou. $
      Development            7500.000 thou. $
      Plant and Equipment   10000.000 thou. $
      Mill Facilities       15000.000 thou. $
      Infrastructure         5000.000 thou. $
      Working Capital        7500.000 thou. $
      Salvage Value             0.000 thou. $

    Investment Schedule    50.00%    50.00%     0.00%
```

Figure 3. Page 1 Information

```
    MINERAL DEPOSIT EVALUATION PACKAGE        PAGE 2
    ===================================

PRODUCTION PERIOD INFORMATION :

    Number of Ore Reserve Categories  2

    Number of Production Investments  1

    Capital Costs :           #1
    Start Time                1.00
    Duration (years)          2.00
    Expl.     (thou. $)       0.000
    Dev.      (thou. $)    2500.000
    P & E     (thou. $)   10000.000
    Mill      (thou. $)       0.000
    Infr.     (thou. $)       0.000
    WC        (thou. $)     750.000
    SV        (thou. $)       0.000
```

Figure 4. Page 2 Information

The variables entered on Page 3 initialize various parameters common to all ore reserve categories. A modification made on this page results in a change across all reserve categories. Figure 5 illustrates the layout of the page. The parameters are mine recovery and dilution, unit operating expenses in terms of mining, milling and overhead costs, annual mill capacity and annual sustaining capital costs. As shown, mine recovery and dilution factors of 100 and 0 percent are given. Total operating costs of $30 per tonne and annual sustaining capital costs of $0.5 million are specified. A mill capacity of 750 000 tonnes per year is assumed.

```
    MINERAL DEPOSIT EVALUATION PACKAGE        PAGE 3
    ===================================

PRODUCTION PERIOD PARAMETER INITIALIZATION

    Mine Recovery           100.00 percent
    Mine Dilution             0.00 percent
    Mill Capacity           750.000 thou. t/y

    Operating Costs :
      Mining Costs           15.00 $/t
      Milling Costs          10.00 $/t
      Overhead Costs          5.00 $/t

    Sustaining Capital Costs 500.000 thou. $/y
```

Figure 5. Page 3 Information

Figure 6 illustrates the layout of information Page 4. There are as many pages of type 4 as there are reserve categories. Each one of these pages is indexed with respect to the reserve category it represents, and contains first of all the same parameters as those specified on Page 3. The value of these parameters may be changed by the user if they differ from those entered on that page. This does not affect the values associated with other reserve categories. Next, the in-situ ore reserve tonnage and the number of contained elements are specified. For each element -- a maximum of five may be specified -- the user must enter the chemical symbol and associated average grade, mill recovery and net smelter return at the minesite. As shown in figure 6, the first ore reserve category consists of 2.25 million tonnes of in-situ material containing copper, zinc and silver, with associated grades, mill recoveries and net smelter returns as indicated. For the sake of being complete, figure 7 lists the information associated with the second ore reserve category.

As shown in figure 8, the user must indicate on Page 5 whether before-tax or after-tax results are desired. Canadian and provincial corporate taxes as well as provincial mining tax models for selected provinces are currently available in the

PRODUCTION INFORMATION - ORE RESERVE CATEGORY 1 :

Mine Recovery	100.00 percent
Mine Dilution	0.00 percent
Mill Capacity	750.000 thou. t/y
Operating Costs :	
Mining Costs	15.00 $/t
Milling Costs	10.00 $/t
Overhead Costs	5.00 $/t
Sustaining Capital Costs	500.000 thou. $/y
In-Situ Reserves	2250.000 thou. t
Number of Elements	3
In-Situ Grades :	

Element	Grade	Mill Rec.	Net Smelter Return
1- Cu	1.500 %	95.00 %	70.00 %
2- Zn	6.500 %	80.00 %	60.00 %
3- Ag	17.500 g/t	85.00 %	95.00 %

Figure 6. Page 4.1 Information

PRODUCTION INFORMATION - ORE RESERVE CATEGORY 2 :

Mine Recovery	90.00 percent
Mine Dilution	10.00 percent
Mill Capacity	750.000 thou. t/y
Operating Costs :	
Mining Costs	25.00 $/t
Milling Costs	10.00 $/t
Overhead Costs	5.00 $/t
Sustaining Capital Costs	600.000 thou. $/y
In-Situ Reserves	5000.000 thou. t
Number of Elements	3
In-Situ Grades :	

Element	Grade	Mill Rec.	Net Smelter Return
1- Cu	2.600 %	95.00 %	70.00 %
2- Zn	5.100 %	80.00 %	60.00 %
3- Ag	11.000 g/t	85.00 %	95.00 %

Figure 7. Page 4.2 Information

package. If after-tax results are specified, a province must be selected, and annual inflation rates must be given for capital costs (which include sustaining capital), revenues and operating costs (which include working capital). Finally, an annual deflation rate must be given for converting the final current dollar cash flow distribution into constant dollars. In this example, the Quebec tax system is to be applied. Capital costs, revenues and operating costs are subject to inflation rates of 5, 5 and 7 percent, respectively. Current dollar cash flows are to be deflated at a rate of 5 percent.

On Page 6, the user must enter the market price forecast associated with each element contained in the ore reserves. All prices are specified in the same currency as that used for capital and operating costs. Figure 9 shows the layout of this page. As seen, long-term price forecasts of $9.00 per ounce and $0.85 and $0.50 per pound are specified for silver, copper and zinc, respectively.

Although ore reserve grades are specified in metric units, price forecasts are quoted in imperial units for the sake of being consistent with the North American market.

AFTER-TAX ANALYSIS (Y/N) :   Y

Province (Que., Ont., B.C.)   Q

ANNUAL INFLATION RATES :

Capital Costs	5.00 percent
Revenues	5.00 percent
Operating Costs	7.00 percent

ANNUAL DEFLATION RATE   5.00 percent

Figure 8. Page 5 Information

MARKET PRICE FORECASTS :

Ag	9.00 $/oz
Cu	0.85 $/lb
Zn	0.50 $/lb

Figure 9. Page 6 Information

3.2 Computation module

When all parameters have been entered to the satisfaction of the user, cash flow computations and economic indicator assessment are initiated with function key F4. Cash flows are determined on an annual basis. Then, economic indicators are assessed from the constant dollar time distribution of cash flows thus obtained. To allow the user to follow the sequence of events, computation steps are displayed on the screen during this phase.

3.3 Output module

As soon as computations are terminated, economic indicators associated with the project are displayed on the screen. Figure 10 illustrates the results associated with the base metal deposit used for illustrative purposes in this paper. These include payback period and DCF rate of return, as well as net present value and present value ratio at discount rates of 0 (undiscounted), 5, 10 and 15 percent. Economic indicators are expressed on the basis of constant dollars.

If a printer is connected to the micro-computer, a cash flow summary report as well as the economic indicators displayed

on the screen may be reproduced permanently. In this case, the number of cash flow
columns and the number of lines per page
must be specified. Default values for
these parameters are set to 8 and 51,
respectively. A title, to be printed as a
header on each page of output, may also be
specified at this time. Printed output
for the hypothetical base metal deposit
example used in this paper is illustrated
in appendix.

ECONOMIC INDICATORS

	Payback Period	6.03 year(s)	

Rate of Return   12.24 %

	Net Present Value ('000 $)	Present Value Ratio
@ 0 percent	32031.094	0.50
@ 5 percent	14287.609	0.24
@ 10 percent	3435.495	0.06
@ 15 percent	-3370.511	-0.07

Figure 10. Economic Indicators

3.4 Built-in utilities

A function key allows current deposit
information to be saved in a file, and as
the program is initiated, a previously
saved data file may be accessed. The user
is prompted for a file name in both of
these cases.

As data is entered interactively, and
immediately before computations are
initiated or data is saved, selected
deposit parameters are validated. If an
invalid parameter is detected at this
stage, a message related to the faulty
variable is displayed in the top left
corner of the screen. Before proceeding
any further, this error must be corrected.

Finally, a function key enables the user
to return from the display of economic
indicators to input/edit mode. At this
time, current deposit information may be
retained, a previously stored data file
may be accessed, or all the deposit information may be cleared for the purpose of
inputing data associated with another
project. By retaining the current deposit
information, input/edit mode is restored
with the cursor located on the same page
and variable it was positioned on before
computations were initiated. Particular
parameters can now be modified before
proceeding to further computations. Sensitivity analysis can thus be carried out
quite conveniently.

4 DESCRIPTION OF THE PCSIM SOFTWARE
  PACKAGE

The PCSIM package performs risk analysis
using a Monte Carlo simulation model.
Essentially, such a model recognizes the
fact that the majority of the parameters
used in the economic evaluation are estimates that have been derived from a limited amount of information. Thus, geological deposit parameters, engineering variables, capital and operating costs, and
market forecasts are considered uncertain,
and are specified in terms of probabilistic estimates. The key variables which can
be specified in such a manner in PCSIM
are:

1) Total preproduction capital costs
2) Total production capital costs
3) Annual sustaining capital costs
4) Total unit operating costs across all
   reserve categories
5) Mine dilution across all reserve
   categories
6) Mine recovery across all reserve
   categories
7) Total in-situ ore reserves
8) Average in-situ grade across all
   elements and reserve categories
9) Mill recovery across all elements and
   reserve categories
10) Market price across all elements and
    reserve categories

The model is kept to its simplest form
by assuming that all variables listed
above are independent from each other.
However, as the above list indicates, mine
recovery and mine dilution are assumed to
be completely correlated from one reserve
category to another. Likewise, average
grade, mill recovery, total operating
costs and market price are assumed to be
completely correlated from one element to
another and from one reserve category to
another.

The package consists of a data
input/edit module, a simulation module,
and an output module. A brief discription
of these modules follows.

4.1 Data Input/Edit Module

PCSIM uses the information initially entered and saved during a PCDEP session.
As the program is initiated, the user is
prompted for the name of the file containing this information. For consistency of
results between the two packages, the
point estimate information created with
PCDEP is considered to be most-likely or
best-guess estimates of all the parameters
associated with the project. In addition,

PCSIM requires supplemental information that quantifies the uncertainty associated with each key variable. This information is entered from the input/edit module, as displayed in figure 11.

As illustrated, a distribution type as well as lower and upper limits are specified for each probabilistic variable. Possible distribution types are uniform, triangular, normal/split-normal and lognormal. The user may also specify that a variable be kept constant by choosing the deterministic (None) type. Uniform and triangular distributions require the specification of lower and upper limits (only lower limit for uniform type), while normal/split-normal and lognormal distributions require the specification of lower and upper confidence limits (only lower limit for lognormal type) of the user's choice. All limits are stated with respect to a standardized most-likely value of 1.0. For instance, preproduction capital costs are assumed to have a split-normal distribution with lower and upper confidence limits of 0.90 and 1.15. In-situ ore reserves are assumed to have a triangular distribution with lower and upper limits of 0.85 and 1.25. Mine dilution and recovery are treated as deterministic estimates.

RISK ANALYSIS PACKAGE
=====================

VARIABLE	Distribution Type	Lower Limit	Upper Limit
Preproduction Cap. Costs	4	0.900	1.150
Production Capital Costs	4	0.900	1.150
Sustaining Capital Costs	4	0.900	1.150
Operating Costs	4	0.850	1.200
Mine Dilution	1	-----	-----
Mine Recovery	1	-----	-----
In-situ Ore Reserves	3	0.850	1.250
Average In-situ Grade	5	0.900	-----
Mill Recovery	4	0.950	1.050
Market Price	2	0.800	-----

Number of Iterations	1001	Confidence	
Discount Rate (%)	10.00	Interval(%)	95.00

Type: 1-None   2-Uniform   3-Triangular
        4-Normal/Split-normal   5-Lognormal

Change any value? (Y/[N]) ---> _

Figure 11. PCSIM Information

Finally, the number of iterations to be carried out in the simulation, the discount rate for the purpose of assessing the net present value and present value ratio, and the confidence interval associated with the lower and upper limits for normal/split-normal and lognormal distribution types must be specified. Figure 11

shows that 1001 iterations are to be performed and that a 10 percent discount rate is to be used. The confidence limits given for preproduction and production capital costs, sustaining capital costs, operating costs, average in-situ grade and mill recovery are such that the estimates are 95 percent confident.

4.2 Simulation module

When all information has been entered to the satisfaction of the user, the simulation module is initiated. First, random values are generated for each probabilistic estimate. The user is informed of the progression of this phase by messages displayed on the screen. As soon as this phase is complete, each set of randomly generated project variables are combined together, as in the PCDEP computation module, to derive a cash flow distribution as well as associated economic indicators. This is repeated for each iteration of the simulation. During the first iteration, all variables are set to their most-likely value.

The economic indicators associated with each iteration are displayed as in figure 10, to allow the user to judge whether the results are meaningful. The iterations may be terminated at any time if the results appear to be erroneous. In this case, program control is returned to the data input/edit module.

4.3 Output module

When all iterations are complete, program control is given to the output module. Here, the user can display the results of the simulation which consists of histograms of the distributions of payback period, rate of return, net present value and present value ratio. These distributions portray the risk associated with the investment project. Figure 12 illustrates the net present value distribution associated with our hypothetical example, as it appears on the screen. These results are based on the deterministic estimates specified in PCDEP and the probabilistic information given in figure 11. The probability of economic loss is represented by the proportion of negative values in the distribution.

The user can also obtain statistics associated with the rate of return and net present value distributions. These are most-likely and average values, overall, lower and upper standard deviations, and

89

probability of economic loss. The statistics associated with the net present value distribution shown in figure 12 are illustrated in figure 13, as they appear on the screen.

If a printer is connected to the microcomputer, histograms and statistics may be reproduced permanently. Here, the class interval or the number of histogram classes must be specified. A title may also be given at this time.

Finally, the user may return to the input/edit module to implement changes to particular simulation parameters. Thus, the effect of changing distribution types or of varying the reliability of project estimates can be analyzed.

```
(thou. $)

-52404.906                                                        1
-48023.695
-43642.484                                                        1
-39261.273 X                                                      2
-34880.063 X                                                      3
-30498.850 X                                                      4
-26117.637 XX                                                     11
-21736.426 XX                                                     11
-17355.215 XXX                                                    13
-12974.004 XXXXXXXXXXXX                                           55
 -8592.792 XXXXXXXXXXXXXXXXXXXXXX                                 104
 -4211.581 XXXXXXXXXXXXXXXXXXXXXXXXXXXXXXX                        138
   169.631 $$$$$$$$$$$$$$$$$$$$$$$$$$$$$$$$$$$$$$$$               161
  4550.842 $$$$$$$$$$$$$$$$$$$$$$$$$$$$$$$$$$$$$                   142
  8932.054 $$$$$$$$$$$$$$$$$$$$$$$$$$$$$$$                        129
 13313.266 $$$$$$$$$$$$$$$$$$$$$$$$$                              104
 17694.477 $$$$$$$$$$$$$$                                        60
 22075.688 $$$$$$$$                                              36
 26456.898 $$$$                                                 15
 30838.111 $$                                                    9
 35219.320                                                       1
 39600.535                                                       1
```

Figure 12. Net Present Value Distribution

```
          RISK ANALYSIS PACKAGE
          =====================

                                      (thou. $)

Most-Likely NPV @ 10.0 %              3435.495

Average NPV                           2489.296

Overall Standard Deviation            11574.061

Lower Standard Deviation              12098.506

Upper Standard Deviation              11012.287

Probability of Economic Loss          41.26 %
```

Figure 13. Net Present Value Distribution
Statistics

5 CONCLUSIONS

Both of these user-friendly software packages are very useful tools for exploration geologists and/or mining engineers. At the user's fingertips in the field or at the head office, the programs can quickly assess the economics of exploration projects or of mine development proposals at the pre-feasibility stage. The availability of sensitivity analysis within the PCDEP package allows a quick response to "what-if-questions" and a convenient method of establishing breakeven conditions. Together, the results of both packages provide a sound quantitative basis for decision-making.

The development of computer software is dynamic. Some secondary but important aspects of economic analysis still need to be incorporated into the packages. For instance, at their current state of development, the packages do not include debt financing considerations and cannot handle foreign currencies. These elements do not present any difficulties and will eventually be included into the packages. The availability of many more mineral taxation systems are being planned. A complete set of Canadian and Australian systems are presently being developed. In the longer term, the correlation between particular project parameters as well as the serial correlation between annual prices, costs and grades need to be addressed in the PCSIM package.

Many aspects of the packages need to be refined as well. The ability of the input modules to accept data under various forms, a more thorough validation of data at entry time, and a wider range of error messages are items that will improve as the packages mature and are used by a wider range of clients.

ACKNOWLEDGEMENTS

The authors wish to acknowledge financial support from MPH Consulting Ltd., St-Joe Canada Inc. and Western Mining Corp. during various development stages of these packages, as well as the assistance of Mr. Denis O'Neill in the conceptualization and programming of the early version of the PCDEP package.

Printed output for hypothetical base metal deposit

Production Summary

Year	1	2	3	4	5	6	7
Mill Feed ('000 t)	0.000	0.000	750.000	750.000	750.000	750.000	750.000

Economic Summary ('000 $)

	1	2	3	4	5	6	7
Revenue	0.000	0.000	49911.558	52407.135	55027.492	59610.260	62590.773
Operating Costs	0.000	0.000	27563.468	29492.910	31557.414	45021.911	48173.444
Operating Profits	0.000	0.000	22348.090	22914.225	23470.078	14588.350	14417.329
Federal Income Tax	0.000	0.000	0.000	1836.716	3820.498	2052.290	2193.393
Provincial Income Tax	0.000	0.000	0.000	264.181	549.514	295.187	315.483
Provincial Mining Tax	0.000	0.000	95.485	669.313	874.146	883.237	1615.711
Total Taxes	0.000	0.000	95.485	2770.210	5244.158	3230.715	4124.587
Preproduction Capital	19687.500	29258.625	0.000	0.000	0.000	0.000	0.000
Production Capital	0.000	0.000	601.073	8240.062	9716.843	809.974	866.672
Sustaining Capital	0.000	0.000	578.813	607.753	638.140	804.057	844.260
Cash Flow (Cur. $)	-19687.500	-29258.625	21072.720	11296.200	7870.938	9743.604	8581.810
Cash Flow (Con. $)	-18750.000	-26538.435	18203.407	9293.412	6167.086	7270.827	6098.932

Production Summary

Year	8	9	10	11	12	Total
Mill Feed ('000 t)	750.000	750.000	750.000	750.000	450.000	7200.000

Economic Summary ('000 $)

	8	9	10	11	12	Total
Renenue	65720.312	69006.328	72456.644	76079.476	47930.100	610740.078
Operating Costs	51545.585	55153.776	59014.541	63145.559	40539.476	451208.083
Operating Profits	14174.727	13852.551	13442.103	12933.918	7390.624	159531.995
Federal Income Tax	2266.306	2285.735	2261.372	2752.565	1491.547	20960.423
Provincial Income Tax	325.970	474.300	469.973	455.696	241.351	3391.654
Provincial Mining Tax	1854.987	1894.974	1825.452	1740.640	904.887	12358.834
Total Taxes	4447.263	4655.009	4556.797	4948.902	2637.785	36710.911
Preproduction Capital	0.000	0.000	0.000	0.000	0.000	48946.125
Production Capital	927.339	992.253	1061.710	1136.030	-18580.581	5771.372
Sustaining Capital	886.473	930.797	977.337	1026.203	646.508	7940.343
Cash Flow (Cur. $)	7913.651	7274.493	6846.259	5822.782	22686.912	60163.244
Cash Flow (Con. $)	5356.271	4689.203	4203.009	3404.460	12632.921	32031.094

Economic Indicators

Payback Period	6.03 year(s)
Rate of Return	12.24 %

	Net Present Value ('000 $)	Present Value Ratio
@ 0 percent	32031.094	0.50
@ 5 percent	14287.609	0.24
@ 10 percent	3435.495	0.06
@ 15 percent	-3370.511	-0.07

Computer Applications in the Mineral Industry, Fytas, Collins & Singhal (eds)
© 1988 Balkema, Rotterdam. ISBN 90 6191 760 3

Le développement de l'informatique à la Division Lac Dufault de Minnova Inc.

Véronique Falmagne
Minnova Inc., Division Lac Dufault, Rouyn-Noranda, Quebec, Canada

RESUME: La micro-informatique est maintenant fermement implantée dans plusieurs mines du Québec, mais cette évolution ne s'est pas faite partout de la même façon et chaque compagnie a choisi l'approche qui lui convenait le mieux. Le cheminement de la Division Lac Dufault est ici présenté comme exemple.

1 INTRODUCTION

Depuis quelques années, l'informatique est devenue chose courante dans l'industrie minière. Cependant, les premiers pas semblent partout avoir été faits par les départements d'administration ou dans le domaine du traitement des minerais. On faisait alors appel à de mini-ordinateurs accompagnés, dans la plupart des cas, d'un programmeur dédié à leur opération.

La micro-informatique, quant à elle, a souvent été introduite dans les mines par les départements de géologie. Ces derniers, comme les administrateurs, étant appelés à traiter un grand nombre de données de façon répétitive, ont très vite pris conscience du potentiel de l'informatique pour augmenter leur productivité. Comme les départements d'ingénierie et de mine ont maintenant emboîté le pas, nous verrons bientôt des mines de dimensions moyennes ou petites qui disposeront de systèmes informatiques complètement intégrés.

2 L'INFORMATIQUE A DIVISION LAC DUFAULT

Dans les années 70, la Division Lac Dufault fut un des pionniers de l'informatique dans les mines avec l'installation d'un système d'échantillonnage automatique en continu et de contrôle de procédé à l'usine de traitement. Le système, qui est toujours en place, est un PDP-8A et est maintenant relié à un IBM-PC. En 1979, la compagnie se dota d'un système 34 d'IBM qui devait servir d'ordinateur central mais qui ne fut utilisé que par l'administration. En 1983, alors que la compagnie exploitait le gisement Corbet, le premier IBM-PC-XT fit son entrée à la mine même. Ce micro-ordinateur devait alors être partagé par les départements d'ingénierie et de géologie et il devint rapidement insuffisant et désuet. Enfin, en 1986, deux personnes des départements d'ingénierie et de géologie reçurent de la direction le mandat d'étudier les besoins de leurs départements et de sélectionner l'équipement requis pour la future mine Ansil. La figure 1 illustre le système envisagé à l'époque. Les trois ordinateurs (Compaq Deskpro 386) ainsi que les traceurs de courbes (Calcomp 1043 GT et HP 7475A), les tablettes graphiques (Calcomp 2500) et l'imprimante (Fujitsu DL 2400) furent achetés tout de suite, mais on décida d'attendre avant de choisir un réseau.

Durant l'année 1987, tout le personnel en ingénierie et géologie a eu la possibilité de se familiariser avec l'équipement et les logiciels existants que ce soit pour l'arpentage, le traitement de texte, les rapports mensuels sur lotus 1-2-3, l'entrée des relevés de carottes, la mécanique des roches, le dessin assisté par ordinateur, etc. Nous avons donc développé un intérêt et des habiletés chez ces utilisateurs dans le but de les faire tous participer au développement du système et d'éviter qu'ils se perçoivent comme des victimes silencieuses d'une technologie envahissante.

La prochaine étape consiste à créer

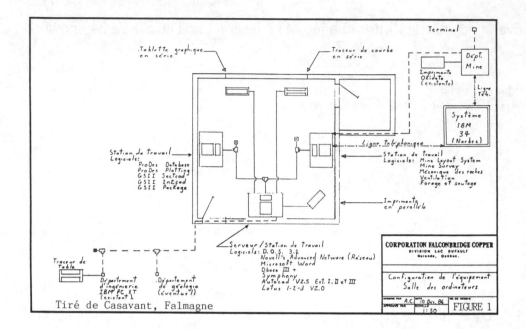

Tiré de Casavant, Falmagne

FIGURE 1

un comité regroupant une personne de chaque département de la Division afin de poursuivre l'intégration totale du système. Celle-ci se fera durant l'année 1988 et devrait être achevée lors de la mise en production du gisement Ansil.

3 DISCUSSION ET CONCLUSION

Plusieurs compagnies, aujourd'hui comme par le passé, ont choisi de confier le développement de leur système informatique à des consultants extérieurs, ou même internes dans le cas de firmes importantes. Ces consultants doivent alors, dans une période de temps relativement courte, rencontrer les intervenants et déterminer les besoins actuels et futurs, dans le but de concevoir un système complet. Leur intervention est ponctuelle et les utilisateurs se retrouvent souvent confrontés à des systèmes et des applications trop compliqués pour leurs besoins et leurs capacités immédiates. Etant donné que ces personnes ne sont pas engagées pour faire de l'informatique mais bien pour accomplir une tâche spécifique, on a souvent vu un ordinateur inactif alors que les utilisateurs potentiels continuaient leur travail manuellement.

A Lac Dufault, nous avons la chance d'avoir un personnel dynamique, conscient du potentiel de l'informatique. Nous disposons dans presque chaque département de personnes familières avec les micro-ordinateurs, ce qui facilitera leur intégration. Nous avons donc commencé notre développement par la base, c'est-à-dire par les utilisateurs, ce qui nous a permis de répandre l'usage de l'informatique mais de façon désordonnée. La tâche du comité mentionné plus haut sera donc d'unifier et de standardiser les applications antérieures et de dicter un plan de développement visant l'intégration totale de la Division.

En conlusion, il existe pour une mine plusieurs façons d'arriver à un système informatique complètement intégré. On peut s'adresser à des consultants, acheter des systèmes déjà existants et généralement très dispendieux, ou bien procéder de l'intérieur et se servir du personnel déjà en place en lui fournissant le temps et la formation nécessaire. A la Division Lac Dufault, c'est cette voie que nous avons emprunté et à date nous pouvons être satisfaits des résultats obtenus.

REMERCIEMENTS

L'auteur tient à remercier la direction et le personnel de la Division Lac Dufault pour leur bonne coopération.

REFERENCES

Bakes, N., Fournier, G., Ingham, K.
1987. Survey of information systems
at Minnova inc., Lac Dufault Division
Rapport interne: Noranda inc.

Casavant, A., Falmagne, V., 1986.
Rapport sur les besoins en informa-
tique pour le projet Ansil, Division
Lac Dufault. Rapport interne:
Minnova inc.

Sanders, D.H., 1983. L'univers des
ordinateurs. Montréal: McGraw-Hill.

Computer Applications in the Mineral Industry, Fytas, Collins & Singhal (eds)
© 1988 Balkema, Rotterdam. ISBN 90 6191 760 3

Microcomputing in international investment negotiations

David Gulley
CERNA, Ecole Nationale Supérieure des Mines de Paris, France
Simon Handelsman
DTCD, United Nations, New York, USA

ABSTRACT: In international investment negotiations for mineral development, computer support is important to the negotiating parties. Financial models are used to evaluate the economic, commercial and financial feasibility of the project. They also can play crucial roles in the refinancing and restructuring of existing operations. In negotiations, there are important intellectual and also social/psychological uses for such models. The central role of such models lies with their ability to caclulate the impact of major variables (such as prices, inflation, and interest rates) on the project's financial and fiscal régime. Sensitivity analyses include the risk of cost overruns and the synergistic effects of variables deviating from forecasts. Computerized financial analysis allows for an instant calculation of the financial impact of major negotiating variables. Computer-generated displays can be very helpful in selling a proposal or tentative agreement to others.

1 INTRODUCTION

The personal computer and electronic spreadsheet have already had profound effects on financial analysis in the developed countries. The new generation of portable but powerful p.c.'s will similarly transform the working environment in many other nations as well. The international nature of mining and energy can involve globe-trotting rounds of negotiation, for which the new generation of portable computers is ideally suited. In this paper, we look at the potential for this technology in the field of mineral investment analysis and negotiations.

When we refer to investment analysis, we mean something a bit different than mine feasibility analysis and valuation. Of course the engineering feasibility study has benefited considerably from modern computer technology, and many of the economic analyses included in a feasibility study can be performed with personal computer spreadsheets. We might mention in this regard the U.S. Bureau of Mines' new Lotus-based cost-estimating system, which is a considerable achievement likely to be of real help to mining engineers. Many mineral economists, the authors included, have performed various real-world mine valuations on personal computers as well. However, bankable feasibility studies require more in the way of computing resources than even the most advanced portable computers. By "mineral investment analysis", we do of course include discounted cash flow (DCF) valuations, as are traditionally envisioned by engineers and business analysts. However, one should also think of other financial analyses supporting negotiations. The first-listed author (Gulley), for example, several years ago built a working capital model for a major U.S. minerals company, which played a key role in negotiation and litigation regarding a Chapter 11 (bankruptcy) petition. Prior to that, he constructed a simulation model to evaluate likely lessee responses to restructured diligence stipulations in energy leases, which may have influenced U.S. legislation. Both of these analyses were performed on what is now considered an "obsolete" (CP/M) portable computer. The point here is that in the present, difficult years for mineral commodities, financial negotiations are likely to involve issues such as financial restructuring, as well as the valuation of a green-fields mineral property.

2 INFORMATION AND NEGOTIATION

Negotiations are both an intellectual and a social process. The role of the computer will therefore naturally encompass both of these aspects. The intellectual process includes both information gathering and information processing. The social process comprises all the arts of persuasion and power, as they relate to the competition and cooperation among parties-at-interest, and also within each negotiation team and its relations with the larger group which it represents.

This is sometimes called the "three-table problem": in a two-party negotiation, the two parties negotiate across the first "table"; but then each of the parties must reach an internal consensus, reflecting the negotiation teams and also the constituency they represent (the other two "tables"). Good negotiation support technologies, such as personal computer software, must be able to address as many of these dimensions as possible.

The uses of personal computing to support information processing are readily appreciated. This improves the teams' ability to create and evaluate options, of course, but it also facilitates the social process wherebye the team can reach a consensus. Having the facility to recalculate the impact of negotiable variables under a variety of scenarios greatly helps all concerned in the negotiation process and contributes to consensus-building, since all parties may assess the same assumptions, and it is easier to develop financial régimes of sufficient flexibility to respond to the various uncertainties of the future. This is particularly useful in an exploration venture; when often the size of the ore body; its grade; and the mining conditions are largely unknown. Constructing a model allows for the development of a range of possible scenarios; and for the design of a fiscal régime able to respond fairly to the alternative scenarios.

However, an assessment of the role of personal computing in negotiations must also describe what the computer does not do, and we must emphasize that it is not a panacea. Particularly for exploration concessions and leases, personal computer analysis is more an illustration of the inherent uncertainty than an accurate measurement of risk. Information processing is less crucial than information gathering. The gathering of information includes each party's probing of other parties in the negotiation, as well as the development of objective information. This, in turn, includes what the other parties say and what information they present; but especially, the information revealed behaviorally, by how far the other parties can be pressed.

A second key feature of many negotiations is time pressure: it is generally in one party's interest to stall as long as possible, so that when the negotiators finally get down to brass tacks, they have surprisingly little time. Years may have been spent deciding the shape of the table or other details; but the real deal will be cut when time has all but run out.

If one finds oneself two days' journey from the nearest telex machine (this has happened to us), there is no way to avail oneself of the equipment at the home office, no matter how wonderous it may be. If a two man-year feasibility study is presented a few scant hours before the decision must reached, one hasn't the time to even read it, much less enter it in a data base. Any agreement reached will still have to be sold "back home", usually to heterogeneous groups with different agendas and varying stakes in the "success" (and that will be defined differently) of the negotiation. This group will wield considerable intellectual

power and presence of mind; but it can be almost guaranteed that crucial people care as much about plebescites as present values. It definitely can be guaranteed that Adam Smith's "economic man" will be interested in self-interest as well as overall project payoff. The key is often persuasiveness -- at all of the "tables" at which one sits. Surveys and experiments of negotiator effectiveness have confirmed this trait (Gulley, Koopman, and Bikerman, 1987).

The point is that for various reasons, intentional or otherwise, financial negotiations and decision-making are seldom entirely rational. This is even the case when the social system is not supposed to be adversarial, as one of us has demonstrated elsewhere (Gulley and Mei, 1985). This view of negotiation puts considerable premium on having accessible technology, quick turnaround, and a self-documenting, persuasive approach to the effort. Portable computers, though no panacea, have obvious merit for exactly these reasons.

3 ISSUES OF EQUITY AND EFFICIENCY

Efficiency refers to the negotiators' ability to maximize the total value of the package they are negotiating. Equity here is used in the neoclassical sense, i.e. not equity investments like common stocks or net worth, but rather fairness and the sharing of the package. The social and intellectual dimensions of negotiations should address both issues.

Negotiators should free themselves from the offer - counteroffer ("positional") school of negotiation. In fact, a back-and-forth haggling over numerical positions is rare in practice, as it should be. The first retreat from such a position is tantamount to signalling that one is playing that game. This is an invitation to others to press their case firmly, to elicit further reductions in demands.

One way around this dilemma is to take initial positions reluctantly and somewhat vaguely (this is one reason serious negotiations come late in the game). We feel, however, that experience shows the importance of creating alternative negotiation bundles, aimed at increasing the size of the pie one is splitting. This masks any retreat from an earlier position, by making a sequence of proposals less immediately comparable. But it is not just a face-saving gesture. An "efficient" negotiation is one in which the terms are structured so as to create as much potential value as possible. Economists will recognize this as the pareto-efficiency criterion. The idea is very simple once it is grasped, but in previous presentations (Gulley, 1987), we have noticed that many people are slow to accept the idea that one can break out of a zero-sum negotiation game ("every dollar I gain is a dollar you lose").

We would like to state the approach very briefly, and refer you to our previous paper cited above, and to more elaborate descriptions (Fisher and Ury, 1981; Raiffa, 1982). We will illustrate the idea with a simple lease negotiation, but emphasize the approach is even more appropriate for cases involving more complex issues.

The simplest negotiation of all involves a lump sum payment. The fact that buyer and seller attach different values to the asset and the cash is what makes such a transaction work. In a typical lease transaction, it may be possible to negotiate a stream of time payments and performance stipulations, which affect the two parties differently. A dollar out of my pocket is not a dollar in your's, because our discount rates, tax rates, present and future liquidity, risk exposure, investment opportunities and expectations are different.

The key to efficient negotiation is the search for packages which exploit these differences. In fact it was probably the commonality of differences among lessors and lessees which have allowed certain lease provisions to become fairly standard.

Table 1, taken from an earlier paper (Gulley, 1987), illustrates the idea. Five different time-series of payments are shown, illustrating how differences in tax and discount rates can affect each party's present value. Space limitations rule-out a full-scale illustration of DCF analysis; and the technique is quite familiar to most readers, anyway.

By listening carefully to the other parties to the negotiation, one can begin to understand what forces drive their behavior. They may be explicit, but even if not, one can form working hypotheses to test. The personal computer can help one create alternatives in which one gives away that which is of more value to others than to one's own team. Similarly, one can search for trade-offs which are particularly attractive to oneself: diligence stipulations and provisions for equity back-ins instead of bonuses or import duties, for example. One is literally limited only by one's imagination, and the ability to make a convincing case.

In summary it is in the search for efficiency that the "intellectual" dimension of negotiation (as referred to in the introduction) is most helped by personal computing. The social role of computing lies in the communicability of the results. Let us begin with the main "table". We feel that negotiators should always strive to present the sincerest case for their position. Over time, this will have an effect on the other side's cynicism. Few people are entirely immune to appeals to fairness; and the climate of negotiation is very important to the overall result. Remarks illustrating the point-of-view of a negotiation team can foster a climate of sincerity and reasonableness. At the same time, a negotiation team will not want to share all of its information.

The second "table" is the negotiating team's own table. Goals must be agreed upon, and the results must be sell-able back home. No matter how sophisticated the analysis, one must be able to couch it in terms immediately understandable to everyone. In particular, the negotiators will need something to anchor them, which they can refer to in the heat of the negotiation without having to take a break in order to digest it.

Graphics capabilities are very helpful in displaying the general approach and general implications of a given proposal. They have the considerable merit of being quickly digested and somewhat vague on details. A particularly useful approach is to create a series of pie-charts of different sizes (for example, net present values), corresponding to more and less efficient proposals; with each party's share shown prominently. These should certainly be shown at one's own table, and may be appropriate for display at the main table.

Our own experience suggests that charts are often more persuasive than actually sitting someone down at the keyboard and doing the cash flows. It is much easier to grasp the implications of royalties, taxes, fees, terms of sale, and other proposals when they are couched in terms of sharing the total value.

CONCLUSIONS

So much has been written on the subjects of negotiations, investment analysis, and personal computing, that we felt a brief paper to be best. After all, we have emphasized the need for easily digestible summaries of the analysis; and would like to hold ourselves accountable to our own standards. It is quite likely the reader is familiar with much of the literature, and the references cited will elaborate on the subject if need be.

People who approach any aspect of life as primarily a "numbers game" do so at considerable peril. All organizational life is social in nature. Moreover, rare indeed is the investment decision in which the numbers are sufficiently solid.

These considerations may relegate the financial analysis to a secondary role. This is so not because the numbers are unimportant -- on the contrary, they are the very heart of the matter. One is tempted to say that personal computing is a victim of its own success:

Having developed facility with the financial calculations, one begins to appreciate that they are only the beginning. Adopting the right stance, proper homework, good team discipline, the search for alternatives to a negotiated agreement, superb timing -- all these and more should be subject to the most rigorous pre-negotiation assessment. These of course are not directly aided by the computer. In a way, the portable computer has if anything accentuated the importance of these -- by doing so very well that which it can do, the portable computer is making it a foregone conclusion that the numbers be properly run, and the financial proposals be as creative as possible.

ACKNOWLEDGEMENT

Professor Gulley's continuing activity on mineral investment analysis and negotiations has been greatly supported by the United Nations, the John D. and Catherine T. MacArthur Foundation, the U.S. National Science Foundation, the Harvard-M.I.T. Program on Negotiation, and his public and private clients. However, the opinions expressed are not necessarily the opinions of any organization, and the authors alone are responsible for any errors or omissions.

REFERENCES

Fisher, R. and W. Ury, 1981, Getting to Yes: Negotiating Agreement without Giving In, New York: Penguinn Press.

Gulley, D. A., 1987, "Some Observations on the Negotiation of Mineral Agreements", EDP Methods in Mineral Exploration and Development (in press).

_____ , C. Koopman, and D. Bikerman, 1987, "The Impact of Arbitrage on Bilateral Agreements", working paper.

_____ , and D. M. Mei, 1985, "The Impact of Decision Models on Federal Coal Leasing", Management Science, v.31, n.12, December, pp.1547-1568.

Raiffa, H., 1982, The Art and Science of Negotiation, Cambridge MA: Harvard University Press.

TABLE 1

PAYMENT STRUCTURE AND PRESENT VALUES, LESSEE AND LESSOR

(in thousands of U.S. dollars)

Year	Schedule 1	Schedule 2	Schedule 3	Schedule 4	Schedule 5
0	1000	500	1000	500	1300
1	100	100	100	100	0
2	100	100	100	100	0
3	100	100	100	100	0
4	100	100	100	100	0
5	100	300	100	300	0
6	100	300	100	300	0
7	100	300	100	300	0
8	100	300	100	300	0
9	100	300	100	300	0
10	100	300	100	300	700
Totals:	2000	2700	2000	2700	2000
PV, lessor:	1614	1709	967	934	992
PV, lessee:	-1419	-1239	-709	-619	-706
	<-- Before Taxes -->		< - - - After Taxes - - - >		

Assumptions

Discount rates: lessor, 10 per cent; lessee, 20 per cent
Tax rates: Schedules 1,2: no taxes;
          Schedules 3-5: 50% except lessor capital gains @ 34%

Computer Applications in the Mineral Industry, Fytas, Collins & Singhal (eds)
© 1988 Balkema, Rotterdam. ISBN 90 6191 760 3

Cash flow vs. cut-off grade criteria for mining planning and production scheduling – Case study on an underground gold mine

J.Q.Rogado, F.H.Muge & F.Freitas
Instituto Superior Técnico, Technical University of Lisbon, Portugal

ABSTRACT: The influence of cut-off grade criterion on the short time cash flows (for instance, in the case of small mines) could be crucial. Yet management decisions to improve
cash flows have strong linkage with the current cut-off grade criterion.
   To tackle with such problems, a package of programs for mine planning and operational
control was developed and is described. The program takes into account mineralogical and
metallurgical characteristics of the mined material.
   The first step is the construction of a typology of the ore to be mined based on its
mineralogical and minerallurgical characteristics. After the definition of the mine preparation works (shafts, drifts, etc..) several mining policies can be adopted. To   each
policy corresponds one time series of run of mine material which will be processed in the
concentrator under the criterion of "maximum conservation". Besides, to each output   of
the concentrator corresponds a cash-flow and several additional economical indicators .
The program is interactive so that the operator can choose the policies he is interested to.
   An example concerning an underground gold mine is presented.

1 PROGRAM DESCRIPTION

The program is the mine planning version
of the software package "SAVARN - System
for the Evaluation and Valorization of
Natural Resources" (C.V.R.M.U.T.L. 1982).
   Figure 1 shows the general flow chart of
the program.

1.1 Typology

This modulus defines typologies using several data analysis techniques (P.C.A.,Corres-
pondence Analysis, etc..).
   The interaction with the operator is high
in this modulus and the outputs can be sent
to the Data Base (D.B.). The typologies are
usually based on sampling data and the attributes or variables can be either quantitative or qualitative. These attributes can
be referred to different mining supports
(core samples, mining blocks, etc...) and
so, this modulus is able to receive input
from the Mine Discretization/Estimation
Modulus.

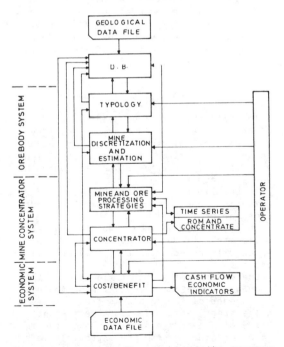

Fig. 1 General flow chart of the program

1.2 Mine Discretization/Estimation

This modulus was designed to model the following mine planning operations:

(i) Definition of the morphology of the orebody using either interactive procedures (e.g. manual interpolation of the hanging and footwalls) or interpolation methods like Indicator Kriging (Soares 1985)

(ii) Definition of the dimensions and orientation of the mining blocks (discretization)

(iii) Estimation of the mining attributes or variables, block by block, using the appropriate geostatistical estimation methods

(iv) Definition of the mine preparation works (shafts, drifts, etc...).

In the case of an open-pit, the package includes software based on an optimal moving cone algorithm (Muge, 1977) in order to define the ultimate pit that maximizes a predefined objective function. The pit can be parametrized in function of the cut-off grade (simple or multiple, constant or variable with depth).

In what concerns underground mining, the operator will have to choose the mining method based on the orebody morphology.

In each case, this modulus is assisted by a graphic terminal that ables the interaction with the operator. The outputs can also be sent to the D.B. .

It is also important to note that this modulus includes a conditional simulation package (Sousa 1983) required for the study of the variability of mining variables.

1.3 Mining strategies

This modulus can operate in two modes: interactive or automatic.

In the interactive mode, the zones to be mined are presented in a graphic display and the operator can choose any criterion to select the mining sequences, obtaining as output, time series of the attibutes he wants to control.

In this mode the operator can alter his criterion upon the analysis of the output time series.

In the automatic mode, two types of criteria are introduced: (i) order relationships (mining constraints); (ii) optimal criteria.

One of the optimal criteria used in this modulus is the total deviation of the average grades of mining blocks from the overall average panel grade (or the average grade of a set of mining panels or the fixed input grade of the concentrator). This means to schedule the operation by removing a certain number of mining blocks at a time,

subject to the geometric (mining) constraints, and such that the total deviation of average block grades from overall average grade is minimum.

A dynamic programming algorithm is used to define the optimal mining sequence(Muge 1979; Ribeiro 1982).

This optimal criterion can also be applied to complex orebodies with several ore characteristics and the deviation of each one of these characteristics can be further constrained by specifying maximum values.In this case, the automatic mode can, optionally, introduce some stationarity criteria such as to exploit the several ore-types in such a rate that each one is stationary (in average and variance or in histogram) till the deflection of the mine (Rogado 1975a).

The outputs of this modulus can also be sent to the D.B..

It is important to note that using this modulus is possible to built the rejection criterion under a global point of view, either working with it independently or interacting with the Ore Processing Strategies modulus and through it with the Concentrator and Cost-Benefit moduli.

1.4 Ore processing strategies

The objective of this modulus is to define the strategy of optimal concentration to the set of ore-types, characterizing the ore processing in terms of grades and recoveries.

The inputs of this modulus are:

(i) the universe of the attributes of the mining blocks generated by the Mine Discretization/Estimation modulus;

(ii) economic data necessary to built the global envelop Cost-Benfit of the mining project.

In order to use this modulus, the ore processing variables, which are the concentrate grade/recovery once fixed the size composition, are affected to each mining block. This means that systematic concentration tests have been done, covering the entire deposit.

The optimal criterion used by this modulus is to obtain the pair (concentrate grade, recovery) that maximizes the GAV (gross added value), assuring the "maximum conservation" principle (Rogado 1975b). This principle consists in obtaining a given concentrate grade with the minimum lost in metal.

Using this principle, the program is able to define the "ideal" working conditions in terms of conservation of the ore reserve.

1.5 Concentrator

This modulus uses a package of models concerning the mineral processing unit operations (e.g. comminution, classification, flottation, etc...). With such models it is possible to simulate several treatment diagrams. Once the treatment diagram is choosen and the estimation of the model parameters is achieved, the modulus can treat any time series produced by the Mining Strategies modulus.

The definition of the treatment diagram and the estimation of the model parameters is carried out with high interaction.Graphical representation can be used.

Connecting the subroutine dynamic processor to the above mentioned package, the processing of the time series yielded by the mining strategy can be easily done. Teh output time series can be stored in the D.B..

1.6 Cost-Benefit

This is a classical economical modulus,able to built up cash-flows and economic indicators. It uses and economic data file.

This modulus can be used interactivelly with any of the above mentioned moduli and its output can also be stored in D.B..

2. CASE STUDY

The methodology was applied to an auriferous quartz vein deposit which has produced during the years 1952 through 1955 and is now being evaluated.

The deposit is located in the Northeastern Portugal. It is part of a quartz vein field intruded in Silurian schists, trending North-Northeast from a hercinian alcaline granite stock. The mineralization changes from W-Sn near the granite to Au-Ag-Pb some 2.5 to 3 Km to the North.

The field was investigated during the late fourties by trenching and sampling , from which results are known of 39 samples. Two veins, inclined some 75° to the West, were selected for development, during which 503 samples were taken and analysed for Au and Ag. Data from ROM and Au-Ag concentrate production is available in a monthly basis. Pb is known to have occurred in both, but in percentages only known as averages for the global production.

Figures 2 to 7 show some outputs of the program.

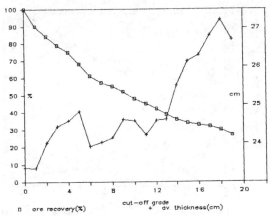

Fig. 2 Orebody recovery and average thickness vs. cut-off grade

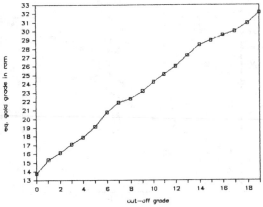

Fig. 3 Equivalent gold grade in the ROM vs. cut-off grade

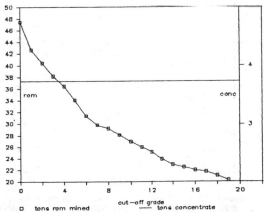

Fig. 4 Tons of ROM to be mined each year and tons of concentrate to be producted yearly in function of the cut-off grade

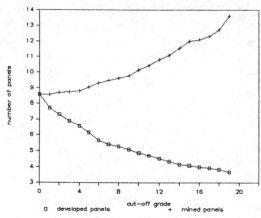

Fig. 5 Number of panels developed and mined each year in function of the cut-off grade

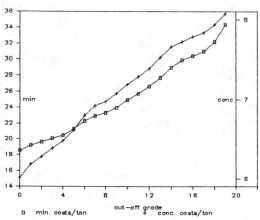

Fig. 6 Mine costs and concentrate costs vs. cut-off grade

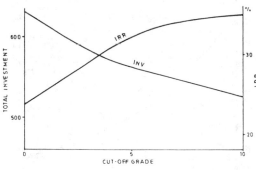

Fig. 7 Total Investment and Internal Rate of Return in function of the cut-off grade

3 COMMENTS

The described software package has some characteristics that must be emphasized:

(i) It allows the establishement of global rejection criteria supported by an economical analysis and so it can be used to ajust the exploitation policies to different conjunctures.

(ii) It is possible to study efficiently the impact of any changement in the mining method or in the concentration.

(iii) It allows the study of alternative mining projects.

REFERENCES

C.V.R.M.U.T.L. 1982. Manual de Gestão do Sistema SAVARN, Lisbon: Centro de Valorização de Recursos Minerais(C.V.R.M.U.T.L.)

Muge, F.H. 1977. Determination of the optimal ultimate pit contour. Técnica 451/452: 261-270.

Muge, F.H., and Pereira, H.G. 1979. Short--term planning in sublevel stoping methods. 16th Application of Computers and Operations Research in the Mineral Industry. Thomas J. O'Neil (ed.) p.323-336.New York: Society of Mining Engineers of AIME

Ribeiro, L. 1982. Dynamic Programming applied to the mining sequence optimization in a sublevel stoping exploitation. 17th Application of Computers and Operations Research in the Mineral Industry. T.B. Johnson and R.J.Barnes (ed.), p. 494-499. New York: Society of Mining Engineers of AIME

Rogado, J.Q.1975a. An optimization method for the mining and beneficiation of ore blocks. International Journal of Mineral Processing. 2: 59-76

Rogado, J.Q. 1975b. Sequential Optimization method for dressing control of ore bodies 13th International Symposium on the Applications of Computers and Mathematics for Decision Making in Mineral Industries. Clausthal.

Rogado, J.Q. 1983. SAVARN - System for the Evaluation and Valorization of Natural Resources. In Encontros sobre Métodos Quantitativos Aplicados às Variáveis Regionalizadas, Lisbon: I.N.I.C.

Soares, A.O. 1985. Conditional Indicator simulation of a multiseam coal deposit. Sciences de la Terre, Ser.Inf., 24: 175--183

Sousa, J., 1983. Aplicação da simulação condicional a uma jazida de ferro. In Encontros sobre Métodos Quantitativos Aplicados às Variáveis Regionalizadas, Lisbon: I.N.I.C.

4 Mineral processing I

Computer Applications in the Mineral Industry, Fytas, Collins & Singhal (eds)
© 1988 Balkema, Rotterdam. ISBN 90 6191 760 3

The simulation of circuit regrind requirements

G.Wonnacott & B.A.Wills
Camborne School of Mines, Redruth, UK

ABSTRACT: Traditional methods for design of middlings regrind circuits require
lengthy and tedious experimental programmes. The "split factor" model proposed
herein, a population balance model, attempts to describe the regrinding of gravity
concentrator middlings from primary grinding density distribution data. The model
provides an optimal start point for a final locked cycle test. This considerably
reduces the quantity of experimentation required for regrind design.

1 INTRODUCTION

Little work has been performed on regrind
simulation, but much research into
various liberation models has been undertaken. Most of these characterise
liberation by:
  1. Particle mass and assay.
  2. Mineralogical structure of the ore.
Unfortunately, in the former case, the
complex nature of resultant equations
precludes an analytical solution for any
but the simplest cases. In the second
type, many restrictive assumptions are
necessary. One such is the decoupling
of the comminution process from the
liberation process, ie: any type of
comminution process will produce the
same degree of liberation for the same
degree of grinding. Clearly this is
physically incorrect.
  Peterson and Herbst(1), describe a
better scheme. This is a multicomponent-
multisize grinding model based on the
classical population balance grinding
models. In this case the grinding model
was applied to three components, pure A,
pure B and locked AB. This model has
since been developed for multicomponent
systems. The model has been validated
using coal and sulphide ores. However,
the model requires a very large parameter
set for complete description of the
process. This set was reduced by making
several restrictive assumptions. Even
with these assumptions the solution is
fairly complex.
  Agar(2-4), has proposed a model for

predicting flotation locked cycle tests.
The model uses "split" factors to characterise the various processes in the
circuit. The split factors represent
the proportion of a mineral recovered to
one of the product streams. The split
factors were related to physico-chemical
conditions pertaining to the process.
One basic assumption made was that each
mineral possessed a characteristic split
factor. Agar doubted whether this was
valid. This assumption became even more
doubtful when applied to recycled middlings
streams.
  The schemes outlined above suffer from
complex mathematics and/or restrictive
assumptions. It was felt that a much
simpler model could be applied to gravity
regrinding design.

1.1 Introduction to the Regrind Model

All gravity process models require information about feed size and density distribution, and usually give the same information
about the predicted products. Hence a
gravity middlings regrind model is required
to predict the mill product density
distribution, given the feed size, feed
density distribution and the product size.
  A model developed by Tucker(5) to predict
density distributions from limited density
data requires use of assay data and restrictive assumptions. The major disadvantage
of assay data is that assays are more
expensive, require more skill to obtain and,
generally, are less accurate than mass data.
Thus a model which only required mass data

for density distributions is more desirable.

The regrinding model under development is a simple population balance model. It consists of a series of simple mass balances on each density class when the material undergoes size reduction.

2 THE SPLIT FACTOR MODEL

The approach adopted was to grind a selected ore to a series of primary grind sizes, and to determine the density distributions at each grind size. The proportion of material in any density class d, after grinding, can be given by the following mass balance:-

$$w_{d(g)} = \sum_{k=1}^{n} w_{k(g-1)} \cdot s_{dk} \qquad ...(1)$$

Here d is the dth density class in a series of n such classes; g is the gth grind size, (the "size number") where the gth is a finer size than the (g-1)th; and s_{dk} is the proportion of density k that appears in density d, after grinding. By definition this is the "split factor" for density k at grind size g-1 into density d at grind size g.

Equation (1) states that the mass of material in each density fraction at the finer size is the summation of the contributions to that density from each of the density fractions in the coarser size. The proportion of density k contributed to density d, at the finer size, is the split factor, s_{dk}.

The "split factor" concept requires some elaboration. Figure (1) shows the definition of split factors schematically. It can be seen that a split factor is the proportion of material contributed from one density to another density.

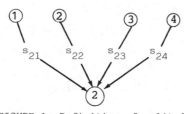

FIGURE 1: Definition of split factors

For example, if an ore is ground from size g-1 to size g, and for simplicity the grind size is an 80% passing size, and there are n density classes, then the following series of mass balances result:

$$w_{1(g)} = s_{11} \cdot w_{1(g-1)} + s_{12} \cdot w_{2(g-1)} + \ldots + s_{1n-1} w_{n-1(g-1)} + s_{1n} w_{n(g-1)}$$

$$w_{2(g)} = s_{21} \cdot w_{1(g-1)} + s_{22} \cdot w_{2(g-1)} + \ldots + s_{2n-1} w_{n-1(g-1)} + s_{2n} \cdot w_{n(g-1)}$$

and so on, to density n. The series of n equations can be represented in matrix form

$$\begin{bmatrix} w_{1(g)} \\ w_{2(g)} \\ \ldots \\ \ldots \\ w_{n(g)} \end{bmatrix} = \begin{bmatrix} s_{11} & s_{12} & \cdots & s_{1n-1} & s_{1n} \\ s_{21} & s_{22} & \cdots & s_{2n-1} & s_{2n} \\ \ldots & \ldots & \ldots & \ldots & \ldots \\ \ldots & \ldots & \ldots & \ldots & \ldots \\ s_{n1} & s_{n2} & \cdots & s_{3n-1} & s_{nn} \end{bmatrix} \cdot \begin{bmatrix} w_{1(g-1)} \\ w_{2(g-1)} \\ \ldots \\ \ldots \\ w_{n(g-1)} \end{bmatrix}$$

which can be summarised by:

$$W_{(g)} = S \cdot W_{(g-1)} \qquad ...(2)$$

Thus to model a regrind process the split factors, s_{dk} for each grinding step must be determined. At the time of writing the effect of grind size on the split factors had not been analysed quantitively.

Once the split factors have been determined the model is a powerful design tool, since the density distribution for any grind size can be determined, without recourse to extensive locked cycle tests.

2.1 Determination of split factors

Comminution population balance models have used two forms of parameter estimation schemes:
1. Direct parameter evaluation.
2. Indirect parameter evaluation.
The former method involves the comminution of an isolated fraction of the population. This isolation can be achieved by one of the following methods:
1. Physically isolating a fraction.
2. Chemically tagging a fraction.
3. Radioactively tagging a fraction.
Analysis of the fractions progeny on grinding yields the parameters. However, all three methods require lengthy test programmes.

Since the aim of the model is to reduce testwork to a minimum, an indirect statistical method of parameter estimation was adopted. Indirect methods, whilst generally mathematically complex, require less information than direct methods, to give the parameter set.

It was initially thought that the split factors could be determined by simultaneous solution. This required grinding the sample to n different sizes, in a sieve series, for n densities. This was the basis for the test programme.

3 EXPERIMENTAL

The experimental work involved taking a sample of tin ore from South Crofty mine, Cornwall, grinding it to n sizes in a root two sieve series, and determining the density distributions at each grind size. Two programmes were undertaken:
1. Simulation of recrushing of DMS middlings, using coarse ore (53mm-2.8mm).
2. Simulation of regrinding of gravity middlings, using fine ore (2mm-125μm).

The former programme showed that, although the regrinding model was applicable, it showed little significant advantage over existing methods. This was due to the large quantities of ore involved for statistical representivity. Thus the results of the coarse ore testwork are not discussed here.

The fine size test programme showed that the regrinding model appeared to be applicable, and has distinct advantages over existing regrind design methods.

3.1 Mineralogy

To enable a clear explanation of the results a brief description of the ore mineralogy is required.

The sample of South Crofty ore is of hydrothermal origin. It is from a vein in granite country rock.

The mineralogy is simple, the vein consists of fine grained blue Tourmaline, in which most of the tin, as Cassiterite, is present. The blue Tourmaline has narrow (-5mm) Quartz stringers running through it. Green Chlorite is present, near the vein sidewalls, often containing Quartz vughs, or pink Feldspars. Minor amounts of sulphides are present in all three vein minerals. The wall rock is granite, consisting of Feldspar, Quartz and Biotite mica. Table (1), below, gives the specific gravity ranges of the minerals present:

TABLE 1: Specific gravity of minerals

Mineral	Density
Tourmaline	2.98 - 3.20
Fluorite	3.00 - 3.25
Quartz	2.65
Chlorite	2.65 - 2.94
Feldspars	2.57 - 2.67
Micas	2.70 - 3.10
Cassiterite	6.80 - 7.10
Sulphides (Typically)	4.00 - 6.20

3.2 Method

A representative sample (95kg) of coarse ore was divided from the bulk sample, and crushed to -2.8mm. A 10kg subsample was taken, and subdivided into 0.5kg samples. Ten samples were ground under standard conditions in a laboratory rod mill for pre-determined times, to give ten different root two sieve sizes (2.8mm to 90um).

A density determination was carried out on the +45μm fraction of each sample, using eight density classes (-2.7kg/l to +3.3kg/l) in 0.1kg/l increments.

To determine the applicability of 80% passing size as a criterion to describe the sample size, a full size analysis was carried out upon each density fraction. The sample size analysis was reconstituted from this data.

A microscopic mineralogical analysis was carried out on each fraction.

3.3 Discussion of results

For simpler analysis, the raw data was converted into mass % of oversize (+45μm) against size number (g). The poor separations achieved by dense liquid analysis of the undersize (-45μm) made undersize data almost useless, hence it was decided not to model this, especially since -45um is practically irrecoverable by most conventional gravity methods.

The mass % data for each density class was plotted against size number. Examination of the plots showed that there were distinct trends in mass % with size. The data for each density were fitted using least sqaures regression, shown in Figures (2)-(4).

The least squares lines fit the data well, with few exceptions. On the basis of mineralogical examination, it can be stated that the few outlier points are due to experimental error, probably in the density determination stage.

It will be noticed that the trendlines fall into two distinct classes:
1. Positive mass trends with size number.
2. Negative mass trends with size number.

The former category represent tailings classes. As size reduction continues, so more gangue is released into these classes.

For example the -2.7kg/l class represents almost pure quartz and pure feldspar, and chlorite/quartz locks. The positive trend exhibited by the +2.8-2.9 fraction is due to this being a "Chlorite concentrate". Reference to Table (1) will show that this is the mean density of chlorite. Microscopic examination of this fraction showed that chlorite was indeed being broken into

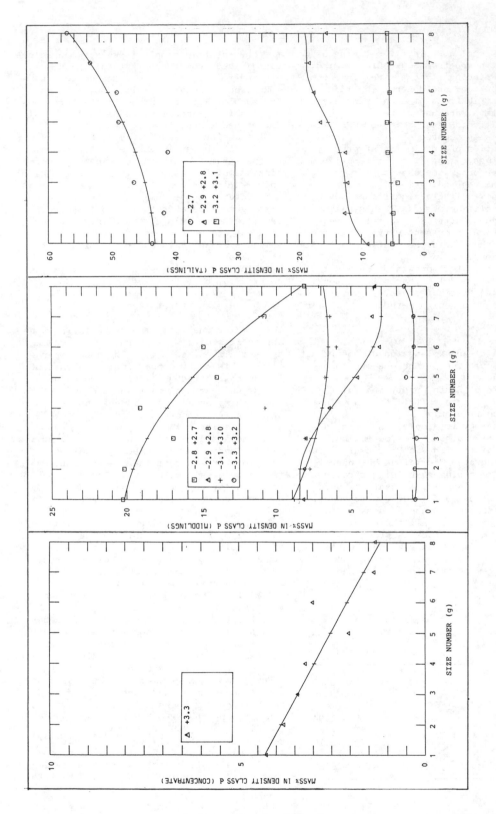

FIGURE 2: Mass% in "Concentrate" against size number.

FIGURE 3: Mass% in "Middlings" against size number.

FIGURE 4: Mass% in "Tailings" against size number.

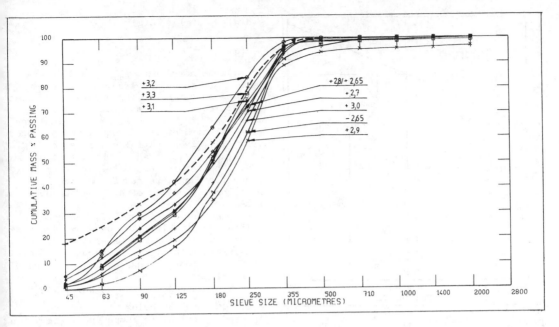

FIGURE 5: Size distributions for each density class, size number (6)

this fraction. The +3.1 -3.2 class represented a Fluorite concentrate and middlings
class. It was noticed that Fluorite was
broken into the class and middlings broken
out resulting in a slight upward trend.

Negative mass % trends represent concentrate, such as the +3.3 fraction, or middlings bands, such as the +2.9-3.0 fraction.
In both cases liberation of dense minerals
from light minerals is occurring, the net
result being a decrease in mass % in density class with size. Examination of the
+3.3 fraction showed that it consisted of
almost pure Tourmaline and Sulphides.
Similarly examination of other fractions
showing negative trends, showed that they
mostly consisted of Tourmaline and Quartz,
or Chlorite, in various degrees of locking.

From the above it can be seen that there
are distinct physical reasons for each
trend in mass % against size number.
These trendlines are important to the parameter estimation scheme, which was being
tested at the time of writing.

The applicability of 80% size as a criterion to describe sample size distribution
was also examined. Figure (5) shows a
typical data set. The size distribution
for each density is plotted for size number
6. The reconstituted data (broken line)
shows that the sample has an 80% size of
250μm. It can be seen that the individual
80% sizes range between 225μm and 300μm.
However, the sample masses were rather
small, thus this range may be due to

experimental error. From this and considering the widespread use of 80% size for
describing sample size range it was decided
to retain its use in the model.

4 PARAMETER ESTIMATION

An attempt was made to estimate the split
factor matrix by least squares simultaneous solution of the density data for all
sizes. This did not work. Thus it is concluded that the split factor matrix is not
constant over the size range being covered.
Hence an estimation scheme which only requires data for two or three sizes was
needed. However, before this could be done
the n squared parameter set, and the split
factor matrix elements, had to be reduced
to about 2n parameters.

4.1 Lumped Parameters

Theoretical consideration of the mass % in
density class versus size trendlines led
to a simplification, from which the parameter set may be obtained. Consider a
typical plot of mass % in density d, for
sizes g-1 to g+1, such as shown in Figure
(6).

Here $x_{d(g)}$ is a convenient way of writing
the gradient, $\Delta w_{d(g)}/\Delta g$. $A_{d(g)}$, $B_{d(g)}$ and
$C_{d(g)}$ are lumped parameters. $A_{d(g)}$ represents the fraction that is broken out of
density d, when size is reduced from g-1 to
g. $B_{d(g)}$ represents the mass of material
broken into density d, on grinding. $C_{d(g)}$

113

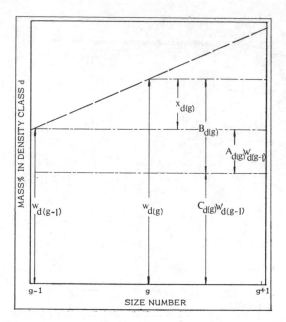

FIGURE 6: Theoretical trendline of mass % in density with size

is the fraction of density d that remains in density d on size reduction. The following relationships can be written, from Equation (1) and Figure (6):

$$A_{d(g)} = \left(\sum_{k=1}^{d-1} s_{kd} + \sum_{k=d+1}^{n} s_{kd} \right) \qquad (3)$$

$$B_{d(g)} = \sum_{k=1}^{d-1} s_{dk} w_{k(g-1)} + \sum_{k=d+1}^{n} s_{dk} w_{k(g-1)} \qquad (4)$$

$$C_{d(g)} = s_{dd} \qquad (5)$$

$$x_{d(g)} = (\Delta w_{d(g)}) / (\Delta (g)) \qquad (6)$$

It can be seen that the three variables are lumped parameters in the classical sense, since each variable lumps many parameters together to reduce the parameter set. Here the reduction is from n squared to 2n independent parameters. The lumped parameters are used to calculate the split factor matrix in conjunction with a few simplifications.

From the above the following relationships can be written:

$$w_{d(g)} = C_{d(g)} w_{d(g-1)} + B_{d(g)} \qquad (7)$$

$$x_{d(g)} = B_{d(g)} - A_{d(g)} w_{d(g-1)} \qquad (8)$$

$$A_{d(g)} = (1 - C_{d(g)}) \qquad (9)$$

It will be noted that $A_{d(g)}$ is dependent upon $C_{d(g)}$ and vice versa. To allow an initial solution the three parameters must be assumed constant over two size reductions (g-1 to g and g to g+1). Simultaneous solution of the 2n equations produced from (7), gives initial estimates of $C_{d(g)}$ and $B_{d(g)}$. $A_{d(g)}$ can be obtained from (9). The initial estimates are then refined for each size reduction step, to give correct values of A, B and C for each size.

A computer program, SPLIT, has been written to do this. The program uses simultaneous solution for sizes 1 to 3, to obtain estimated lumped parameters. An iterative algorithm using the following constraints:

1. $C_{d(g)}$ and $A_{d(g)}$ sum to unity, ie:
2. Equation (8) holds.

is used to calculate lumped parameters for each size, using the previous size's parameters as initial estimates. This has been found to give more meaningful results, than calculating new initial parameters for each size. Typical results are given below, in Table (2):

TABLE 2: Typical lumped parameters

d	$A_{d(2)}$	$B_{d(2)}$	$C_{d(2)}$
1	0.21683	0.39106	0.78316
2	0.62122	0.59810	0.37878
3	0.14285	0.86484	0.85715
4	0.21688	1.09199	0.78311
5	0.49132	3.40796	0.50868
6	0.75046	9.59327	0.24953
7	0.28501	4.64084	0.71499
8	0.30299	14.26412	0.69701

The lumped parameters for tailings and concentrate density classes have been found to be fairly constant over the size range covered. However there are considerable differences in parameters for middlings band fractions with size. This is due to increasing liberation of heavy from light minerals, as size number increases, thus A tends to increase with size for these fractions.

Once the lumped parameters have been obtained the split factors must be calculated.

TABLE 3: Calculated split factors for size reduction 2-3

+3.3	+3.2	+3.1	+3.0	+2.9	+2.8	+2.7	-2.7
0.7887	0.1708	0.0019	0.0008	0.0014	0.0039	0.0000	0.0046
0.0452	0.3749	0.0195	0.0007	0.0016	0.0064	0.0000	0.0055
0.0018	0.1225	0.8624	0.0369	0.0024	0.0086	0.0000	0.0083
0.0015	0.0041	0.0126	0.7883	0.0505	0.0109	0.0000	0.0110
0.0043	0.0110	0.0045	0.0304	0.5062	0.1421	0.0000	0.0345
0.0142	0.0341	0.0119	0.0133	0.1676	0.2469	0.0834	0.1483
0.0030	0.0076	0.0030	0.0029	0.0076	0.0604	0.7178	0.0891
0.1415	0.2750	0.0843	0.1267	0.2626	0.5207	0.1987	0.6987
Σ 1.0000	1.0000	1.0000	1.0000	1.0000	1.0000	1.0000	1.0000

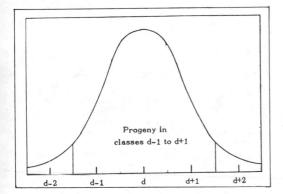

FIGURE 7: Normal distribution of progeny

TABLE 4: Calculated masses and errors, for split factor matrix in Table (3)

Class	Actual mass%	Calcd. mass%	Error %
+3.3	3.417	3.447	0.86
+3.2	0.876	0.885	1.03
+3.1	5.262	5.305	0.82
+3.0	7.447	7.508	0.82
+2.9	7.718	7.778	0.77
+2.8	12.560	12.708	1.18
+2.7	18.656	18.797	0.75
-2.7	44.665	44.996	0.74

4.2 Split Factors

Calculation of the split factor matrix (nxn parameters) from the 3n lumped parameters requires an initial simplifying assumption. This assumption is that the split factors are normally distributed around the diagonal as shown in Figure (7). Using this assumption gives an initial solution for the split factor matrix. This initial solution is improved iteratively by the program SPLIT, using the constraints:

1. Sum of column split factors must be unity
2. Sum of $s_{dk} w_{k(g-1)}$ must be $B_{d(g)}$,
   note that the contribution of d to d is not included in this sum.

The error resultant from not fulfilling these two constraints is subtracted from the column/row elements in proportion to their magnitude, it being assumed that the error in split factors is dependent upon the size of the split factors.

Occasionally the program converges upon split factors which give large mass errors. These mass errors are at equilibrium. If this occurs, then a routine forces convergence by altering the value of $A_{d(g)}$

slightly. This has been found to work satisfactorily. Here the assumption is that $A_{d(g)}$ has been calculated slightly incorrectly.

As with the calculation of the lumped parameters, the split factor matrix from the preceding stage is used as an initial estimate for the next stages calculation. This is a physically meaningful procedure, since the material is the same for all stages. Thus the matrix for the following stage is dependent upon the preceding stage's matrix.

A typical split factor matrix is given in Table (3), for a size reduction from 1.4 to 1.0mm. Table (4) shows the error in calculating mass in density fraction, compared with the actual mass, ie: it is a guide to the errors in convergence.

The minor split factors were found to vary quite considerably with size, however the major split factors show distinct trends with size. Two major trends are:

1. Diagonal split factors decrease with size number.
2. The split factor referring to the contribution of material from density d to the tailings fraction increases with size number.

The physical reasons for these trends are fairly obvious: as size decreases, so liberation increases, thus the quantity of material leaving a fraction will increase, thus the diagonal split factor will decrease. Since the bulk of material is quartz and other low density minerals, the proportion of mineral entering the -2.7 fraction will increase with size number, and hence the split factors to this fraction will increase. It will be noted that the diagonal split factors for middlings fractions are small, as would be expected.

Generally SPLIT will converge to within 1% of the true values of mass % in density class. Errors up to 2% are acceptable, since these are well within the bounds of experimental error involved in dense liquid tests. It must also be remembered that the model provides an optimum starting point for a locked cycle test, and as such is not required to give an optimal solution by itself, and a high degree of accuracy is not required.

However, SPLIT, fails to converge to less than 2.5% with the size reduction from 1 to 2 and n-1 to n. This is due to poor values of lumped parameters, since they are being estimated from two size reductions only. Thus two additional tests would be required, if these two size reductions are to be simulated correctly.

5 CONCLUSION

The model has been tested using a tin ore, a parameter estimation scheme evolved, and split factor matrices have been calculated for all sizes. The split factors obtained appear to be physically meaningful, on the basis of mineralogical observation. The split factor matrix appears to exhibit a definite relationship with size. At the time of writing this relationship had not been tested. The predictive capacity of the model is to be tested prior to this paper being presented, and some detail changes in the parameter estimation program may be made.

Finally, it can be concluded that once the model is developed into a software package, the designer of gravity regrind circuits will have a powerful time saving design tool.

REFERENCES

(1) Peterson, Herbst. 1985. Estimation of kinetic parameters of a grinding-liberation model. Int. Journ. of Min. Proc. 14: 111-126.
(2) Agar, Kipkie. 1978. Predicting locked cyle flotation test results from batch data. C.I.M. Bull. 799: 119-125.
(3) Agar, Stratton-Crawley, Bruce. 1980. Optimizing the design of flotation circuits. C.I.M. Bull. 824: 173-181.
(4) Agar, Stratton-Crawley. 1982. Bench scale simulation of flotation plant performance. C.I.M. Bull. 848: 93-98.
(5) Lewis, Tucker. 1986. Computer method for simulating relative density distribution analysis with application to gravity separations. Trans. I.M.M. Sect. C 95: 35-40.

Computer Applications in the Mineral Industry, Fytas, Collins & Singhal (eds)
© 1988 Balkema, Rotterdam. ISBN 90 6191 760 3

Instrumentation and control of a continuous 0.6-M SAG mill circuit

A.L.Mular, T.Bond & J.F.Kingston
Department of Mining and Mineral Process Engineering, University of British Columbia, Vancouver, Canada

G.D.Farnell
Machinery Development, Dominion Engineering Works Ltd, Lachine, Quebec, Canada

ABSTRACT: A 60-cm semiautogenous grinding (SAG) mill was designed for continuous wet operation in closed circuit with a vibrating screen. The circuit is fully instrumented and automated. Part of the control system incorporates Foxboro SPEC 200 controllers around the pump box to maintain level. The rest of the control system involves an ACTION model BC2 basic direct digital controller that handles all other inputs/outputs and contact closure sensing. The success of the control system is attributed to the uniqueness of specially designed mass flow metering systems. Various control strategies are possible. Employed to date has been control of a mill load setpoint by cascading the deviation from load setpoint as the setpoint to the fresh feed rate "controller". Mill feed water is ratioed to total flow rate of solids fed to the mill. The performance of the circuit is being correlated with the performance of full scale SAG circuits in British Columbia to develop models, to assess the feasibility of scale up and to investigate the effects of design variables on responses such as circuit throughput.

1 INTRODUCTION

Where ore types are appropriate, semiautogenous grinding and/or autogenous grinding (AG) has replaced conventional grinding. Secondary/Tertiary crushing plants are thereby eliminated so that capital expenditures are less and overall comminution costs can be lower (Barratt 1982). To establish the suitability of an ore for SAG circuits, pilot scale tests with 1.7 to 1.8 M diameter mills are often conducted. At least one test per ore type in a deposit has been necessary. Often, each test involves a large amount (up to 50 tonne) of sample, so that the testwork can be expensive and may discourage the use of SAG or AG circuits. However, MacPherson (1977) has employed a continuous 45 cm dry Aerofall mill circuit closed with screens and an air classifier to determine whether ores are suitable for SAG milling. The amount of sample per ore type is of the order of 60 kg or less. Thus an established SAG mill scale up procedure that employs small circuits is highly desirable.

The purpose of this paper is to describe a small 60 cm wet continuous SAG/AG mill pilot circuit assembled at the University of British Columbia to study the feasibility of scale up, to assess the influence of design variables, to develop mathematical models and to evaluate the effects of alternative control strategies on circuit performance.

2 DESCRIPTION OF CIRCUIT

Figure 1 is a simplified flow diagram of the circuit, which highlights major equipment, primary sensors and analog loops associated with the level control system around the pump box.

2.1 Major equipment

Major equipment components include the feed hopper/feeder assembly, a 60-cm mill and mill drive assembly, the pump box/pump system and a vibrating screen.

The feed hopper/feeder assembly supplies the mill with fresh feed and consists of an Eriez 40A vibrating chute feeder with a 30N vibrator, an Eriez VFT/115 vibration intensity control box and a fresh feed mass flowmeter described later. The vibrator is actuated in response to the demand for fresh feed solids. Water is fed to the mill via a 1.25 cm PVC line.

The 60-cm mill was designed and built by

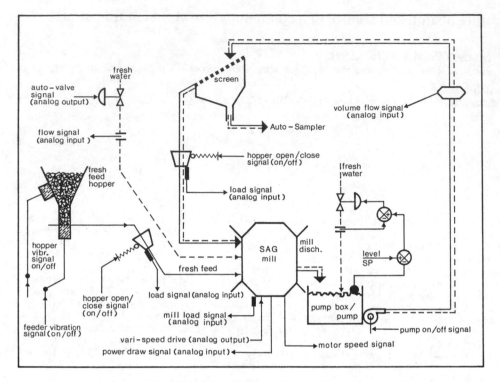

Figure 1: Simplified circuit diagram

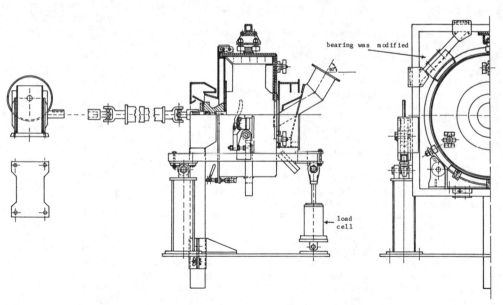

Figure 2: Sectional View of 0.6 M Mill

Dominion Engineering Works Ltd (Farnell 1981). The mill is 25 cm in length, has a diameter of 60 cm and operates via a variable speed DC drive in the range 0 to 58 rpm. Figure 2 is a sectional view which shows important details of construction. The mill outer shell has a peripheral track that rides on two horizontal rollers mounted on a vertical frame which is supported by a horizontal frame that rests on two posts and a load beam transducer. The horizontal frame is pinned at the top of the posts but can move up and down on the load cell in response to a change in mill load. Axial movement of the mill is restricted by air thrust bearings and roller guides attached to the vertical frame. The mill discharge end assembly is removable to permit access to mill contents. The discharge grate (1 cm slots) is attached to an end plate/trunnion that permits the release of pulp from the mill. The vertical frame and hence the mill can be rotated with respect to the horizontal frame to assist in discharging the mill contents. The mill is driven by a 3 HP variable speed DC motor complete with gear reducer and a DC motor speed controller (Ratiotrol) to manipulate mill speed. The shaft of the reducer attaches to a torquemeter that connects via a drive shaft to a spline bearing mounted on the mill discharge assembly. The spline bearing prevents horizontal forces from affecting the load beam output.

Mill product discharges to a 75 cm by 40 cm steel pump box sloped at the bottom. The discharge end is deepest at 70 cm from the top. Pump box slurry is fed to the vibrating screen by means of a 6.35 cm vac-seal centrifugal pump whose speed is set manually by adjusting a variable diameter pulley. The discharge rate is proportional to pump speed and pump box level. Thus water to the pump box is manipulated to maintain a constant level so that discharge rate is constant for a particular pulley setting. Water addition lines are 3.1 cm PVC.

Mill discharge is sized on a 60 cm by 90 cm Universal vibrating screen which has 0.625 cm square openings and is angled at 25° to the horizontal. Undersize is diverted to a drum thickener, while oversize is transmitted back to the mill via a specially designed mass flowmeter. Screen efficiency is very high.

2.2 Primary sensors

Primary sensors include a screen oversize mass flowmeter, a fresh feed mass flowmeter, a load beam to detect mill weight, a clarostat to measure pump box level, a shaft mounted torquemeter to measure mill power draw, a mill shaft counter to measure mill speed, a magnetic flowmeter to measure pump discharge rate and dp cells with orifice plates for pump box water and feed water flow rate measurements. The water flow rates are manipulated by means of control valves.

Mass flowmeters were specially designed for the circuit to accommodate flows of 0 to 4 kg/min. The meters dispense measured weights of solids (batches) from a weighing bin at constant time intervals (batch times). Figures 3(a) and 3(b) show components of the fresh feed and oversize mass flowmeter weighing bin assemblies. Referring to Figure 3(a), the cycle of operation starts with the vibrating feeder filling

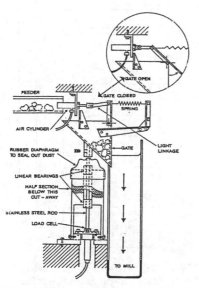

Figure 3(a): Fresh feed flowmeter

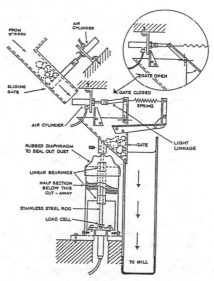

Figure 3(b): Oversize flowmeter

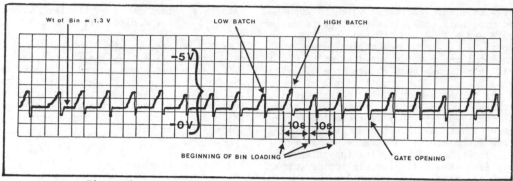

Figure 4: Mass flowmeter load cell output at 1034 gm/min

the weighing bin until the weight of ore, as measured by a load cell, reaches a batch setpoint. At this juncture the vibrating feeder is stopped and a final batch weight is taken. Next, the bin gate opens so that ore particles discharge onto the fresh feed chute leading to the mill feed trunnion. The gate closes, the empty bin is weighed (for calculating the net weight of ore) and the system is left idle for the duration of the batch time. At the beginning of the next interval the vibratory feeder is restarted and the cycle is repeated. A digital control computer (see control section) employs a machine language subroutine which actuates the feeder vibrators (on or off) and the air valve that drives the air cylinder to open or close the gate. The load cell signal (proportional to bin weight) is amplified, filtered and converted to a digital signal for use in the subroutine. The fresh feed mass flow setpoint (either user-supplied or supplied by a controller) is converted to a desired average batch weight, while the batch time is a user input that is constant. To maintain the desired average, the subroutine compensates for overweight or underweight batches by subtracting or adding weight, respectively, to the batch weight setpoint to be used in the current cycle.

Figure 4 shows the amplified output of the fresh feed mass flowmeter load cell, where the mass flow setpoint is 1034 gm/min (equivalent to a desired batch weight of 172.3 gm at 10 second intervals). As ore enters the weighing bin, the output increases until the weight is greater than or equal to 172.3 gm (i.e., output at 1.24 volt). After a 1.4 second delay, the gate opens as noted by the rapid drop in output. Note how each time there is a low batch weight relative to setpoint, then the next batch weight is increased to compensate.

The screen oversize mass flowmeter depicted in Figure 3(b) differs in that a sliding pneumatic gate stops the flow of oversize into the weighing bin to permit the measurement of a batch weight. The sliding gate is left open until the weight of material in the bin reaches a user-determined batch weight setpoint. Then the gate is closed by means of an air cylinder/valve assembly, a final batch weight is taken, the bin gate is opened to dispense oversize and then is closed. At this point the sliding gate is opened and the whole cycle is repeated. Every 30 seconds the weight of the batches are summed and an average oversize mass flow rate is calculated. A flow chart of the machine language subroutine employed for the oversize is shown in Figure 5 below.

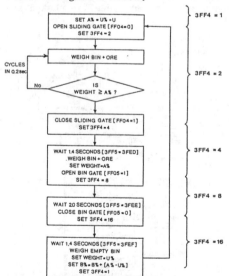

Figure 5: Flowchart of machine language subroutine

120

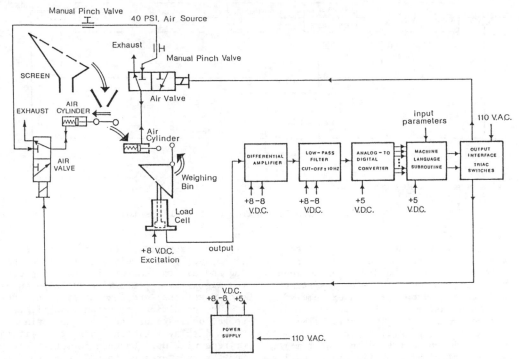

Figure 6: Major components of oversize mass flowmeter system

Figure 6 shows the major components of
the complete oversize mass flowmeter system.

The mill weight detector is an electronic load beam with a capacity of 500 kg and
a sensitivity of 11.85 A/D units per kg.
The detector has been extremely reliable
and rugged.

A Shinkoh torquemeter is employed to detect net power draw. It has a sensitivity
of -5.556 A/D units per kg and a capacity
of 30 kg-m. The output is filtered and
decreases with increasing torque.

A Foxboro orifice plate/differential pressure transmitter assembly is employed to
measure feed water flow rate. Manipulation of water is by means of an I/P signal
converter that actuates an air-to-open
control valve.

A counter mounted on the mill drive shaft counts the number of revolutions, where
a machine language subroutine keeps track
of the number of mill shaft turns.

Pump box level is detected by means of
a styrofoam float attached to a Clarostat
potentiometer. When level rises, an arm
attached to the float rotates and causes
the potentiometer to output a larger current. Pump box water is measured with an
orifice plate/dp cell assembly and manip-

ulated by means of an I/P converter that
actuates an air-to-open control valve.

A Foxboro magnetic flow transmitter/
converter measures pump discharge rate.

3 DESCRIPTION OF CONTROL SYSTEM

Automatic control (Bond and Mular 1982,
1985; Kingston, 1987) of the 0.6 M circuit involves a Foxboro SPEC 200 analog
control system which maintains pump box
level at setpoint and an ACTION BC2 digital control computer which performs all
other control functions.

3.1 Analog control

Figure 7 illustrates the analog control
loops involved in the control of pump box
level. The Clarostat signal, proportional
to level, is compared with a level setpoint by a proportional-integral level controller. The deviation from setpoint is
outputted as the remote setpoint (cascad-
ed) to a proportional-integral pump box
water flow rate controller. The latter
compares the remote setpoint to the water
flow rate signal from the orifice plate/dp

121

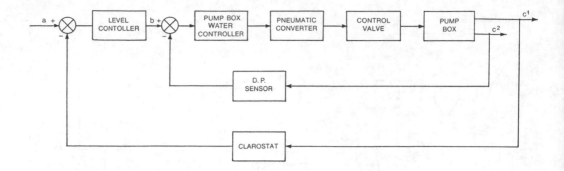

a = PUMP BOX LEVEL SET POINT
b = PUMP BOX WATER SET POINT
c¹ = PUMP BOX LEVEL
c² = PUMP BOX WATER FLOW RATE

Figure 7: Analog control loops for level

cell assembly and generates a deviation that becomes the output which manipulates an automatic valve in a direction to reduce the deviation to zero. The resulting cascade control loop performs extremely well. Both master and slave loops employ the PI controller mode in accord with:
$$c(t) = KE(t) + (K/I)E(t)dt + c$$
where $c(t)$ is controller output at time t, c is the controller bias signal (actuating signal when $E(t) = 0$), K is proportional gain constant, I is integral time constant and $E(t)$ is deviation from setpoint at t. Controllers are SPEC 200 series made by Foxboro Ltd. The package consists of two display stations and a nest of cards with power supply. Cards are: an input that converts measuring device output signals (4-20 ma) to DC voltage (0-10 v) for use with the SPEC 200 system, two square root extractors with one to determine the square root of the dp cell output signal and one to linearize the level sensor output signal, two PI controllers with one serving as the level controller and the other as the pump box water flow controller and an output that converts controller 0-10 v DC signals to 4-20 ma signals used by final control elements.

3.2 Digital Control

Digital control functions are performed by an ACTION Model BC2 digital control computer. Functions involve measurement of circuit physical parameters (i.e., direct sensing of mechanical contact closures, measuring time periods, converting analog signals to digital format), computation of resultant data (i.e., performing mathematical and logic functions as well as data

storage), control of external devices (i.e., activating relays, generating analog and digital outputs by executing control algorithms, performing timing/sequencing functions) and communication of the status, conditions or parameters (i.e., operational status and conditions via a video display, printing data, transmitting or receiving data between peripherals). Some key features of the BC2 computer are: a Z80 microprocessor operating at 2.5 MHz, Z80 industrial BASIC language interpreter in ROM (8K), 16K of RAM and 4 K of EPROM sockets with on-board EPROM programmer, built-in cassette interface for storage/retrieval of programs and data, 24 hour real-time clock, six vectored hardware interrupts, digital input/output of 32 general purpose sense inputs, 32 general purpose flag outputs, 8 relay outputs with on-board electromechanical relays and 8 lite outputs with on-board lamps. The ACTION computer runs the control program which executes control algorithms and supervises real time operation. Use is made of both machine language subroutines and ZIBL (Z80 Industrial Basic Language). The control program samples input signals, executes control algorithms, stores and operates on data and actuates final control elements.

Essential features of the BC2 control computer are shown in Figure 8.

3.2.1 Input signals sampled

At time intervals specified by the user, seven measurement signals are sampled and converted to digital form by an A/D converter (except for mill speed which is in digital form already). These are feed

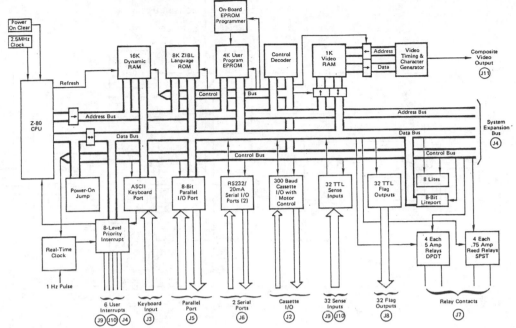

Figure 8: BC2 control computer block diagram

water flow rate (orifice plate/dp cell assembly), oversize mass flow rate (load cell), fresh feed mass flow rate (load cell), mill power draw (torquemeter signal), pump discharge flow rate (magnetic flowmeter) and mill speed (shaft counter). Mass flow rates of feed and oversize are sampled as part of machine language subroutines. On/Off states of various digital relays are sensed by means of a built-in function when ZIBL is employed so that programming is simple.

3.2.2 Control algorithm execution

A control program, written as ZIBL statements and machine language subroutines, supervises the real time operation of the computer. The program executes the control algorithms and produces control commands. Machine language timing subroutines determine when to initiate an algorithm and output a control command, while machine language utility subroutines can be either called explicitly or are interrupt driven. The advantage of machine language subroutines is that the computer executes them at high speed, so that a long series of tasks are performed rapidly. For example, the control of fresh feed rate re-

quires turning on and off two digital relays and sampling the fresh feed bin weight. These functions are rapidly performed in a specific order and for specific time intervals in a machine language subroutine. In fact, the fresh feed rate control is an interrupt driven subroutine to allow the computer to execute the fresh feed control algorithm as a high speed background operation while ZIBL commands are being executed.

Figure 9 shows, in a schematic context, the digital control loops involved in the control program along with primary sensors and final control elements employed.

The mill weight and feed water controllers employ a velocity form of the PI control mode written as:

$$D2 - D1 = K(1 + T/t)E2 - KE1$$

where D2 is controller output at sample time n, D1 is controller output at sample time n-1, K is proportional gain constant, T is sampling interval, t is integral time constant, E2 is deviation from setpoint at sample time n and E1 is deviation from setpoint at sample time n-1. This form is an advantage since it does not need initialization and it protects against integral wind-up.

The feed water setpoint is found from:

$$W2 = (Y(D + Q) + W1)/2$$

123

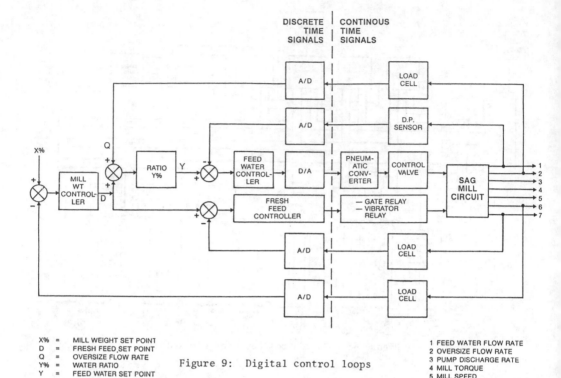

X%	=	MILL WEIGHT SET POINT
D	=	FRESH FEED SET POINT
Q	=	OVERSIZE FLOW RATE
Y%	=	WATER RATIO
Y	=	FEED WATER SET POINT

Figure 9: Digital control loops

1 FEED WATER FLOW RATE
2 OVERSIZE FLOW RATE
3 PUMP DISCHARGE RATE
4 MILL TORQUE
5 MILL SPEED
6 MILL WEIGHT
7 FRESH FEED RATE

where W2 is feed water setpoint at sample time n, W1 is feed water setpoint at sample time n-1, Y is desired water to solid ratio, D is mill weight controller output at sample time n and Q is oversize flow rate at sample time n.

The fresh feed controller uses a proportional mode in the form:

$$F2 - F1 = K(E2 - E1)$$

where F2 is controller output at sample time n, F1 is controller output at sample time n-1, K is proportional gain constant, E2 is deviation from setpoint at sample time n and E1 is deviation from setpoint at sample time n-1.

The control loops illustrated in Figure 9 have been extensively employed in grinding tests at UBC with substantial success, although alternative control algorithms may be employed. For example, the circuit could be controlled to a power draw setpoint (Mular, 1986) via a cascade arrangement as shown in Figure 10. Other alternatives, such as controlling to a mill load setpoint by cascading to power draw which cascades to fresh feed rate, are readily studied along with more modern approaches to control (Flintoff, 1986; Hales and Herbst, 1987).

3.2.3 Data storage

Every 30 seconds the fresh feed flow rate, oversize flow rate, mill torque, mill weight, feed water flow rate, mill speed, pump discharge flow rate and the elapsed

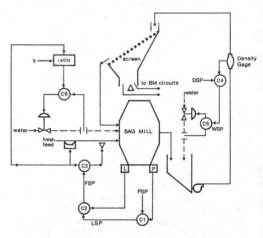

Figure 10: SAG circuit control

time are sampled and saved. Every one minute the control program instructs the computer to average the values over the last
minute (current value plus the value 30
seconds earlier divided by two) and store
the result in RAM. This data will be used
later to calculate the power draw, the steady state feed rate and other values of
interest.

3.2.4 Actuating final control elements

After execution of control algorithms, control commands in the form of discrete signals undergo D/A conversion via a machine
language subroutine when the signal is required in analog form. For example, the
feed water control element is a control
valve. Hence the feed water controller
output must first be converted to an analog signal and subsequently converted to a
pneumatic signal to actuate the control
valve.

Control commands for the fresh feed control elements are on/off information to
digital relays. Their activation/deacti-
vation is accomplished by means of a built
in function in the computer and does not
require any programming.

4 CONCLUSIONS AND FUTURE WORK

A small, continuous SAG circuit, comprised
of a feed hopper/feeder assembly, a 60 cm
SAG mill/drive unit, a pump/pump box system, and a vibrating screen, has been instrumented. Primary sensors include a screen oversize mass flowmeter, a fresh feed
mass flowmeter, a load beam to detect mill
load, a Clarostat to measure pump box level, a torquemeter to determine mill power
draw, a mill shaft counter for measurement
of rpm, a magnetic flowmeter to determine
screen feed slurry flow rate and orifice
plates/dp cells to detect water flow rates.
All sensors have proven reliable.

The control system includes analog and
digital systems. To control pump box level, SPEC 200 controllers, which operate in
PI mode, are employed in a cascade arrangement. An ACTION BC2 digital control computer performs all other control functions.
The digital control system is flexible and
and readily permits the evaluation of alternative control strategies. To date,
mill load has been controlled to a setpoint
by cascading the mill load deviation as
the setpoint to the fresh feed rate "controller". Mill water is ratioed to total
mill feed rate.

The 60 cm circuit has been employed for
more than 300 grinding tests and functions
extremely well either in fully autogenous
or in semiautogenous mode. When operated
as an autogenous mill circuit, feed rates
at steady state tend to be low for hard
ore types. In the absence of the specially designed mass flowmeters, it would be
difficult to maintain steady state operation.

The volume of sample required for a single test is approximately one half of a
30 USGal concentrate drum (about 90 kg of
a sample that has a bulk density of 100
lb/ft^3). An actual test can be completed
within a 4 hour period, although density
determinations and size analyses (including the mill contents) at key circuit
points may take an additional day.

4.1 Mathematical modelling

Data acquired from more than 300 tests are
being analyzed to develop a mathematical
model of the mill, where ore type, mill
speed, ore load setpoint at constant mill
volume loading, weight of ball charge, ball
size distribution and feed water-to-ore
ratio have been studied in accord with experimental design methodology (Bacon and
Mular, 1968).

4.2 Scale up studies

At present, the 60 cm circuit is operated
under a standard set of conditions to determine an operating autogenous work index.
Full scale SAG circuits, including Highland Valley Copper (formerly Lornex), Afton
Mines, Similkameen Mines and Island Copper
Mine, have been sampled in accord with established procedures (Mular and Larsen
1985). Operating autogenous work indexes
for each circuit have been determined.
Feed belt samples were taken during sampling and sent to the CMP (coal and mineral
processing) laboratory at UBC for standard
tests in the 60 cm mill to correlate with
full scale runs.

4.3 Effect of design variables

The effect of design variables, such as
grate open area, grate spacing and configuration, screen opening, lifter height/
spacing, on the steady state feed rate at
a fixed volume loading has not been experimentally quantified for SAG mills. Such
studies have been initiated at UBC and
are expected to play a role in the scale
up of SAG mills.

5 ACKNOWLEDGMENTS

Financial support for this project has been provided by Dominion Engineering Works Ltd., Lachine, Quebec, for which we are grateful.

6 BIBLIOGRAPHY

Barratt, D. J. 1982. Factors which influence selection of comminution circuits. Ch 1, in Mular, A. L. and G. Jergensen (eds) Design and installation of comminution circuits, Littleton, CO, SME-AIME.

MacPherson, A. R. 1977. A simple method to predict autogenous grinding mill requirements for processing ore from a new deposit. SME-AIME Trans. 262:236-240.

Farnell, D. G. 1981. Instruction manual for 60 cm SAG mill. Montreal, Dominion Engineering Ltd.

Bond, T. and A. L. Mular 1985. Operations manual for 0.6 M SAG circuit. Vancouver, Univ. of B. C.

Bond, T. and A. L. Mular 1982. Microcomputer controlled mass flowmeter. Vancouver, Univ. of B. C.

Kingston, J. F. 1987. Control and instrumentation of a continuous 60 cm SAG mill circuit. M. Eng. report, Vancouver, Univ. of B. C.

Mular, A. L. 1986. Automatic control of mineral processing circuits. Ch 40 in Somasundaran, P. (ed), Littleton, CO, SME-AIME.

Flintoff, B. C. 1986. An introduction to advanced control methods. Sudbury, Ont., Falconbridge control seminar.

Hales, Lynn B. and John A. Herbst 1987. The use of expert systems in concentrator control. Salt Lake City, UT, SME-AIME ann. gen. meet.

Bacon, D. W. and A. L. Mular 1967. Statistical design of experiments and process optimization in metallurgical engineering. Kingston, Ont., Queen's Univ.

Mular, A. L. and C. Larsen 1985. Grinding circuit sampling for steady state modelling. Laguitton, D. (ed), The SPOC manual. Ottawa, Ont., CANMET.

Computer Applications in the Mineral Industry, Fytas, Collins & Singhal (eds)
© 1988 Balkema, Rotterdam. ISBN 90 6191 760 3

Computer applications in the mineral industry: The role and contribution of CANMET

D.Laguitton & R.Pilgrim
Mineral Sciences Laboratories, CANMET, Ottawa, Ontario, Canada

ABSTRACT: One of the primary goals of the Canada Centre for Mineral and Energy Technology (CANMET) is to support R&D performed by industry in order to improve the economic performance and productivity of industry. Since the early days of computer applications, this has been achieved by undertaking in-house projects and supporting industrial and academic efforts in process simulation. As computer technology evolved towards microcomputers, the transfer of software has become a key element of the global technology tranfer from CANMET to industry. The early 80's have seen CANMET become the chief supplier of software for the evaluation and optimization of mineral processes as a result of the SPOC project (Simulated Processing of Ore and Coal). The commercial diffusion of Artificial Intelligence products, especially of expert system shells has opened a vast domain of computer applications for process diagnosis, operator training and process control. This new computer technology has been incorporated in the R&D projects supported by CANMET and the impact of these early efforts is already being felt in the industry.

The paper reviews CANMET's contributions to computer applications in the mineral industry, outlines the various mechanisms by which technology transfer can be achieved and provides information on how to benefit from existing and up-coming projects.

1 INTRODUCTION

Since 1970, scientists at the Canada Centre for Mineral and Energy Technology (CANMET) have been researching and developing safer and more efficient ways to extract, process and utilize Canadian energy and mineral resources. In so doing, CANMET fulfills its three primary goals:
- providing information to the Minister for policy-making related to non-renewable resources;
- serving government social objectives for health, safety and the environment;
- supporting R&D performed by industry in order to improve the economic performance and productivity of industry (Anon.1986b).

The present paper reviews how the third mission of R&D support is achieved in the area of computer applications in the mineral processing industry. The various means of R&D support are first discussed with a historical retrospective on how they have been used in the last fifteen years. Next, the SPOC effort that has constituted the decisive turning point in CANMET's contributions to computer applications in mineral processing is presented. Finally, the latest R&D projects in expert system prototyping for the mineral industry under the CAMP project (Computer Aided Mineral Processing) are outlined.

2 R&D SUPPORT MECHANISMS AT CANMET

In order to achieve its support mission in Research and Development, CANMET has systematically used the vast array of funding programs provided by the Government of Canada (Anon. 1987b). These programs vary from tax incentives to direct financial assistance for research and development, technology support through procurement, through institutes, through regional development agreements, energy programs and other national or international initiatives.

Three funding mechanisms are discussed here in more details as they have constituted the chief source of funding by CANMET, in the area of computer applications to the mineral industry, namely contracts, Research Agreement

Proposals and Industrial Research
Assistance Programs sponsored by CANMET.

2.1- Contracts

In 1972, the government established a
contracting-out policy which directs
that government requirements for
mission-oriented science and technology
are to be contracted-out to the private
sector, preferably to Canadian industry
unless a department can justify
intramural or foreign performance.

The policy applies to present and
new requirements in all scientific and
technological activities in the natural
sciences and to human science and
technology requirements in the field of
urban, regional and transportation
studies. Although the contracting-out
policy is primarily a procurement
policy, it is intended to promote the
development of a Canadian industrial
R&D capability.

The contracting-out policy was
expanded in 1974, to cover unsolicited
science and technology proposals (UP)
which are submitted by industry and
which fall within the mission of a
government department. The unsolicited
proposals program, managed by Supply
and Services Canada, permits the
government to respond quickly to sound,
unique proposals from the private
sector in support of government science
missions. Part of the UP program is a
fund that provides bridge financing for
proposals which are accepted from the
point of view of sponsorship,
scientific merit and uniqueness, but
which cannot be funded from the
sponsoring departments's current
appropriations.

The unsolicited proposals program
was expanded in 1985 to allow projects
that aim at demonstrating new
technologies in operational settings
within the government. This program
expansion is specifically targetted at
small Canadian businesses. Table 1
lists the contracts that were awarded
by CANMET in the area of computer
applications in the mineral processing
industry since this discipline appeared
in CANMET's programs. In this table,
the contracts that resulted from
unsolicited proposals are identified by
UP in the year column.

Table 1. Research contracts funded by
CANMET for computer applications in
mineral processing.

RECIPIENT	YEAR	TITLE
Consultant (R.E.Hamilton)	1976	Definition of an experimental design to optimize a continuous grinding circuit consisting of a ball mill and classifier in closed circuit for a fine grained complex Pb/Zn/Cu pyrite ore of the New Brunswick type
U.Laval (Y.Berube)	1976	Definition of an experimental design to optimize a continuous grinding circuit consisting of a ball mill and classifier in closed circuit for a fine grained complex Pb/Zn/Cu pyrite ore of the New Brunswick type
U.Laval (D.Hodouin)	1976	The design of an experimental grinding circuit for a fine pyrite of the New Brunswick type
U.Laval (D.Hodouin)	1976	A new approach for the analysis of thirty-two continuous grinding experiments involving three different materials
		Report on the sampling campaign performed on the grinding circuit of Heath Steele Mines Ltd. (N.B.) during March 1977
		Modelling of twelve continuous grinding experiments on a New Brunswick sulphide ore
U.Laval (M.D.Everell)	1977	Development of a simulator of the Heath-Steele Mines grinding circuit Modelling of grinding and classification of fine-grained suphide ores of New Brunswick
		Modelling of Heath Mines hydrocyclones and ball mills

U.Laval   1978 Development of a
(M.D.Everell)   classification model
                  for the simulation of an
                  industrial circuit
                  processing fine-grained
                  sulphide ores

                  Modelling of the Heath
                  Steele Mines rod mill and
                  evaluation of a combined
                  rod mill/grinding
                  circuit simulation

U.Toronto 1980 Sampling methodology for
(H.W.Smith)   industrial mineral and
                coal slurries

U.Laval   1980 Material balance program
(M.D.Everell)   for process evaluation and
                 modelling in mineral and
                 energy processing plants

McGill U.   1980 Study of modelling
(J.Finch)   methodology illustrating
                 principles and benefits in
                 mineral energy processing
                 plants

U.Alberta   1980 Normalization of mineral
(L.R.Plitt)   and coal process models
                 and review of the
                 executive structure and
                 software methodology of
                 flowsheet simulation
                 programs

UBC   1980 Mathematical models of
(A.L.Mular)   comminution and flotation
                 for mineral and coal
                 processing plants

Noranda   1981 Ball mill modelling
Inc.
(Res. Centre)

Brunswick   1981 Workshop on sampling
Mining and
and Smelting

U.Laval   1981 Workshop on material
                 balance

U.Toronto 1982 Supplement to chapter 2 -
(H.W.Smith)   SPOC Manual

UBC   1982 Simulated Processing of
(A.L.Mular)   Ore and Coal (SPOC)
                 project - phase 3
                 feasibility

U.Western 1982 Further development of a
Ont.   unique computer software
(SACDA)   system for computing
                energy balances

U.Alberta 1983 Mathematical modelling of
(L.R.Plitt)   auto-medium cyclones

Devco,   1983 Demonstration of the SPOC
Sydney,N.S.   methodology in the coal
                 processiing industry

U.Laval   1983 Workshop on process
(G.Barbery)   evaluation, simulation and
                 optimization

Quasar   1983 On-line demonstrations of
Systems Ltd.   process simulation
                 software

General   1983 Development of a system
Cybernetics   to produce audiovisual
Ltd.   material from a computer

U.Western 1983 Development of a simulator
Ont.   (UP) of the Morgensen sizer
(J.Beecksman)

Kilborn   1984 The development of a
(Sask) Ltd.(UP) prototype for the
(D.Madge)   balancing of mass and
                 energy in a potash
                 crystallizer circuit

Les   1984 Computer aided design of
Industries   crushing plants
Welfab Can.
Ltd.

Kilborn   1984 Review of SPOC project
(Sask)Ltd.

Industries 1984 Computer Aided design of
Piedmont Ltee. crushing plants

IOTA   1985 Conversion of SPOC
Consulting   software to the
Ltd. Ottawa   micro-computer

Consultant 1985 To provide consultation
(F.Flament)   services for the SPOC
                project

Lavalin-   1986 Pilot application of
Cosigma   (UP) expert system technology
Inc.   to process control

Canada   1986 Development of a
Cement   prototype expert system
Lafarge Ltd.   in a grinding circuit

EMR-   1987 To acquire a 8-10 minute
Communication   videotape presentation in
Topek   English and French on the
Production   subject of "Simulated
                 Processing of Ore and
                 Coal" SPOC

Kilborn   1987 To develop a model for the
(Sask) Ltd.(UP) analysis of a potash
(D.Madge)   compaction circuit

2.2 Research agreement proposals (RAP)

The Research Agreements Program is designed to take advantage of the capabilities of a variety of disciplines and technologies, available in Canada, to facilitate the better use of our mineral and energy resources and to provide information on Canada's landmass. Applications are considered on the basis of the potential significance that the results of studies will have on the Department's objectives (Anon.1987a). In this context, CANMET's objectives in mineral processing are defined as: to develop technology to assist Canadian industry in the assessment and more efficient extraction and refining of mineral resources and to minimize energy consumption and enviromental impact in mineral processing as well as to provide policy support.

Table 2 lists the research agreement proposals funded by CANMET in the field of computer applications in the mineral industry since 1972.

Table 2. Research agreement proposals funded by CANMET for computer applications in mineral processing.

RECIPIENT	YEAR	TITLE
Laval (A.Galibois)	1972	Application of mathematical models for cost analysis, planning and decision making
UBC (A.L.Mular)	1972	Analysis, optimization and control of mineral processes
UBC (A.L.Mular)	1973	Grinding circuit simulation from Bond grindability tests
U.Alberta (L.R.Plitt)	1973	Mathematical modelling of hydrocyclone classifiers
(U.Laval) (A.Galibois)	1974	Developpment de modeles mathematiques pour accroitre la productivite dans l'industrie miniere
Queen's U. (D.J.Spottiswood)	1974	Simulation of the flotation process
UBC (A.L.Mular)	1975 1976 1977	Digital simulation of the Brenda concentrator
U.Alberta (L.R.Plitt)	1977	Size and mineral density effects on hydrocyclone performance
Queen's U. (J.G.Patterson)	1978	Process simulation of an ore processing plant
U.Alberta (L.R.Plitt)	1978	Size and mineral density effects on hydrocyclone performance
U.Laval (M.D.Everell)	1978 1979 1980	Dimensionnement de circuits industriels de comminution
UBC (M.S.Davies)	1979	Dynamic modelling and estimation of mineral flotation circuit
Laurentian U. (B.H.Kaye)	1980 1981 1982 1983	Shape and size distribution of crushed coal
U.Laval (D.Hodouin)	1982	Elimination par cyclonage des tres fines particules d'un minerai
U.Laval (G.Barbery)	1984 1985	Etude du comportement hydrodynamique de cellules flottation agitees mecaniquement
U.Laval (D.Houdouin)	1985 1986	Pilotage automatique des circuits de broyage
UBC (A.L.Mular)	1986 1987	The estimation of full scale sag mill performance from a 0.6M sag mill test circuit
U.Alberta (L.R.Plitt)	1987	Laboratory simulation of coal flotation
U.Laval (G.Barbery)	1987	Calibration des modeles cinetiques de la flottation au laboratoire
U.Laval (D.Hodouin)	1987	Application de la theorie des systemes flous a la commande automatique de la flottation des minerais

From the examination of the above Tables 1 and 2, it is apparent that the list of Canadian research centers active in computer applications in the mineral processing industry and publishing in scientific and technical journals is fairly well represented by the list of recipients of CANMET funds for R&D.

2.3 Industrial research assistance program (IRAP)

The Industrial Research Agreement Program of the National Research Council (NRC) is a federal government program established to further the economic development of Canada through encouragement of R&D and increased utilization of science and technology. Through IRAP, the scientific, technical and R&D management resources of NRC and other research organizations of the federal government are able to assist industry to increase the calibre and range of industrial research and development across Canada and to foster the use of available technology. The program is divided in several elements that provide various forms and levels of assistance ranging from simple advice on industrial technology to short term or small risk studies and longer term, high risk major projects involving technology transfer (Anon. 1985a). The contributions to major projects involving technology transfer incorporate NRC's former Program for Industry/ Laboratory Projects (PILP) and is designed to assist Canadian companies to undertake projects that take advantage of technology in the form of scientific and engineering knowledge and resources existing within government laboratories. The program has been recently extended to include technology originating within Canadian Universities.

The program supports such projects from the conceptual stages through to prototype or pilot plant development within the company in order to provide a clear indication of whether this technology can result in a commercially viable product or service.

Although these projects are not funded by CANMET, the sponsorship of proposals that are most directly in line with the department's research programs can play an important role in the preparation, documentation and support they receive. Two recent IRAP projects in the area of computer applications in the mineral industry have been sponsored by CANMET. One of them addresses in-situ bioleaching of uranium and is carried out jointly by a team from The University of Waterloo and from Denison Mines. The other addresses process control and expert systems for oil and sand extraction and is carried out by a team from Syncrude, the University of Alberta and CANMET.

3 IN-HOUSE R&D ON COMPUTER AIDED MINERAL PROCESSING AT CANMET

The first computer applications in mineral processing at CANMET, can be found in the work of Kelly & Pilgrim (1972) who, in the early 70's, were active in grinding circuit simulation and sponsoring outside research on the subject. Computing power was then available through the remote card readers and batch printers of the Computer Science Center of the Department of Energy, Mines and Resources (EMR).

By the late 70's, time-sharing computing had become quite common and a variety of low-cost terminals were installed to distribute the computing power of the main frames into the offices of the mineral dressing section of CANMET. This was reflected in further computer applications in-house by Klymowsky & Cameron (1979) Laguitton & Wilson (1979) and the sponsoring of contracted research on process simulation as listed in the tables of contracts and research agreements.

Then, in 1980, the mineral dressing section of CANMET started a six year project entitled SPOC (Simulated Processing of Ore and Coal). The objective, as detailed in the first progress report by Laguitton & Sirois (1981), was to provide the Canadian mineral and coal industries with an easily usable computer methodology for process evaluation and optimization.

The project sponsored numerous contributions from industry and universities (Fig.1), through contracts and resulted in the release of an engineering manual divided into 18 chapters under separate covers (Anon. 1985b). The methodology was also embodied in a score of computer programs written in Fortran and initially executable on the Department CDC Cyber mainframe computer. They cover several essential aspects of mineral process evaluation and optimization, namely : sampling, material balance calculations for metallurgical inventories, model calibration by least-squares techniques of various complexities and process simulation also known as predictive material balance computation.

The microcomputer technology which had been in a state of innovative turmoil in the early 80's could not be used advantageously during the first years of the project, because the wide spectrum of incompatible hardware and software was not a safe support for a project with strong emphasis on technology transfer. Fortunately, by 1984, the market forces converged towards a much smaller number of microcomputer types, the IBM-PC became a pseudo-standard device, as witnessed by the number of IBM-PC clones. Powerful

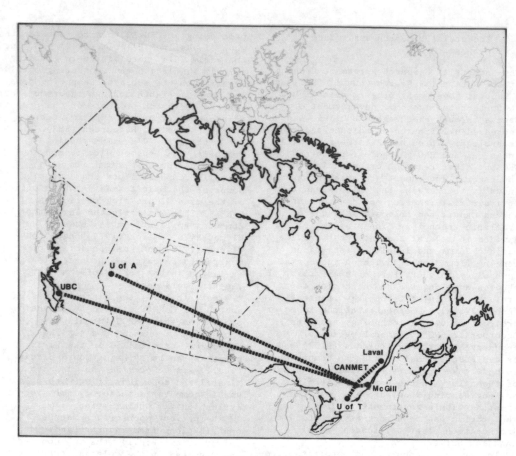

FIG. 1 SPOC INITIATION (1980)

Fortran compilers became available for scientific users, and it became obvious that a major break-through in technology transfer would be possible if the entire collection of SPOC programs was converted to the personal computer format. This was accomplished in 1985, and by April 1986, CANMET was able to release the entire collection of manuals and software. The needs were real and the number of requests for software grew rapidly. More than 130 organizations have so far acquired SPOC software in Canada and abroad (Fig.2 and 3) and some testimonials have been published in scientific literature (Reid 1982; Carmola 1982).

Besides distributing software and manuals upon request, CANMET continued providing interested users with training workshops on various aspects of the SPOC methodology. This method of technology transfer was found to be essential since both the theory of process computation as well as the practice of using computers as a tool were not in the curriculum of most of to-day's practicing engineers and require familiarization and demystifi-

cation. This situation is changing rapidly as the first computer-age graduates in mineral engineering arrive in the field, many of them already owning a personal computer. The original assumption that the SPOC methodology would result in industrial initiatives in data collection and model upgrading seems as reasonable as ever and as such, obsolescence and emulation of similar applications are, for SPOC, another measure of success.

The year 1986 has seen considerable effort in-house, to take full advantage of the PC user-friendliness to access and execute the SPOC programs. Such features had not been used in the hasty conversion from mainframe to microcomputer in 1985. A few of these improvements are discussed below.

3.1 Current and future improvements of the SPOC software

Full screen data entry and type-in-the-box formats highlighted by the use of colors are the most visible improvements made over the old unformatted sequential input

132

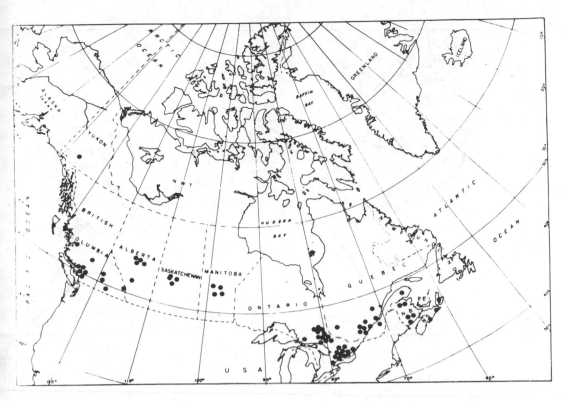

FIG. 2   SPOC SOFTWARE IN CANADA (1987)

files. The user can selectively enter or modify any field of an input-data file by simply moving the cursor to the desired area of the screen. Help facilities provide instant on-screen definition of the required data.

Flowsheet drawing by simple cursor movements is another major area of improvement. Not to be confused with the high resolution CAD systems available on the market, the graphic capabilities added to most SPOC programs allow the automatic numbering of the streams and units of the flowsheet at drawing time as well as after unit or stream deletions. When the user has satisfactorily completed the flowsheet definition, the stream connection matrix that numerically represents the flowsheet in subsequent computations is internally generated. This permits the automatic call to the required subroutines for process simulation in the order required by the sequential modular approach (Laguitton 1987). In the case of material balance computation by the MATBAL and BILMAT programs, the selection of the mass conservation equations which constituted one of the main difficulties in using the programs can be done automatically from the stream connection matrix.

The output files are also being redesigned to take full advantage of graphic displays, screen paging and movable windows.

Although most new software related to the SPOC methodology is expected to be developed outside CANMET, some key areas will remain in the direct line of activities at CANMET. Sponsoring of various model updates or of new model developments is likely to continue as already done for the mass and energy balance of a potash crystallizer by Kilborn and a simulator of a Mogensen sizer at the University of Western Ontario.

Modifications are being made to the material balance methods to include the case of leaching circuits where material to be accounted for appears in the solid and the liquid phases. Preliminary work in this direction can be found in a master thesis by Cimon (1987).

These upgrades and others confirm the microcomputer as the ideal hardware for the transfer of the SPOC software. Assistance to industrial applications of the methodology through contracts or on a cost recovery basis will also be considered case by case.

133

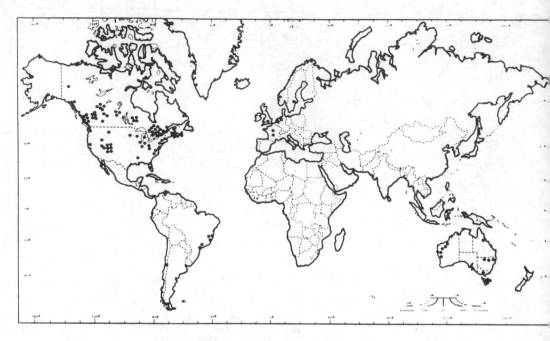

FIG. 3 SPOC SOFTWARE WORLDWIDE (1987)

3.2 Expert system applications in mineral processing at CANMET

In 1986, two complementary projects have been started at CANMET to explore the applications of expert systems in mineral processing. One project covered the application of shells using conventional hardware and the second explored the benefits of specialized tools. As usually recommended by the experts (Anon.1986a), a prototyping approach has been chosen whereby successive prototypes are developed and assessed. The alternative would be the design of massive knowledge bases whose structure could be faulty from the beginning or made obsolete by new technological developments.

These activities in Simulation of Process Expertise (SPEX) are being developed as a complement to those in simulation (SPOC), within CANMET's R&D efforts in Computer Aided Mineral Processing (CAMP), as illustrated in Fig.4.

3.2.1 Development of a prototype expert system to advise the operator of a large control system in a zinc refinery.

This project, results from an unsolicited

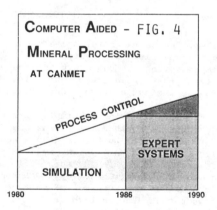

proposal by Cosigma- Lavalin and aims at assessing the potential of an expert system in the control room of the Canada Electrolytic Zinc (CEZ) plant in Valleyfield, Quebec. The CEZ plant is a very complex sequence of roasting, leaching and electrolysis processes that involve considerable process delays and delayed feed-back effects in the recycle streams. Such a plant requires a large experience by the supervisors to reach pseudo-steady state and optimal recovery at maximum throughput. This type of expertise is well suited to be captured in an expert system. Conventional material balance and simulation algorithms are very difficult to apply in this type of process. As a preliminary target, it has been decided to focus on the leaching plant.

The project whose main contributors are CANMET, LAVALIN and CEZ, started in November 1986 and is due to be completed by April 1988. The first version of the prototype was delivered to the plant for assessment in October 1987. The inference engine is provided by Personal Consultant (Texas Instruments), the process data entry by the operator is simulated by a Lotus 1,2,3. electronic sheet and the on-line process data-base is represented by DBASE-3 (Ghibu 1987). The hardware considered for the first prototype is an IBM compatible using a 80386 processor.

3.2.2 Development of a prototype expert system as operator adviser in a cement factory.

This project has brought together CANMET, Canada Cement Lafarge Inc., an AI consultant UNISYS (formerly SPERRY Corp.) and a University group GRAIIM (Groupe de Recherche sur les Applications de l'Informatique a l'Industrie Minerale, Universite Laval).

This project complements the LAVALIN-CEZ project by following the intensive and specialized approach to expert system prototyping, and using the most advanced tools and expertise. The system was developed on a TI-Explorer LISP machine, using the KEE shell (INTELLICORP). The domain expertise being captured was that of a clinker grinding circuit operator. At this stage of the cement manufacture, clinker discharged from the raw material cooking kiln is blended to additives and ground in a closed ball- mill circuit. The character- istics of the final product are highly dependent on grinding circuit variables and require very expert decisions by the operator. This expertise has been encapsulated into a first prototype completed in April 1987 (Vanderstichelen 1987).

3.2.3 Process control and expert system for oil sand extraction

A process control system soliciting process state information from multiple sensors, which optimizes product quality and quantity in the oil sand extraction process is the major objective of this

IRAP project sponsored by CANMET. When implemented, the system will adjust process conditions in response to both feed-forward and feed-back information. Work started in late 1987 and reports on developments will appear in scientific literature over the next 3 years.

By sponsoring these projects, CANMET aims at mastering the most recent techniques of expert system development and at being able to continue its role of leader and adviser for the numerous requests for information on AI applica- tions that are bound to originate in the mineral industry, as expert systems penetrate every sector of expertise. The knowledge implemented in any particular system is often confidential but the lessons learned by structuring knowledge according to different techniques, languages and machines will be most important in developing future systems.

4 CONCLUSION

The improvement of the economic performance and productivity of the Canadian mineral processing industry is one of the primary goals of the Canada Centre for Mineral and Energy Technology (CANMET). Very few areas of R&D are better adapted to support that goal than the development of computer tools for process evaluation and optimization.

CANMET has been active in this field since the early 70's and especially since 1980 with the SPOC project which has provided a widely accepted methodology for computer aided mineral processing. Besides a continued effort in developing microcomputer based software for material balance and simulation, the activities of CANMET in this area have been extended to include expert system prototyping as a complement to conventional computer applications. In order to fulfill its mission, CANMET has systematically used contracts, Research Agreement Proposals and Industrial Research Assistance Programs as means to support R&D in the private sector and in universities across the country.

5 DIRECTORY

The following addresses should be used to obtain further information on the various programs of support for technology development discussed in this paper.

Contracts : Science Programs Branch

RAP : Supply and Services Canada
Place du Portage, Phase 3,-
12C1. Ottawa, Ontario K1A 0S5
Coordinator, External research
Programs. Department of
Energy, Mines and Resources,
580 Booth St., 20th floor
Ottawa, Ontario K1A 0E4

IRAP : IRAP Laboratory Network
NRC Montreal Road, Building -
M-55. Ottawa, Ontario K1A 0R6

CANMET : Office of Technology Transfer
(programs) CANMET-OTT
555 Booth St., 3rd floor
Ottawa, Ontario K1A 0G1

CANMET : The Mineral Technology
(SPOC) Information Officer
CANMET-TID
562 Booth St.
Ottawa, Ontario K1A 0G1

6 REFERENCES

Anon. 1986a. AI presentation kit and
symposium proceedings. Texas
Instrument. AI Satellite Symposium
II. P.O. Box 181153, Austin, Texas
78718.

Anon. 1986b. CANMET Review 1984-1985.
CANMET Report 85-10. Minister of
Supply and Services Canada 1986,
Cat.M 38-13/85-10

Anon. 1987a. Information Guide on
Research Agreements in the Natural,
Physical and Social Sciences and
Engineering. External Research
Programs, Energy, Mines and
Resources Canada, 20th floor, 580
Booth St., Ottawa K1A 0E4.

Anon. 1985a. IRAP NRC/Industrial
Research Assistance Program. NRC
Montreal Road, Building M-55,
Ottawa, K1A 0R6.

Anon. 1985b. SPOC Manual. CANMET
Special Publication 85-1 Catalogue
M38-16/1985E. Canadian Government
Publishing Centre, Supply and
Services Canada, Ottawa, K1A 0S9

Anon. 1987b. The Government of
Canada's Support for Technology
Development, 1987. Minister of
Supply and Services Canada 1987,
Cat. ST31-13/1987.

Carmola,R., M.R.Hoover, J.H.Kim & S.J.
Klode 1983. Computer simulation of
synthetic fuels feed preparation
circuits. Proceedings of the first
conference on use of computers in the
coal industry. Wang Y.J. &
R.L.Sanford (eds). AIME 177-186.

Cimon, D. 1987. Application des
techniques de bilan matiere a une
usine de traitement d'un minerai
d'or. These de Maitrise, Universite
Laval.

Ghibu, C. & Dupuis, D. 1987. Expert
system prototype as an advisor to the
operator of a modern electrolytic
zinc refinery: preliminary results.
Proceedings of the CANMET Symposium
on Expert Systems. CANMET, SP87-4,
CANMET-TID 555 Booth St., Ottawa,
K1A 0G1

Kelly, F.J. & R.F. Pilgrim 1972.
Simulation of a grinding circuit and
encountered problems. Trans. SME/AIME
252.

Klymowsky I.B. & W.H.Cameron 1979.
Industrial application of
mathematical modelling to the
treatment of fine-grained sulphide
ores. Proceedings of the 11th
Annual Meeting of the Canadian
Mineral Processors, Ottawa.

Laguitton, D. & M.Bilodeau 1987.
Computer aided mineral processing.
Proceedings of the Copper 87
Conference. Vina del Mar, Chile,

Laguitton, D. & L.L.Sirois 1981.
SPOC-project; status and prospects
after one year of joint effort;
Proceedings of the 13th Annual
Meeting of the Canadian Mineral
Processors,Ottawa.

Laguitton, D. & J.M.D. Wilson 1979.
Material balance: a step towards
modelling of fine-grained sulfide
ores treatment. Proceedings of the
11th Annual Meeting of the Canadian
Mineral Processors, Ottawa.

Reid, K.J., K.A.Smith, V.R.Voller &
M.Cross 1982. A survey of material
balance computer packages in the
mineral processing industry.
Proceedings of the 17th APCOM
Conference. Johnson,T.B. & R.J.
Barnes (eds). AIME 41-62.

Vanderstichelen, J., F.Flament,
D.Laguitton & D.Hodouin 1987.
Prototyping of an expert system for
trouble shooting of clinker
grinding mills. Proceedings of
Symposium on Artificial
Intelligence in Minerals and
Materials Technology. The
University of Alabama, Tuscaloosa.

Computer Applications in the Mineral Industry, Fytas, Collins & Singhal (eds)
© 1988 Balkema, Rotterdam. ISBN 90 6191 760 3

USIM: An easy to use industrial simulator for mineral processing plants

A.Broussaud, P.Conil, G.Fourniguet & J.-C.Guillaneau
BRGM, Orléans, France

ABSTRACT : USIM is a steady state mineral processing plant simulator, designed especially for use by process designers and engineers. Three families of USIM versions are
available : USIM industrial versions, customized for the design and the optimization of
a given projected or existing plant, the USIM exploratory version, for the early stages
of a project and pilot plant experiments, and USIM hydro versions, devoted to hydrometallurgical and chemical processes. The function of USIM and the practical ways of
entering data and making simulations are presented. The methodology for building
accurate industrial versions is explained and more advanced uses of USIM are mentioned.

INTRODUCTION

A steady state mineral processing plant
simulator is basically software, capable
of predicting plant operation according
to the characteristics of the ore feed
and of the circuit. Fig. 1 shows the typical function of such a simulator.

Steady state simulation can be a very
effective approach for plant design
since it makes it possible to consider a
great number of hypotheses for flowsheets, choices of equipment, etc., and
to quickly arrive at the configuration
for a nearly optimal industrial plant.

However it is not only a tool for
designing new plants, it is also an
operational aid for decision-making in
operating plants :
. to adapt the flowsheet to changes in
  the ore or in the concentrates market,
  or just to improve it,
. to choose the settings or the operating parameters of certain devices.

The basic elements of a simulator are
usually :
. simulation software, per se, which
  makes possible communication between
  user and simulator and coordination of
  calculations. As this is the only
  element visible to the user, it is
  often called the simulator,
. mathematical models for unit operations, buried inside the simulator as
  subroutines. They constitute however,

the vital core since the validity and
the limitations of the simulation
results depend entirely on the quality
of the models,
. auxilliary software which is generally
  used only for calibrating models.

The authors have been promoting the
practical use of steady state simulation
for plant design and optimization for
several years (ref. 2 to 4 and 6 and 7).
They have formed a library of tested
operational models, mostly based on

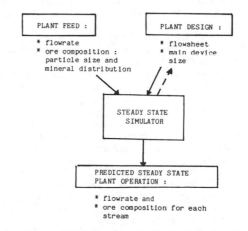

FIGURE 1. Typical function of a direct
steady state simulator.

various published concepts, and a methodology to select and calibrate the models adapted for each given real application (4). Their approach made it possible (2) for a barite plant to improve recovery by about 3 % and to adapt the flowsheet to a change in production objectives, eg., production of chemical quality barite in place of drilling mud quality. More recently, (6) a simulator was built for a phosphate ore processing plant in Senegal, the models having been calibrated from experimental data collected directly in the plant ; then the reliability of the simulator was confirmed when a major modification of the flowsheet took place. The simulator is now used to evaluate a priori any new ideas for modifying the operative settings or the flowsheet.

The authors' earlier applications were supported by main frame computer based simulators, at first a home made simulator (2), and later an adapted version of MODSIM (3) (8). However these softwares were designed for and could only be used by specially trained users.

To speed up the introduction of simulation, the authors decided to make it easy to use by people on sites, and thus created USIM, a microcomputer simulator designed especially to be operated by process engineers and technicians, and not only by trained specialists.

Therefore, USIM satisfies a number of constraints :
- data input and results display only involve symbolic graphics and words which are clear to the practician. Flowsheets are presented with the traditional graphic appearance. All visible numbers have a physical meaning such as flowrates, particle size and mineral distributions, inside diameter of mill, length, rotation speed, mass of grinding media...
- data are entered on clear simple questionnaires. The user can at any time come back and correct any data easily (on the same screen-page or on a previous page) and numerous safeguards practically eliminate the risk of error due to lack of training in computer or keyboard use,
- data, which concern plant flowsheets, plant devices, and plant feed, are stored in a way which enables the display, reuse, or modification of old data, quite simply,
- software development method ensures that USIM is robust.
- USIM runs on PC compatible microcomputers. The optimum required hardware costs approximately less than 6000 US$.

Three families of USIM versions are delivered :
- USIM industrial : a personalized simulator dedicated to a given ore. It contains models of physical processing unit operations selected from the authors' model library and adapted in accordance with the ore and/or the process(es) used or planned. For processing industry or engineering.
- USIM exploratory : non-personalized version for exploratory studies of physical beneficiation processes, or exercises at universities. It contains models of crushers, rod and ball mills, cyclones, classifiers, spirals, shaking tables and jigs, flotation banks, thickeners, band filters.
- USIM hydrometallurgy : an adapted version for hydrometallurgical or chemical processes e.g. liquid-liquid extraction, extraction by resins or by activated carbon (CIL, CIP or CIC), soluble salts flotation, phosphates washing...

This version is capable of predicting plant operation according to the characteristics of the feed composition (up to four phases and six components) and of the circuit.

USIM HYDRO may be used by process designers and engineers in all industries where exchanges between phases are taking place.

USIM can be delivered in English or in French and is documented by a detailed "user's manual"

HOW TO SIMULATE A PLANT WITH USIM :

The best way to explain how USIM works is to show a typical application. The figures presented here are screen pictures or hard copies.

Data input

The USIM user enters the simulator by the main menu shown on fig. 2. Options 1 to 5 are for data entry and display. Data to enter for current use of a USIM simulator consist of :
- a plant flowsheet,
- a description of plant devices : geometric sizes, main settings and operation parameters,
- a description of plant feed : ore and water flowrates, ore particle size and mineral distribution.

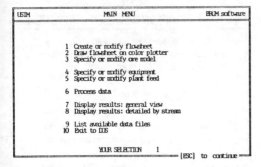

FIGURE 2. USIM simulator : main menu.

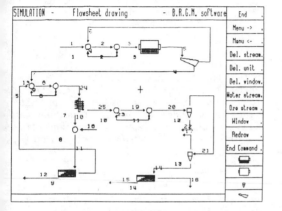

FIGURE 3. Interactive flowsheet drawing.

These data are stored in files organized in a modular form, so that the user can generate data libraries consisting of a set of flowsheets, a set of plant feeds and a set of plant devices. Old data can at any time be easily displayed, reused or modified.

The **plant flowsheet** is entered or modified with option 1 of the main menu. This option is fully graphic and is icon driven (fig. 3). A cross, moved by the keyboard arrows, is used to select functions in the side menu (on the right side) and to locate devices and streams on the drawing. The side menu displays only three or four device symbols at a time but a library of 25 symbols is available on-line and can be displayed by scrolling. Flowsheets created are not only stored as drawings but their physical meaning is also analysed and error messages are displayed if they are not understandable.

Option 2 is used to generate a color plot of the flowsheet on a sheet of paper or on transparencies (fig. 4).

Plant devices and settings are described by option 4. For example, fig. 5 shows parameters which have to be entered for some common devices. Default values are proposed to the user.

GRAVIM FILE : FLOWSHEET

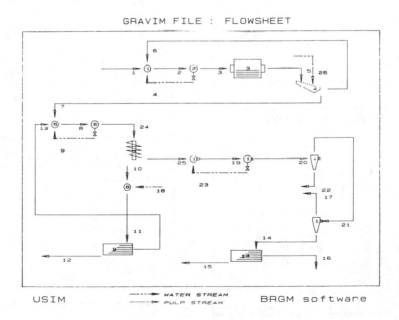

FIGURE 4. Flowsheet print-out.

USIM — SPECIFICATION OF EQUIPMENT PARAMETERS — BRGM software

UNIT # 3
ROD MILL

NUMBER OF MILLS IN PARALLEL	1.00000
MILL DIAMETER (m)	2.13000
MILL LENGTH (m)	3.05000
ROD CHARGE WEIGHT (t)	20.00000
CRITICAL SPEED FRACTION	0.70000
ROD SPECIFIC GRAVITY (g/cm3)	7.80000
NUMBER OF THE PHYSICAL PROPERTY WI	1.00000

[ESC] to continue

FIGURE 5. Specification of equipment parameters for a rod mill.

Plant feed is described by option 5 of the main menu. Each feed stream description mainly consists of a solid flowrate, a water flowrate, a particle size distribution, and a particle type distribution for each particle size class (fig. 6).

Particle size classes, mineral names and physical porperties and particle types (liberated mineral species and/or locked particles), are indicated in a ore model file which can be easily created or modified with option 3 (fig. 7 and 8). However, for a current use of USIM industrial versions, option 3 is not used, since convenient ore model files are delivered which were created when the models were calibrated.

Calculation

Once all the data related to a situation have been entered and stored in files (the names of which are chosen by the user), USIM can simulate the corresponding steady state operation of the plant.

The user selects option 6 of the main menu, and first fills the screen page

USIM — DESCRIPTION OF ORE MODEL — BRGM software

PARTICLE SIZE FRACTIONS: UPPER LIMITS (micron) in decreasing order

31500 20000 12500 8000 5000 3150 2000 1250 800
500 315 200 125 80 40 20 10

MINERALS: NAME	DENSITY	PROPERTIES				
WOLF	7.100	8.000	0.000	0.000	0.000	0.000
SCHE	6.200	8.000	0.000	0.000	0.000	0.000
GANG	2.700	15.000	0.000	0.000	0.000	0.000
.	.	.	.	.	.	.

[ESC] to continue

FIGURE 7. Ore model : specification of particle size classes and mineral names and physical properties.

shown in fig. 9. Thus, he indicates the files in which are the data which describe the situation to be simulated : flowsheet, device parameters and settings, ore model and plant feed. He can also mention a number of allowed iterations and a convergence criteria, although there are default values for these.

The USIM algorithm is iterative. At each iteration, the output streams of each device are calculated from the device input streams by the device model. During the calculation, USIM displays the number of the iteration performed.

At the end of each iteration, all calculated streams are compared with those calculated by the previous iteration, and convergence is achieved if the difference is smaller than the convergence criteria. Typical convergence time is in the range of a few minutes.

USIM — STREAM INITIALIZATION — BRGM software

PARTICLE SIZE DISTRIBUTION FOR STREAM # 1
ORE FLOWRATE(t/h) 14.00 WATER FLOWRATE(t/h) 1.00

SIZE micron	WEIGHT %	GRADE WOLF	GRADE SCHE	GRADE GANG	SIZE micron	WEIGHT %	GRADE WOLF	GRADE SCHE	GRADE GANG
31500	8.20	2.25	0.24	97.51	200	0.55	2.38	0.24	97.38
20000	26.60	2.14	0.16	97.70	125	0.70	2.35	0.25	97.40
12500	19.20	2.15	0.16	97.69	80	0.80	2.20	0.19	97.61
8000	16.30	2.20	0.19	97.61	40	0.60	2.22	0.20	97.58
5000	8.40	2.22	0.20	97.58	20	1.10	2.20	0.15	97.65
3150	6.60	2.24	0.25	97.51	10	1.60	2.34	0.26	97.40
2000	4.50	2.30	0.19	97.51					
1250	2.10	2.25	0.29	97,46					
800	1.30	2.34	0.26	97.40					
500	0.80	2.36	0.23	97.41					
315	0.65	2.30	0.25	97.46					

[ESC] to continue

FIGURE 6. Feed stream description.

USIM — DESCRIPTION OF ORE MODEL — BRGM software

PARTICLE TYPES: MINERAL CONTENTS

	WOLF	SCHE	GANG	
1	1.000	0.000	0.000	.
2	0.000	1.000	0.000	.
3	0.000	0.000	1.000	.
.	.	.	.	.

REAGENTS: NAMES

[ESC] to continue

FIGURE 8. Ore model : particle types description.

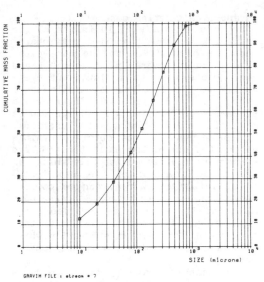

FIGURE 9. Information input for calculation.

```
USIM              DATA PROCESSING              BRGM software


        FLOWSHEET FILE        GRAVIM

        ORE MODEL FILE        GRAVIM

        INITIALIZATION FILE   GRAVIM

        EQUIPMENT FILE        GRAVIM

        CONVERGENCE CRITERION (%)   0.0010

        MAXIMUM NUMBER OF ITERATIONS  50
                                        [ESC] to continue
```

FIGURE 9. Information input for calculation.

```
USIM                   GLOBAL RESULTS              BRGM software
      ore   water
ST. FLOW. FLOW.    WOLF        SCHE         GANG
#   t/h   t/h  GRADE RECOV. GRADE RECOV. GRADE RECOV.  GRADE RECOV.
1  14.00  1.00  2.20 100.00  0.19 100.00  97.61 100.00   .    .
2  17.42  7.46  1.90 107.70  0.16 107.79  97.93 124.81   .    .
3  17.42  7.46  1.90 107.70  0.16 107.79  97.93 124.81   .    .
4   0.00  6.12  0.00   0.00  0.00   0.00   0.00   0.00   .    .
5  17.42  7.46  1.90 107.70  0.16 107.79  97.93 124.81   .    .
6   3.42  0.34  0.69   7.70  0.06   7.79  99.25  24.81   .    .
7  14.00 12.12  2.20 100.00  0.19 100.00  97.61 100.00   .    .
8  14.22 26.41  2.58 119.45  0.22 118.78  97.19 101.13   .    .
9   0.00 13.34  0.00   0.00  0.00   0.00   0.00   0.00   .    .
                                          [ESC] to continue
```

FIGURE 10. Global presentation of results.

```
USIM                  DETAILED RESULTS              BRGM software

                    SIZE DISTRIBUTION IN STREAM #   7
          ORE FLOWRATE (t/h)  14.00 WATER FLOWRATE (t/h)  12.12

SIZE PASS.        GRADE           SIZE PASS.        GRADE
micron  %    WOLF  SCHE  GANG     micron  %    WOLF  SCHE  GANG

31500 100.0  0.00  0.00  0.00      200 66.3  1.66  0.14  98.20
20000 100.0  0.00  0.00  0.00      125 52.6  2.00  0.17  97.83
12500 100.0  0.00  0.00  0.00       80 42.0  2.42  0.21  97.37
 8000 100.0  0.00  0.00  0.00       40 28.7  2.93  0.25  96.82
 5000 100.0  0.00  0.00  0.00       20 19.1  3.41  0.30  96.29
 3150 100.0  0.00  0.00  0.00       10 12.5  4.64  0.41  94.95
 2000 100.0  0.00  0.00  0.00
 1250 100.0  0.52  0.05 99.43
  800  98.9  0.75  0.07 99.18
  500  90.2  1.03  0.09 98.88
  315  78.0  1.33  0.11 98.55
                                          [ESC] to continue
```

FIGURE 11. Detailed description of the composition of a stream.

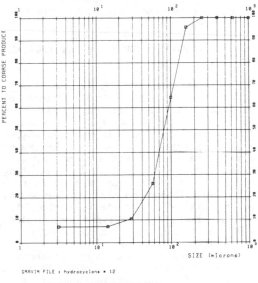

GRAVIM FILE : stream # 7

—□ SIZE DISTRIBUTION IN STREAM # 7

FIGURE 12. Size distribution curve print-out.

GRAVIM FILE : hydrocyclone # 12

—□ CLASSIFICATION CURVE FOR HYDROCYCLONE 12

FIGURE 13. Partition curve print-out.

Results display

Option 7 displays on several screen pages the solid and water flowrates and the mineral content of all the streams. This is a general view of plant operation (fig. 10). For more details option 8 allows a full description of any stream, including particle size distribution and mineral distribution in each particle size class (fig. 11).

Graphic output for size distribution curves of streams (fig. 12) and classifier (hydrocyclone...) partition curves (fig. 13) are also provided.

Advanced calculation control

It is usual to run USIM in the simple way which has been described previously. But USIM also offers to the more experienced user a variety of powerful and comfortable ways to follow, and eventually to speed up the calculation by a proper initialization of intermediate streams. For example, a first run can be made with a small number of iterations. If convergence is not obtained, values calculated during the last iteration are stored in an intermediate file which has the exact standard of the plant feed file. This file content can be comfortably observed with option 5 of the main menu and it can also be modified : the intermediate file can be used as a starting point for another simulation which will probably require less iterations to converge.

For another example, to simulate a situation close to a previously simulated situation, plant feed can be specified by option 5, not in a new file but in a result file which already contained an equilibrium situation close to the one simulated. Thus the number of iterations and the convergence time will be reduced.

USIM INDUSTRIAL VERSIONS : HOW THEY ARE BUILT

From the user, there is no visible difference between a USIM exploratory version and a USIM industrial version. In both cases, the software design results in a combination of real power and apparent simplicity. However, a USIM industrial version has been customized, tailored for a given projected or existing plant. Therefore it can be used to predict the influence on plant performances, of variations of flowsheet, device size and settings, feed rate and composition, with an operational accuracy, but only for a given project or plant.

Contrasting with the user's comfort, a complex methodology is involved to build a USIM industrial version and has been proven successful (7). Figure 14 shows the main stages of this advanced methodology.

Experimental data must be collected carefully, taking into account the usual sampling and measurement difficulties. Steady state material balances of pilot or industrial situations are basic information for model calibration and validation. If available plant or pilot data are too inaccurate, accurate

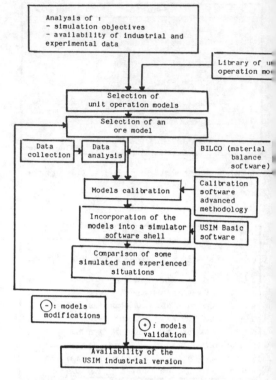

FIGURE 14. Typical methodology to build a USIM industrial version.

results from calibrated models, cannot be expected nor model validation. However, material balancing software, such as the BILCO software developed by the authors (5) is used to extract the most accurate information from available plant data, and help to overcome some measurement deficiencies.

Model calibration itself is a complex stage witch includes both ore model calibration and unit operation model calibration. The ore model calibration consists in calculating particle size distributions and particle type distributions for each particle size class of the feed streams. As has already been shown, (ref. 4) the BILCO material balance software can be used for this purpose, to fit the mineral model with the chemical composition.

The unit operation model calibration remains the main task. It consists of finding values of parameters which are not directly visible by the USIM user (unlike devices dimension and settings), but govern the accuracy of the models. Some of these parameters are directly

142

derived from specific experiments, such as those accounting for breakage function, residence time distributions... The other parameters are obtained by a specific calibration software, CALI, developed by the authors, which reads USIM data files and injects optimal parameters into USIM parameter files.

Generally, the development of a USIM industrial version requires about a few weeks or a few months, depending upon the complexity of the process.

USIM EXPLORATORY VERSION

A plant flowsheet and a rough estimation of device sizes are often required at early stages of a mining project (prefeasability study or even earlier). At any time in the development of a project, USIM exploratory can provide this information, based on the limited available data.

A key step of a project is often a pilot plant experimentation, for which the USIM exploratory version is an aid in defining a flowsheet, and in choosing devices and settings. The USIM approach avoids testing unpromising situations and can therefore save time and money. The authors have experienced that at this stage, USIM is often more effective than the engineers' intuition.

Unit operation models included in the USIM exploratory version are based on simple proven concepts. Rod and ball mill models are based on the energetic approach (ref. 1 and 9), including efficiency factors. But, apart from mill diameter, length, rotation speed and grinding media mass, the only parameter required from the user is a measured (or guessed !) Work Index. Eventually, different work indices can be used for different minerals to account for differences of grindability. The flotation model is a simple kinetic model and the user is asked, apart from the number and size of cells, for a foreseeable half flotation time for each mineral. The half flotation time of a given mineral can vary or not from one bank to another. Tables with usual values are provided for guesses.

USIM HYDROMETALLURGY VERSION

The basic elements described above are globaly kept for this version dedicated to the hydrometallurgical and chemical processes. As can be seen in the main menu (fig. 15), the different steps of a

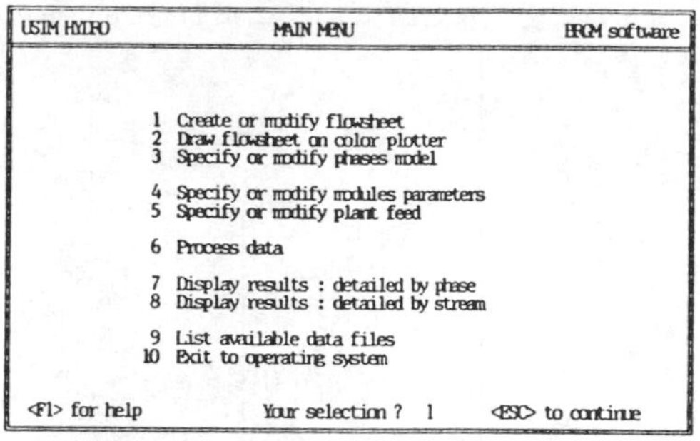

FIGURE 15. Main menu of the USIM-HYDRO version.

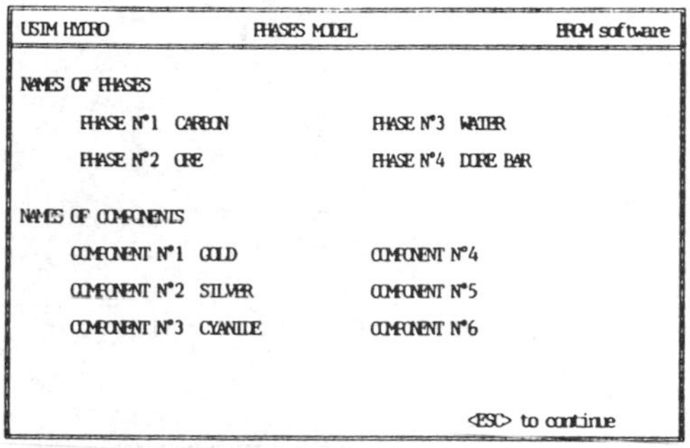

FIGURE 16. Model of phases : example of Gold process.

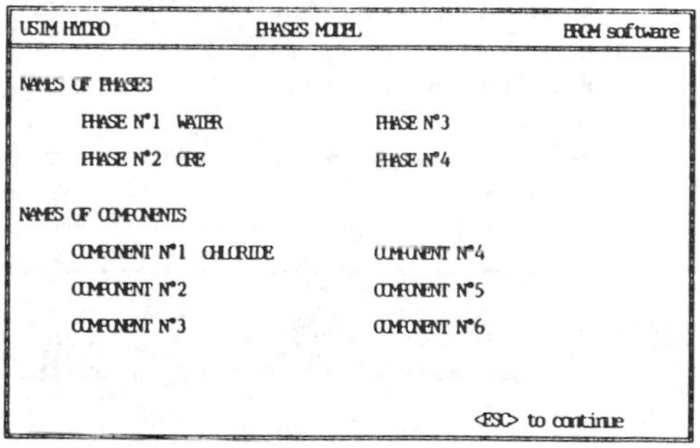

FIGURE 17. Model of phases : example of Phosphate process.

simulation are the same. However, the manipulated concepts and the models are totally different ; for example for a gold ore beneficiation process the feed will be described (fig. 16) by four phases (activated carbon, ore, water and dore bar) and three components (gold, silver and cyanide), for the removal of

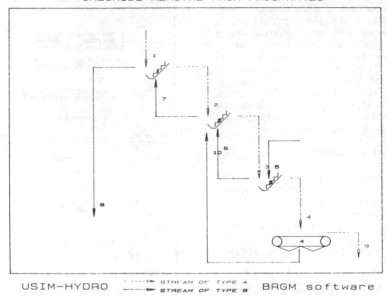

USIM-HYDRO ----▷ STREAM OF TYPE A / ———▶ STREAM OF TYPE B   BRGM software

FIGURE 18. Flowsheet plotting : example of phosphate process.

USIM HYDRO	PARAMETERS OF THE MODULES	BRGM software
	MODULE # 1 WASHING (Screw classifier)	
Volume (in cubic meter)		3.2500
Number of the washed phase (1 to 4)		2.0000
Number of the cleaning phase (1 to 4)		1.0000
Number of the exchanged component (1 to 6)		1.0000
Kinetic of exchange (inverse of a time : 1/h)		4.0000
Number of the property corresponding to the porosity (1 to 5)		2.0000
		<ESC> to continue

FIGURE 19. Parameters of the Modules : example of Phosphate process.

USIM HYDRO		RESULTS BY PHASE			BRGM software		
			PHASE N°2 : ORE CONCENTRATIONS OF THE COMPONENTS				
N° STREAM	PARTIALS FLOW.	N°1 CHLORIDE	N°2	N°3	N°4	N°5	N°6
1	10000.000	3200.00					
2	10000.000	2217.51					
3	10000.000	1309.30					
4	10000.000	619.91					
5	0.000	0.00					
6	0.000	0.00					
7	0.000	0.00					
							<ESC> to continue

FIGURE 20. Results by phase example of Phosphate process.

choride from a phosphate ore by counter-current washing (fig. 17), two phases are considered (water and ore) and one component (chloride).

The models, called after the flowsheet drawning (fig. 18), calculate the output streams of all the devices (columns, filters, exchangers...) for each phase and the variations of the concentrations of one or more component between two or more phases (fig. 19).

The results may be displayed through two forms (with the options 7 and 8) either by phase for all the streams (fig. 20) or by stream for all the phases (fig. 21).

USIM HYDRO is a tool for the design and the optimization of the hydrometallur-gical or chemical process. It enables the formulization of the know how of process designers or engineers and the quick evaluation of a large number of new possibilities for a process at the levels of the flowsheet, of the quantity and the quality of the phases or of the operating parameters.

144

```
USIM HYDRO          RESULTS BY STREAM          BRGM software

              CONTENT OF PHASES FOR THE STREAM # 2
                         CONCENTRATIONS OF THE COMPONENTS
                   N°1    N°2   N°3   N°4   N°5    N°6
  PHASES  FLOWRATES  CHLORIDE
  WATER   1200.000  11629.90
  ORE    10000.000   2217.51



  TOTAL  11200.000

                                        <ESC> to continue
```

FIGURE 21. Results by stream : example of Phosphate process.

SPECIAL VERSIONS

As described above, USIM is a **direct** simulator : it works like a real plant. Plant and plant feed are defined and USIM simulates the behaviour of the material in the plant when steady state is obtained. However, USIM structure and models are such that it has been possible to add another interactive option to use USIM as a **reverse** simulator. With this option, the user indicates which parameters he wishes to optimize to get a given situation characterized by expected flowrates and flow compositions. Parameters can be, for example, mill diameter and length, hydrocyclone diameter, or solid percentage for a given pulp stream... Generally, USIM can optimize 3 or 4 parameters at a time, possibly up to ten. One original feature of this approach is that two devices can be optimized at a time (mill and cyclone for example). This option has already been developed and tested by the authors, although its commercial availability is not yet decided in 1987.

CONCLUSION

The authors feel that the USIM software helps in the introduction of simulation as an operational tool for plant design and optimization.

All the versions of USIM are designed to be used directly by process designers or engineers. USIM helps to analyse the impact of each parameter of the process and so to take the right decisions for the design or the optimization of a plant.

REFERENCES

[1] BOND, F.C., JEN-TUNG-WANG. 1952. A new theory of comminution. Trans. AIME, Vol. 193, p. 163-198.

[2] BROUSSAUD A., BRACHET C., FOURNIGUET G., AUDOLI H., LAPLACE G. 1985. Juin. Prédiction par simulation de l'influence de modifications de flowsheet sur les performances de l'usine de flottation de barytine de Chaillac (INDRE - FRANCE). Proceedings of the XVth IMPC, Cannes, France.

[3] BROUSSAUD A., ALBERA F. 1986. Selection of a model for the B.R.G.M. pilot plant rod mill. Vth IFAC, Symposium on Automation in Mining, Mineral and Metal Processing, (TOKYO - JAPON).

[4] BROUSSAUD A., ALBERA F. 1987. Models of selective grinding and mineral liberation in rod and ball mills for mineral processing plant simulation. Annual meeting, SME-AIME (DENVER, CO, U.S.A.) to be published in Process Mineralogy, vol. 7.

[5] BROUSSAUD A., CLIN F., CROISE G., FOURNIGUET G. 1987. Oct.. Méthodologie d'échantillonnage et bilans matière cohérents dans un pilotage de minerai sulfuré. Congrès annuel de la S.I.M. (MARRAKECH -MAROC), to be published in "Industrie Minérale".

[6] CONIL P., BROUSSAUD A., NIANG S., DELUBAC G. 1987. Oct.. Utilisation d'un simulateur pour l'aide à l'évaluation d'hypothèses d'évolution d'un atelier de concentration à l'usine de phosphate de Taïba (SENEGAL). Congrès annuel de la S.I.M. (MARRAKECH -MAROC) to be published in "Industrie Minerale".

[7] BROUSSAUD A. 1988. Advanced computer methods for mineral processing : their function and potential impact on engineering practices. XVIth Int. Min. Proc. Congress, Stockholm.

[8] KING R.P. Revision august 1986. MODSIM : a modular method for the design, balancing and simulation of ore dressing plant flowsheets. University of the Witwatersrand, Department of metallurgy and materials engineering, report number GEN/2/83.

[9] ROWLAND C.A. 1982. Selection of rod mills, ball mills, pebble mills and regrind mills. Design and Installation of Comminution Circuits, SME-AIME, Mular, Jergesen, (eds).

Computer Applications in the Mineral Industry, Fytas, Collins & Singhal (eds)
© 1988 Balkema, Rotterdam. ISBN 90 6191 760 3

Process control with Bailey at Selbaie (Abstract)*

Jim Lauder
Les Mines Selbaie, Joutel, Quebec, Canada

ABSTRACT. The Selbaie operation use a Bailey controls Network 90 distributed process control system. Modules, containing powerful micro-computers, communicate with each other over an Ethernet bus. Various areas of the mill are linked to one another and to operator interface units via a high-speed communication loop. Thus, any module can receive or send data to any other module in the same cabinet or across the entire mill. Likewise the operator can control any loop of the process from a local or remotely located console. Also installed at Selbaie is the Network 90 Data Management System (LAN-90 DMS). It provides an interface to 'live' process data from the Network 90 Distributed Process Control System and the PC-90 Programmable Controllers. LAN-90 is highly user configurable and user friendly. These features allow current process data to be incroporated in management reports which in turn reflect the most up-to-date operation of the mill and process.

*Paper not received on time for publication.

5 Orebody modelling – Geostatistics

Computer Applications in the Mineral Industry, Fytas, Collins & Singhal (eds)
© 1988 Balkema, Rotterdam. ISBN 90 6191 760 3

Orebody modelling – A case study for a polymetallic tabular deposit

P.C.Satchwell
Bauer, Calder & Workman, Inc., Washburn, N.Dak., USA

C.W.Pelley & K.R.Notley
Queen's University, Kingston, Ontario, Canada

ABSTRACT: The use of computer-assisted orebody modelling techniques to assist with the definition and reserves calculation of an essentially "two-dimensional" underground orebody is described. The different stages of the modelling process are considered, including input of diamond drilling data, definition of ore intersections, statistical analysis, orebody contouring, block model development, two-dimensional block kriging, estimation of block and global ore reserves, and the development of grade-tonnage curves.

INTRODUCTION

Orebody modelling may be defined as the definition of an orebody with respect to its size, shape, grade and other physical parameters. The model should provide a mathematical and visual representation of the deposit which is of direct use in the design and planning of its optimum extraction.

This paper describes the development of computer-assisted modelling techniques for a stratiform polymetallic Cu-Zn-Ag deposit located in western Canada. Use of a contouring package was made to assist with the visual understanding of the orebody. The study was primarily concerned with mathematical aspects of the orebody model for use at the project feasibility stage. This involved the establishment and statistical analysis of a diamond drilling database. It was followed by the creation of a two-dimensional block model which formed the basis of the reserves estimation process.

Computers have been used extensively in recent years for modelling at large open pit mines, and are likely to play an increasingly important role in small underground deposits. A computerized block model based on diamond drilling data provides a "mineral inventory" for the deposit (Barnes, 1980), and a tool which may be used during mine feasibility and design studies. For the model to be used for meaningful ore reserves estimation it must take into account the natural variability of the mineralization, and the deposit geology.

In a period of low base metal prices and high project risk it is essential to obtain the best possible reserves estimate for available orebody information. It is also required to quantify the accuracy of the reserves estimate for a given level of risk. For these reasons geostatistical estimation techniques were employed for the study described. These are very calculation intensive and may not be realistically employed without the assistance of a computer.

2 GEOLOGY

The deposit studied was 0-8 meters in thickness, and dipped at around 30 degrees to the north. The mineralized zone measured approximately 1200 meters by 175 meters. Host rocks were metamorphosed carbonaceous and calcareous sediments of 500 MY age, consisting of phyllites and schists. Though a minor fold structure was detected at its southern end, the orebody was essentially tabular in nature.

The mineralogy of the sulphide layer consisted of pyrrhotite, chalcopyrite and sphalerite. These occurred disseminated in the host rock matrix, as small veinlets oriented with foliation, and in fractures. The footwall contact was generally sharp, with transitional hangingwall contacts and patchy hangingwall mineralization. There were no distinct mineralization zones reported in drill logs, other than some oxides near the surface at the up-dip end of the orebody.

3 ESTABLISHMENT OF OREBODY DATABASE

The establishment of a complete and accurate orebody database is an essential first step to the modelling process. It requires the input of data from diamond drilling, the definition of ore-waste contacts, and in many cases the digitization of the available geological interpretation.

3.1 Input of diamond drilling data

Data from 58 diamond drill holes was available for this study, from two separate drilling campaigns. Initial drilling consisted of closely spaced short holes through the oxide zone into sulfides. These were in most cases rejected from the database, to avoid clustering of information, and because in some cases the hangingwall was not picked up.

One approach to data input is the development of a "customized" template for the deposit. For small orebodies this could be achieved using a PC based spreadsheet program. It is necessary that the software employed be capable of producing files in ASCII format which may be read by later application programs. An interfacing program is required to rigorously check for input errors and missing data.

A second approach, which was used for this study, is the development of a menu based interactive program for data input. Each borehole was identified by its number, and located by its collar coordinates. Dip and bearing were entered for the collar, and wherever down the hole survey data was available. Checks for data error were carried out during input, and included tests for incorrect data type (character when expecting number) and for unlikely or impossible values (for example, borehole inclinations of greater than 90 degrees).

In the simple tabular orebody studied it was not required to code rock or mineralogical zone types at the data input stage. In many cases, however, this will be essential. In a two-dimensional orebody it should be possible to outline different rock and mineralogical zone types within the model by digitizing the geologists' interpretation.

The program developed presented data in an "edit mode" for final checking and correction prior to filing. For grade data the appropriate "from..to" distance down the borehole was entered. Checks were carried out for reversed or non-contiguous down hole sample interval distances. The user was alerted to gaps in sampling, and grade values for each metal were input for each sampled intersection.

3.2 Definition of orebody contacts

The definition of the orebody in diamond drilling intersections is a vital stage, as it translates raw sample data into the composited values required in the orebody model database. In large massive deposits this may be achieved "automatically" by a computer program which calculates composite assay values within pre-determined intervals which correspond to block thickness in the model. This method causes the inclusion of waste material in blocks at the edges of the defined ore or mineralogical zone limits, but in a large orebody with many blocks this may not be too significant at the initial estimation stage.

In a thin tabular orebody, however, block thickness is determined by the orebody hangingwall and footwall positions, and in effect every block is an "edge block". A computer program was developed which allowed the program user to manually define orebody limits in the intersection.

In many cases it was obvious from visual inspection of grade values where to put the hangingwall and footwall. For the studied orebody it was assumed that mechanized mining methods were to be employed. As such, the concept of a minimum mining height was considered in the ore definition philosophy. Geostatistical estimation methods attempt to quantify the natural variability of ore. It was not permitted, therefore, to include hangingwall waste in samples composited over the mining height where the orebody was thin. Mining dilution must instead be considered at the planning stage.

In some boreholes it was difficult to assign a hangingwall due to the diffuse nature of the mineralization. It was decided, therefore, to use "equivalent grades" to assist with ore definition. Zinc and silver grades were converted into "copper equivalent" values using "recovery factors" which incorporated the expected metal prices and metallurgical recoveries. In this way the value of all three metals were considered.

Equivalent grade was calculated by the following expression:

$$EG = \%Cu + \frac{(\%Zn \times RFZn)}{RFCu} + \frac{(oz/ton\ Ag \times RFAg)}{RFCu}$$

Where RFCu,Zn

$$= \frac{Price\ (c/lb) \times Recovery\ (\%) \times 2204}{1000000}$$

$$RFAg = \frac{Price\ (c/tr.oz) \times Recovery\ (\%)}{10000}$$

3.3 Creation of sample database

The program developed for this study presented the grade values for each sampled intersection in a borehole at the computer terminal. The user input possible hangingwall and footwall contacts as down hole distances. The list of assays was re-displayed, showing the selected contacts, and the weighted average grades of each metal, equivalent grade, and the true-thickness of the intersection were calculated. The program user was able to try out different positions until a satisfactory result was obtained.

In the studied orebody the bearing of angled drill holes was at 90 degrees to the orebody strike. Simple trigonometric relationships could therefore be employed to calculate true thickness values for each sample composite. Mathematically more rigorous approaches are available when this is not the case (Royle, 1979).

Trigonometry was also employed to calculate the three-dimensional coordinates of each hangingwall and footwall intersection point. These coordinates were then calculated for the mid-point of each defined ore intersection. Correction was made for down hole deviation where indicated by camera survey data. Table 1 shows the portion of the diamond drill database used for modelling the orebody. It includes the prices and recoveries used in calculation of equivalent grade, together with the easting, northing and elevation coordinates of each composite sample mid-point.

BHNO	OBEAST	OBNRTH	OBELEV	TT	GR1	GR2	GR3	GRPM	EQGR
NG-3	14900.719	2073.205	821.159	2.77	0.14	0.05		0.04	0.18
74-1	14995.121	1960.924	913.068	4.27	3.89	0.00		0.26	3.99
NG-2	15000.289	2160.908	774.814	1.53	0.75	0.77		0.18	1.24
NG-1	15000.969	2074.824	812.819	3.30	3.69	1.51		0.38	4.67
NG-4	15100.801	2064.957	817.033	6.60	4.24	1.54		0.48	5.27
NG-33	15101.352	2232.020	697.419	6.20	1.16	0.65		0.16	1.58
NG-32	15105.141	2146.635	756.834	3.60	5.93	1.70		0.57	7.09
NG-13	15196.531	2216.180	718.160	3.55	4.38	2.11		0.62	5.78
NG-5	15200.543	2064.852	834.548	6.00	1.66	2.12		0.33	2.95
NG-8	15201.684	2158.152	756.530	3.78	1.27	1.84		0.37	2.43
NG-14	15207.730	2295.187	668.342	3.58	6.73	1.94		1.05	8.21
NG-9	15288.172	2168.484	765.695	5.50	2.23	2.54		0.42	3.79
NG-21	15295.598	2400.475	612.449	4.62	0.94	0.75		0.20	1.43
NG-15	15304.387	2314.584	670.501	4.87	2.17	1.72		0.39	3.27
NG-11	15305.074	2226.490	727.255	7.30	3.83	4.73		0.69	6.70
NG-7	15306.715	2130.182	802.046	5.30	2.45	3.10		0.56	4.37
NG-12	15372.820	2225.607	745.851	1.93	1.88	2.22		0.51	3.30
NG-36	15380.988	2572.061	519.724	1.53	0.14	0.53		0.06	0.45
NG-22	15394.570	2500.576	573.545	2.39	3.35	1.08		0.38	4.09
NG-16	15401.457	2333.830	673.521	6.00	2.24	2.18		0.36	3.58
NG-10	15402.176	2172.539	784.443	6.00	0.66	0.22		0.16	0.84
NG-20	15416.941	2408.783	628.160	7.39	3.53	2.94		0.52	5.35
NG-26	15493.988	2580.900	540.753	2.59	2.03	1.10		0.35	2.77
NG-28	15498.711	2656.074	489.120	8.31	1.97	1.08		0.29	2.68
NG-17	15499.250	2346.340	685.933	4.93	0.53	0.70		0.19	0.99
NG-23	15499.992	2496.998	591.895	5.51	3.27	2.77		0.50	4.99
NG-41	15501.199	2722.379	448.983	7.90	0.28	0.19		0.05	0.40
NG-19	15508.371	2426.367	632.540	3.63	0.98	1.93		0.30	2.16
NG-18	15604.398	2356.096	701.482	1.90	0.31	0.02		0.20	0.40
NG-24	15608.449	2498.650	609.124	6.31	0.63	0.35		0.18	0.89
NG-27	15612.289	2688.893	494.571	7.38	4.82	2.74		0.60	6.56
NG-30	15614.898	2760.312	444.748	4.54	3.69	2.63		0.60	5.37
NG-25	15627.762	2583.246	564.168	3.34	0.46	0.36		0.13	0.71
NG-29	15693.551	2759.773	466.223	8.00	3.24	2.04		0.44	4.53
NG-40	15699.590	2682.520	513.990	3.00	0.50	0.85		0.13	1.02
NG-38	15704.051	2833.377	425.279	6.90	6.05	3.53		0.94	8.36
NG-50	15705.828	2950.324	372.746	5.26	0.74	0.77		0.12	1.21
NG-42	15750.039	2767.277	471.674	7.38	1.39	1.72		0.30	2.45
NG-43	15803.699	2833.967	441.250	6.90	0.99	2.37		0.23	2.38
NG-49	15803.699	2992.100	380.747	6.60	0.92	0.48		0.13	1.23
NG-48	15803.703	2921.109	405.102	5.69	1.61	2.17		0.30	2.92
NG-45	15903.641	2990.270	389.990	2.38	0.76	0.29		0.18	0.99

PARAMETERS USED FOR CALCULATION OF EQUIVALENT GRADE

METAL	PRICE	RECOVERY (%)
CU	65.00	80.00
ZN	38.00	75.00
AG	605.00	75.00

Table 1: Database of orebody intersections

4 OREBODY CONTOURING

One of the methods which may be employed to achieve a better understanding of a deposit is contouring. These may also be a significant aid to the mine design process. A computer program was developed for this study which interfaced the sample database with the DI3000 contouring package (Precision Visuals, 1983).

This package employed a sophisticated bivariate interpolation method (Akima,

1978) to provide values at points on a regular grid. A contour map of footwall elevation was used in a separate study for location of underground development at the mine design stage.

Figure 1 shows an example of a two-dimensional contour map of equivalent grade times thickness for the deposit studied. A useful visual representation of an orebody may also be obtained using a three-dimensional contouring subroutine, such as that shown in Figure 2, for orebody true-thickness.

153

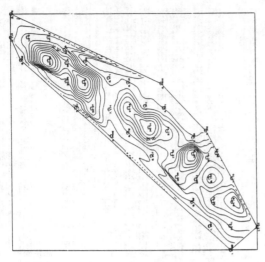

Figure 1: Contour map of equivalent grade times true thickness

CLASS INTERVAL		FREQUENCY %		HISTOGRAM
-1.61 TO	-1.34	1	2.38	2.38	I**
-1.34 TO	-1.07	0	0.00	2.38	I
-1.07 TO	-0.80	1	2.38	4.76	I**
-0.80 TO	-0.53	1	2.38	7.14	I**
-0.53 TO	-0.26	0	0.00	7.14	I
-0.26 TO	0.02	0	0.00	7.14	I
0.02 TO	0.29	1	2.38	9.52	I**
0.29 TO	0.56	2	4.76	14.29	I****
0.56 TO	0.83	2	4.76	19.05	I****
0.83 TO	1.10	2	4.76	23.81	I****
1.10 TO	1.37	3	7.14	30.95	I*******
1.37 TO	1.64	3	7.14	38.10	I*******
1.64 TO	1.91	2	4.76	42.86	I****
1.91 TO	2.18	3	7.14	50.00	I*******
2.18 TO	2.45	3	7.14	57.14	I*******
2.45 TO	2.72	5	11.90	69.05	I************
2.72 TO	2.99	5	11.90	80.95	I************
2.99 TO	3.27	4	9.52	90.48	I**********
3.27 TO	3.54	2	4.76	95.24	I****
3.54 TO	3.81	2	4.76	100.00	I****

MEAN OF SAMPLES = 14.348 VARIANCE = 760.255
STANDARD DEVIATION 27.53

FROM 42 SAMPLES

Figure 3: Histogram of logarithmic data for copper grade x thickness

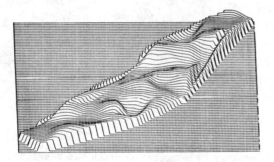

Figure 2: 3D Contour map of orebody true thickness (vertical scale exaggerated 10 times)

5 STATISTICAL ANALYSIS OF OREBODY DATA

Geostatistical methods of ore reserve estimation employ the concept that orebody parameters are spatially correlated (David, 1977). Inter-dependence of samples is measured by the variogram, which measures the average squared difference between sample values a given difference apart. The converse function, covariance, measures the similarity of sample values, and is employed in the kriging estimation method.

An important consequence of the "volume-variance" relationship in geostatistics is that samples must be of the same size if they are to be represented as point values for kriging. For the two-dimensional orebody studied all sample composites had different lengths. It was therefore necessary to include the thickness dimension in the variables analyzed.

Geostatistics employs probability theory to estimate the error of an estimate. It was therefore necessary to determine whether sample values in the orebody could, when considered as independent variables, be approximated by standard statistical distribution models. A program was developed which calculated and presented histograms of sample data values. An example is given in Figure 3.

Another approach used was a probability plot. The straight line shown over much of the graph in Figure 4, for example, indicates that the copper grade times thickness data for the studied deposit could be approximated by a lognormal model.

Copper and equivalent grade times thickness values were found to be approximately lognormal, and true thickness was normally distributed. Coefficients of correlation between grade times thickness and true thickness were found to be 0.55, 0.7 and 0.8 for Cu, Zn and Ag respectively.

The manual calculation of variogram values is very tedious, even for a small deposit with few boreholes. The minimum recommended number of boreholes for use of geostatistics is 30 (David, 1984a), which generates 435 combinations of sample pairs. For each sample pair the separation distance must be calculated from three dimensional coordinates, and

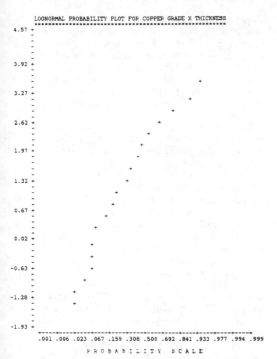

Figure 4: Lognormal probability plot for copper grade x thickness

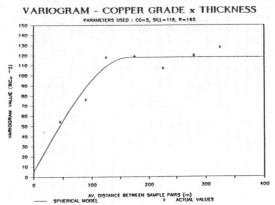

Figure 5: Variogram of copper grade times true thickness

the square of each difference in grade value obtained. A computer program was therefore developed to generate variogram plots.

It employed a three-dimensional cone search technique for sample pairs. The maximum sample separation distance was set by the user, together with a step distance increment. All sample pairs whose separation distance fell within this increment were included in the calculation of the average variogram value for that step. The average separation of sample pairs within each step increment was calculated, and a variogram plot obtained.

An example of such a plot is given in Figure 5. Sample data from an orebody can often be approximated by a spherical variogram model. A visual estimate of sill, range and nugget effect was made, and the spherical model generated using these parameters. Variogram model parameters were obtained for true thickness and for each grade times thickness parameter.

Variogram parameters can vary with direction, the phenomenon of "geometric anisotropy" (David, 1977). The program developed included the facility to vary the bearing and inclination of the sample search cone, and to specify an angular

tolerance. In the case studied there was no obvious geometric anisotropy, and in general an angular tolerance of 180 degrees was allowed in a cone search carried out in the plane of the orebody.

6 CREATION OF THE BLOCK MODEL

The representation of an orebody by a model containing discrete geometric blocks, usually of fixed dimensions, is becoming increasingly common in the mining industry. Orebody data may be rapidly and efficiently stored or accessed, and the use of blocks is consistent with the representation of production units later in the life of the mine.

In the two-dimensional orebody studied, thickness was one of the spatially correlated variables estimated by geostatistical methods. The block model was thus reduced to two dimensions, and block corners were represented by points on a regular grid. This is illustrated in Figure 6. The size of blocks in any model must be realistic for the density of

available drilling information. A reasonable size for estimation blocks is not less than one third of the drilling grid size (David, 1984a).

A 50 meter square grid was used to model the studied orebody. This was horizontal, and oriented with orebody strike. The computer program employed to assist with block model creation transformed sample locations into the block model coordinate system.

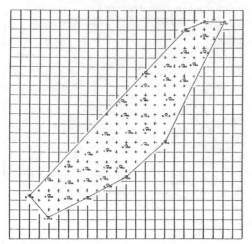

Figure 6: Block model showing block corners as grid locations, with orebody bounded by convex hull

The gridding routine from the contouring package was used to interpolate elevations of ore intersection mid-points onto the corners of the grid model. This routine also created a "convex hull" which joined the outermost data points (diamond drill holes) in the deposit, as shown in Figure 6.

Since boreholes which intersected waste were not included in the database, the convex hull formed lateral limits for the orebody in the model. A data file was created containing only those grid points which fell within these limits. Blocks with less than four boundary grid points were "flagged" as partial blocks. This approach was only possible due to the simple geology and stratiform nature of the deposit. In other cases digitized lateral orebody and zone limits should be employed.

7 ORE RESERVE ESTIMATION

The geostatistical method known as kriging provides an estimate of the value of a point or block which has the least estimation error. This error can be quantified, and is unbiased in the sense that there is no systematic over- or under-estimation of the kriged values. It has been well described in numerous publications (David, 1976, 1977; Journel & Huijbreghts, 1978; Rendu, 1978).

7.1 Estimation of block reserves

For the deposit studied simple linear

kriging was carried out on 50 x 50 meter ore blocks for true thickness and grade thickness values. A kriging computer program was developed which accessed the diamond drill database of composite sample intersections, and grid data from the block model. It searched for the nearest samples within variogram range of each block center.

Sample to sample and sample to block covariance matrixes were calculated. To achieve the latter an array of nine discrete points was established in the "mid-plane" of each block. The elevation of these points was derived from those interpolated onto grid corners during the block model creation stage.

The sample to sample covariance matrix has the highest values on its diagonal, which allowed a very efficient algorithm to be employed for solution of a system of linear equations (Davis et al. 1980). The solutions to the kriging equations were the set of linear weights applied to influencing samples, such that the estimate with minimum error was obtained for the block.

This error was calculated by multiplying each sample weight by its sample to block covariance, and subtracting this value from the block to block covariance associated with 50 meter square blocks. Small blocks have high block to block covariance, thus the error in estimation increases as block size decreases. The value of partial blocks at the orebody limits were estimated, but with lower accuracy.

The estimation error for grade values was obtained from those associated with true thickness and grade thickness. It is obtained in the following manner (David 1984a):

$$\sigma_G^2 = \frac{\sigma_{GT}^2}{GT^2} + \frac{\sigma_{TT}^2}{TT^2} - (2 \times r \times \frac{\sigma_{GT}}{GT} \times \frac{\sigma_{TT}}{TT})$$

where σ_G^2 = estimation variance of grade
σ_{GT}^2 = estimation variance of grade times thickness
σ_{TT}^2 = estimation variance of true thickness
GT = kriged grade thickness value
TT = kriged true thickness value
r = correlation coefficient between TT and GT for all orebody samples

Block tonnages were obtained by multiplying the grid area by the kriged true thickness value, factored by the

^^

BLOCK #	BLOCK TONNAGE	TRUE THICKNESS		
42	10603.35	4.70		

CU.GR	ZN.GR		AG.GR	EQ.GR
3.69	0.03		0.00	3.75

(PARTIAL BLOCK - NO INDIVIDUAL BLOCK CONFIDENCE LIMITS AVAILABLE)
^^

BLOCK #	BLOCK TONNAGE	80.0% CONF. INTERVAL		TRUE THICKNESS	80.0% CONF. INTERVAL	
		LCL	UCL		LCL	UCL
112	34604.65	30108.66	39100.65	3.41	2.97	3.85

CU.GR		ZN.GR			AG.GR		EQ.GR	
3.39		0.65		0.00	0.20		3.83	
2.88	3.99	0.29	1.48	0.00 0.00	0.07	0.58	3.07	5.03

^^

Table 2: Example of block reserves showing confidence limits for complete blocks

average dip of each block. Since the block model was oriented with average orebody strike, a reasonable estimate of block dip was obtained from considering elevation differences between block corners. Tonnages of partial blocks were crudely estimated on the basis of how many of the 9 discrete points used for covariance calculation fell within the block area.

Precision statements could be made for kriged block values, as sample values in the deposit were approximated by statistical distribution models. Upper and lower confidence limits were established for each 50 meter square block in the model, for a desired confidence interval. Since true thickness was a kriged variable, it was also possible to measure the error associated with block tonnage. An example of block reserves calculated for the study is shown in Table 2. Confidence limits were not calculated for edge blocks.

7.2 Estimation of global reserves

The global reserves for the modelled orebody were calculated simply by summing the individual block and partial block tonnages. This result was probably conservative due to the use of the convex hull to define lateral ore limits. Average grade for each metal was obtained by weighting each block value by block tonnage. Estimated block values are also spatially correlated in an orebody. It was not permitted, therefore, to take a simple weighted average of block estimation variances and apply them to the deposit as a whole.

The average error associated with a global ore reserves estimate can be obtained by a "polygonal approximation" (David, 1984b), as it is similar to that which would be obtained if the polygon method had been used. A computer program developed for global reserves estimation utilized this principle.

The points on the grid model were considered to discretize the orebody. For each point on the grid the nearest sample was found. For each sample in the database, the grid points which found it nearest were used to discretize its polygon. The estimation variance for a polygon can be calculated knowing the variance of its sample with the discrete points it contains, and the variance of these points themselves.

The average estimation error for the deposit was then obtained by weighting each individual estimation variance by the number of grid points associated with that polygon (David, 1984b). Upper and lower confidence limits were obtained for each global reserve estimate, including tonnage. These are shown in Table 3. It can be seen that as the required confidence in an estimate is increased (risk decreased) the quality of the estimate is reduced for the available sample data.

8. ESTIMATION OF MINABLE RESERVES

The final stage of orebody modelling at the feasibility stage is the determination of what part of the global reserve may actually be mined. Certain parts of the orebody may be rejected for physical reasons, such as inaccessibility for the chosen mining method, or for geotechnical reasons. However, the most important factors will be economic, and relate to the application of a cut-off value.

Serious mistakes can be made with the use of grade-tonnage curves in a feasibility study (David, 1972). The greatest error results in application of a cut-off

PARAMETER	GLOBAL VALUE	LOWER 90.0% CONF. LIMIT	UPPER 90.0% CONF. LIMIT
GLOBAL TONNAGE	4647458.00	3868483.00	5426433.00
TRUE THICKNESS	5.18	4.31	6.04
CU.GRADE	2.32	1.67	3.23
ZN.GRADE	1.46	0.81	2.62
AG.GRADE	0.33	0.11	1.03
EQ.GRADE	3.25	2.16	5.07

GLOBAL RESERVES STATEMENT AT 80.0% CONFIDENCE INTERVAL

PARAMETER	GLOBAL VALUE	LOWER 80.0% CONF. LIMIT	UPPER 80.0% CONF. LIMIT
GLOBAL TONNAGE	4647458.00	4132609.00	5162306.00
TRUE THICKNESS	5.18	4.60	5.75
CU.GRADE	2.32	1.87	2.89
ZN.GRADE	1.46	0.99	2.15
AG.GRADE	0.33	0.16	0.70
EQ.GRADE	3.25	2.47	4.34

Table 3: Global ore reserves calculated at two different confidence intervals

grade to sample values, which is the basis of the polygonal method. If this is done, and the cut-off is greater than the mean of samples, then the tonnages of normally and lognormally distributed deposits are systematically over-estimated.

The "volume variance" relationship recognized in geostatistics provides a mathematical explanation for this. Ore is mined in volumes which will correspond to a planned production unit. An economic cut-off should not be applied to sample values unless the deposit is to be mined out by diamond drill coring.

8.1 Cut-off applied to kriged blocks

A computer program was developed which calculated minable reserves for the polymetallic deposit studied. Cut-off was applied to true thickness, individual grade parameters, and to equivalent grade. In each case the tonnages of kriged blocks above cut-off were summed, and the weighted average for each orebody parameter calculated. This was achieved over a range of cut-off values selected by the program user.

A grade tonnage curve produced by applying a cut-off to equivalent grade is shown in Figure 7. The variation of average true thickness and individual grade values for a variable equivalent grade cut-off is shown in Figure 8. These

graphs were generated using the data in Table 4.

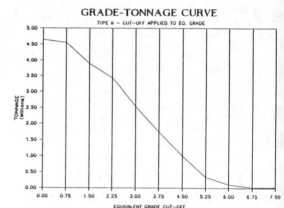

Figure 7: Grade tonnage curve produced by application of equivalent grade cut-off kriged blocks

It should be appreciated that an estimate of minable reserves obtained this way is biased, even though kriging was used for blocks. In a small underground orebody such as that studied a 50 meter square block will contain many

production scale blocks. Some will be of appreciably higher grade than the kriged value, and some will be rejected by grade control as waste.

CUT-OFF	# BLOCKS	TONNAGE	TR.TH	CU.GR	ZN.GR	AG.GR	EQ.GR
0.00	170	4647458.00	5.18	2.32	1.46	0.33	3.25
0.75	123	4559476.00	5.18	2.37	1.48	0.34	3.31
1.50	107	3892949.00	4.76	2.62	1.65	0.37	3.67
2.25	93	3437814.00	4.69	2.80	1.75	0.40	3.91
3.00	69	2546743.00	4.67	3.10	1.95	0.45	4.35
3.75	45	1741162.00	4.59	3.43	2.07	0.50	4.76
4.50	25	1004866.62	4.56	3.85	2.12	0.56	5.23
5.25	9	349953.50	4.61	4.23	2.75	0.66	6.00
6.00	3	107450.87	4.17	4.98	2.54	0.76	6.68
6.75	1	23874.76	3.58	6.73	1.43	1.05	7.93
7.50	1	23874.76	3.58	6.73	1.43	1.05	7.93

Table 4: Grade and tonnage data for equivalent grade cut-off

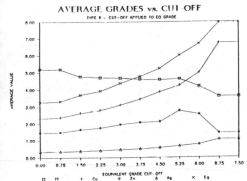

Figure 8: Variation of grade when equivalent grade cut-off is applied to kriged blocks

Kriged blocks with grade below cut-off will likewise contain some material in a production unit size which is ore. In a small underground operation this may not be accessed by mining. In these cases the loss of ore from kriged ore blocks will not be compensated by the gain from kriged waste blocks, and a shortfall of tonnage occurs.

8.2 Cut-off applied to unlocated blocks

It is possible, if sample data is approximately normal or lognormal, to provide an unbiased estimate of the minable reserve from an orebody (David, 1972). The proportion of material above a cut-off value is related to partial areas under normal and lognormal probability distribution curves (David, 1977, 1984a).

For example, the tonnage above cut-off for a lognormally distributed cut-off grade parameter can be expressed by the following:

$$T_c = \text{Global tonnage} \times (1 - F [z])$$

where T_c = Tonnage above cut-off
F = function for area under normal probability curve

$$z = (1/B) \times \log (x_c/\overline{x}) + (B/2)]$$

where
x_c = cut-off value
$\overline{x}$ = mean value of samples
B^2 = variance of blocks of a given size

Calculations of this type were carried out by a computer program for the case study. The user input the size of blocks to be considered, and the parameter to which cut-off should be applied. The variance between blocks was calculated by subtracting the average variance of points within a block of that size from the variogram sill value.

This method allows grade-tonnage curves to be produced for block sizes which are being considered as production units. These blocks, however, are unlocated in the deposit, and the results are of no direct use for mine production scheduling. This information is useful, though, in estimating the probable grade and tonnage of material falling within certain grade ranges. It can be applicable to mill design, and the planning of stockpile and materials handling requirements.

9 CONCLUSIONS

The use of computers to assist with orebody modelling brings the advantages of speed and accuracy of calculation into a critical area of mining. It also allows more sophisticated calculation intensive ore reserve estimation techniques to be used. Provided the data from the orebody can be approximated by certain mathematical models, the level of risk associated with block and global reserves estimates can be quantified.

It should always be remembered, though, that the computer is only a tool, which may produce results which are incorrect, false or unrealistic. However, when combined with the interpretation and judgement of experienced geologists and engineers the computer can provide invaluable assistance to the orebody modelling process.

REFERENCES

Akima, H. 1978. A method of bivariate interpolation and smooth surface fitting for irregularly smoothed data. ACM Transactions in Mathematical Software, Vol. 4, No. 2 : 148-159

Barnes, M. 1980. Computer-assisted mineral appraisal and feasibility. Soc. Mining Engineers of AIME : 167pp.

David, M. 1972. Grade-tonnage curve - use and misuse in ore reserve estimation. Trans. IMM, Sect A : A90-A93.

David, M. 1976. The practice of kriging. Advanced Geostatistics for the Mining Industry. Reidel Pub. Co. : 31-48.

David, M. 1977. Geostatistical Ore Reserve Estimation. Elsevier Sci. Pub. Co : 364pp.

David, M. 1984a. Geostatistical Mineral Reserve Estimation Course. McGill University,, Montreal

David, M. 1984b. Applied advanced geostatistical ore reserve estimation. Gamma Geostat International, Montreal (draft).

Davis et al. 1978. A fast method for the solution of a system of simultaneous equations - a method adapted to a particular problem. Math. Geol. Vol 12, No. 4 : 369-374.

Journel + Huijbreghts. 1978. Mining Geostatistics. Academic Press : 600pp.

Precision Visuals. 1983. DI3000 Contouring Syatem - Users Guide.

Rendu, J-M. An introduction to geostatistical methods of mineral evaluation. SAIMM, Monograph No. 2

Royle, A. 1979. Plane projections of tabular orebodies for evaluation purposes. Trans. IMM,   : A87-A91.

Computer Applications in the Mineral Industry, Fytas, Collins & Singhal (eds)
© 1988 Balkema, Rotterdam. ISBN 90 6191 760 3

Dealing with outlier high grade data in precious metals deposits

A.G.Journel
Applied Earth Sciences Department, Stanford University, Calif., USA

A.Arik
Mintec, Inc., Tucson, Ariz., USA

ABSTRACT: In precious metals deposits, a minute percentage of outlier data may contribute a disproportionate amount of the ounces recovered. Such outlier data usually correspond to specific high grade mineralizations which should be mapped, either deterministically or probabilistically, prior to their average grade estimation. The indicator kriging (IK) algorithm is shown to be such a two-step estimation procedure. Application to local grade estimation at the Sunbeam gold property is presented.

KEYWORDS: outlier data, reserves estimation, indicator kriging, epithermal gold deposit

1-INTRODUCTION

In precious metals deposits the proper handling of outlier high grade is crucial to ore reserves estimation. By outlier high grade datum, it is understood a datum much higher than the median or mean of other data, and an actual reliable value corresponding to a known mineralization, not an assay error or a misplaced decimal point! Such outlier data may represent only a minute percentage of the total data set but can contribute a large percentage of the total in situ quantity of metal.

Traditional practice consists in either setting back the outlier data to a maximum limit or to cut them out altogether. In other situations these outlier data are unchanged and treated equally to any other data, but the final ore grade estimate is cut down by a "dilution" factor. Such practice when backed by actual production figures from the same site can be reasonably accurate, however a problem arises when no prior production experience exists. More recently, geostatisticians have introduced reserves estimation procedures based, either on an interpolation of the logarithms of grades, or on some "lognormal shortcut" modeling of the probability distribution of the unknown grade, see David (1977). These lognormal techniques have yet to receive wide acceptance either because they are delicate to apply or more

possibly because they do not appeal to geological intuition.

In most precious metal deposits, the high grade data correspond to specific mineralizations clearly distinct from the predominant and usually more pervasive low to median grade mineralization. These high grade mineralizations are usually very limited in their spatial extent (veinlets, fracture or fault planes, breccia pods, etc.), but can contribute a large proportion of the ore reserves.

The ideal would be to delineate all such high grade mineralizations and limit the extrapolation of high grade data to these zones. But, such detailed geological mapping is not always possible from exploration holes, particularly if these holes carry little or no lithological information. Short of an extensive and accurate mapping the idea is to predict the volume proportion of such high grade occurrences and estimate their average grades using only relevant and neighboring high grade data. Conversely, these high grade data should not be used to estimate the average grade of the complement proportion of low to median grade occurrences. This two-step estimation procedure would limit the spatial influence of high grade data while accounting for the possibility that not all high grade zones have been encountered by the sparse drilling available. Because the influence of "outlier" data is limited to their class or mineralization, they need not be set back or cut out.

2-THE CONSEQUENCES OF DELETING EXTREME DATA

Consider a lognormal histogram modeling the spatial distribution of gold assays within a mine site. The gold assays are composited on equal lengths and the density of drilling is assumed to be uniform throughout the site. The lognormal model has been chosen for it yields closed analytical expressions for reserves parameters after truncation by a cut-off grade, see Journel and Huijbregts (1978, p. 481), not because the authors believe that any asymmetric distribution should be lognormal.

The lognormal histogram of Figure 1 is fully defined by its two parameters, mean: $m = 10$ ppm and variance: $\sigma^2 = 900$ ppm². Note the high coefficient of variation $\sigma/m=3$, not an uncommon figure for gold grades defined on composites of 3 to 10 feet.

The proportion of data above 100 ppm is found to be .0114, i.e. slightly over 1%, and the average grade of these 1% outlier data is 196 ppm. This is more than 19 times the overall mean $m = 10$ ppm and more than 62 times the overall median grade $M=3.16$ ppm.

The contribution to the variance of these 1% outliers is approximately: $(196 - m)^2 \times .0114$, i.e. a relative contribution of : $394/\sigma^2 = .438$. The 1% highest data thus contribute 44% of the total variance. This simple remark indicates that any variography based on all the data available will be rendered meaningless by the disproportionate influence of a few highest data.

In most situations these 1% highest data would be indeed "outliers" corresponding to a different mineralization. Such outlier mineralization may correspond to a specific mode in the data distribution but would be transparent to most lognormal probability plots, for the latter are rarely plotted beyond the 95th percentile, or if they are, deviations beyond that percentile are not considered significant (and wrongly so).

The contribution of the 1% highest data to the total quantity of metal is striking:

$$196 \times .0114/m = 2.24/10 = .224$$

In other words, one percent of the data informs over 22 percent of the total in situ quantity of metal.

Consider now a selective operation with an economical cut-off close to the median

Figure 1: Lognormal distribution of gold data

mean = 10 ppm
median = 3.16 ppm

variance: 900 ppm
coefficient of variation: 3

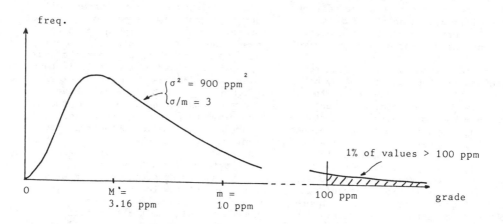

1% of the data are greater than 100 ppm and are considered as outliers. They average 196 ppm, and contribute 22.4% to the overall mean m=10 ppm.

162

datum value M = 3.16 ppm. Such an operation would recover about half the in situ tonnage, which is not an unrealistic figure. The actual ore grade is found to be 18.7 ppm. However,
- if the 1% highest data are cut out, that figure is estimated to be:

[(18.7 x .5) - (196 x .0114)] /
(.5 - .0114) = 14.56 ppm,
i.e. 22% underestimation

- if the 1% highest data are set back to 100 ppm, the ore grade would be estimated at:

[(14.56 x .4886) + (100 x .0114)] / .5 =
16.51 ppm,

still a solid 12% underestimation about equal to the profit margin possibly expected from such a mining venture!

The solution to the problems posed by outlier data is not to reject them but to treat them as coming from a separate high grade mineralization.

3-ESTIMATION OF TWO CO-EXISTING POPULATIONS

Consider the case of two types of mineralization within the same orebody: a predominant low-grade, disseminated, pervasive mineralization (type-1); and a localized, high-grade, fracture or vein controlled mineralization (type-2) cutting across the orebody in a certain direction (see Figure 2).

Figure 2: Two co-existing grade populations

The high grade, fracture filling, type-2 mineralization is imbricated into the predominant, pervasive type-1 mineralization.

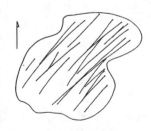

The type-2 mineralization may represent only a minor percentage of the total volume of the mineralized rock in the deposit; but, because of its high grade nature, may constitute a major portion of the total quantity of recoverable metal. In Figure 2, the type-2 mineralization appears as a series of parallel to subparallel, fracture-controlled veins which can be defined as a large-scale, stock-work type mineralization with a NE trend. These high-grade structures will yield, by definition, high grade assay values, but fewer in number than the numerous assay values obtained from the type-1 mineralization. Further, the high grade values from the type-2 mineralization will exhibit good continuity along its NE strike, but they will have little or no correlation with the neighboring lower values of the type-1 mineralization. Conversely, the grades within the pervasive type-1 mineralization will have good spatial correlation which will possibly reflect an isotropic nature (i.e., similar in all directions).

The type-2 mineralization is so interlaced with the pervasive type-1 mineralization that its accurate mapping is not possible at the scale of the exploration drill holes. An alternative to such a detailed mapping is the evaluation of the probability for any location x within the deposit to belong to either population.

Define an indicator, function of the location x, equal to 1, if x belongs to type-1, equal to zero otherwise:

$$I(x) = \begin{cases} 1, \text{ if } x \text{ belongs to type-1} \\ 0, \text{ if } x \text{ belongs to type-2} \end{cases} \tag{1}$$

At each uninformed location x, the previous indicator I(x) can be seen as a binary random variable whose mean (more exactly conditional expectation) can be estimated by a weighted linear combination of the surrounding indicator data $I(x_\alpha)$, $\alpha = 1,\ldots,n$:

$$I^*(x) = \sum_{\alpha=1}^{n} a_\alpha I(x_\alpha) \in [0,1] \tag{2}$$

$$= \text{Prob}^* \{I(x) = 1 \mid n \text{ data}\}$$

$$= \text{Prob}^* \{x \in \text{type-1, given the n data}\},$$

$$\text{with}: a_\alpha \in [0,1], \text{ and } \sum_{\alpha=1}^{n} a_\alpha = 1$$

The n weights a_α can be taken proportional to the inverse distances of x to the data locations x_α, or better they can be provided by a kriging system using the indicator data variogram thus capitalizing on possible patterns of anisotropy of these indicator data. For example, from Figure 2 it appears that NE would be the direction of maximum continuity for the indicator data.

The complement value $[1 - I^*(x)]$ provides an estimate for the probability that x belongs to the type-2 mineralization:

$$1 - I^*(x) = \tag{3}$$
$$\text{Prob}^* \{x \; \epsilon \; \text{type-2, given the n data}\} \; \epsilon [0,1]$$

There remains to estimate the grade of each mineralization type if it were to prevail at location x. The type-1 grade will be interpolated from the n surrounding grade data of type-1, using either a traditional inverse distance weighting scheme, or better a kriging system capitalizing on the pattern of spatial correlation within type-1 as characterized by a type-1 grade variogram.

$$[z(x) \mid x \; \epsilon \; \text{type-1}]^* = \sum_{\alpha_1 = 1}^{n_1} b_{\alpha_1} \cdot z(x_{\alpha_1}), \text{ with} \tag{4}$$

$z(x_{\alpha_1})$, $\alpha_1 = 1, \cdots, n_1$: grade data of type-1, and : $n_1 \leq n$

b_{α_1} : weight applied to the datum $z(x_{\alpha_1})$, and usually:

$$b_{\alpha_1} \geq 0 \;, \; \sum_{\alpha_1 = 1}^{n_1} b_{\alpha_1} = 1 \;.$$

Similarly, the type-2 grade at location x will be interpolated from the surrounding type-2 grade data. Because the type-2 grade data are scarce, one may have to extend the search neighborhood for such data beyond that used for the indicator estimation:

$$[z(x) \mid x \; \epsilon \; \text{type-2}]^* = \sum_{\alpha_2 = 1}^{n_2} b_{\alpha_2} \cdot z(x_{\alpha_2}) \tag{5}$$

with possibly n_2 such that:

$n_1 + n_2 > n$

The two grade estimates (4) and (5) are then recombined with probabilities $I^*(x)$ and $[1-I^*(x)]$, respectively, to provide the estimated grade at location x:

$$[z(x)]^* = \tag{6}$$

$$I^*(x) \cdot [z(x) \mid x \; \epsilon \; \text{type-1}]^* +$$

$$[1 - I^*(x)] \cdot [z(x) \mid x \; \epsilon \; \text{type-2}]^*$$

3.1 Remarks

- The previous algorithm corresponds to a two-step estimation, an indicator or probability estimation followed by a grade estimation. If the information available is dense enough, geological mapping may replace the first step of indicator estimation.

- If the type-2 high grade data show no spatial auto-correlation, the grade estimate (5) reduces to an equal-weighted mean of the n_2 surrounding type-2 data available over the site.

- A common application of this two-step estimation algorithm corresponds to the case where type-2 represents waste or values below detection limit set to zero. The type-2 grade estimate is then zero everywhere and the overall grade estimate (6) reduces to its type-1 component weighted by $I^*(x)$.

$$[z(x)]^* = I^*(x) \cdot [z(x) \mid z(x) > 0]^* \tag{7}$$

see Parker et al. (1979) for application to a sedimentary uranium deposit.

4-GENERALIZATION TO K POPULATIONS

Consider the case where K mutually exclusive and exhaustive populations co-exist over a given site: at any location x, one and only one population prevails.

Again short of a detailed geological mapping, the probability that a given uninformed location x belongs to population k may be estimated by the weighted proportion of neighboring data of type k:

$$[I_k(x)]^* = \sum_{\alpha=1}^{n} a_\alpha I_k(x_\alpha) \; \epsilon \; [0,1] \tag{8}$$

$$= \text{Prob}^* \{x \; \epsilon \; \text{type } k, \text{ given the n data}\} \;,$$

$$\text{with} : I_k(x) = \begin{cases} 1, \text{ if } x \; \epsilon \; \text{type } k \\ 0, \text{ otherwise} \end{cases}$$

$$\text{and} : \sum_{k=1}^{K} I_k(x) = \sum_{k=1}^{K} [I_k(x)]^* = 1$$

The latter relation is ensured if the weights a_α do not depend on k and add up to one:

$$\sum_{\alpha=1}^{n} a_\alpha = 1$$

The grade of each mineralization type if it were to prevail at location x is estimated using the sole neighboring data of the same type.

$$[z(x) \mid x \; \epsilon \; \text{type } k]^* = \sum_{\alpha_k = 1}^{n_k} b_{\alpha_k} \cdot z(x_{\alpha_k}) \tag{9}$$

These K type-specific grade estimates are then recombined into an estimate of the overall grade:

$$[z(x)]^* =$$
(10)

$$\sum_{k=1}^{K} [I_k(x)]^* \cdot [z(x) \mid x \in \text{type } k]^*$$

4.1 Remarks

- The series of K independent type-specific grade estimations is suboptimal if there exists correlation across population boundaries. Even if two populations are mineralogically different, they should be pooled if their grade ranges are similar and their grade values are spatially inter-correlated.
- The algorithm requires that each grade datum $z(x_\alpha)$ be unequivocally classified into one of the K populations. This requirement may not be met if there are too many populations, or if the grade assays $z(x_\alpha)$ are not informed by any lithological or mineralogical description, which is unfortunately not an uncommon case with rotary, reverse circulation holes. Also composite data may intersect several populations. At the limit, one may have to define the populations on classes of grades.

5-GENERALIZATION TO K CLASSES OF GRADES

The indicator kriging algorithm (IK) corresponds to the previous approach where the K populations are actually K mutually exclusive and exhaustive classes of grades. Because classes of grades are naturally ranked from low to high, a cumulative indicator is defined:
(11)

$$I(x; z_k) = \begin{cases} 1, & \text{if the grade } z(x_k) \text{ is } \leq z_k \\ 0, & \text{otherwise} \end{cases}$$

with: z_k being the upper bound of the kth class $]z_{k-1}, z_k]$.

At any uninformed location x, the series of (K-1) indicator values $I(x; z_k), k=1,\ldots,K-1$ is estimated by a weighted linear combination of the corresponding indicator data:
(12)

$$[I(x; z_k)]^* = \sum_{\alpha=1}^{n} a_\alpha(k) \cdot I(x_\alpha; z_k) \in [0,1]$$

$$= \text{Prob}^* \{Z(x) \leq z_k, \text{ given the } n \text{ data}\}.$$

In the IK practice, these (K-1) indicator estimates are obtained by (K-1) kriging systems using (K-1) different indicator variograms (see the Sunbeam case study hereafter and also Journel (1983), Davis (1984), Verly and Sullivan (1985)). Alternatively these indicator krigings may be replaced by more traditional weighting algorithms, accounting for the fact that, as the cut-off increases, the indicator data become less spatially correlated.

The (K-1) indicator estimates (12) provide a discretized estimation of the probability distribution of the unknown grade z(x). They must therefore verify the following order relations:
(13)

$$I^*(x; z_k) \geq I^*(x; z_{k+1})$$

$$I^*(x; z_k) \in [0,1]$$

The difference between two contiguous indicator estimates $[I^*(x; z_k) - I^*(x; z_{k-1})]$ gives an estimate of the probability that x belongs to that grade class. There remains to estimate the corresponding class mean value.

The simplest alternative consists of estimating the K class means by the arithmetic average of data falling in the corresponding grade classes. Let $[m_k(x)]^*$, $k = 1,\ldots,K$ be the class mean estimates. An estimate of the grade at location x is then:
(14)

$$[z(x)]^*_E =$$

$$\sum_{k=1}^{K} [m_k(x)]^* \cdot [I^*(x; z_k) - I^*(x; z_{k-1})]$$

with by convention:

$$I^*(x; z_0) = 0, \text{ for the first class}$$

$$I^*(x; z_K) = 1, \text{ and: } z_K = \max z, \text{ for the last class}$$

5.1 Remarks

- The estimate (14) appears as an estimate of the (conditional) expectation of the unknown z(x), hence its name "E-type estimate".
- The accuracy of the E-type estimate (14) or for that matter any other estimate of the unknown z(x) is directly given by the (K-1) estimated probabilities (12). In particular, the probabilities of exceedance:

$1 - I^*(x;z_k) =$
$\text{Prob}^* \{Z(x) > z_k, \text{ given the n data}\}$

can be retrieved for each location x and isopleth values of these probabilities can be contoured.
- As mentioned before, ideally the thresholds z_k should separate geological or mineralogical populations, in which case the number K can be small from 2 to 4. Short of any geological information, the K populations correspond to mere grade classification, and K should not exceed 10, with for example the z_k's being the 9 deciles of the overall grade histogram.

6-THE SUNBEAM MINE CASE STUDY

The Sunbeam mine is located in central Idaho within the Yankee Fork Mining District, a gold mining district since the mid 1800's. The mine site lies within a volcano-tectonic graben consisting of rhyolitic lava, ash-flow tuff and tephra. The fracture zones within the rhyolite masses were favorable locations for epithermal gold-silver mineralization.

Since the late 1960's, over 120,000 feet of drilling has been completed on the property by three different mining groups. Most of these drillings consisted in rotary, reverse circulation holes yielding little information for detailed geologic control. In addition, some 17,000 feet of channel sampling were taken from road cuts; this information will prove valuable for the short distances indicator variography, see hereafter Figure 3.

At a large scale two types of mineralization have been found to coexist within the deposit:

(i) a predominant, disseminated, low grade mineralization in zones of altered, intensely fractured rocks adjacent to intersections of faults and fractures.
(ii) a higher grade mineralization found in altered pyroclastic, brecciated rocks along fault and fracture planes.

Although important for grade control, the transitions between these two types of mineralizations are at a much smaller scale than that of the rotary drilling spacing (100 to 200'). In other words, local reserves estimation has little or no geological control to rely upon.

Table 1 gives the statistics of the 10 foot composite gold grades available over one area of the mine site, called the North pit. For proprietary reasons all grade figures have been multiplied by a constant; this factor does not affect the spatial correlation (variogram) structures, nor the coefficient of variation.

The in situ mean grade (m) is .024 oz/t with an extremely high coefficient of variation (σ/m) of 5.2. The grade range spans from a detection limit of .001 oz/t to a high of several ounces per ton. 51.6% of the data are above the .009 oz/t cut-off grade, which corresponds roughly to the median of the data distribution. The data above the cut-off contribute 90% of the in situ ounces of gold. On the other hand, the .033 oz/t cut-off grade, which corresponds to the economical cut-off, leaves only 10.5% of the data above it. These data contain a large proportion (61%) of the in situ ounces of gold averaging .139 oz/t. The variability of these 10.5% data is still very high, showing a coefficient of variation of 2.6.

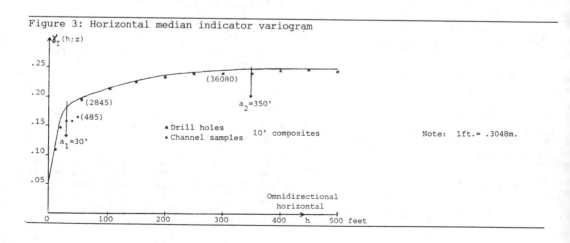

Figure 3: Horizontal median indicator variogram

7-INDICATOR VARIOGRAPHY

Initially, 9 decile cut-offs were selected defining 9 indicator data values, see relation (11). The corresponding 9 indicator variograms were run. Only those cut-offs for which the indicator variogram shapes were notably different were retained; they are the four, non-zero cut-offs shown in Table 1.

Figure 3 gives the omnidirectional horizontal variograms for the median indicator (cut-off .009). The crosses indicate the experimental values based on drill hole indicator data at spacing 100' with some zones in-filled at 50'. The dots indicate the experimental values deduced from channel sampling performed along road cuts. In parentheses are given the number of pairs of data contributing to some experimental variogram values.

These experimental values have been fitted by the sum of a nugget effect (discontinuity at the origin of amplitude $C_0 = .05$) and two spherical structures of respective ranges 30' and 350'. Note that a spherical variogram model is defined as:

$$Sph_a(h) = \begin{cases} \frac{3}{2}\frac{h}{a} - \frac{1}{2}(\frac{h}{a})^3 \text{ , for } h \leq a \\ 1 \text{ , for } h \geq a \end{cases}$$

The model indicates that at 30' horizontal interdistance about 75% of the total indicator variance of .25 is already accounted for. Any accurate geological control at Sunbeam would require information at spacing 10 to 20'; with only the available 50 to 100' spacing, probabilistic-type estimates are relevant.

Table 2 gives the parameters of the indicator variogram model for each of the four cut-offs. Beyond .033 oz/t the pure nugget model indicates that a location x close to a high grade datum is not more likely to yield a high grade than a more distant location.

7.1 A test panel

Figure 4 gives the relative configuration of a test panel of size 40' x 40' x 20' and its data search neighborhood of size 300' x 300' x 60' spanning 3 benches, the bench containing the panel and the benches above and below. Forty closest 10'composites were retained within that search neighborhood, of which six intersected the panel. The 40 data have been ranked by increasing distance to the center of the panel, see Figure 4.

Table 3 gives the equal weighted statistics of the 40 data. The largest datum .519 oz/t contributes by itself 39% of the data mean .0333. The composite just below that outlier on the same hole is valued only .003 oz/t. Clearly such an outlier datum should not be extrapolated as far as the other more median-valued data.

Yet interpolation algorithms, such as inverse distance weighting or ordinary kriging, which do not account for the data values in determining the weights will give inappropriate weight to such outliers, failing to recognize their specificity.

Table 4 lists the inverse distance squared (IDS) weights: the outlier datum .519 oz/t receives 33% of the weight resulting in an unrealistically high IDS estimate for the panel: .19 oz/t.

Table 1: Sunbeam N-pit: 10 feet composite statistics

Cut-off	% above	% of Au above	m	σ/m
0.0	100.0	100	0.024	5.2
0.009	51.6	90	0.041	4.1
0.015	31.6	80	0.060	3.6
0.020	21.0	72	0.081	3.2
0.033oz/t	10.5	61	0.139oz/t	2.6

Sample size: N=3500 composites

10% of the data inform over 60% of the total quantity of metal and most of the recoverable ore.

Conversely, the indicator kriging algorithm considers separately 5 classes of grades defined by the 4 cut-offs .009, .015, .20, and .033 oz/t. For each cut-off, the series of 40 IK weights are derived from an indicator kriging system built from the corresponding indicator variogram model (Table 2). These IK weights are listed in Table 4. As the cut-off increases the IK weight attributed to the outlier datum decreases from a high 34% to an equal weight of 1/40 = 2.5%. The probability that the grade at a point x exceeds the highest cut-off 0.033 is simply equal to the proportion of such high grade data, irrespective of their proximity to the center of the panel.

The IK estimation is performed for a certain number of locations x discretizing the panel. At each cut-off, the average of the various point-IK estimates provides an estimate of the volume proportion above that cut-off within the panel, see Table 5. For example 81.8% of the panel is estimated to be above the first cut-off 0.009, while only 17.5% of that panel will have grade above 0.033 oz/t.

There remains to reconstitute the overall estimated grade for the test panel The five local class means shown in Table were estimated by the average of data falling into the corresponding class. Whe the number of data within any particular class was below five, a default class mean corresponding to the entire bench was used -which amounts for that class mean to extend the search neighborhood to the entire bench containing the panel being estimated.

For this particular test panel, the E-type estimate of the panel grade is:

Table 2: Parameters for the indicator variogram models

Cut-off z	$m_I^* = 1 - F(z)$	C_0	C_1	C_2	a_1	a_2	
0.009	0.52	0.05	0.13	0.07	30	350	horiz.
						250	vertical
0.015	0.32	0.10	0.08	0.04	30	150	Isotropic
0.020	0.20	0.08	0.08		100		Isotropic
0.033	0.11	0.90	0	0	Pure nugget		

$m_I^* =$ proportion above cut-off

$\gamma(\rho, h_3) = C_0 + C_1 Sph_{a_1}(\sqrt{\rho^2 + \lambda_1^2 h_3^2}) + C_2 Sph_{a_2}(\sqrt{\rho^2 + \lambda_2^2 h_3^2})$

$\rho = \sqrt{h_1^2 + h_2^2}$: isotropic horizontal distance

h_3 : vertical distance

Table 3: Statistics of the data found in the search neighborhood

	mean	CV (σ/m)	max	min
40 data	0.0333	2.44	0.519	0.001
39 lowest data	0.0208	1.18	0.076	0.001

Block Estimates $\begin{cases} \text{IDS weighting: .19 } oz/t \\ \text{E-type (IK): .04 } oz/t \end{cases}$

Figure 4: Location map of the test panel and the 40 data locations

Next to its locations, the datum distance ranking to the center of the panel is given. Refer to Table 4 for the data values.

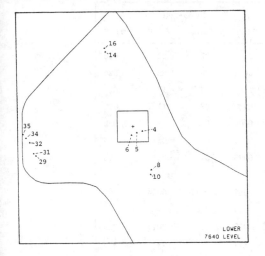

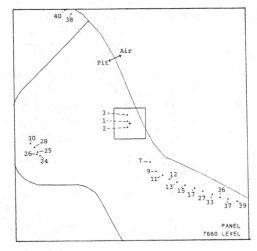

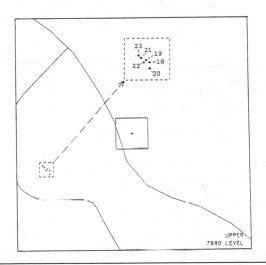

(.182 x .004) + (.065 × .0125) + (.578 x .028) + (.175 x .125) = .040 oz/t,

a value much smaller than the .19 oz/t yielded by the IDS algorithm.

The influence of the outlier grade datum 0.519 oz/t has been limited to its sole class of influence which accounts for only 17.5% of the panel tonnage. Had there been less than 5 data higher than 0.033 oz/t that outlier value 0.519 would have been even more diluted into the bench default

class mean. In other situations, an outlier high grade located far from the panel would be essentially ignored by the IDS algorithm; in the IK approach it would contribute by increasing the proportion of grades within the panel which belong to the last grade class.

In short, the IK algorithm limits the extrapolation in space of outlier data but compensates by allowing a probability for not having encountered all high grade spots.

Table 4: Weights applied to the 40 data

Rank	Distance ft.	Datum value	IDS weights	IK Weights 0.009	0.015	0.020	0.033
1	5.3	.030	.456	.35	.27	.22	.025
2	6.3	.519	.327	.34	.26	.22	.025
3	12.0	.022	.091	.11	.14	.17	.025
4	17.8	.030	.041	.08	.08	.11	.025
5	20.9	.027	.030	-.0009	.033	.78	0.025
6	27.5	.003	.017	-.0084	.017	.051	.025
7	56.7	.004	.004	.0095	.017	.021	.025
.	.	.	.	.	.	.	.
20	122.6	.001	.0009	.0045	.0055	.0054	.025
.	.	.	.	.	.	.	.
30	132.4	.070	.0007	.0018	.0030	.0015	.025
.	.	.	.	.	.	.	.
40	171.6	.013	.0004	.0042	.0084	.0095	.025

The forty data are ranked by increasing distance to the center of the test panel. Note the outlier datum .519 oz/t and the corresponding weights.

Table 5: Determination of the E-type estimate for the panel grade

Cut-off grade: z_k	.009	.015	.20	.033
Percent tons $> z_k$	81.8	75.3	75.3	17.5
% tons in class	18.2	6.5	0	57.8	17.5
Local class mean	.004	.0125	.020	.028	.125
Class contribution	.0007	.0008	0	.0162	.0219

$$\text{SUM} = Z_E^*(x) = 0.0396 \, oz/t$$

The IK algorithm provides estimates of the tonnage proportion of each class of grades within the panel. The contribution of each class to the panel grade is the product of the class tonnage proportion by the local class mean.

CONCLUSIONS

Extreme high grade data, when proven not to be measurement errors, are not bothersome outliers to be handled casually. The profitability of the mining operation may

170

rely in great part on the existence of such high grade spots, however small their extent. Could one imagine Mr. de Beers discarding all cores containing gem quality stones?

Outlier data often corresponds to specific high grade mineralization(s) contained within a predominant, more pervasive, low grade mineralization. Ideally, a prior geological mapping would delineate these different mineralizations, and grade interpolation should be limited to within each specific mineralization. When the information available does not allow such detailed deterministic mapping, the probabilistic alternative consists of estimating the probability of occurrence of each mineralization at each unsampled location. Given that a location belongs to a specific mineralization, the corresponding grade is estimated using only the neighboring data of the same type. This formalism allows restricting the influence (extrapolation) of high grade data to either their mineralization of origin or to their class of grades.

The IK formalism amounts to replace (or approximate) the previous distinct mineralizations by a series of classes of grades: indicators of classes are kriged (interpolated) to provide the proportion of each grade class within any panel, then the corresponding class mean is estimated using the data that fall into the same class.
The overall panel mean grade is then reconstituted by adding the contribution of each class/mineralization.

Application of indicator kriging to the Sunbeam mine reserves valuation improved on inaccuracies of traditional inverse distance squared (IDS) estimates. Such local inaccuracies may have severe consequences on a pit design. Relative to the traditional IDS reserve estimates, the IK approach decreased the estimated ore tonnage, increased the ore grade leaving unchanged the total ounces of gold. It thus improved the global economic feasibility.

ACKNOWLEDGEMENT

The management of Sunbeam Mining Corp., and A.J. Silva in particular, are gratefully acknowledged for having supported that study and allowed publication of its results.

REFERENCES

David, M. 1977. Geostatistical ore reserve estimation. New York: Elsevier. p.364.
Davis, B. 1984. Indicator kriging as applied to an alluvial gold deposit. In Geostatistics for Natural Resources Characterization. Reidel. Part 1, p.337-348.
Journel, A.G. 1983. Non-parametric estimation of spatial distributions. Math. Geol. 15,3:445-468.
Journel and Huijbregts. 1978. Mining Geostatistics. London: Acad. Press. 600 p.
Parker, H.M., Journel, A.G., and Dixon, W. 1979. The use of conditional lognormal probability distribution for the estimation of open pit ore reserves in stratabound uranium deposits: A case study. In Proceedings of the 16th International APCOM Symposium. p.133-148.
Verly, G. and Sullivan, J. 1985. Multi-gaussian and probability krigings. Mining Engineering. June:568-574.

Computer Applications in the Mineral Industry, Fytas, Collins & Singhal (eds)
© 1988 Balkema, Rotterdam. ISBN 90 6191 760 3

Pour une meilleure diffusion de la géostatistique dans l'industrie minière

Georges Verly
Geomines Ltd, Montreal, Quebec, Canada

RÉSUMÉ: La géostatistique est maintenant reconnue dans l'industrie minière comme une
méthode fiable d'estimation des réserves d'un gisement. Elle est cependant sous-utilisée
car d'une part, trop peu de personnes sont assez compétentes en la matière et d'autre
part, un investissement informatique substantiel est nécessaire pour la pratiquer. Il en
résulte que bien souvent seules les importantes compagnies minières se permettent
d'utiliser la géostatistique tandis que les compagnies de moindre taille se limitent à
des méthodes plus traditionnelles, mieux connues et plus abordables. Le seul remède à
cette situation est une plus grande accessibilité de la géostatistique aux niveaux de la
méthodologie et de l'informatique. Ce papier est une discussion des divers points à
considérer pour atteindre cet objectif d'accessibilité.

1 INTRODUCTION

L'estimation des réserves de minerai d'un
gisement est une étape essentielle dans son
étude de faisabilité. Il existe de nombreuses méthodes d'estimation des réserves dont
la géostatistique, la seule qui est une
approche probabiliste. Il est reconnu
depuis plusieurs années que la géostatistique possède de nombreux avantages sur les
autres méthodes et plusieurs centaines de
gisements ont ainsi été estimés avec
succès. Malgré cela, la géostatistique
n'est pas nécessairement la première méthode choisie lors d'une estimation des réserves. Deux facteurs expliquent en partie
cette situation: 1) le nombre limité de
personnes qui comprennent les concepts
géostatistiques et se sentent comfortables
lorqu'il faut les utiliser, 2) le coût
élevé d'acquisition d'une expertise en
géostatistique pour une petite entreprise.

Il n'y a qu'un seul remède à cette situation: diffuser davantage la géostatistique
et la rendre plus abordable. Ce papier
retrace tout d'abord l'historique de la
géostatistique et fait ensuite le point de
la situation actuelle. Des moyens sont
ensuite proposés afin de diffuser davantage
la méthode dans l'industrie minière.

2 HISTORIQUE

La géostatistique est une branche des
statistiques appliquées à la reconnaissance
et à l'estimation des phénomènes naturels.
Il semble que Matern l'utilisa le premier
dans le secteur de la forestrie (Matern,
1960). Ce fut cependant Matheron qui, face
à des problèmes d'estimation minière, vit
le plein potentiel de cette méthode et la
développa afin de l'appliquer dans ce nouveau domaine (Matheron, 1962 et 1965).
Matheron fut suivi dans cette tâche par de
nombreux chercheurs et, dès la fin des
années 70, plusieurs recueils de géostatistique étaient disponibles (David, 1977;
Journel et Huijbregts, 1978; Rendu, 1978;
Clark, 1979). Aujourd'hui, la géostatistique est appliquée non seulement dans le
domaine minier mais aussi dans les domaines
du pétrole, de la géotechnique, de l'hydro-
logie, de la géochimie et de l'environnement. En somme, presque tous les phénomènes naturels peuvent se prêter, d'une
manière ou d'une autre, à un traitement
géostatistique.

La géostatistique fut très vite appliquée en Europe. Son acceptation en Amérique
du Nord fut cependant plus lente et ne se
fit qu'à partir du milieu des années 70.
Plusieurs facteurs ont été invoqués pour
expliquer ce fait, notamment un manque de
littérature anglaise et des concepts mathématiques et probabilistes difficiles à com-
prendre. Il semble aussi que la majorité
des gisements exploités au cours des années
60 et 70 avaient des minéralisations en fer
ou métaux de base peu variables et les

méthodes d'estimation existantes étaient considérées satisfaisantes (Parker, 1984).

Cette situation changea complètement à partir de 1976 lorsque commença une intense exploration pour l'uranium. Les cibles uranifères étaient en effet petites, erratiques et difficiles à délimiter. On fit alors appel à la géostatistique afin de diminuer le nombre de forages nécessaires à la délimitation et à l'estimation de ces gisements (Sandefur et Grant, 1976; Parker, 1984).

En 1979, l'exploration pour l'uranium fut brutalement abandonnée. Elle fut remplacée par l'exploration pour les métaux précieux qui continue encore aujourd'hui. La distribution spatiale des métaux précieux étant très erratique, on fit encore appel à la géostatistique pour les estimer. De nouvelles méthodes d'estimation furent mises au point (Journel, 1983; Sullivan, 1984; Lemmer, 1984; Verly, 1984; Marcotte, 1987). Ces méthodes ont été utilisées et sont encore utilisées de nos jours avec succès (Verly et Sullivan, 1985; Froidevaux et al., 1986; Guibal, 1987; Schofield, 1988).

3 AVANTAGES ET DESAVANTAGES

La géostatistique est une approche probabiliste à l'estimation des réserves minières qui tient compte de:
- la continuité spatiale de la minéralisation, donc de la nature même du gisement;
- la taille et géométrie des blocs miniers à estimer;
- la taille et répartition spatiale des échantillons.

Parmi les principaux avantages de la méthode, on retrouve les possibilités de:
- mesurer la variabilité spatiale de la minéralisation et d'en tenir compte dans les calculs de réserves (variogramme);
- obtenir des valeurs estimées optimales associées à des intervalles de confiance (krigeage et variance d'estimation);
- calculer le gain de précision dû à des échantillons additionnels avant qu'ils ne soient prélevés;
- déduire la distribution des teneurs de blocs miniers à partir de la distribution des teneurs d'échantillons (relation de krige), ce qui est nécessaire pour le calcul des courbes tonnage/teneur;
- estimer les teneurs locales dans un réseau de blocs miniers, ce qui permet de recalculer facilement les réserves lorsqu'il y a modification de la teneur de coupure ou de la taille des blocs;
- simuler un gisement permettant ainsi l'analyse de l'impact des variabilités locales des teneurs sur la production (simulation conditionnelle);
- estimer une variable à partir de l'information correspondant à cette variable et de l'information correspondant à une deuxième variable (cokrigeage).

Les avantages de la géostatistique ne s'obtiennent malheureusement pas gratuitement. Les concepts probabilistes de la méthode sont difficiles à assimiler et requièrent une bonne expérience pour être appliqués correctement. Un bon support informatique est aussi indispensable. Enfin, une étude par méthode géostatistique est en général plus longue que par une autre méthode, l'information disponible étant analysée de façon plus poussée. Il en résulte que les compagnies qui se lancent pour la première fois dans une étude géostatistique doivent faire un investissement substantiel. En effet, il faut faire appel aux services d'un spécialiste (interne ou externe), développer ou acheter des programmes géostatistiques, et allouer le temps nécessaire au bon déroulement de l'étude. Un fois la procédure installée cependant, il n'est guère plus cher de mettre à jour les réserves d'un gisement par traitement géostatistique que par une autre méthode.

4 SITUATION ACTUELLE

La plupart des grandes compagnies minières reconnaissent les avantages de la géostatistique et ont décidé de l'utiliser soit conjointement avec d'autres techniques, soit comme une seule méthode d'estimation. Certaines compagnies ont même intégré complètement leurs études de faisabilité en réunissant sous la même équipe géologues, géostatisticiens, ingénieurs miniers et analystes financiers (Parker, 1984).

La situation est malheureusement différente dans les petites compagnies minières. En effet, si toutes ont entendu parler de la géostatistique, peu l'utilisent de façon courante probablement à cause de l'investissement initial que cela implique.

L'importance du calcul des réserves dans le cadre d'une étude de faisabilité est indéniable et toutes les compagnies sont prêtes à investir davantage pour obtenir des calculs plus précis. Il faut cependant que ce ne soit pas prohibitif. Les grandes compagnies sont avantagées à ce point de vue puisque l'investissement initial d'acquisition d'une expertise en géostatistique peut être amorti sur divers projets ou sur le même grand projet pendant plusieurs années. Il n'en va pas de même pour les plus petites compagnies qui bien souvent doivent amortir leur investissement sur un même projet de moindre envergure.

5 SUGGESTIONS POUR UNE MEILLEURE DIFFUSION

La géostatistique devrait être considérée
dès qu'un calcul des réserves est à faire.
Pour cela cependant, il faudrait qu'elle
soit plus accessible que maintenant, ce qui
n'est possible qu'au travers d'une
meilleure diffusion de la méthodologie et
des logiciels. -

5.1 Niveau méthodologie

Parker aborda le sujet de la diffusion il y
a quelques années et recommanda notamment
(Parker, 1984):
   - la publication de livres décrivant en
détails les étapes de base d'une étude
géostatistique;
   - la publication de recueils d'études de
cas dont la méthodologie est clairement
exposée.

Il semble que ces recommandations aient
été suivies puisque plusieurs publications
de ce genre sont maintenant disponibles ou
vont l'être dans un proche avenir (Matheron
et Armstrong, 1987; David, 1988;
Srivastava et Isaaks, 1988).

Il faudrait cependant que la diffusion
commence au niveau des universités. Ainsi,
une nouvelle génération de géologues et
d'ingénieurs miniers serait au courant des
méthodes géostatistiques et plus généralement des méthodes géomathématiques. À
noter qu'il existe actuellement des cours
dans ces matières. Ils ne sont cependant
pas disponibles dans toutes les institutions et ne sont pas toujours considérés
comme très importants. En conséquence, il
y a encore aujourd'hui des diplômés univer-
sitaires qui ne savent pas calculer
correctement un histogramme ou qui sont
prêts à estimer les réserves d'un gisement
par méthode polygonale sans être au courant
des dangers de cette méthode.

5.2 Niveau informatique

En 1984, Parker considérait qu'il existait
sur le marché de très bons logiciels com-
merciaux intégrant graphisme, planification minière et géostatistique. Il
s'agissait alors de logiciels coûteux
fonctionnant sur grands ordinateurs.
Depuis cette époque, cependant, il y a eu
une révolution informatique. Des micro-
ordinateurs toujours plus puissants et
moins coûteux sont apparus sur le marché et
une foule de logiciels allant de la gestion
de données au dessin de cartes de contours
et de sections sont devenus disponibles.
Toutes les compagnies minières ont maintenant un ou ou plusieurs micro-ordinateurs;
de plus en plus elles possèdent une table
traçante et une table numérique. Il en résulte que la plupart des compagnies, même
les plus petites, ont maintenant les ressources nécessaires aux traitements de
données qui, il y a quelques années, exi-
geaient un investissement de plusieurs centaines de milliers de dollars en matériel
et logiciel informatiques.

Cette popularité de la micro-informatique
dans le monde minier est l'occasion de
démocratiser la géostatistique en concevant
des logiciels fonctionnant sur
micro-ordinateurs et respectant la
philosophie de la micro-informatique,
c'est-à-dire: prix abordable, facilité
d'utilisation, graphisme impeccable et
bonne performance au niveau des calculs.
À noter qu'il y a déjà eu du travail de
fait à cet égard (Kark, 1985; Froidevaux,
1988).

Le reste de cette section traite des qualités que devrait idéalement posséder un
logiciel de géostatistique. Des recommandations sont faites dans le contexte d'un
utilisateur qui occasionnellement doit
faire une estimation des réserves et est
prêt à considérer cette méthode. Il est
donc supposé que cette personne a de connaissances de base en géostatistique ou est
prête à les acquérir.

5.2.1 Géostatistique de base

Le but d'un logiciel de géostatistique
minière est l'estimation des réserves de
gisements. Un tel logiciel doit donc
permettre de passer aisément au travers de
toutes les étapes nécessaires à une estimation géostatistique standard, c'est-à-
dire:
   - travail préparatoire;
   - analyse des données;
   - analyse structurale;
   - estimations locales et globales,
courbes tonnage/teneur.

Travail préparatoire

Cette première étape consiste à récolter
toute information pertinente au niveau géo-
gique et au niveau minier. Il s'agit d'une
étape importante car ces informations dé-
terminent plusieurs aspects de la méthodologie à suivre. Il est ainsi nécessaire de
connaître les différents types de roches,
les principaux aspects structuraux et les
limites du gisement. Au niveau de l'échantillonnage, il est important de savoir si
plusieurs types d'échantillons (chips,
carottes de différentes longueurs) ont été
récoltés. Il faut aussi connaître la taille

et la géométrie des blocs ou chantiers à estimer ainsi que la localisation des développements miniers.

Du point de vue informatique, cette étape se traduit par la gestion d'une grande variété de données. Afin de gérer efficacement ces données, les points suivants sont suggérés:

- que des banques de données flexibles soient utilisées pour stocker l'information;

- que la saisie des données puisse se faire du clavier, par lecture d'un fichier ASCII, par l'intermédiaire d'une table numérique ou d'un écran graphique;

- que de puissantes fonctions de gestion puissent relier les différentes banques de données non seulement entre elles mais aussi avec la quincaillerie informatique; par exemple, il devrait être possible d'attribuer un code à tous les échantillons localisés à l'intérieur d'une zone simplement en pointant cette zone sur l'écran;

- que des fonctions spécialisées telles que calcul des coordonnées des échantillons ou calcul de localisation des galeries soient incorporées;

- que toutes les sorties graphiques nécessaires, notamment sections et cartes de localisation, soient intégrées.

Analyse de données

Cette étape constitue le premier contact statistique avec les échantillons. Diverses statistiques locales et globales sont calculées afin de détecter toute particularité de la minéralisation pouvant affecter la méthodologie de l'estimation. Parmi ces particularités, on retrouve un manque d'homogénéité de la minéralisation, des valeurs très erratiques, ou une très grande proportion de stérile. Cette étape est très fastidieuse car elle nécessite de nombreuses manipulations sur différents ensembles de données. Afin de minimiser le temps perdu dans ces manipulations, il est recommandé:

- d'intégrer au système une série de fonctions statistiques pouvant travailler directement sur des sous-ensembles des banques de données;

- d'inclure la possiblité d'obtenir des statistiques tenant compte du suréchantillonnage de certaines zones (en anglais, "declusterized statistics");

- d'incorporer de puissantes fonctions permettant rotation, projection et calcul des composites;

- d'inclure un ou plusieurs algorithmes permettant d'approximer l'enveloppe du gisement à partir d'intercepts minéralisés;

- d'offrir toutes les sorties graphiques pertinentes, incluant dessins d'histogrammes, de nuages de corrélation et cartes de contours.

Analyse structurale

Cette troisième étape est en fait la première reliée à un concept purement géostatistique: le variogramme. Le variogramme est une fonction qui quantifie la variabilité (ou inversement la continuité) de la minéralisation. Il s'obtient en deux temps: calcul du variogramme expérimental et modélisation.

La plupart des programmes de calcul de variogrammes sont adéquats en ce qui a trait aux résultats obtenus. Certains permettent même le calcul simultané de plusieurs types de variogrammes et ont des capacités de modélisation interactive.

L'analyse structurale est cruciale dans une étude géostatistique. Afin de la faciliter au maximum, il est recommandé:

- que tous les types de variogrammes traditionnels soient disponibles, notamment les variogrammes standards et relatifs;

- que plusieurs variogrammes robustes, par exemple des variogrammes de l'indicateur ou de différentes transformées, puissent être calculés dans le cas de minéralisations erratiques;

- que les variogrammes puissent être calculés dans toutes les directions, le long des sondages, globalement à l'intérieur de zones prédéfinies ou dans des voisinages glissants;

- qu'une analyse de sensibilité du variogramme aux valeurs extrêmes puisse être faite interactivement;

- que la modélisation du variogrmme soit interactive avec plein emploi du graphisme sur écran;

- qu'ils y aient des programmes permettant de vérifier la justesse du modèle choisi;

- que toutes les sorties graphiques nécessaires soient incorporées.

Estimations locales et globales

Les estimés locaux sont obtenus par krigeage, une technique donnant accès à des estimations optimales ainsi qu'à leur degré de précision. Pour ce faire, le krigeage considère la continuité de la minéralisation, la taille et la géométrie des blocs à estimer ainsi que la localisation des échantillons.

Comme pour le variogramme, la plupart des programmes de krigeage donnent des résultats adéquats. Quelques points à considérer pour faciliter cette tâche sont:

- l'intégration du ou des programmes de krigeage au système de banques de données;
- la possibilité d'effectuer tous les types de krigeages courants;
- la possibilité de kriger simultanément plusieurs zones irrégulières avec des paramètres différents;
- l'incorporation de techniques permettant de vérifier les résultats, notamment en bordure du gisement;
- l'intégration d'un programme qui, partant du modèle de blocs krigés, calcule les réserves à l'intérieur d'enveloppes minéralisées en tenant éventuellement compte des développements miniers;
- l'incorporation de sorties graphiques adéquates.

5.2.2 Autres méthodes d'estimation

Il pourrait être profitable pour une meilleure diffusion de la géostatistique d'offrir des algorithmes géostatistiques et des algorithmes plus traditionnels tels que l'inverse des distances ou la méthode des sections. Ceci permettrait de comparer facilement les différentes méthodes et, éventuellement, pourraient rassurer ceux qui en sont à leur première expérience géostatistique.

5.2.3 Module de formation

Il est souvent utile soit dans un but éducationnel, soit dans un but de vérification, d'effectuer un exercice géostatistique sur un exemple très simple, par exemple le calcul des poids de krigeage dans une situation particulière, ou encore le calcul d'une variance d'estimation. Il est recommandé que de tels exercices puissent se faire interactivement avec l'ordinateur.

Il est aussi recommandé que tous les logiciels de géostatistique soient accompagnés de plusieurs études de cas détaillées au niveau de l'utilisation des programmes et de la méthodologie employée. Une analyse des résultats obtenus serait aussi très utile.

5.2.4 Performance

L'utilisateur ne s'attend pas à ce qu'un système de logiciels lui donne automatiquement une estimation des réserves. Il s'attend cependant à ce que toute manipulation de données se fasse facilement et sans perte de temps. Pour lui donner satisfaction, il faut donc tous les programmes - gestion de données, graphisme et calculs géostatistiques - soient non seulement

performants mais aussi totalement intégrés entre eux.

L'aspect performance des algorithmes de géostatistiques a été longuement travaillé durant les années 70. En conséquence, il existe maintenant des algorithmes efficaces pour la plupart des calculs géostatistiques. Par contre, les performances aux niveaux de la gestion, du graphisme ainsi que l'intégration pourraient être améliorées. Il est intéressant de noter que ce problème n'est pas géostatistique mais purement informatique.

La plupart des programmes de géostatistique sont codés en FORTRAN, ce qui constitue probablement une barrière à l'intégration totale de la gestion, du graphisme et des calculs. Il est donc suggéré qu'une étude soit menée afin de déterminer le langage de programmation, le système de gestion de données et la librairie graphique qui permettraient une telle intégration. Il est à noter qu'une récente étude parrainée par le gouvernement canadien et le gouvernement de l'Ontario recommandait le langage C et le système d'exploitation UNIX (CANMET, 1987).

Il n'est pas discuté ici du choix de l'ordinateur, celui-ci ayant déjà été fixé par l'industrie minière qui utilise en grande majorité des ordinateurs IBM PC ou compatibles.

Flexibilité

La distribution spatiale de la minéralisation d'un gisement est extrêmement complexe et il est impossible de concevoir une série de logiciels parfaitement adaptés à toutes les situations. Afin que l'usager ne reste pas bloqué devant un problème imprévu, il est suggéré d'accroître la flexibilité d'un système compilé (et donc cadenassé) en y ajoutant:
- des codes source permettant d'effectuer certains calculs géostatistiques en dehors du système;
- des codes objet permettant d'intégrer des applications particulières au reste du système sans pour autant qu'il y ait divulgation d'algorithmes secrets.

5.2.6 Facilité d'emploi

L'aspect externe d'un produit compte beaucoup lorsqu'il est évalué par une tierce personne. Il est donc recommandé que des logiciels de géostatistique destinés à la diffusion soient visuellement beaux et faciles d'emploi. La saisie des paramètres devrait se faire "plein écran" avec possibilité de corrections. Les valeurs des paramètres devraient être proposées par le programme. Les textes explicatifs affichables sur écran devraient non seulement expliquer les paramètres du programme mais aussi certains concepts géostatistiques en donnant des références adéquates.

6 CONCLUSION

Après une longue période de gestation durant laquelle elle n'était appliquée que dans quelques centres de recherche, la géostatistique est maintenant acceptée par l'industrie minière comme une méthode valable d'estimation des réserves d'un gisement. Cette acceptation est attribuable à deux facteurs: 1) une diffusion de la méthodologie aux moyens de séminaires, de livres et de publications scientifiques, 2) un nombre croissant d'estimations des réserves de gisements pour lesquelles la géostatistique a donné un rendement supérieur aux autres méthodes. On remarque toutefois que malgré cette acceptation, la géostatistique reste une méthode coûteuse que seules les compagnies relativement importantes se permettent d'utiliser. Il y a en effet peu de professionnels assez compétents en la matière pour répondre à la demande et les programmes disponibles sur le marché sont relativement chers. La communauté géostatistique a reconnu ce problème et une tendance actuelle est de diffuser davantage la méthode en publiant des livres sur la pratique de la géostatistique et des recueils d'études de cas.

Il est recommandé dans cet article d'attaquer le problème au niveau universitaire en offrant davantage de cours de géostatistique et de géomathématique. Il est aussi suggéré de profiter de la révolution micro-informatique pour améliorer les logiciels existants et éventuellement en réduire le prix. Il est probable que seules de telles mesures de démocratisation permettront une plus grande diffusion de la géostatistique dans le monde minier.

RÉFÉRENCES

CANMET 1987. Computer program specifications for the Ontario mining industry. DSS file no. 14SQ.23440-6-9009.

Clark, I. 1979. Practical geostatistics. London: Applied Science Publishers Ltd. 129 pp.

David, M. 1977. Geostatistical ore reserve estimation. New York: Elsevier. 364 pp.

David, M. 1988. Handbook of applied advanced ore reserve estimation. New York: Elsevier. 198 pp. (à être publié).

Froidevaux, R. et al. 1986. Estimating and classifying gold reserves at Page-Williams C-Zone: a case study in non parametric geostatistics. Dans M.David et al. (eds.), Ore reserve estimation, methods, models and reality. Canadian Institute of Mining and Metallurgy. pp.280-298.

Froidevaux, R. 1988. Geotool box software (à être publié).

Guibal, D. 1987. Recoverable reserves estimation at an Australian gold project. In G.Matheron & M.Armstrong (eds.), Geostatistical case studies. Dordrecht, Holland: Reidel. pp. 149-168.

Journel, A.G. 1983. Nonparametric estimation of spatial distributions. In Journal of the International Association of Mathematical Geology, vol. 15, no. 3. pp. 445-468.

Journel, A.G. & C.J.Huijbregts 1978. Mining geostatistics. New York: Academic Press. 600 pp.

Kark, M.J. 1985. A software approach to linear geostatistics. Thèse de maîtrise. Applied Earth Sciences Dept., Stanford University.

Lemmer, I.C. 1984. The mononodal cutoff. Thèse de doctorat. Dept. of Applied Earth Sciences, Stanford University.

Marcotte, D. 1987. Le krigeage multigaussien. Thèse de doctorat. Dépt. de Génie Minéral, École Polytechnique de Montréal.

Matern, B. 1960. Spatial variation. Meddelanden Fran, Stateus Skogsforsknings-institut, Band 49, no. 5. 144 p.

Matheron, G. 1962-1963. Traité de géostatistique appliquée, vols. 1 & 2. Paris: Technip.

Matheron, G. 1965. Les variables régionalisées et leur estimation. Paris: Masson.

Matheron, G. & M.Armstrong. 1987. Geostatistical case studies. Dordrecht, Holland: Reidel. 248 pp.

Parker, H.M. 1984. Trends in geostatistics in the mining industry. Dans G.Verly et al. (eds.), Geostatistics for natural resources characterization, Part 2. Dordrecht, Holland: Reidel. pp. 915-934.

Rendu, J.M. 1978. An introduction to geostatistical methods of mineral evaluation. South African Inst. of Mining and Metallurgy Monograph, Johannesburg. 84 pp.

Sandefur, R.L. & D.C.Grant 1976. Preliminary evaluation of uranium deposits, a geostatistical study of drilling density in Wyoming solution fronts. Int. Symposium on Uranium Exploration, I.A.E.C., Vienna.

Schofield, N. 1988. Ore reserve estimation at the Entreprise Gold Mine, Pine Creek, Northern Territory, Australia, Part 1: Structural and variogram analysis, Part

2: The multigaussian kriging model. Canadian Institute Mining Journal (à être publié).

Srivastava, M. & E.Isaaks 1988. Geostat with Moh and Ed. Oxford Press (à être publié).

Sullivan, J. 1984. Non-parametric estimation of spatial distribution. Thèse de doctorat. Dept. of Applied Earth Sciences, Stanford University.

Verly, G. 1984. Estimation of spatial point and block distributions: the multigaussian model. Thèse de doctorat. Dept. of Applied Earth Sciences, Stanford University.

Verly, G. & J.Sullivan 1985. Multigaussian and probabilty krigings - application to the Jerritt Canyon deposit. Mining Engineering Journal. pp. 568-574.

Computer Applications in the Mineral Industry, Fytas, Collins & Singhal (eds)
© 1988 Balkema, Rotterdam. ISBN 90 6191 760 3

Indicator approach to two-dimensional problems: Theory and example

Miguel Armony
BP Mineração, Rio de Janeiro, Brazil

ABSTRACT: A point-recoverable reserve estimation method is presented for bi-dimensional type deposits using Journel's indicator approach. Two new density functions are defined: ore tonnage and metal weight per unit area. The results and formulae obtained apply especially to underground deposits with no clear geological limits, and where only chemical analysis can be used to draw boundaries between ore and waste, or between levels of profit. Theoretical and numerical inconsistencies arising from a classical geostatistical approach are presented for the case of a real base metal deposit. Ore reserve calculations redone using density functions provide straightforward results and solve the previous contradictions.

1 THEORY, FORMULAE AND USER'S GUIDE

1.1 Introduction

This work intends to apply Journel's indicator methodology to deposits where a two-dimensional treatment can (and sometimes must) be given. If the deposit is underground, additional mining constraints are imposed such as minimum thickness simultaneously with cutoff grade. In some cases there are veins or layers having detectable hanging walls and footwalls. But many times no clear geological boundaries can be found and the hypotheses must be drawn only on quantitative data. Then, layer, hanging wall, footwall, and the whole orebody are modeled in space following grade distribution. The resultant spatial surface is then projected onto a chosen plane.

In this way thickness is transformed into a parameter and the problem becomes a bi-dimensional one. In this type of underground deposit, cutoff grade is applied first; in a second stage positive holes are analysed for their accumulation (grade x thickness).

1.2 The deposit

Let us consider an underground deposit D

projected onto a given plane S, and let us take the area A' within S. The area A' is a projected one, and must be corrected to the real area A by dividing it by a factor ≥ 1. The application of the theory can be done directly on A without any loss of generality.

The area A can be seen as the union of a very large number k of samples with the same very small unit area. A sample is said to be mineralized if it fulfills the following conditions:

a priori $z(x) \geq zo$ (z = grade, zo = cutoff grade)

a posteriori $p(x) \geq po = eo \cdot zo$ (p = accumulation, eo = minimum thickness);

so for each sample the mineralized area is given by:

$$a(x_j) = (A/k) \cdot i(x_j; po)$$

with the conditional distribution function

$$Fx(p|(N)) = Prob\{p(x) \geq po \cap z(x) \geq zo | (N)\} / Prob\{z(x) \geq zo | (N)\};$$

The recoverable sample weight is given by:

$$t(x_j) = (A/k) \cdot d(x_j) \cdot e(x_j) \cdot i(x_j; po)$$

where d = specific gravity

e = thickness .

The recoverable weight of the metal content in a sample is simply:

$$q(x_j) = (A/k) \cdot d(x_j) \cdot p(x_j) \cdot i(x_j;po);$$

the samples with a grade below zo, or for which the accumulation is below po will not be recovered.

Note about notation: in reality the correct notation is d(x;po), e(x;po) , q(x;po) etc. instead of simply d(x) etc.. But for simplicity the accumulation cutoff will be indicated only for the function i(x;po).

1.3 Area proportion

The total mineralized area in A is given by:

$$A(po) = (A/k) \cdot \sum_{J=1}^{K} i(x_j;po).$$

The proportion of mineralized area is then:

$$P(A;po) = (1/k) \cdot \sum_{J=1}^{K} i(x_j;po)$$

which is the average of the indicators; if a panel (or other area) b A is to be evaluated from n available samples, the indicator average P(b;po) can be estimated by some weighted average using the data available:

$$P^*(b;po) = \sum_{K=1}^{m} \lambda_K \, i(x_k;po), \qquad (1)$$

for example by traditional geostatistics (indicator variogram and kriging).

1.4 Point recoverable ore tonnage and quantity of metal

Total ore tonnage is given by:

$$T(A;po) = (A/k) \cdot \sum d(x_j) \cdot e(x_j) \cdot i(x_j;po) ;$$

calling

$$f(x_j;po) = d(x_j) \cdot e(x_j) \cdot i(x_j;po) \qquad (2)$$

then $f(x;po) = \begin{bmatrix} d(x) \cdot e(x), & p(x) \geqslant po \\ 0 & , p(x) < po. \end{bmatrix}$

Note that f is the density function giving ore weight per unit area.

For the estimation of the point recoverable ore tonnage for a given panel (or area) b from only n available samples, the f(b;po), ore weight density average, can

be estimated by some weighted average using the data available:

$$f^*(b;po) = \sum_k \lambda_k \, f(x_k;po) \qquad (3)$$

for example by traditional geostatistics (variogram and kriging). Tonnage of b ore is estimated by:

$$T^*(b;po) = b \cdot f^*(b;po) . \qquad (4)$$

In the same way metal content is given by:

$$Q(A;po) = (A/k) \cdot \sum d(x_j) \cdot p(x_j) \cdot i(x_j;po);$$

calling $g(x_j;po) = d(x_j) \cdot p(x_j) \cdot i(x_j;po)$

with $g(x;po) = \begin{bmatrix} d(x) \cdot p(x), & p(x) \geqslant po \\ 0 & , p(x) < po \end{bmatrix}$

where g represents the metal weight per unit area, hence a new density function. Thus, an estimation of b metal content is given by:

$$g^*(b;po) = \sum_k \lambda_k \, g(x_k;po) \quad , \text{ and} \qquad (5)$$

$$Q^*(b;po) = b \cdot g^*(b;po) \quad , \qquad (6)$$

Q being the total of point recoverable quantity of metal.

As before, classical geostatistics can be used for panel or area b estimation from n data available.

1.5 Average grade and thickness

It is easy to conclude that the average grade of a panel b estimated from n point data available may be taken as

$$m^*(b;po) = Q^*(b;po) / T^*(b;po) . \qquad (7)$$

For average thickness inside an ore zone the variogram can be built for the thickness density e(x).i(x;po), which will behave like the f function variogram. Then panels can be estimated by linear kriging.

When specific gravity is taken as a constant the ore thickness can simply be written as:

$$e^*(b;po) = T^*(b;po) / (P^*(b;po) \cdot b \cdot d) \qquad (8)$$

and for total panel average thickness:

$$e^*(b;po) \geqslant f^*(b;po)/b + (1 - P^*(b;po)) \cdot e_m \qquad (9)$$

where e_m = minimum mining width.

Note that when d = constant, thickness means ore weight and accumulation means quantity of metal, having d as a proportion factor.

1.6 Continuous treatment

Note that density functions depend on the measure (support) a. When $a \to 0$ we have:

$$dt = d(x) \cdot e(x) \cdot i(x;po) \cdot da$$
$$dq = d(x) \cdot p(x) \cdot i(x;po) \cdot da$$

and a continuous treatment can be given to the problem yielding the same final formulae.

1.7 Multiple cutoffs case

Given $1+1$ cutoffs $zo < z_1 < \ldots < z_1$ and consequently $po < p_1 < \ldots < p_1$, with the same minimum mining width eo, for each cutoff p_i, conditioned to z_i, the proportion of the mineralized area is given by

$$P(A;p_i) = (1/k) \cdot \sum_{j=1}^{K} i(x_j;p_i).$$

The cutoffs can be based on mining parameters, h-scattergram analysis or other practical or theoretical reasons.

For an area $b \subset A$ the proportion of area above p_i conditioned to z_i, will be the weighted average

$$P^*(b;p_i) = \sum_{K=1}^{m} \lambda_K i(x_k;p_i)$$

if only n samples are available.

And, in the same way for cutoff z_{i+1}, and correspondently p_{i+1}:

$$P^*(b;p_{i+1}) = \sum_{\ell=1}^{m} \lambda_\ell\, i\,(x_1;p_{i+1}).$$

The quantity

$$\Delta P^*_{i,i+1}(b) = P^*(b;p_{i+1}) - P^*(b;p_i),$$

is the difference between the proportion of the area above p_{i+1} submitted to the prior condition $z \geq z_{i+1}$ and proportion of area above p_i submitted to the different condition $z \geq z_i$; even the thickness of each sample may not be the same for both conditions; indeed

$$e\,(x_j;z_i) \geq e\,(x_j;z_{i+1}).$$

So P gives the proportion of point recoverable mineralized area the grade of which lies between z_i and z_{i+1}. Within each interval $(i,i+1)$ the estimation for grade and thickness can be done by simply taking the average of accumulation and

thickness for the samples fulfilling the condition $z_i \ z \ z_{i+1}$; one can choose an arithmetic mean, a logarithmic mean or a Sichel mean, or the median, the middle point of the interval, the most probable value or any other value. The choice will depend on the data distribution, the type of estimation required and the problem that is to be solved. For each interval the estimated grade will be given by:

$$m^*_{i,i+1}(b) = p^*_{i,i+1}(b) \ / \ e^*_{i,i+1}(b).$$

And straightforward for point recoverable ore tonnage and quantity of metal within the interval:

$$\Delta T^*_{i,i+1}(b) = \Delta P^*_{i,i+1}(b) \cdot b \cdot e^*_{i,i+1}(b)$$
$$\Delta Q^*_{i,i+1}(b) = \Delta P^*_{i,i+1}(b) \cdot b \cdot p^*_{i,i+1}(b)$$

and, consequently:

$$\Delta T^*(b;po) = \sum_{i=0}^{\ell-1} \Delta T^*_{i,i+1}(b)$$

estimator for point recoverable tonnage for cutoff po conditioned to zo in $b \ A$.

And

$$\Delta Q^*(b;po) = \sum_{i=0}^{\ell-1} \Delta Q^*_{i,i+1}(b),$$

estimator for point recoverable quantity of metal for cutoff po conditioned to zo in $b \subset A$. Consequently

$$m^*(b;po) = Q^*(b;po) \ / \ T^*(b;po),$$

estimator for grade.

1.8 User's guide

Let suppose, a priori, that the two dimensional deposit presents no clear geological distinction between poor, waste and high grade areas. Chemical analysis is then the only guide for drawing internal boundaries. This means that the deposit, or the part of it under study, has no other cutoffs than numerical: objective parameters (mining or economic parameters) or subjective (like apparent natural cutoffs). This chapter refers to the objective parameters and intends to present some guidelines for usage of its several items and formulae.

Formula (1) can be used for every bidimensional deposit even if an open pit is required. In some cases, like precious metal deposits with highly skewed values, indicators for area proportion may be the unique way to obtain robust results.

183

Usage of formula (1) can provide, for point recoverable areas:
- probability or risk contour maps
- proportions inside panels.

In principle, formulae (3) to (9) can be used when the functions $f(x;po)$ and $g(x;po)$ present clear and robust variograms. The question is when and how to use it, what this usage leads to, or in other words, what kind of answers will be given by each formula:
- formula (4) gives the point recoverable ore tonnage for any area. Inside a totally mineralized area it means the total area tonnage;
- formula (6) gives the total point recoverable quantity of metal within any area;
- contours can be drawn for g for boundaries of mineralized areas and levels of profit;
- formulae (3) to (9) replace usage of subjective cutoffs.

The supposition of natural cutoffs based only on chemical analysis often leads to variograms that exclude the non-mineralized holes. In the kriging procedures, mineralized areas, panels tonnage and quantity of metal will be overestimated. To avoid this, the non-mineralized holes are included in the kriging calculation in many occasions. However, the wrong variogram is being used!

1.9 Multiple cutoff case

Ni, Zn, Cu and many other major metal deposits present well behaved f and g variograms, so that the formulae above work. But the f and g functions normally don't work for gold and other precious metals. No robust (if any) variograms can be modeled for their metal weight density function g even if the ore weight density function may present some structure. In these cases the multiple cutoff method allows us to use only indicator variograms. For each interval, average values are chosen for accumulation and thickness without cutting out the so called "outliers". If there is a large amount of sample data the number of cutoffs can be increased. This reduces the problem of high values just to a small proportion of each area or panel.

2 ORE RESERVE ESTIMATION IN A BASE METAL
  DEPOSIT USING DENSITY FUNCTIONS

In the first chapter two new density functions were defined: ore weight and quantity of metal per unit area. Here a direct application of these functions is presented in a real case, a base metal sulphide deposit located in the State of Minas Gerais, Brazil as it was done by BPM-Brazil. It will be shown how density functions usage can help to solve the theoretical contradictions and numerical inconsistencies resulting from classical geostatistics methods.

2.1 The deposit

The deposit under study generally lies between a serpentinite hanging wall and a chert footwall but it never fills the whole space between them neither orthogonally to the layer, nor in the extension. Sometimes the ore is remobilized giving rise to several branches within the serpentinite or the chert.

In the first drilling campaign a more or less regular grid of 100 m x 100 m. could be attained. In the second stage, new holes were added, most in the richest part of the deposit. In total 108 ddh was drilled as can be seen in Figure I(a).

Because of the lack of clear geological boundaries in the three dimensions some definition had to be given to what was to be called ore. One way to do this is to search for a numerical cutoff.

Examining the data available, a natural cutoff grade was assumed for ore. A second cutoff, for thickness (2 metres minimum width), was imposed as a mining constraint. Mineralized intervals within each hole were chosen in the following way:
* the whole interval grade must be not below cutoff;
* the samples in the extremes of the interval must not have a grade equal or above the cutoff.

It is important to mention that the samples size was not constant. Most samples vary between half metre to one metre, but some of them were as small as 30 centimeters or as long as 1.50 metres. For the purpose of this discussion, the vertical projection of the deposit will be used directly. This deposit strikes in an assumed north-south direction. The referred thickness will always be the true thickness. Real grades will be multiplied by a factor and given in fictitious units (uBm). Specific gravity is assumed constant as its variation is very small. It will be taken to be equal to 1. It was shown that density functions are related to ore and metal weight instead of thickness and accumulation. By taken $d = 1$ the numerical values become the same for both sets of variables. This allows comparisons to be made between the results of the two methods. For this reason, the f function will be called "thickness" and g function will be called "accumulation". Two different cutoff grades will be used successively in both methods: 1.00uBm and 1.14uBm. While using also the mining parameter imposed of 2.00 metres minimum thickness, the accumulation cutoffs will be 2.00muBm and 2.28muBm.

In all cases it was decided not to dilute the thickness to avoid the destruction of possible thickness mathematical structure.

All variables presented well behaved spherical variograms, and except for the indicators, a high nugget effect. Some of them are shown in Figures IX to XII.

2.2 Ore reserve evaluation methods

The hypothesis of a natural cutoff:

The hypothesis of a natural cutoff grade is used very often for evaluation of geological reserves, when no clear geological boundaries can be defined for ore. This implies the supposition that there are two different independent populations, and that the behaviour of one does not affect the mathematical structure of the other. This is the real underlying hypothesis. In this way, every hole where some length was found with a Bm grade not below cutoff belongs to the "ore" population; and every hole where no sample was found the grade of which is equal or above the cutoff belongs to the "waste" population. The validity of this hypothesis will not be discussed; the intention here is just to make clear what the hypothesis really means and to use it in a coherent way. So the universe under study includes the set of holes where at least one piece of core is found to be equal or above the cutoff grade.

Usage of density functions:

In this case only strictly physical measured quantities are used. The f function gives sample recoverable ore weight; the g function gives the quantity of metal recoverable from the sample. Both will be zero if the sample is not to be mined and they assume the measured values if the sample will not be below cutoff. The universe will be the whole set of samples, with functions i, f and g taking values according to the probability distribution given by:

$$\text{Prob}\left[p(x) \geqslant p_0 \cap z(x) \geqslant z_0 \mid (N)\right] / \text{Prob}\left[z(x) \geqslant z_0 \mid (N)\right].$$

Irrespective of the method chosen, the tasks can be summarized in the following steps:
  * search for the mathematical structures of each sample universe - the variograms;
  * estimate the limits of the mineralized zone;
  * estimate the grade and thickness distributions inside the mineralized zone - in this case by kriging procedure.

2.3 Ore evaluation using natural cutoff

Search for mathematical structure:

from the original 108 holes, 91 was found to belong to ore population and were used for building experimental variograms for thickness and accumulation. Figure I(b) presents a map with hole locations showing the population to which each hole belongs. The variograms presented a clear spherical structure as can be seen in Figure IX.

Estimate the limits of the mineralized zones, and distribution of values inside them:

is it possible to define the limits of the mineralized zones using coherently the mathematical structure found? There is a variogram modeled from 91 mineralized holes. Kriged points can be calculated in a regular grid for thickness and accumulation. By using some plotting software isothickness and isopower contours may be drawn. As all grades are not below 1.00uBm, the contour of 2.00muBm accumulation will give the limits of the mineralized zone. Is this true? Of course not! Waste holes were not taken into account. So, the limits must be defined before drawing contour maps, otherwise absurde results will be obtained as shown in Figure II. There, the mineralization and the ore thickness advance strongly into obvious waste area. The second possibility is a simple manual intervention: supposing a 100 m x 100m grid, one can draw by hand a geometrical contour. For example, using Matheron's formula for the relative variance of the estimation of the surface (David,1977). Such a contour is shown in Figure III(a).

Distribution inside the defined mineralized area can be given by kriging: blocks for mine planning purposes or point kriging in a regular grid for contours and

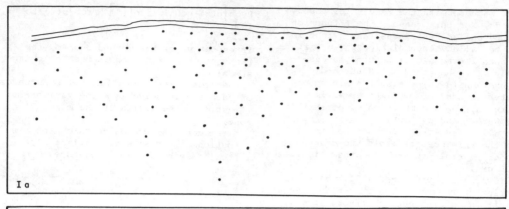

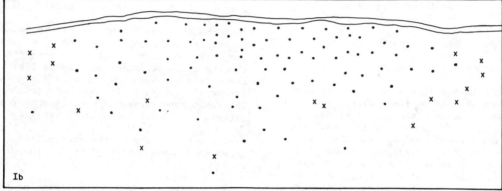

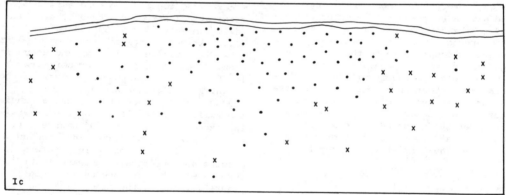

Figure I

a) Hole locations
b) Positive and negative holes using natural cutoff
c) Positive and negative holes using density functions

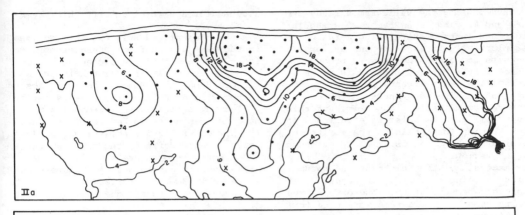

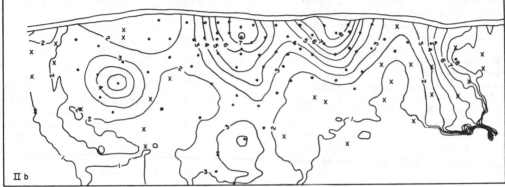

Figure II

Contour maps for 1 uBm cutoff using the 91 mineralized holes
a) Accumulation
b) Thickness

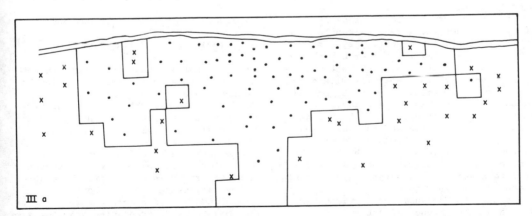

Figure III-a

A possible hand made delimitation of ore area using a 50x50 grid

isocurves. Using the variogram already defined and kriging in a 25 m x 25 m regular grid, then interpolating using a reasonable plotting software, the curves for thickness and accumulation were produced. But see what a bizarre picture is shown in Figure III(b) with strange physical "step function" realizations where no corrections for border effect can help. For this reasons all the 108 holes were included in point kriging for obtaining the limits of the mineralized zone, using the previous modeled variograms.

By extension, also thickness and accumulation isovalue contours were produced. It means that because of the absurdity of the results, the ore variograms were used for waste (thus wrong) holes, or the wrong variograms were used for the right holes. Here lies a strong inconsistency!

But that was the way found to solve this practical problem. The results were pretty good and were coherent with the known values (Figure IV) so they could be kept.

Assuming the contour line of 2.00muBm accumulation as the limit of the mineralized area, it will be shown for illustration purposes, a few 100 m x 100 m kriged blocks inside this defined mineralized area. For that, only the 91 mineralized holes were used along with their correspondent variogram as previously modeled. The results for block thickness, accumulation and grade (accumulation divided by thickness) can be seen in Figure V or in Table I.

When the change is made to the 1.14uBm cutoff grade, the mineralized intervals must be rebuilt according to the new conditions. Their length must be smaller or equal to the previous. It was found, actually, that two holes had this interval reduced to zero.

What variograms were to be used? As the intervals change, no more work can be done with the 1.00uBm variograms! They must be rebuilt for 1.14uBm. The former hypothesis of a "natural" cutoff at 1.00 unit must be preserved, so the two holes with zero thickness and accumulation were to be included in the new variogram calculations, otherwise the hypothesis would be changed during the calculation and that also leads to practical contradictions at least in the regions near to these two holes.

As in the previous case, good spherical variograms were obtained for the 91 holes, as can be seen in Figure X.

Note that the 1.14uBm cutoff variogram shows a smaller nugget effect and, for thickness, a lower variance, although a shorter range is presented. Does it mean that 1.14uBm is a number as good as

1.00uBm for a natural cutoff? This seems to be true.

A new mineralized area was then defined in the same way as for 1.00uBm cutoff. Now it is the region inside a contour of 2.28muBm accumulation. Point kriging was performed on all 108 holes to avoid absurd pictures, and the isocurves were drawn by a plotting software, so reproducing the same incoherencies as before. Then, inside the new mineralized zone, using the 91 holes already defined, the same 100 m x 100 m square blocks are kriged.

The results can be seen in Figure V and in Table I. Now, a surprise:
* three blocks have higher thickness for 1.14 than for 1.00;
* four blocks have lower grade for 1.14 than for 1.00;
* five blocks have higher accumulation for 1.14 than for 1.00.

Of course these figures are unacceptable for a coherent mine planning.

Thus the work was restarted. The same hypothesis of a natural cutoff at 1.00uBm was assumed. However, a different method for 1.14uBm cutoff (kriging the decrements in thickness and accumulation from 1.00 to 1.14) was used yielding coherent results.

2.4 Ore evaluation using density functions

As no clear geological boundaries appear along the extension of the vertical deposit, the mining cutoffs will be applied. This means that the variograms for f, g and i functions for all the 108 holes have to be calculated. In Figure I(c) shows the positive and the negative holes.

The f and g functions are given by:

$$f(x_j;2.00) = e(x_j) \cdot i(x_j;2.00) \text{ and}$$

$$g(x_j;2.00) = p(x_j) \cdot i(x_j;2.00)$$

Remember that in this study these functions have numerically the same value as thickness and accumulation.

All the three variables have presented good spatial structure as can be seen in Figures XI, and could be modeled in a spherical scheme.

Note that the nugget effects is lower than before. Contour maps were produced using these models, just straightforward without any change on hypothesis.

Both indicator and accumulation contour maps may be used for drawing the limits of the mineralized area, see Figures VI where thickness contour is also presented. Indicator shows probabilities of mineraliza-

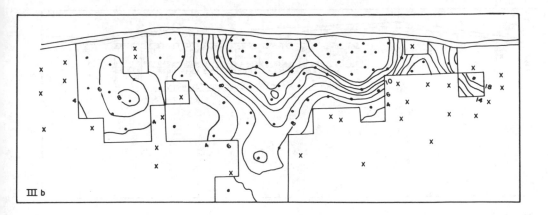

Figure III-b

Accumulation isocurves using the 91 mineralized holes, inside the hand made ore area

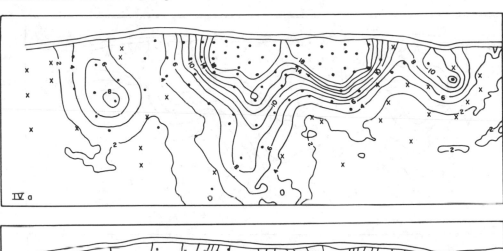

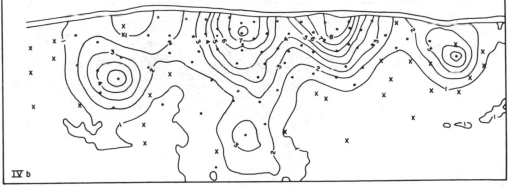

Figure IV

Contour maps using all the 108 holes
a) Accumulation
b) Thickness

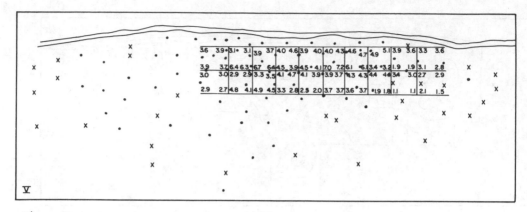

Figure V

Block grades and thicknesses using natural cutoffs:
a) Grade at 1u cutoff   b) Grade at 1.14u cutoff
c) Thickness at 1u cutoff   d) Thickness at 1.14u cutoff

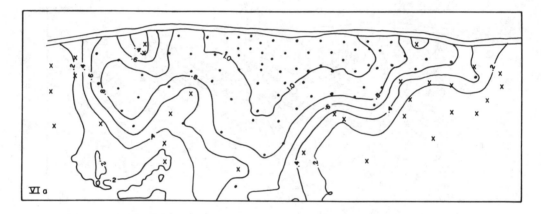

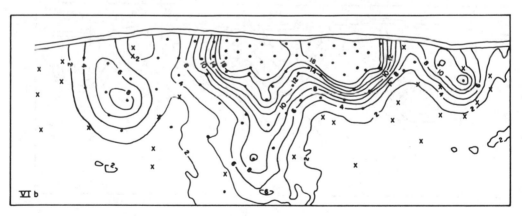

Figure VI

Contour maps using density functions for 1uBm
a) Indicator
b) Accumulation

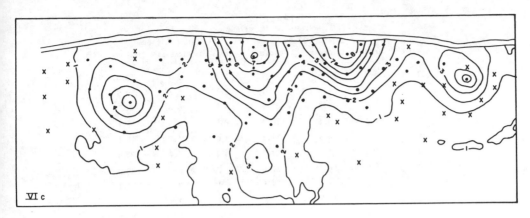

Figure VI-c

Contour maps for thickness using density functions for 1uBm

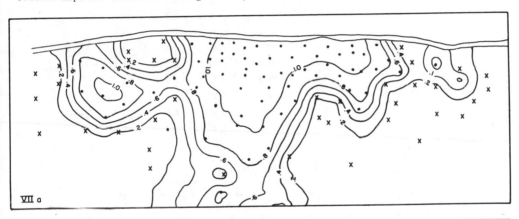

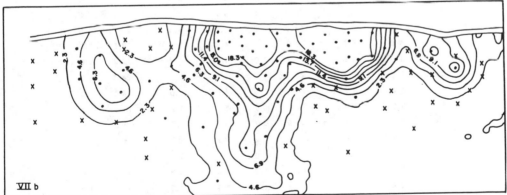

Figure VII

Contour maps, cutoff 1.14uBm, using density functions for 1.14uBm
and 2.28muBm
a) Indicator
b) Accumulation

x means waste holes

TABLE I
Natural cutoff method

block	grade u	Cutoff 1.00 uBm thickness m	accum. mu	grade u	Cutoff 1.14 uBm thickness m	accum. mu
1	3.6	3.9	14.1	3.9	3.7	14.3
2	3.1	6.4	20.1	3.1	6.3	19.6
3	3.6	6.7	23.6	3.7	6.4	24.1
4	4.0	4.5	18.1	4.6	3.9	17.1
5	3.9	4.5	17.5	4.0	4.1	16.1
6	4.0	7.0	27.9	4.3	7.2	29.6
7	4.6	6.1	27.6	4.7	6.1	28.8
8	4.9	3.4	16.5	5.1	3.2	16.1
9	3.9	1.9	7.5	3.6	1.9	6.8
10	3.3	3.1	10.2	3.6	2.8	10.0
11	3.0	2.9	8.6	3.0	2.7	8.1
12	2.9	4.8	13.9	2.9	4.1	11.9
13	3.3	4.9	16.1	3.5	4.5	15.7
14	4.1	3.3	13.5	4.7	2.8	13.1
15	4.1	2.5	10.2	3.9	2.0	7.8
16	3.9	3.7	14.4	3.7	3.7	13.6
17	4.3	3.6	15.5	4.3	3.7	15.9
18	4.4	1.9	8.4	4.4	1.8	7.9
19	3.4	1.1	3.7	3.0	1.1	3.3
20	2.7	2.1	5.7	2.9	1.5	4.4

TABLE II
Mining parameters method

	grade u	Cutoff 1.00 uBm thickness m	accum. mu	ind.	grade u	Cutoff 1.14 uBm thickness m	accum. mu	ind.
1	3.7	3.9	14.4	1.0	3.7	3.8	14.1	1.0
2	3.1	6.5	20.2	1.0	3.1	6.2	19.2	1.0
3	3.6	6.7	24.1	1.0	3.7	6.4	23.7	1.0
4	4.0	4.5	18.0	1.0	4.4	4.0	17.6	1.0
5	3.9	4.5	17.6	1.0	4.0	4.2	16.8	1.0
6	4.0	7.1	29.4	1.0	4.1	7.0	28.3	1.0
7	4.6	6.2	28.5	1.0	4.7	6.0	28.3	1.0
8	5.0	3.3	16.5	0.9	5.1	3.2	16.3	0.9
9	4.0	1.7	6.8	0.7	4.3	1.5	6.5	0.5
10	3.6	2.4	8.6	0 6	3.9	2.1	8.2	0.4
11	3.0	2.9	8.7	1.0	3.1	2.7	8.4	1.0
12	2.7	4.8	13.0	1.0	3.0	4.3	12.9	1.0
13	3.4	5.0	17.0	1.0	3.6	4.5	16.2	1.0
14	4.3	3.2	13.8	1.0	4.6	2.9	13.3	1.0
15	4.1	2.2	9.0	1.0	4.4	2.0	8.8	0.8
16	3.9	3.6	14.0	1.0	4.0	3.3	13.2	0.6
17	4.3	3.6	15.5	0.9	4.3	3.5	15.1	0.9
18	4.7	1.7	8.0	0·6	4.9	1·6	7.8	0.6
19	3.6	0.6	2.2	0.4	4.0	0.6	2.2	0.3
20	2.7	1.4	3.8	0.3	3.0	1.2	3.6	0.2

tion and have lower relative variances than accumulation.

These three variables were kriged for the same blocks as before; some differences arose from this kriging procedure: boundary blocks have lower thickness and accumulation, internal blocks have higher accumulation and thickness, all grades being higher than before. This picture seems to be a more realistic one. Indicator value means the proportion of ore inside each block: 1.00 means 100%, 0.27 means 27% and so on.

Now, if mining engineers ask for the new cutoff grade 1.14, the same procedure must be followed for the new intervals. All formerly mineralized holes where accumulation is below 2.28muBm are assigned zero: that means there is no recoverable ore in these holes. All other holes have an equal or reduced interval. In these conditions, density functions being $f(x; 2.28)$ and $g(x ; 2.28)$, and the indicator $i(x; 2.28)$, the variograms are produced for them using all the 108 holes, as before (Figure XII).

New contour maps are drawn (see Figure VII); now the 2.28muBm can be assumed as the mineralization limit. Kriging for the same blocks as before yields the results shown in Figure VIII and Table II. Note that no discrepancies appear for thickness, accumulation or grade or indicator values.

The results are coherent and acceptable and there is no need for new assumptions or changes on basic hypothesis.

3. TWO NOTES

3.1 Contour maps

An isovalue curve means that every point inside the area limited by this curve has a value greater or equal (or less or equal, depending on the case) the value of the isocurve. Contour maps are important (and impressive) at management level and can give, at a glance, an idea about what is happening in the area. A set of blocks with little numbers printed inside or a lot of figures printed on a computer listing have not this ability. Using graphical facilities, areas can be contoured, colored or divided by levels of risk or levels of profit pointing out, for example, the limits of richer or poorer regions, suggesting a mining sequence, etc.

But let us now analyse and understand the way a contour map is build to determine how far they can be reliable. Contour maps

are normally drawn by some software using a mechanical or electrical equipment called plotter. Generally, before drawing, a regular grid must be built. It may be built directly by the software using a chosen method (closest point, weighted average, linear trend, kriging, etc.), but also pre-established grids are accepted. In a second stage a polynomial interpolation between these points is used for drawing the contours. What happens in the case of this base metal deposit? There is a grid already estimated by using kriging techniques. But take care! These are not measured, but estimated points with a high variance associated with each other. It is different from airborne or ground geophysics where each point in the grid is really measured. It can be seen, then, that because of software limitations the real measured points do not enter in the interpolation procedure. Moreover, the points were estimated in a 25 m x 25 m grid as it makes no sense to krige in a smaller scale because of the original grid size - but a polynomial interpolation between these points was allowed to draw a continuous contour!

When may a polynomial interpolation be accepted? Let take again the geophysical examples. For instance, magnetic field distribution. The resultant magnetic field at a point is due to a very large number of magnetic bodies, whose individual fields can be described by a polynomial. So, the resultant field can also be represented by a polynomial. While kriging is philosophically an opposite to the polynomial approach (Matheron,1970)! In conclusion, isocurves may be used for a visualization of spatial distributions, and even for establishing regional limits, but it is not to be surprised if some points inside an isoregion show discrepant values. Contouring geometrical blocks calculated by kriging is much more secure and reliable.

Some words about associated errors: variance is very high for each estimated point and a correct contour on these values shows a region where each point is within a certain range of associated error. It is not a value that can be associated to something greater than a point. Knowledge of point variance can be important in cases like pollution by radiation where no point may pass a certain treshold, but it has no importance in mining problems. So a contour map for point associated errors gives a completely false idea about error distribution. Contour must be done around kriged blocks.

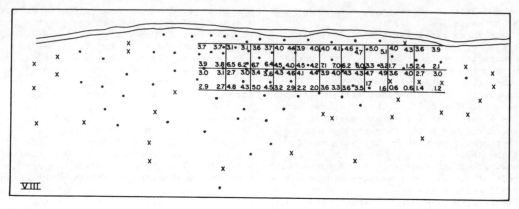

Figure VIII
Block grades and thicknesses using mining parameters

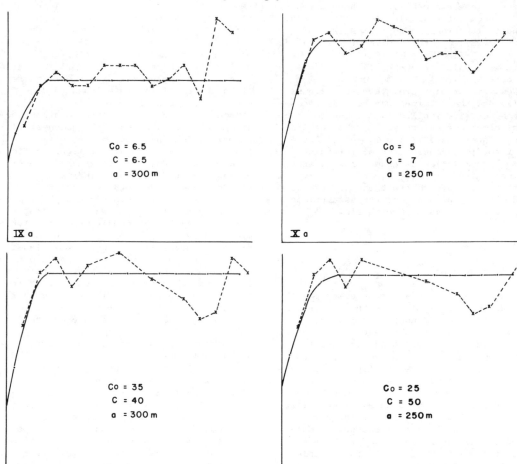

Figure IX
Variograms using natural cutoff 1u
a) Thickness
b) Accumulation

Figure X
Variograms using natural cutoff 1.14u
a) Thickness
b) Accumulation

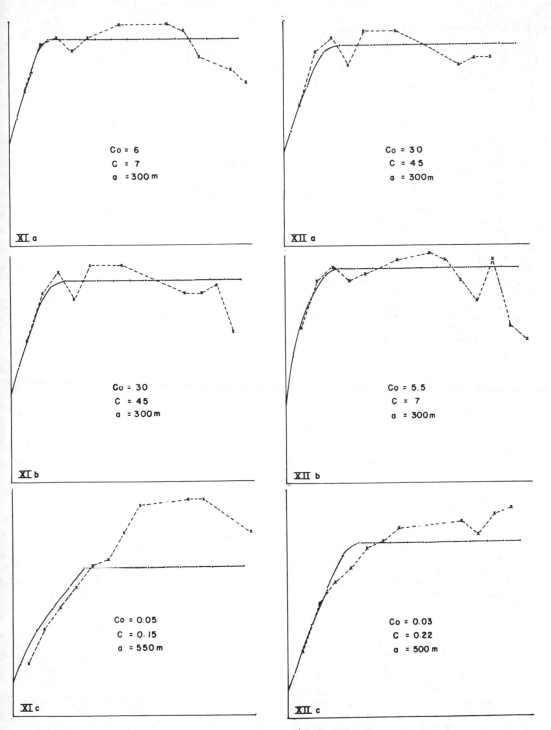

Co = 6
C = 7
a = 300 m

XI a

Co = 30
C = 45
a = 300m

XII a

Co = 30
C = 45
a = 300 m

XI b

Co = 5.5
C = 7
a = 300 m

XII b

Co = 0.05
C = 0.15
a = 550 m

XI c

Co = 0.03
C = 0.22
a = 500 m

XII c

Figure XI
Variograms using density functions for 1u
cutoff. a) Thickness;b)Accumulation
c) Indicator

Figure XII
Variograms using density functions for 1.14u
cutoff. a) Accumulation;b) Thickness;
c) Indicator

195

3.2 Mathematics & Physics

Every mathematical assumption has a physical meaning; and the reverse is also true: every physical hypothesis can be represented mathematically. So accumulation means metal weight, the polygons method calls for step function. Natural cutoff assumes that there are geological differences in the field. Mining cutoff divides the deposit into "to be mined" and "not to be mined" samples.

When a model is changed in the middle of the calculation, there is also a change in the physical meaning of what is being done and the results become less coherent. An ore reserve calculation is a long sequence of not-so-precise features: sample collecting, drill-hole deviation, layer dip and azimuth definition, true thickness calculation, geological layer definition, sample transportation, preparation and analysis, and so on. It is very important that in each phase, procedures and values must be as precise as possible, and so in the ore reserve calculation the best - and most coherent - hypothesis must be assumed.

BIBLIOGRAPHY

Armony, M. 1987. Contributions on indicator approach for two-dimensional problems - (internal paper). Rio de Janeiro:BPM.

Armony, M. 1987. Ore reserve estimation in a base metal deposit   using density functions (internal paper). Rio de Janeiro:BPM.

Internal report on a base metal deposit in Brazil. 1986. Rio de Janeiro: BPM.

David, M. 1977. Geostatistical ore reserve estimation, p. 218-221. Amsterdam: Elsevier.

Journel, A. G. 1983. Non-parametric estimation of spatial distributions. Math. Geology. 15-3:445-468.

Journel, A. G. 1984. Indicator approach to toxic chemical sites (report). Project No. CR-811235-02-0. Las Vegas: Environment Monitoring Systems Laboratory.

Journel, A. G. 1985. Recoverable reserves estimation the Geostatistical approach. Mining Engineering. June:563-568.

Matheron, G. 1970. La theorie des variables regionalisees, et ses applications. Les Cahiers du Centre de Morphologie Mathematique de Fontainebleau. "le Krigeage Universel". Fascicule 5: chapitre 4.

Computer Applications in the Mineral Industry, Fytas, Collins & Singhal (eds)
© 1988 Balkema, Rotterdam. ISBN 90 6191 760 3

Modified block modeling – A different approach to ore deposit modeling

Khosrow Badiozamani & Foad Roghani
Morrison-Knudsen Engineers, Inc., Boise, Idaho, USA

ABSTRACT: Geology and separation of ore and waste are two fundamental factors in modeling epigenetic deposits. Epigenetic deposits are those deposits where there is a time lag between the deposition of the ore and the host rock. Geologic features such as lithology, facies, and faults are important contributing factors in ore deposition and should be taken into consideration during data categorization and modeling. Separation of ore and waste should be handled at the block level. Even though the conventional block modeling technique is claimed to take geology into consideration, this is not accomplished to the extent necessary to produce a representative model and to take advantage of all the available information. For example, two sand channels belonging to two depositional systems cannot be differentiated and modeled as such. In conventional block modeling ore and waste are not structurally separated. In another words, the exact location of geologic units containing ore and waste is lost due to the modeling. A new modeling technique developed by MKE, called "modified block modeling" (MBM), takes these geologic features into consideration during data categorization and continues using related information throughout the modeling process. MBM differentiates between ore and waste through blocks of variable height representing individual units of ore or waste separated by the geologic features. This technique has been applied successfully to various oil sand deposits throughout Canada during the past three years and it is currently being used at various locations for geologic modeling, long and short range mine planning and daily operations.

INTRODUCTION

The value of ore in a mineral deposit is heavily dependent on the position of the ore in the mineralized zone and the value of ground surrounding it. As a result, during mineral evaluation, modelers have tried to incorporate the concept of "area of influence" in their evaluation. Numerous modeling techniques such as Polygonal, Triangulation and Inverse Distance Weighting were developed to accommodate this intuitive weighting process (Popoff 1966). Traditional statistical processes also have played an important role in ore modeling (Hazen 1967; Agterberg 1974; Barnes 1980).

The major difficulty with the above estimating techniques is their inability to fully account for the regionalized variations. The occurrence of high grade and low grade sections and lower grade fringe areas makes ore grade a "regionalized variable" (Royle 1980). Using empirical techniques developed by Krige, Matheron (1963) developed mathematical concepts to account for the regionalized variables. This branch of ore reserve estimation has been applied extensively to various mineral deposits (David 1970 1977; Huijbregts 1973; Journel 1976).

Geostatistics has been used extensively in conventional block modeling. However, to quote Ramani and Stanley (1979), "the selection of a method to develop an ore body depends on geological considerations; exploration and/or sampling methods; availability, reliability, and volume of data; specific purpose of the estimation; and the requirements of accuracy. We would like to add to this list: the complexity of the deposit and mineralization stage as well.

Geologic modeling and separation of ore and waste in epigenetic deposits play a major role in an ore body evaluation. The separation process is not an easy task and requires detailed analysis and sophisticated tools. One of the tools used in modeling lithologic units and ore-waste separation is the conventional block modeling technique where, based on the cutoff grade, blocks are separated into ore and waste zones. Due to arbitrary selection of blocks and cutoff grades, this modeling technique is not capable of providing a true separation between the ore and the waste zones where such separation is necessary. Conventional block modeling falls short of ideal conditions in ore reserve estimation of such deposits as Canadian oil sands. In this type of deposit it is essential to separate lithologic units as well as geologic facies. Conventional block modeling is too coarse for finer separation of facies.

Arney and Peruzza (1972) introduced a new modeling method called "Zone Interpolation" to account for facies changes in a unit. Their method, called for generation of a fixed thickness block model. The zone interpolation modeling technique was modified later on to accommodate "variable thickness" blocks (Yildirim 1985).

Usage of variable thickness block provided a vehicle for representing the finer geologic units and the facies variations in a model. Yildirim provides an extensive discussion of concepts for developing a computer program to model irregular blocks. In this paper Yildirim recommends dividing individual zones at each drill hole into a fixed number of subzones of equal thickness. As a result, the thickness of subzones vary from one drill hole to the next, but all of the subzones of a zone within a drill hole are of equal thickness. The equal thickness subzones reduce the effectiveness and advantages of this technique.

To eliminate this problem we present the concept of variable subzone thickness even at the drill hole level. We have refined the concept of the variable block model to

a "Modified Block Model" so that an accurate representation of the ore and waste can be obtained with minimal effort by the geologist using Morrison-Knudsen Engineers' computer software.

CONVENTIONAL BLOCK MODEL

Conventional block modeling techniques follow a four step process, i.e. input of geologic and assay data, compositing of the assay data into predetermined benches, designation of block size and finally estimation of the ore content for each block. To provide a comparison between the conventional block and modified block models we need to provide a more detailed discussion of the compositing, block size definition, and estimating process.

Block modeling technique for a disseminated ore deposit is based on division of the ore bearing region into a set of horizontal benches with a given height. The size of each bench and the size of the block is dictated by the mining equipment used. Assay data along each DDH is then composited by weight averaging along the length of the DDH and within each bench. The weighted average value so obtained is assigned to the center of each segment of the DDH intersecting the bench (Figure 1). The weighted assay values, the bench height, and the blocks are basically independent of the geologic units.

Following assignment of the DDH assay data to each bench, an interpolation or extrapolation technique such as inverse distance or geostatistics is used to estimate the grade or any other desired variable value for each block. The estimation process continues for all uniform thickness blocks. Again, this process does not directly consider the effect of the geological variations. These variations may only be represented by variations in the assay data.

MODIFIED BLOCK MODELING -- BASIC CONCEPT

Modified Block Modeling follows the same four steps of conventional block modeling process; however, the compositing, block size specification, bench definition and ore estimation are all dependent on the geological factors such as lithology, facies, and grade in the deposit. Yildirim (1985) gives an excellent introduction to

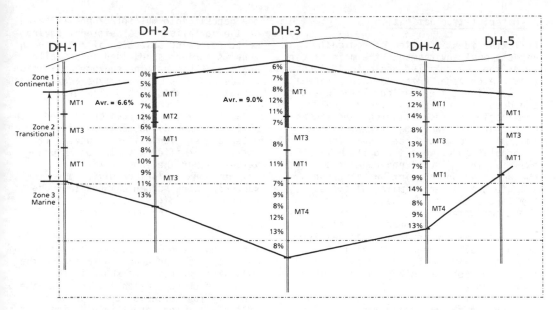

Figure 1 - Compositing by bench using conventional block modeling approach, without geologic consideration

the variable block modeling techniques and discusses in detail the steps involved. However, at the subzone level, he recommends dividing each zone into a number of equal thickness subzones which results in a possible combination of unrelated features, therefore losing the integrity of the detail geology and effectiveness of this modeling technique.

The main reason for dividing an orebody into variable block heights is to interpolate between the related information and exclude unrelated units during modeling. This process will increase the accuracy of estimation by properly separating ore from waste. Such separation of ore and waste is impossible by conventional block modeling techniques. In the following sections we will discuss in more detail the steps involved in the modified block modeling process.

A. Correlation

The first step in developing a modified variable block model is division of the mineralized body into correlatable or related zones on a regional basis. In this case a geologic unit such as formation or group is divided into more related or correlatable units based on depositional environments. Such division reduces the possibility of interpolation among unrelated units or extensive grade variation caused by changes in porosity or permeability. For example, a group or formation can be divided into marine, transitional, and non-marine or continental zones. If additional information is available each of these zones can be further divided to improve geologic control prior to modeling. For instance, a transitional unit can be divided into Delta Plain, Delta Front, and Prodelta, (Fisk et al. 1954; Colman et al. 1980; Weimer et al. 1979,1985). If some of the zones are not present in a given area, their stratigraphic location should be marked and a zero thickness should be assigned to them to preserve the continuity of the sequence.

B. Material type definition

Following identification of regional depositional environments the main task is to define the detailed geologic features within each zone. Prior to interpolation the user identifies various lithologies and depositional facies. The unique combination of each lithology and facies constitute what we name "material type" (MT) which forms the basic unit for interpolation between the drill holes. For example, in a fluvial environment a channel sand and a point bar sand constitute two different material types. Similarly, two channel sands with different characteristics, or two sand bodies belonging to two different channels, are considered as two separate material types. During modeling, regardless of the algorithm used, only similar material types are used for interpolation within each subzone.

C. Zone and subzone definition

If information is scant a zone may just be represented by a formation. If detailed information is available more zones may be defined by breaking a formation into its finer depositional units such as marine, transitional and continental. Division of

the deposit into multiple zones is based on the variation in depositional environment. Following selection of zones, each zone is further subdivided into smaller units or subzones. The subzone definition can be based on material type and/or grade variations within each drill hole.

For example, a zone can be divided into a fixed number of subzones based solely on material type variations. However, if there is significant variation in grade values within the zone it will be advantageous to further divide the subzones where sudden changes in grade occur. Subdivisions based on grade variation improves the grade estimation for the blocks.

Figures 2 and 3 illustrate the effect of material type and grade variation on the estimation process. As shown in the schematic cross-section in Figure 2, there are only four material types present (MT1-MT4), which result in division of the zone 2 into four subzones. Due to extensive and sudden grade changes in DH-3 and DH-4, the subzones are further divided into four smaller units to separate high and low grade material (Figure 3). This process improves grade interpolation and further aids in separation of ore and waste units.

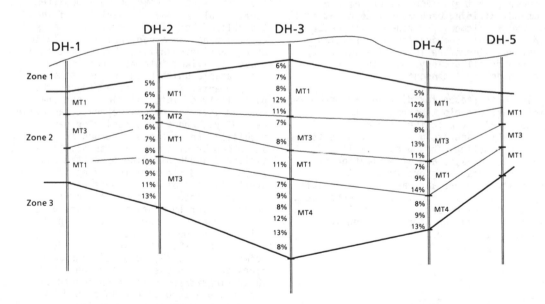

Figure 2 - Zone and subzone designation based on material type (MT).

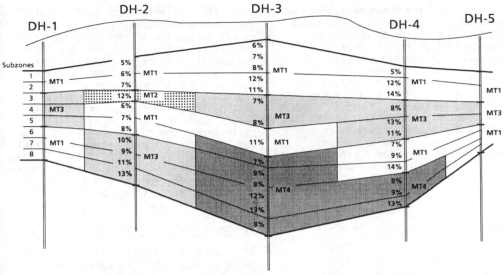

Figure 3 - Subzone designation based on material type and grade variation greater than 3%.

Figure 3 also demonstrates the extent of influence of each grade or material type around each drill hole. Figure 4 shows the material type designation for each subzone at each drill hole. The main advantage of subzone interpolation is its ability to account for sudden changes in lithology, facies or grade both in vertical as well as in horizontal direction. The interpolation process will consider the material types present in the area. As shown in Figure 3, the grade interpolation is performed within each subzone and among the same material types. As an example, in subzone 3, material type 3 is present in four out of five drill holes (Figure 4). Drill hole 2 in the center contains material type 2. The block value interpolation for material type 3 will use all four drill holes for estimation. However, blocks under the influence of DH-2 will not be assigned any value from material type 3; rather, they will be assigned values from material type 2.

SUBZONE 1	SUBZONE 2	SUBZONE 3	SUBZONE 4
DH-1 o 1 DH-4 o 1 DH-2 o 1 DH-3 o 1 DH-5 o 1	DH-1 o 1 DH-4 o 1 DH-2 o 1 DH-3 o 1 DH-5 o 1	DH-1 o 3 DH-4 o 3 DH-2 o 2 DH-3 o 3 DH-5 o 3	DH-1 o 3 DH-4 o 3 DH-2 o 1 DH-3 o 1 DH-5 o 3
DH-1 o 3 DH-4 o 1 DH-2 o 1 DH-3 o 4 DH-5 o 1	DH-1 o 1 DH-4 o 1 DH-2 o 3 DH-3 o 4 DH-5 o 1	DH-1 o 1 DH-4 o 4 DH-2 o 3 DH-3 o 4 DH-5 o 1	DH-1 o 1 DH-4 o 4 DH-2 o 3 DH-3 o 4 DH-5 o 1
SUBZONE 5	SUBZONE 1	SUBZONE 1	SUBZONE 1

Figure 4 - Plan view distribution of material type with subzones.

201

As evident from these schematic diagrams the modified variable block models allow for detailed separation of ore and waste with substantial thickness variations as well as abrupt changes in lithology or facies in horizontal and vertical directions. Figures 5, 6, and 7 depict the difference in block designation among the conventional, variable, and modified block models. If desired, it is possible to have equal thickness subzones throughout the property; but, the uneven thickness subzones provide a greater flexibility and better differentiation of ore and waste units.

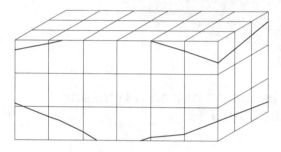

Figure 5 - Conventional block model with fixed blocks

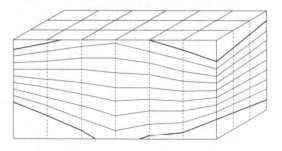

Figure 6 - Variable block model with equal thickness subzone blocks.

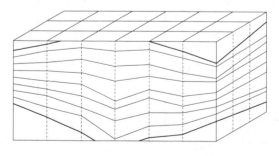

Figure 7 - Modified block model with variable thickness subzone blocks.

Modified variable block model provides the vehicle necessary to interpolate grade values across the property regardless of the lateral grade variations; provided interpolation is performed within the same material type. Yildirim (1985) advises against grade interpolation between high and low grade material due to the possibility of generating intermediate values where they are not warranted. We believe that the modified variable block model process described above will eliminate the possibility of misinterpolation, and can be freely used to generate intermediate values from low and high grade material as long as they belong to the same material type.

D. Compositing of sample data

After zone and subzone designation and prior to interpolation, various analytic values such as ore grade have to be calculated for each subzone. Generally, sample intervals do not coincide with the top and bottom of each subzone, or multiple samples may be available for each subzone. If the subzone and the sample intervals coincide, the value of the sample is assigned to the subzone. If there is more than one sample per subzone, the samples are composited and the composite value is assigned to the subzone. Composite value is weight averaged by the thickness of the samples or by a product of thickness and density in situations where there is a substantial density difference between samples. The weight averaging by thickness and density is performed for elements whose value is reported as weight percent such as grade, and not for elements which are based on volume percent such as porosity.

E. Interpolation and block generation

Upon completion of the compositing process, different interpolation algorithms can be used to estimate each element. For example, grade can be modeled using geostatistics and thickness can be modeled using an inverse squared distance method.

Depending upon the mining conditions and mining equipment selected, the subzones may be combined and assigned to different benches. A bench may pass through a number of zones or subzones. It is also possible that a bench may divide a subzone into two parts, where each part will be assigned to a different bench.

Even though the subzones are assigned to one bench or the other, their identity and associated information remain intact. This will allow separate reporting of each subzone information and material type during mining process, which can result in a more accurate estimation of the grade delivered to the plant.

APPLICATION TO CANADIAN OIL SANDS

Morrison-Knudsen Engineers, Inc., as part of its EAGLES (Engineering, Analytic, Geologic and Land Engineering System) has developed a set of software which allows automatic generation of subzones using lithologic, facies, and grade variations. This system assists the users in modeling and mine planning using the MBM approach. The system has been successfully applied to various oil sand deposits in the Province of Alberta, Canada and other ore deposits. In its application to the oil sand deposits, the user has various options to choose from during the modeling process. For example, in some applications the deposit was divided into three zones i.e., Clearwater Formation, McMurray Formation, and Devonian Limestone (Flach et al. 1985). Figure 8 shows a schematic cross section of these units. The McMurray units were further divided into marine, tidal, mixed fluvial-estuarine, and continental zones, as shown in Figure 8. Based on the lithology and specific depositional environment of each litho-facies, each of the units within the McMurray formation, were assigned to a different material type. For example, the sand units from the tidal channel and the sand flat were assigned to two different material types. As a result, we were able to differentiate between similar litho-logic units of the same zone based on their depositional environment. Diff-erentiation of this sort, allows us to recognize a number of material types with very distinct ore bearing characteristics within the same zone. During the modeling, interpolation was performed between similar material types, eliminating the mixture of unrelated materials and allowing the separation of ore from waste. The top of the McMurray formation was modeled as the marker structure. Depending on the amount of information available, and spatial correlation between various elements, different interpolation tech-niques were used to model various char-acteristics of each subzone.

Where the detailed geologic information was not available, the McMurray formation was divided into a number of correlatable zones such as upper ore zone, center reject and lower ore zone. Each of these zones were divided into a number of subzones based on the grade variations and maximum number of samples in each subzone. The interpolation was then performed within each subzone according to proc-edures described earlier.

The automatic generation of subzones by computer and the ability to review and update the subzones graphically expedites the process tremendously. The geologist has more time to analyze the model and establish the distinct material types in the deposit. The geologist also has the ability to limit the areal extent of each material type graphically and according to his interpolation.

EAGLES also allows modification of the blocks individually or collectively on the graphic terminal at a later date, when information from additional drilling or actual mining becomes available. This function provides the modeler with a tool to update the model on a local basis at any time and without a need to re-run the entire model.

CONCLUSION

Conventional block models cannot ad-equately represent geologic variation and abrupt depositional changes in a deposit. As a result, the ore and waste separation and grade calculations are not accurate, especially in epigenetic deposits such as Canadian oil sands. The new modified block model described in this paper, provides the necessary mechanism for separation and recognition of depositional changes to the finest detail desired. The MBM allows division of geologic units into smaller and more closely associated units based on geologic, facies or grade variations. The result of finer separation is the elimin-ation of unlike material and increase in reliability of estimation. We believe the new MBM provides a much better and more accurate representation of the deposit which will result in a more accurate ore reserve estimation. EAGLES software developed specifically for this purpose automatically generates the subzones and therefore expedites and simplifies the process.

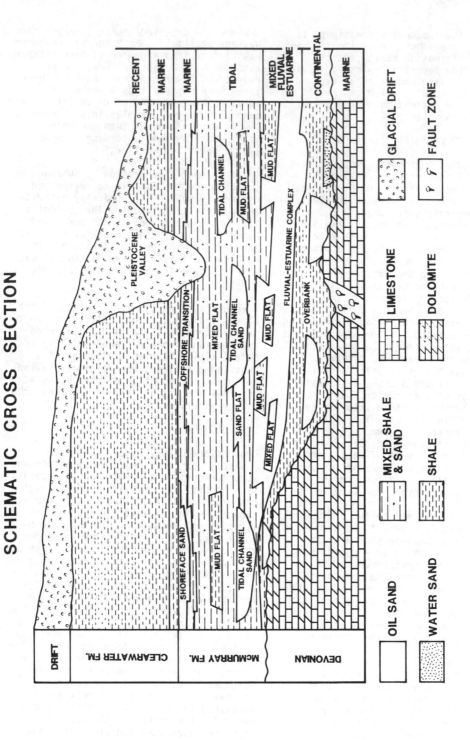

Figure 8 – Schematic Cross Section of Canadian Oil Sand

ACKNOWLEDGEMENT

We are very grateful to many individuals
who have helped us in formulation of the
ideas and development of our software
during the past few years. Especially we
are indebted to Messers. Ian McLachlan and
Desmond Milner from Syncrude Canada, and
Mr. George Hawes from Esso Resources
Canada, and Mr. Erdal Yildirim from
Canadian Occidental Petroleum for their
insight and discussions with regards to
problems associated with oil sand modeling.

Even though the discussions with the
above individuals have been most helpful
to us, the ideas presented in this paper
are those of the authors and do not necessarily reflect or agree with their ideas.

REFERENCES

Agterberg, F.P. 1974. Geomathematics,
  Amsterdam: Elsevier.
Arney, T.O. and Perruza, A. 1972. Block
  value model development for highly
  irregular ore bodies. The Council of
  Economics of AIME: San Francisco.
Barnes, M.P. 1980. Computer-assisted
  mineral appraisal and feasibility, New
  York: Society of Mining Engineers.
Coleman, J.M. and Prior, D.P. 1980.
  Deltaic sand bodies, Amer. Assoc. Pet.
  Geol., Continuing Education Course Note
  Series #15.
David M. 1977. Geostatistical ore reserve
  estimation, Amsterdam: Elsvier.
Fisk, H.N.; McFarlan Jr. E.; Kolb, C.R.;
  Wilbert Jr., L.J. 1954. Sedimentary
  framework of the Mississippi Delta,
  Jour. Scd. Pet. 24: 76-79.
Flach, P.D., Mossop, G.D. 1985.
  Depositional environment of lower
  Creteceous McMurray formation, Atabasca
  oil sands, Alberta. Am. Assoc. Pet.
  Geol. 69: 1195-1207.
Hazen Jr., S.W. 1967. Some statistical
  techniques for analyzing mine and
  mineral-deposit sample and assay data.
  Bulletin 621, US Bureau of Mines.
Huijbregts, C. 1973. Regionalized
  variables and applications to
  quantitative analysis of spatial data.
  Proc. NATO Advanced Study Institute,
  display and analysis of spatial data.
  London: Wiley.
Journel, A. 1976. Ore grade distribution
  and conditional simulation-two
  geostatistical approaches. Guarascio,
  M.; David M. and Huijbregts C. (ed.).
  Advanced geostatistics in the mining
  industry. Netherland: Reidel.

Matherton, G. 1963. Principles of
  geostatistics. Econ. Geol. 58: 1246-1266.
Popoff, C.C. 1966. Computing reserves of
  mineral deposits: Principles and
  conventional methods. Information
  circular 8283, US Bureau of Mines.
Ramani, R.V. and Stanley, B.T. 1979.
  Ore-body modeling. Weiss, A.(ed.).
  Computer methods for the 80's p. 245-252.
  New York: Soc. Min. Eng. Port City Press.
Royle, A.G. 1980. Why geostatistics? In
  Geostatistics. New York: McGraw-Hill.
Yildirim, E. 1985. A Block-value modeling
  application in tar sands. AOSTRA seminar
  on the advances in petroleum recovery
  and upgrading technology. Edmonton.

6 Mineral processing II

Computer Applications in the Mineral Industry, Fytas, Collins & Singhal (eds)
© 1988 Balkema, Rotterdam. ISBN 90 6191 760 3

Mass balance calculations for plants and laboratories

M.Rémillard & M.Thérien
CRM, Ministère de l'Energie et des Ressources, Quebec, Canada

ABSTRACT: Data from two industrial plants are treated by mass balance package BILMAT. The first application concerned a metal mining producer and the second, an industrial mineral producer. Also LOTUS 123 software can easily be adapted for laboratory test mass balances using an IBM-PC computer.

1 INTRODUCTION

The present paper deals with two different BILMAT program applications. The first one concerns a copper-zinc-gold-silver producer based in North-western Quebec and the second, concerns the implementation of a hydrocyclone for the recovery of iron chips in a Quebec granite cutting company.

Afterwards, we deal with some LOTUS 123 software applications for laboratory test mass balances using an IBM-PC.

2 BASE METAL AND P.M. PRODUCER

Noranda Inc, Mattagami division, which operates mines in North Western Quebec, produces copper and zinc concentrates from a sulphide ore. Small amounts of gold (0,34 g/t) and silver (20 g/t) are present. At the time of the study, only 30% of the silver content was recovered in the copper concentrate. The main objective of this study is to evaluate the metallurgical behaviour of silver in the flotation circuits, to increase its recovery in the copper concentrate and therefore to improve its final economic value. It is then worthwhile to identify the silver bearing minerals and to determine the critical flotation stages where they are lost. The methodology proposed is based on the coupling of a mineralogical study and a detailed mass balance calculation in the flotation circuit performed using the BILMAT mass balance package (Hodouin, 1980, 1981, 1985a).

The results are discussed with respect to the distribution of silver in the circuit. The recoveries obtained at the various separation stages for the minerals present in the ore are also discussed.

2.1 Description of the flotation circuits

The Cu-Zn mill (4000 metric tons per day capacity) received, at the moment of this study, two different sulphide content ores from Mattagami Lake and Norita Mines.

Metal contents of the mill feed usually averages 0,9% Cu, 5,0% Zn and 0,03% Pb. The sulphide mineralization is fine-grained. Mattagami Lake ore has a talcose gangue. The main contaminant of the Norita ore is pyrite.

After mixing and crushing the two ores, the product is ground in two parallel grinding circuits to 83% minus 200 mesh and directed after a classification stage to the copper flotation circuit. Copper and zinc flotation circuits (figures 1 and 2) are similar. A rougher concentrate is produced and cleaned in three stages. The second and third cleaner tails are reground prior to returning to the first cleaner. The first cleaner tails go through a cleaner scavenger which tails are discarded out of the circuit.

The copper rougher concentrate grade may vary widely (2-6% Cu) depending on the initial content of talc in the feed. The talc makes an abundant and persistent froth that is not suitable for the production of a clean rougher concentrate. This mineral is further depressed in the cleaners stages. Most of the depressant is added in the first cleaners. A poor concentrate grade is obtained when over-frothing conditions prevail. A good quantity of zinc can also follow the copper rougher concentrate. It is then depressed using SO_2 in the copper cleaning circuit and discarded to the zinc roughers.

The pH control in the copper roughers is obtained with soda ash. Lime is added in the zinc rougher circuit where a zinc rougher concentrate of about 30% zinc is obtained.

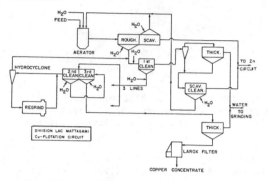

Figure 1: Flowsheet of the Cu-flotation circuit

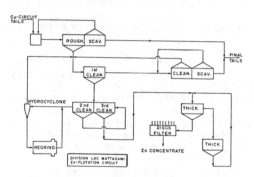

Figure 2: Flowsheet of the Zn-flotation circuit

2.2 Experimental techniques

Twenty streams of the Cu-Zn flotation circuits were sampled and analysed for pulp solid percentage, particle-size distribution, chemical content and precious metal grades in particle-size intervals. Figure 3 summarizes the different information levels which are generated by the stream samples and the corresponding process variables which are measured. A secondary sample is used for base metal contents (Cu, Pb, Fe, Zn), Au, Ag, S and also for Sb, Te, Se, As, Bi elements which could be involved in complex minerals containing precious metals. Another sample is taken for wet sieving on a 325 mesh screen. The oversize fraction is dried and sieved on

the ROTAP sieving column and all the undersize fraction mixed and split into 40 g samples for cyclosizer analysis. More details concerning the cyclosizer analysis can be obtained in the Hodouin article (1986).

Polished sections of some samples were first examined under an optical microscope to identify silver bearing minerals. They are galena, stützite and tetrahedrite (observed in the copper concentrate). The galena is mainly liberated in the copper concentrate and 38 grains were analysed with the microprobe. The silver content varies from 0 up to 0,9% (average content of 0,2%). Three stützite grains (Ag_3Te_2) were analysed and present an average silver content of 56,5% (very close to the theoretical composition). Tetrahedrite was only observed in the copper concentrate as unliberated grains always associated with galena. Six grains (as determined by microphobe analysis) average 20,1% Ag.

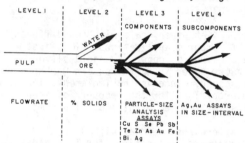

Figure 3: Information levels and process variables

2.3 Mass balance calculations

The variables which have been measured in the flotation circuit are not independent from each other, because the mass conservation constraints link them together. For example, the flowrate of silver which enters a bank should be equal to the sum of the concentrate and tail silver flowrates. However this is generally not verified when using the experimental data because of the dynamic process fluctuations and of the measurement errors. This problem becomes very puzzling when the number of measured variables is large as in the present case. There are many ways to calculate the distribution of the ore and the metal contents in the flowsheet. An optimal solution to this problem is proposed in the modern automatic mass balance packages. The BILMAT program (Hodouin, 1980, 1981, 1985a) is well adapted to the present situation where data are distributed into four different levels of information as presented in figure 3.

2.3.1 Basic principles

The BILMAT mass balance technique consists in the simultaneous estimation of all the process variables of the flowsheet (flow-rates, particle-size distributions, as-says...), such that the estimated values are as close as possible to the measured values, according to their assumed accuracy, while maintaining the laws of mass conservation. This can be formulated as follows:

$$\text{MINIMIZE } J = \sum_{\substack{\text{ALL INFORMATION} \\ \text{LEVELS}}} + \sum_{\substack{\text{ALL CIRCUIT} \\ \text{STREAMS}}} (Z_j - \hat{Z}_j)^2 \frac{1}{\sigma_j^2} \quad (1)$$

under the constraints $H(\hat{Z}_j) = 0$ (2)

where:

Z_j: observed value of the jth variable measured with a variance σ_j^2
$\hat{Z}_j$: the estimated value
H: the system of mass conservation equations

   H can be written throughout the flow-sheet for the various components consi-dered (ore, water, particles, metals con-tents, metal content by particle-size, particle-size). Precious metals are mea-sured at two different information levels (ore and particle-size intervals). This creates, in the system H, a supplementary set of coherency equations expressing that the sum of the gold or silver present in the size fractions should be consistent with the ore head grades. This constraint is now included in the latest BILMAT ver-sion (Hodouin, 1985b).

2.3.2 Networks

The program automatically builds the sys-tem of equations $H(\hat{Z}) = 0$ inasmuch as the flowsheet is described by networks expres-sing the material circulation and conser-vation in the circuit. Theoretically, each component should have its own network but, in the present case, many components have the same. Only three different net-works have to be considered:

1) for pulp and water
2) for ore and its metals contents
3) for particles
   The first one is shown in figure 4, where each branch corresponds to a stream and each node to a process unit (or a set of process units).
   The second one is obtained by removing water streams (figure 5) while the third one (figure 6) is obtained by deleting two

nodes in the second one. These nodes in-clude grinding stages wherein there is no conservation of particles with respect to their size property and to their chemical content in the particle-size intervals.
   The largest network (pulp-water) con-tains 11 nodes, 20 pulp streams and 9 wa-ter streams. The total number of process variables Z (ore, water, 14 size classes, 12 chemical elements, 2 metal grades in particle-size intervals) is 1129 and the total number of equations in H is 583. This shows the need for a suitable compu-tation algorithm to solve this huge mathe-matical sets of equations (1) and (2).

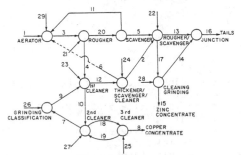

Figure 4: Pulp and water network

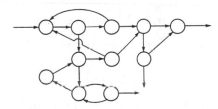

Figure 5: Solids and chemical elements network

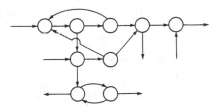

Figure 6: Particle size classes network

2.3.3 Prior evaluation of errors varian-ces

The weighting factors $1/\sigma_j^2$ (the least square criterion (1)) play an important role, since they allow the corrections $(Z_j - \hat{Z}_j)$ to be large when the measu-rements are imprecise and vice-versa. It is not necessary to have an exact evaluation of the variances but at least to be able to classify the data according to their degree of accuracy. The systematic approach proposed by Gy (1979) is useful even though it cannot completely solve the problem of the evaluation of measurement statistical properties. The random events that occur in measurements by sampling can be divided into fundamental, segregation, operation and analysis errors.

The fundamental error can be evaluated, after Gy, by:

$$\sigma^2 = [\frac{1}{M_s} - \frac{1}{M_1}] \; [\frac{1 - a_1}{a_1}] \; [(1 - a_1)\rho_c + a_1\rho_g)] \; lfgd^3 \quad (3)$$

where:

M_s = sample mass
M_1 = total mass from with the sample is taken
a_1 = the weight fraction of the valuable mineral
ρ_c = specific gravity of the valuable mineral
ρ_g = specific gravity of the gangue
l = a liberation factor
f = a particle shape factor
g = a particle-size distribution factor
d = 95% passing particle size

This error accounts for particles heterogeneity and must be evaluated for the initial sampling of the stream as well as for the subsequent subsampling. In equation 3, σ is the coefficient of variation, i.e., the relative standard deviation. This formula is also applied, in simplified forms, to the percentage of solids in the pulp and to the particle-size distributions (Smith, 1985). The segregation error is not evaluated.

More details concerning operating errors and posterior adjustment of errors variances can be obtained from Hodouin (1986).

2.4 Results of mass balance computation

The program print out of the mass balance results gives for each process variable the measured value, the estimated value, the relative percentage adjustment and the measurement coefficient of variation as a percentile. Figure 7 and table I present the ore flowrates and pulp percentage of solids for all the streams (only measured and estimated values are shown). We also obtain the particle-size distributions, the overall chemical assays (table II) and the gold and silver grades (table III) of the particle-size intervals, (for only the

6 first streams of the copper flotation circuit). The global results considering only the feed, the 2 concentrates and the tails are summarized in table IV.

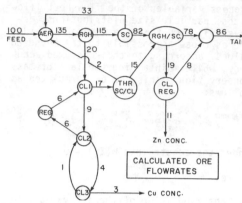

Figure 7: Ore distribution in the flotation circuit

Table I: Solid flow rates and pulp percent solids obtained from BILMAT

STREAM	FLOWE RATES CALCULATED	PULP % SOLIDS	
	CALC	MEAS	CALC
FRESH FEED*	100	62.7	63.1
CU.SC/CL.T.	15	25.3	25.3
CU.RO/SC.F.	135	44.1	42.4
CU.RO. C.	20	28.9	28.9
CU.SC. T.	81	44.1	44.1
CU.SC/CL.C.	2	15.0	15.0
CU.CL2 T.	6	11.3	11.3
COPPER CON.	3	33.2	33.2
CU.REGRIND	6	11.1	11.1
CU.CL2 F.	9	21.5	21.5
CU.SC. C.	33	21.4	22.3
CU.CL1 T.	17	17.7	17.7
ZN.RO/SC.T.	78	33.0	32.8
ZN.CL. T.	8	5.0	4.8
ZINC CONC.	11	31.4	31.4
TAILINGS	86	21.0	21.2
ZN.RO/SC.C.	19	-	24.5
CU.CL3 F.	4	45.6	45.7
CU.CL3 T.	1	-	19.2

* Reference value

Table II: Chemical assays for streams 1 to 6 (BILMAT print-out), %

	FRESH FEED		CU SC CL T		CU RO SC F	
CHEMICAL ASSAY	MEAS	CALC	MEAS	CALC	MEAS	CALC
COPPER	1.50	0.75	0.30	0.32	0.99	0.91
ZINC	5.82	6.15	5.96	5.89	6.20	6.05
IRON	28.70	29.05	25.90	25.68	28.00	27.98
LEAD	0.09	0.07	0.07	0.08	0.09	0.09
SULPHUR	19.30	19.26	19.20	18.42	19.10	19.14
ANTIMONY	28.00	27.87	26.00	27.38	35.00	27.79
ARSENIC	210.00	114.92	103.00	102.80	115.00	112.84
SELENIUM	77.00	68.02	80.00	83.19	67.00	77.73
TELLURIUM	20.00	21.67	20.00	19.52	20.00	25.06
BISMUTH	28.00	29.69	40.00	39.57	38.00	37.15
GOLD (g/t)	0.34	0.35	0.34	0.35	0.49	0.49
SILVER (g/t)	18.60	18.92	24.20	24.19	30.20	27.94

	CU RO C		CU SC T		CU SC CL C	
CHEMICAL ASSAY	MEAS	CALC	MEAS	CALC	MEAS	CALC
COPPER	3.42	3.33	0.14	0.13	2.22	2.23
ZINC	5.27	5.30	6.18	6.33	4.49	4.48
IRON	25.60	25.88	30.80	29.70	22.20	22.19
LEAD	0.21	0.20	0.04	0.04	0.27	0.27
SULPHUR	19.80	20.00	19.90	19.07	19.00	18.93
ANTIMONY	33.00	30.83	24.00	27.36	39.00	39.26
ARSENIC	103.00	104.92	122.00	118.63	175.00	175.14
SELENIUM	110.00	105.55	74.00	60.58	160.00	161.88
TELLURIUM	47.00	34.55	20.00	19.26	62.00	61.06
BISMUTH	55.00	51.56	95.00	24.92	63.00	62.97
GOLD (g/t)	1.10	1.01	0.20	0.22	1.60	1.62
SILVER (g/t)	55.00	54.22	13.00	11.40	80.00	81.40

Table III: Silver grades in particle-size intervals for streams 1 to 6 (g/t)

SIZE INTERVALS	FRESH FEED		CU SC CL T		CU RO SC F	
	MEAS	CALC	MEAS	CALC	MEAS	CALC
+ 35	5.30	4.76	0.00	0.00	6.00	4.76
- 35+ 48	5.30	5.07	0.50	0.50	6.00	5.72
- 48+ 65	8.20	6.05	5.60	6.67	8.00	7.49
- 65+100	10.40	5.98	9.00	9.98	11.80	8.84
-100+150	16.60	11.22	13.60	10.85	18.20	14.96
-150+200	18.20	13.89	17.60	15.55	17.00	19.48
-200+270	21.00	14.81	16.60	16.25	20.20	21.89
-270+325	21.00	10.07	17.80	19.65	21.40	14.30
CYCL1	23.40	7.80	22.00	21.19	19.00	12.89
CYCL2	25.40	15.52	22.40	21.34	17.60	21.06
CYCL3	23.00	17.69	17.60	17.14	15.60	22.00
CYCL4	23.80	20.63	16.80	18.07	22.40	27.11
CYCL5	22.00	24.66	20.60	19.62	27.20	30.72
-CYCL5	43.80	32.15	24.20	30.68	56.60	44.91

SIZE INTERVALS	CU RO C		CU SC T		CU SC CL C	
	MEAS	CALC	MEAS	CALC	MEAS	CALC
+ 35	0.00	0.00	4.50	4.76	0.00	0.00
- 35+ 48	36.00	5.36	4.50	4.94	0.00	0.00
- 48+ 65	36.00	11.34	5.20	5.75	57.00	57.39
- 65+100	36.00	24.42	4.70	5.14	57.00	57.12
-100+150	36.70	43.11	11.20	9.23	57.00	55.38
-150+200	57.00	70.47	6.70	8.57	56.00	51.61
-200+270	57.00	57.44	7.30	8.80	60.00	55.69
-270+325	55.00	30.10	6.00	7.74	53.00	52.83
CYCL1	75.00	44.02	2.40	2.83	64.00	61.61
CYCL2	67.00	54.72	8.40	8.91	69.00	66.09
CYCL3	60.00	60.62	10.60	8.98	75.00	69.38
CYCL4	57.00	57.70	10.80	9.95	87.00	85.00
CYCL5	62.00	66.55	11.60	10.54	80.00	72.64

Table IV: Measured and calculated Cu, Pb, Zn, Au and Ag content in circuit feed and products (% or g/t)

Stream	Metal	Measured Content	Estimated Content	Adjustment %
Feed	Cu	1.50	0.75	101
	Pb	0.09	0.07	31
	Zn	5.82	6.15	5.4
	Au (ppm)	0.34	0.35	3.8
	Ag (ppm)	18.60	18.92	1.7
Copper Concentrate	Cu	19.20	19.10	0.5
	Pb	0.75	0.76	0.9
	Zn	2.83	2.81	0.9
	Au	4.26	3.98	7.0
	Ag	203.00	190.86	6.4
Zinc Concentrate	Cu	0.47	0.40	1.6
	Pb	0.05	0.03	90
	Zn	54.00	53.13	1.6
	Au	0.50	0.46	10.0
	Ag	37.50	39.53	5.1
Final Tails	Cu	0.12	0.13	4.8
	Pb	0.04	0.05	18.4
	Zn	0.61	0.59	3.1
	Au	0.24	0.21	11.0
	Ag	8.80	10.29	14.5

In comparison to the raw data, a major advantage of the balanced data is obviously their consistency with respect to the mass conservation law. This can be illustrated by comparing, for instance, the reconstituted silver grades from particle-size intervals grades and the overall silver grades using either the raw or the balanced data. The results appear in figure 8. The balanced data produces points necessarily located on the graph diagonal, unlike the raw data. This is obtained at the expense of large adjustments in some cases: Cu, Pb, As in circuit feed, Te in rougher concentrate, Bi in scavenger tails, Au and Ag in 35/48 class of rougher concentrate, Ag in 48/65 class of rougher concentrate, Ag in 270/325 class in circuit feed, etc...

2.5 Discussion of results

The silver distribution in the circuit will be discussed in conjunction with the metallurgical behaviour of the other metals and minerals. Figure 9 gives the silver distribution assuming 100% in the feed. The major part of silver (47%) is lost in the tailings, while 31% and 22% accompany the copper and zinc concentrates respectively.

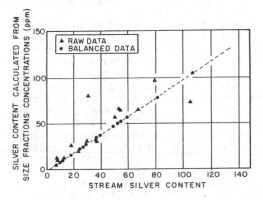

Figure 8: Comparison of stream silver content and reconstructed stream silver content, for raw and balanced data

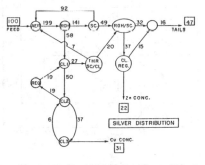

Figure 9: Silver distribution (%) in the flotation circuit

2.5.1 Coupling of silver and minerals distributions

In order to relate the silver distribution to the silver-bearing minerals distribution, it is possible in a first step to evaluate the concentration of the streams with respect to the three identified silver-bearing minerals (galena (G), stützite (S) and tetrahedrite (T)). Assuming that G and T are respectively the only Pb and S bearing phases, their proportion can be calculated in each stream, with the help of the mass balance results. Using the microprobe silver content determined in G S and T (respectively 0,2, 56,5 and 20%) it is possible to determine the amount of stützite in the 19 streams by sharing the silver for these 3 phases. We observe the amount of stützite, to be too small or even negative. However, this calculation correlated the Te content. By progressively assuming lower Ag content in the tetrahedrite, it is possible to obtain reasonable stützite and Te contents in all circuit streams. A 7% to 9% Ag content seems to be more probable for tetrahedrite than the 20% Ag content found by microprobe in the copper concentrate. Figure 10 compares the amount of stützite predicted by this silver balance approach and the Te content as adjusted by the mass balance program. Discarding the results for the copper concentrate cleaner 3, there is a reasonable correlation between predicted (ppm) content of stützite and the balanced Te contents. The experimental points are slightly above the theorical curve corresponding to stoichiometric stützite, suggesting that Te should appear in other secondary phases.

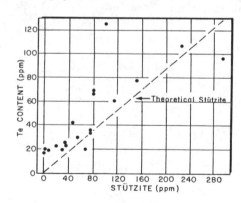

Figure 10: Comparison of streams stützite content as predicted from Ag balance and streams balanced Te contents

A second approach to the determination of the silver-bearing species is to assume that the Ag content in galena measured by the microprobe is correct and to explain the remaining silver by its association to Sb and Te (i.e. to T and S). This can be

performed by a multilinear regression technique of the type: (ppm of Ag uncombined to galena) = a x (ppm Sb) + b x (ppm Te). The regression performed on the 19 streams values leads to coefficients a and b which correspond to a tetrahedrite containing 6% Ag and a stützite containing 40% Ag. The latter figure is quite compatible with the measured 56,5%, considering that Te appears also in other phases than stützite. This approach confirms that the tetrahedrite is poorer, with respect to silver, than observed in the microprobe analysis.

2.5.2 Silver recovery

Table V summarises the results of the two previous sections on the silver behaviour in the flotation circuit. In order to rebuild the stützite amount present, a 7% silver grade in tetrahedrite particles is assumed, despite the fact that a richer tetrahedrite phase is observed in the final copper concentrate. This is acceptable since the richer tetrahedrite observed in the copper concentrate account for less than 3% of the overall tetrahedrite in the ore. Stützite, the richer silver-bearing mineral, is recovered in the two concentrates but accounts only for 31% of the total silver present in the ore. The second richer silver-bearing mineral, tetrahedrite and the silverous galena report to the tails. The latter also reports in part to the copper concentrate (34%), but accounts only for about 10% of the total silver present in the ore.

Table V: Summary of Ag and Ag-bearing minerals distribution in the flotation circuit

Ag and Ag-bearing minerals Distribution (%)	Feed	Cu concentrate	Zn concentrate	Tails
Silver	100	31	22	47
Galena	100	34	4	62
Tetrahedrite	100	5	4	91
Stützite	100	49	37	14
Silver Content (ppm)	19	203	38	11

Proportion of Ag bearing minerals (%)	Feed	Cu concentrate	Zn concentrate	Tails
Galena	85	95	78	85
Tetrahedrite	14	2	8	15
Stützite	1	3	14	0
Silver distribution in minerals (%)				
Galena	10	10	2	14
Tetrahedrite	59	8	7	86
Stützite	31	82	91	0

3. GRANITE INDUSTRY APPLICATION

A hydrocyclone has been installed in a granite cutting plant in Quebec. Before setting this apparatus in the granite circuit, its behavior had to be simulated to verify its efficiency. In order to develop equations to simulate the hydrocyclone, sampling of the pulp was necessary. The latter is a suspension of granite dust and iron chips in water. Data obtained was then balanced by the BILMAT mass balance package (Hodouin, 1980, 1981).

Plant experiments consisted of operating the hydrocyclone in a closed-circuit illustrated in figure 11. Three series of measurements were made. For each series, feed, overflow and underflow pulp samples were taken. Pulp flow rate of the underflow was also measured. Samples were then analysed for solids' content and particle-size distribution.

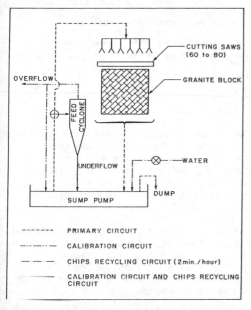

Figure 11: Flowsheet of hydrocyclone circuit

Before executing the mass balance in order to adjust experimental data, the variance of sampling errors was calculated. The results of these calculations have shown that operational errors were more important than fundamental errors of sampling.

Mass balance computations were then executed for each of the three series of measurements. These computations were performed with the BILMAT mass balance package with its release in APL language.

The hydrocyclone circuit was represented as the network consisting of one node and three branches; one input and two output branches (figure 12).

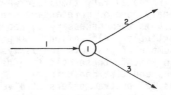

Figure 12: Solids and particle size classes network

Tables VI and VII present an example of the results. The BILMAT program not only adjusted experimental data but also calculated missing data which was not measured (overflow and feed flow rates). Furthermore, we can see that the percentage of solids and particule-size distributions were not appreciably modified by the program.

Calculations were then computed to simulate the behavior of the cyclone.

Table VI: Macroscopic results

STREAM	PULP FLOW RATES		WATER FLOW RATES	
	Obs.	Cal.	Obs.	Cal.
FEED	.00	20.65	.00	8.13
OVERFLOW	.00	19.07	.00	7.96
UNDERFLOW	1.58	1.58	.00	.17

STREAM	SOLID FLOW RATES		MASS FRACTIONS OF SOLIDS	
	Obs.	Cal.	Obs.	Cal.
FEED	.00	12.52	60.70	60.63
OVERFLOW	.00	11.11	58.20	58.26
UNDERFLOW	.00	1.41	89.10	89.11

Table VII: Microscopic results

Particle Size Analysis (%)

Branches Names

	FEED		OVERFLOW		UNDERFLOW	
	MEAS	ESTIM	MEAS	ESTIM	MEAS	ESTIM
- 20+ 28	0.40	0.40	0.00	-0.09	4.27	4.27
- 28+ 35	2.12	2.18	0.43	0.25	17.41	17.33
- 35+ 48	1.99	2.01	0.33	0.26	15.91	15.81
- 48+ 65	1.92	1.85	0.37	0.51	12.23	12.40
- 65+100	2.38	2.30	0.90	0.95	12.78	12.97
-100+150	3.11	2.90	1.77	1.82	10.92	11.43
-150+200	5.37	5.41	4.87	4.86	9.93	9.71
-200+270	4.47	4.33	4.50	4.55	2.53	2.58
-270+400	8.41	8.62	9.57	9.49	2.10	1.78
-400	69.82	70.00	77.27	77.40	11.91	11.70

4. LOTUS 123 APPLICATIONS

At the laboratory scale, some calculations are needed to establish distribution of elements in the various test products. These calculations must, from time to time, be repeated many times per week. Occasionnaly, one or more assays can be wrong. It is then necessary to repeat these particular calculations. Depending on the number of products and assays involved in a test, these calculations can be tedious and time consuming.

Lotus 123 program offers us a very flexible tool for laboratory scale calculations. For simple mass balance, it is generally possible to build a program including some particular conditions (product names and weights, assays, particular products combinations, etc.) in less than one hour or so. From a virgin table we insert product names, weights and assays for the test considered, Lotus 123 makes the mass balance for you.

You have a doubt concerning two analyses! The combined grade calculated and the head show discrepancies! Should you think two assays results are inverted, you change the numbers and Lotus 123 gives you the new mass balance. You then ask your analyst to make new analysis for the two samples to check the first assays.

If you have 5 products and 15 analyses, you can imagine the work needed to calculate this material balance if a few corrections are needed.

We show two examples for cyanidation tests. The first deals with a batch test made in a Winchester bottle (table VIII) and the second for a CIP test (table IX). Both tests dealt with a gold ore.

Table VIII: Batch test with a Winchester bottle: gold distribution and content of products

Project: ....   Date: ....
Sample: ....   Cyanidation Test ..... Hours

Test no. ....

Hours	Assays pp (ppb)	Calc.	Units	Unit of each sample taken — Liquid cc	To add	Cumul.	Total units	Distr. %
1:00	1.90		1900.00	20.00			1900.00	55.27
4:40	2.90	2.94	2900.00	20.00	38.00	38.00	2938.00	85.47
24:00	3.10	3.20	3100.00	20.00	58.00	96.00	3196.00	92.98
28:40	3.00	3.16	3000.00	20.00	62.00	158.00	3158.00	91.87
47:40	3.00	3.22	3000.00	20.00	60.00	218.00	3218.00	93.62
Filtrate	3.00		2670.00	890.00		218.00	2888.00	84.02
1e wash	580.00*		284.20	490.00			284.20	8.27
2e wash	2.00*		16.00	8000.00			16.00	0.47
Residue	498.00*		249.25	500.50			249.25	7.25
							3437.45	100.00

	Weight, g	Assays, ppm	Units	Distr. %
Solid	500.50	0.50		
Soln. (final)	1000.00			

* Note: These analysis are in ppb.

Table IX: CIP test on a gold ore: analyses and metal distribution

Hours	Assays (ppb) Results	Calc.	Units	Liquid soln. Add.	Units Add.	Cumul.	Total	Distr. %
	2.95		158.356				158.36	62.77
	3.30	3.301	177.144	20.0	0.0590	0.0590	177.20	70.24
	3.38	3.382	181.438	20.0	0.0660	0.1250	181.56	71.96
	3.53	3.534	189.490	20.0	0.0676	0.1926	189.68	75.18
	3.55	3.555	190.564	20.0	0.0706	0.2632	190.83	75.64
	3.53	3.536	189.490	20.0	0.0710	0.3342	189.82	75.24
	3.58	3.588	192.174	20.0	0.0706	0.4048	192.58	76.33
	3.50	3.509	187.880	20.0	0.0716	0.4764	188.36	74.66
	3.52	3.530	188.954	20.0	0.0700	0.5464	189.50	75.11
	3.63	3.642	194.858	20.0	0.0704	0.6168	195.48	77.48
	3.14	3.153	168.555	20.0	0.0726	0.6894	169.24	67.08
	3.72	3.734	199.690	20.0	0.0628	0.7522	200.44	79.45
	3.66	3.675	196.469	20.0	0.0744	0.8266	197.30	78.20
	3.78	3.797	202.910	20.0	0.0732	0.8998	203.81	80.78
	3.83	3.848	205.594	20.0	0.0756	0.9754	206.57	81.87
	3.83	3.850	205.594	20.0	0.0766	1.0520	206.65	81.91
	3.78	3.801	202.910	20.0	0.0766	1.1286	204.04	80.87
	3.83	3.852	205.594	20.0	0.0756	1.2042	206.80	81.97
	3.58	3.604	192.174	20.0	0.0766	1.2808	193.46	76.68
	3.73	3.755	200.226	20.0	0.0716	1.3524	201 58	79.90
	4.18	4.207	224.382	20.0	0.0746	1.4270	225.81	89.50
	4.23	4.258	227.066	20.0	0.0836	1.5106	228.58	90.60
	4.02	4.050	215.794	20.0	0.0846	1.5952	217.39	86.16
	4.30	4.331	230.824	20.0	0.0804	1.6756	232.50	92.15

	Weight g	Analysis g/t	Units	Distr.	
Solid	44000.0	0.450	19.80	7.85	Project no:
Soln. (final)	53680.0	4.331	232.50	92.15	Sample: Test no.
Calc. head		5.734	252.30	100.0	Date: Batch no.:

5. CONCLUSION

A very large number of data were available on the studied Cu-Zn flotation circuit. Sometimes, specially for the trace elements, the measured values were inaccurate. The mass balance program BILMAT was found to be a powerful tool to process such a large data set when the information is distributed on various interconnected levels. The process variables (flowrates, chemical contents, particle-size distributions and precious metals grades in size-intervals) were estimated by the program and resulted in a consistent set of data which can be confidently used for subsequent metallurgical calculations such as metals distributions, flotation banks recoveries and minerals contents.

The main silver-bearing minerals in the ore were identified as galena, tetrahedrite and stützite. From a combination of mass balance calculations and microprobe studies it was evaluated that approximately 10% of the silver is in the galena, 60% in the tetrahedrite and 30% in the stützite. Regression studies based on balanced data indicate that the tetrahedrite analysed with the microprobe is a Ag-rich phase, not representative of the average tetrahedrite in the ore which should contain less silver. Galena reports mainly to the tails (60%), tetrahedrite almost completely to the tails (90%), while stützite is equally shared in both concentrates. In order to increase silver recovery in the copper concentrate, in which presently only 30% of the silver is recovered, an attempt should be made to recover more lead in the copper-concentrate. The mass balance indicates a strong coupling between lead-silver metallurgical performances due to the presence of silver in galena and to the textural associations of tetrahedrite and galena observed on the micrographs.

The BILMAT package has also been useful in the context of a material balance application within the granite industry.

Lotus 123 provides a versatile tool to establish a material balance for a laboratory test. It avoids tedious and time consuming calculations. Examples of flotation and cyanidation tests, demonstrate its easy applications.

6. ACKNOWLEDGMENTS

Susan Sawyer-Beaulieu of the Lac Mattagami division of Noranda Inc. supervised the data acquisition process, Bernard Kieller of the Centre de Recherches minérales conducted the micrographic and microprobe analyses and Magella Bilodeau of CANMET, Energy, Mines and Resources Canada, formerly with GRAAIM, Laval University, participated to the BILMAT package application. We also thank Rémi Tremblay of the Centre de Recherches minérales, for technical advise for this article and Daniel Hodouin and Michel Garon respectively from Laval University and Minnova Inc., division Opémiska.

REFERENCES

Hodouin, D. and Everell, M.D. 1980. A hierarchical procedure for adjustment and material balancing of mineral processes data. Int. J. Miner. Process. 7: 91-116.

Hodouin, D., Kasongo, T., Kouamé, E. and Everell, M.D. 1981. BILMAT: an algorithm for material balancing mineral processing circuits. CIM Bulletin: 123-131.

Hodouin, D. and Flament, F. 1985a. BILMAT - Computer program for material balance data adjustment. The SPOC Manual, chap. 3.1, Edit. D. Laguitton, CANMET, SP85-1/3.1 E, Energy, Mines and Resources Canada.

Hodouin, D., Garon, M., Rémillard, M. and Thérien, M. 1986. Assessment of precious metals distribution in Lac Mattagami flotation plant by computer mass balance calculation, Paper 25, CMP, Ottawa.

Hodouin, D. and Vaz Coelho, S. 1985b. Mass balance calculation around mineral processing units using particle size and composition analyses. Int. J. Miner. Process., 33 pages, presently in print.

Gy, P.M. 1979. Sampling of particulate materials: Theory and Practice. Elsevier Scientific Publishing Company, Amsterdam.

Smith, H.W. 1985. Sampling methodology for ore and coal process evaluation and modelling. The SPOC Manual, chap. 2, Edit. D. Laguitton, SP85-1/2E, Energy, Mines and Resources Canada.

Hodouin, D., Gelpe, T. and Everell, M.D. 1982. Sensitivity analysis of material balance calculations. Powder Technology, 32: 139-153.

Hodouin, D., Bazin, A. and Trusiak, A. 1984. Reliability of calculation of mineral process efficiencies and rate parameters from balanced data. Control '84, J.A. Herbst Editor, AIME-SME/TMS, New York: 133-144.

Computer Applications in the Mineral Industry, Fytas, Collins & Singhal (eds)
© 1988 Balkema, Rotterdam. ISBN 90 6191 760 3

Dynamic simulators for the mineral processing industry

D.Hodouin
GRAIIM, Département de Mines et Métallurgie, Université Laval, Quebec, Canada

ABSTRACT: Although steady-state simulation is now recognized as a valuable technique for mineral processing plant design and tuning, dynamic simulation is still in its infancy. However since time-variations of the process variables play an important role in the operating performances of production units, valuable information can be generated by dynamic simulation. The specific structure of dynamic simulators is described and their applications are reviewed. Three examples of dynamic simulators are discussed. The first one deals with responses of flotation circuits to disturbances and to control variable manipulation. In the second one, the analysis of transient regimes in CIP processes is used for the design and optimization of gold adsorption circuits. Finally, grinding circuit simulators are used for filter and controller design.

1 INTRODUCTION

Simulation techniques of mineral processes are now widely used for computer aided design and optimization of production units (Hodouin, 1985). Simulators are computer programs of extremely variable structure and containing process models of very different levels of complexity. Some make use of mass balance equations, along the flow-sheet to be simulated, coupled with some constant coefficients expressing the transformation or separation efficiencies of the various equipment. At the other end, very specific and detailed simulators can be found which describe the complexity of the physico-chemical mechanisms involved in the processes. In their majority, simulators are designed for steady-state operation of the production circuit and, as a consequence, are not able to describe process dynamics and assist in the development of automatic control strategies. The objective of this paper is to discuss the nature of the new information that is brought by dynamic simulators in comparison to the steady-state ones and to illustrate their potential applications.

2 TIME-VARIATIONS OF MINERAL PROCESS VARIABLES

Time variations of the variables of mineral processing units occur for a number of reasons. First, when the process itself is not continuous, but rather a batch or semi-batch operation, the process variables can never reach equilibrium values. This is the case of discontinuous filtration, batch leaching, adsorption on carbon, ion-exchange extraction, carbon or resin elution...

A second fundamental cause of the unsteady behaviour of mineral processing operations is the time-variation of the chemical, mechanical, geometrical, mineralogical characteristics of the treated ores. This type of disturbance is very significant when the orebody heterogeneity is strong or when the concentrator processes ores from different mines.

Additional factors to variations of process variables can be identified in relation to the process itself: disturbances of material transportation systems (bins, feeders, conveyors, pumps...) and of actuators (reagent feeders, valves, motors...), wear of equipment, natural instabilities of the processes (material recycling in pelletizing plants, pump surging and rod or ball mill pulsed discharge in comminution plants...) and finally manual or automatic adjustments of the manipulated process variables.

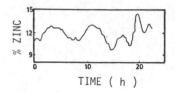

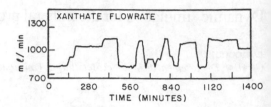

Figure 1: Zinc assay as a function of time in the feed to a flotation circuit.

Figure 2: Xanthate addition rate as a function of time at the feed of a flotation circuit.

Because of all these random or deterministic time events, the performance of a true process can be significantly different from what have been established or expected assuming steady-state operation. Typical examples of time-variations of process variables are now given to illustrate the importance of deviations from equilibrium states in mineral processes plants.

Figure 1 gives the variation of the zinc assay in the feed of a lead-dezincing flotation circuit (Bazin and Hodouin, 1985a) during a 24 hours period. The magnitude of the variation is 40% of the nominal value and can be considered as a very significant disturbance. For the same circuit, Figure 2 shows the variation of one of the manipulated variables, the

xanthate addition rate (Bazin and Hodouin, 1985b), as it is selected by the operators to keep the circuit operation in an acceptable range of performance. Again, variations as high as 50% of the average value are observed.

Figure 3 shows the variations of two disturbances (ore hardness index and particle size corresponding to 80% passing) and one manipulated variable (ore feed rate) as measured during a three-month period at the feed of a grinding circuit. Again, the magnitude of variations of the operating conditions of the circuit is considerable and very rarely allows the grinding circuit to operate under steady-state conditions. For another grinding circuit, Figure 4 gives the histogram of the percentage of

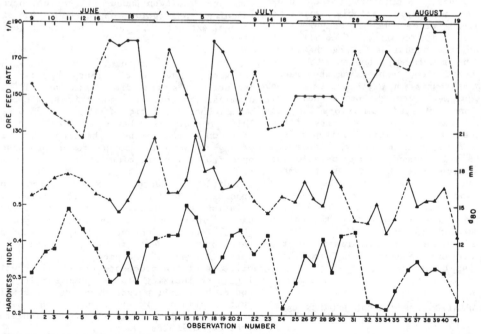

Figure 3: Variation of the ore characteristics of a rod mill feed.

220

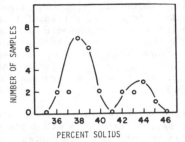

Figure 4: Frequency distribution of the percentage of solids in a hydrocyclone overflow of a grinding circuit.

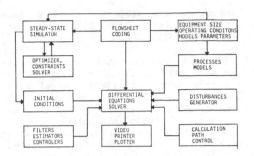

Figure 6: Architecture of a dynamic simulator.

solids in random samples of hydrocyclone overflow, taken during a period of one hour. This result is typical of cyclic behaviors related to pump-sump instabilities. As a result the hydrocyclone is almost never in its average operating condition which theoretically corresponds to 40% of solids in the overflow.

The last example illustrates the variation of the slurry flowrate as measured with an ultrasonic flowmeter at the feed of a CIP circuit during a period of one hour (Hodouin and Carrier, 1987b). Very short but intense disturbances can be noticed in Figure 5.

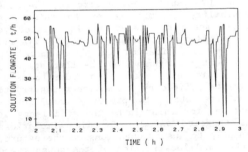

Figure 5: Variation of the slurry feedrate to a CIP circuit.

3 ARCHITECTURE OF DYNAMIC SIMULATORS

A usual structure of a dynamic simulator is depicted in Figure 6. As any other mineral processing simulator, it contains an executive which sequentially performs the computation, following a path in the flowsheet of the circuit, and calls for the various models of the equipment incorporated in the flowsheet. The models which are used are close to the models used in steady-state simulators where mass balance differential equations are generally written before setting the derivatives of the process variables to zero. Also, the computation evolution during the program execution is not very different in dynamic and steady-state simulators since, in both cases, iterative calculations are performed. In the dynamic case the computation increment corresponds to a time step, while in the steady-state case it corresponds to the substitution of the previously simulated values of the recycling streams in the new sequence of computation which is run through the flowsheet. The natural convergence of the sequential modular technique of calculation in steady-state simulation corresponds approximatively to the transient behaviour for going from an initial state to a new equilibrium state, although no time-scale or disturbances are generated.

Furthermore a steady-state simulator is frequently imbedded in the dynamic simulator to find initial conditions corresponding to an equilibrium state of the circuit. Ideally this module has to be connected with an optimizer (or constraint solver) which allows a rapid tuning of the flowsheet for finding a steady state corresponding to suitable production objectives and states of the simulated circuits.

An important module in the computer program is the disturbances generator. It is essential that very flexible time-functions can be plugged to any process variable of the circuit. These functions can be deterministic (impulse, step, pulse train or cyclic functions) or stochastic (pseudo-binary-random-sequences, auto-regressive and moving average processes, randomly accuring deterministic disturbances, white noises...) and should be easy to implement for generating deterministic variations of manipulated variables, disturbances of ore properties and process operating conditions and finally sensor and actuator noises.

It is also important that the structure of the simulator is flexible enough for an easy incorporation of filters, model estimators and controllers, directly in the program or in an external program communicating with the main simulation program. Conventional PID control loops should at least be directly available in the structure by a suitable definition in the coded flowsheet.

Finally, since dynamic simulators generate much more numerical results than steady-state ones, they must be equipped with graphical facilities allowing a fast appraisal of the variation of the process variables.

4 APPLICATIONS OF DYNAMIC SIMULATION

Dynamic simulators are built for handling the variations of the process variables with the objective to apply the generated information to practical purposes. The applications of dynamic simulation can be divided into four categories: characterization of mineral circuit dynamics, design of control strategies, computer-assisted in-plant decision and personnel training.

4.1 Characterization of circuit dynamics

The very first application is to characterize the dynamics of a flowsheet in reference to its operating performances. The following aspects of the dynamic performances of a mineral processing complex unit have to be considered: the circuit efficiency, the operation stability, the process state observability and the circuit controllability.

The efficiency of a complex unit (metal recovery, throughput, operating costs...) depends upon its average behaviour (steady-state level of operation), but also largely upon the transient regimes around it. As an example the transient regimes of a CIP circuit generated by the carbon transfer from tank to tank correspond to a considerable increase of the gold content of the slurry directed to the mill tailings. Another example can be taken from comminution operation. Although the variations of the grindability of an ore, as depicted in Figure 3, generate corrective actions by the operators, it has been evaluated, on

this circuit, that in fact the final product is only a very short period of time at the set point value corresponding to the optimal liberation degree. During the periods of overgrinding, efficiency of grinding is lost, and during the periods of undergrinding, efficiency of subsequent separation is lost.

The stability of an operation (tendency to reject or amplify disturbances) depends upon the structure of the disturbances and upon the structure of the flowsheet. For example it is not warranted that a flotation circuit design, which corresponds to a maximum metal recovery, is the best one for natural damping of the disturbances or the easiest to operate. Simulation can help answering questions such as: Is it better to have a small or a large number of recycling streams? In what situation the unavoidably tuning errors of the operators have the smallest impact on the stability and, consequently, on the efficiency? In a grinding circuit it is generally admitted that multistage classification improves steady-state efficiency. Dynamic simulators can be used to answer other questions as: are the operating instabilities, which could be generated by multistage arrangements, detrimental to the average efficiency of the grinding and separation circuits?

The observability of a circuit, i.e. the possibility to measure or to infer its actual state of operation, is related to its structure and to the sensor availability and localization in the plant. A flowsheet arrangement may be theoretically very efficient, however, in order to maintain its efficiency, it may be necessary to know its state to properly made interventions for maintaining the efficiency at its targeted value. Dynamic simulations can help to answer questions on the nature, number and localization of the sensors and to select the most observable flowsheet structure.

The controllability, i.e. the possibility to master circuit operation by manipulation of control variables, depends upon the observability of the system and on the process variables which have to be manipulated to maintain the circuit in the state where the efficiency is maximum. This is related to the flowsheet structure and to the equipment needed to activate the necessary process variables (reagent feeders, variable speed pumps, automatic values...). Dynamic simulation can help to handle this type of problems of selection of the manipulated variables, a preliminary step to the design of control strategies.

4.2 Design of control strategies

After a flowsheet has been designed with suitable levels of efficiency, stability, observability and controllability, the design of the control strategies can be assisted by dynamic simulation. There are two broad classes of control strategies: the conventional empirical monovariable feedback loops involving PID regulators and the model-based mono- or multivariable control strategies (Hodouin, 1985). A simulator can be first used to test the efficiency of conventional PID loops for disturbance effects rejection. More specifically, it can be used to select the couples of controlled/manipulated variables for each loop, to determine the impact of the loops interaction and finally to tune the gains of the loops for an acceptable level of performance, evaluated by stability, overshoot and time response criteria.

When process models are available, and this should obviously be the case if a dynamic simulator is assumed to be used, it is possible to use this prior knowledge of the process to develop more efficient control strategies than the PID loops. The model can be known from fundamental principles, or determined by experimentation on the process; it can be considered as constant or adapted to the process on a real-time basis (adaptive strategies). The model, whether it is identified off-line or on-line, can be used at two different levels in the control loops: filtering and controlling (Figure 7).

The information coming from the sensors can be used with maximum efficiency to observe the state of the production unit, if the data is made coherent with what is known of the processes, i.e. with their mathematical models. Model-based filters can be designed and tuned with dynamic simulators.

Model-based controllers select the values to be given to the manipulated variables using the filtered information on the state of the process and the performance criterion selected for the circuit operation. The performance index generally contains requirements such as: minimum quadratic variations around set points under the constraints of minimum quadratic variations of the control variables and eventually some additional requirements as maximum throughput or minimum costs. Many ap-

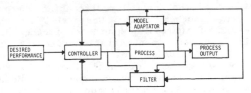

Figure 7: Organization of a model-based control strategy.

proaches to this problem have been developed in the recent years (Foulard et al., 1987). However they are not frequently implemented, and dynamic simulators can be of great help to test their efficiency and robustness.

Finally the model adapter is also an important component in the control loop since it allows tracking of the process model which can vary significantly due to the time-variations in the ore or the process itself. The model estimator is based on an analysis of the differences between the process outputs and the model predicted values. Many strategies have been developed for model identification (Landau and Dugard, 1986) and require a strong assessment by dynamic simulation in order to be properly managed. Finally dynamic simulation is helpful to test simultaneously the behavior of these three modules of a control strategy and evaluate the robustness of the overall automatic control process (Figure 7).

4.3 Computer-aided in-plant decision

A plant dynamic simulator, running in parallel to the actual process, can be helpful to support decisions of operators about the modifications to be brought to the operation of a circuit, when some offset is detected in its behaviour. The simulator is used as a predictor helping to select the best values to be assigned to the manipulated variables. This is in fact some kind of manual control assisted by predictive simulation, which can be applied when automatic control loops are too far from normal operating conditions and inefficient to correct the observed deviations from set-points.

For instance, simulation in parallel to the process can give answer to questions such as: what will happen immediately and in long term if the water addition to a grinding circuit is decreased or the number of parallel hydrocyclones increased? How much time should the operator wait for the effect of the applied modification, before trying another one? What will be the result of increasing xanthate, frother and air feed rate in a flotation circuit? Is there a solution which gives a better response to the circuit? What is the amount of xanthate which should be added? and where? Since no elution tank is available to extract the carbon of the first tank of a CIP circuit, what will be the effect of accumulating carbon in the first tank? For how long is it reasonable to do that? Is it necessary to add more fresh carbon in the last tank to compensate for the efficiency decrease? etc. Many other examples of the usefulness of in-plant simulators could be given, assuming obviously that they have been carefully calibrated on recent data acquired from the circuit.

4.4 Personnel training

In the same way as flight simulators are used to train aircraft pilots, mineral processes simulators could be used to train students, engineers and operators. The intuitive knowledge of the plant dynamics is frequently incorrect, and as a consequence the attitude of operators can be misadapted to industrial problems. Manipulation of fictitious situations on simulators should help the personnel to be aware of the time response magnitudes of the processes as well as of the direction and magnitude of the variable variations, during transient regimes and equilibrium states. Dynamic behaviors with overshoot, inverse response, time delays are generally not well understood by operators, who certainly would have to learn from simulated dynamic situations.

5 EXAMPLES OF DYNAMIC SIMULATION: 1. FLOTATION

Modeling the flotation mechanisms is a very complex task because of the nature of the superficial phenomena which are involved in the formation of the bubble-particle aggregates and of the three-phase froth. The first difficulty arises in the description of the properties of the particles (mineralogical composition of the surface and the volume of the particles), of the adsorption of the reagents on the surfaces and of the hydrophobicity of the resulting particles. A second difficulty is in the description of the interaction of the bubbles and the particles in relation to the bubble sizes, particle properties and hydrodynamics of the material in the cell. A third diffi-

culty is the description of the froth structure, of its exchanges with the pulp (entrainment and drainage) and of the mechanism of overflowing into the concentrate launder.

The simplest models consider the flotation process as a perfectly mixed first order reaction which describes the apparition of the valuable mineral in the concentrate launder (Flament et al., 1985). The water which is directed to the concentrate, also "floated" using a first order kinetic mechanism, is responsible for gangue entrainment into the final concentrate (Lynch et al., 1981). At the other end of the models spectrum, attachment and detachment phenomena can be taken into account in the pulp as well as in the froth (Bascur and Herbst, 1982). Intermediate approaches make use of distributed kinetic constants to take into account size factor, surface factors and liberation degree (King, 1975).

The simplest model allows a qualitative description of the process dynamics with respect to feed concentrations and flowrate variations. However the utilization of much more complex models is required to describe the dynamics with respect to the manipulated variables such that the reagents feed rates (collector, frother, air...) and to the disturbances of the volume and surface properties of the particles. However these models are, for some aspects, very speculative since it is extremely difficult or impossible to measure the parameter values.

5.1 Dynamic simulation of a flotation cell

In this first example a prototype of a dynamic model of a unique cell has been built for testing model structures, evaluation of dynamic responses and selection of an adequate model for incorporation in a dynamic simulator of a whole flotation plant. The population of particles is classified with respect to size and mineral composition, allowing the definition of mixed particles. The cell is divided into two zones: the pulp and the froth. The attachment reaction which occurs in the pulp for each particle class i is considered of first order with respect to concentration of free particles and to free bubble surface. Water is entrained by the bubbles-particles aggregates. The entrainment of the free particles with the water is a function of the particle mass. The reaction which occurs in the froth is mainly a washing kinetic process, related to water drainage which selectively drains particles from the froth to the pulp. The concentrate which overflows is assumed to have the average properties of the froth, an assumption which is obviously not true but efficient, taking into account that very little is known on this flotation aspect. The model in the dynamic simulator is the assembly of the mass and volume balance for the various elements: free and attached particles, water and air.

Typical simulated dynamic responses of the cell to disturbances of the input variables are given in Figures 8 and 9 for a 10% step variation of the air flowrate. The observed responses are the consequence of the modelling hypotheses and of the numerical values selected for the model parameters of a fictitious ore and have not been experimentally verified. The objec-

tive is here to show the interest to have a simulator to predict the flotation process dynamic behaviour, which is not obvious to predict by intuitive reasonings.

5.2 Dynamic simulation of a flotation plant

Although the type of model presented above can be extented to a complete flotation plant, its calibration is very difficult. For representative industrial dynamic results we have used a simpler cell model using first order kinetic approximation of the whole cell (pulp plus froth) and perfect mixing behaviour with backmixing from cell to cell within a bank of cells.

The simulated circuit (Figure 10) is the lead-dezincing unit of a Zn-Pb-Cu flotation plant (Bazin and Hodouin 1985a). The

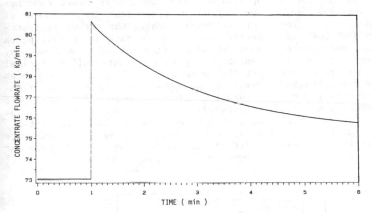

Figure 8: Dynamic response of the concentrate flowrate to a step change in the air flowrate.

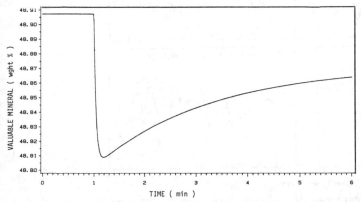

Figure 9: Dynamic response of the valuable mineral content in the concentrate to a step change in the air flowrate.

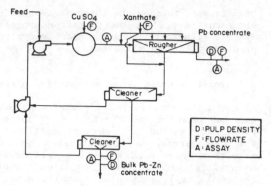

Figure 10: Flowsheet of a lead-dezincing flotation circuit.

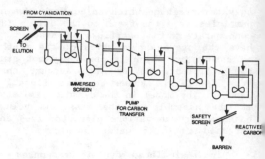

Figure 12: A CIP process flowsheet.

variation of the kinetic constants for the flotation of the various minerals (sphalerite, galena, chalcopyrite, pyrite) with respect to xanthate flowrate has been determined using steady-state data. The simulator responses observed for the circuit concentrate (bulk Pb-Zn concentrate) are compared to the measurements in the plant for the same variations of the solids and water feed rate, the feed concentration and the xanthate flowrate. Results appearing in Figure 11 show that some trends are well represented by the simulator, which is, however, unable to take into account all the disturbances occuring in an industrial unit.

6 DYNAMIC SIMULATION EXAMPLE: 2. CIP PROCESS

The carbon-in-pulp (CIP) process for gold extraction is a pulsed counter-current operation where the slurry flows continuously from tank to tank in a series of perfect mixers, while the carbon is transferred from tank to tank in a semi-batch mode at time intervals of about one day length. A dynamic simulator of the typical circuit depicted in Figure 12 has been developed using a first order adsorption-desorption kinetic model for the gold transfer from the cyanided solution to the activated carbon (Menne 1982, Nicol et al. 1984, Hodouin et al., 1987a). Tanks are modelled as perfect mixers of constant volumes and the pumps for carbon transfer

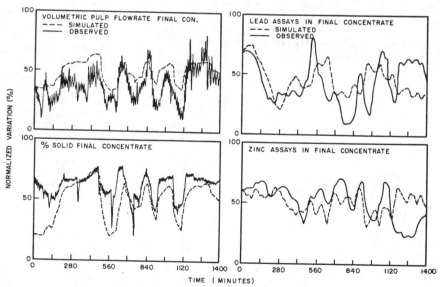

Figure 11: Compared dynamic responses of the simulator and the flotation circuit of Figure 10.

226

are assumed to deliver constant volumetric flowrates.

The dynamic simulator was calibrated using plant data. Figure 13 shows a comparison of the simulated and observed solution gold contents, when the slurry feed to the first tank is suddenly bypassed to the second tank (Hodouin and Carrier, 1987b). Assuming that the modelling equations are sufficiently accurate for describing the whole circuit behavior, as confirmed by other dynamic experiments, the simulator can be used to predict the recovery of CIP circuits for different design or different operating strategies. The gold recovery is directly related to the gold content in the solution leaving the last tank. The simulator can predict the evolution of this process variable as shown in Figure 14. During the carbon transfer period which lasts almost half of the time, over a cycle of 24 hours, the gold in

solution dramatically increases and leads to gold losses.

For a correct conception of a CIP circuit, it is essential to take into account the intrinsic unsteady behaviour of the process. For instance, the dynamic simulator has been applied to the determination of the gold recovery as a function of the mass of carbon in the tanks (Figure 15) and the size of the tanks (Figure 16) and to the tuning of a given operation with respect to the carbon transfer strategy (Carrier and Hodouin 1987).

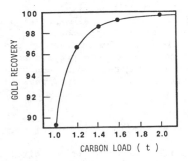

Figure 15: Simulated gold recovery as a function the carbon mass in the tanks.

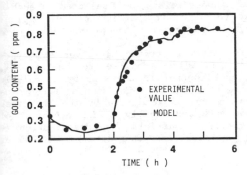

Figure 13: Simulated and observed gold contents in the second tank when the feed is by-passed to the second tank.

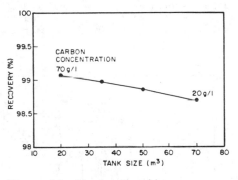

Figure 16: Simulated gold recovery as a function of the tanks size.

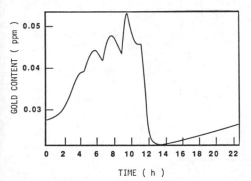

Figure 14: Simulated variation of the gold content in the last tank.

7 DYNAMIC SIMULATION EXAMPLE: 3. GRINDING

The grinding dynamic simulator presented here is based on a phenomenological description of the comminution process, coupled with stochastic elements for the simulation of disturbances and measurement noises.

227

The grinding mill model is based on chemical reaction engineering concepts such as perfect mixers in series, interacting tanks with residual volume for transport dynamics and a first order grinding reaction in each mixer characterized by a rate factor $S_j{}^k$ and a fragment distribution $b_{ij}{}^k$ for each size class j and grindability class k. Other elements are described by simpler models such as single mixers for the sumps, pure delays for the pipes and zero order classifications models based on semi-empirical concepts for the hydrocyclones. The detailed mathematical formulation of these models is given by Dubé et al. (1987).

Stochastic functions such as random input variables, actuators noises, process noises and sensor noises are automatically programmed by a noise generator preprocessor. Random variables are generated by empirical ARMA equations driven by centered gaussian white noises.

7.1 Simulation of a pilot grinding circuit

Test runs performed in a pilot circuit composed of a rod mill (90x180cm), a ball mill (90x180cm) and a hydrocyclone were simulated using the above described computer program. The models of the three pieces of equipment were calibrated using standard techniques (Gupta and Hodouin, 1985). Then the disturbances of the feed variables, as measured during the test run,

were applied to the simulator. Result appear in Figure 17 showing the comparison of the simulated and observed percentage of solids of the circuit product (hydrocyclone overflow). The discrepancy between the two time-variations arise from the unknown disturbances (fresh ore grindability and size distribution), the modelling approximations (no interaction between grinding kinetics and slurry density, no specific gravity effects in the hydrocyclones) and the measurement noises.

7.2 Simulation of control strategies for an industrial grinding circuit

The two-stage grinding circuit of Figure 18 was simulated using industrial data for the calibration of the models of the two mills, the hydrocyclones and the sump (Hodouin et al. 1980). The rod mill is simulated as a series of ball mills. The variances of the white sequences, which drive the ARMA generators of the disturbances and simulate the sensors noises, are selected in such a way to be representative of the typical signals monitored in the plant.

The control strategy involves a proportioner to adjust the water to the rod mill, a level control in the sump by a variable speed pump, a PID regulation of the hydrocyclone solids feed rate by the ore fresh feed rate and finally a PID regulation of the product particle size distribution (% of particles finer than 37 μm) by the water

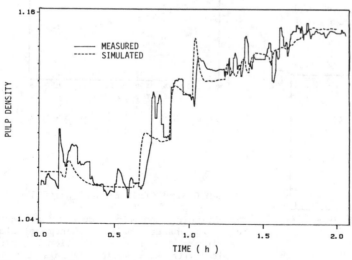

Figure 17: Observed and simulated pulp density of a pilot grinding circuit product.

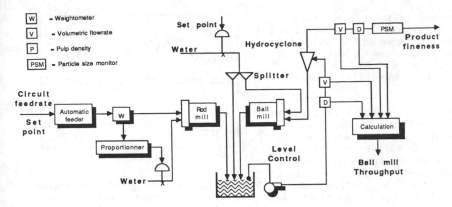

Figure 18: Two-stage grinding circuit.

to the ball mill circuit. The PID regulator gains were tuned directly on the simulator and Figure 19 shows a typical variation of the two controlled variables, when the circuit is submitted to random variations of the fresh ore size distribution and to a simultaneous step decrease of the grindability (Hodouin et al., 1987c). The measured values of the two controlled variables are very noisy, as can be seen in Figure 19 and are preconditioned by autoregressive first order filters before being fed to the PID algorithms. This study shows that this type of control strategy is efficient in a highly disturbed operation,

when variations of the percentage of particles finer than 37 μm in the range 60.5 $\pm$ 1.5% are acceptable for the plant.

7.3 Design of model-based grinding circuit filters

To improve the information given by the sensors, model-based filters have been designed with the help of the grinding circuit simulator described in the two previous examples. As for model-based controllers, the model used in the filter can be phenomenological or empirical

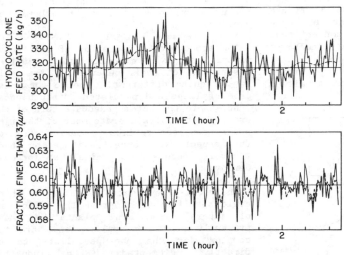

Figure 19: Variations of the controlled variables (hydrocyclone feed rate, product fineness) as simulated for the circuit of Figure 18.

(transfer functions calibrated on experimental data). The two approaches were considered, the first one for a laboratory grinding circuit (Dubé and Hodouin, 1986) and the second one for the previously described two-stage industrial grinding circuit.

The phenomenological approach requires a complex rewriting of the models to make them compatible with usual filtering structures as the Kalman filter. However it gives a description of the state of the circuit, in reference to physico-chemical concepts and allows the estimation or prediction of multi-size particle distributions. The empirical approach is more directly applicable to filtering because of the simple linear equations used for describing the overall process, but requires a more elaborated model identification step.

Both approaches need some model parameter adaptation procedure to properly track the effects of major disturbances. In the phenomenological approach a grindability index accounting for the comminution efficiency variation has to be updated at each time-increment. In the empirical case, the process variable mean values must be continuously adapted.

Figures 20 and 21 show an example of the simulated results obtained at the design stage of a filter based on an empirical model (Lanthier and Hodouin 1987). The two filtered variables are the mill throughput and the percentage of particles finer than 74 μm delivered by the circuit of Figure 18. It should be stressed that this type of filter is better than the conventional one, because the filtered values contain, in addition to the measurement values, corrections coming from the predicted output variables of the circuit submitted

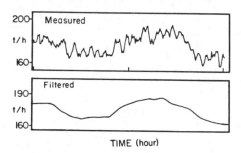

Figure 20: Measured and filtered values of the mill throughput.

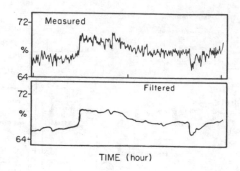

Figure 21: Measured and filtered values of the product size characteristic.

to known values of the input variables. The utility of dynamic simulator is specially important to determine the correct weighting, in the filter, of the sensors and the model informations.

8 CONCLUSION

The paper gives some information on the usefulness of dynamic simulators of mineral processing plants. From the examples given the following conclusions can be drawn:

- For not truly continuous processes, as the CIP gold extraction process, it is essential, even at the design stage, to consider the time-variations of the variables during operating cycles. The methods based on steady-state assumptions predict erroneous recovery values and should be avoided.

- The steady-state operation of continuous mineral processing circuits is an idealized description of their behaviour. The reality is quite different. Due to the presence of large and persistent disturbances, inherent to the heterogeneous nature of the ores, the circuits operate most of the time under transient regimes. Consideration of this aspect using dynamic simulation leads to more accurate appraisal of process performances.

- The model-based control strategies, which could significantly improve the efficiency of mineral processing plants but have presently very few applications, are too complex for the non-specialists to be directly implemented in the industry. Dynamic simulation would be of great help in the understanding and designing of these tools.

- Finally dynamic simulators can be used to train plant personnel for a better understanding of the process dynamics and of the consequences of the manipulation of the control variables. When simulators can be run in-plant, they can help immediately the operators to take the right decision about the settings of the manipulated variables.

Although the use of dynamic simulators is more specific than the use of steady-state ones, it is expected that in the future they will follow the same trajectory and will become more and more present in the universities, the research centers and the plants.

ACKNOWLEDGEMENTS

Results presented in this paper issue from various research projects. The author would like to acknowledge the contribution of C. Bazin (Brunswick Mining and Smelting), F. Flament and J. Paradis (GRAIIM) for the flotation simulation, the contribution of C. Carrier (Cambior, Mine Yvan Vézina) for the CIP process simulation, and the contribution of Y. Dubé (Université du Québec à Trois-Rivières), R. Lanthier (GRAIIM) and J.C. Guillaneau (BRGM, France) for the grinding simulation. Financial support for these projects was obtained from CRM (Energie et Ressources Québec), CANMET (Energie, Mines et Ressources Canada) and the Ministère de l'Enseignement supérieur et de la Science du Québec.

REFERENCES

Bascur, O.A., Herbst, J.A., 1982. Dynamic Modelling of a flotation Cell with a View Toward Automatic Control. XIVth Int. Mineral Process. Cong., Toronto: Can. Inst. Min. Met.

Bazin, C. and Hodouin, D. 1985a. Analysis, modelling and simulation of the lead dezincing circuit of a complex sulfides flotation unit. Proceedings of Inter. Symp. Complex sulfides. San Diego: Soc. Min. Eng. of AIME.

Bazin, C. and Hodouin, D. 1985b. Modelling of a Pb-Zn flotation circuit with a view toward automatic control. Proceedings of the 23th Can. Miner. Processors Meeting, Ottawa: Energy, Mines and Resources.

Carrier, C., Hodouin, D. and Courchesne, M. 1987. Dynamic simulation of the CIP gold recovery process. Proceedings of the Int. Symp. on Gold Metallurgy. Met. Soc. of the Can. Inst. of Min. and Met. Pergamon Press: p. 309-326.

Dubé, Y. and Hodouin, D. 1986. Design of an adaptive filter for a laboratory grinding circuit. Proceedings of the 5th IFAC Symposium on Automation in Mining, Mineral and Metal Processing. Pergamon Press.

Dubé, Y., Lanthier, R. and Hodouin, D., 1987. Computer-aided dynamic analysis and control design of for grinding circuits. CIM Bulletin, October.

Flament, F., Laguitton, D., Ter Heijden, E., Chassat, P., 1985. Plant Simulators for Mineral Dressing. The SPOC Manual, chapt. 6, Ottawa: CANMET, Energy, Mines and Resources Canada.

Foulard, C., Gentil, S. and Sandraz, J.P. 1987. Commande et régulation par calculateur numérique. Paris: Eyrolles.

Gupta, V.K., Hodouin, D. and Spring, R. 1985. FINDBS-program for breakage and selection functions determination in the kinetic model of ball mills. The SPOC Manual, chap. 7.2. Ottawa: CANMET, Energy, Mines and Resources, Canada.

Hodouin, D., McMullen, J. and Everell, M.D. 1980. Mathematical simulation of the operation of a three-stage grinding circuit for a fine-grained Zn/Pb/Cu ore. Proceedings of Int. Symp. Particle Technology. Amsterdam: Fédération européenne de génie chimique.

Hodouin, D. 1985. L'ordinateur pour la conception et l'automatisation des opérations minéralurgiques. Comptes-rendus du Colloque OPTIMINE, Rouyn-Noranda, Québec, Mars 1985, Editions du Gouvernement du Québec.

Hodouin, D., Carrier, C., Barbery, G. and Courchesne, M. 1987a. Evaluation of the performance of a CIP gold recovery process. Proceedings of the 25th Canadian Mineral Processors Meeting, Ottawa: Energy, Mines and Ressources: p. 644-677.

Hodouin, D., Carrier, C. 1987b. Influence des régimes transitoires sur les performances d'un circuit d'extraction de l'or par le procédé charbon-en-pulpe. Comptes rendus du Colloque "De l'or et des profits", Rouyn-Noranda, Novembre 1987, Editions du Gouvernement du Québec.

Hodouin, D., Dubé, Y. and Lanthier, R. 1987c. Stochastic simulation of filtering and control strategies for grinding circuits. Int. J. Mineral processing. In press.

King, R.P. 1975. Simulation of flotation plants. Soc. Min. Eng. AIME Transactions, V. 258: p. 286.

Landau, I.D. and Dugard, L. 1986. Commande adaptative, aspects pratiques et théoriques. Paris: Masson.

Lanthier, R. and Hodouin, D. 1987. Empirical models for multivariate filtering of

closed grinding circuits real-time data.
Proceedings of Int. Symp. The mathematical modeling of metals processing operations, TMS/SME of AIME.

Lynch, A.J., Johnson, N.W., Manlapig, E.V.,
Thorne, C.G., 1981. Mineral and Coal
Flotation Circuits: Their Simulation and
Control. Amsterdam: Elsevier Sc.Pub.Co.

Menne, D. 1982. Predicting and assessing
carbon-in-pulp circuit performance.
Proceedings of the XIVth Int. Mineral
Processing Conf. Toronto: Canadian Inst.
Min. Met.

Nicol, M.J., Fleming, C.A. and Cromberge,
G. 1984. The adsorption of gold cyanide
onto activated carbon. II. Application of
the kinetic model to multistage circuits.
J. of S. Afr. Inst. Min. Metc., V. 84: p.
70-78.

Computer Applications in the Mineral Industry, Fytas, Collins & Singhal (eds)
© *1988 Balkema, Rotterdam. ISBN 90 6191 760 3*

Column flotation simulation

R.del Villar, J.A.Finch, J.B.Yianatos & A.R.Laplante
Department of Mining and Metallurgical Engineering, McGill University, Montreal, Quebec, Canada

ABSTRACT : A flotation column circuit simulator written in BASIC (IBM-PC) is presented. It comprises a scale-up model of a flotation column and a networking routine. The circuit is represented by arrangements of three types of ideal nodes: junctions, splitters and flotation columns. The experimental parameters required are the collection rate constants and the solids carrying capacity of the system. The use of the simulator for preliminary design of column flotation circuits is illustrated through a case study. Emphasis is on the impact of recycle and bias water rate on circuit capacity.

1.- INTRODUCTION

Flotation columns are receiving considerable attention as alternatives to conventional mechanical flotation machines, particularly for cleaning applications. Most testing to assess the metallurgical potential for columns is conducted in pilot-scale units, typically 5 cm in diameter and up to 10 m high. An objective of our work has been to obtain data for scale-up from these same pilot-scale units.

2.- THE SCALE-UP MODEL

The general approach is to model recovery separately in the two zones of the flotation column, the collection (or recovery) zone and the cleaning (or froth) zone, and then to combine into an overall column model. At present the more complete model is that of the collection zone.

2.1.- Collection zone - recovery model

Assuming flotation is a first order process, recovery is estimated from the flotation rate constant, k, particle retention time, T_p, and mixing characteristics (i.e. the vessel dispersion number, N_p) by (Levenspiel 1964) :

$$R = 1 - \frac{4a \exp(0.5/N_p)}{(1+a) \exp(0.5a/N_p) - (1-a)\exp(-0.5a/N_p)}$$

where $a^2 = (1 + 4kT_pN_p)$ (2)

Rate constant: This value must be experimentally determined. It is obtained from pilot-scale tests by determining R as function of T_p. Since a typical pilot unit, 5 cm x 10 m approaches plug flow, a plot of $\ln(1-R)$ vs T_p yields a slope equal to k.

Complications which may arise are the need to divide into fast and slow-floating fractions (Lynch et al. 1981, Dobby and Finch 1986), or the need to dilute the feed pulp to avoid overloading the column (Espinosa-Gomez 1987).

The technique of estimating k involves a compromise because the froth zone is present. Until satisfactory methods of decoupling the collection and froth zones are developed this compromise will remain.

Particle retention time: For particles less than about 100 μm, retention time in a countercurrent bubble swarm can be estimated by (Dobby et Finch 1985, Yianatos, Finch and Laplante 1986) :

$$\frac{T_p}{T_L} = \frac{U_L}{U_L + U_p}$$ (3)

The liquid residence time, T_L is estimated as (Yianatos, Espinosa and Finch 1986) :

$$\tau_L = \frac{A \, H \, (\, 1 - \epsilon_g)}{10 \, Q_t} \tag{4}$$

The relative particle/liquid velocity U_p is estimated from the general equation proposed by Masliyah (1979):

$$U_p = \frac{g \, d_p^2 \, (\, 1 - \epsilon_g)^{2.7} \, (\, \rho_p - \rho_{sp})}{18 \, \mu_L \, (1+0.15Re^{0.687})} \tag{5}$$

Particle mixing characteristics: The mixing characteristics are described in terms of a particle vessel dispersion number N_p :

$$N_p = \frac{D_p}{(\, U_L + U_p \,) \, H} \tag{6}$$

The dispersion coefficient D_p has been related empirically to column diameter d_c by Dobby and Finch (1985):

$$D_p = 0.063 \, d_c \tag{7}$$

Refinements to eq (7), for example, to include the effect of gas rate, are continuing (Laplante, Yianatos and Finch 1988).

2.2.- Cleaning zone - carrying capacity

Whilst a detailed model of the froth zone is lacking it is appreciated that a column does have a maximum concentrate solids rate (carrying capacity) and this is probably dictated by the froth (Espinosa-Gomez et al. 1988).

The carrying capacity should be determined experimentally. This involves measuring concentrate solids rate as a function of increasing feed solids rate till a maximum is reached. The carrying capacity parameter used in the simulator is mass of solids per unit time per unit of column cross-sectional area, C_a (e.g. g $min^{-1}cm^{-2}$). The simulator recovers particles according to eq.(1) till the carrying capacity is reached. Some authors (King 1972, Sutherland 1977) have approached the bubble loading problem differently.

From extensive pilot and full-scale results, Espinosa-Gomez et al. (1987), suggest

$$C_a = 0.068 \, (d_{80} \ast \rho_p) \tag{8}$$

The effect of column diameter on C_a appears to be minor (Espinosa-Gomez et al.

1987). This model should be used only if experimental carrying capacity is not available. The program gives the option of selecting the model or inputting experimental values

3.- THE SIMULATOR

The simulator consists of a circuit input section, a data input section, an executive program and a final results printout section. The flowchart is presented in Figure 1.

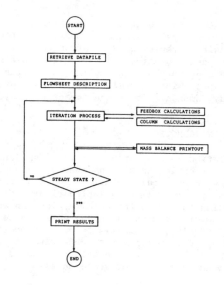

Figure 1. - Simulator Flowchart

3.1.- Circuit input

The circuit flowsheet is retrieved from a previously created file. No modifications are allowed at this stage. The circuit configuration (flowsheet) is coded using a network of nodes and streams (Del Villar et al. 1986). Three types of ideal nodes are used, junctions, splitters and flotation columns. After setting the general characteristics of the circuit (number of nodes, streams, etc), information about the different nodes must be entered as prompted by the computer. Input data can be modified at the end of each node description. This information is then saved on a disk as a sequential file. The program allows for multiple feeds and final products, several stages of columns, recycles to a previous stage, etc.

Column stages consist of two units: a feed box (junction) and a flotation column. The streams associated with these two units are:

Feed box,

    entering : – feed, either fresh or product from a previous stage
            – water, used when streams are recycled, to simulate launder water additions
            – recycled streams from downstream column stages

    leaving   : – feed to the stage column

Flotation column,

    entering : – feed from a box
            – wash water

    leaving   : – concentrate
            – tailings

Splitters may be used to feed parallel column stages, whereas extra junctions could be used to merge products from different stages into an overall circuit product.

3.2.– Data input

Data are retrieved from a previously created datafile and are displayed on the screen for possible modification.

The information is organised in six sections:

    a) fresh feed characteristics
    b) mineral characteristics
    c) column design characteristics
    d) operating variables
    e) carrying capacity parameters

3.3.– Executive program

Once all data have been inputted, the program moves to the calculation section. The sequential modular method is used – i.e. material is routed through the circuit using unit simulators (junctions, splitters and flotation columns).

Figure 2 shows the flowchart of the column flotation subroutine. Concentrate mineral flowrates are calculated and mineral recoveries and grades for each stream are then calculated.

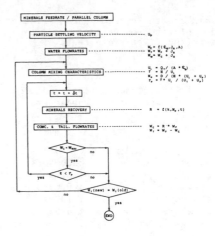

Figure 2.– Flowchart of column flotation subroutine

Water recovery to the concentrate is estimated from gas flowrate assuming a liquid fraction at the lip level (Yianatos, Espinosa-Gomez and Finch 1986); tailings water content is calculated from the feed water and the bias rate. Wash water addition is then determined by balance. The method of estimating water in the overflow will be modified as more experimental data become available.

3.4.– Final results

During program execution, mass balance results for the different species and mixing characteristics of the columns are displayed on the screen to monitor the progress of the simulation. Once the convergence criteria are satisfied (i.e. steady state is achieved), the printout subroutine is called. First, all the input data (as in the datafile or after modification by the user) are presented. Then, simulation results are printed. Four types of results are provided: mixing parameters for all columns, and recoveries, flowrates and grades for all stages, species and streams.

4.– A CASE STUDY

The simulator was used for the preliminary design of a column flotation circuit to clean a copper concentrate. Table 1 summarizes the stream characteristics and some of the operating conditions used during the pilot-scale tests.

lumn baffled into equivalent number of base units. The 1 m dia. base unit was selected since the performance of this size is well documented.

Table 1. Pilot-scale test conditions and experimental parameters

Pilot-scale test conditions

Feed copper grades	: 8-9.5 %
Feed pyrite grade	: 30-40 %
Superficial gas rate	: 0.25-3.6 cm/s
Superficial bias rate	: 0.05-0.6 cm/s
Froth depth	: 100-130 cm
Nominal retention time	: 6-26 min
Tailings flowrate	: 600-2800 cm³/min

Experimental parameters

Carrying capacity	: 1.97 g/min/cm²
Rate constants,	
chalcopyrite	: 0.080 min⁻¹
pyrite	: 0.010 min⁻¹
gangue	: 0.002 min⁻¹

Model parameters required by the simulator were determined using a pilot-scale flotation column (internal diameter 5.08 cm, height 1050 cm). The flotation rate constants were determined by the standard procedure using a diluted feed pulp (10% solids) to avoid overloading.

5.1.- Simulator validation

Simulator predictions for various pilot-scale test conditions were compared to actual test results. Good agreement was obtained for both recoveries (Figure 3) and grades. Similar satisfactory agreement was reached for other parameters, e.g. retention time and pulp densities.

5.2.- Use of the simulator for preliminary design

Column size, circuit configuration and the impact of some operating variables were explored.

Column size: As column size (diameter) influences retention time and mixing characteristics, the adopted approach was to take a 1 m diameter * 10 m high collection zone column as a base unit and vary the size of a column stage by varying the number of base units in parallel. The physical reality is a large diameter co-

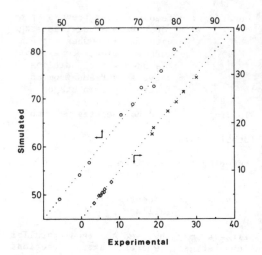

Figure 3. - Simulator validation: mineral recovery; O: chalcopyrite; X : pyrite; ◇: gangue

Figure 4 shows the effect of the number of parallel base units on metallurgical performance. Three column designs were simulated: 4, 6 and 8 parallel base units. The grade-recovery curve is generated by selecting different number of equivalent stages, up to 4, as per the circuit insert.

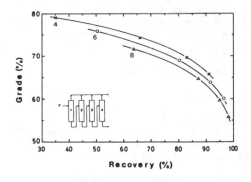

Figure 4. Effect of number of parallel base units (column size) on metallurgical performances; column diameter: X=4; O=6, Δ=8

The circuit with 8 parallel base units gives the poorest grade-recovery results because the decreased flow through each base unit results in a higher vessel dispersion number. The 6-parallel base unit column was chosen as the standard column stage.

Open vs Recycle: The effect of recycling the concentrate from the third and fourth stages of the series circuit is illustrated in Figure 5. Recycling decreases retention time in all stages ahead of the recycle source lowering recovery but increasing grade.

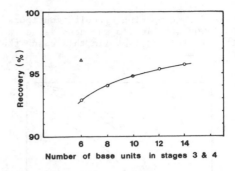

Figure 6A.- Effect of number of units on recycled circuit performance: recycle conc. 4 to feed box 3; △: open circuit

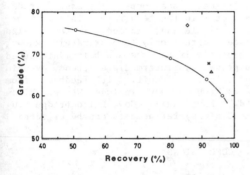

Figure 5.- Effect of recycling on circuit performance; O: open circuit (4 stages in series), ◇: recycling 3rd,4th stage conc. to first stage, X: recycling 4th stage conc. to first stage, △: recycling 4th stage conc. to 3rd stage

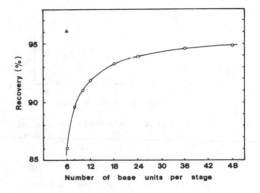

Figure 6B.- Effect of number of units on recycled circuit performance; recycle conc. 3 & 4 to feed box 1; △: open circuit

A greater selectivity is achieved upon recycling because a reduced τ has more impact on the species with lower rate constants. Also circulating loads mean increased flowrate through the base units and thus decreased vessel dispersion number. To regain the recovery level of the open circuit, larger columns are required. This is explored in Figures 6A and 6B. In Figure 6A, the 4th stage concentrate is recycled to the 3rd column. By increasing the number of base units in stages 3 and 4 to about 14, the original (open) circuit recovery is retrieved. If both 3rd and 4th stage concentrates are recycled to the 1st column, then all stages must be increased to 48 base units. This is a formidable increase in stage size. Clearly circuits with recycle must be designed very carefully.

Effect of bias rate: Figure 7 shows the impact of bias water on recovery. Assuming a target of 94% recovery, bias rates less than 0.1 cm/s are required to reach the target in 4 stages. To maximize recovery (or capacity), bias rates should be kept at a minimum value compatible with cleaning action. There is evidence that $0 < J_B < 0.05$ cm/s is adequate (Espinosa-Gomez 1987) yet lack of a direct measure of J_B makes operating with such low bias rates difficult.

6.- DISCUSSION

Circuit simulators are simply mass balancing routines with rules for splitting minerals at the various stages. There is no denying the facility the simulator

237

offers for comparing circuit configurations. The simulator's weakness is in the splitting rules, i.e. in this case the flotation model.

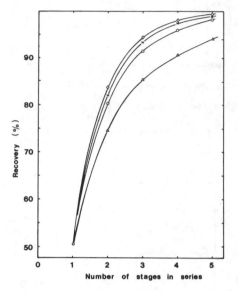

Figure 7.- Effect of bias rate on circuit recovery; $\triangle$: J_B = 0.30, O : J_B = 0.10, X : J_B = 0.05, $\diamond$: J_B = 0.01

The simulator uses a single flotation rate constant to describe the behaviour of each species. This is generally inadequate, as variables such as particle size and grade afects the rate constant. A distributed flotation rate constant is normally used. A number of distributions have been proposed (Dowling et al. 1984). Some could easily be integrated into the model, such as a discrete distribution of non-floating, slow and fast floating components, which has been found particularly effective (Bourassa and Barbery 1986).

Another limitation of the simulator is the assumption that the flotation rate constant and froth recovery can be directly scaled-up. This is not the case for mechanical flotation, where scale up of the froth phase kinetics and froth removal rate is far from completely understood (Kaya and Laplante 1986). In the column, preliminary indication are that froth removal is not affected by column diameter. To preserve the rate constant when scaling up the column, a necessary condition is a similar bubble size at similar gas rates. Frother dosage is the principal governing factor but a notable secondary factor is the relative column cross-sectional area to sparger surface area (Xu 1987).

Specific to column flotation is the attempt to estimate mixing using eq. (7). The model shows D_p is proportional to d_c. The dependence of D_p on design and operating variables has recently been reviewed (Laplante et al. 1988). The estimation problem was alleviated here by adopting a 1 m x 10 m column as a base unit and scaling up the number of columns required. Mixing in columns of this size is reasonably well known, and well estimated by the model. An effort was also made to avoid $N_p > 1$ for any column stage since there are no data available in this range.

Refinements to the simulator are always possible. However it is important not to lose sight of the objective, which is preliminary scale-up. The most critical parameter is the feed rate. This is usually variable and, indeed, at the design stage often only approximately known, especially if columns are being introduced to process an intermediate stream, even in an existing circuit. With feedback from users (which now include Mt.Isa Mines Ltd., Bougainville Co., Falconbridge Ltd. and Cominco) refinements can be expected.

7.- CONCLUSIONS

A simulator for scale-up of flotation columns has been developed. The simulator uses a model of recovery in the collection zone and an estimate of (froth) carrying capacity. The recovery model uses an experimentally determined flotation rate constant and a prediction of particle retention time and mixing characteristics based on column geometry. Flotation rate constants and carrying capacities must be supplied.

In a previous communication (Yianatos et al. 1987), a simulator was described based on two alternative configurations (cleaning and scavenging). The simulator has been made more flexible by using a networking routine (Del Villar et al. 1986). This permits circuits of arbitrary configuration to be tested and also allows for multiple feed and products.

A case study was used to illustrate the impact of column size, recycling and bias rate. Recycle improved selectivity, but requires considerably more column capacity. Minimum bias rate was shown to be an advantage in maximizing recovery.

The simulator can be used at the preliminary design stage and also at later stages to refine the design as more data become available.

8.- LIST OF VARIABLES

Note: units (e.g. cm or m, s or min) are those used by the simulator, based on common usage)

A = column cross-section, m^2
D_p = dispersion coefficient, m^2/s
d_p = particle size, μm
d_c = column diameter, m
d_{80} = conc 80% passing size, μm
g = acceleration due to gravity
H = height of collection zone, m
J_g = superficial gas rate, $cm^3/s/cm^2$
J_B = superficial bias rate, $cm^3/s/cm^2$
k = flotation rate constant, min^{-1}
N_p = vessel dispersion number
Q = volumetric flowrate, m^3/min
R = recovery, fraction or %
Re = Reynolds number
t = time
U_L = net interstitial liquid velocity, cm/s
U_p = particle settling velocity relative to the fluid, cm/s
W = mass flowrate, kg/min

Greek letters

ϵ_g = fractional gas holdup
ϵ_L = fractional liquid holdup, $1 - \epsilon_g$
ϵ_c = fractional concentrate holdup at the top of the column
τ = mean residence time any phase, min
ρ = specific gravity
μ = viscosity

Subindices

f = feed
c = concentrate
g = gas
L = liquid
p = particle
t = tailings
sp = suspension

9.- REFERENCES

Bourassa, M. and G. Barbery, 1986, "Erreurs experimentales et calibration des modeles cinetiques de flottation au laboratoire", 88th Annual General Meeting of CIM, Montreal

Del Villar, R., J.A.Finch and A.R.Laplante, 1986, "Network configuration program", Int. Mine Computing, Sept.-Oct., pp.24-27.

Dobby, G.S. and J.A.Finch, 1985, "Mixing characteristics of industrials flotation columns", Chem. Eng. Sci. 40 (7), 1061.

Dobby, G.S. and J.A.Finch, 1986, "A model of sliding time for flotation size bubbles", Can. Inst. Min. Metall. Bull. 79, (889), 89

Dowling, E.C., R.R.Klimpel and F.F.Aplan, 1984, "Model discrimination in the flotation of a porphyry copper ore", Annual Meeting of AIME, Los Angeles.

Espinosa-Gomez, R.G., 1987, PhD Thesis, McGill University.

Espinosa-Gomez, R.G., J.B.Yianatos, J.A. Finch and G.S.Dobby, 1987, "Flotation column carrying capacity: particle size and density effects", submitted to Minerals Engineering, July.

Espinosa-Gomez, R.G., J.B. Yianatos, J.A. Finch and N.W. Johnson, 1988, "Carrying capacity limitations in flotation columns", Symp. Column Flotation, AIME (Ed..K.Sastry), Phoenix, January.

Laplante, A.R., J.B. Yianatos and J.A. Finch, 1988, "On the mixing characteristics of the collection zone in flotation columns", Symp. Column Flotation, (Ed. K.Sastry), AIME, Phoenix, January.

Levenspiel, O., 1962, Chemical Reaction Engineering, Ed. John Wilcy, New York.

Lynch, A.J., N.W. Johnson, E.V. Manlapig and C.G. Thorne, 1981, Mineral and Coal Flotation Circuits, Vol 3, Developments in Mineral Processing, Elsevier.

Kaya, M. and A.R.Laplante, 1985, "Investigation of batch and continuous flotation kinetics in a modified Denver laboratory cell", Can. Met. Quart., Vol.25, No.1, pp.1-8.

King, R.P., 1972. Flotation research work of the N.I.M. Research Group, and the Dep. of Chemical Engineering, Univ. of Natal, J.S. Afr. Inst. Min. Metall., 72(4):135-145

Masliyah, J.H., 1979, "Hindered settling in a multi-species particle system", Chem. Engng. Sci., 34, pp.1166-1168.

Sutherland, D.N., 1977, "An appreciation of galena concentration using a steady-state flotation model", Int.J.Min.Proc., Vol. 4, pp.149-162

Xu Manqiu, 1987, M.Eng. Thesis, McGill University.

Yianatos, J.B., J.A. Finch and A.R. Laplante, 1986, "Apparent hindered settling in a gas-liquid-solid countercurrentcolumn", Int. J. Min. Proc. 18, 155.

Yianatos, J.B., R.G. Espinosa-Gomez and J.A. Finch, 1986, Seminar on Column Flotation, McGill University, May 9-10.

Yianatos, J.B., R. del Villar, J.A. Finch, A.R. Laplante and G.S. Dobby, 1985, "Preliminary flotation column design using pilot scale data", COPPER 87, Vina del Mar, Chile.

Computer Applications in the Mineral Industry, Fytas, Collins & Singhal (eds)
© 1988 Balkema, Rotterdam. ISBN 90 6191 760 3

Coal preparation process control R&D

A.I.A.Salama & M.W.Mikhail
Coal Research Laboratories, CANMET, Energy, Mines & Resources Canada, Edmonton, Devon, Alberta

ABSTRACT: An industry-wide survey of Canadian coal preparation plants was conducted, with two main objectives: i) identify industry needs and problem areas, and ii) generate industry feedback as an input to FPL's planning and overall strategy. The paper presents the main results of the survey along with the strategy for FPL's coal preparation process control R&D program.

1 INTRODUCTION

During the past three years the demand for coal has decreased which coupled with an increase in production capacity, has resulted in an oversupply on world market. To overcome reduced prices and a shrinking market, the coal industry is challenged to optimize their operations to reduce the cost of production. The coal cleaning cost can be reduced by improving fine coal recovery and in general by maximizing plant yield. These approaches can be helped by the utilization of process control and by new developments in computer technology.

Application of process control techniques and schemes in coal preparation can achieve : flexibility in operation, improved plant performance and product yield, better utilization of manpower, tighter quality control and minimization of production cost. Successful implementation of process control techniques and schemes is expected to improve coal preparation plant recovery by as much as 5% (Bradley et. al. (1981))

The mandate of CANMET (Canada Centre for Mineral and Energy Technology) is to address needs and provide new technologies to industry to enhance the role and contribution of minerals and energy to the Canadian economy, by means of mission-oriented R&D which is supportive of R&D performed by industry. As part of the close interaction and contacts required by this mandate, an industry survey was conducted recently (1986) in conjunction with field visits. Discussions were held with plant operational management to determine coal industry interest in process control development and priorities regarding specific plant circuits.

2 COAL PREPARATION AND COMPUTER APPLICATIONS

In the last twenty years an enormous progress in the area of computer technology has been witnessed. The rapid advances in computer hardware and the less rapid but significant advances in computer software have led to development of the minicomputer, microcomputer and microprocessor. Significant computer control development in mineral processing has already been achieved. Although progress has been made in the area of computer applications in coal preparation, more concentrated effort is needed to utilize the full potential of computer technology development.

The computer applications in coal preparation can be viewed as off-line and on-line applications. However the development in one area could be utilized in the other. Considerable progress has been reported in the area of off-line computer applications, especially in the areas of process modelling (static models), performance evaluation and/or prediction, flowsheet analysis and/or optimization and statistical analysis (Fig 1). However, little effort has been reported in on-line computer applications, especially in the areas of process control, model reference control, process optimization and adaptive control.

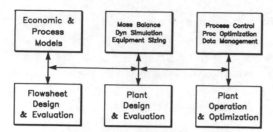

Fig.1 Coal preparation computer applications

3 CANADIAN COAL PREPARATION PROCESS CONTROL DEVELOPMENT

Most Canadian coal preparation plants were commissioned during the past two decades with minor provision for utilization of process control. Most of these plants include manual sequential start-up and shut-down with analog instrumentation for some process variables. The fast growing and developing computer technology in the 1970's and 1980's had a strong impact on the utilization of process control in industrial applications. Consequently the technology of programmable logic controller (PLC) found its way into recently built plants.

In 1981, the Fuel Processing Laboratory (FPL) of CANMET, began the development of a distributed process control scheme for the various circuits in its comprehensive pilot plant facilities at Devon, Alberta (Salama et. al. (1985)). The pilot plant is being used to develop, test and evaluate various process control strategies and on-line instrumentation before recommending them for industrial use. Part of the effort in process control R&D at FPL is carried through research contracts, two of which were on control systems for heavy medium and froth flotation circuits. The heavy medium control system is designed to regulate the medium density at a prescribed set value (Salama et. al. (1985)). The developed system is installed in the pilot plant and was successfully tested. This system will be the focus of further development and improvement. The design of a computerized process control system for the flotation cells has been completed. At present, in-house expertise is being utilized to implement the sequential start-up and shut-down of the pilot plant.

4 CANADIAN INDUSTRY COAL PREPARATION PROCESS CONTROL SURVEY

To maintain the relevance of FPL's process control development program, an industry-wide survey was conducted, with two main objectives :
- identify industry needs and problem areas, and
- generate industry feedback as an input to FPL's planning and overall strategy.

A questionnaire aimed at fulfilling the above objectives was prepared and hand delivered or mailed to twelve coal washeries across Canada; seven responses were received. Careful and thorough analysis of the response received indicate that six out of the seven respondents are in favour of more process control R&D activities. However, one operator noted that present economic restraints are creating an unfavourable atmosphere for any new development or changes in their plant. The interpretation and discussion of the questionnaire main points are presented in the next subsections.

4.1 Circuit Priorities

Responses received indicate the froth flotation circuit is a top priority in reference to plant benefits, followed by HMB/HMC and thermal dryer.

4.2 Problematical Circuits

The froth flotation circuit was identified by the survey as the most problematical circuit in the plant and stands to greatly benefit from implementation of process control. This is in agreement with the previous findings.

4.3 On-Line Analyzers

An area which is important for successful and effective implementation of process control schemes is the development and availability of reliable on-line analyzers (ash and/or sulphur). The responses received reflect regional needs (i.e. eastern Canada needs sulphur analyzer while western Canada needs ash analyzer). A critical study on the evaluation and availability of on-line monitors will be carried by the author in the the near future and the findings will form a separate report.

4.4 Programmable Logic Controllers

Out of the seven plants surveyed, six are utilizing PLC's which indicate a good support for the utilization of PLC's among Canadian coal preparation plants and is a reflection of the effectiveness and reliability of PLC's in monitoring and control.

5 CANADIAN COAL PREPARATION PROCESS CONTROL R&D: FUTURE DIRECTIONS

5.1 Plant Automation

5.1.1 Distributed Process Control

A Direct Digital Control (DDC) system utilizes a central process control computer to perform most of the functions of local controllers. The DDC thus offers a mean of simplifying many of the analog controls : complex analog control loops are replaced by simple DDC loops and control calculations are handled digitally. The one serious disadvantage that limits the usefulness of DDC systems in industry is related to overall system backup. For example, if the central control computation should fail, the central control function is lost and no back-up remains.

The disadvantage that the control functions of all local controllers are handled by a central process computer can be eliminated by introducing the concept of a distributed system (Fig. 2). In the distributed system the control functions are spread between two or three or more control computers or microprocessors (Reed Marchant (1980) and Willie (1983)), which are easily interfaced with a supervisory digital computer. The various distributed microprocessors are interconnected by data highways and can talk to the supervisory computer. The distributed system is in a sense the digital version of analog control which utilizes microprocessors instead of analog control elements. The distributed microprocessor supervisory control has the back-up feature and the communication advantage of DDC. Moreover, this system offers greater reliability, more flexibility, greater ease in service and simpler operation. These advantages indicate that microprocessor based distributed process control systems will be used for coal preparation process control applications in the future.

5.1.2 Programmable logic controller

In general the control commands of the process are generated by the computer but some control units such as programmable logic controllers (PLC's) are used instead of or in conjunction with the computer. The PLC has proved to be reliable, robust and gives the system the following capabilities for control: sequential start-up and shut-down , interlocking of plant (i.e. failure in a unit of equipment will cause the up-stream flow to be halted), emergency shut-down, detection of alarm conditions and monitoring ability. Also there are software available for these devices to be interfaced with existing and known minicomputers and personal computers. The versatile CRL pilot plant at Devon, Alberta was redesigned in 1983 with the installation of some new equipment. The flexibility in operation covers 15 flowsheet options to serve research and industrial needs. This large number of flowsheet options needed the utilization of PLC. As a result CRL installed a medium size Allen Bradley AB PLC 2/30 system to handle the sequential start-up and shut-down, alarm and monitoring features. The next stage is to install an Allen Bradley AB-Advisor colorgraphic system to display the status of all motors and all analog quantities of any particular option.

The availability of PLC's with enlarged capabilities (computation and storage) at reasonable price compared to computers provides an extra advantage of operating in a standby mode which provides reliability and flexibility.

5.1.3 Utilization of Modern Control Theory

From a coal preparation process control perspective, limited control has been implemented; a concentrated effort is required to utilize the full potential of the theoretical results reported in modern control theory. Most coal preparation circuits are accessible for formulation in the frame of modern control theory, which will facilitate the utilization of computer (Roesch et. al. (1976), Allgood et. al. (1982) and williams and Meloy (1983)).

5.1.4 Optimization and Adaptive Techniques

Modern control theory provides the vehicle for the analysis and optimization of most physical processes by utilizing a digital computer. The level of optimization is governed by the hierarchy of the coal preparation facility (Roesch

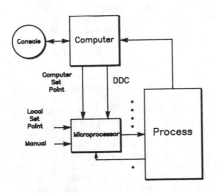

Fig.2 Distributed process control

et. al. (1976)), however we can classify the optimization levels as:

Level 1 - Simple regulation : this can be viewed on the single loop level and deals with local regulation involving simple, physical phenomena such as level control, flow control, speed control etc.

Level 2 - Micro optimization : this is on the circuit (HMC, jig, WOC, thickener etc.) level and concerns optimizing the efficiency of an elementary circuit, in relation to technical production criterion (e.g. yield maximization) or to an internal or economic consideration (e.g. reduction of energy consumption).

Level 3 - Macro optimization : the optimization is viewed here in a more wider context where the elementary circuits are considered to be included in the whole production operation. It is therefore necessary to optimize the process operation in relation to an internal criterion (production quality) which depends on external factors such as competition and the laws of supply and demand.

In most situations the above optimization levels can be achieved utilizing dynamic, static, linear, nonlinear and distributed parameter system formulation. These optimization problems can be solved utilizing the existing techniques (linear, nonlinear, quadratic, mathematical programming and maximum principle and dynamic programming approaches) which determine the optimum decisions with respect to some economic benefits.

The modern control theory has serious disadvantage of assuming that process parameters (such as residence time etc.) are known, and can not easily handle random disturbances. In many practical situations some of the process parameters are not known. Moreover, many coal preparation circuits such as flotation or HMC have inherent process delays (time elapsed between the initiation of a control signal and circuit response). In these situations the adaptive control approach provides an effective means in dealing with unknown parameters and system delays. This type of control can be achieved by utilizing a microprocessor-based controller to review the performance of the control loop, to determine the quality of control achieved against a predetermined standard, and to calculate the new tuning parameters needed to maintain specified performance. In this way, without creating process upsets, the controller is maintaining loop performance in an 'expert' manner - in other words, in the same way as would an instrumentation engineer when he 'tunes the loop'. This type of self-tuning control adapts not only to load dependent dynamic gain changes, but also to variations in time related and/or random dynamic characteristics. Such applications include : (a) pH - control, (b) level - dynamics changes with plant throughput and (c) flow - pumps and valves wears, pipes foul up etc. Experience with these controllers indicates saving of thousands of dollars per year in reagent costs and in energy expended. In the future more emphasis and utilization of these controllers in coal preparation process control will be implemented.

5.1.5 On-Line Ash and/or Sulphur Analyzers

On-line ash and/or sulphur analyzers are essential monitors for effective product quality control and process control schemes. A summary of the international manufacturers of on-line ash analyzers is presented in Table 1 (Kamada et. al. (1986)). A critical review of the available ash and/or sulphur monitors will be prepared by the author in a separate report.

5.1.6 Expert Systems

In the last two decades we have witnessed a dramatic development in computer technology with respect to reduction in size and increased computation speed, accompanied by development of efficient computer algorithms. This dramatic development coupled with the fascination of people with the prospect of intelligent artifacts have speeded up the introduction of Artificial Intelligence (AI) technology. This technology is in its infancy due to incomplete information and knowledge of how the human brain functions and processes information; however some success has been reported in the areas of medicine, chemistry and mining. Based on the current rate of advancement of computer technology it is anticipated that AI will find its way into many technical and industrial applications (e.g. robotics).

Artificial Intelligence (AI) has two different products: models of human cognition and intelligent artifacts. Expert systems (ES) belongs to the latter. Therefore ES can be defined as : "a knowledge-based system that emulates expert thoughts to solve significant problems in a particular domain of expertise". Intelligent artifacts are produced to solve problems and this is the main reason for building ES. Artifacts can be classified as general or domain-dependent or as special or domain-specific. ES's are domain-specific

Table 1 - International manufacturers of ash monitors

Brand or manufacturer	Country	Radiation source	Manner of radiation	Ash content range (%)
CENDREX (DSM)	Netherlands	Soft X-ray	Backscattering	3-15 <60
SIMCAR	United Kingdom	170-Tm(γ)	Backscattering	2-30
GUNSONS SORTEX (Phase 3)	United Kingdom	238-Pu(γ)	Backscattering	5-20
HUMBOLDT-WEDAG (KHD)	F.R. Germany	241-Am(γ)	Backscattering	3-10
ASHSCAN	Australia	241-Am(γ) 137-Cs(γ)	Transmission	8-12 22-35
COALSCAN (PP)	Australia	226-Ra(γ)	Transmission	-----
AMDEL	Australia	241-Am(γ) 137-Cs(γ)	Transmission	-----
NUCOALYZER	United States	252-Cf(n)	Transmission	<30
MICHIGAN I.T.	United States	109-Cd(γ)	Transmission	45-80
EMAG	Poland	241-Am(γ)	Backscattering	3-30
AZUK, EAZ	Soviet Union	90-Sr(β) 241-Am(γ), 137-Cs(γ)	-------------- --------------	----- -----

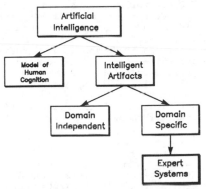

Fig.3 Artificial intelligence and expert systems

category (Fig. 3).

An ES is essentially composed of a knowledge base, a data base, an inference engine and some support software (Fig. 4). The central part is the inference engine.

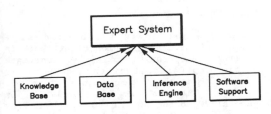

Fig.4 Expert system components

The knowledge base holds whatever information found appropriate to solving problems in our chosen domain. It is specific to the particular application. The data base is the work area of the system. The support software provides the interface to the environment. The inference engine provides the motive power to the system. Its functions determine what data it needs to solve the problem at hand, to get this data via the support software, to lodge it in the database, to employ the contents of the knowledge base to draw inferences and to record these in the database. It exercises these functions repeatedly until it can do, or need do, no more (Sell(1985)). The main desirable characteristics of ES's are :

- perform well on difficult problems
- easy to implement
- interact well with humans
- able to explain themselves
- real-time pace
- easy to modify and augment the knowledge base
- inference machine works irrespective of the theory

An application of ES in process control is the area of alarm management. When an alarm or out of limit signal is received by the system, the inference engine uses its rules to decide which process measurements are important for an analysis of the situation. These can then be scanned faster than others, effectively focusing on the data needed for diagnosis. A parameter that may not normally be measured could be called up if necessary.

Structuring of rules allows the appropriate ones to be found without searching the entire database. The rule altered by the data base might direct the inference to the next rule to consider or alternatively it could direct it to search all rules relating to a certain aspect of the system such as a tank or temperature indicator. With the capabilities of present systems, analysis and diagnosis can be completed at a speed fast enough to prevent potentially hazardous situations.

It is estimated that over half of current development work on expert systems is in the field of alarm management. It is anticipated that, perhaps 3-5 years, they will be embodied in control systems, not just as an add-on, in the same way that the VDU (visual display unit) has become the main operator interface.

5.2 Circuit Priorities

5.2.1 Flotation Cell

Western Canadian coals are friable in nature and as a result large proportions (20-60%) of fines (-0.5mm) are present in the run-of-mine. Efficient separation in the fines circuit is essential for an acceptable overall plant recovery. Froth flotation is the most common process used to beneficiate coal fines and is used in 10 out of 12 preparation plants in Canada.

Although simple in concept, coal flotation is a complex process that is affected by many variables among them :
- particle size and size distribution
- size and population density of bubbles
- coal rank
- chemical and hydrodynamic conditions
- surface conditioners and collectors
- reagents dispersion state
- pH of pulp phase
- turbulence and fluid flow patterns in cell.
- height and structure of froth layer
- flotation mechanisms (collision, attachment, detachment)

There are strong interactions between the various parameters that influence the performance of the flotation process. Consequently it is generally acknowledged that the flotation section of a coal preparation plant is the most difficult to operate and control to achieve optimum recovery. This conclusion is supported by the response obtained in the survey.

Modest developments in flotation modelling and control from Australia, USA and Canada have been reported in the literature (Envirotech Report (1981), Hammoude and Smith (1980), Lyman et. al. (1979), Lynch et. al. (1978) and Williams and Meloy (1983)) which indicate that a concentrated effort is still needed to develop means to improve coal recovery.

This effort should concentrate on :
- better understanding of flotation mechanisms and development of effective, reliable and realistic dynamic models to simulate these mechanisms.
- utilization of modern control theory to develop effective control strategies.
- utilization of available self-tuning and adaptive control algorithms (Hammoude and Smith (1980)).

5.2.2 HMB/HMC

Considerable progress has been made in the development of heavy medium circuit process control in Australia, USA, UK, South Africa, France and Canada (Carr et. al. (1982), Lambert and Mentzer (1986), Lyman et. al. (1982) and Moyers et. al. (1983)). In all of the developed systems the objective was to regulate the medium density and in some cases the viscosity. Most of the developed control systems utilize classical proportional-integral (PI) and proportional-integral-derivative (PID) controllers. The performance of PI and PID controllers are acceptable to some degree. However, a concentrated effort is needed in advanced digital control utilizing :
- pulse amplitude modulation scheme
- pulse frequency modulation scheme
- pulse modulation scheme
- self-tuning and adaptive scheme
to improve the performance of the control system in terms of the response time and efficient tracking of the set points.

5.2.3 Jig

It appears that most of the effort on the development of process control systems for jigs is taking place in Germany (Carr et. al. (1982)). The jig bed depth is continuously controlled based on the ash content of the clean product. It may be advantageous to investigate the feasibility of controlling the air pressure window based on the yield maximization subject to ash level in the products. The German development could be introduced to jigs operating in Canadian preparation plants.

5.2.4 Thermal Dryers

The thermal dryer is an expensive component of a preparation plant, both in capital and operating costs. One estimate of the capital cost of the equipment is $54-$73 per year per tonne of water removed (Moyers et. al. (1983)). In general 70 to 80% of the total thermally dried tonnage is dried using fluidized bed dryers. For coal dried from 14 to 5%

246

moisture, assuming a fuel cost of $1 per million Btu and an electricity cost of 5¢ per kWh, these two components of operating cost amount to $0.61 to $0.90 per tonne of dried coal.

To minimize both fuel and electricity consumption in a fluidized bed dryer, the gas temperature entering the drying chamber should be as high as possible without causing the coal to burn, to lose its volatile matter, or to be oxidized appreciably. Furthermore, the moisture-laden gases must leave the drying zone at a temperature sufficiently high that they are not cooled below the dew point before they reach the scrubber. The instrumentation and control system should properly control the combustion rate and gas flow to maintain drying conditions that will result in safe, efficient, reliable operation at the desired degree of product dryness.

6 CRL COAL PREPARATION PROCESS CONTROL R&D : FUTURE DIRECTIONS

In 1981 CRL began a process control program to automate its 10 tph pilot plant facility at Devon, Alberta. At that time it was realized that the distributed process control scheme is superior to the direct digital control scheme, especially with regard to the backup feature and flexibility in development as was discussed earlier. Moreover the pilot plant is being used to develop, test and evaluate process control strategies and on-line instrumentation before recommending them for industrial use.

Part of the effort in process control R&D at CRL is carried through research contracts, two of which were on control systems for heavy medium and froth cell circuits. The heavy medium cyclone project has been successfully completed. This system will be the focus of further development and improvement especially in reference to control algorithms. The design of a computerized process control system for the flotation circuit in CRL's pilot plant is completed. The implementation stage is underway.

Most of the circuits are manually controlled i.e. setting of cutpoints, feed rates etc. are adjusted manually. At present, in-house expertise is being utilized to implement the sequential start-up and shut-down for the 13 operating options of the pilot plant utilizing the Allen Bradley PLC 2/30 programmable logic controller (existing).

Based on the analysis of the responses received FPL future efforts will concentrate on :

- Sequential start-up and shut-down
- Instrumentation
- Monitoring and report generation
- Simple control loops
- Process optimization, adaptive control and parameter identification
- Process modelling
- Expert systems

in relation to the following circuit priorities:
- Flotation
- Water recovery
- Jig
- WOC
- Spiral

The sequential start-up and shut-down of the FPL pilot plant, interlocking, alarm and monitoring are handled by an Allen Bradley AB PLC 2/30 (existing) (Fig. 5). The logic programming of the different operating options has been started and it is expected that this task will be completed in 1987. The next step in our development is the installation of of an AB-Advisor colorgraphic system for flowsheet, process variables, motor status displays and process variables trending. Storage of plant or run data, number crunching and data manipulation, report generation and higher level control are handled by a process control minicomputer HP 1000-A700 (existing). Communication between AB PLC 2/30 and HP 1000-A700 is established via a fiber optics link.

Most of the instrumentation, except for on-line ash and/or sulphur analyzers and per cent solids, are reliable and available. During the course of development of process control systems for different circuits in the pilot plant the required instrumentation and control loops will be evaluated and implemented. The dedicated microprocessors designed to

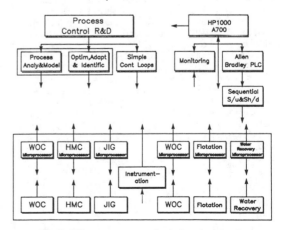

Fig.5 FPL process control developments

Table 2 - Process control development methodology

Area	Methodology	Availability
Sequential start-up and shut-down	Logic ladder diagram programming	National
Instrumentation		National and/or International
Monitoring and report generation	Software programming	National
Simple control loop	Software and hardware programming	National
Process optimization	Modern control theory Adaptive control schemes Parameter identification schemes	National and/or International
Process modelling and simulation	Mathematical models Empirical models Probabilistic models Computer simulation	National and/or International
Expert systems	Modelling of human cognition Intelligent artifacts Utilization in coal preparation	National and/or International

control and optimize a particular circuit is required to communicate with the HP 1000-A700.

The major areas of process optimization, adaptive control, parameter identification, process modelling and simulation and expert systems in coal preparation are necessary to achieve better and effective control. The research carried at FPL and the utilization of international developments could complement each other in achieving a comprehensive and efficient process control scheme (Brown and wieckowski (1986), Finlayson (1983), Salama and Mikhail (1986), Salama (1986), Wizzard et. al. (1983) and Zigmond et. al.(1983)). To reinforce a future strategy for CRL development in coal preparation process control it is useful to present the areas of development with the methodology of achieving it and the availability of such methodology and technology (national and/or international) related to it. This information is given in Table 2.

7 CONCLUSIONS AND RECOMMENDATIONS

As a result of preparation plants visits, discussions and survey responses, the following main R&D directions emerge:

- continuing need for accurate and reliable process monitors particularly on-line ash and/or sulphur monitors.
- process control strategies for coal preparation circuits to be pursued, especially to maximize the recovery of flotation circuit.
- to stabilize and improve the moisture control on plant products.
- development of effective computer models (dynamic) which can be used as a reference for the process operation.

Regarding on-line ash and/or sulphur monitors, it is recommended to evaluating existing ash monitors and if not satisfactory establish a joint project with the Canadian Atomic Energy Establishment to develop ash and/or sulphur monitors (stream, slurry and froth applications).

The in-house process control development must continue and be implemented on a step-by-step approach depending on the availability of funds. The effort should concentrate on developing effective process control strategies for the different circuits to maximize the recovery. As a direct response to the industry needs, the flotation circuit automation is a top priority. The international developments and any fundamental work carried out by Canadian universities could be utilized depending on the applicability to Canadian coal preparation plants.

The development of effective computer models (dynamic) is being and will continue to be pursued at FPL. This development will enhance the circuit operation, produce better prediction/simulation and could be used to train coal preparation engineers. Any recent fundamental and/or overseas developments will be incorporated in FPL development.

In conclusion, it must be noted from the survey that the industry is striving to achieve high recovery to minimize the effect of difficult economic and operating conditions.

REFERENCES

Allgood, G.O., Canright, G.S., Brown Jr, C.H. and Hamel, W.R. 1982. Dynamic modeling and simulation of froth flotation and vacuum filtration units. Transactions ISA: 21: 3: 45-53.

Bradley, N.C., Allgood, G.O. and Moyers, J.C. 1981. Potential economic benefits from process control of coal preparation plants. ORNL - 5736.

Brown, T.S. and Wieckowski, A.M. 1986. Process simulation of coal preparation plants. The 10th Int Coal Prep Congress, September 1-5, 1986; Edmonton, Canada; 1: 217-231: paper 3.3.

Carr, K.R. et al. 1982. State-of-the-assessment of coal preparation plant automation. ORNL - 5699.

Chessel, T. 1983. Process control - what's new and what's coming. Can Chem Process: November, 1983: 16-17, 19.

Control systems in coal preparation plants. 1981. EPRI CS - 1880, prepared by ENVIROTECH.

Finlayson, N. 1983. Australia develops process control system for coal preparation plant. Australian Coal Miner: 5: 2: 46-51.

Hammoude, A. and Smith, H.W. 1980. Experiments with self-tuning control of flotation. Proc. 3rd IFAC Symposium on Automation in Mining, Mineral and Metal Processing, Aug 18-20, 1980; Montreal, Canada: 213-218.

Kamada, H., Kawaguchi, H. and Onodera, J. 1986. On the coal blending process control by on-line ash monitors. Proc 10th Int Coal Prep Congress, September 1-5, 1986; Edmonton, Canada: 1: 245-266: paper 3.5.

Lambert, J.L. and Mentzer, P. 1986. Automatic washing density regulation in a dense medium preparation plant"; Proc. 10th Int Coal Prep Congress, September 1-5, 1986; Edmonton, Canada; 1: 286-299: paper 4.2.

Lyman, G.J., McKenzie, C.K., Leach, K.R., Lynch, A.J. and Bateman, A.J. 1979. The automatic control of coal flotation circuits. The 8th Int Coal Prep Congress, May 21-26, 1979; Donetsk, USSR 2: 5-18.

Lyman, G.J., Askew, H., Wood C.J. and Davies, J.J. 1982. Dynamic modelling of dense medium cyclone washing circuits. Proc. Aust I.M.M.-North West Queensland Branch Mill Operators' Conference.

Lynch, A.J., McKenzie, C.K., Leach, K.R. and Bateman, K.W. 1978. Modelling of central Queensland coal flotation circuits for control purposes. Proc. Aust I.M.M. Conference - North Queensland.

Moyers, J.C. et al 1983. Coal preparation plant automation"; Noyes Data corporation, Mill Road, Park Ridge, New Jersey 07656.

Reed Marchant, G. 1980. The 80's-microprocessors, distributed control and advanced instrumentation for the mineral industry. Proc 3rd IFAC Symposium on Automation in Mining, Mineral and Metal Processing, August 18-20, 1980: Montreal :15-24.

Roesch, M., Ragot, J. and Degoul, P. 1976. Modelling and control in mineral processing industries. Int J of Mineral Processing: 3: 219-246.

Salama, A.I.A. and Mikhail, M.W. 1986. Optimization of coal or mineral circuits. Proc. 3rd Conference on The Use of Computers in The Coal Industry. July 28-30, 1986; Morgantown, West Virginia; 169-176.

Salama, A.I.A. 1986. Yield maximization in coal blending. Proc. 10th Int Coal Prep Congress, September 1-5, 1986; Edmonton, Canada; 1: 196-216: paper 3.2.

Salama, A.I.A., Mikhail, M.W. and Shaw, D.L. 1985. Development of a computer control system for a heavy medium circuit. Proc. 13th Annual Mining and Metallurgy Industries Symposium and Exhibit, ISA, May 15-17, 1985; Salt Lake City, Utah: 12: 49-59.

Salama, A.I.A., Mikhail, M.W. and Mikula, R.J. 1985. Coal preparation process control. CIM Bulletin: 78: 881: 59-64.

Sell P.S. 1985. Expert systems - a practical introduction"; London, UK, MacMillan Publishers LTD.

Williams, M.C. and Meloy, T.P. 1983. Dynamic model of flotation cell banks-circuit analysis. Int J Mineral Processing: 10: 141-160.

Wizzard, J.T., Killmeyer, R.P. and
    Gottfried, B.S. 1983. Computer program
    for evaluating coal washer performance.
    Mining Engineering: 35: 3: 252-257.
Wyllie, R.J.M. 1983. Distributed process
    control: today's prep plant philosophy.
    World Coal: February, 1983: 34-37.
Zigmond, R.D., Ramani, R.V. and Frantz,
    R.L. 1983. Computer program for the
    analysis of coal preparation plant
    economics. Mining Engineering: 34: 12:
    1688-1696.

Computer Applications in the Mineral Industry, Fytas, Collins & Singhal (eds)
© 1988 Balkema, Rotterdam. ISBN 90 6191 760 3

Simulation techniques in mineral liberation analysis

Gilles Barbery & Robin Pelletier
Université Laval, Quebec, Canada

ABSTRACT: Simulation techniques are used in liberation analysis studies in order to test, demonstrate, or calibrate reconstruction procedures to transform primary measurements results, obtained by image analyzers on sections or test lines, into three dimensional information. The difficulties encountered in these Monte Carlo simulations are presented. In the case of simple problems, some elementary but powerful integral geometry theorems can be used. Examples of errors in the simulations are illustrated. Simulation of spherical caps composite particles are carried out, which demonstrate the application of these integral geometry results. Simulations are carried out to demonstrate particle composition distribution reconstruction methods. A new method based on the calibration of the original ore texture is presented.

1 INTRODUCTION

The measurement of liberation, or rather, particle composition distribution, has only been developing very slowly over the past years. With the advent of automatic image analyzers, based on Scanning Electron Microscopes or Electron Microprobes, the primary measurement has become readily accessible. Information produced by these equipments is a degraded version of reality. Particles are observed on sections, and measurements are made on a linear basis, or on an areal basis. It has been known for many years that this primary measurement is biased, and that it cannot be used without serious limitations. A complete review of developments till 1984 was presented by Barbery (1985 b). The derivation and implementation of mathematical procedures for the correction of primary results has been a stumbling block to various researchers. In most cases, primary results, obtained on lines or on sections, are presented as true volumetric information, without proper warning to the user of the results. The problem is most challenging, since there cannot be a complete solution. The lack of standard materials against which it could be possible to calibrate equipments and methods is also a difficulty.

Due to these limitations, some researchers have attempted to simulate the operation of image analyzers on computers, using Monte Carlo techniques. The present paper will illustrate some of the difficulties and errors that can be made (and that have actually been made!), and procedures that can be used to minimize them. A demonstration will be made of a correct simulation procedure, and the results will be applied to a new method for particle composition distribution reconstruction in 3-D from 2-D or 1-D information.

2 MONTE-CARLO SIMULATION TECHNIQUES IN LIBERATION STUDIES

Barbery (1974) has presented an application of Monte Carlo simulations to the determination of particle shape factors for the reconstruction in 3-D of particle size distribution measured on random lines. Some of the concepts given in the present paper were already applied. The simulation of image analyzers operation by Monte Carlo simulations requires the definition of various random orientations and positioning (random orientation of a 3-D particle, random orientation and positioning of a section plane or a test line). The type of random orientation must first be defined. Coleman (1969) has presented a classification of random lines in 3-D space and identified that the randomness used in image analysis, and in stereology in general, is uniform randomness (μ-randomness). A random plane in this context has a direction which is perpendicular to a random line, which in turn can be obtained as a point of uniform probability on the sphere of radius one. The equation of a random plane is thus:

$$x\cos\theta\cos\phi + y\,\sin\theta\cos y + \sin\phi + d = 0$$

where θ is drawn at random from 0 to 2π, and $\cos\phi$ is drawn at random from -1 to +1. Any other method used for the definition of random planes will lead to preferential orientations.

This method has been used for Monte Carlo simulations of composite particle characterization by linear image analyzers by Barbery and Huyet (1977) and Barbery et al. (1978), and for shape analysis of particle sections by Durand et al. (1978).

As was stated in an early book on Monte Carlo simulation (Hammersley and Hanscomb, 1964), "optimism and utilitariamism are required in simulation"; the only test of randomness is "whether or not the random numbers yield an unbiased estimate or a reliable answer to the problem". Under such circumstances, it is most useful to test simulation techniques to cases for which the results are known analytically. For particles of simple shape and texture, such results are readily obtained.

Integral and stochastic geometry must be used for the derivation of theoretical solutions. Standard books are available in this field (Kendall and Moran, 1963, Santalo, 1976, Solomon, 1978). The results that are valuable in the present case are limited in number, but they are very powerful in rejecting wrong simulation procedures; they should be used to analyze very carefully the results and the claims that are made. Some of the results are routinely used in image analyzers:

$$S_s = L_l = V_v$$

which corresponds to the Delesse and Rosiwal equations using Underwood's notations (1970),

$$S_v = 2\,P_l$$

The probability of observing composite particles can also be quantified; it is equal to the ratio of the mean projected height of the interfaces to the particles mean projected height for areal measure-

ments; it is the ratio of areas of interface to particles surface area for linear measurements, see Kendall and Moran (1963) and Stewart and Jones (1977). For simple geometric features, Underwood (1970) gives a table of mean projected heights.

In the case of linear intercepts, a very stringent criterion is that of Crofton's second theorem (1963), for <u>convex</u> particles and phases:

$$\frac{\pi E(1^4)}{3E(1)} = \frac{E(V^2)}{E(V)}$$

where 1 is a random line and V is feature volume. For monodisperse particles or phases, the ratio is equal to particle or phase volume.

Some simulations of composite particle population that have been made by various authors in recent years do not always appear to provide the proper results. Exemple of such simulations are given in Broussaud et al. (1984), Jones and Horton (1979) and Moore and Jones (1980), Hill et al (1987), Sepulveda et al. (1985) and Lin and Miller (1987).

The results of Broussaud et al. (1984) are based on the simulations of random spherical particles with various internal textures. The results agree well on elementary checks (composite volumetric ratio, apparent locked factor), but in some cases fail on the more difficult Crofton theorem. In one case, and for the case of spherical caps composites, random orientation was achieved by describing a plane by:

$$ax + by + cz + d = 0$$

and drawing a, b, and c at random between -1 and +1! some bias can be shown on phase volume as calculated from Crofton theorem.

Jones and Hornton (1979) and Moore and Jones (1980) give their results for the distribution of intercepts by number as a fonction of linear grade. Their results are in good agreement with theory for the parameters that can be calculated, especially apparent composite intercept number.

Hill et al. (1987), in attempting to derive a method for particle composition distribution for linear and areal measurements carry out simulations of spherical caps particles.

Their results appear to be in error even for the elementary results of free sections of each phase.

Sepulveda et al. (1985) and Lin and Miller (1987) have developed a most interesting software package, PARGEN, which enables to simulate complex particles sectioning by random planes and analysis by areal or linear measurement. The results presented for simple cases do not always satisfy theoretical results. For example, from their table 4 in Sepulveda et al. (1985), it is possible to test Crofton theorem, and to compare calculated particle volume to theoretical values. A significant disagreement can be observed: mean volume of 3144.5, instead of 5000 as theoretically simulated.

Since all the authors mentioned developed their simulation packages in order to demonstrate and test, and even calibrate reconstruction methods for particle composition distribution, the presence of such errors is annoying, since it casts doubts on the validity of the procedures that are derived.

3 EXAMPLE OF A CORRECT SIMULATION METHOD FOR SPHERICAL CAPS COMPOSITE PARTICLES

In the present study, simulation of spherical caps composite particle was carried out in order to demonstrate the checks that can be made. Figure 1 presents the random orientations that were used. The plane separating phases is located at a distance do from the origin, and defines the volumetric phase ratio in particles. Random orientations such as defined above were used. Planes of section are drawn at random perpendicular to xOy and a z value taken uniformly between -1 and +1. Surfaces of section are calculated for

the particle and each phase. For linear intercepts, a line parallel to Ox is drawn in the plane of section, at a random distance from the center of the circle of section; this distance takes a value uniformly distributed between -1 and +1 for y. Intercepts lengths for particle and each phase are calculated.

Results are provided in terms of tables of results and figures. Figure 2 gives the results of intercept and areal grade distribution for composite particles with uniform 0.5 volumetric concentration. The results can be compared to theoretical values in Table 1.

The procedure can be used to generate data for the testing of various reconstruction methods.

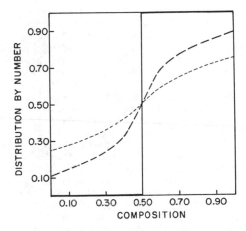

Fig. 2 Distribution of sections and intercepts grade for 0.5 spherical caps composite particles (volume, sections, intercepts)

Table 1. Comparison of simulation and theoretical volumes

	true	calculated
Mean particle volume (from Crofton theorem)	4.18879	4.17643
Mean phase A volume (from Croften theorem)	2.09439	2.07771
Volumetric phase ratio (based on area measurements)	0.5	0.4980
(based on linear measurements)		0.4998
Probability of observing composite sections	0.7854	0.7806
composite intercepts	0.5000	0.5043
		(20.000 simulations)

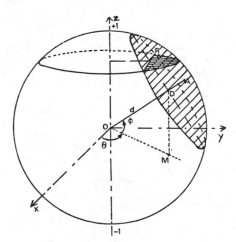

Fig. 1 Randomization procedure used in the simulation

4 RECONSTRUCTION OF OVERALL PARTICLE POPULATION PARAMETERS

Barbery (1985b) has presented a review of various methods put forwards till 1984 to reconstruct particle composition distribution. A development was made in 1985 in order to correct some errors in the model that had been developed by researchers at BRGM, France (Barbery et al. 1985, Barbery, 1985a). In the present application the following demonstration can be made. A rectangular distribution of particles (by number), as a function of volumetric grade, is generated. All particles have the same volume. The primary results of the simulations are given on Figure 3. The results for mean particle volume and mean phase A volume (based on Crofton theorem) were 4.183882 and 2.807417 respectively. Using Barbery's corrected method (1985a), an incomplete Beta function can be fitted to the distribution of volumetric grade, to give values of parameters a and b of an incomplete Beta function that can be calculated by: $a/(a+b) = 0.5$ (measured volumetric grade) and $(a+1)/(a+b+1) = 2.807417/4.183882 = 0.671$. The resulting values of a and b are a=b=0.9692, very close to the theoretical values of a=b=1, characteristic of rectangular distributions. In the present case, the agreement between theoretical values and calculated parameters is excellent. It should be noted that the assumptions used in the derivation are very limited: particle and phase convexity only, in order to apply Crofton theorem. No assumption is made on particle shape or particle internal texture.

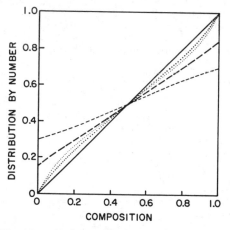

Fig. 3 Distribution of composite features, number basis (volume, sections, intercepts, 3-D reconstruction from sections and intercepts)

Applications of this reconstruction method to real cases is more limited, due to the poor sampling statistics of the fourth order moment of particle and phase intercept lengths. The assumption of uniform particle volume has also been found to be too restrictive for some applications. A demonstration will be given here of a new method that has been derived in order to solve some of these inefficiencies.

The general principle of the method is to apply the methodology which has already been used by Barberv (1987a), and Barbery and Leroux (1986 and 1987) in order to predict particle composition distribution by modelling texture and particle production. It is based on powerful concepts used in integral geometry, as developed by Gilbert (1962), Kellerer (1986) and particularly by Davy (1984). A paper has been prepared on the method (Barbery, 1987b), which will only be summarized here. The following assumptions are made:

- particle shape can be quantified by random intercept distribution, measured in image analyzers
- particle composition distribution results from the random application of particles on a 3-D texture
- texture quantification can be made using its covariance function, which can be measured on image analyzers for the original material, if available
- independance of texture and particle generation is assumed.

Under these circumstances, Davy (1984) has developed relationships relating $A=E[V_a(V-V_a)]/E(V)$, $B=E[S_a(S-S_a)]/E(S)$ and $C=E[L_a(L-L_a)]/E(L)$ to integrals of functions containing the texture covariance and the particle shape descriptors (Barbery, 1987b). Predicting methods for particle composition distribution take into account particle descriptors and texture covariance (Barbery, 1987a, Barbery and Leroux, 1986 and 1987); the application to reconstruction in 3-D from measurements made in 1-D or 2-D requires the measurements of C or B in image analyzers, and the calibration of covariance functions. Since one value is quantified, only texture covariance functions with one parameter can be used, such as Poisson polyhedra or Boolean processes with Poisson polyhedra as primary grains (Barbery and Leroux, 1987). 3-D reconstruction then results from the methods that have already been published (Barbery and Leroux, 1987, Barbery, 1987a), and especially the use of the MODLIB software (Barbery et al. 1986).

In the present case, during the simulations that are described above, calculations are made of B and C. For the case described in Figure 3, the following values were obtained: B=0.5069507, C= 0.1884044. Calibration of a texture based on Poisson polyhedra gives the following parameter: $\lambda_3 = 0.143$ when texture is calibrated on C and $\lambda_3 = 0.1624$ when it is calibrated on B. Calibration made on the theoretical value of A gives $\lambda_3 = 0.200$. Simulated results for primary results and reconstructed distribution using this new method are given on Figures 3 and 4.

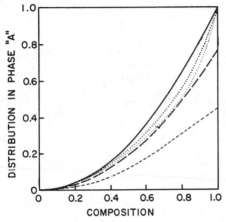

Fig. 4 Distribution of phase A as a function of feature grade (same symbols as Figure 3)

The method presented here has been applied to various simulation results (Barbery, 1987b), including those of Broussaud et al. (1984), as well as to experimental measurements made in Australia on the QEM*SEM instrument (Barbery and Sutherland, 1988). The advantages of the method lie in the simplicity of the measurements: only covariances are measured, which have improved statistics compared to fourth order moments as used in the original method. In the actual measurements, only particles on sections or traverses which in fact reveal apparent composites are taken into account, as shown in B and C. Particle shape is used and results from the quantification which is made during the measurements. The constraint on uniform particle volume is removed; the only limitation is that particle composition and particle volume distributions are independent. The limitation on convexity of phases in particles is removed. Particle texture results from the modelling of ore texture, and of particle size and shape. The dif-

ferences between theoretical and reconstructed particle composition distributions that are obvious on Fig. 3 and 4, are mainly related to the total lack of fit of the Poisson polyhedral texture to composite particles which are spherical caps! Even under these circumstances, the agreement between reconstructed areal measurements and theoretical values is considered very good.

5 CONCLUSIONS

The values and difficulties of using Monte Carlo simulation techniques in liberation analysis are presented. These techniques are valuable in the production of simulated experimental results in conditions where theoretical volumetric distributions are known. The application of such techniques to stochastic geometric processes is prone to errors, which are usually related to difficulties in randomizing orientations and positioning. It is strongly recommended to check simulated results against powerful integral geometry theorems, such as Crofton theorem. The demonstration made for spherical caps composite particle illustrates this point. The examples given for reconstruction method for overall population characteristics proves the validity of the method and the simulation. A new method is illustrated, which consists in estimating the parameters of the original ore texture from the results of the 1-D or 2-D analysis of broken fragments. The method removes most of the difficulties that have been presented, and gives results which are more than adequate for reconstruction in practice.

REFERENCES

Barbery, G. 1974. Determination of particle size distribution from measurements on sections, Powder Technology, 9: 231-240.
Barbery, G. 1985a. Nouvelle méthode pour caractériser la libération minérale par analyses d'images à une et à deux dimensions: théorie et exemple d'application, Proc. XVth Int. Mineral Processing Congr., CEDIM, France, I: 20-30.
Barbery, G. 1985b. Mineral liberation analysis using stereological methods: a review of concepts and problems, Applied Mineralogy, AIME, New-York, 171-190.
Barbery, G. 1987a. Random sets and integral geometry in comminution and lib-

eration of minerals, Mineral and Metallurgical Processing, 96-102.

Barbery, G. 1987b. Particle composition distribution reconstruction from image analysis measurement: method of fundamental parameters, submitted to Powder Technology, 1987.

Barbery, G., Bloise, R., Bodziony, J., Gateau, C. and Kraj, W. 1985. Measurement of mineral liberation of comminuted ores, Archiwum Gornictwa, 30: 297-310.

Barbery, G. and Huyet, G. 1977. Mineral liberation analysis: theoretical study and computer simulation, paper presented at AIME Annual Meeting, Atlanta.

Barbery, G., Huyet, G., Bloise, R. and Laurenceau, H. 1978. Theoretical study and simulation of the reconstruction of the volumetric composition of composite particles derived from measurements carried out on sections, Proc. 2nd. Europ. Symp. on Quant. Analysis of Microstructures, Riederer Verlag, Stuttgart, 212-223.

Barbery, G. and Leroux, D. 1986. Prediction of particle composition distribution after fragmentation of heterogeneous materials: influence of texture, Process Mineralogy VI, AIME, New-York, 415-430.

Barbery, G., Leroux, D. and Guillaneau, J.C. 1986 MODLIB, computer software for IBM-PC, for the prediction of particle composition distribution after breakage of heterogeneous materials, Laval University.

Barbery, G. and Leroux, D. 1987. Prediction of particle composition distribution after fragmentation of heterogeneous materials, accept. for public. in Int. J. Mineral Processing.

Barbery, G. and Sutherland, D.N. 1988. QEM*SEM experimental study of mineral liberation in the Mount-Lyell ore: results and interpretation, to be present. at AIME Ann. Meet., Phoenix, Az, Jan. 1988.

Broussaud, A., Fourniguet, G. and Brachet, C. 1984. Libération Minérale - simulation et évaluation de modes de mesure par analyse d'images, Rapport BRGM 84 RDM 024 MIN, 1984.

Coleman, R. 1969. Random paths through convex bodies, J. Appl. Prob., 6: 430.

Davy, P. 1984. Probability models for liberation, J. Appl. Prob., 21: 260-269.

Durand, M.C., Barbery, G., Groslier, J. and Hucher, M. 1978. Stéréologie de quelques polyèdres simples, Proc. 2nd Europ. Symp. on Quant. Analysis of Microstructures, Riederer Verlag, Stuttgart.

Gilbert, E.N. 1962. Random subdivision of space into crystals, Ann. Math. Statist., 33: 958-972.

Hammersley, J.M. and Hanscomb, D.C. 1964. Monte Carlo Methods, Methuen, London.

Hill, G., Rowlands, N. and Finch, J. 1987. Data correction in two-dimensional liberation studies, present. at AIME Ann. Meet., Denver, Feb. 1987.

Jones, M.P. and Hornton, R. 1979. Recent developments in the stereological assessment of composite (middling) particles by linear measurements, Proc. XIth Commonwealth Min. and Met. Congr., IMM, London, 113-122.

Kendall, M.G. and Moran, P.A.P. 1963. Geometrical probability, Griffin, London 1963.

Kellerer, A.M. 1986. The variance of a Poisson process of domains, J. Appl. Prob., 23: 307-321.

Lin, C.L. and Miller, J.D. 1987. Effect of grain size distribution and grain type of linear and areal grade distribution, Process Mineralogy VI, AIME, New-York, 1987, 405-413.

Moore, S.W. and Jones, M.P. 1980. Random linear probes through composite spheres that have multiple planar phase boundaries, Trans. I.M.M., 89: C190-C193.

Santalo, L.A. 1976. Integral geometry and geometric probability, Addison Wesley, Reading, Mass.

Sepulveda, J.E., Miller, J.D. and Lin, C.L. 1985. Generation of irregularly shaped multiphase particles for liberation analysis, Proc. XVth Int. Mineral Processing Congr., GEDIM, France, I: 120-132.

Solomon, H. 1978. Geometric probability, Soc. for Ind. and Appl. Math., Philadelphia.

Stewart, P.S.B. and Jones, M.P. 1977. Determining the amount and composition of composite (middling) particles, in Proc. XIIth Int. Mineral Processing Congr., DNPM, Sao Paulo, Brasil, III: 91-116.

Underwood, E.E. 1970. Quantitative Stereology, Addison-Wesley, Reading, Mass.

7 Geotechnical applications

Computer Applications in the Mineral Industry, Fytas, Collins & Singhal (eds)
© 1988 Balkema, Rotterdam. ISBN 90 6191 760 3

A finite element computer model for stress analysis of underground openings

K.S.Chau, H.S.Mitri, F.P.Hassani & M.J.Scoble
Department of Mining and Metallurgical Engineering, McGill University, Montreal, Quebec, Canada

ABSTRACT: A finite element model for linear elastic analysis of underground openings has been developed for use on microcomputer. The model is designed to serve as a useful analysis tool for the mining engineer to conduct a 2-dimensional analysis of multiple-material domain surrounding a mining excavation for the purpose of predicting areas of potential failure and critical stress. The model accommodates any arbitrary geometry of the opening and allows for a variety of loading conditions and combinations which may be encountered in a mine environment such as in situ stress, gravity loading and boundary tractions. Material failure is checked by means of Drucker-Prager, Mohr-Coulomb and Hoek and Brown criterion.

The model utilizes a 4-node quadrilateral isoparametric element and employs a mesh generating routine to facilitate node and element data preparation. Element stiffness matrix as well as consistent load vectors are evaluated numerically using the Gauss quadrature rulle. The discretized equilibrium equations are solved by a banded Gauss solver routine.

The use of the model is demonstrated by a numerical example of an underground tunnel opening subject to in-situ stress loading.

1 INTRODUCTION

Most mining engineering problems involving stress calculations are so complicated that closed form solutions are difficult to derive and that numerical methods have to be employed. The finite element method is among the numerical techniques most suitable for mining problems. In fact, its ability to handle materials of anisotropic properties under a variety of loading conditions makes it superior to other numerical methods available for stress and displacement analysis. The finite element method has a wealth of applications in mining engineering. Analyses involving stress distribution around underground tunnel openings of various shapes and roof loading conditions have been carried out; see for example Duffant and Feuga (1985). Also, problems of roof stability in longwall faces (Vervoort et al. 1987) and stability of inter panel pillars for longwall mining (Fama and Wardle 1987) have been tackled by the finite element method. Other applications include the determination of relaxed zones boundaries around underground caverns to provide a guide for rock-bolting design (Lee and park 1985).

Many general purpose finite element programs have been developed over the past two decades or so, such as SAPIV, NASTRAN, ABAQUS, ADINA and NONSAP. Such programs are designed to run on mainframe computers such as CDC, AMDAHL, CRAY and IBM and they usually require considerable effort in input data processing. Unfortunately, mainframe computers are not often easily accessible at mine sites. Also, little effort has been paid towards the development of special purpose programs for mining engineering applications.

The purpose of this paper then is to present a finite element model for use by the mining engineer as a tool to predict the response of rock mass around underground openings. The model is designed to run on personal computers so that it can be used at mine sites where mainframe computers are unlikely to be available. The system also includes a preprocessor to help generate the data and a postprocessor to help analyze the results.

2 PROBLEM FORMULATION

The two-dimensional, quadrilateral iso-parametric element developed by Irons (1966) is used. As shown in Fig. 1, the element is defined by four corner nodes with two degrees-of-freedom per node making a total of eight degrees of freedom per element. The element nodes as well as the sides are numbered counterclockwise. The displacements (u,v) and the position (x,y) of any point within the element are found by inter-polation of its nodal displacements (u_i,v_i) and nodal coordinates (x_i, y_i) respectively using the shape functions N_i, where $i = 1$ to 4. Thus,

$$u = N \cdot \bar{u} \qquad (1)$$

$$x = N \cdot \bar{x} \qquad (2)$$

where: $u^T = (u,v)$ $\qquad (3)$

$$x^T = (x,y) \qquad (4)$$

$$\bar{u}^T = (u_1,v_1,u_2,v_2,u_3,v_3,u_4,v_4) \quad (5)$$

$$\bar{x}^T = (x_1,y_1,x_2,y_2,x_3,y_3,x_4,y_4) \quad (6)$$

and N is the shape function matrix.

The shape functions N_i ($i=1$ to 4) are based on Lagrange polynomials and are expressed in terms of the natural coordi-nates (s,t) shown in Fig. 1. The func-tions are given by Cook (1981).

The strain-displacement relationship at any point within the element can be written as

$$e = B \cdot \bar{u} \qquad (7)$$

in which B is the strain-displacement matrix and is 3 x 8. The vector of strain components e contains e_x, e_y, γ_{xy}. For convenience of subsequent computation, the order of B is raised to 4 x 8 by shifting its third row to the fourth one and filling row 3 with zeros.

The total stress at a point due to external loads and initial stresses can be calculated from

$$\sigma = D \cdot e + \sigma^0 \qquad (8)$$

where $\sigma^T = (\sigma_x, \sigma_y, \sigma_z, \tau_{xy})$ is the stress vector, and σ^0 is the vector of initial stresses in the same directions. The stress-strain elasticity matrix D for an orthotropic material making an angle with the global X-axis of the cartisian frame of reference (Fig. 2) is given by

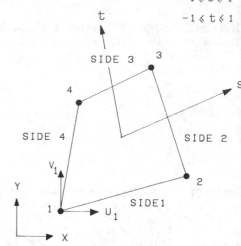

$$-1 \leqslant s \leqslant 1$$
$$-1 \leqslant t \leqslant 1$$

Fig. 1 Quadrilateral Isoparametric Element

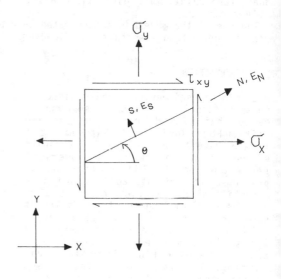

Fig. 2 Orthotropic Material in N-S System

$$D = T^T \cdot \bar{D} \cdot T \qquad (9)$$

where T is a transformation matrix from the local N-S system of the material and the global X-Y coordinate system (Cook,

260

1981). The matrix $\bar{D}$ will take one of two distinct forms depending on the case being analyzed. For the plane strain

case, $\bar{D}$ is given by the inverse of the strain-stress relations, i.e.

$$
D = \begin{bmatrix}
\dfrac{1}{E_N} & \dfrac{-\nu_{NS}}{E_N} & \dfrac{-\nu_{NT}}{E_N} & 0 \\[2mm]
 & \dfrac{1}{E_S} & \dfrac{-\nu_{ST}}{E_S} & 0 \\[2mm]
\text{Symm.} & & \dfrac{1}{E_S} & 0 \\[2mm]
 & & & \dfrac{1}{G_{NS}}
\end{bmatrix} \quad (10)
$$

where E_N, E_S = moduli of elasticity in N and S directions respectively,
G_{NS} = shear modulus in NS plane,
ν_{NS} = Poisson's ratio in S direction due to a strain component in N direction and vice versa for ν_{SN}.

For the case of plane stress ($\sigma_z = 0$), the stress-strain matrix for orthotropic materials is given by

$$
\bar{D} = \begin{bmatrix}
\dfrac{1}{E_N} & \dfrac{-\nu_{NS}}{E_N} & 0 & 0 \\[2mm]
 & \dfrac{1}{E_S} & 0 & 0 \\[2mm]
\text{Symm.} & & 1 & 0 \\[2mm]
 & & & \dfrac{1}{G_{NS}}
\end{bmatrix} \quad (11)
$$

The element stiffness matrix can be derived from the principle of virtual work which leads to the following expression.

$$
K = \int_V B^T . D . B \, dv \quad (12)
$$

For 2-dimensional elements dv = t.dxdy and a relationship between dxdy and dsdt must be found since B is obtained in terms of s and t. Such a relationship is furnished through the determinant of the Jacobian i.e. dxdy = $|J|$ dsdt. Furthermore, the resulting integral becomes a complex function of (s,t) and numerical

integration must be used. A four point (2 x 2) Gauss quadrature scheme has been employed to evaluate Eq. (12) numerically. The locations of integration points are shown in Fig. 3 and are given by (Cook 1981)

$$
\begin{aligned}
s_1 &= t_1 = -0.57735\ 02691\ 89626 \\
s_2 &= t_2 = +0.57735\ 02691\ 89626
\end{aligned} \quad (13)
$$

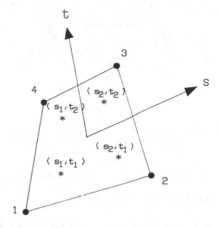

Fig. 3 Distribution of Integration Points

The element load vector P is derived from the virtual work principle and is composed of several load vectors. These are vectors due to concentrated nodal loads P^n, gravity forces P^g, boundary tractions P^b and initial stresses P^o i.e.

$$
P = P^n + P^g + P^b + P^o \quad (14)
$$

Except for P^n, which is calculated by simply lumping the nodal force contributions, the load vectors are evaluated by Gauss quadrature rule of numerical integration. Both gravity and initial stress load vectors are calcualted from a 2 x 2 integration scheme whereas the boundary traction vector is evaluated using 2 integration points only along the loaded boundary. Both pressures and shears are accounted for and are assumed to vary linearly along the boundary as shown in Fig. 4.
Finally the global equilibrium equations are assembled as follows:

$$
P = K . u \quad (15)
$$

which are solved for the nodal displacements by means of a banded Gauss elimination solver.

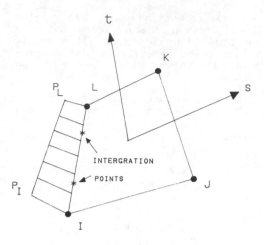

Fig. 4   Integration Scheme for Boundary
          Load Vector

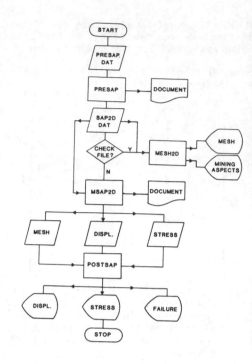

Fig. 5   Flowchart of MSAP2D System

3 COMPUTER MODEL MSAP2D

A numerical model called MSAP2D (Microcomputer Static Analysis Package for 2-
Dimensional problems) has been developed
for use on personal computers equipped
with mathematical coprocessor and fixed
disk.

Fig. 5 shows a flowchart of the basic
structure of MSAP2D system. The system
is composed of four modules, which are
highlighted by rectangular boxes on the
figure. The first two modules, program
PRESAP and program MESH2D, represent the
preprocessor and the interface of the
system.   The third module, program
MSAP2D, is the core processor of the
system.   The fourth module, program
POSTSAP, is the postprocessor.

PRESAP is a mesh generation program for
the core program MSAP2D. It reads a
simple datafile and generates the
required datafile for program MSAP2D. In
addition, PRESAP also outputs a document
file which gives general information of
the input data.   The data for program
PRESAP includes geometry of the structure, loading conditions, mesh grading
requirements, boundary conditions,
properties of materials and required
mining sequences.

MESH2D is a graphic program for checking the output file from program PRESAP,
which is also the input file for program
MSAP2D. Geometry of the structure, mesh
grading, boundary conditions, mining

sequences as well as material types of
the structure can all be graphically
displayed and checked.

MSAP2D is the main finite element
program of the system. it is a simplified two-dimensional version of the more
general, well known SAPIV program written
by Bathe et al. (1973).   The present
version was first   produced by CANMET-MRL
(Toes and Yu, 1974) on a CDC 6000 mainframe. Displacements of nodal points as
well as stresses at prescribed locations
are outputted. The stress output locations include the centroid and the mid-
points of the four sides of each element.

POSTSAP is a graphic program designed
to interpret the output from program
MSAP2D. Nodal displacements are displayed in vectorial form. Major principal stresses and minor principal stresses
of all elements are also plotted in
vectorial form. Moreover, failure criteria have been incorporated into the
program to allow parameter studies to be
performed in order to check the safety
level of an excavation design.

The solution procedure is described as
follows.

1. Prepare the datafile for the program PRESAP (e.g. PRESAP.DAT) which identifies the problem layout, boundary conditions, material properties and loading conditions as well as the required number of subdivisions.

2. Using PRESAP.DAT, run PRESAP. This will generate the required datafile for the core program, namely SAP2D.DAT. Also, will be available is an information output for documentation purposes.

3. The user has the option of making a visual check of SAP2D.DAT using MESH2D or bypassing this stage to run the core program MSAP2D.

4. Having run MSAP2D, three output files will be produced for further use by the postprocessor in addition to a documentary output file for the user.

5. The final step in the solution procedure involves the use of the postprocessor along with the three datafiles generated previously by MSAP2D. This enables checking stresses, displacements and material failure. The latter is carried out by means of Mohr-Coulomb, Drucker-Prager or Hoek and Brown failure criterion.

4 PRE- AND POST-PROCESSORS

4.1 PRESAP

The mesh generation scheme employed in program PRESAP originates from the work done by Zienkiewicz and Phillips (1971). It is based on using the shape functions of the 8-node quadratic isoparametric element to subdivide a quadrilateral, called the superelement, into any number of subdivisions to create a finite element mesh. The superelement is defined by 4 corner nodes, 4 side nodes and 4 sides which can be straight or curved depending on the location of the side node. Fig. 6 illustrates the generation of a nonuniform 4 x 4 finite element mesh from one superelement. In this example, the physical problem domain is taken as one superelement. In practice, however, the physical domain may need be subdivided into several superelements thus making a superelement model.

4.2 MESH2D

Undoubtedly, the chances of human error occurrence while preparing data for moderate size problems are not little. Even with the help of a preprocessor which requires relatively little information,

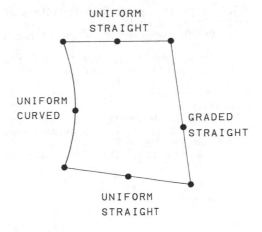

(a) SUPERELEMENT

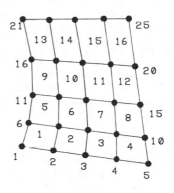

(b) MESH

Fig. 6 Mesh Generation Scheme

it would still be worthwhile reviewing the generated finite element grid and other relevant aspects by means of a graphical processor. MESH2D has been developed for use on personal computer in an interactive, user-friendly mode. The main benefits of MESH2D are:

1. Check element shapes and connectivity.

2. Identify the element and node numbering schemes of the generated grid for, naturally, the original node numbers defining the superelement model are not retained.

263

3. Check the number and locations of materials selected.

4. Check the opening shape and location.

5. Zoom-in facility. Not infrequently when the finite element mesh is so crowded that attempting to identify the element or node numbers is virtually impossible.

6. Ease of documentation. With the use of a pen-plotter, all relevant information can be plotted.

Fig. 7 displays the flowchart of program MESH2D.

4.3 POSTSAP

The numerical output of any practical problem solved by the finite element method is often long and quite tedious to examine by hand. A postprocessor routine which is capable of transforming such an output into graphical representations of displacement vectors and principal stresses is an essential component for the computer model to be efficient and easy to use. For this reason, POSTSAP has been developed for the MSAP2D system; see flowchart on Fig. 8. The basic functions of POSTSAP are:

1. Plot the nodal displacements.

2. Plot the principal stresses at the element's centroidal point as a pair of perpendicular lines.

3. Check material safety level at the centroid of the element and classify it as one of three categories: safe, critical or failed.

The safety level of a material is evaluated by calculating the stress at the element center and comparing it with the strength of the material. The stress of the material is obtained from the output of MSAP2D while the strength of the material is calculated based on a failure criterion. Whenever the stress of the material exceeds its strength, the corresponding element will be filled with red colour on the screen thus indicating material failure. If the stress of the material is close to its allowable strength, the corresponding element will be painted in yellow colour on screen to signify that the element while still safe is near failure. Otherwise, the element will be painted in green colour. However, for hardcopy output, these safety levels are illustrated by different hatching patterns instead of solid colours.

Three failure criteria are considered in program POSTSAP. The first two crite-

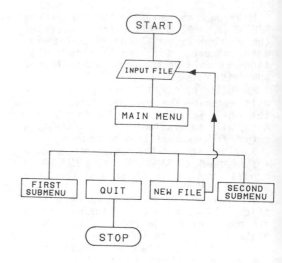

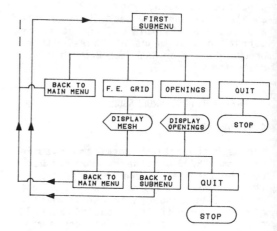

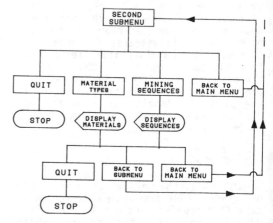

Fig. 7 Flowchart of MESH2D

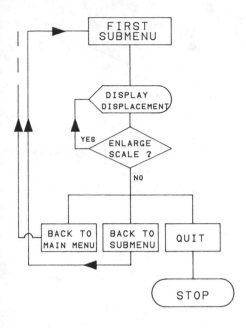

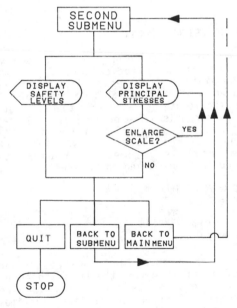

Fig. 8 Flowchart of POSTSAP

ria are Mohr-Coulomb and Drucker-Prager and their input parameters include the cohesion c and the internal friction angle of the material (Chen and Saleeb 1982). The third failure criterion considered is Hoek and Brown's (1980) whereby the required parameters include the uniaxial compressive strength, and m and s values of the material.

Mohr-Coulomb failure criterion has long been used by geotechnical engineers especially in soil mechanics. The criterion states that shear failure occurs at a point along the failure plane when (Fig. 9)

$$\tau_{max} = c + \sigma_N \tan \phi \qquad (16)$$

where τ_{max} is the shear strength at that point along the failure plane, σ_N is the normal stress at that point acting perpendicular to the failure plane.

In POSTSAP, the shear stress τ along the failure plane is calculated first and is then compared with the shear strength τ_{max}. The ratio τ_{max}/τ represents the factor of safety, see Table 1.

Drucker-Prager failure criterion is a modification of Von Mises failure criterion commonly used for ductile materials. The criterion states that the material will fail when

$$K' = a\, J_1 + (J_2')^{0.5} \qquad (17)$$

where J_1 is the first stress invariant, J_2' is the second stress invariant of deviatoric stress and a and K' are material constants.

The sign convention for this failure criterion is that tensile stress is considered to be positive and is illustrated in Fig. 10. By letting the left hand term of Eq. 17 equal to K such that $K = aJ_1+(J_2')^{0.5}$, the safety level of the material can be evaluated by the index value K'/K; see Table 1.

According to Hoek and Brown (1980) the rock fails when the stress level reaches the state such that:

$$\sigma_1 = \sigma_3 + \sqrt{m\sigma_c \sigma_3 + s\, \sigma_c^2} \qquad (18)$$

where σ_1 = major principal stress
σ_3 = minor principal stress
σ_c = uniaxial compressive strength of the intact rock
m = material constant which depends on rock type
s = material constant which depends on the degree of rock fragmentation.

265

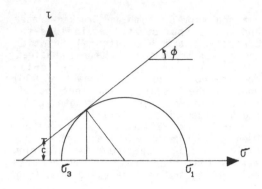

Figure 9   Mohr-Coulombe failure
criterion

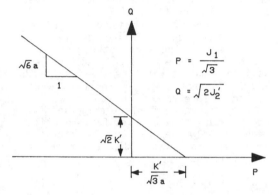

$$P = \frac{J_1}{\sqrt{3}}$$

$$Q = \sqrt{2J_2'}$$

Figure 10   Drucker-Prager failure
criterion

The sign convention for this failure
criterion assumes that compressive stress
is positive. By letting the right hand
term of Eq. (18) equal to $_{max}$ the safety
level of the material in this case can be
determined by the ratio $_{max}/$; see
Table 1. The failure envelope is plotted
in Fig. 11.

Table 1   Indices for assessing material
safety level

Criterion	Index	1.5	<1.5 >1.0	<1.0
M-C	$\tau_{max}/$ $^\tau$safe	safe	critical	failed
D-P	K'/K	safe	critical	failed
H&B	$\sigma_{max}/$ $^\sigma$safe	safe	critical	failed

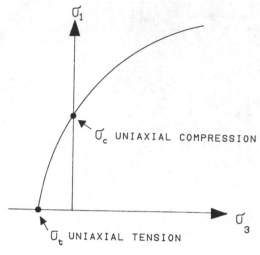

Figure 11   Hoek and Brown failure
criterion

5   NUMERICAL EXAMPLE

Consider an example of a long underground
tunnel of an elliptical cross section.
The width of the opening is 4 m and its
height is 2 m, thus making a ratio of 2:1
of the length of major axis to the length
of minor axis. The opening is subjected
to in-situ stress of 20 MPa vertically
and 6.67 MPa horizontally. The rock
material is assumed isotropic, its
properties are given as follows.

E = 70000 MPa
ν = 0.2
g = 0.028 MN/m^3
c = 16 MPa
ϕ = 35^0

It is required to find the stresses and
deformations around the tunnel opening
and determine its safety level. The
boundary of the domain is taken 10 times
the size of the opening away from the
opening centre. Due to double symmetry
of both loading and geometry, one quadrant of the opening is considered. The
tunnel is simulated by confining the
outer boundary by rollers on the sides
and hinges at the corners, then applying
initial stress loading throughout the
domain. Fig. 12a shows the superelement model which, after discretization,
produces a mesh of 525 elements as shown
in Fig. 12b. A blown-up picture of the

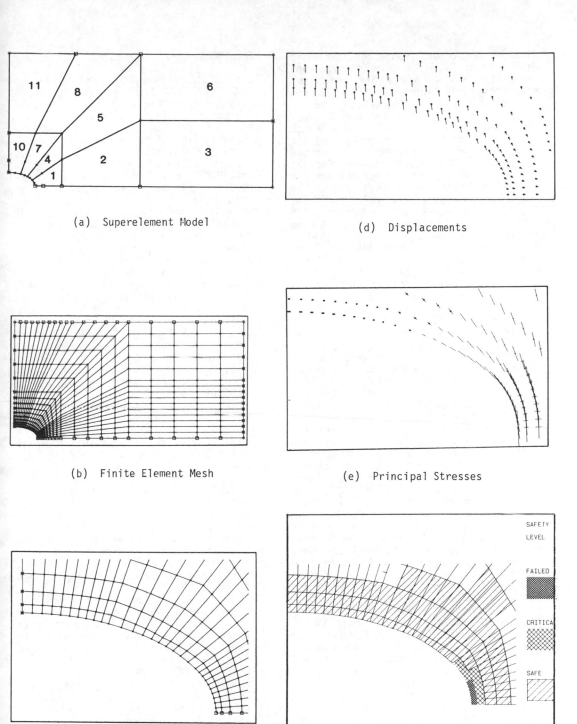

(a) Superelement Model

(d) Displacements

(b) Finite Element Mesh

(e) Principal Stresses

(c) Zoom-in Feature

(f) Safety Levels

Fig. 12 Data and Results of Numerical Example

finite element grid around the opening is displayed in Fig. 12c. The displacement and principal stress plots are shown respectively in Figs 12d and 12e. It can be seen that because of the higher vertical pressure there is a tendency to further ovalize the tunnel opening which causes high compressive stress along the major axis and tensile stress along the minor axis. The computed values for the maximum compressive and maximum tensile stress concentration factors around the opening are respectively 4.5 and 0.26. These concentration factors compare with 4.67 and 0.33 obtained from closed form solutions by Obert and Duvall (1967). Better correlation is possible if a finer mesh were used. According to Mohr-Coulomb failure criterion, this results in a failure pattern of the tunnel opening as shown in Fig. 12f.

6 CONCLUSIONS

A finite element computer model called MSAP2D has been developed. The model consists of four modules namely PRESAP, MESH2D, MSAP2D and POSTSAP. The two processing units PRESAP and MESH2D serve to reduce data preparation time for users and to perform extensive error checking of input data graphically. The core program MSAP2D is capable of analyzing 2-dimensional, linear elastic structures subject to static loading conditions. The fourth module, program POSTSAP, is a postprocessor designed to display graphical pictures of displacement and stress outputs from MSAP2D. Also, the safety levels against rock failure can be examined graphically using Mohr-Coulomb, Drucker-Prager or Hoek and Brown failure criterion.

The system is by no means completed and further developments are certainly recommended to better simulate the behaviour of underground openings. Of most importance is the introduction of non-linear analysis capability into the system. Other features like dynamic analysis and the effect of rock joints are also useful options for future work.

REFERENCES

Bathe, K.J., Wilson, E.L. and Peterson, F.E. 1973. SAP IV: A structural analysis program for static and dynamic response of linear systems.

Chen, W.F. and Saleeb 1982. Constitutive equations for engineering materials. Elasticity and modeling. Vol. 1, John Wiley & Sons.

Cook, R.D. 1981. Concepts and applications of finite element analysis. 2nd ed., John Wiley & Sons.

Duffaut, P. and Feuga, B. 1985. Towards a theory of tunnels in heterogeneous formations, Numerical analysis of some cases of heterogeneity around tunnels. Proc. 5th Int. Conf. Num. Meth. in Geomech., Nagoya, April, 1039-1048.

Fanna, M.E. and Wardle, L.J. 1987. Numerical analysis of coal mine chain pillar stability. Proc. 6th Int. Cong. Rock Mech., Montreal, 859-864.

Hoek, E. and Brown, E.T. 1980. Underground excavations in rock. Instn. of Mining and Met., U.K.

Irons, B.M. 1966. Engineering applications of numerical integration in stiffness methods. AIAAJ, Vol. 4, No. 11: 2035-2037.

Lee, C.I. and Park, Y.J. 1985. A numerical analysis for underground rock cavern in Samrangjin pumped-storage power plant project. Proc. 5th Int. Conf. Num. Meth. in Geomech., Nagoya, April, 1085-1092.

Obert, L. and Duvall, W.I. 1967. Rock mechanics and the design of structures in rock. John Wiley & Sons.

Toews, N.A. and Yu, Y.S. 1974. SAP2D documentation 2-D linear elastic finite element computer system. Mining Research Laboratories, CANMET, Energy, Mines and Resources Canada. Report 74/138.

Vervoort, A., Thimus, J.F., Brych, J., Crombrugghe, O.D. and Lousberg, E. 1987. Verification by the finite element method of the influences on the roof conditions in longwall faces. Proc. 6th Int. Cong. Rock Mech., Montreal, 1311-1316.

Zienkiewicz, O.C. and Phillips, D.V. 1971. An automatic mesh generation scheme for plane and curved surfaces by isoparametric coordinates. Int. J. Num. Meth. Eng., Vol. 3, 519-528.

Computer Applications in the Mineral Industry, Fytas, Collins & Singhal (eds)
© 1988 Balkema, Rotterdam. ISBN 90 6191 760 3

Three hinge beam analysis for buckling of stratified rock strata

D.S.Cavers
Hardy BBT Ltd, Burnaby, BC, Canada

Raj K.Singhal
Surface Mining Laboratory, CANMET, Devon, Alberta, Canada

ABSTRACT: A three hinge model for analyzing buckling movements of rock beams or slabs has been developed for microcomputers using Turbo Pascal. The work was part of a study funded by Canada Centre for Mineral & Energy Technology (CANMET) which developed new methods for analysis and design of footwalls in mountain surface coal mines. The model analyzes the buckling zone as discrete blocks with pin connections. Pore pressure, self weight and external forces are included. The program is capable of analyzing a wide variety of geometries including plane and curved strata. The algorithm used for the calculations is discussed. The program uses a spreadsheet-like screen editing method of data input which facilitates rapid trials. The number of stability conditions which can occur is much larger than in conventional stability analyses. These are discussed in the paper.

1 INTRODUCTION

This paper discusses the analysis of buckling of strata on rock slopes by approximating the buckling zone as a three hinge beam with 2 discrete moving blocks. Buckling may occur on rock slopes in which one or more slabs are approximately parallel with the rock face. In general, the failure mechanism requires that the slope dips more steeply than the angle of internal friction along the discontinuity forming the slabs involved. Typically, buckling may occur in sedimentary rocks where slabs are separated by bedding planes but may occur in other rock types if thoroughgoing discontinuities (joints, faults) are parallel to the slope face. Failure involves buckling of the slab near the toe of the slope followed by transla-tional failure of the slab above the buckle. Figure 1 depicts some of the possible failure modes on plane and curved slopes.

The use of 3 hinge methods is reason-able for slabs which are cut by cross joints, thus forming blocks. The present model does not allow for the formation of new tensile cracks between blocks.

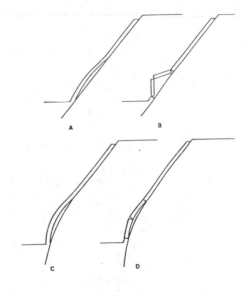

Fig 1 Possible modes of buckling failure
a) Flexural buckling of plane slope.
b) Three hinge buckling of plane slope.
c) Flexural buckling of curved slope.
d) Three hinge buckling of curved slope.

PHOTO 1: Overall view of major buckling failure of plane strata with a roll at the toe.

PHOTO 2: Closeup of roll at toe of Photo 1.

Buckling failure modes are of particular importance in the design of high footwall slopes, such as occur in Western Canadian coal mines. The work described in this paper is based in part on work undertaken as part of a study funded by CANMET to develop new methods for the analysis and design of footwalls in mountain surface coal mines (Cavers and Baldwin, 1986).

The application of 3 hinge methods in this paper is directed specifically to slopes but similar methods have also been applied to the stability of stratified roofs in underground openings (Chugh, 1977) and to problems in structural engineering. The equations developed can be easily modified for use in other applications.

Three hinge beam analysis offers one approach to the general problem of analyzing the stability of curved or plane strata on a slope. Other approaches which may be taken include Euler (elastic) buckling criteria (Cavers and Baldwin, 1986 and Cavers, 1981) and analysis of the stress conditions in curved strata (Corbyn, 1978). The latter approach is non-conservative for some geometries as discussed later.

Photos 1 and 2 illustrate a failure in curved strata of the type which may be analyzed using program CURVBUK, discussed in this paper. The failure comprised 2 sandstone layers separated by weaker material. During mining, a roll in the seam was uncovered and failure was initiated at the roll.

Program CURVBUK is designed to facilitate the analysis of footwall slopes, particularly those containing rolls.

Information regarding the interrelationship of dip, shear strength, roll geometry, bedding thickness, water pressures and bolting forces can be obtained.

The development of the equations follows from work described in Cavers (1978) and Cavers (1981). The present program CURVBUK is coded in Turbo Pascal (C) and runs on IBM-PC(C) and PC compatibles.

2 EQUATIONS

Figure A-1 (Appendix A) shows freebody diagrams for the blocks. The failure zone is divided into two blocks separated by three cross joints which form the "hinge" points on which movement occurs. The following assumptions are made:

1. Individual blocks are rigid. It has been found by model testing and confirmed by the program, that for slopes in which cross jointing is frequent, that an individual failing block may be cut by internal cross joints. The cross joints cutting an individual critical block remain closed until failure is well developed. Thus the critical block length may be longer than the individual blocks formed by cross joints actually present in the slab.

2. Cross joints forming blocks 1 and 2 are perpendicular to the local dip of the footwall slab. This is a reasonable assumption in most sedimentary strata.

3. The slope surface and the discontinuity bounding the slab are parallel. This assumption may be relaxed for certain modes of operation of the program

in which the geometry is completely described by the user.

4. Water forces are parallel to the slope are neglected except that they may be input explicitly as external forces.

5. Lateral release at the sides and crest of the slab is assumed to exist.

6. The slope is analyzed at the point of incipient failure. The forces between the blocks and the underlying slope are assumed to be zero.

Failure under a three hinge mechanism involves the centre hinge point (Point B, Figure 2) moving outward from the slope. Support for the slab above the roll above is reduced and failure occurs providing that the footwall slab is not in equilibrium (Figure 2).

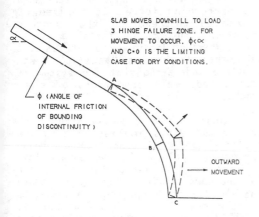

Fig 2: Failure mode for 3 hinge model

The derivation of the equations governing the three hinge block failure mode is given in Appendix A. The results of the equations give the values of P_n and Q_n (normal and shear forces) at the block hinge points at the point of limiting equilibrium.

To determine whether or not the slope is stable, the program applies an outward perturbing force to the center hinge point between the two blocks. The possible results are discussed in Section 4.

It should be noted that as failure starts and the centre hinge point moves away from the slope (Figure 2), the failure may be extremely rapid and can be dangerous. Very few slopes have been found poised in a state of partial failure.

The values of shear and normal forces at the hinge points can be checked to determine whether shear failure is

possible on a hinge point. Normally this is not a problem, but could occur for cases of high lateral loading, either by high water pressures or by lateral loading.

3 PROGRAM CURVBUK

CURVBUK is coded in Turbo Pascal and solves the system of equations given above. The program allows the complete geometry to be input with all variables defined or, for certain cases involving semi-circular rolls, the program will do many of the geometrical calculations automatically.

Figure 3 summarizes the modes of failure analyzed by the program. These are as follows:

1. Mode 1: Uniform thickness T with 2 semi-circular blocks. Moment arms, block weights and water forces are generated automatically by the program.

2. Mode 2: Completely general. Block thickness need not be uniform and the roll need not be semi-circular. This mode requires all water forces, moment arms, etc. to be calculated and input by the user.

3. Mode 3: Assumes that block 1 is planar and block 2 is semi-circular. Much of the input data is calculated automatically as in mode 1. The results of simple model testing and numerous runs of the program on different geometries show that this is frequently the critical mode of failure.

The program uses a "spreadsheet" style interface (Figure 4 – at end of paper) in which the cursor is moved using the arrow keys to allow inputting or changing the various input variables. This allows making repeated runs while changing 1 or 2 variables to find the critical case. The program jumps the bar type cursor over cells for which the results are calculated automatically under the mode in use. The program also originally had standard question and answer and input file interfaces but these were found to be cumbersome in comparison to the spreadsheet style. It may be noted that the coding to implement this interface is considerable (some 800 lines).

Output from the program goes to a printer or, using DOS redirection facilities, to a file. The results are also displayed on the screen.

The process of analyzing a slope using 3 hinge buckling methods requires finding the critical geometry by varying the lengths and geometries of the two blocks. This may also involve varying thickness.

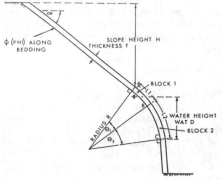

MODE 1 - CIRCULAR ROLL. BOTH BLOCKS ON ROLL.
(THIS MAY NOT BE THE CRITICAL CASE).

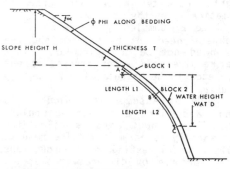

MODE 2 - COMPLETELY GENERAL

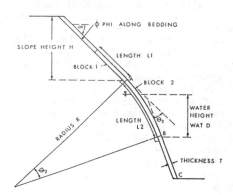

MODE 3 - PLANE SLAB (BLOCK I) ABOVE CIRCULAR ROLL.
ACCESSIBLE ONLY FROM METHOD 2 INPUT USING
FULL SCREEN EDITING.

Fig 3: Modes analyzed by program. Many of the input parameters are calculated automatically for modes 1 and 3.

The process is analogous to finding the critical failure surface for conventional limit equilibrium slope stability analysis using, for example, Bishop's method. In the present program the search for the critical block geometry is not automated, although it would be relatively simple to do this for modes 1 and 3.

The program differentiates between four cases of stability. The differentiation is made on the basis of the response of force P to a perturbing force applied to the center hinge point. These cases are as follows:

1. Slope is stable: This implies that force P_A is less than zero, and that when the perturbing forces is applied, P_A increases to zero and then becomes positive. Thus, to put the block system into a state of limiting equilibrium, the slope has to have a "tensile" force applied, indicating that the system is stable under all conditions of slope height.

2. Slope is conditionally stable - P_A is minimum: This implies that P_A is positive, and that when the perturbing forces is applied, P_A decreases. This indicates that at least this value of P_A (or the corresponding slope height above the roll) must be present in order to produce stability. This condition is normally encountered when water pressures are included, and results from requiring a certain minimum resisting moment to counteract the outward water pressure thrust. This situation can be dangerous if there is cohesion present in the slope, and the actual value of P_A existing in the field is less than that determined by analysis. Under such conditions, the slope would fail, possibly unexpectedly.

3. Slope is conditionally stable - P_A is maximum: This indicates that when the perturbing force is applied, P_A decreases. The value of P_A is the maximum which the roll can support.

4. Slope is unstable: The analyzed slope is unstable under all conditions, and will not support any driving force (P_A) whatsoever. P_A may be negative.

The different cases result from the different loading conditions which are possible, combined with the restoring moments that exist on the blocks due to their finite thickness. The desireable design conditions are for complete stability or for P_A being a maximum for conditional stability.

4 DISCUSSION

Once facility is gained with the input parameters, a given slope geometry may be rapidly analyzed, although the results are somewhat more complicated than for conventional limit equilibrium analyses. The authors believe that the program can also be adapted, by using mode 2, to analysis of other situations such as stratified mine roofs.

One of the early results obtained with the program was to show that a failure mode involving lifting of the plane slab above the roll (similar to mode 3 in Figure 3) was frequently the critical failure mode. This result, which has been confirmed by simple model testing, is in contrast to earlier work by others (for example, Corbyn, 1978), who implicitly assumed that the failure would be confined to the semi-circular portion of the roll. On the basis of this finding, the earlier work may be non-conservative for some slope geometries.

It may be noted that the four different stability cases discussed in Section 3 range from stable to unstable with two conditionally stable cases. One of these conditionally stable cases involves P_A being a maximum value. If the slope above the roll is not high enough or the bounding discontinuity is too strong, this minimum force, P_A, will not be applied and failure will result. For this case, higher slopes or weaker bounding discontinuities tend to increase stability. Slopes which exhibit this characteristic tend to be planar or have flat rolls with low angular dip change. Slopes in this condition are considered to be dangerous. P_A may also temporarily decrease due to slope freezing or similar effects.

For stable conditions, the value of the perturbing force which brings the roll to a condition of limiting equilibrium provides a measure of the "safety factor". Low values of perturbing force indicate that the roll is close to being unstable and that minor outward forces, for example, water pressures, could precipitate failure.

5 EXAMPLES

Cavers and Baldwin (1986) present stability tables calculated using CURVBUK. Independent variables in these tables include roll radius, slab thickness, and angular length of roll. Table 1 shows an example of these stability charts for slope dip ( ) above the roll of 40 degrees.

In Table 1, which is calculated for dry conditions, an unbracketed number is a value for P_A and represents the maximum driving force which may be applied to the top edge of the roll before failure. Bracketed negative numbers indicate that the slope is stable under all values of P_A and the slope would be at limiting equilibrium for the value of perturbing force indicated. This perturbing force was analyzed assuming L1=5T (T=thickness).

Figure 5 shows an example of calculation of the minimum height of slope above a semicircular roll of 8 degrees. This case is based on an actual operating coal mine. As discussed earlier, the minimum slope height is required to provide a restoring moment on the blocks in the roll. This is a dangerous method of operation, even if the slope height is present, since the normal force on the upper hinge point is dependent on the shear strength of the discontinuity bounding the slab. Nevertheless, the results may be used to provide some basis for vertical spacing of drains during mining past the roll.

Figure 6 shows the effect of anchoring the nose of a roll in terms of the allowable slope height above the roll and considering the effect of different water pressures. A second alternative would be to anchor the plane slab above the roll; however, this area may be inaccessible by the time that the roll is recognized.

The analysis shown in Figure 6 provides the restraining force required to prevent outward movement of the roll shown. Where anchors are used to provide this force, the depth of the anchors and the stability of the thicker slab formed by anchoring must also be determined. The use of anchors which are too short may lead to failure of the thicker beam particularly where the roll radius is large.

CURVBUK - 3 HINGED CURVED STRATA BUCKLING - INPUT

TITLE   TEST RUN

MODE   3

GEOMETRIC DATA

ALPHA	45.000	T	3.000	
RADIUS	345.000	L1	45.000	
THETA1	0.000	L2	72.257	CALCULATED IN MODE 3
THETA2	12.000			

FORCE DATA
CALCULATED(MODE=3)

GAMMAW	9.810	W1	3.375E+003	XW1B	14.849		
GAMMAR	25.000	W2	5.419E+003	XW2B	25.329		
WATD	23.000	U1	0.000E+000	XU1B	0.000	BETAU1	0.000
PHI	38.000	U2	3.182E+003	XU2B	62.508	BETAU2	34.561
C	0.000						
CALCULATED		D1		XD1B	0.000	BETAD1	0.000
XPAB	3.000	D2	0.000E+000	XD2B	0.000	BETAD2	0.000
XQAB	45.000						
XP3B	10.539						
XQ3B	71.730						

[F1] PRNT,SAVE   [F2] CALC INPUT   [F4] OUTPUT [F5] CALC PRNT [F6] DOS

Figure 4:   Input Screen for CURVBUK. A bar type cursor is moved between data items using the arrow keys.   The vaidity of items is checked at entry time.   The cursor jumps across items which are calculated by the program under a particular mode.

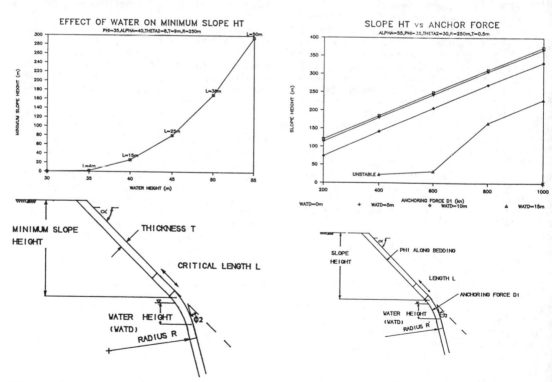

Figure 5: Effect of water on minimum slope height above a roll

Figure 6: Restraining force required to prevent movement of a roll

6 ACKNOWLEDGEMENTS

The authors wish to acknowledge the assistance of CANMET in funding the most recent studies. The assistance of Mr. George Headley, Mrs. Rikki Raison and Hardy BBT Limited in the preparation of this paper is gratefully acknowledged.

7 REFERENCES

Cavers, D.S. (1981): Simple Methods to Analyze Buckling of Rock Slopes, Rock Mechanics, N14, pp 87–104.

Cavers, D.S. and Baldwin, G. (1986): Report on Development of Design Criteria for Steeply Dipping Footwall Stability in Mountain Regions, 2 vols., DSS file number 075Q.23440-5-9221. (Prepared for CANMET)

Cavers, D.S., et. al (1986): Design Methods for Open Pit Coal Mine Footwalls; in Geotechnical Stability in Surface Mining, Raj K. Singhal, Ed. (Rotherdom, A.A. Balkema Publishers).

Chugh, A.K. (1977): Stability Analysis of a Jointed Beam; Int. Jour. for Numerical and Analytical Methods in Geomechanics, V.1, pp 323–341.

Corbyn, J.A. (1978): Stress Distribution in Laminar Rock During Sliding Failure, Int. J. Rock Mech., Min. Sci. and Geomech. Abstr., V. 15, pp 113–119.

Table 1. Stability Tables for Three Hinged Buckling

Alpha = 40 Degrees
Radius = 10 m
Value of P_A or (Perturbing Force) (KN)

THETA 2 DEG	T=0.5m	T=1m	T=2m	T=3m	T=6m	T=9m
10.0	(−20)	(−65)	(−227)	(−482)	(−1788)	(−3873)
15.0	(−23)	(−72)	(−241)	(−503)	(−1832)	(−3956)
20.0	7955	(−77)	(−252)	(−521)	(−1876)	(−4018)
25.0	1105	(−81)	(−261)	(−536)	(−1908)	(−4072)
30.0	545	6297	(−268)	(−548)	(−1935)	(−4114)
35.0	321	2053	266210	(−555)	(−1953)	(−4151)
40.0	227	1106	15887	(−558)	(−1962)	(−4168)
45.0	168	722	5773	48744	(−1961)	(−4175)

Radius 50m

THETA 2 DEG	T=0.5m	T=1m	T=2m	T=3m	T=6m	T=9m
5.0	(−33)	(−91)	(−281)	(−569)	(−2015)	(−4320)
10.0	13201	(−128)	(−355)	(−681)	(−2247)	(−5678)
15.0	2242	16858	(−417)	(−778)	(−2440)	(−4975)
20.0	1194	4820	47042	(−856)	(−2606)	(−5232)
25.0	849	2782	13688	56754	(−2733)	(−5426)
30.0	667	1945	7500	21112	534700	(−5581)
35.0	543	1500	5040	12169	102900	(−5676)
40.0	431	1157	3651	8096	47206	226160
45.0	348	910	2774	5837	28267	97298

Radius 100m

THETA 2 DEG	T=0.5m	T=1m	T=2m	T=3m	T=6m	T=9m
5.0	(−53)	(−132)	(−363)	(−694)	(−2302)	(−4764)
15.0	2368	8556	65488	(−1102)	(−3103)	(−5997)
25.0	1260	3394	11073	26028	226420	(−6867)
30.0	1029	2668	7780	16303	84447	362810
35.0	856	2120	5896	11766	48107	144920
40.0	712	1728	4629	8822	32384	83705
45.0	575	1393	3641	6863	23347	55434

Radius 300m

THETA 2 DEG	T=0.5m	T=1m	T=2m	T=3m	T=6m	T=9m
5.0	25917	(−319)	(−715)	(−1211)	(−3295)	(−6262)
10.0	6199	19811	99187	471910	(−4600)	(−8247)
15.0	4208	11037	34172	77008	589390	(−9917)

Table 1: Example of stability chart for dip above roll of 40 unbracketed numbers are maximum values for P (Figure A-1) while bracketed numbers represent the perturbing force to bring a stable slope to an unstable condition

APPENDIX A

DERIVIATION OF EQUATIONS

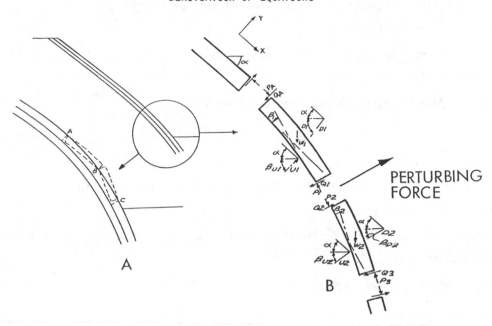

DEFINITIONS OF SYMBOLS

P_A	Normal force on upper hinge of Block 1
P_1, P_2, P_3	Normal forces acting on block hinge points
Q_A	Shear force corresponding to P_A
Q_1, Q_2, Q_3	Shear forces acting on block hinge points
W_1, W_2	Weight of Block 1 or 2
D_1, D_2	External forces applied to Blocks 1 and 2
U_1, U_2	Water pressure forces
X_{WIB} etc.	Moment arms of force about stated point (here the moment arm of W_1 about point B).
β $U1$ etc.	Angle between horizontal and named force (here U_1)
θ_1, θ_2	Curvature of Blocks 1 and 2
α	Dip of slope.

BLOCK 1:

Equilibrium in the X Direction:

$$-P_1\cos\theta_1 - Q_1\sin\theta_1 + P_A + W_1\sin\alpha - D_1\cos(\beta_{D1}+\alpha) + U_1\cos(\beta_{U1}+\alpha) = 0 \tag{1}$$

Equilibrium in the Y Direction:

$$P_1\sin\theta_1 - Q_1\cos\theta_1 - Q_A - W_1\cos\alpha - D_1\sin(\beta_{D1}+\alpha) + U_1\sin(\beta_{U1}+\alpha) = 0 \tag{2}$$

Moment Equilibrium About B:

$$-P_A X_{PAB} + Q_A X_{QAB} + W_1 X_{WIB} + D_1 X_{DIB} - U_1 X_{IB} = 0 \tag{3}$$

BLOCK 2:

Equilibrium in the X Direction:

$$P_2\cos\theta_1 + Q_2\sin\theta_1 - P_3\cos(\theta_1+\theta_2) - Q_3\sin(\theta_1+\theta_2) + W_2\sin\alpha - D_2\cos(\beta_{D2}+\alpha) + U_2\cos(\beta_{U2}+\alpha) = 0 \tag{4}$$

Equilibrium in the Y Direction:

$$-P_2\sin\theta_1 + Q_2\cos\theta_1 + P_3\sin(\theta_1+\theta_2) - Q_3\cos(\theta_1+\theta_2) - W_2\cos\alpha - D_2\sin(\beta_{D2}+\alpha) + U_2\sin(\beta_{U2}+\alpha) = 0 \tag{5}$$

Moment Equilibrium About B:

$$P_3 X_{P3B} - Q_3 X_{Q3B} - W_2 X_{W2B} - D_2 X_{D2B} + U_2 X_{U2B} = 0 \tag{6}$$

By Symmetry: $P_1 = P_2$ \hfill (7)

$\qquad\qquad Q_1 = Q_2$ \hfill (8)

SUBSTITUTING EQS. (7) and (8), REARRANGING AND EXPRESSING
IN MATRIX FORM YIELDS:

$$[A] \begin{Bmatrix} P_1 \\ Q_1 \\ P_A \\ Q_A \\ P_3 \\ Q_3 \end{Bmatrix} = \{B\} \tag{9}$$

[A] MAY BE SIMPLIFIED FOR INVERSION BY SUBSTITUTING EQS. (3)
AND (6) INTO EQS. (2),(4), AND (5)

SOLVING EQ. (3) IN TERMS OF Q_A YIELDS:

$$-X_{PAB}P_A + X_{QAB}Q_A = B_3 \quad \text{OR} \quad Q_A = \frac{B_3 + X_{PAB}P_A}{X_{QAB}} \tag{10}$$

SOLVING EQ(6) IN TERMS OF Q_3 YIELDS:

$$X_{P3B}P_3 - X_{Q3B}Q_3 = B_6 \quad \text{OR} \quad Q_3 = \frac{X_{P3B}P_3 - B_6}{X_{Q3B}} \tag{11}$$

REWRITING EQ. (2):

$$P_1\sin\theta_1 - Q_1\cos\theta_1 - \frac{X_{PAB}P_A}{X_{QAB}} = B_2 + \frac{B_3}{X_{QAB}} \tag{12}$$

REWRITING EQ. (4):

$$P_1\cos\theta_1 + Q_1\sin\theta_1 - P_3\left[\cos(\theta_1+\theta_2) + \frac{X_{P3B}}{X_{Q3B}}\sin(\theta_1 + \theta_2)\right]$$

$$= B_4 - B_6\frac{\sin(\theta_1+\theta_2)}{X_{Q3B}} \tag{13}$$

REWRITING EQ. (5)

$$-P_1\sin\theta_1+Q_1\cos\theta_1+P_3\left[\sin(\theta_1+\theta_2)-\frac{X_{P3B}\cos(\theta_1+\theta_2)}{X_{Q3B}}\right]=B_5-\frac{B_6\cos(\theta_1+\theta_2)}{X_{Q3B}} \tag{14}$$

THE ABOVE SYSTEM OF 4 EQUATIONS (EQS. (1), (12), (13), (14))
WITH 4 UNKNOWNS CAN BE EXPRESSED IN MATRIX FORM AS:

$$[A]\begin{Bmatrix} P_1 \\ Q_1 \\ P_A \\ P_3 \end{Bmatrix} = \{B\}$$

WHERE:

$$A = \begin{bmatrix} -\cos\theta_1 & -\sin\theta_1 & 1 & 0 \\ \sin\theta_1 & -\cos\theta_1 & \frac{-X_{PAB}}{X_{QAB}} & 0 \\ \cos\theta_1 & \sin\theta_1 & 0 & -\cos(\theta_1+\theta_2)-\frac{X_{P3B}}{X_{Q3B}}\sin(\theta_1+\theta_2) \\ -\sin\theta_1 & \cos\theta_1 & 0 & \sin(\theta_1+\theta_2)-\frac{X_{P3B}}{X_{Q3B}}\cos(\theta_1+\theta_2) \end{bmatrix}$$

$$B = \begin{Bmatrix} -W_1\sin\alpha + D_1\cos(\beta_{D1}+\alpha) - U_1\cos(\beta_{U1}+\alpha) \\[1ex] W_1\cos\alpha + D_1\sin(\beta_{D1}+\alpha) - U_1\sin(\beta_{U1}+\alpha) - \frac{W_1X_{W1B}}{X_{QAB}}-\frac{D_1X_{D1B}}{X_{QAB}}+\frac{U_1X_{U1B}}{X_{QAB}} \\[1ex] -W_2\sin\alpha + D_2\cos(\beta_{D2}+\alpha) - U_2\cos(\beta_{U2}+\alpha)-(W_2X_{W2B}+D_2X_{D2B}-U_2X_{U2B})\frac{\sin(\theta_1+\theta_2)}{X_{Q3B}} \\[1ex] W_2\cos\alpha + D_2\sin(\beta_{D2}+\alpha)-U_2\sin(\beta_{U2}+\alpha)-(W_2X_{W2B}+D_2X_{D2B}-U_2X_{U2B})\frac{\cos(\theta_1+\theta_2)}{X_{Q3B}} \end{Bmatrix}$$

279

Computer Applications in the Mineral Industry, Fytas, Collins & Singhal (eds)
© 1988 Balkema, Rotterdam. ISBN 90 6191 760 3

Optimization of fragmentation in underground blasthole mining

Yves Lizotte, Ajit Singh & Reginald Maclachlan
Department of Mining & Metallurgical Engineering, McGill University, Montreal, Quebec, Canada

ABSTRACT

A research project on fragmentation in underground blasthole mining has been initiated. The purpose of the study is to optimize fragmentation. Factors influencing size distribution and amount of oversize material, as well as consequential costs related to fragmentation are reviewed.

The optimization approach used involves an iterative integration of three computer modules. The first module is a commercially available program for predicting relative blast fragmentation as a function of drilling and blasting parameters. The second module, developed in-house, is a Load-Haul-Dump simulation program using both deterministic and stochastic elements and is well suited for assessing fragmentation effects on haulage production. The third module, integrates drill, blast and fragmentation parameters from the first module and performance parameters from the second module for a range of scenarios, to evaluate total system economics.

1 INTRODUCTION

The concept of optimization of fragmentation is not new. It is recognized that increasing drilling and blasting costs by some increment will not necessarily increase overall operating costs. Beyond recognizing this fact is the necessity to recognize and assess the inter-relationships between all appropriate cost components. Examination of fragmentation in underground mining leads the authors to re-stating that **"optimization of rock fragmentation implies minimization of total mining system costs"**. This can be demonstrated by examining the influence of fragmentation on each unit mining operation.

Several authors have stated that because not all costs increase proportionately to fragmentation, a minimum-cost fragmentation level could be achieved. The concept was initially put forward during the 1960's in the context of open-pit mining where the inter-relationships are more readily apparent. Increased mechanization and new bulk mining methods in underground mining arising during the 1970's inspired other authors to perceive the concept applicable to underground mining. However, the establishment of precise inter-relationships and difficulties related to comprehensive model formulations, measurements and computation of generalized solutions have hindered the practical application of the concept for underground mining.

The attraction of striving for the optimization of fragmentation in underground mining is apparent when one considers the greater cost of mining, on a unit product basis, compared to surface mining. A review of all the cost components associated with fragmentation is firstly warranted. It is also necessary to examine the factors which determine fragmentation. Further it is appropriate to re-examine the concept in the light of recent advances made in all the associated fields. This will demonstrate that a number of hurdles have been overcome and work can now be pursued to link these technological advances. The proposed computer model for underground blasthole mining is subsequently detailed.

2 THE COST OF FRAGMENTATION

In order to characterise the discussion of fragmentation related costs, a simplified conventional blasthole mining operation is used. Figure 1 schematically illustrates the operations involved, including the flow of broken ore from the stope to beneficiation processes in a mining operation using both trackless and rail equipment, crushing and hoisting. Where appropriate, note is made of other popular or possible ore flows.

Drilling and blasting costs are the first operations to link to fragmentation. However, these costs are (nearly) the only ones which increase with an increase in fragmentation. A first indication of the difference in strategies with respect to fragmentation is the wide gamut of equipment and techniques applied in blasthole, hard rock mining; ranging from fan drilling with 10cm holes on a tight pattern to parallel drilling patterns with holes up to 20cm for what could be regarded as **similar** mining conditions. A comprehensive examination could thus include costs associated to the development, layout and dimensioning of stopes.

The expenditures on ground support is in part linked to drilling and blasting practices. The geotechnical characteristics of the ore and wall rocks can cause varying degrees of dilution which can add large oversized fragments to the blasted material.

The size distribution and maximum particle size of the blasted rock will affect flow characteristics in drawpoints. High stress concentrations are induced around the drawpoints by the ore flow. Various costs are specifically linked to the size distribution:

1. Large blocks create arching which jams drawpoints and requires hazardous and time consuming techniques to resolve.
2. Blocks too large to be dropped into ore passes will require secondary blasting. This involves secondary blasting costs and delays during which time the drawpoint is occupied.
3. The effects that size distribution has on the loadability of the material. "Free-flowing" muck signifies faster loading time, smaller particle sizes signify easier digging and less equipment wear. Excess fines may lead to packing, and in some ores to cementation.

The size distribution will affect Load-Haul-Dump equipment wear and travel time. Secondary blasting costs must include LHD operating time and delays at drawpoints. Fragment size affects ore pass design, operation and wear. The impact of large fragments induces additional wear of rail cars.

The expenditure associated with crushing will be inversely proportional to drilling & blasting. It is less obvious that hoisting, surface transport and storage costs can be numerically expressed as functions of fragmentation, while a qualitative relationship may be apparent. Assuming the physical process ends at mineral liberation, the total energy input can be added and include grinding costs.

A conventional mining operation is selected for the purpose of illustrating fragmentation costs. Fragmentation optimization is equally important for new materials handling systems employing continuous loading devices and conveyors, their operation being sensitive to fragment size. Three cost categories are distinguished:

1. Costs which can be directly measured and linked to the fragmentation process. An abundance of industrial data is available for this purpose, organized in traditional well categorized costs of each unit operation for specific operating conditions.
2. Costs which were impossible to determine in the past without a great deal of field work. These costs can now be assessed due to technological break-throughs in associated areas.
3. Costs for which causal relationships to fragmentation can be stated but for which numerical functions can only be inferred.

3 FRAGMENTATION

Rock blasting is a complex phenomenon in which several parameters, difficult to evaluate dictate the outcome. These can be divided into four categories: site geometry and drilling parameters, explosive type, loading and firing sequence, and rock parameters. Geological factors are beyond the control of the blasting engineer, however with an intimate knowledge of the geological factors it is possible to better evaluate the controllable variables.

In the past, blasting was done on a trial and error basis. Based on experience a particular blast design was selected. Frequently this process back-fired, resulting in uneconomical, unsafe and undesirable results. After many years of working in the same geological environment, a blaster could eventually achieve the desired results.

Fragmentation is a difficult characteristic to quantify. In the general sense it refers to the fineness of the blasted product. The degree of fragmentation is influenced by the size of the material before blasting which is a function of the in-situ joint spacings and orientations. Improved fragmentation can be achieved by:
1. Increasing the powder factor by either increasing the drill hole diameters or closing the drilling patterns. Explosives with higher energy may also be used. This increase in powder factor has a limit, after which energy is wasted resulting in excessive flyrock, throw and high levels of vibration.
2. The existing powder factors can be used more efficiently in some operations Careful layout of the blast patterns, design of charges, control of the front row and the delaying sequence may lead to more efficient utilization of the explosive.

Although the science of blasting has developed to a great extent, there is no foolproof method of realistically assessing the resulting fragmentation.

For the purpose of optimizing the blasting operations, attempts have been made to evaluate and to predict the fragment size distribution in a blasted muckpile. Commonly the approach has been to derive size distribution equations, or to modify existing equations of comminution known to the field of mineral processing. Small scale blasts were conducted in an attempt to validate these models. Commonly used models are summarized in table 1.[2,3] Sophisticated prediction models have been developed on the basis of the Kuznitsov and the Rosin-Rammler equations [4,5]. This is essentially a one dimensional approach and uses the total available explosive energy and the average rock properties.

These evaluation methods have been severely handicapped by the high cost of sizing the fragmentation components of full scale field blasts. Visual techniques for blast fragmentation evaluation have been resorted to as a direct consequence. One such procedure, involving the photographing of blasted heaps in the course of their being excavated, was critically examined in an attempt to establish the precision of determining the oversize fragment content of a muck heap[6]. This work will be extended in the context of this project for assessing fragmentation in drawpoints. The potential for size segregation in blasted material will also be assessed.

4 TECHNOLOGICAL ADVANCES

A number of recent technological advances in various fields allow the authors to believe that fragmentation optimization is now a more attainable goal. The previous section outlined how fragmentation can now be predicted and measured. Large scale commercial software packages are now also available for improved overall blast design[4,7]. A better understanding of the influence of rock characteristics has been achieved[3,8,9]. Improved models for estimating drilling and blasting costs can now be applied[10].

The most significant technological advance is the increased use and availability of computer software and hardware. The problem can now be examined knowing that extensive data gathering, monitoring, data manipulation and lenghty iterative procedures can be computer-assisted. A variety of simulation models have been developed which enable fragmentation effects on materials handling systems to be ascertained[11,12]. The theory of bulk solids flow has been applied to blasted rock and ore passes in underground mines[1,13].

The present research is supported by the recent renewed interest in the optimization of fragmentation, for open-pit[14] as well as underground mining[15,16]. However, these studies rely on simplified approaches based on historical data interpretations and calculating a limited number of alternatives.

5 COST MINIMIZATION

The principal objective of optimum blasting in the overall mining environment can be defined as achieving a degree of fragmentation which allows minimization of the overall operating costs. The drilling and blasting is critical because it affects a host of other interrelated mining operations. The principal cost components are traditionally divided into loading, hauling, crushing and grinding. Figure 2 illustrates the relative size of these costs for a typical operation (excluding grinding), for a fixed fragmentation. Figure 3 shows the typical variation of unit mining costs with the degree of fragmentation. Generally as the degree of fragmentation goes up, drilling and blasting costs increase while all other costs decrease.

Optimization of the complete system may be achieved by an iterative approach using the concept of optimum blasting and by developing appropriate cost allocation procedures. It is essential to be able to predict the size distribution of fragments resulting from a drilling/blasting operation and to establish relationships between the degree of fragmentation and the relative costs of the other operations.

For a fixed degree of fragmentation, total costs are computed from the very fundamental equation of the total mining costs. This basic equation can be set up as:

$$C = C_{Dr} + C_{B1} + C_L + C_H + C_D + C_{Cr} \qquad (1)$$

where,

C_{Dr} = cost of drilling
C_{B1} = cost of blasting
C_L = cost of loading
C_H = cost of hauling
C_D = cost of dumping
C_{Cr} = cost of crushing

In this equation the direct costs of fragmentation are the drilling and blasting costs. For a given degree of fragmentation, the blasting cost can be determined. This cost is a function of the price of the explosive being used, the charge weight, cost of accessories and the labour costs. The minimum diameter of the blast holes for a particular powder factor is a straightforward calculation. The drilling cost (C_{Dr}) is given by:

$$C_{Dr} = \frac{V_R/B \times S}{D_{rate}} \times C_{(O\&O)} \qquad (2)$$

where,

V_R = volume of rock blasted
B = burden
S = spacing
D_{rate} = rate of drilling, distance/hr.
$C_{(O\&O)}$ = drill owning and operating cost in $ per hr.

The last four items of cost in equation 1, i.e, the loading, hauling, dumping and crushing costs depend upon the size distribution of the blasted muckpile. These costs are harder to quantify. The cost of loading and hauling decrease as the degree of fragmentation increases. The decrease in the loading cost is limited to a stage where no further reduction in loading time is achieved by finer fragmentation. Indeed loading costs may rise due to fines causing muck packing and reduced ease of flow to the digging bucket. A program has been written to determine hourly LHD costs[10]; the simulation program described below must be executed to determine the cost per ton associated to a de-

gree of fragmentation.

The cost of dumping can be measured indirectly by the delays caused due to bridging at the crusher or jamming of ore passes. This can considerably influence the performance of the primary crushing system and hence its economics. These can be reduced to some extent by resorting to secondary blasting. To determine the full extent of secondary blasting it is also necessary to use the simulation module since removal of oversize material affects LHD productivity.

The primary crushing cost is mostly an energy related factor and can be estimated from the comminution equations, e.g., Bond's index equation which directly relates the energy required to break the rock to a particular size.

Figure 1 suggests that a number of additional costs are associated to the degree of fragmentation. These cost could conceivably be determined from several years operating experience of an existing operation. Fragmentation affects equipment wear and subsequently equipment life. This can now only be assessed in relative terms. This is a recognized limitation of the global applicability of the present procedure.

6 ESTIMATION OF FRAGMENTATION

The Fragmentation program uses a set of basic parameters that need to be estimated. Initial parameters need not be very precise because the values are modified in the iterative process. Table 2 shows the input variables required for this program. These parameters can be determined from past experience at a particular mine or by using any of the various empirical relations available. One such formula currently used was developed by C.J. Konya[17] for burden determination:

$$B = 0.38D_e \frac{SG_e}{SG_r} \qquad (3)$$

where,

B = burden in meters
D_e = diameter of the explosive in mm.
SG_e = specific gravity of the explosive
SG_r = specific gravity of the rock

The stemming is normally approximately 0.7B and in weak and fissured strata this figure should be appropriately adjusted. At this stage the weight of the charge can be calculated.

$$CH = (HD - S) CLD \quad (4)$$

where,
CH = Charge weight
HD = Hole depth
S = Stemming
CLD = Charging loading density

The rock strength is estimated by experience. For very weak fissured strata this factor is taken as 1 whereas for strong and massive rocks it is 10. Once all the input variables are known they are used in the fragmentation module which gives the resulting size distribution. Sample outputs from this program are presented in figure 4 and table 3. If a satisfactory fragment size distribution is obtained then the LHD simulator can be started. Otherwise, the spacing and burden, the charge diameter or the charge weight are adjusted to achieve the required size distribution of the fragments.

The model used to predict fragmentation is commercially available. The predicted size ranges are close to actual field blasting conditions but the model tends to exaggerate the amount of the larger sizes while showing a smaller amount of fines. The predicted size distribution can be slightly modified on the assumption that muck from the collar region and back sloughing can account for some of the oversized muck. For example a scaling factor of 0.2 can be applied to the coarser fragmentation and 0.8 to the finer muck. After this conditioning of the size distribution from the fragmentation module, it is inserted as part of the information for the LHD simulation module.

7 THE LOAD-HAULAGE-DUMP SIMULATOR

A computer simulation program using both deterministic and stochastic elements has been developed prior to this research on fragmentation[11,12]. Three significant elements make this LHD simulator appropriate for studying the effects of fragmentation:
1. The program computes LHD travel times along haulage level circuits. Equipment interference along paths and at loading and dumping points can be realistically assessed. This is computationally facilitated by using deterministic simulation for estimating travel times from equipment performance.
2. Loading times are expressed as stochastic variables. A load time distribution can be stipulated directly as a function of the material characteristics.

3. A stochastic variable which is associated to the probability of encountering oversize material (requiring special handling and secondary blasting) is included in the program.

These three basic features enable the output from the Fragmentation module to be used for assessing the appropriate load and dump distributions and the probability of encountering oversize. Figure 5 illustrates the typical results achieved for simulating a LHD system for a fixed haulage distance with varying number of machines and different probabilities of encountering material requiring secondary blasting. Prior results have also demonstrated that the effects of oversize also vary as a function of haulage distances, enforcing the conclusion that optimization of fragmentation must be assessed for each specific case. The difference in this approach and prior published procedures is that the haulage costs not only vary as a function of the mean fragment size but also as functions of the size distribution and the percentage of large size material. The LHD simulator is further improved to include other effects of fragmentation by including the probabilities of complete drawpoint and ore pass hangups. The cost evaluation program can be used to determine equivalent annual costs and unit costs for various projected LHD economic lives.

The iterative procedure thus consists in using the Fragmentation and Simulation modules together with cost evaluation programs for assessing the total costs, including capital equipment, for a given scenario. The sequence of program use will depend on the particular case study, e.g. an operating mine or the development of a new mine. The studies may be assisted by automated information gathering of LHD performance and determination of activity times. The application of judgement based on knowledge from operating experience will be considered as a means of reducing the possible number of permutations involved.

8 CONCLUSIONS

It is now possible to assess optimum fragmentation in underground mining for the purpose of total mining cost minimization. A set of computer programs can be used to fully determine the influence of fragmentation on materials handling operations. However, there are a number of costs and parameters which are related to fragmentation, including stope layout, equipment useful life and wear. These factors limit the general applicability of the proposed procedure. Research work remains to be done to fully model the rock fragmentation (by explosives) process and gain a better understanding how fragmentation affects equipment and overall mine productivity.

Acknowledgements

The financial support of this research by NSERC (Canada) is gratefully acknowledged. Mr. Ross Hill is thanked for the research related to this project he has done.

References

1. Just, G. D., Rock fragmentation and the design of underground materials handling systems, CIM Bulletin, February 1980, pp. 45-51.

2. Just, G.D., The Application of Size Distribution Equations to Rock Breakage by Explosives, National Symp on Rock Fragmentation, Adelaide, February 26 - 28, 1973, pp 18 - 23.

3. Kondos, P.D., Rock Structure: An Important Factor In Forecasting Blast Fragmentation, M.Eng. Thesis, McGill University, 1983, pp. 328.

4. Jorgenson, G.K., Chung, S.H., Blast Simulation Surface and Underground With the SABREX Model, CIM Bulletin, August 1987, pp. 37-41.

5. Zeggeren, F. van, Chung, S.H., A model For The Prediction of Fragmentation, Patterns and Costs in Rock Blasting, Explosives Research Laboratory, Canadian Industries Limited McMasterville, P.O., Canada

6. Singh, A., Photographic Evaluation of Blast Fragmentation, M.Sc. Thesis, McGill University, Canada, 1983, pp. 200.

7. Favreau, R.F., "Blaspa" a Practical Blasting Optimization System, 6th Conference on Explosives and Blasting Techniques, 1980, Tampa.

8. MacLachlan R.R., Scoble M.J., Techniques For The Evaluation of Rock Mass Structure in Blast Design, Proceedings of The second Mini Symposium On Explosives And Blasting Research, Society Of Explosives Engineers, Atlanta, Georgia, Feb.13-14, 1986, pp. 132-144.

9. Ismail, M.A., Gozon, J.S., Effects o discontinuities on fragmentation by blasting, International Journal of Surface Mining, Vol. 1 (1987), pp. 21-25.

10. Borquez, G.V., Estimating drilling and blasting costs - An analysis and prediction model, E&mJ, January 1980, pp. 83-89.

11. Hill, R.D., A Micro-computer Based Underground Mine Haulage Simulation Program, M. Eng. Thesis, McGill University, 1987, pp. 147.

12. Hill, R.D., Lizotte, Y., Application of Computer Simulation to Evaluate Cost of Drawpoint Blockage, CIM-AGM, 1987.

13. Hambley, D.F., Design of ore pass systems for underground mines, CIM Bulletin, January 1987, pp. 25-30.

14. Nielson, K., Sensitivity analysis for optimum open-pit blasting, Proceedings of The Eleventh Conference on Explosives and Blasting Technique, Society of Explosives Engineers, January 1985, San Diego, California.

15. Nilsson, D., Optimization of fragmentation in an Underground Mine, Proceedings of the annual meeting, Society of Mining Engineers of AIME, Feb, 1985

16. Nielson, K., Optimum Fragmentation in Underground Mining, 19th APCOM Symposium, AIME, New York, 1986, pp. 746-753.

17. Hemphill, G.B., Blasting Operations, McGraw Hill Book Company, 1981, pp. 258.

Equation	Author(s)	General form of equation
1	Bond	$W_i = \dfrac{W\sqrt{P}}{10}$ or $P = 100(\dfrac{W_i}{W})^2$
2	Gates-Gaudin-Schumman	$Y = (\dfrac{X}{k})^m$
3	Rosin-Rammler	$Y = 1 - \exp(-bX^n)$
4	Gaudin-Melov	$Y = 1 - (1 - \dfrac{X}{X_0})^r$
5	Da Gama	$P_s = aW^b(\dfrac{S}{B})^c$
6	Just	$Y = \exp 4.8(\dfrac{X}{D})^g$, and $q - \dfrac{1}{F}(\dfrac{d}{D})^2 = \dfrac{R^2}{F}$

Where,

W_i = work index
W = energy input in kw hrs/ton
P = size 80% of the product passes
Y = fraction of the material finer than size X
X = characteristic partical size
k = size modulus
m = distribution mudulus
b = constant in Rosin-Rammler equation
n = distribution parameter
X_0 = maximmum particle size
r = distribution parameter
P_s = percent cummulative undersize weight of size "S"

a, b and c = parameters in Da Gama's equation, site specific to rock type, blast pattern and explosive used
S = size of the fraction in the Da Gama equation
B = burden
q = fragmentation gradient
F = fragmentation factor
d = actual charge depth
D = optimum depth for the weight of explosive detonated
R = charge depth ratio

Table 1. Commonly used size distribution equations in rock fragmentation by blasting.

```
Diameter of the explosive............................_____
Burden................................................_____
Spacing..............................................._____
Stemming............................................._____
Sub-drill............................................._____
Bench height........................................._____
Hole Depth..........................................._____
Pattern of drilling (square or staggered)............_____
Number of rows Blasted..............................._____
Explosive strength (ANFO =100)......................._____
Rock strength........................................_____
Explosive charge /hole..............................._____
```

Table 2. Input data specifications for fragmentation program

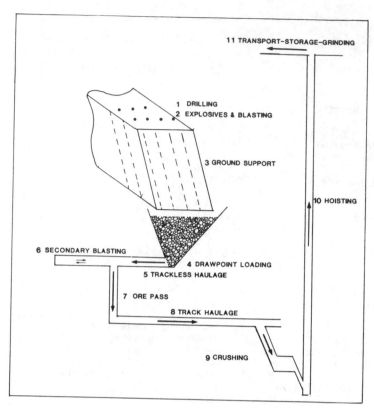

Figure 1. Costs associated to fragmentation in blasthole mining

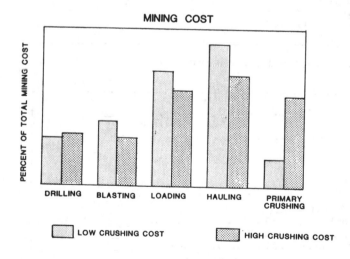

Figure 2. Relative cost of various mining components for a typical operation

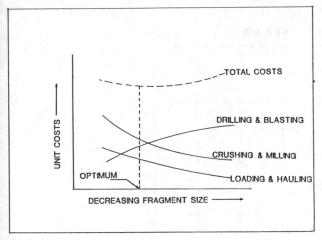

Figure 3. Typical costs of mining operations and optimum blasting definition

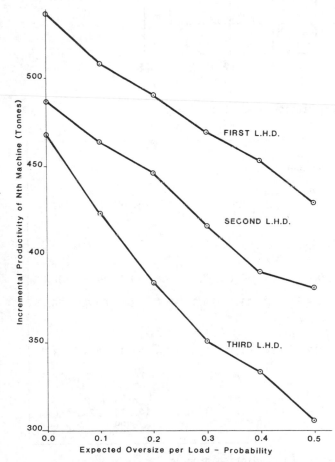

Figure 5. Load-Haul-Dump productivity versus expected amount of oversize (11)

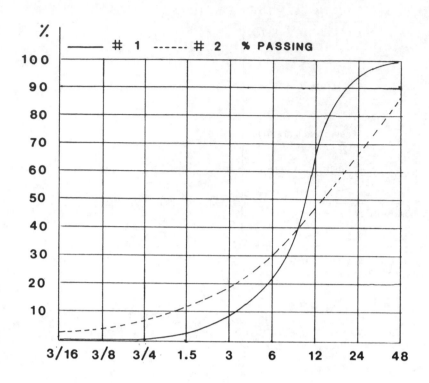

Figure 4. Typical results of fragmentation prediction program.

	Screen	Size							
(in.)	3/16	3/8	3/4	1/5	3	6	12	24	48
(mm)	4.8	9.5	1.9	3.8	7.6	15.2	31.0	61.0	122.0
Entry No 1									
% Passing									
Entry # 1	0.0	0.2	0.6	2.1	7.2	23.2	60.7	96.3	100.0
Entry # 2	2.2	3.9	6.8	11.5	19.3	31.3	48.3	68.5	86.8
% Fraction									
Entry # 1		0.2	0.4	1.5	5.1	16.0	37.4	35.6	3.7
Entry # 2		3.9	2.9	4.8	7.8	12.0	16.9	20.2	18.3

	Entry No. 1	Entry No 2
Average size - in (mm)	10.189 (258.8)	12.775 (324.5)
Fragmentation index	1.820	0.809

Table 3. Typical results of fragmentation prediction program.

Computer Applications in the Mineral Industry, Fytas, Collins & Singhal (eds)
© *1988 Balkema, Rotterdam. ISBN 90 6191 760 3*

Computer applications in geomechanical stope and pillar design at Noranda Minerals Inc.

W.F.Bawden, D.M.Milne, W.Hrastovich, P.Germain & H.Kanduth
Department of Geomechanics, Centre de Recherche Noranda, Pointe Claire, Quebec, Canada

ABSTRACT: The science of applied rock mechanics forms the basis of any engineered mine design. The rapid escalation in microcomputing power during the past ten to fifteen years has been instrumental in the realization of rock engineering as a practical mine design tool.

Geomechanical stope and pillar design incorporates a number of interrelated areas, each of which relies to a greater or lesser degree on the practical use of computers. These key areas are listed below:

1. Geomechanical database
2. Laboratory testing systems
3. Numerical stress analysis models
4. Empirical mine design programs
5. Blast analysis
6. Field instrumentation systems

Although specific cases of these areas may require the generation of dedicated software, there exists public domain and/or very low cost software that can be adapted or used directly to perform most of the above functions. All of the necessary software   can be run on powerful but low cost modern microcomputer hardware.

This paper describes how the Centre de Rechrche Noranda has used such packages to fulfill most of its software requirements. Brief examples of the utility of the programs, and their benefits and limitations are given.

1. INTRODUCTION

The use of computers in mine engineering offices has increased substantially during the past few years. This has been especially true with operations that have an in-house geomechanics team working as part of the mine engineering group.

The Mining Technology Division (MTD) of the Centre de Recherche Noranda makes extensive use of computer facilities. The heart of the MTD system is a Microvax II minicomputer supporting 16 users, including a modem and a central graphics station. Individual users are generally supplied with PC's (e.g. IBM-XT equivalent), hardwired to the Microvax. A limited number of portable computers are also used. The latter are easily transportable to field operations which may be lacking suitable hardware.

The Microvax is run using the UNIX operating system which incorporates multi-user capabilities. While most engineering software is written in professional Fortran, some of the more specialized software is written in Basic. The more powerful C language is not yet being used at the MTD, largely due to lack of training and familiarity.

Additionally to the already described hardware the MTD also has two older HP9816 microcomputers. These units are now dedicated to data acquisition and processing systems in the laboratory. One unit acts as a dedicated data acquisition/processing system for MTD's Bemek index test and shearbox test systems, while the second unit is used as a back-up as well as for other more specialized testing and software development work.

Most of the Noranda Minerals Inc. mine engineering offices operate with IBM-AT or equivalent computers. More recently, some operations have been changing to newer 32-bit microcomputers. It is believed at the MTD that much more extensive use of computers is warranted at operating mine engineering offices. As more technology, such as numerical mine simulation, is transferred to the mine operators as a day to day design tool, computer use will rapidly become more intense.

The potential utility of sophisticated computing systems in the mine engineering office, however, goes well beyond the limited area of geomechanics. In order to realize the potential, much more extensive database systems, which are dedicated to mine engineering needs, will be required. Such systems should include the complete geomechanics database plus complete data on stope and pillar geometries, dilution levels, blasting, surveying, etc. Such comprehensive databases, combined with rapid advances in the numerical modelling field, will result in the need for again increased computing capabilities both at the operations and at central technical groups. If the recent trend in computing hardware continues, the need for such increased computing power should present no problem.

There is no question that the increased "effective" use of computers at mining operations can result in significantly enhanced productivity for technical groups such as engineering, geology and surveying. As improved databases, etc., are developed and the use of the computer becomes commonplace in the mine engineering office, the realized benefits are certain to increase manyfold beyond our present assumptions.

This paper describes the present extent of computerization and its use in geomechanical stope and pillar design at the Centre de Recherche Noranda. More specifically, the paper describes how readily available and relatively inexpensive software can be used to perform reasonably complex design analyses. It goes without saying that such analyses must be tempered with good common sense and sound engineering practice.

2. GENERAL GEOMECHANICS DATABASE

The key to any rational geomechanics design is realistic input for the properties of the rockmass. A rockmass is a highly complex medium, making it very difficult to quantify the mechanical properties. Due to the highly variable nature of a rockmass, a large number of measurements are required to obtain statistically valid properties. The necessity for a large database makes this an area where readily available commercial software packages for database management are of value.

Two principal areas of a geomechanics database include the laboratory and the field database. These are discussed in the following sections.

2.1 Laboratory testing systems

Key input in geomechanics mine design are the rockmass properties. Laboratory testing can be used to determine the properties of intact rock as well as the strength characteristics of joints and discontinuities.

In an effort to make laboratory testing more efficient and to handle the large amount of data generated by testing, the MTD has successfully integrated two testing systems with a computer. One system is a rock index test machine and the other is a direct shearbox. The systems are discussed below.

2.1.1 Rock index test machine

The MTD uses an inexpensive index test apparatus (the Bemek Rock Tester) for the determination of the basic mechanical properties of rock. Low cost of testing is possible by employing such methods as a 4-point bend test to determine the modulus of elasticity and the Poisson's Ratio, a point load test to determine the point load index and the uniaxial compressive strength, and a Brazilian test for the indirect tensile strength. The advantage of the Bemek is that it allows a statistically more valid, albeit relatively unsophisticated determination of properties. The increased use of computer modelling for the study of rock excavation has created the need for such statistical data.

In the case of Noranda, tests have been carried out for most operations for mine design and numerical modelling studies. The work invariably involved testing of a large number of core samples from various rock types, the results of which were then transferred onto an HP9816 computer for the performance of statistical analyses.

The transfer by hand of the data to the computer was a lengthy and arduous task, greatly diminishing the efficiency and the cost effectiveness of the apparatus. In addition, it was realized that preparatory work, such as sample numbering and description, as well as documentation, could be streamlined. Integration into a computer system was the obvious solution to the shortcomings. An integrated system was designed, comprising the following components:

1. Bemek testframe with hydraulic pump.
2. Electronic bridgebox-balancing unit for the pressure transducer and two strain gauges.
3. HP3421A data acquisition/control unit, in effect an A/D converter transferring the data to the computer.
4. HP9816 computer with printer and plotter peripherals, controlling the test parameters, acquiring the data, performing calculations and statistical analyses, and providing printed reports and graphs.

In-house software was developed to control the tasks, the key elements of which are i) sample numbering and description, ii) test control and iii) statistics and graphs. The results of the computerization have been very satisfactory. The unit cost of testing has been reduced considerably and the actual execution of the tests has been made easier. The reduced cost of testing will encourage operations to have more tests carried out and thus create better input data for numerical and empirical modelling.

2.1.2 Shear testing device

Shear testing is another area where a data acquisition system can save substantial time and energy, as well as enhance the precision of the resulting statistical analysis.

The PHI-10 portable shearbox was obtained to determine laboratory shear strength values for rock discontinuities. The apparatus was designed so that the data is fed via a data acquisition unit directly to the computer. Easy to use software was developed that reads data directly from the shear test, stores it on a diskette and allows the user to manipulate the data from one or more tests to produce several kinds of plots. The program is written in Basic 2.0 and is designed to run on the HP9816 computer in conjunction with an HP3421A data acquisition/control unit and the Roctest PHI-10 shearbox signal amplifier.

The program is divided into two components, the data acquisition and the data manipulation routine. During the Data Acquisition Phase, readings from the instrument's Linear Variable Displacement Transducer (LVDT) and two pressure transducers are recorded in accordance with the sample mode defined by the user. Three sampling modes are available:

- DeltaDispl (readings initiated by change in shear displacement)
- DeltaShear (readings initiated by change in shear loading)
- DeltaTime (readings initiated by change in time)

The Data Manipulation Phase is divided into two separate stages. The first stage allows the user to work with a single data file and perform operations such as reading and storing data files, plotting normal and shear loads against displacement or putting out the raw data listing. The second stage lets the user plot and compare the results of a series of tests. It is in this stage that the peak and residual Phi angles are calculated. Figure 1 gives an overviw of the general flowchart of the software.

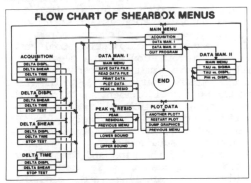

Figure 1. Flowchart of shearbox menus.

2.2 Field database

Accurate field data is essential for any realistic mine design. Field data is also required to relate intact properties to those of a fractured rockmass.

A large number of joint measurements are required to delineate joint sets in the rock. Computer programs are available to record and analyze such data and represent it graphically. Rockmass classification is done throughout the mine to estimate the variability and magnitude of the rockmass strength and modulus. This data can be recorded on a computer along with the mine geometry. Another invaluable piece of field data is a record of dilution, ground falls and stope or pillar failures, all of which can be used to calibrate numerical and empirical modelling techniques. Also, background data on the type of ground support and on blasting practices can prove very successful.

To develop a comprehensive field database, it rapidly becomes obvious that a great deal of data will be generated, and that the data must be easily cross- referenced. Commercially available software packages can be ideal for the task, and the MID is working towards integration of all the above data into a single comprehensive database. A database of this magnitude requires input from both the MID and the operations. An approach for handling such a database is described below.

Each mine will be equipped with a suitable P.C. and will be able to communicate via modem with the Research Centre where the information will be centralized. Operators will be able to send information to an electronic mail box where it will be filtered, reorganized and transferred into a central database which is accessible to all the operations. For the mines it will represent a comparative tool and source of information which, hopefully, will stimulate communication between all Noranda mining operations. For the MID it will be a tool to identify trends and needs among operations and to help modify our research accordingly.

The database will comprise the following subdirectories: general description of the mine and its mining method(s); geomechanics laboratory and field data; drilling and blasting practices; mining equipment; ventilation. From the geomechanics point of view the data will serve as input for numerical and empirical mine design tools.

3. NUMERICAL STRESS ANALYSIS

There are three basic types of numerical stress analysis models which are commercially available. These include i) finite element models (FEM), ii) boundary element models (BEM) and iii) distinct element models (DEM). All of these models are used to calculate the flow of stress around mine openings. Each has particular inherent strengths and weaknesses which are mentioned here briefly.

Computing hardware requirements vary widely for the various programs, depending on the type of model and the required sophistication. There is also variation between the models with respect to the ease of data input and the interpretation of the data output. All these parameters must be carefully considered prior to software selection. The present software requirements at the MID can be summarized as follows: simple elastic stress analysis to derive predicted stresses, displacements and failure zones at Noranda Minerals Inc. hard rock underground mines; elastic stress analysis incorporating effects of multiple materials and key geological structures to derive predicted stresses, displacements and failure zones at Noranda Minerals Inc. hard rock underground mines; stress analysis, including large displacements, discrete block behaviour, key block analysis, etc.; simple elasto-plastic stress analysis to give time dependent stress and displacement at Noranda Minerals Inc. soft rock (potash) mine.

The MID has obtained commercial software at a reasonably low investment that will satisfy most of the objectives mentioned above. Where suitable comerical software was not available it has been developed under contract for Noranda. In summary, total software investments have been minimal while derived benefits in terms of helping solve day to day operating problems have been significant.

3.1 Finite element codes

Finite element technology today can be considered highly advanced. Commercial modelling packages are available that are capable of handling very sophisticated problems (e.g. elastic, non-linear, plastic, thermo-elastic/plastic, etc.). However, this software is not yet widely used in the mining industry despite its advanced technology level and the potential benefits that could be gained. The main reason for this is the time consuming nature of the data input and the large demands on computer hardware.

The MID has both the 2-D and 3-D finite element codes available. Both are suitable for use on an IBM-AT or equivalent microcomputer. Naturally, the allowable problem size is dictated by the available computing power. Additionally, the MID has a 2-D visco-plastic Finite Element Model (FEM) "GEOROC", which was obtained from CANMET. This model is used for mine simulations at Noranda Minerals Inc. Central Canada Potash operation. Another package, Viscot, is commercially available through the U.S. Bureau of Mines and J.F.T. Agapito and Associates Inc. of the United States. There is evidently a good supply of commercially available finite element models that are capable of satisfying most present day mining requirements. The problem with this technology rests more with affordable computing power and input/output requirements.

3.2 Boundary element codes

Boundary element type codes include both so-called "boundary element analysis" (BEA) and "displacement discontinuity" (DD) codes. The mathematical basis for such models is thoroughly discussed in the standard text by Crouch and Starfield. For most practical purposes the models perform the same tasks as those of the finite element models (FEM) discussed earlier. One key difference between the FEM and the boundary element model (BEM) is that the latter needs only the discretization of the excavation boundary, resulting in much simpler data input than that for the FEM's. There are a number of BEM's available in the public domain, two of which are used by the MID as described below:

1. A 2-D BEM, given in the standard text by Hoek and Brown, coupled with the ubiquitous joint program "DISCO", developed by P. Choquette, Universite Laval.

The program is a simple 2-D, single material, linear elastic model which is useful for the 2-D analysis of inclined or horizontal transverse sections.

Figure 2 is a graphical representation of the maximum principal stress calculated by this program.

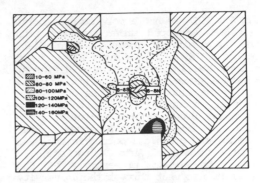

Figure 2. Stress contour of a mine cross section.

2. A linear elastic 2-D displacement discontinuity model "Mintab", obtained from CANMET and downloaded to run on an IBM-AT or equivalent computer.

This program allows modelling of the longitudinal section of tabular orebodies. The model is strictly linear elastic but does allow modelling of stope sequencing. Mathematically this model assumes that the orebody is a thin slit, which makes the program most applicable to rather thin, tabular orebodies.

Near the openings only stresses perpendicular to the section are accurate. Therefore, principal stresses cannot be calculated and failure criteria cannot be incorporated. For this reason the model is used primarily for stress analysis through key pillar areas. The Mintab model also allows the inclusion of backfill. However, the value of this is questionable since it is strictly a linear elastic model (i.e. small displacement). A capability to determine "off-reef" stresses is also included in the program. Figure 3 shows the result of a Mintab simulation on a longitudinal section of an orebody.

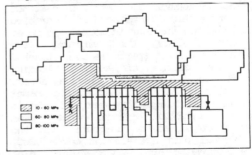

Figure 3. Typical result of a Mintab simulation on a longitudinal section of an orebody.

295

3. The MID has had a more comprehensive 2-D BEM developed under contract with the University of Toronto. The model, called "MINE 5", is linear elastic and allows the inclusion of five different material types, discrete joints, faults, etc., and also includes the ubiquitous jointing program of Choquette and the Hoek and Brown failure criteria. The data is input using a spread sheet format, and graphical output is achieved via the commercial Auto-CAD package.

4. The MID has presently a 3-D BEM under development through the University of Toronto. The objective of the program is to develop a basic linear elastic 3-D BEM model with sophisticated "solid geometry" input and output so that the input can be verified and the output interpreted. The reason why the model is being developed is the large number of cases, including complete orebodies or parts of orebodies, where none of the models discussed above give suitable results due to geometric boundary constraints.

3.3 Distinct element codes

The development of distinct element or block model numerical codes has been led by the work of P. Cundall. The initial models were largely of academic and teaching interest and were considered useful primarily for obtaining an improved qualitative understanding of the behaviour of jointed rock masses.

Development in this area, however, has been expanding rapidly. A 2-D distinct element model is now commercially available through Itasca Consultants, Minneapolis, and a 3-D version is under development. Additional work is underway cooperatively between Itasca and the Norwegian Geotechnical Institute to incorporate Barton's constitutive joint behaviour in the 2-D distinct element code. New methods of describing rock joint behaviour in normal and shear loading allow the physical and conducting joint apertures to be tracked in the course of complex loading histories, i.e., during closure, shear, dilation, shear reversal, cyclic normal loading, with asperity wear, etc. This joint behaviour model has been incorporated as a sub-routine in the universal distinct element code (UDEC) for studying the changes in joint geometries that can occur in jointed reservoirs under drawdown, and in rockmass surrounding tunnel excavations, with and without internal or external

fluid pressure.

Naturally, problems remain to adequately define joint properties behind an exposed face.

3.4 Finite difference code

One of the earliest numerical techniques to receive extensive computerization was the finite difference method. Most of this early work was done in the field of hydrogeology. Later, extensive developments in the finite element technique largely overshadowed the finite difference method.

The finite difference technique has recently been revived in a novel numerical model developed by Itasca, called FLAC which stands for Fast Lagrangian Analysis of Continua. The technique appears quite practical and useful, especially since it allows for large displacements. The model has the advantage of modelling plastic flow, so that the post peak failure behaviour of pillars can be modelled. Also, the interaction of rockbolts and other support can be taken into account.

The code is available commercially and several mining corporations, including Noranda, have purchased it. At the time of writing, however, the MID has not had time to thoroughly test the model's capabilities in practical mining situations. It is hoped to have this done by the time this paper is presented in March 1988.

3.5 Modelling summary

Numerical modelling packages are available, which can fill almost all the requirements of stress determinations at Noranda Minerals Inc. mines. Limitation still exists in obtaining programs which can handle three dimensional geometry in a usable fashion. Some mine geometries can only be simulated realistically with true 3-D geometry which has traditionally been very time consuming to input and resulted in very cumbersome output. The shortcomings in data output, however, are rapidly being overcome with improved computer graphics.

Another area which is in need of improvement in the field of mine modelling is the description of the mining medium. Rock is a complex material, and the properties of this material control the results of the computer stress and failure analysis. Due to the difficulties in analytically determining realistic rockmass properties, more empirical techniques are often used. The empirical method most frequently used at the MID is described in the following section.

4. EMPIRICAL STOPE DESIGN

An empirical stope design technique, called the Mathews Method (see reference) has been developed to analyse the stability of open stopes. The technique is graphical and plots two factors against one another.

The first factor, the stability number 'N', accounts for the properties of the rockmass and the existing induced stress acting on the stope. The second term is the shape factor 'S' which accounts for the geometry of the stope surface. This term is simply the area of the surface divided by the perimeter. The two factors then plot on regions of a graph which are defined as either stable, potentially unstable or potentially caving.

The technique is simple to use and provides a useful check against numerical analysis results. The analysis is easily performed by hand, but it is planned to integrate it into a comprehensive computing package which will also include a rock mechanics database and a mine design package. The empirical method requires the induced stresses around stopes, which can be estimated accurately only with the numerical modelling programs discussed in the preceding section. Also, laboratory determined rock strength values and structural data gathered in the field, as discussed in Section 2, are required in the analysis.

Since the technique is empirical, it is constantly upgraded for each operation as the database for case histories is expanded. All the factors discussed above point out the value of an overall database where background geomechanics data, numerical and empirical design techniques and case histories can be accessed from one single system.

5. FIELD INSTRUMENTATION

Instrumentation plays a very important role in geomechanical mine design. Geomechanics is not an exact science on account of the difficulty in quantifying the properties of the rockmass.

Therefore, it is always necessary to verify the validity of a particular design, generally by monitoring changes in stress or the competence of rock in response to mining activity.

In general, instrumentation is designed to measure movement, whereby small movement is interpreted as changes in stress and larger measurements are considered ground dislocations associated with discrete structural elements. In the mining industry instrumentation is not used to the extent that would be desirable. The reason is the expense associated with its purchase and installation, but more importantly, the expense of sending men underground to read the instrument output data, a task which is labour-intensive and time consuming.

Research at the MID is directed towards making the monitoring process more efficient. One of the possibilities investigated is the installation of an underground cable network, possibly fiber optic, and permanently wiring the instrumentation into the network. The data can be fed directly to a computer, either located underground or on the surface, for immediate data interpretation. This approach is very attractive for mines which have an existing network or are planning the installation of one for the purpose of microseismic monitoring, allowing the instrumentation signals to be "piggy-backed" onto the system.

In addition to making the monitoring system more efficient and cost-effective, such an approach would allow the computerized integration of the data input into existing databases where it is used for the calibration of the various numerical models.

6. BLAST DESIGN AND ANALYSIS

Blast design and analysis is a very important aspect in mine design. In the process of blasting ore in a mine, considerable damage can be done to the surrounding rock. This damage changes the properties of the rock and, if ignored, will cause the geomechanical design to be overly optimistic. The results could be serious stability problems. Therefore, mine excavations must be designed on rock quality after blasting.

Blast design has traditionally been regarded as an empirical art. Considerable work has been done in the past few years in an attempt to make this art more rational. Efforts have included both the development of numerical models to assist in the design of blasts for particular situations, and research into monitoring blasts to evaluate their efficiency and effectiveness. The most important research in the latter has been conducted by the Julius Kruschnitt Mining Research Centre, Australia, over the past ten years.

The MID is active in both research areas, but with special emphasis on the area of blast monitoring. The monitoring hardware and much of the signal processing software is very similar to that used for microseismic monoitoring, another area where the MID is active. Consequently, efforts in blast and microseismic monitoring have been complimentary. Research in blast monitoring by the MID is discussed in a companion paper presented at this conference by Mr. D.L. Sprott, titled "Computer Processing of Blast Vibration Records."

It is believed at the MID that considerable improvement in blasting practice is possible through improved design and monitoring. Computer usage in both these areas will continue to be of importance. It is also believed that many of the geomechanical problems will be minimized with improved blasting practices.

7. CONCLUSION

The key components of a geomechanics mine design have all been discussed in this paper. They include a laboratory and field data base to quantify the properties of the rockmass, numerical and empirical models to predict the ground behaviour relative to the induced stress, and field instrumentation to verify the accuracy of the results. Blast design and analysis was included because of the profound effect blasting can have on the properties of the rockmass.

All of the components of a geomechanics mine design can be, and in many instances, have been computerized. This paper is an attempt to show at what stage the MID is in computerizing its geomechanics design work and what the goals are with regard to developing an integrated database and mine design package.

REFERENCES

Barton, N. & Makurat, A. 1987. Modelling rock mass conductivity changes in disturbed zones. 29th U.S. Symposium on Rock Mechanics, Tuscon, 29 June to 1 July 1987.

Crouch, S.L. & Starfield, A.M. 1983. Boundary element methods in solid mechanics. George Allen & Unwin, Pub.

Matthews, K.E., Hoek, E. & Wyllie, D.C. 1980. Prediction of stable excavation spans for mining at depths below 1,000 meters in hard rock. CANMET DSS serial no. OSQ80-00081, April 1981.

Computer Applications in the Mineral Industry, Fytas, Collins & Singhal (eds)
© 1988 Balkema, Rotterdam. ISBN 90 6191 760 3

Vers un système expert pour le choix du soutènement en galeries

H.Baroudi & R.Revalor
Cerchar, Nancy, France

J.P.Piguet
Ecole des Mines, Nancy, France

RESUME : Depuis 1985, une recherche a été entreprise en vue de l'élaboration d'un système expert pour le choix du soutènement en galeries. Une phase d'étude de faisabilité
a conduit au choix d'un moteur d'inférence et à la mise au point d'un prototype sur micro-
ordinateur (compatible PC), comportant environ 150 règles intéressant le soutènement par
boulonnage et par cintre.
Le projet s'est poursuivi par l'enrichissement de la base de connaissances grâce aux
résultats d'une enquête menée systématiquement dans des galeries minières. Le système
expert dans sa forme actuelle permet de recourir aux modèles de calculs analytiques
couramment utilisés par les experts.
Un des avantages les plus intéressants de cette approche, l'aspect évolutif, sera mis à
profit pour adapter dans le futur le système au choix du soutènement dans les ouvrages
de Génie Civil.
L'article analyse les choix qui ont été faits lors de l'élaboration du système ; il présente l'outil dans son état actuel ainsi que son application au choix du soutènement
dans quelques galeries où ce problème était posé.

La France produit annuellement 16 millions de tonnes nettes de charbon dont 90 % à partir d'exploitations souterraines.

Pour réaliser ce résultat, 78,6 kms de galeries sont creusés chaque année dont 55,4 kms sont des galeries au charbon qui découpent les futurs panneaux exploités.

Or, depuis un certain nombre d'années, une évolution caractéristique se dessine : le coût de ces creusements pèse de plus en plus lourd dans le prix de revient global des exploitations. Cette tendance est liée :

* aux investissements de plus en plus importants réalisés dans ces chantiers en liaison avec leur mécanisation
* à la nécessité de creuser et de maintenir ouvertes des sections de galeries plus importantes
* à l'obligation de soutenir les chantiers dès la phase de creusement en mettant en oeuvre des soutènements de plus en plus performants.

L'amélioration des performances des soutènements des galeries avait toujours été un souci constant de l'industrie charbonnière française, en particulier compte-tenu des conditions de plus en plus difficiles rencontrées dans les chantiers à grande profondeur. Néanmoins, depuis 4 ans, sous l'impulsion du CERCHAR (*) et des mines concernées, la recherche dans ce domaine s'est intensifiée en se focalisant sur 2 points :

* rassembler les différentes méthodes de choix de soutènement existantes et les adapter aux conditions actuelles régnant dans les chantiers
* intégrer ces méthodes dans un outil d'aide à la décision, type Système Expert.

L'article rassemble les principaux résultats tirés de cette étude: il présente le système expert qui a été conçu pour être appliqué au choix du soutènement des galeries minières.

1 LA PROBLEMATIQUE DU CHOIX DU SOUTENEMENT DANS LES TRAVAUX SOUTERRAINS

Dans les travaux souterrains, la décision de mettre en oeuvre, dans un chantier donné, un procédé de soutènement (ou de renforcement) est souvent guidée par l'avis "d'experts". Ceux-ci sont explicitement (ou implicitement) reconnus comme experts, de par l'étendue de leur expérience vécue dans le domaine.

(*) CENTRE D'ETUDES ET RECHERCHES DE CHARBONNAGES DE FRANCE
B.P n° 2 60550 Verneuil en Halatte, France

Ils peuvent fonder leur avis sur un ensemble assez disparate de considérations théoriques et pragmatiques comme par exemple :
* des informations concernant la géologie, les conditions hydrogéologiques, les données géomécaniques...
* la géométrie, la durée de la vie, la fonction dévolue à l'ouvrage
* les matériels et matériaux disponibles, l'expérience et les habitudes des opérateurs

* les résultats de calculs, etc
Il faut souligner deux particularités importantes dans le raisonnement des spécialistes appliqué à "des objets naturels" (comme les terrains).
1. Il se nourrit souvent d'appréciations qualitatives et doit beaucoup à l'observation
2. Les propriétés caractéristiques de ces objets sont toujours dispersées et malaisément prévisibles ce qui conduit à travailler avec un niveau d'incertitude parfois grand.
En conséquence, on fait largement appel, dans ce domaine, au raisonnement par analogie avec des situations jugées comparables, rencontrées antérieurement et ailleurs.
Aussi, les choix fondamentaux concernant le soutènement (y compris son dimensionnement et la détermination des modalités techniques de mise en oeuvre) sont sous l'influence partagée de l'empirisme (plus ou moins formalisé) et de l'approche scientifique mathématisée.
On retrouve cette dualité dans les méthodes de choix et de dimensionnement du soutènement en galerie. Certaines méthodes sont exclusivement empiriques et si peu formalisées qu'elles peuvent être non écrites (ainsi les diverses "Règles de l'Art" comme le fameux "Art des Mines"). D'autres, au contraire, reposent sur la conduite de calculs complexes à partir d'une modélisation mathématique des terrains, du soutènement et de leur comportement mécanique.

Méthodes empiriques élaborées

* Méthodes statistiques directes : Elles consistent à utiliser, pour prévoir le comportement d'un ouvrage et le soutènement le mieux adapté, des lois statistiques établies à partir de données recueillies dans un grand nombre d'ouvrages analogues. Par exemple, une formule proposée dans les charbonnages allemands (GROTOWSKY, 1977) donne une estimation de la convergence dans les voies influencées par une exploitation en fonction de facteurs comme la profondeur, l'épaisseur de la veine, etc.
La valeur de cette convergence est mise en correspondance avec différentes modalités de soutènement (par cintres métalliques) conformément à l'expérience.
* Classifications géotechniques : L'expérience en matière de creusement de tunnels et galeries a permis de repérer les paramètres géotechniques importants pour leur stabilité (fracturation naturelle des roches, résistance, présence d'eau etc.). En les combinant, plusieurs auteurs ont proposé des méthodes qui permettent d'affecter le terrain rencontré dans une catégorie. A chaque classe de terrain correspond un type de soutènement (BARTON, 1974), (BIENIAWSKI, 1976), (DEJEAN et RAFFOUX, 1978).
Ces méthodes sont très diverses, leur application plus ou moins commode, et surtout elles sont dédiées à des chantiers particuliers pour lesquels elles ont été conçues (par exemple mines de charbon, ou mines métalliques, ou galeries de génie civil etc.). Les projeteurs ont souvent la prudence d'en utiliser au moins deux.

Méthodes de calculs

Ce sont les différentes formes de modélisation des terrains et du soutènement qui distinguent les méthodes de calcul employées (PIGUET, 1983). Ainsi, le massif rocheux peut être assimilé à un milieu continu, ou, au contraire, être étudié comme un assemblage de blocs (HELIOT et ARCAMONE, 1987). Dans le premier cas, le choix de la loi rhéologique (élasticité, plasticité, visco-élasticité etc...) détermine la complexité des calculs. D'autres hypothèses, plus directement inspirées de la description concrète du site, reviennent par exemple à considérer les bancs du toit d'un ouvrage de section rectangulaire creusé en terrain stratifié, comme des poutres (ou des dalles) en flexion (TANG, 1985). Selon les cas, les extrémités peuvent être supposées encastrées, ou sur appuis déformables. En terrain homogène, on cherche parfois à déterminer le contour d'une cloche d'éboulement potentielle et à calculer ses conditions d'équilibre.
Le mode de calcul peut être soit analytique (résolution explicite des équations) soit numérique (comme par la méthode des éléments finis, par exemple), (figure 1).
De manière générale, les grandeurs résultant de ces calculs sont des forces, des états de contraintes (comparés le cas échéant à des critères de rupture), des déformations, des déplacements.
Le développement de ces méthodes de calculs au cours des dernières années a été notoirement lié au progrès des moyens informatiques utilisés. Mais la technique de programmation est alors de type "algorithmique" classique, ce qui ne permet pas de prendre en compte commodément tous les aspects empiriques de la démarche qui conduit à choisir un soutènement.

C'est pourquoi il nous a paru intéressant, pour ce problème, d'explorer les possibilités de l'approche informatique dite "heuristique", associée avec le développement récent de l'Intelligence Artificielle et des Systèmes Experts.
De là est né le système expert SOUT, présenté dans la suite de cet article.

Modèle des éléments finis

Hypothèses : • massif continu

Entrées : • paramètres du modèle rhéologique
          • forme
          • σ_v , σ_H

Sorties : • contraintes
          • déformations

Modèle convergence-confinement

Hypothèses : • massif continu homogène
             • galerie circulaire
             • σ_v , σ_H

Entrées : • paramètres du modèle rhéologique
          • $\sigma_v = \sigma_H = \sigma_0$

Sorties : • raideur de soutènement

Modèle "poutre"

Hypothèses : • conditions d'encastrement
             • poutres élastiques

Entrées : • caractéristiques des poutres
          • L . H

Sorties : • contraintes
          • déformations

Fig. 1 : Principes de quelques modèles de calcul

2 SOUT : VERS UN SYSTEME EXPERT POUR LE CHOIX DU SOUTENEMENT EN GALERIES

2.1 Système Expert : Principes et réalisation

Un système expert est un programme informatique capable d'utiliser une grande masse de connaissances et de reproduire des stratégies de raisonnement pour résoudre des problèmes dans un domaine spécialisé.

Cette connaissance est formalisée sous forme de règles de production.

Compte-tenu de ces deux entités, un système expert comprend classiquement une base de connaissances et un moteur d'inférences qui permet de l'exploiter de manière à conduire un raisonnement sur un problème posé.

Les connaissances proviennent d'experts humains et/ou de connaissances livresques.

L'architecture de base d'un tel système est schématisée sur la figure 2 (BONNET, 1986).

Compte-tenu de la complexité du problème posé, la réalisation d'un Système Expert fait appel à une démarche en plusieurs phases (WATERMANN 85) :

1. Maquette de démonstration : au cours de cette phase, une première base de connaissances est réalisée à partir d'une trentaine de règles. Elle permet de tester la faisabilité du projet et de cerner les problèmes liés à son développement.

2. Prototype de recherche : à ce stade est abordé le développement règle par règle de la base de connaissances qui est ensuite testée et validée sur un certain nombre de cas judicieusement choisis. A la fin de cette phase, la base de connaissances contient 200 à 300 règles.

3. Produit final : la base de connaissances demeure pratiquement en l'état : Après amélioration et optimisation des interfaces avec l'utilisateur, le produit est mis en exploitation pour tests industriels.

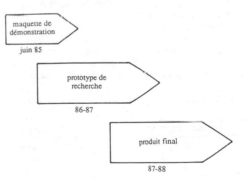

Fig. 2 : Architecture de base d'un système expert

La figure 3 situe dans le temps les différentes phases de développement de SOUT. La réalisation du prototype de recherche, stade actuel du développement, est décrite dans ce chapitre.

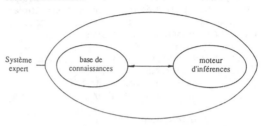

Fig. 3 : Phases de développement de SOUT

2.2 Représentation des connaissances dans SOUT

La base de connaissances de SOUT est constituée de règles et de faits.

2.2.1 Les faits

On distingue :

1. Les faits permanents du domaine, équivalents au contenu d'une base de données. Dans SOUT ce sont essentiellement quelques valeurs de résistances moyennes de roches. Ce qui permet de conduire un raisonnement par défaut dans le cas où l'utilisateur ne connaît pas ces valeurs.

2. Les faits propres à un problème à résoudre, ou données, c'est la mémoire de travail du système expert.

Exemple : " La largeur de la galerie est 5.50 m".

Les faits ne sont pas introduits dans le système en langage naturel. Les recherches en intelligence artificielle sont en partie orientées dans ce sens. Il faut donc, d'une part, structurer les faits et d'autre part respecter la syntaxe imposée par le moteur d'inférences.

Dans SOUT un fait peut être :

1. Un triplet : (attribut, comparateur, valeur)
Exemples : "Le terrain est stratifié
    (type de terrain, = , stratifié) ;
    "La longueur du pilier est inférieure à
    60 m"
    (largeur du pilier, < , 60).

Ce formalisme permet de manipuler un ou plusieurs paramètres (ou attributs) relatifs à un seul objet (terrain, pilier, galerie..), implicite dans notre cas.

2. Un prédicat : la logique des prédicats (GHALLAB 87) est indispensable dans les problèmes où les paramètres prennent des valeurs correspondant à plusieurs objets, c'est le cas pour décrire le toit d'une galerie dans un terrain stratifié (figure 4) constitué de plusieurs bancs (les objets) auxquels on associe les paramètres : nature, épaisseur, résistance..

Les paramètres sont appelés prédicats, relatifs à la variable "banc-X", X étant le numéro du banc :
Notations : nature (banc-X), épaisseur (banc-X)...
Le moteur d'inférences va instancier la variable "banc-X" par les valeurs banc 1, banc 2... banc N.

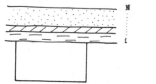

1, 2...N bancs = objets
variable : bancs_x
prédicats : nature, résistance

Fig. 4 : Description d'un toit stratifié

2.2.2 Les règles

La syntaxe des règles de production est définie par

SI conditions ALORS conclusion.

"condition" peut être une conjonction, disjonction ou la négation de faits. La "conclusion" consiste en général , à ajouter un fait dans la base de faits,dans le cas où les conditions de déclenchement de la règle sont satisfaites .

Exemple :
Règle 1 : SI largeur du pilier est comprise entre
    5 et 60m
    ALORS il y a effet stot ET état de
    contrainte élevé.

On distingue trois classes de règles spécifiques aux tâches effectuées par le système expert :

1. Le diagnostic : par analogie avec le diagnostic médical ou le diagnostic de pannes, et à partir de la base de faits, on cherche à caractériser l'environnement de la galerie et en particulier le degré de stabilité des terrains.

Exemples :
Règle 2 : SI effet stot
    OU effet pilier
    ALORS zone fortement surcontrainte.
Règle 3 : SI le terrain est homogène
    ET la résistance de la roche = Rc
    ET la contrainte initiale verticale est
    inconnue
    ET la profondeur = h
    ET la densité des terrains = d
    ET Rc < 3 x d x h
    ALORS risque d'écaillage

2. La conception : il s'agit de l'utilisation des résultats du diagnostic pour choisir et dimensionner le soutènement.

Exemples :
Règle 4 : SI la zone est fortement surcontrainte
    ET le cintre choisi est un TH29
    ALORS intercadre = 0.6 m.
Règle 5 : SI il y a risque de cisaillement
    ALORS l'utilisation de l'ancrage
    ponctuel n'est pas recommandée.

3. Gestion de la consultation : certaines règles permettent de gérer le dialogue avec l'utilisateur (envoi de messages...)

Exemple :
Règle 6 : SI afficher (il est recommandé que le
    diamètre de la cartouche soit d'environ
    4 mm plus petit que le diamètre du trou)
    ALORS conseil-réparti-résine.

Pour tenir compte du raisonnement incertain, au début du développement de SOUT, nous avons étudié la possibilité d'utiliser les coefficients de vraissemblance pour représenter les connaissances. Ainsi :

Règle : SI le terrain est stratifié
    ALORS le boulonnage à ancrage réparti est
    mieux adapté (0.8) qu'un ancrage ponctuel
    (0.3).

Nous nous sommes heurtés aux problèmes classiques d'attribution des coefficients (ici 0.8 et 0.3 sur une échelle de 0 à 1) et de leur propagation au cours du raisonnement, d'autant plus que l' algèbre proposée par le moteur ne permet pas de les contraindre à l'intérieur de certaines limites. Cette possibilité a été abandonnée.

2.3 Acquisition des connaissances

L'acquisition des connaissances s'est alimentée à 3 sources (figure 5) :

a) L'analyse de textes bibliographiques et de monographies d'experts nous a conduit à extraire des pseudo-règles en identifiant le concept "SI... ALORS". Après une phase de validation "manuelle", la pseudo-règle a été érigée en règle et introduite dans la base.

c) Pour disposer d'une connaissance actualisée, une enquête a été réalisée sur des galeries récemment exploitées.

Des relevés ont été réalisés de manière à caractériser globalement une galerie du point de vue de son environnement naturel et d'exploitation, de son soutènement et de son comportement.

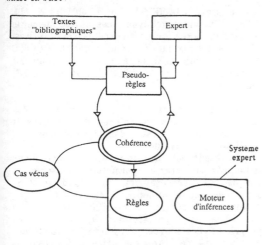

Fig. 5 : Acquisition des connaissances

Cette démarche nous a demandé un temps relativement long du fait que pour affiner les conditions d'application d'une règle, il nous a fallu souvent fusionner plusieurs textes ou plusieurs auteurs, en passant par d'éventuelles contradictions ou difficultés d'interprétation.

Nous illustrons sur la figure 6 à partir d'un exemple les différentes étapes d'élaboration des règles à partir d'un texte bibliographique.

b) Conformément à ce qui a été développé précédemment, il est apparu essentiel de faire intervenir un expert du domaine afin de disposer d'une base de connaissances optimisée.

L'expert choisi a été interviewé selon des techniques d'entretiens adaptées (WATERMAN, 1985), (MAHE, 1987). Son intervention s'est située à deux niveaux :

* au début de l'élaboration de la base, contrôle et validation des connaissances tirées de l'analyse bibliographique. Certaines de ces connaissances ont été précisées et même amplifiées, surtout en ce qui concerne les règles de mise en oeuvre d'un soutènement donné.

* après réalisation d'une base de connaissance conséquente (une centaine de règles), confrontation du système et de l'expert sur quelques cas réels, préalablement choisis.

Ce travail s'est étalé sur 6 mois, ponctués par une dizaine de séances d'entretien d'une demi-journée chacune.

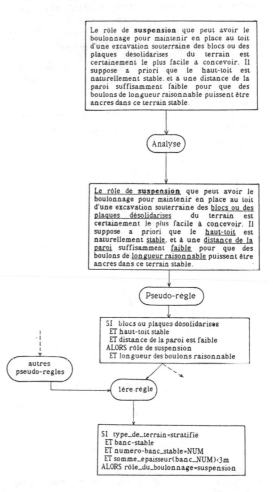

FIG. 6 : Formalisation de la connaissance "bibliographique"

Plus de 30 galeries minières ont ainsi été suivies : Une base de données permettant de gérer cette information a été constituée et sera interfacée à terme avec la version "PRODUIT FINAL" du Système Expert.

2.4 Structuration de la base de connaissances

La connaissance a été structurée en blocs de règles. Il n'y a pas d'intéraction entre deux blocs, en revanche ils partagent la même base de faits au cours de la consultation.

On distingue un bloc "boulonnage" et un bloc "soutènement porteur", correspondant aux types de soutènement pris en compte par SOUT, ainsi qu'un bloc "général" qui aide l'utilisateur à choisir entre un soutènement par boulonnage ou un soutènement porteur (figure 7).

Cette structuration sert à charger en mémoire de l'ordinateur un bloc de règles à la fois au cours d'une consultation.

En outre, cette organisation permet une actualisation aisée et d'autres blocs, correspondant à d'autres modes de soutènement, pourront être ajoutés ultérieurement

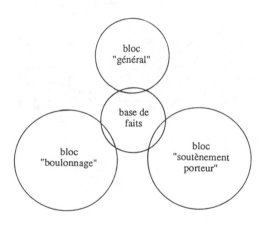

Fig. 7 : Structure de la base de connaissances

2.5 Le moteur d'inférences

Le moteur d'inférences utilisé est M1 de FRAMENTEC sur micro ordinateur compatible IBM/PC. M1 répond aux contraintes imposées par la représentation des connaissances.

M1 utilise le chainage arrière comme mode de déclenchement des règles. Etant donné un but fixé par l'utilisateur ou implicite dans le système, le moteur d'inférences essaie de le prouver en examinant les règles concluant sur ce but, ce qui l'amène à vérifier de nouveaux sous-buts en menant un dialogue avec l'utilisateur pour demander , entre autres, les informations nécessaires sur le problème.

Les buts principaux de SOUT sont le mode de soutènement et les paramètres relatifs à chaque mode.

Le système expert SOUT tel qu'il a été présenté précédemment a été utilisé pour le choix du soutènement dans certaines galeries minières.Nous présentons en annexe les principales lignes de la consultation correspondant à une galerie rectangulaire dans une mine de charbon située à forte profondeur

3 - DEVELOPPEMENTS EN COURS

Les phases "MAQUETTE" et "PROTOTYPE DE RECHERCHE" étant achevées, ces développements concernent la réalisation du produit final. SOUT est en cours d'implantation sur un COMPAQ DESKPRO 386 (compatible IBM PC/AT) sous un nouveau moteur d'inférences : NEXPERT OBJECT de NEURON DATA.

Ce moteur offre une représentation des connaissances , des mécanismes d'inférences et des interfaces avec l'environnement et l'utilisateur très évolués :

* représentation des connaissances type objets structurés ou "frame" permettant d'associer à chaque objet les propriétés qui le définissent (exemple : banc =objet ; nature, épaisseur, résistance = propriétés), ainsi que les sources d'informations permettant d'évaluer ces propriétés. Ces sources d'informations pourront être une base de données associée ou les résultats de l'exécution d'un programme externe de calcul.

* mécanismes d'inférences évolués, chainage avant et arrière, permettant en particulier, lors d'une consultation, de choisir entre différentes hypothèses générées automatiquement par le moteur.

* interfaces très performantes permettant en particulier, en cours de consultation, de suivre de manière très précise le raisonnement suivi par le système expert , de faire appel à des procédures de calcul externes et d'exploiter leurs résultats, de consulter une base de données associée.

Dans cette phase de réalisation du "PRODUIT FINAL", l'objectif est la conception d'un système à base de connaissances (SBC) regroupant les différents types de connaissances en la matière, cités précédemment (figure 8). L'architecture d'un tel système est schématisée sur la figure. Ce produit final qui, compte tenu des qualités de NEXPERT OBJECT, devrait allier performances et souplesse d'utilisation, sera opérationnel vers le milieu de l'année 1988.

4 - CONCLUSIONS

Afin d'aider au choix du soutènement des galeries minières, un Système Expert "SOUT" a été conçu et réalisé.

* Le "PROTOTYPE DE RECHERCHES" contient 200 règles couvrant le domaine du soutènement par boulonnage et du soutènement porteur.

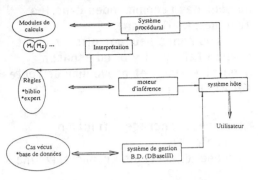

Connaissance en matière
de soutènement en galerie

Fig. 8 : Architecture du système à base de connaissances

* Ces règles sont issues d'une bibliographie très exhaustive et d'interviews d'expert du domaine ainsi que de l'expérience tirée d'une enquête conduite récemment dans plus de 30 galeries minières.

* Cette connaissance est amplifiée dans le "PRODUIT FINAL" du Système Expert "SOUT" par adjonction de modèles de calcul appelés en cours de consultation.

* Ce "PRODUIT FINAL", qui permet une représentation très élaborée de la connaissance et présente une grande souplesse d'utilisation, est en cours d'achèvement.

* D'une manière générale, ces travaux nous ont montré que l'approche par Système Expert pouvait trouver des applications intéressantes dans le domaine minier. Dans le domaine du choix de soutènement, les avantages de cette approche entrevue au début de l'étude , (représentation, actualisation, capitalisation et transmission des connaissances largement fondées sur l'empirisme) n'ont pas été démentis.

REFERENCES

BARTON N, et al.1974. Engineering classification of rock masses for design of tunnel support . Rock mechanics, vol 6 n°4 .

BIENIAWSKI Z.T.1976. Rock mass classification in rock engineering. Symposium on rock engineering , Johannesbourg

BONNET A., HATON J.P, TRUONG J.M .1986. Systèmes experts - vers la maitrise technique . Paris:InterEditions.

DEJEAN M. , RAFFOUX J.F.1978. Choix des paramètres d'un soutènement par boulonnage . Revue de l'Industrie Minérale, N° spécial boulonnage .

GHALLAB M., FARRENY H. 1978 .Eléments d'intelligence artificielle . Paris : Hermès .

GROTOWSKY U. 1977 . Nouvelles techniques de soutènement en voies . 6ème conférence sur les pressions de terrains , Banff .

HELIOT D., ARCAMONE J. Modèles numériques appliqués aux massifs rocheux fracturés . Revue sciences de la Terre , Série Informatique (à paraitre) .

MAHE H., VESOUL P. 1987. Aquisition des connaissances et adaptation à l'utilisateur : outils et méthodes . 7ème journées internationales , les Systèmes Experts et leurs applications, Avignon : EC2.

PIGUET J.P. 1983.Modélisation en mécanique des terrains et son application à l'exploitation minière .Thèse doctorat es sciences, Ecole des mines de Nancy , France .

TANG D.H.Y., S.S. PENG .1985. Reinforcement analysis and design of mechanical roof bolting systems in horizontally bedded mine roofs. Int. Jal. Mining Engineering, 3, 1-25.

WATERMAN D.A.1985. A guide to expert systems. New-York : Addison Wesley .

ANNEXE : EXEMPLE DE CONSULTATION

Le cas traité correspond au choix du soutènement dans une galerie rectangulaire située dans une mine de charbon à forte profondeur.

Les instructions émanant de SOUT et les réponses de l'utilisateur sont précédées par "M1 >" ou ">>".

1. Faits généraux :

...

M1 > Quelle est la largeur de la galerie ?

>> 6.5.

M1 > Quelle est le type de terrain au toit ?

>> stratifié.

...

2. Sélection du type de soutènement :
SOUT demande des informations sur la configuration d'exploitation et des facteurs naturels .

...

M1 > La zone étudiée se trouve-t-elle en zone de faille? (c.a.d qu'on traverse une faille)

>> non.

...

M1 > Quelle est la largeur du pilier ?

>> 60.

SOUT envoie un message à l'utilisateur :

M1 > Les dimensions du pilier ne sont pas critiques. Nons considerons que la zone est hors influence d'un stot .

...

M1> Les regles de pressions de terrains n'empechent pas l'emploi du boulonnage . Un soutenement porteur est egalement possible.

M1> soutenement_par_boulonnage = yes : regle-4

SOUT permet d'expliquer son raisonnement :

M.1> list regle-4.
  regle-4:
  if type_d_exploitation = taille
  and hors_influence_d_une_faille
  and zone_vierge
  and pas_de_terrain_de_faible_
    resistance
  then soutenement_par_boulonnage.

M.1>show zone_vierge.
    zone_vierge = yes : regle-19.
M.1>list regle-19.
  regle-19:
  if effet limite d exploitation = non
  and effet stot = non
  and effet pilier = non
  then zone_vierge.

...

3. Dimensionnement du boulonnage.

  * type de boulons :
      ...

M1> Quelle est l'ecartement moyen des fissures ?
    >> 1.

M1> Avez vous fait un calcul de stabilite des bancs du toit,par le modele "poutre" ?
    >> non.

...

M1> Les boulons a ancrage ponctuel ne sont pas recommandes dans les zone de cisaillement.
    Les bancs peuvent glisser entre eux, du fait qu'il y a un espace annulaire vide , et provoquer de fortes deformations .

...

M1>mode_d_ancrage = friction :
    regle-87.
    mode_d_ancrage = reparti_resine :
    regle-82.

L'utilisateur intervient pour choisir le type d'ancrage, suivant les indications données par SOUT.

M.1>set choix_ancrage =
    reparti_resine.
  >> réparti-résine

  * Longueur des boulons :
      ...
M1> Le C.E.Tu (cf Etienne) recommande une longueur de 2.1 m .
      ...
  * Densité des boulons .
  * Orientation .
  * Mise en oeuvre :
      ...

M1> Quelle est le type de materiel de boulonnage utilise ?
    - manuel (semi stoper)
    - mecanique .
    >> manuel.

M1> La mise en oeuvre est manuelle, le temps de pose des boulons risque d'etre superieur au temps de prise de la resine. Il est recommande d'utiliser
...

Computer Applications in the Mineral Industry, Fytas, Collins & Singhal (eds)
© 1988 Balkema, Rotterdam. ISBN 90 6191 760 3

Application of the blast simulator BLASPA to predict and minimize the coal damage induced by blasting in strip mining operations

R.F.Favreau
Collège Militaire Royal de St.Jean, Quebec, Canada

D.P.Lilly
Du Pont, Wilmington, Del., USA

ABSTRACT: Blast induced coal damage in a strip mining operation is a serious technical problem. Blast simulations on the model BLASPA help to design blasts that do not induce coal damage but yet break the rock overburden. The paper explains the principles involved.

1 INTRODUCTION

The objective of operating a coal strip mine is to retrieve coal laying beneath the rock overburden. The main activity of such a mine, however, is the engineering and actual removal of the rock overburden, as required to expose the coal. In most cases, overburden removal involves blasting with explosives. Explosives are used so as to break and move the rock. The problem is that the coal seams are in contact with the rock, so there is a risk that the breaking and moving action of the explosions may act not only on the rock but on the coal as well. When this occurs, the coal may be damaged or even lost. It is therefore important that blasts may be designed in such a way that the rock be successfully broken and moved, but the coal not be damaged. The paper explains how blast simulation may help to achieve this, both by explaining how the explosions may damage the coal and by assisting to quantitatively design blasts that minimize such damage.

2 VARIOUS KINDS OF COAL DAMAGE

Coal damage resulting from the blasting action of explosives may be of various kinds. The principal kinds of coal damage encountered in mines are reviewed below.

2.1 "Chilling" the Coal

Blast induced "chilling" refers to coal that has been fragmented by the action of the explosion. "Chilled" coal can often be recognized because it appears as fines or crushed material, with many cracks dis-tributed throughout a volume of the coal. Another indication of chilling is when the digging equipment, either dragline or shovel used to move the broken rock or the dozer used to clean and level the top surface of the exposed coal, dips down into the soft fines causing an uneven floor on top of the coal seam. The depth and intensity of the chilling often display a regular pattern associated with the original lay-out of the pattern of shot holes. The depth of "chilled" coal may extend from a few inches up to as much as several feet in extreme cases.

2.2 "Shooting" the Coal

Even more serious than chilling the coal is a situation where the blast "shoots" the coal. Blast induced "shooting" refers to coal (usually in the zone near the edge of the seam closest to the vertical high-wall face) which physically gets moved forward during the blast and is totally lost under the spoil pile (see fig.1). In an extreme case, many horizontal feet of the entire coal seam will be pushed out together with the blasted overburden, moving forward with it and therefore becoming a part of the spoil pile. As much "shot" coal ends up underneath the rock in the spoil pile, it is therefore irretrievably lost. In the case of a thick coal seam, the shot coal may include only the top corner of the seam (fig.1(b)). If before the shot there was an extended ledge of coal ahead of the high-wall face, then the coal in this extended ledge is especially susceptible to being shot out (fig 1(c)); such a ledge may be left in place because debris rock from the high-wall had fallen on the coal before the latter was dug, thereby making retrieval hazardous.

2.3 Other Kinds of Coal Damage

Other kinds of coal damage are possible. For example, if explosive is detonated in contact with a narrow local coal whose exploitation is not intended, then ignition of the coal may conceivably occur. Another example is the situation found for some harder coals, where blasting of the exposed coal seam has to be carried out in order to loosen the coal and make it easier to dig. In such a case, blast induced overbreak of the coal at the back of the bench of coal (see fig.2) may weaken the toe of the high-wall of the next shot, thereby causing hazardous instability of the latter. Such instability often causes debris to fall from the high-wall unto the coal during the time lapse between blasting and digging of the coal, which may make it unsafe to dig the coal right up to the toe of the high-wall, leaving a ledge of coal that may be lost in the next round as explained in part (2.2) above.

3 IMPORTANCE OF COAL DAMAGE

The economic implications of coal damage (ref.1) may be very significant. "Chilling" creates cracks in the coal seam; partial "shooting" of the coal leaves large vertical openings in the seam. During mucking, and during the period between exposure of the coal by removal of the overburden and actual retrival of the coal itself, the action of the mucking equipment and of other equipment driving on the surface of the exposed coal all contribute to drive overburden impurities into the cracks in the coal seam. This reduces the cleanliness quality of the coal and increases the level of ashes produced when the coal is burnt at the power plant. Cleaning of the coal is possible, but this adds to its cost. In a serious case of chilling induced contamination, complaints from the end-user may lead to a downgrading of the quality of the coal, with the possibility of a reduction in the selling price or in the application of contractual penalties.

The consequences of "shooting" of the coal are usually even more dire because it usually leads to the complete loss of a percentage of the total coal extracted from the mining operation. In the case of a thin coal seam with a thick layer of overburden, the economic implications of wastage of a percentage of the coal originally in the seam may be very serious. In the case of a thick seam, shooting also leads to the complete loss of a percentage of the coal; furthermore, blast induced overbreak in a coal seam being blasted may lead to a ledge of unretrieved coal due to high-wall instability (see section 2.3 above), the latter often also being totally lost to the mine because of the blast of the next shot. Thus even in a thick-seam operation, loss of coal may be serious. Many mines have found (ref.1) that they are losing from 3 to 12 percent of the coal that they have uncovered because of blasting damage. When a mine has removed 40 to 100 or more feet of overburden at a cost of $3 to $5 per ton of coal, each ton uncovered becomes precious.

Coal that has been "shot" into the spoil pile presents a further inconvenience; it may spontaneously or accidentally be ignited, sometimes leading to long lasting spoil pile fires. In thick-seam operations where the coal itself is blasted, coal overbreak under the toe of the high-wall of the next shot leads to high-wall instability whose safety hazards are objectionable to the workers and their union.

For all these reasons, the control of coal damage is of prime importance to the coal strip-mining industry.

4 ASPECT OF BLAST SIMULATIONS RELATED TO COAL DAMAGE

The principles underlying the blast simulation model BLASPA have been explained elsewhere (refs. 2, 3, 4, 5); some of its practical applications to assist design blasts for the coal mining industry can be found in refs. 6,7. This section of the present paper will therefore review mainly those aspects of the computer model BLASPA which relate specifically to addressing the problem of coal damage. Fig.3 shows the chief mechanisms involved during a blast. These may be described as follows: Following the passage of the detonation head from one end of the charge to the other, at the well known Velocity of Detonation, the explosive is converted into explosion gases at very high pressure. This very high pressure causes a rapid but limited expansion of the borehole diameter, the limit being dependent on the ability of the competent rock to oppose the force of the gases; rocks of high Young's modulus Y resist expansion better than rocks of low Y value. This first gas expansion creates strong compressive waves which travel through the rock mass (fig. 3 a); rock being strong in compression such waves cause little damage to the rock. On arrival at the free faces, however, the compressive waves are reflected in tension and travel back into the rock mass (fig.3b); rock being weak in tension, such reflected tensile waves initiate rock breakage, causing a large number of small cracks that weaken the previously competent rock. Such weakening of the rock allows gas expansion to resume, thereby causing additional fragmentation to the rock. The manner whereby this second kind of fragmentation occurs is shown in Fig.3c, where a crack front is seen to advance from the region of the borehole towards the free faces under the action of a semi-static stress field; ahead of the crack front the rock is only weakened by the earlier passage of the reflected tensile waves, behind it it is fully broken by the added effect of the semi-static stress field. When a blast has been well designed, the crack front will advance all the way to the free faces, whereupon the whole rock bench will burst out as shown in fig. 3d; during the burst-out, each fragment of rock will have acquired its own particular value of rock burst-out velocity. The movement of all the rock pieces will eventually lead to the final muck-pile shape.

Equations have been found (e.g. refs. 3, 5) for the various mechanisms outlined in fig.3, and these equations are built into the simulator BLASPA so that the latter can quantitatively predict the intensities of the various stress waves and stress field, as well as the rock burst-out velocities, involved in a blast. Such simulated values of stresses and velocities take into account the full thermochemistry of the explosives, starting with their basic ingredients, as well as the rock properties and the full geometry of the blast design. Values of stresses and movement velocities are not restricted to average values; on the contrary, a simulation evaluates stress levels and velocities at every location within the bench. As an example, fig.4 maps out contours of local values of stress levels T associated with the reflected tensile waves described in fig. 3b.

The manner in which a T stress contour map, such as that simulated in fig. 4a, can be used in blast design is as follows: The map of contours of fig. 4a includes many stress levels T. Of these, one is of particular importance; it may be called the T_o stress level, and it corresponds to the minimum stress level required to break the rock. Thus in fig 4a, $T_o = 226$, and contours have been simulated corresponding to stress level T_o, stress levels T less than T_o, and stress levels greater than T_o. The map of fig. 4a depends on all of 1 the properties of the explosives involved, 2 the rock properties, as well as 3 the geometry of the blast If any of these is changed, then the whole map in fig. 4a will shift. For example, in fig. 4b the burden has been increased, and the T_o and each other contours have shifted to the left. The interpretation of figs.4a and b is as follows. Rock subjected to stress level less than T_o does not even begin to break, and so cannot be expected to ever be excavated; rock subjected to stress level T_o just begins to break, while rock subjected to T greater than T_o is more completely broken. Thus the T_o contour decides the limit of possible excavation for that particular blast design. Following this thinking, fig. 4a suggests that for the blast design in that figure the rock bench will fully excavate all the way down to the top of the coal as well as all the way back to the line of the hole and even beyond; fig. 4b, on the other hand, suggests that excavation will be difficult at the back near the coal, and toe can be expected. Fig. 4c, where the burden has been reduced and the length of the charge shortened, suggests that a large block in the zone of the stemming will not be broken, although it may be expected to just fall down on top of the muck-pile due to natural faults in the rock mass. In this manner, blast simulations predict the expected back break (Fig. 4a), toe (Fig. 4b) and large blocks (Fig. 4c), and so they may be used to design blasts that reduce back-break, or prevent toe or large blocks. For example, if the blast design of Fig.4b was in use, a first simulation identifies the expected toe problem, and a subsequent series of simulations can be carried out to remove toe by seeking on the computer blast designs that move the T_o contour far enough to the left at least to intersect the line where the hole meets the top of the coal. Adequate movement and easy digging is also predictable from the simulated values of local rock movement burst-out velocities, as shown in fig. 5a (easy digging) and fig. 5b (hard digging).

5 APPLICATION OF SIMULATIONS TO PREDICT COAL "CHILLING"

Section 4 has explained how rock breakage is initiated only if the rock is stressed at least to the minimum breakage level T_o;

309

rock stressed below the level T_o is left undamaged. In the same manner, one may surmise that for each coal type there exists a stress level, which we may call T_c, below which it is left undamaged and above which it will begin to crack i.e. be chilled. Simulations on BLASPA can predict the maximum stress level that reaches the coal at the rock/coal interface, as shown in fig.6. In fig. 6a, for example, the maximum stress level reaching the coal is $T_c = 375$, the lower end of this T_c contour just barely touching the coal near the rock free face; all other rock stresses reaching the coal are less than T_c. In fig. 6b, on the other hand, changes in blast design are such that the T_c contour is seen to "penetrate" into the coal; thus it may be surmised that some of the coal will be "chilled" if the blasting procedure of fig. 6b is used at a mine where the value of $T_c = 375$. If at a given mine, the actual blasting procedure (say that of fig.6b) is known to cause "chilling" of the coal, then simulations can be used to prevent such "chilling"; this is done simply by varying the blast design, on the computer, till blast parameters are found which simulate a T_c contour that does not penetrate into the coal (fig. 6a), or for more safety, one which stays well out of the coal (fig. 6c). In practice, T_c is easily evaluated by the use of simulations. This is done by searching through the mine's past records for a blast procedure known to have caused "chilling"; this past blast procedure is then simulated to seek a stress contour which does penetrate into the coal, and the value of stress of this contour is then identified with T_c from the fact that "chilling" was known to occur from that particular blast procedure.

6 APPLICATION OF SIMULATIONS TO PREVENT COAL CHILLING BUT INSURE ROCK EXCAVATION

The previous section has explained how BLASPA may be used to identify and prevent coal damage through simulations of the T_c contour; in essence, different blast designs are simulated till one is found which does not send the T_c contour into the coal. Sections 4, on the other hand, has explained how successful rock breakage and excavation requires that the contour of minimum rock breaking stress level T_o must fully include the volume BxH (see fig.4), where B is the burden and H is the thickness of overburden. Depending on the relative values of T_o for the rock involved and T_c for the coal at the mine, the prevention of coal chilling may not be compatible with successful rock excavation. For example, in fig. 7a, $T_c = 375$

while $T_o = 226$; the simulations of fig. 7a show that for these relative values of T_c and T_o the blast procedure found not to cause "chilling" is indeed compatible with successful excavation in that the T_c contour does not reach the coal while the T_o contour does reach beyond the line of the hole at the top of the coal. In Fig. 7b, on the other hand, $T_c = 315$; while $T_o = 226$; the simulations of fig. 7b show that in order to pull the T_o contour out of the coal, the back-fill had to be increased to such an extent that the T_o contour no longer reaches to the point where the line of the hole intersects the top of the coal, thereby predicting toe and hard digging. For the same values of T_c and T_o, fig. 7c shows that reduction of back-fill sufficient to prevent toe is expected to cause "chilling" of the coal. This dilemma can, in practice, be overcomed by the blast design of fig. 7d, where a top charge is shot first, thereby not damaging coal but leaving toe, but the latter is then broken by a small delayed bottom deck charge. The delay between bottom and top decks must be sufficiently long that an effective "free face" be created by the latter at the T_o contour in the region of the toe; otherwise, the lower deck will also particpate in the intensity of the stresses reaching the coal (see fig. 7e) and T_c will indeed penetrate the coal and cause "chilling" of the latter.

At first glance, one may be surprised at the relative values of $T_c = 315$ and $T_o = 226$, as shown in say fig. 7b. One might have expected the coal breaking tension T_c to be less than the rock breaking tension T_o. In fact, three arguments militate in favour of the reasonableness of the relative magnitudes of T_c and T_o in fig.7. These are as follows:

1 The sresss levels T_o and T_c are associated with tensile waves of high amplitude but short duration, as explained in section 4. Such stresses are known to be more effective at fracturing brittle materials. It is well known that rock is more brittle than coal. Hence it is not unreasonable that T_c be higher than T_o.

2 The stress levess labelled T_c in figs. 6 and 7 are actually values simulated in the rock itself. It is proposed in the paper that when the stress level T_c in rock reaches the coal, the threshold for coal "chilling" is reached. But in fact, in a case such as that of Fig. 7c where T_c is shown as "penetrating" into the coal, there is wave transmission from

the rock to the coal and the impedance mismatch between them is expected to reduce the stress level in the coal itself, while partial reflection may in fact reinforce somewhat the stress level in the rock itself. Thus the actual stress level inside the coal which chills the latter may be lower than the actual value of T_c used in figs. 6 and 7. Simulations of the wave intensity transmitted into the coal is possible; in fact work is ongoing to develop a user-friendly computer subroutine to do this. However, computer costs of such a subroutine will increase by comparison with those for the subroutine presently used to calculate T_c in the manner of figs. 6 and 7. In fact, the procedure to prevent coal chilling is completely justified, since simulations are used to keep T_c outside the coal; thus any inaccuracy in T_c when the latter is simulated inside the coal is of little practical significance since blasting procedures whose simulation suggest that T_c enters the coal should not be used anyway.

3 If a blast broke the coal much more easily than it breaks the rock, then no coal strip mining operations could possibly use blasting to remove overburden. This is contrary to the known fact that huge amounts of coal have been extracted this way. Hence T_c, as used in fig. 6 and 7, must indeed be greater than T_o. So far, all simulation studies done to prevent "chilling" of the coal have indicated that T_c is greater than T_o.

7 APPLICATION OF SIMULATIONS TO PREDICT "SHOOTING" OF THE COAL

Sections 5 and 6 have explained how the simulator BLASPA may be applied to design blasts that prevent "chilling" the coal. But, as discussed in section 3, the consequences of "shooting" the coal may be even more dire than those of "chilling" it, since shooting may result in the total loss of some of the coal while chilling just increases its impurity content. Simulations have proven to be a tool to address the problem of "shooting" the coal. To understand how this may be done, let the physics of "shooting" the coal be examined more closely. Referring to fig.8, the reason why a blast may "shoot" the coal would seem to be as follows: If a blast is to successfully move the rock, then the latter should acquire burst-out velocity U_r all the way down to the top of the coal. On the other hand, the ground below the coal seam does not move, so its velocity

is zero in fig.8. Through the action of friction at the interfaces, then, the rock velocity U_r just above the coal creates a forward force F_1 which tries to move the coal, while the non-moving material below creates a force F_2 which tries to prevent forward coal movement. The deciding factor as to whether the coal does or does not move depends therefore on the relative magnitude of F_1 as compared with that of F_2. The magnitude of F_1 in turns depends on the velocity U_r of the rock just above the coal seam. Hence there must be, for each rock and coal types, a threshold value of rock velocity U_r below which the coal does not move and above which the coal does move; let this critical velocity be called U_c. As the coal which may most easily be induced to move is that at the face, the value of U_r most susceptible of "shooting" the coal is that in the rock at the face just above the coal seam; this is in fact also the highest value of U_r anywhere along the rock/coal interface. As explained in fig. 5, simulations with BLASPA supply for any given blasting procedure the value of rock burst-out velocity U_r of any rock fragment in the bench above the coal. Hence if a given blast procedure does produce U_r greater than U_c in the rock near the face just above the coal, as simulated in fig. 9a, then "shooting" of the coal may be expected. In that case, a series of simulations can be carried out with different blasting parameters till a blast design is found on the computer for which U_r in the rock near the face just above the coal is equal to or less than U_c (see figs. 9b and c), whereupon "shooting" of the coal is no longer expected to occur.

In practice, U_c is evaluated at a given mine by searching amongst past blasting records for a blasting procedure known to have caused "shooting" of the coal. This particular blast procedure is then simulated to find out the value of U_r near the face just above the coal seam. This value of U_r is then the first estimate for U_c, the rock movement velocity that can cause "shooting" of the coal. Subsequent simulations of blasting procedures observed just barely to "shoot" the coal usually lead to a more accurate evaluation of U_c.

Normally, there are two values of U_c, namely a lower one for the first row of a blast, where the coal is more free to move since the coal in the pit has already been removed, and a higher one for back rows to be used when the front row coal is known not to "shoot". The effective value of U_c for the front row may be increased if a pile of clay is stacked up against the coal face before the blast.

8 CONCLUSIONS

Coal mine operators have explained to the authors of this paper that coal damage was an important problem to them. Many mines have reduced or even illiminated coal damage from blasting by trial and error in the pit. In so doing, however, it is often possible that the new blasting procedure does not damage the coal, but it may not very effectively break the rock nor give easy digging. In particular, if a mine wishes to cast, it will often be found that trial and error in the pit will not find the optimum between good casting and damaging the coal. The paper has explained how simulations on BLASPA may be used to determine the values of T_o for the rock, T_c and U_c for the coal. Once these are known, further simulations can be done to insure that each one of the following criteria be respected:

8.1 The T_o contour be such that all the rock in the zone BxH of the bench is successfully broken.

8.2 The T_c contour be such that it does not penetrate into the coal seam, insuring that there be no "chilling" of the coal.

8.3 The value of rock movement velocity U_c be a safe distance above the top of the coal seam.

When all of these criteria are met simultaneously, the blast design is expected to break and excavate all the rock overburden, yet without causing coal damage. In practice, it is found that all three criteria can be satisfied more easily at mines where the rock and coal characteristics are such that T_c is substantially greater than T_o. Yet, even when T_c is not much greater than T_o, simulations can still help to design blasts that break all the rock but cause no coal damage (e.g. see fig. 7d).

Blast design using simulations on BLASPA and the above three criteria bring a new level of technical competence to the coal mining industry. Because it is a quantitative approach, it usually leads to quality of blast results closer to optimum than those achieved by the method of trial and error in the pit. The subroutines that simulate T_o, T_c and U_c may be used in conjunction with other subroutines of BLASPA, such as that used to improve shovel productivity (ref.1), or that used to design optimized cast blasts (Ref. 6). When this is done, blast design passes from an art to a true engineering endeavour, resembling say the way that engineers design bridges.

REFERENCES

Favreau. R., Lilly, D. "Computer Aided Blast Design for the Coal Industry", presented at the 1987 Indiana Mining Institute Annual Meeting, Evansville, Indiana, April 1987.

Favreau, R. "BLASPA, a Practical Blasting Optimization System", presented at the 6th Annual Symposium of the Soc. of Explosives Engineers, Tampa, U.S.A. Feb. 1980.

Favreau, R. "Blasting Simulations - Present and Future", presented at the Annual Symposium of the Canadian Institute of Mining, Ottawa, Canada, April 1984.

Favreau, R., Williams, R. "A Review of the History and Present Status of Blast Modelling", presented at the Special Session of the SEE on Computer Blast Modelling, Miami, Feb. 4, 1987.

Favreau, R. "Generation of Strain Waves in Rock by an Explosion in a Spherical Cavity", J. of Geophysical Research, 74, 4267, 1969.

Favreau, Mahan and Schmidt, "Quantitative Evaluation of Cast Blasting by Airborne Surveys and Computer Simulations", presented at the Pennsylvania Blasting Conference, Pennsylvania State University, Nov. 14-15, 1985.

Favreau, Lilly, "The Use of Computer Blast Simulations to Evaluate the Effect of Angled Holes in Cast Blasting", presented at the 3rd. Conference on the Use of Computers in the Coal Industry, West Virginia University, July 1986.

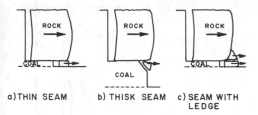

a) THIN SEAM   b) THISK SEAM   c) SEAM WITH LEDGE

FIG.1: "SHOOTING" THE COAL

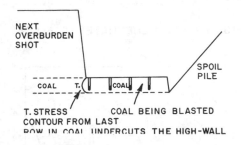

NEXT OVERBURDEN SHOT

SPOIL PILE

COAL   T.   COAL

T. STRESS CONTOUR FROM LAST   COAL BEING BLASTED
ROW IN COAL UNDERCUTS THE HIGH-WALL

FIG. 2: DESTABILIZING THE HIGH-WALL BY BLAST IN COAL

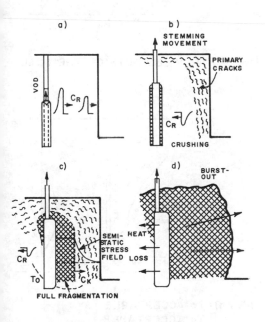

a)

VOD C_R

b)

STEMMING MOVEMENT

PRIMARY CRACKS

C_R

CRUSHING

c)

C_R

SEMI-STATIC STRESS FIELD

T_0 C_K

FULL FRAGMENTATION

d)

BURST-OUT

HEAT LOSS

FIG.3: MECHANISMS OF THE BLASTING PROCESS

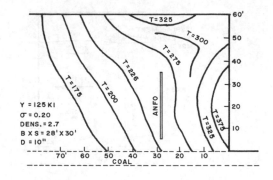

T=325   T=300

T=275

T=226

T=200

T=175

ANFO

T=375

T=325

Y = 125 KI
σ = 0.20
DENS. = 2.7
B X S = 28' X 30'
D = 10"

COAL

FIG. 4(a): MAP OF STRESS CONTOURS

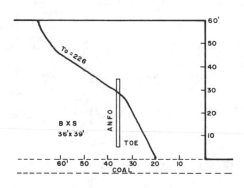

T_0 = 226

ANFO

B X S
36' x 39'

TOE

COAL

FIG. 4(b): PREDICTION OF TOE

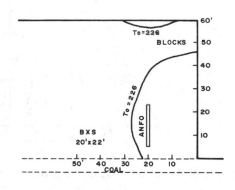

T_0 = 226

BLOCKS

T_0 = 226

ANFO

B X S
20' x 22'

COAL

FIG. 4(c): PREDICTION OF BLOCKS

313

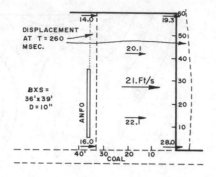

FIG. 5 a): HARD DIGGING

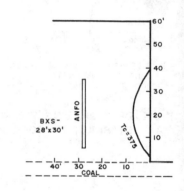

FIG. 6 b): Tc PENETRATES THE COAL

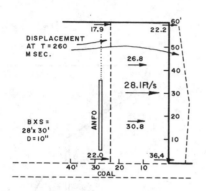

FIG. 5 b) EASY DIGGING

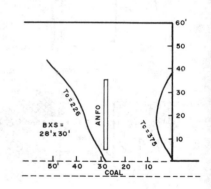

FIG. 6 c): Tc WELL OUTSIDE THE COAL

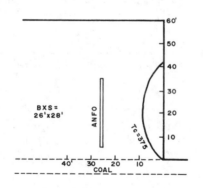

FIG. 6 a): Tc JUST TOUCHES THE COAL

FIG. 7 a): Tc ACCEPTABLE
To ACCEPTABLE

314

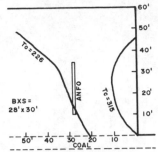

FIG. 7 b): Tc ACCEPTABLE
To NOT ACCEPTABLE

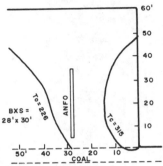

FIG. 7 c) To ACCEPTABLE
Tc NOT ACCEPTABLE

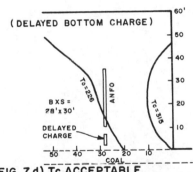

FIG. 7 d) Tc ACCEPTABLE
To NOT ACCEPTABLE

(DELAYED BOTTOM CHARGE)

(UNDELAYED BOTTOM CHARGE)

FIG. 7e) Tc UNACCEPTABLE
To ACCEPTABLE

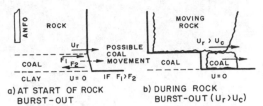

a) AT START OF ROCK
BURST-OUT

b) DURING ROCK
BURST-OUT (Ur > Uc)

FIG. 8 : MECHANISMS OF "SHOOTING"
THE COAL

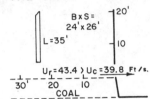

FIG. 9 a): Ur > Uc SOME "SHOOTING" OF
THE COAL

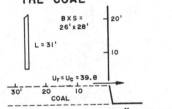

FIG. 9 b): Ur = Uc POSSIBLE "SHOOTING"
OF THE COAL

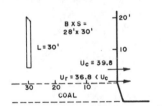

FIG. 9 c): Ur < Uc NO "SHOOTING"
THE COAL

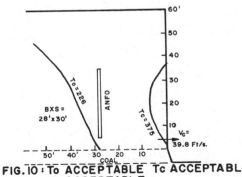

FIG. 10 : To ACCEPTABLE Tc ACCEPTABLE
Uc ACCEPTABLE

315

8 Mine design

Computer Applications in the Mineral Industry, Fytas, Collins & Singhal (eds)
© 1988 Balkema, Rotterdam. ISBN 90 6191 760 3

The application of computer aided drafting and design at INCO Ltd's Creighton Complex

W.R.McBride
INCO Ltd, Copper Cliff, Ontario, Canada

ABSTRACT: Technological change has resulted in an evolution of the Mines Engineering Office. Slide rules and hand drawn plans have been replaced by computer workstations and Computer Aided Drafting and Design. The initial CADD system, originally purchased for Creighton Mine, has been expanded to include three separate mines within the Sudbury basin. As well the original software package is continually being upgraded. CADD has contributed to some meaningful productivity gains. Repetitive drafting tasks have been reduced. More drawings are constantly being added to the CADD database. This augmented database provides for an increased number of drawings that can be rapidly revised to produce a new drawing.

1 INTRODUCTION

Mining at INCO Limited has evolved from a labour intensive industry into a highly mechanized, automated industry. This evolution has been made possible by an increased use of bulk mining methods as well as by an increased use of technological advances in the design and operation of mining equipment. Remote control equipment goes where no equipment went before. Remote controls allow one man to perform tasks that previously required two men. Programmable Logic Controllers tied into computers convey and hoist the ore, monitor the ventilation system, control the mine pumps, and monitor the sandfill pouring with minimal human involvement.

This technological evolution is also apparent within the Mines Engineering Department. Micro-computer workstations are widely used. Computer terminals at the minesite are tied into the central mainframe in Copper Cliff for access to the databases contained within. Computer Aided Drafting and Design software packages are in use at all of the mines within the Sudbury basin.

Creighton Mine has been on the forefront of this computerization. The first micro-computer was a Radio Shack TRS80 purchased in October, 1980. In April, 1984 Creighton Mine purchased the first CADD system in the Mines Engineering

Department of INCO Limited. Since then the TRS80 has become obsolete and been replaced while the CADD system has been substantially upgraded and expanded. Today the CADD system consists of Holguin's ADC400 Version 3.05 software running on a Hewlett-Packard HP1000 model A900 mini-computer. The Creighton CADD system now services three separate mines that are collectively referred to as the Creighton Complex.

This paper examines the application of CADD within the Creighton Complex. The purpose, the hardware, the software, the implementation, and the benefits to date are detailed. Possible future applications and benefits are also examined. Productivity enhancement is paramount to the survival of the mining industry. A centralized computer system using CADD software provides for productivity enhancement by reducing repetitive drafting tasks, by using the same basic drawings and modules at all mines, and by improved drafting quality and consistency making the drawings easier to understand by those using them in the field.

2 HISTORY

The Creighton Complex consists of Creighton Mine, Copper Cliff South Mine, and Crean Hill Mine. Creighton Mine is the largest of the three with production scheduled to be 9700 tons per day in 1988. The first ore shipment from Creighton, produced by an open pit mining method, was made in 1901. This ore was brought up a ramp by horse drawn carts. Mining has come a long way from those early days.

The mining methods have evolved to meet economic as well as safety concerns. Most recently this evolution has resulted in a bulk mining method known as Vertical Retreat Mining. This method is cheaper, safer, and less labour intensive. The mining equipment incorporated into the VRM method is also undergoing a transformation. Remote controls allow access to all of the ore without exposing men to unsafe conditions. Continuous mining systems are being devised that are more suited to bulk mining. Work areas are mined out quicker. More emphasis is placed on developing unmined areas as quickly as possible to replace the mined out areas.

In order to keep up with the demands of this fast paced industry the Mines Engineering Department invested in micro-computers to initially aid in compiling reports, schedules, and budgets. The high acceptance of these early micro-computers led the department to investigate other areas of computerization. Included in this investigation was a search for an applicable CADD package. Three and a half years after the introduction of micro-computers the first CADD system was purchased for Creighton Mine.

The system chosen was based on Holguin software and a Hewlett-Packard mini-computer. The CADD system as it initially was purchased only serviced the Mines Engineering office at Creighton Mine. The system has now been expanded to include all of the Engineering and Geology offices at all of the mines within the Creighton Complex. As well the Maintenance Department personnel at Creighton Mine make extensive use of CADD and as such have their own terminals.

3 PURPOSE

CADD was purchased to address the following needs: (a) simplify drawing files and allow rapid retrieval of requested drawings, (b) reduce the number of hard copy drawings being carried by the Mines Engineering Department, (c) simplify the updating of multiple copies of drawings, (d) reduce the time required to produce a drawing and (e) provide the capability for future electronic data transfer.

The CADD package chosen was a relatively low cost system from Holguin and Associates. The Holguin CADD software that was purchased is that generally used by architectural or engineering firms, and by contractors. This system is as well being used by J. S. Redpath of North Bay. Otherwise it sees minimal use in the mining industry. The alternative to the Holguin CADD is a more costly fully featured system such as the Intergraph CADD system in use at INCO Limited's General Engineering Department in Copper Cliff.

The advantage of the lower cost Holguin CADD is that it uses a network of workstations tied into a Hewlett-Packard mini-computer. The multiple work-station concept is deemed more desirable for our needs since it means every person in the Engineering office is capable of using CADD. The alternative to this is to have one or two fully featured work-stations with only a couple of individuals employed full time to create and modify drawings. The desire to maintain job enrichment coupled with the large number of drawings in existance resulted in the decision to have everybody capable of using CADD.

4 HARDWARE

The initial system that was purchased in April, 1984 consisted of a Hewlett-Packard HP1000 model A900 mini-computer, a HP 7914 hard disc drive, a colour graphics terminal, three black and white graphics terminals, a digitizer, a plotter, and a printer. The terminals were "dumb" terminals dedicated to the mini-computer for CADD useage. Refer to Figure 1.

The wish to have the ability to work on report generation, budgets, schedules, and CADD using only one computer led to the purchase of two HP 150A micro-computers with built in terminal emulators that allow them thereby to be used as full featured workstations as well as terminals. Since then the concept of one computer that can do it all at each persons desk has grown. Presently the micro-computers at Creighton Mine are used as stand alone computers, are tied in as terminals for the mini-computer, and are tied in as terminals for the company mainframe.

The demands on the system have grown since its initial purchase and as such the system has grown. The system now covers three separate mines and consists of a

HP1000 model A900 mini-computer, two HP7914 disc drives, a HP7937 disc drive, three colour terminals, five black and white terminals, twelve HP150 micro-computers, two HP Vectra micro-computers, one IBM AT micro-computer, five Compaq 286 micro-computers, five Compaq 386 micro-computers, four digitizers, four plotters, and one printer. The system is all connected together by switchmux multiplexers and high speed modems. The data transfer rates are set for 9600 baud and 4800 baud. Refer to Figure 2.

The HP1000 model A900 mini-computer has 3.8 megabytes of internal memory and operates at a speed of three million instructions per second (3 MIPS). Two of the disc drives are HP7914 with a disc storage capacity of 132 megabytes each. One of the HP7914 disc drives has a 65 megabyte tape backup. The other disc drive is a HP7937 with a disc storage capacity of 570 megabytes. The CADD program is contained within the HP7937 disc drive. As well the drawings and modules are stored in the larger HP 7937 disc drive. The HP7914 disc drives are used for continuous backup of the drawings. As well a tape backup of the drawings is done monthly and these tapes are stored off site.

A system as large as that at Creighton is best serviced by one person. With the expansion of the system to two other mines came the implementation of such a system manager. This one person is responsible for maintaining the system hardware, ordering and installing new equipment, and ordering supplies. As well this person is responsible for addressing software problems, training the CADD users, and ensuring management's policies regarding CADD and computers are followed.

5 SOFTWARE

The CADD software was supplied by Holguin and Associates. The licences were purchased through Systemhouse Ltd. in Ottawa. Two CADD workstations were licenced initially for the Mines Engineering office in April, 1984. Two additional licences were purchased for the Mines Geology office in June, 1984. The licencing agreement was expanded in May, 1986 to cover two workstations at Copper Cliff South Mine and again in November, 1986 to cover two workstations at Crean Hill Mine. The November, 1986 licence purchase actually allows 24 workstations to be on line at any given time. This licence purchase also addressed complaints by the CADD users of having to queue to get into the system.

The initial software program as supplied by Holguin was known as "CEADS-CADD". In March, 1985 this was upgraded to a new version containing about 50% more features and had a new name "CEADS-CAD 2000". Since then the CADD package has been upgraded and renamed through "ADC400 version 2.01" to the most recent release in October, 1987 "ADC400 version 3.05".

The Holguin CADD is a 2-D menu driven Computer Aided Drafting and Design System. A master menu appears on the screen as soon as the CADD program is initiated. This master menu accesses the sub-menus used to create or modify a drawing. Refer to Figure 3. The function keys control 14 functions that can be used in the creation of a drawing. Refer to Figure 4. The heirachy of CADD menus combined with the function keys control hundreds of commands and thousands of steps of execution.

Each CADD workstation is capable of working independent of the other workstations. The user calls up the master drawing and works on the drawing within his own unique workspace. The master drawing remains on the disc for others to access and work with at another workstation. When a drawing is completed the user can assign a new name and save the drawing independent of the master. Power failure or exiting the system does not delete the drawing being worked upon. Upon initializing the CADD system the option to "RESTORE DRAWING IN WORKSPACE" exists independently for each workstation.

Each drawing that is saved onto the disc drive is assigned to the CADD library. The task of maintaining an organized meaningful library listing rests with the users. They are responsible to assign a name to the drawing being saved that describes the drawing to others referring to the library. The users are also responsible for advising the system manager which drawings are obsolete so that these obsolete drawings can be deleted. In order to better organize the library of drawings within the CADD system each of the three mines have been assigned a two letter code to be used at the beginning of their drawing names to place those drawings in separate areas of the library listing. Drawings saved on the disc drive can be called up onto a workstation one at a time and overlayed to create a composite drawing. Composite drawings can be useful when interpolating between different levels of the mine to get a feel for the big picture. This is as close to 3-D as one can get with the ADC400 CADD.

Two of the more powerful options offered by Holguin's CADD are layers and subsets. These options allow the sorting or grouping of data. Each drawing can be divided into 256 layers of which 20 layers can be "Active" at any given time. Each layer can be further divided into an infinite number of subsets. Common elements in a drawing can be grouped into a layer. Layers can then be turned off or on dependent on the need to see the elements contained within the layers. Through the use of layers and subsets as much of the drawing as required can be plotted, transferred to a new drawing, or edited to create a new drawing.

The Holguin CADD software also includes a training module composed of two series of drills; the basic courses (41 drills) and the advanced courses (39 drills). The new user can advance through the training drills at his own pace. All of the drills need not be completed before usable drawings are being created.

6 IMPLEMENTATION

The first objective with the new CADD was to develop a system for the storage and retrieval of the then current drawings. The drawings used most often had to be stored first. What information to put on which layer had to be defined in order to produce drawings with varying amounts of information from the original. The layering of the drawings had to be compatible with the current needs as well as compatible with the future needs. Most of all the system had to be uncomplicated to encourage the maximum useage of CADD.

Before any work with the existing drawings could start the initial users had to be trained. In April, 1984 two supervisors and four planners began the self training drills. Each trainee was to spend 25 per cent of his time on the drills. Within two weeks, although the drills had not been completed, usable mine layouts were being created. Training of the CADD users is the most important means of encouraging the use of the CADD system. Even personnel that have used CADD for an extended time period can benefit by performing a particularly difficult series of tasks as part of an advanced training program. The majority of the Creighton Complex engineering and geology personnel have now been trained in the CADD system operation. During the last summer vacation period a one week training

program was given to eight beginners and a one week program was undertaken by five other advanced CADD users. All software training to date has been handled in house with the majority being of a self taught nature.

The next step after the startup was to determine which layer was to contain what information. Layers 1 to 19 were determined to contain mining, survey notes, grids, and mining proposals; layers 20 to 29 were reserved for geology data; layers 50 to 59 were for services, ventilation, drainage, air and water distribution lines, and electrical distribution lines; and layers 59 to 256 were reserved for future use. The descriptions of the subsets and the names to be used for these subsets were also determined after the system startup. These guidelines were then maintained in the digitizing that followed.

The existing mine level plans as well as the more commonly used drawings were then digitized using these layering and subset guidelines to maintain consistency. The digitizing of all existing drawings at Creighton is an incredible task. Such a large quantity of drawings can not be digitized overnight. The last two summer vacation periods have been utilized for digitizing drawings. All of the mine level plans now exist in the CADD system memory with some still requiring the plotting of survey plugs and other such details before they can be deemed complete. Over 2,000 drawings have been incorporated into the CADD system memory for reproduction or use at a latter date. Another 700 drawing modules have been placed into the system for use in the mine layouts.

7 PRESENT DAY USE

Most of the drafting work in the Mines Engineering office is the creation of mining layouts. Refer to Figure 5. These layouts can be stope and development layouts, raise sections, or drill layouts for I.T.H. or longhole drilling. The layouts are layered according to convention to show as little or as much detail as possible. The co-ordinate grids are created by the computer and thereby are much more accurate than those done by hand. The layouts also are composed of modules to accelerate the creation of a completed drawing. Refer to Figure 6. A module is simply a drawing that is common to a great many drawings and is saved as an individual item in the CADD system memory. A module can be a north arrow, a co-ordinate grid, boreholes, chutes, manways, vent raises, written mining procedures, and even written survey notes.

The desire to maintain consistent drawing conventions and to have user friendly instructions for the creation of mine layouts led to the compiling of a reference manual by all of the Creighton mine CADD users. Many hours were spent in user meetings assigning the tasks and troubleshooting problem areas. As well many phone calls to the Holguin "Hotline" were made to better define the features of the ADC400 CADD. The results are a comprehensive CADD reference manual geared towards the creation and updating of mine layouts. Once trained in the fundamentals of the Holguin CADD system the beginner need only refer to his reference manual in order to make a mine layout that is equal to any created by advanced CADD users.

In the Mines Engineering environment there is also the need to update the same drawing but in a different scale. The CADD system gives the ability to plot an excavation or drawing in any scale and thereby quickly update the other scale drawings for the same item. Another ability of CADD is to superimpose one drawing on top of another to study the interaction between openings.

All surveying at Creighton is done using Topcon GTS-3B survey instruments with detachable data loggers to record the survey data underground. The data loggers latter transfer the data to an IBM-AT micro-computer for calculating and plotting the results using AutoCAD. The plot presently requires digitizing to update the Holguin CADD drawings. No interface is in place for the Holguin CADD system as of the writing of this paper. Software for this interface is on order from Holguin and is anticipated to allow automatic survey data transfer to the Holguin CADD drawings without the intermediate steps of plotting and digitizing. An additional benefit of this interface will be to use CADD to define openings for evaluation using Numerical Modeling. Presently AutoCAD is used to define the openings for Numerical Modeling.

8 BENEFITS

If no data base of drawings exist to cover a required layout the time required to produce that layout is about the same using CADD as by hand. The advantage of CADD is that subsequent revisions are much faster. It is estimated that a drawing can be created two to three times faster as long as you can draw on older drawings and modules as components of the new drawing. Where drawings are continually revised and added to, as is the case with many mine drawings, even higher productivity improvements are possible. Automatic updating of different scale drawings of the same item can be done with little effort or time required.

In 1978 the Mines Engineering office at Creighton Mine employed 32 people. By 1987 the number of employees was 22 people. Yet the amount of time spent underground for each Mines Engineering employees at Creighton Mine has increased by thirty per cent. It is apparent that computerization of the office workload and the use of CADD has contributed to the engineering staff spending more of their time underground.

The accuracy of the layouts using CADD reduces the potential for mistakes. The co-ordinate lines are at exact intervals. Once the existing openings are digitized then tracing errors are substantially reduced since reproduction of the digitized openings will always be exactly the same every time. Improved drawing quality and consistency means that the layouts are easier for the operators to understand.

Another benefit of CADD using a mini-computer is multiple access to all drawings by all users. The same modules are in use at all three mines in the Creighton Complex. The same drawing conventions regarding layers and regarding drawing names are also in place at all three mines. More time is spent creating meaningful drawings and less time is spent on duplication of effort.

Multiple access also applies to the ability for anybody to log onto the CADD system without waiting. The profusion of terminals within the complex and the licencing agreement providing for infinite workstations that can use CADD at any given time means that nobody should have to wait to perform their work. No special allotment of time has to be made for individuals to use the CADD system. It is thereby easier and desirable for the engineering personnel to spend the morning underground examining a problem and to spend the afternoon creating a layout that solves the problem.

9 FUTURE BENEFITS

The downloading of the survey information

from the data logger into the Holguin CADD system is an absolute necessity. The automatic update feature when combined with the survey information downloading will eliminate many manhours of duplicated effort.

A borehole database exists in the Copper Cliff mainframe. The development of an interface program that allows access to this database means that the boreholes would automatically be plotted in their exact location on the mine layouts. Another benefit of accessing this database would be the ability to use the Holguin CADD system to plot the borehole information on sections and plans as part of the ongoing evaluation of ore reserves. The existing program for plotting boreholes is contained on the mainframe and as such is remote and inflexible for our needs.

The HP1000 mini-computer has the potential to become a Management Information System with any of the terminals or micro-computers tied into the system capable of accessing the database. Word Processing, Spreadsheet, and Database Management software are installed on the mini-computer available for all to use. Once the use of CADD becomes second nature to all of the engineering and geology employees, these other software packages will be put into use.

10 CONCLUSIONS

The introduction of a sophisticated CADD system has been well accepted within the Creighton Complex. The goal of training everyone in the office in the use of CADD is close to being achieved. The benefits of using CADD are obvious for all to see. The reduction in time required to do a layout allows the personnel involved to spend more time underground examining the problems first hand.

Ongoing training and the resisting of the urge to manually complete a task are necessities for the successful implementation of a CADD system. Management must rigidly enforce the use of CADD for drawing creation. Management must recognize the need for and allot the necessary time to have an ongoing training program.

A smaller computer with a stand alone CADD package can be used to create mine layouts but many of the productivity gains achieved through a multi-user system are not recognized. The CADD system has to be available when someone needs it. The access by all users to any drawing within the database without delay is a must for maximum efficiency. The CADD system combined with the other software packages available simultaneously make the mini-computer a powerful device for mine planning.

FIGURE 1.
CREIGHTON MINE ENGINEERING DEPARTMENT
CADD SYSTEM HARDWARE SETUP - INITIAL - APR/84

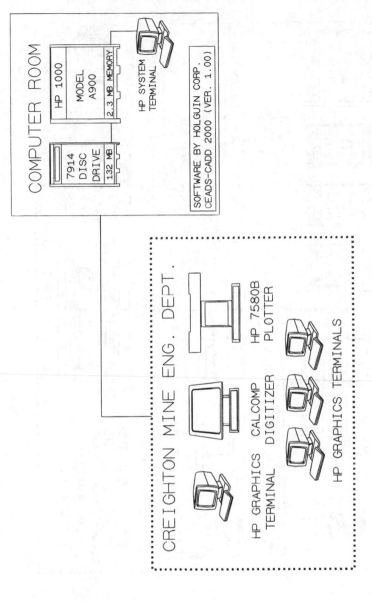

COMPUTER ROOM

7914 DISC DRIVE
132 MB

HP 1000
MODEL A900
2.3 MB MEMORY

HP SYSTEM TERMINAL

SOFTWARE BY HOLGUIN CORP.
(CEADS-CADD 2000 (VER. 1.00)

CREIGHTON MINE ENG. DEPT.

HP GRAPHICS TERMINAL

CALCOMP DIGITIZER

HP 7580B PLOTTER

HP GRAPHICS TERMINALS

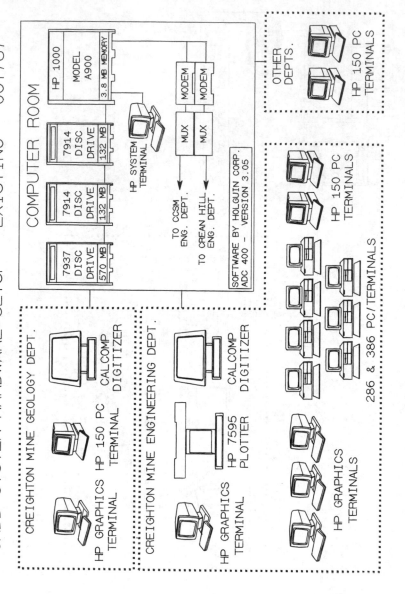

FIGURE 2.
CREIGHTON MINE ENGINEERING DEPARTMENT
CADD SYSTEM HARDWARE SETUP – EXISTING – OCT/87

COMPUTER ROOM

HP 1000
MODEL A900
3.8 MB MEMORY

7937 DISC DRIVE 570 MB

7914 DISC DRIVE 132 MB

7914 DISC DRIVE 132 MB

HP SYSTEM TERMINAL

MODEM
MODEM

MUX
MUX

TO CCSM ENG. DEPT.

TO CREAN HILL ENG. DEPT.

SOFTWARE BY HOLGUIN CORP.
ADC 400 – VERSION 3.05

OTHER DEPTS.

HP 150 PC TERMINALS

HP 150 PC TERMINALS

286 & 386 PC/TERMINALS

CREIGHTON MINE GEOLOGY DEPT.

HP GRAPHICS TERMINAL

HP 150 PC TERMINAL

CALCOMP DIGITIZER

CREIGHTON MINE ENGINEERING DEPT.

HP GRAPHICS TERMINAL

HP 7595 PLOTTER

CALCOMP DIGITIZER

HP GRAPHICS TERMINALS

FIGURE 3.

CADD MENU

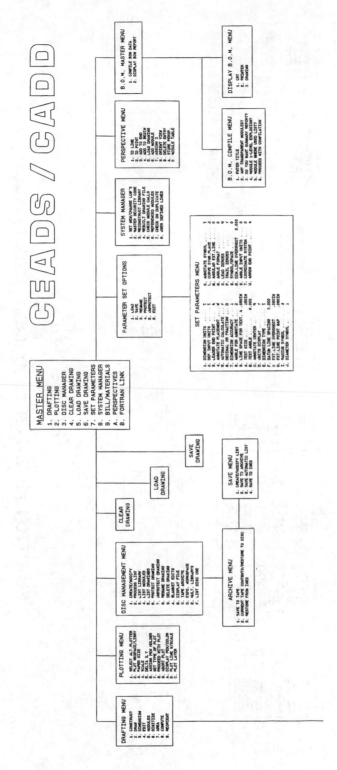

FIGURE 3. (CONTINUED)

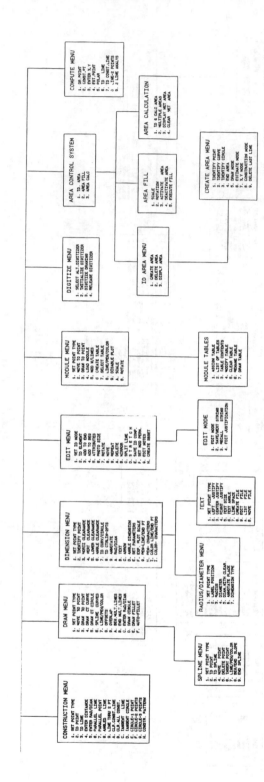

FIGURE 4.

FUNCTION KEYS FOR CADD (ADC 400) PROGRAM

F 1 EXIT	F 2 MENU	F 3 ZOOM	F 4 RDISPLAY	F 5 RSCALE	F 6 MOVE	F 7 SKIP	F 8
F 8/F 1 DELETE	F 8/F 2 MODES	F 8/F 3 RUBR LIN	F 8/F 4 SNP(A/G)	F 8/F 1 PEN/LAY	F 8/F 2 TOG CONST	F 8/F 3 LAYERS	SHIFT

FIGURE 5. : TYPICAL MINING LAYOUT

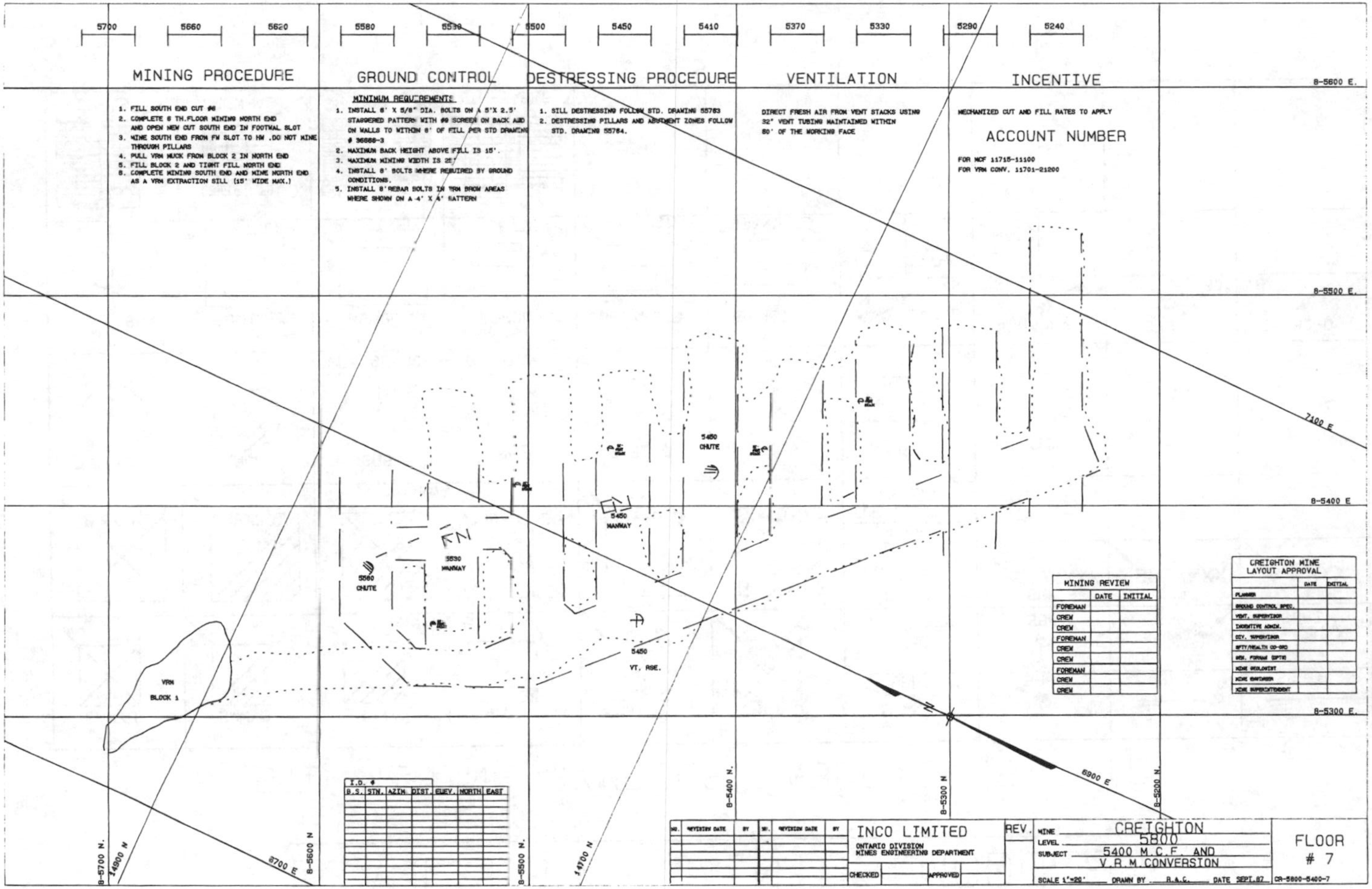

FIGURE 6.

EXAMPLES OF MINING MODULES

STD MINING SYMBOLS TABLE 2 (SMS-T2)

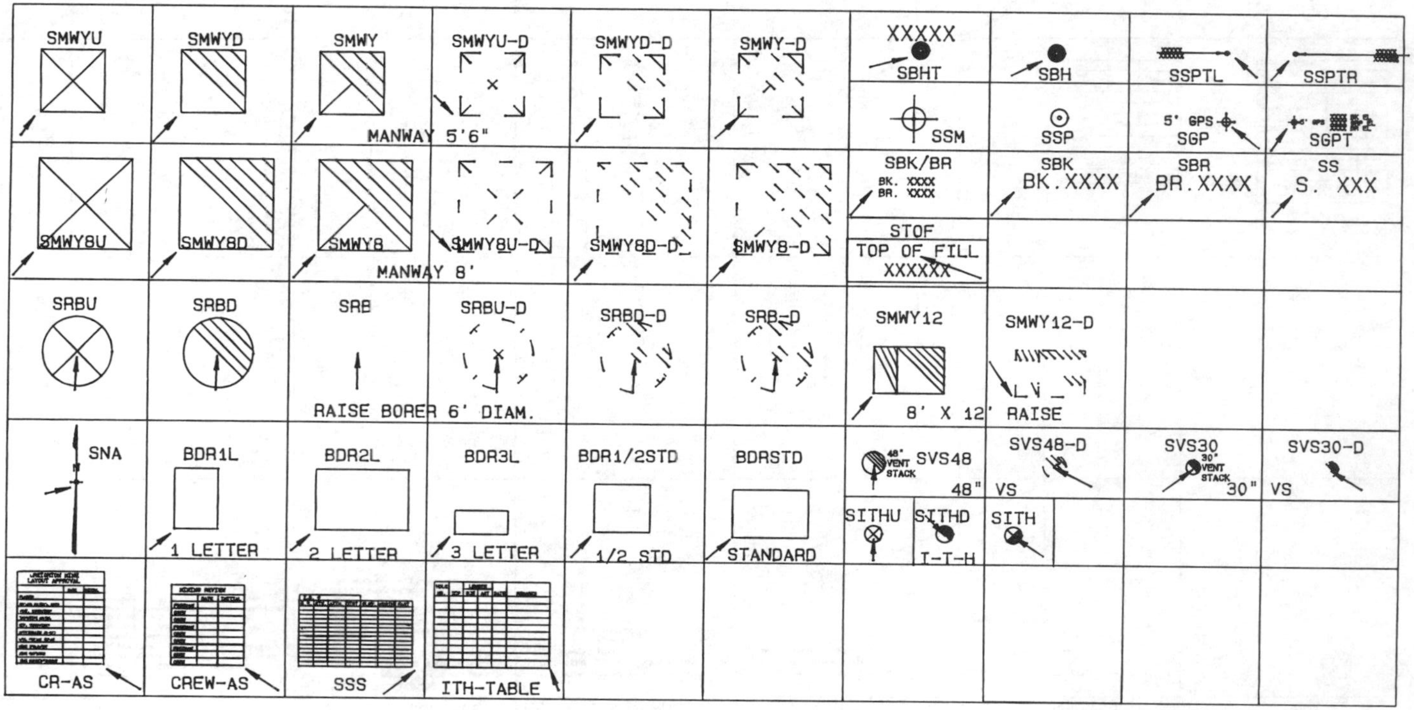

Computer Applications in the Mineral Industry, Fytas, Collins & Singhal (eds)
© 1988 Balkema, Rotterdam. ISBN 90 6191 760 3

Beyond optimization in open pit design

J.Whittle
Whittle Programming, Melbourne, Australia

ABSTRACT: A method for reducing the number of economic variables involved in open pit design to one major and one minor factor is presented. It is shown that a family of optimal open pits can be generated for a range of values of the major factor. This permits the rapid production of mining schedules and discounted cash flows for a range of economic scenarios, and thus facilitates rational risk analysis in the choice of the best pit. An example is given.

1 THE VALUE OF A PIT

For a particular ore body, there can be any number of different open pit designs. From these, we try to find the optimal design, that is the pit with the highest value. To calculate that value, we take into account a large number of physical and economic factors. These can include:

  the amount of ore
  the grade
  the recovery fraction
  the amount of waste
  the cost of blasting ore
  the cost of blasting waste
  the cost of trucking
  the cost of processing ore
  the administrative costs
  the capital costs
  etc.

If any of these factors change, then the value of each of the possible pits will change and a pit chosen as optimal may no longer be so. Note that some of these factors are themselves dependent on the pit design, in particular its size.

The number and complexity of these factors appear to make the examination of optimal pits for all possible scenarios quite impossible. However, as we shall show in this paper, the factors can be reduced to one, or, at most, two, and a useable set of optimal pits can be produced. This set of pits allows objective examination of things such as economic risk, as part of the design process.

2 PIT OPTIMIZATION

If all the factors relevant to the value of a pit are known, then methods for finding the optimal pit outline are well established, and suitable computer programs are readily available. For example Lerchs and Grossmann published a method in 1965.

The first step is to produce a block model of the deposit, giving the amount and grade of ore, the amount of waste and possibly the amount of air in each block in the deposit. A block value is then calculated for each block. This is the value of the block once it has been uncovered, and consists of the return obtainable from any ore in the block, less the cost of mining and processing the block.

In the case of a block containing ore, the value will (generally) be positive. In the case of a block consisting entirely of waste, the value will be negative. Pure air blocks have zero value.

In theory, an optimal pit can be generated by repeatedly searching for incremental pits consisting of combinations of blocks which have a positive total value and which obey the slope constraints. Such incremental pits are 'worth' mining and, as each is found, it is added to the pit being generated and thus increases its value. From then on the blocks involved are treated as air blocks.

By definition, each time we add a positive increment to the pit we prove that the previous pit was not optimal, because we now have a pit with a higher value.

Conversely, when no further positive incremental pits exist, the pit we have is optimal.

The simplest case of an incremental pit consists of a single ore block together with the inverted cone of waste blocks which must be removed to uncover it. This led to the development of the 'Moving Cone' method of optimization. However, incremental pits can also consist of two or more ore blocks with cones which are not positive when taken individually, but which share some waste blocks and are positive when taken together. Unfortunately, an exhaustive search for all possible combinations of this type is beyond the power of even the largest of today's computers and generally the moving cone method does not find the true optimum.

The Lerchs-Grossmann method takes a different approach to the problem but always finds the optimal pit. An explanation of this method is beyond the scope of this paper.

There are some problems with the Lerchs-Grossmann method, but in most cases they are more apparent than real.

Firstly, in its simple form, the method takes no account of such essential requirements as haul roads and minimum mining widths. In practice, however, it is possible to overcome most of these difficulties by including extra constraints and/or by adjusting pit slopes to the average required, including any haul roads or berms.

Secondly, the list of blocks is not itself a pit design. More work has to be done, usually by hand, to smooth out the jagged edges caused by the blocks and to put in the detail of haul roads, etc. However, given the starting point provided by the method, the time taken for this is an order of magnitude less than that required for a full design from the beginning. Also, if we want to compare different designs produced under different assumptions, then the difference between two optimal block lists will usually closely approximate the difference between the two corresponding detailed designs. Thus, for comparisons, it is not necessary to do the detailed manual work at all.

Finally, some of the economic factors can depend on the pit design, particularly its scale. (This applies to any pit design method.) If this is the case, then it may be necessary to estimate the factors initially, produce an optimal block list, recalculate the factors and repeat the process until no further change is necessary. Two, or at most three, iterations should be more than adequate for this.

In summary, the Lerchs-Grossmann method provides a quick way of producing very good open pit designs for a given set of economic and pit slope assumptions.

3 THE FACTORS

For the purposes of assessing the value of a block, some of the technical and economic factors mentioned in an earlier section are added together and all we are concerned with is their total. Consequently, the number of independent factors involved in pit design can easily be reduced to seven with no practical loss of generality. These are:

> The tonnes of rock in the
> block (ore and waste)   (ROCK)
> The tonnes of ore in the block (ORE)
> The units of product in the
> block   (METAL)
> The fraction of the product
> which can be obtained
> from the ore   (RECOVERY)
> The price obtainable for a
> unit of product less any
> delivery costs   (PRICE)
> The cost of mining and
> removing a tonne of rock   (COSTM)
> The extra cost of processing
> a tonne of ore   (COSTP)

The value of a block, once it is exposed, is calculated by the formula:

$$(METAL \times RECOVERY \times PRICE - ORE \times COSTP) - ROCK \times COSTM$$

where the expression in brackets is omitted if it is negative.

The value of a particular pit design is the sum of the values of the blocks contained within it.

We should expand on two of these factors:

COSTM - This is the cost of mining and removing a tonne of waste.

COSTP - This is what it costs to mine and process a tonne of ore over and above what it would cost to mine it as waste. It is not just the processing cost.

COSTM and COSTP must both include their fair share of any variable overheads. COSTP should not carry all the overheads on the basis that 'ore pays for the mine'. Truly fixed overheads, which are rare, should not be included for the purposes of optimization. They should merely be subtracted from any value obtained for the pit.

When we design an optimal pit, by whatever method, we create a list of blocks such that the total of the block values is the maximum possible while still obeying the slope constraints.

4 THE NATURE OF THE FACTORS

The values of ORE, METAL and ROCK for each block describe the deposit. If we do not know these values to some acceptable degree of accuracy, no quantitative evaluation of the deposit is possible. We will regard them as fixed.

In contrast, although we may know the current values of PRICE, COSTM and COSTP, we need their values at some time in the future in order to design a pit which might take some years to mine out. Whilst COSTM and COSTP will probably be affected approximately equally by inflation, our knowledge of PRICE in the future can be very uncertain indeed.

If we are to plan a pit sensibly, we need to consider a range of optimal pits covering the likely range of values of PRICE, COSTM and COSTP. We will then be in a position to plan a mining strategy which will minimise the risk due to changes in economic circumstances.

In theory we could compute an optimal pit for every combination of values of PRICE, COSTM and COSTP, but this would not only present a massive computing task, it would also result in a massive quantity of data to be assessed.

We need to reduce the number of factors.

5 REDUCING THE NUMBER OF FACTORS

Note that if, for a particular set of factors, we divide the value of each block in the model by a constant positive value, we have merely changed the units of currency. The size and shape of the optimal pit for that model will be unaffected.

In other words, it doesn't matter whether we assess the value of blocks in dollars, pesos or ounces of gold.

If we choose to divide the value by COSTM, the cost of mining a tonne of waste, then the formula for the value of a block becomes:

$$(METAL \times RECOVERY \times [PRICE/COSTM] - ORE \times [COSTP/COSTM]) - ROCK$$

We now have only two economic factors,

neither of which is affected by general economic inflation.

[COSTP/COSTM] is only likely to change significantly with time if new mining or processing methods are introduced or if there is a very significant change in a particular cost component. Let us redefine this as CRATIO (cost ratio).

[PRICE/COSTM] is now the only significant unknown. In the form shown, its meaning is not obvious, but if we invert it to get [COSTM/PRICE] and cancel out the dollars, it becomes a number of units of product divided by a number of tonnes. It is therefore a 'grade' and is, in fact, the amount of product which must be sold to pay for the mining of a tonne of waste. Let us redefine this as MCOSTM (the 'Metal Cost of Mining').

The modified formula for the value of a block now becomes:

$$(METAL \times RECOVERY/MCOSTM - ORE \times CRATIO) - ROCK$$

CRATIO is relatively stable with time, so for sensible strategic planning, we now need to consider a range of optimal pits covering the likely range of values for only one unknown (MCOSTM) rather than three. This is a very significant improvement.

6 THE SET OF OPTIMAL PITS

Let us assume that CRATIO is fixed and that we have generated an optimal pit (Pit A in Figure 1) for a particular value of MCOSTM. This pit contains all the ore blocks which are worth mining, together with the waste blocks which must be removed to expose them, subject to the required pit slopes. Because this is an optimal pit, to mine any ore block which lies outside the pit will require the removal of extra waste costing more than the ore block is worth. Conversely, to omit any waste block will also involve omitting one or more ore blocks with a total value which is greater than any saving on the mining of waste.

If we decrease MCOSTM and re-calculate the block values, then reference to the modified formula for a block value shows that the value of every block in the model will either increase or stay the same. Therefore all of Pit A is still worth mining and will be included in any optimal pit. In addition there may be ore blocks outside Pit A which are now worth mining. Pit B consists of Pit A plus these additional blocks and their attendant waste. Therefore, each time MCOSTM is decreased, the corresponding optimal pit will always include the previous pits. In consequence, optimal pits generated for a range of values of MCOSTM will form a nested set.

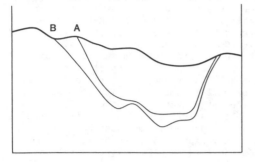

Fig.1 Nested pits

The Lerchs-Grossmann pit optimization method can be extended, as is suggested in the original paper, to generate this nested family of pits.

The approach is to set up the block values using the lowest required value of MCOSTM and then to obtain the optimal pit. This pit is the largest in the family and all blocks outside it can now be excluded from further consideration. We then set up the remaining block values using the highest required value of MCOSTM and again obtain the optimal pit. This is the smallest pit in the family and all blocks inside it can be excluded from further calculations. We can then calculate the intermediate pits using interpolated values of MCOSTM until we have obtained the required resolution. As this process continues, we deal with ever decreasing numbers of blocks so that, although the whole calculation takes longer than a single optimization run, it takes much less time than would be required if each optimal pit was calculated from the beginning.

The end product of the computer run consists of a block model, where each block is labelled with the highest value of MCOSTM, the metal cost of mining, for which it is mined. This model can then be analysed in a number of different ways.

Software has been written which will carry out the above multiple optimization. It will handle:

> Multiple ore parcels per block
> Multiple processing methods and ore
> types in any combination
> Slopes which change with both
> position and direction
> Mining costs which vary with
> position
> Processing costs which vary with
> position.

7 ANALYSIS

From the block model which details the set of nested pits, a range of different reports and graphs can be prepared.

Bench by bench tonnages and grades for each of the nested pits can be printed.

Graphs of optimal pit volume, ore tonnage, waste tonnage and average grade can easily be prepared for the range of values of MCOSTM considered. Alternatively, if the cost of mining waste (COSTM) is regarded as known, these values can be plotted directly against product price.

Simple 'best' and 'worst' mining schedules can be printed for a series of economic scenarios. Each gives tonnages and grades for each period together with cash flow and discounted cash flow. Each is based on the pit which is closest to optimal for the economic scenario in question.

Ore tonnages and grades can be reported either by processing method or by ore type.

All the above is done for a particular value of CRATIO. In practice, one would repeat the exercise for values, say, 20% above and 20% below the expected value so as to get some feel for the sensitivity in this direction. Note that the optimal pits generated in this way will not necessarily nest with the initial set.

8 THE SIZE OF THE BLOCKS

When setting up a block model of a deposit the block dimensions are usually chosen so as to be convenient for ore body delineation and/or mining scheduling. This can lead to a model, including waste, of anything from 100,000 to 10,000,000 blocks.

Experience in the use of pit optimization has shown that it is often possible to increase the block size and reduce the number of blocks used for optimization purposes with little or no effect on the final design, provided that the smaller blocks are combined correctly. This can dramatically reduce the computation involved. A model of 50,000 to 100,000 enlarged blocks is usually quite adequate for pit optimization and this can easily be handled on a small mini-computer.

9 EXAMPLE

A small low-grade gold deposit containing both reef and stockwork structures has been used as an example.

A combined operation using both milling and heap-leaching was proposed, with the program allocating parcels of ore between the two on economic criteria. Milling was given a CRATIO of 9.35 and its recovery was estimated at 0.95. Heap-leaching was given a CRATIO of 3.20 and its recovery was estimated at 0.65. Only one ore-type ('ORE') was used.

A nested set of 21 pits was generated covering a range of MCOSTM from zero to 0.2 g/tonne in steps of 0.01. The zero case corresponds to infinite gold price and causes all ore to be mined. The 0.2 case corresponds to a gold price of $7.50/g ($233/oz) with $1.50 as the mining cost.

```
Pit 13 (D)     which is optimum for
======
                an MCOSTM of            0.080

              ------------------------------------
              Processing  Ore
                 Method   Type        Cut-off
                   MILL   ORE            1.64
                   HEAP   ORE            0.39
              ------------------------------------

    Bench       Method      Tonnes      Metal  Grade

     14 Rock                1695294
                   MILL      122034     287227  2.35
                   HEAP      521596     428903  0.82

     13 Rock                1840436
                   MILL      270864     770671  2.85
                   HEAP      829542     729337  0.88
                   .           .          .      .
                   .           .          .      .

      8 Rock                  49248
                   MILL        4617       7966  1.73
                   HEAP       23085      23395  1.01
                            ---------  --------- -----
Totals:    Rock             7875710
                   MILL     1357851    4261826  3.14
                   HEAP     3608851    3084629  0.85
                            ---------  --------- -----
```

Fig.2 Bench by bench details of a particular optimal pit

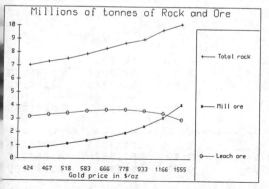

Fig.3 Tonnages relative to gold price assuming a waste mining cost of $1.5/tonne

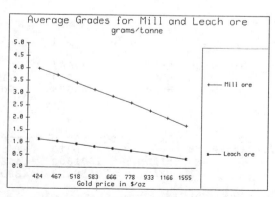

Fig.4 Average grades of Mill and Leach feed relative to gold price assuming a waste mining cost of $1.5/tonne

Usually a smaller range of MCOSTM would be covered and about 40 pits would be generated, but the numbers chosen are better for illustrative purposes.

Figure 2 shows the bench by bench details of one of the pits. Please note that the cut-offs and grades are in grammes per tonne. The cut-off given for milling is not the marginal cut-off but the grade below which it is more profitable to heap-leach the ore. The cut-off given for 'HEAP' is the marginal cut-off for that method. These cut-offs are calculated by the program from the cost-ratios, the recovery fractions and MCOSTM.

The 'D' in brackets by the pit number is the character (0-9,A-Z) used to identify this pit in simple printed sections and plans produced by the software.

Figure 3 shows the rock, milling and heap-leach tonnages as functions of the gold price, assuming a mining cost of $1.50 per tonne. Figure 4 shows the milling and heap leach average grades under the same circumstances. Note that each gold price corresponds to a different optimal pit.

Figure 5 shows the print-out for the worst case schedule for a scenario where the price of gold is $18.00/g ($560/oz), the mining cost is $1.50 per tonne, the maximum rock and milling per period are 2,000,000 and 350,000 tonnes respectively.

Price of metal	($/UNIT) :	18.00
Cost of mining	($/TONNE) :	1.50
Maximum mining per period	(TONNES) :	2000000
Discount rate	(% per period) :	10.00
Pit number	:	13 (D)

Processing Method	Ore Type	Recovery Ratio	Processing Cost	Cut-off		Processing Method	Maximum Tonnes
MILL	ORE	0.950	14.03	1.71		MILL	350000
HEAP	ORE	0.650	4.80	0.41		HEAP	1000000

Worst case schedule, with each bench completed before the next is started.

Period	Method	Tonnes	Metal	Grade	Cash-Flow	Discounted Cash-Flow
1 Rock		2000000			-3000000	-3000000
	MILL	199782	489767	2.45	5573076	5573076
	HEAP	677474	545277	0.80	3127866	3127866
					5700942	5700942
2 Rock		1927916			-2891874	-2602687
	MILL	350000	991251	2.83	12041650	10837485
	HEAP	856021	711943	0.83	4220832	3798749
					13370608	12033547
.		.	.	.	.	.
.		.	.	.	.	.
5 Rock		1232687			-1849031	-1213149
	MILL	251520	718518	2.86	8759094	5746842
	HEAP	743935	609797	0.82	3563743	2338172
					10473807	6871865
Totals: Rock		7875710			-11813565	-9966215
	MILL	1501302	4489728	2.99	55718586	44686127
	HEAP	3559123	2892575	0.81	16759336	13782299
					60664357	48502211

Fig.5 Worst case mining schedule for a particular economic scenario

Heap-leaching is effectively unlimited. The (real) discount rate is set to 10%.

The mining cost fixes the processing costs via the CRATIO values used during optimization. The gold price and mining cost fix MCOSTM at 0.0833. The nearest pit in the family to this has an MCOSTM of 0.08 and is the one shown in figure 2. Because the match is not exact, the cut-offs and allocation between milling and heap-leaching are different, however the total tonnes of rock and ore agree with the tonnages shown for Pit 13 in Figure 2.

The best case schedule for this scenario, with inner pits always mined out first, is also printed but is not shown here. The total discounted cash-flow is $49,706,262.

It should be emphasised that these schedules are both extremes which would never be used in practice. Nevertheless, any real schedule would lie somewhere between the two and they thus represent a working range, particularly with regard to cash-flow.

10 CONCLUSION

We have shown that the economic factors which determine the shape and size of an optimal pit can be combined and scaled in such a way as to reduce their number to two. One of these, the ratio between processing and mining costs, is relatively stable with time. The other, the amount of product which must be sold to pay for the mining of a tonne of waste or 'The Metal Cost of Mining', constitutes the main economic unknown when planning for the future.

We have shown that a family of pits can be generated, where each is optimal for a different value of the metal cost of mining. This set of pits forms an excellent data base for the analysis of future possibilities, and for rational assessment of commercial risk.

REFERENCES

Lerchs, H. & Grossmann, I.F. 1965. Optimum Design of Open-Pit Mines. Trans. C.I.M. Vol LXVIII, pp 17-24.

Computer Applications in the Mineral Industry, Fytas, Collins & Singhal (eds)
© 1988 Balkema, Rotterdam. ISBN 90 6191 760 3

Role of the computer in managing and applying geotechnical data to surface mine design

N.H.Wade & F.Naderi
Monenco Consultants Ltd, Calgary, Canada
D.Nikols
Alberta Research Council, Edmonton, Canada
(Formerly TransAlta Utilities Corp., Calgary)

ABSTRACT: In large open cast operations, effective management of considerable volumes of specialized data generated by various earth science disciplines is necessary to ensure optimum pit design and equipment utilization. The development of a computerized geotechnical data base for a coal strip mine is described and the interrelationship between the data base and the geological computer model is discussed, highlighting the advantages as well as shortcomings of the systems approach. Geotechnical input to mine development is outlined including computer models for determining safe pitwall and spoilpile slopes, classifying the mine area into sub-areas of relatively consistent geotechnical characteristics and assessing formation diggability.

1 INTRODUCTION

With the increasing size and complexity of equipment and depth of surface mining operations, present day mine planning requires the collection and evaluation of greater volumes of numerical data from a wider variety of engineering disciplines than ever before. Mining engineers, geologists, geotechnical specialists, geophysicists and hydrogeologists not only have to cope with new and changing technology but also each has to possess some knowledge of the professional requirements of the other disciplines. This paper outlines the role played by geotechnical engineers in the planning and operational phases of surface mining, describes, for a strip mine in the Alberta plains coal province, an independently developed systems approach to mine planning incorporating a computerized geotechnical data base and illustrates the interdisciplinary nature of the planning procedure.

2 BASIC AIMS OF MINE PLANNING

The principal aim of mine planning, once a site has been selected, is to design mine layouts and schedules that allow operations to be optimized, resource recovery maximized and costs minimized.

The planning must be based on a practical, economic and pragmatic approach to operations and problem solving that minimizes risk to the safety of mine personnel. The disciplines involved in mine planning have a continuing role in the development of a project, from the conceptual stage through feasibility studies to the end of the project life. To ensure that the mining system is designed to best advantage, the nature of the deposit must be thoroughly assessed in terms of geological structure, geotechnical characteristics and groundwater regime. As reiterated by Jeffreys et al. (1986), the planning team must have a fundamental understanding of the deposit to be mined. Contributions to this understanding are made by:

1. Geophysicists - geophysical logs of drillholes, reflection/refraction surveys utilizing surface seismic techniques;

2. Geologists - dimensions of the deposit, depth of cover, number and thickness of coal seams or other ore formations; type and thickness of partings, interburden, underburden, overburden; dip and strike of bedding, faults, major discontinuities; degree of folding; occurrence and nature of weak planes, clay seams, joint sets and other defects; geologic model of the deposit;

3. Geotechnical Engineers - physical characteristics of all materials to be

excavated, both insitu and after excavation, as well as the pit floor strata including parameters for classification, strength, density, slake durability, flowability, stickiness, diggability, trafficability, slope stability;

4. Hydrologists - drainage basin size and characteristics; maximum-average-minimum rates of precipitation; infiltration and evapotranspiration rates, hydrographs and maximum flood estimates for stream courses;

5. Hydrogeologists - groundwater regime, groundwater recharge, water quality, drainability, potential leachate generation/migration, hydraulic conductivity;

6. Mining Engineers - characteristics of the coal chemical/physical properties such as natural moisture, heat value, combustion properties, ash content, specific gravity and hardness.

All disciplines must appreciate, of course, the reliability of the available information and establish the confidence with which the data can be used.

In chronological order, the various stages of mine development which involve planning commitment are generally as follows:

1. conceptual mine design based on limited data and evaluation;

2. various preliminary design phases and intermediate reappraisal stages as more information becomes available;

3. final design for economic feasibility and project commitment;

4. detailed mine planning for optimum equipment usage and production requirements;

5. short-term mine plan (1 to 5 yr) for forecasting profits;

6. medium term plans (5 to 10 yr) for equipment justification;

7. long term plans (10 to 30 yr) to identify type and number of equipment required;

8. periodic updated plans during mine operation to accommodate any changes in geologic stratigraphy observed in excavations, experience with equipment operation, market fluctuations, etc.

Concomitant with the experience of others involved in planning of surface mines (see for example papers by Jeffreys et al, 1986; Mallett et al, 1987; Moelle and White, 1987; Kim, 1987; Scoble et al, 1986 and Scoble et al, 1987), the writers consider that an interdisciplinary or systems approach to planning is necessary for large mines. In addition, the use of computerized data

bases and specialty computer programs is required to effectively evaluate and manage the large volumes of data generated. A list of commonly used programs for earth science data management and analysis with a brief description of each is illustrated in Table 1.

3 GEOTECHNICAL ASPECTS OF SURFACE MINING

A brief outline of geotechnical involvement in specific mine development activities is described in the following paragraphs.

3.1 Deposit evaluation

1. Organize the exploration program in conjunction with mining, geological and hydrogeological specialists;

2. Prepare detailed logs of materials encountered; ascertain formation properties such as penetration resistance and strength by in-situ testing; obtain representative undisturbed samples of the principal strata affecting pitwall and floor stability for laboratory classification and strength testing; install piezometers to determine the existing groundwater regime and location of aquifers;

3. Determine, in conjunction with geologists, the structural geology of the deposit including formation dip and strike; the presence, location and orientation of faults; the extent of jointing and especially the presence of discontinuities such as weak clay seams, slickensided joints and shear zones.

3.2 Ore evaluation

Determine the geotechnical properties of the coal sequence or other ore such as density, natural moisture, strength, etc., and assess the consistency of these parameters throughout the deposit.

3.3 Pit stability

1. Develop probable failure mechanisms and carry out appropriate stability analyses to establish safe pitwall slopes;

2. Monitor excavated pit slopes in the operating mine to warn of impending failure or, on the other hand, ensure overly conservative slopes are not being excavated;

Table 1. Typical earth science computer programs used in surface mine design.

Program Title	Applications
Computer aided drafting (CAD)	interactive system for digitizing and plotting maps of the mine area.
Mine evaluation and design system (MEDSYSTEM)	geological modelling, estimating ore reserves, calculating excavation quantities, generating strata contours and thickness isopachs.
CYCLE	calculation of truck haul cycles utilizing truck rimpull curves to evaluate truck performance on road segments.
Hydrologic engineering center (HEC) programs	modelling of surface water flow in order to design drainage systems.
SLOPE II	limit equilibrium analysis to evaluate the stability of pitwall and spoil slopes. Program can accommodate complicated surface profiles and stratigraphic conditions, piezometric or groundwater data, and externally applied loads including seismic loading.
Slope Analysis Preprocessor (SLAP)	interactive system for generation of models to be used in slope analysis.
DATABASE	depository for geotechnical, hydrogeological and geological data on physical and chemical characteristics of ore, overburden, interburden and underburden, as well as groundwater regime.
Digital groundwater models (PRICKETT-LONNQUIST & U.S. Geological Survey)	Two- and three-dimensional models to predict hydrostatic heads behind excavated slopes for use in slope stability analyses and to estimate seepage inflow to the pit.
SURF II	plotting package for piezometric heads.
ROADS, COCO	used in the design of mine haul roads.
Geostatistical methods (Kriging technique)	used to interpret exploration and drill hole data to generate a three dimensional geologic model of the mine and an estimate of ore reserves.

   3. Conduct back analysis of failed slopes to confirm material parameters and groundwater levels assumed during design;
   4. Assess the potential for pit floor heave due to excessive hydrostatic pressures in the underlying formations or stress relief from excavation unloading.

3.4 Formation diggability

Assess the ease with which the overburden materials and ore can be excavated with the proposed mining equipment and designate those areas in the mine where blasting is required.

3.5 Vehicular trafficability

Determine from geological/geophysical logs, seismic surveys and geotechnical data the presence of fine grained or slakable formations which could soften in the presence of water and thus hamper traffic mobility on benches, ramps and roadways; design mitigative measures to improve trafficability on such materials; compute safe bearing pressures for critical formations.

3.6 Mine infrastructure

Design roads, embankments, ditches and provide recommendations for building foundations.

3.7 Spoil stability

Provide design recommendations and safe slope angles for spoil dumps (containing dry mine spoil, ash from thermal plant, etc.) and slurried tailings empoundments.

3.8 Land reclamation

Establish longterm stability, settlement potential and potential for groundwater contamination from leachate generation; design mitigative measures as required for safe end use of reclaimed land.

4 GEOTECHNICAL DATA REQUIREMENTS

To carry out the tasks described in Section 3, the geotechnical engineer requires a reliable assessment of the geological conditions at the mine site including the lithology and stratigraphy of the deposits which form the pitwalls and the mine floor. Accordingly, he leans heavily on the expertise of the geologist to obtain these and other data such as details on the presence, location and orientation of major faults, the extent of jointing, the occurrence of other structural defects such as weak clay seams, brecciated zones and slickensides, and, in glaciated regions, the depth and extent of glacial disturbance in the pit formations. By utilizing a computerized geological model and subroutines with X-Y plotting capability, the geologist is able to readily provide representative geological sections at any azimuth and thickness isopachs or surface contours for individual stratigraphic units. By incorporating the proposed cut geometry obtained from the mining engineer and the groundwater regime developed by hydrogeologists into the geologic sections, the geotechnical engineer, after assigning material properties to the individual strata, can compute the factor of safety against failure for the proposed pitwall slopes. Slopes with an adequate safety factor are established by iteration. Material properties required for stability computations include insitu density, cohesion and friction angle, values for which are obtained from laboratory tests on representative specimens recovered from core holes. A summary of the data required for slope stability analysis and other geotechnical computations is given in Table 2.

Geotechnical input to surface mine development is further illustrated in the following example of interdisciplinary planning at Highvale Mine.

5 HIGHVALE MINE - A CASE HISTORY

5.1 Background

Highvale Mine is located just south of Wabamun Lake about 70 km west of Edmonton, Canada. Maximum coal reserves exploitable by conventional surface mining techniques have been estimated to be about 750 million tonnes which will be mined from fourteen separate pits by stripping up to 3.7 billion bm^3 of overburden. The general layout of the current mine is illustrated on Figure 1.

About 12 million tonnes of coal per annum are presently mined from three pits designated 02, 03 and 04. All the output from the mine is consumed by two mine mouth thermal generating plants. This surface mining operation constitutes one of the largest in Canada and, with depths

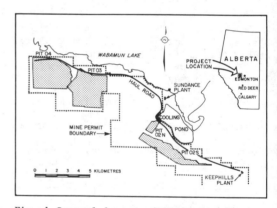

Fig. 1 General layout of Highvale mine

to coal currently up to 30 m increasing ultimately to 60 m, reliable geotechnical assessment of the overburden characteristics has become a necessary part of mine design and planning.

The overburden materials at Highvale are composed of gently dipping sequences of mudstones, siltstones, sandstones and coal of Upper Cretaceous and Tertiary age. Overlying this succession are Pleistocene deposits consisting of clay till containing ice thrusted blocks of

Table 2. Parameters required for geotechnical analysis.

Analysis	Geotechnical data requirements
Pitwall stability	geologic sections, water table, cut geometry, weight of excavator (dragline, bucketwheel or bucketchain excavator); material properties of principal strata including in-situ density, cohesion, friction angle, uniaxial compressive strength.
Classification of geotechnical areas	geologic sections, strata thickness isopachs, contours of top of ore, cut geometry.
Diggability assessment	geologic sections, strata thickness isopachs, sonic transit times from geophysical logs for individual strata, uniaxial compressive strength, specific cutting resistance from O&K wedge tests.
Trafficability assessment	geologic sections, strata thickness isopachs; material properties of critical strata including natural moisture, Atterberg Limits, grain size, uniaxial compressive strength; water table; rut depths from vehicular traffic obtained during field trials, vehicle weight and number of passes to form rut.
Bearing capacity computations	geologic sections, water table, uniaxial compressive strength of critical formations.
Spoilpile stability analysis	bulk density, cohesion, friction angle and uniaxial compressive strength of spoil material and foundation strata; water table; spoil height.
Settlement potential of reclaimed land	bulk density, coefficient of consolidation, and compression index of landfill material; water table; magnitude of superimposed loads.

Tertiary rock. Glaciation has resulted in varying degrees and depths of bedrock disturbance, evidence of which includes minor faulting, folding, fracturing, crushing and shear plane formation.

The coal is generally won from five of the six seams present, which vary from one to three metres in thickness. The overall coal and interburden thickness between Seams 1 and 6 is almost constant at 14 m throughout the mine area. The strata immediately above the coal are characterized by a mudstone and siltstone sequence between 4 and 8 m thick, containing a number of thin calcareous and sideritic bands. Overlying this sequence is a massive, poorly indurated sandstone enveloping a number of hard, calcareous/ferruginous bands of variable thickness to about 1 m. Overburden is currently stripped in all three pits by draglines operating in both the chopcut mode (excavating above the dragline bench level) and bankcut mode (excavating below the bench), the extent of each mode depending on the relevant geotechnical constraints and the depth to coal in the

particular pit. Prestripping by scraper is presently carried out to recover and stockpile topsoil and subsoil for reclamation purposes. The typical mining method used at Highvale Mine is shown on Figure 2 (Wade & Peterson, 1986).

5.2 Planning approach at Highvale

Within the last few years as coal production and mine depths increased, the need for a co-ordinated approach to mine planning was recognized. Although some detailed procedures are still being developed, the general interdisciplinary planning procedure currently being used is illustrated by the model given on Figure 3. A brief description of the respective activities is outlined in the following paragraphs.

1. Field work - Whether required for extension of an existing pit or opening a new pit, the first step involves gathering data in the field. Subsurface

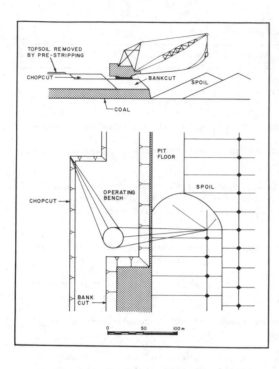

Fig. 2 Mining method at Highvale

exploration is carried out to obtain new information or augment existing data on the geology of the overburden and the coal deposit, groundwater regime, as well as geotechnical properties of the principal strata to be excavated. In setting up the drilling program, the requirements of the geotechnical engineer, the coal geologist, the mining engineer and the hydrogeologist are accommodated to the fullest possible extent consistent with the available budget. The number, type and spacing of drillholes are selected on the basis of interdisciplinary co-ordination meetings prior to program commencement as well as from studies of aerial photographs, the results of seismic surveys and structural mapping of existing rock outcrops and adjacent pit excavations.

Holes designated as Geotechnical are usually cored to allow the recovery of representative rock cores for laboratory classification and strength testing. Undisturbed samples of unconsolidated overburden are obtained from Becker, auger or rotary holes. Inclinometers and slip indicators are installed in holes close to active mining areas to monitor

ground movement and potential slope instability which could jeopardize equipment operation and coal recovery.

Mining or geological holes, on the other hand, are generally cored through the coal zone only to recover coal samples and obtain information on seam thickness, parting frequency and coal quality. Piezometers are installed for geotechnical/hydrogeological requirements to determine the groundwater regime, most often in selected holes drilled primarily for other purposes but sometimes in rotary holes specifically required by the hydrogeologist to complement existing data. Samples required by reclamation specialists are obtained from auger holes. All holes are geophysically logged to provide a complete set of geophysical logs such as sonic, natural gamma, neutron, caliper, focused density, gamma-gamma and resistivity.

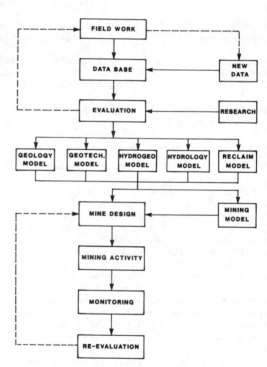

Fig. 3 Highvale mine planning model

2. Data base – Upon completion of the field work, the data obtained are entered into a computerized data base by the respective disciplines. Hard copies are also retained by each discipline. The

Table 3. Number of geotechnical tests on overburden in data base

Test	Till	Displaced Rock	Siltstone/ Mudstone	Sandstone	Bentonite Seams	Hard Bands
natural moisture	703	324	283	222	43	27
Atterberg limits	132	57	156	54	35	7
bulk density	130	78	102	66	16	5
compressive strength	105	22	52	93	4	19
grain size	24	2	10	20	–	–
triaxial	31	33	78	34	5	2
direct shear	12	23	72	13	21	3

geological data set contains information on formation lithology, stratigraphy and structure. A register of the vital statistics of all piezometers is maintained by the hydrogeologist. Geophysical logs are recorded in digital form on magnetic tape and are available for further processing as required. The geotechnical data set contains the results of laboratory and insitu tests which are depth corrected to match the geophysical log and labelled with the strata code assigned by the geologist. Typical material data included in the data set are illustrated in Table 3.

An example of triaxial test data stored in the data base is illustrated by the computer print-out reproduced in Table 4.

The discipline data sets for each drillhole are provided with a header sheet which includes drillhole data such as hole number, co-ordinates, ground elevation etc., insitu tests carried out in the hole, geophysical logs run, instrumentation installed (e.g. piezometers, inclinometers), tests run on samples from the hole including tests on groundwater, and any other pertinent data from the hole.

TABLE 4. Typical Triaxial Data Stored in Data Base

Drill Hole Number: HV 783 411

14–Nov–1986   Geotech. lab test-Triaxial

DEPTH FROM	TO	INTcm	STRAT CODE	SAMPLE NO.
19.07	19.67	60	80	RC5

Consold undrained   Y
Consold drained
Unconsold Undrained

	Spec 1	Spec 2	Spec 3	Units
Initial Water Content	12.90	12.30	11.50	%
Dry Density	1998.00	1986.00	1997.00	kg/m^3
Bulk Density	2255.00	2231.00	2227.00	kg/m^3
Effective Cell Pressure	400.00	600.00	800.00	kPa
Pore Pressure at Max. P(1)-P(3)	205.65	452.58	693.55	kPa
Sigma 1 at Max. P(1)-P(3)	2891.44	2385.81	2561.25	kPa
Sigma 3 at Max. P(1)-P(3)	194.36	147.41	172.14	kPa
Axial Strain at Max. P(1)-P(3)	1.98	2.48	3.28	%
Pore Pressure at Max. P(1)/P(3)	398.18	579.75	789.59	kPa
Sigma 1 at Max. P(1)/P(3)	1807.64	1567.76	2108.97	kPa
Sigma 3 at Max. P(1)/P(3)	1.82	20.25	10.36	kPa
Axial Strain at Max. P(1)/P(3)	0.96	1.03	1.75	%

Remarks: Clay shale, silty, hard, coal stringers.

3. New data – New or additional data which may be included in the data base consist of, for example, results of special laboratory tests such as the O & K Wedge test for assessing formation diggability, additional field testing such as field pumping tests to assess the hydraulic conductivity of aquifers or CBR testing for haul road design or fill-in data obtained from additional drilling programs carried out after evaluation of the initial drilling results.

4. Evaluation – Once the data base is established, evaluation of the data can commence to determine the adequacy, consistency and confidence limits of the data, assess whether additional data are required, either by further exploration or research, and allow models to be created by the discipline involved. Of particular significance to highwall stability is the occurrence, continuity, dip and strength of weak planes, especially in formations below the dragline bench. Data on the presence, orientation and spatial continuity of a given seam are obtained from the geological model of the mine and geophysical logs of drill-holes in the area being considered. An evaluation of the weak seam strength is made by plotting the results of direct shear tests from the geotechnical data set on the Mohr diagram as shown on Figure 4.

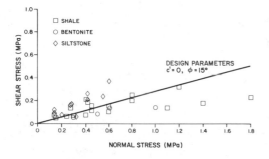

Fig. 4 Strength parameters for weak seams

5. Research – Geotechnical research studies carried out at Highvale have included instrumented field tests to determine the variation in ground bearing pressures exerted by the tracks of an electric shovel during different operating modes (Monenco, 1984).

Other studies involved assessment of overburden diggability characteristics for Bucketwheel Excavator (BWE) operation (Wade et al., 1987); back analyses of highwall failures (Monenco, 1983a); investigation of vehicular trafficability on clay till, glacially disturbed Tertiary formations and mine spoil (Monenco, 1983b); evaluation of stabilization measures for mine spoil to provide adequate foundation support for a Mine Services Building (Wade, 1983); settlement of small structures on reclaimed land (Sonnenberg et al., 1987) and assessment of pit slope stability using the proprietary SPDR method (Das et al., 1987).

6. Geotechnical models – As with models produced by other disciplines, a number of geotechnical models are created after evaluation of the parameters in the data base. Such models include those for classifying the mine area into sub-areas having uniform geological characteristics, analyzing the stability of highwall and spoil pile slopes taking into consideration operational constraints, grouping areas with similar geotechnical properties and assessing formation diggability.

The highwall stability model illustrated on Figure 5 depicts the types and sources of data required to carry out bankcut stability computations. An example of the resulting stability curves for different overburden lithologies and dragline loadings is given on Figure 6.

Input to the stability program includes the local groundwater level obtained from the hydrogeological model. However, this hydrostatic level requires modification behind the highwall to account for face drainage and, in fine grained sediments, for piezometric head reduction due to stress relief from excavation unloading. The latter component, not generally considered in analysis of long term slope stability, is sometimes significant during the short term (e.g. one to two months) provided the coefficient of consolidation of the highwall material is less than about 40 m^2/year (Brown, 1983) and the formation is non-gaseous, i.e. not subject to gas exsolution upon unloading (Coffey & Partners, 1985; Wade, 1986).

Using the stability curves, target elevations for the dragline bench from the mine model and stratigraphy from the geology model, each pit is subdivided into areas of relatively consistent geotechnical characteristics, strata thicknesses and cut geometry. These subdivisions, termed geotechnical areas, are then incorporated into mine planning computations to determine material balance quantities, schedules for equipment operation and coal production, and mine operation costs. A plan showing typical geotechnical areas is included on Figure 7.

In the BWE diggability study, the geotechnical model adopted is shown on Figure 8. As indicated on the figure, geotechnical material parameters from the data base were correlated to obtain empirical relationships which could be used to modify published diggability criteria to evaluate the types of data available from Highvale (see Table 5). The modified criteria were then used to prepare isopach maps of the mine denoting difficult-to-dig areas and areas requiring blasting. The modified criteria, reproduced in Table 6, relate diggability to sonic transit time, unconfined compressive strength and cutting resistance determined by the O & K wedge test (Wade et al, 1987). Also shown in Table 6 are similar diggability criteria developed by O'Regan et al (1987) for the Goonyella Mine in

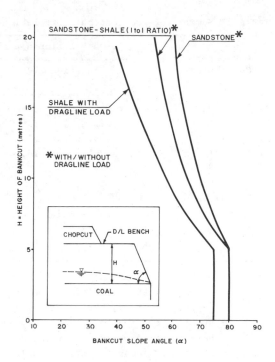

Fig. 6 Bankcut stability curves

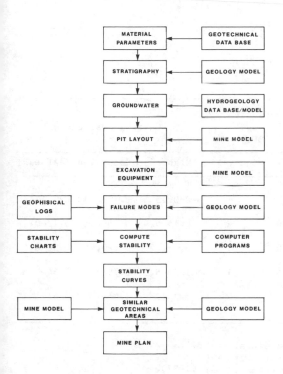

Fig. 5 Geotechnical model – highwall stability

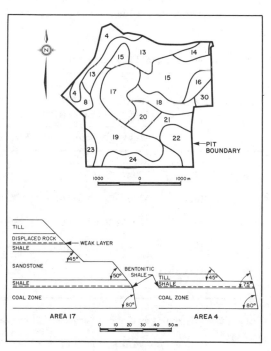

Fig. 7 Typical geotechnical areas

TABLE 5. Summary of geotechnical correlations

Parameters Correlated	Relationship obtained		Remarks
	Equation	Coefficient of Correlation	
1. F_a vs Q_u	$F_a = 0.445\ Q_u^{0.74}$	0.83	Highvale data
2. F_a vs $I_s(50)$	$F_a = 0.11 + 0.89\ I_s(50)$	0.98	Highvale data
3. F_a vs S_t	$F_a = 4.36 \times 10^{13}\ S^{-5.35}$	-0.76	Highvale data
4. F_a vs I_d	$F_a = 0.018\ I_d - 0.85$	0.86	Highvale data
5. Q_u vs S_t	$Q_u = 8.4 \times 10^{10}\ S^{-3.88}$	0.79	Data after Carroll (1966) and Highvale data
6. $I_s(50)$ vs Q_u	$I_s(50) = 0.042\ Q_u$	0.88	Data after Franklin (1972)
7. RQD vs F_f	$F_f = 18.21 - 0.151\ RQD$	-0.88	Data after Deere (1966)
8. F_a vs Fl_s	$Fl_s = 117.3\ F_a - 5.42$	0.85	Highvale data
9. RQD vs S_t	$RQD = 122.56 - 0.086\ S_t$	-0.33	Highvale data

Where:

F_a = specific cutting resistance per area of cut from O&K wedge test (Orenstein & Koppel, undated), MPa

Fl_s = specific cutting resistance per length of cutting wedge (O&K wedge test), kN/m

Q_u = Uniaxial Compressive Strength, MPa

$I_s(50)$ = Point Load Index (Broch & Franklin, 1972), MPa

S_t = Sonic Travel Time from geophysical logs, micro-sec/m

I_d = Slake Durability Index (Franklin & Chandra, 1972), %

RQD = Rock Quality Designation (Deere et al, 1966), %

F_f = fracture frequency, fractures/m

TABLE 6. BWE diggability criteria

Class	Digging Resistance	Sonic Travel Time (micro-sec/m)		Compressive Strength (MPa)		Cutting Resistance (MPa)		
		Highvale	Goonyella	Highvale	Goonyella	Highvale	Goonyella	Kozlowski
1	Easy	500+	667+	0-2	–	0-0.15	0.15-0.45	0-0.17
2	Moderate	420-500	500-667	2-10	–	0.15-0.4	0.45-0.6	0.17-0.36
3	Hard	340-420	400-667	10-20	10	0.4-1.25	0.6-0.75	0.36-0.54
4	Marginal	260-340	333-500	20-30	15	1.25-5.25	0.75-1.0	0.54-0.8
5	Undiggable	0-260	0-333	30+	–	5.25+	1.0+	0.8+

References: o Highvale criteria – after Wade et al. (1987).

o Goonyella criteria – after O'Regan et al. (1987).

o Kozlowski data from field measurements on operating BWE – after Kozlowski (1980).

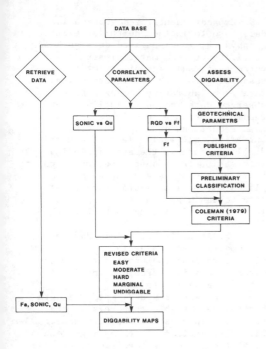

Fig. 8 Geotechnical model – diggability assessment

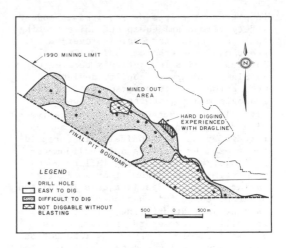

Fig. 9 BWE diggability map

Australia as well as cutting resistance values for various diggability ratings as determined by Kozlowski (1980) from field measurements on an operating BWE. Additional confidence in this approach was achieved when the dragline at Highvale experienced hard overburden digging in the area designated on Figure 9.

7. Mine Design – Mine design at Highvale draws upon the discipline models illustrated on Figure 3 and encompasses the following aspects for the current operating pits:

1. general plan for each pit showing the annual progression of overburden stripping and coal removal;
2. annual production schedules for each pit showing coal and overburden stripping quantities, and equipment hours and units;
3. annual capital and operating costs;
4. annual reclamation quantities;
5. blasting requirements for overburden and coal;
6. drainage schemes for both surface water and mine water;
7. cut orientation;

8. haul roads, ramps, dragline walk roads and other infrastructure;
9. electrical requirements and modifications to supply lines or routes;
10. geotechnical constraints to mining, e.g. potentially unstable areas where more than 5 m of glacially disturbed material occurs below the planned dragline bench level.

Current planning activities at Highvale involve mine plans in various stages of development. Short range plans for the operating pits, for example, are in the Mining Activity category on Figure 3. For proposed new pits and long-term planning, mine design is in a conceptual or preliminary stage and thus planning consists primarily of data collection and evaluation.

8. Mining activity – Geotechnical involvement in mining activity at Highvale includes input to design of haul roads, ramps, dragline walk roads; recommendations for foundation design of infrastructure buildings; carrying out trafficability and bearing capacity studies; collecting data on existing highwall and spoil pile slopes, groundwater levels, etc. to confirm assumptions made during design.

9. Monitoring – At Highvale, safe operation of the mine on a day to day basis is the responsibility of the mining contractor who monitors highwall and

in-pit spoil slopes as required to ensure safety of men and equipment while meeting coal production targets. However, a program of geotechnical and geological data collection is routinely carried out by mine planners to confirm design assumptions. This program consists of geologic mapping of freshly exposed highwall cuts, measuring representative highwall and spoil heights/slope angles for different geotechnical areas, recording piezometric levels and inclinometer readings, and determining failure modes when significant highwall or spoilpile failures occur. Specific data collection programs conducted when conditions warrant include drilling, test pitting, installing piezometers, inclinometers, slip indicators or other instrumentation, laboratory testing and preparing detailed slope failure reports.

After evaluation, the collected data are either entered in the respective data bases or, if significant anomalies are evident, utilized in stability computations, or in updating stability curves, geotechnical areas or other models. The revised geotechnical and geological models are then incorporated into long and short range mine design. Photographs and audio-video tapes with a commentary on salient features depicted during successive mine cuts are used to aid data evaluation.

6 FUTURE PLANNING REFINEMENTS

The interdisciplinary approach to mine planning used for Highvale mine has been outlined with emphasis being given to the geotechnical component. The procedure, however, is still being refined and revised as more of the existing data are being added to the database. Potential refinements currently being considered include:

1. Subdivide existing geologic strata codes to account for different lithologies occurring within presently designated strata. For example strata code 60 is assigned to sandstone units whereas in some areas of the mine this unit contains numerous thin mudstone seams which are of significance in stability computations.

2. Develop diggability criteria for different equipment having different digging capability and correlate diggability rating with strata code and formation thickness. Confirm diggability criteria developed with field observations and measurements.

3. Expand computerized geotechnical data base to include depth-correlated geophysical logs; derived data such as stability curves, material strength parameters, and diggability criteria; measured data such as stress-deformation data from laboratory strength tests; observed data including inclinometer readings, failure modes and piezometric head reductions due to excavation unloading; all available geotechnical drillhole data in the area including information from foundation investigation holes at Keephills and Sundance power plants, as well as exploratory penetration holes in glacial lake deposits located outside the current ultimate pit limits.

4. Obtain additional field data on piezometric head reduction due to stripping operations to better assess groundwater levels in overburden immediately behind highwall cuts.

5. Enter more raw data into the data base using portable computers in the field to minimize transcribing time. Such data include for example geotechnical logs, groundwater levels, and description and location of samples taken.

7 ACKNOWLEDGEMENT

The authors wish to thank TransAlta Utilities Corporation, owner of Highvale Mine, for permission to publish this paper and are grateful to Mr. D.G. Bacon, Director of Fuel Supply, for his review comments. Mine planning for Highvale is carried out by Monenco Consultants Ltd. and the systems approach currently being used was encouraged and supported by TransAlta.

REFERENCES

Broch, E.B., and Franklin, J.A. 1972. The point load strength test. Int'l J. Rock Mech. and Mining Sc., Vol. 9, No. 6.

Brown, A. 1983. The influence and control of groundwater in large slopes, Proc. 3rd Conf. Stability in Surface Mining, Vancouver, B.C.

Carroll, R.D. 1966. Rock properties interpreted from sonic velocity logs. J. SMFD, ASCE, 92.

Coffey & Partners Pty Ltd. 1985. Lochiel coal deposit 1984 field program. Report prepared for The Electricity Trust of South Australia, Vol. 12, Aug.

Coleman, J.F. and Fitzhardinge, C.F.R. 1979. The geotechnology of excavation equipment selection with particular emphasis on bucket wheel excavators. Int'l Conf. on Mining Machinery, Brisbane, July.

Das, B., Serata, S. and Nikols, D.J. 1987. Application of the Stress-Property-Deformation Relation (SPDR) integrated method to the surface mine slope stability problem. 89th Annual General Mtg. of CIM, Toronto, May.

Deere, D.U., Hendron Jr., A.J., Patton, F.D. and Cording, E.J. 1966. Design of surface and near-surface construction. Eighth Sym. on Rock Mech., Univ. of Minnesota, Sept.

Franklin, J.A. and Chandra, R. 1972. Slake durability test. Int'l. J. Rock Mech. and Mining Sc., Vol. 9, No. 3, May.

Jeffreys, J.T., Brett, J., Aspinall, T., White, B., Stevens, C. and Johnson, M. 1986. Mine planning. Australasian Coal Mining Practice - Monograph 12, pub. by The Australasian Institute of Mining and Metalurgy, Australia, June; 174-188.

Kim, Y.C. 1987. Geostatistics applied to short-term mine planning. Int'l. J. of Surface Mining, Vol. 1, No. 1; 73-82.

Kozlowski, T. 1980. Technika prowadzenia robot odkrywkowych. W.G. Katowice (In Polish).

Mallett, C.W., Duncan Fama, M.E., and Seedsman, R.W. 1987. Geomechanical data acquisition and numerical modelling for coal mine design. Coal Power '87, The AusIMM Annual Conference, Newcastle, N.S.W., Australia, May, 7-15.

Moelle, K.H.R. and White, B. 1987. A systems approach to the planning of coal resource investigation and mining development. ibid; 65-78.

Monenco 1983a. Highvale mine highwall failures at pits 03 and 04. Report prepared for TransAlta Utilities Corp. by Monenco Consultants Ltd., Oct.

Monenco 1983b. Highvale mine trafficability study. ibid; Dec.

Monenco 1984. Highvale mine shovel test pit. ibid; Oct.

O'Regan, G. Davies, A.L. and Ellery, B.I. 1987. Correlation of bucket wheel performance with geotechnical properties of overburden at Goonyella mine, Australia. Proc. Intl, Sym. on Continuous Surface Mining, Ed. by Golosinski & Boehm, Trans Tech Publications, Fed. Rep. of Germany, 381-396.

Orenstein & Koppel Canada Limited, Undated. Pamphlet outlining procedure and equipment for carrying out wedge splitting tests on samples of geologic formations.

Scoble, M.J., Hadjigeorgiou, J. and Lizotte, Y. 1986. An integrated stability assessment system for surface mine design. Proc. Int'l Sym. on Geotechnical Stability in Surface Mining, Ed. R.K. Singhal, Pub. A.A. Balkema, Rotterdam, Nov.; 51-60.

Scoble, M.J., Hadjigeorgiou, J. and Nenonen, L. 1987. Development of an excavating equipment selection expert system based on geotechnical considerations. Proc. 40th Canadian Geotechnical Conference, Regina, Sask., Oct.

Sonnenberg, R.G., Thomson, S., Scott, J.D. and Sego, D.C. 1987. Small structures on reclaimed land study. Final Summary report for TransAlta Utilities Corp. prepared by the University of Alberta, Edmonton, Sept.

Wade, N.H. 1983. Stabilization of open-pit mine spoil by dynamic compaction. Proc. 36th Canadian Geotechnical Conference, Vancouver, B.C., June.

Wade, N.H. 1986. Review of Lochiel geotechnical studies. Report prepared for The Electricity Trust of South Australia, Jan.

Wade, N.H. and Peterson, T.W.P. 1986.
Highwall monitoring, instrumentation
and stability analysis at Highvalee
coal mine, Alberta. Proc. Int'l Sym.
on Geotechnical Stability in Surface
Mining, Ed. R.K. Singhal, Pub. A.A.
Balkema, Rotterdam, Nov.; 373-384.

Wade, N.H., Ogilvie, G.M. and Krzanowski,
R.M. 1987. Assessment of BWE Digga-
bility from Geotechnical, Geological
and Geophysical Parameters. Proc.
Int'l Sym. on Continuous Surface
Mining, Ed. by Golosinski & Boehm,
Trans Tech Publications, Fed. Rep. of
Germany; 375-380.

Computer Applications in the Mineral Industry, Fytas, Collins & Singhal (eds)
© 1988 Balkema, Rotterdam. ISBN 90 6191 760 3

Computerized dragline mine planning

L.H.Michaud & P.N.Calder
Queen's University, Kingston, Ontario, Canada

ABSTRACT: A computerized dragline mine planning package that can be adapted to all mining environments, including mountainous terrain, is being developed at Queen's University. This package differs from most others presently available on the market in that it avoids taking a 'Black Box' approach to mine planning. All decision making, such as positioning the dragline and the placement of the spoil, will be left to the engineer.

With this package, the computer is used as a tool to speed up the time consuming aspects of the design work such as calculating and tabulating all waste, coal and spoil volumes and drawing and updating plans and sections. 3D images of the mine can also be generated at any stage of the mining operation. The computer also regulates the planning operations by ensuring that digging and spoiling stay within the operating limits of the dragline. This paper outlines the general approach being taken to develop this computerized dragline mine planning package.

1 INTRODUCTION

Although several computer packages have been developed to simulate dragline coal mining operations, most tend to be limited to fairly regular coal structures where the total mine area can be represented by one or a few generalized geological cross sections. Often a 'Black Box' approach is used to determine the 'best' mining parameters. This means that the user cannot follow the logic of the program and has no means of checking the answers.

These factors make it impossible to adapt the packages to simulate coal mining in areas with irregular geology such as in a mountainous environment. Here the topography is rugged and the coal seams are folded and faulted, causing substantial variations in seam thicknesses, seam location and overburden depths over short distances across the mine area. In these cases it is impossible to generate typical cross sections for the mine area and planning must therefore be carried out on a full set of closely spaced sections.

Although Truck/Shovel mining is more commonly used in mountainous terrain, there are situations that lend themselves well to dragline mining. Where applicable, this can result in lower overall production costs. The key to success is to minimize the rehandled material and this in turn is a result of good planning. Accurate volume estimates must be produced and the engineer must be able to quickly evaluate different mining scenarios in order to determine the best alternative.

2 PACKAGE DEVELOPMENT

In order to develop a computerized dragline mine planning package that can handle any geology, a more general approach to the planning process is necessary. This can be achieved by developing a program that follows the same basic steps as those taken when planning is carried out manually. The engineer will control the planning at every stage. All actual decision making, such as the determination of dragline movements, placement of the spoil and specification of the cut dimensions will be carried out by the engineer. In this way the package avoids taking a 'Black Box' approach to the planning process. All planning operations will be carried out interactively at the computer terminal.

The computer is used as a tool to speed up the time consuming aspects of pit planning. A computer graphics system can be used as an interactive tool to draw and update plans and sections at any stage in the mine design process and to generate 3D images of the mining operation. The computer will also provide the capability for the rapid calculation, display and tabulation of coal and waste volumes and stripping ratios.

These features allow the engineer more time to calculate different scenarios such as changes in dragline dimensions or alterations to the mine design parameters, thereby optimizing the dragline operation.

The ability to view the mine in 3D at any stage in the planning process can also greatly aid in the conceptualization of the planning process at a mine site or in an educational setting and can be a useful method of presenting the plan to others.

3 CASE STUDY

For the development of this design package at Queen's the case study that is being used is a coal mine in mountainous terrain. The area consists of multiple coal seams dipping into a hillside at a 15 to 20 degree angle (see Figure 1). The coal seams also dip down in both directions along the hillside (see Figure 2). A mining operation that incorporates the removal of the two upper seams along the hillside is being simulated. This case study has been chosen because it offers enough variation in the overburden depth and the seam position that several dragline mining techniques are required to mine it. These include the use of simple side casting, double pass mining and the construction of extended benches, advance benching and chopdown benches. Because of the double pass mining, spoil room is very limited and rehandle of waste material is necessary in order to avoid the mine becoming spoil bound, resulting in the loss of otherwise mineable coal.

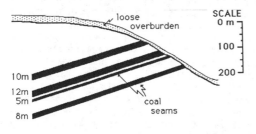

Fig. 1 Case Study: Cross Section of the mining area

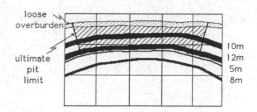

Fig. 2 Case Study: Longitudinal Section of the mining area

4 COMPUTER PLANNING PROCESS

The steps required to create a dragline mine plan can be outlined as follows:
   1) Determining material and dragline design parameters.
   2) Inputting the mine geology.
   3) Determining the location of the working sections.
   4) Determining the approximate mine limits.
   5) Producing a plan view and the working sections.
   6) Determining the approximate cut limits and locating them on the plan and sections (spoil placement is not considered).
   7) Determining the mining sequence.
   8) Developing the range diagrams.
   9) Producing the final sections and plan view.

Geotechnical characteristics of the material and design parameters for the dragline that influence the planning process are listed in Table 1. These parameters can be entered interactively by the user or through an input data file. Dragline parameters can be chosen from a library of dragline models available with the package or they can be input independently.

Geological contour maps or sections specifying the topography, overburden contacts and the top and bottom of the seams will be digitized and a 3D geological model developed. There are several good packages presently available on the market to generate 3D geological models from drillhole data. It will be possible to incorporate such a system into the package in order to generate the required model to begin the planning process.

Table 1. Design Parameters

MATERIAL CHARACTERISTICS
Swell Factors
Highwall Angles
Safety Berm Width
Spoil Pile Angle

DRAGLINE PARAMETERS
Maximum Reach
Maximum Spoil Height
Maximum Digging Depth
Maximum Chopdown
Minimum Pad Width

Because of the complexity of a multiple coal seam model the geology can best be viewed by cutting sections across the mine area. It will be possible to generate sections at any angle by inputting the start and end point on a plan view at the terminal. The location of these sections will be dependent on the rate of variation of the geology and, because the sections will be used to develop cut dimensions, on the dragline parameters. The spacing should be less than the maximum dragline reach so that the influence of spoiling from one dragline location will spans three sections.

The initial rough ultimate mining limit can be determined at this stage. An approximate limit, generated from the maximum digging depth and maximum acceptable stripping ratio will be automatically generated and displayed on the sections and on the plan at the terminal. This limit can be altered by the user or accepted. It can then serve as a guide to determining the individual cut limits and overall mining sequencing. Upon acceptance, the ultimate limit will also be incorporated into the 3D mining model. As with any operation, altering the mine limit on a section will result in the plan view and the 3D model also being updated.

5 DETERMINING DRAGLINE CUT DIMENSIONS

The next step in the planning process is to generate individual cut limits. Working on each section, these are determined by taking the ultimate pit limit, maximum cut depth and the minimum and maximum permissible cut width into consideration. A cut limit can be entered at the terminal by positioning the cursor at the desired highwall crest and entering the point. The

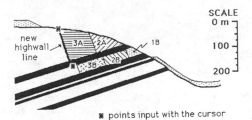

* points input with the cursor

1) Input highwall crest with the cursor.
2) Highwall is drawn to the maximum depth.
3) Input the highwall toe.

Fig. 3 Inputting the dragline mining cut limits

highwall will be automatically drawn at the predefined angle, extending down to the maximum digging depth of the dragline. If loose overburden is present, a safety berm will also be incorporated. The user can then input the desired highwall toe and the cut bottom will be drawn. This process is illustrated in Figure 3. The cut dimensions, stripping ratio and the waste and coal volumes will be calculated and displayed on the screen. The highwall location can then be altered by cursor control or accepted. Note that, at this stage, these are only idealized cut limits as spoil room requirements have not yet been considered.

Once the cut dimensions have been accepted for each section, the user can input the initial dragline position for each cut. A minimum width dragline pad will be displayed which can be altered by inputting a new pad crest. If a filled dragline pad

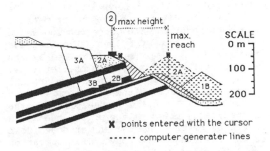

* points entered with the cursor
...... computer generater lines

- Showing the dragline reach and maximum spoiling height.

Fig. 4 Cut 2A: Inputting the dragline pad and spoil pile

355

is required, the pad slope will be automatically drawn and the required fill material calculated and displayed. (See Figure 4.)

The cut limits and dragline position can be smoothed by working between sections and the plan view. This operation will be aided by the ability to view any three sections or a section and the plan simultaneously at the terminal.

Upon completion of this stage the dragline positions can be used to develop a dragline movement sequence between sections and between cuts along the length of the mine.

6 SPOIL PLACEMENT

After the initial cut limits have been calculated, the placement of spoil can be considered and range diagrams generated. All spoil placement will be directed by the user and only the spoiling limits controlled by the computer. Spoil can be placed anywhere within the maximum dragline reach along the path of dragline movement between sections and can extend to a height equal to the maximum spoiling height of the dragline. When viewing a section at the terminal these limits will be indicated as is illustrated in Figure 4. On the same screen a plan view will show the maximum spoil limits for the area influenced by that section and the total available spoil volume will be displayed.

Within these limits waste material can be used to build the required dragline pad. It can also be placed into the previously excavated cut, or can be spoiled behind the dragline anywhere within the maximum dragline reach.

Required spoil room for each cut will be calculated by digitizing the cut or partial cut to be removed. The corresponding waste and coal volumes will then be calculated from the in-situ volumes multiplied by the swell factor and displayed at the terminal. The spoil pile can then be drawn on the section by inputting the spoil crest with the cursor. The sides of the pile will be automatically drawn at the predefined angle of repose of the material and extended down to the existing topography. This operation is illustrated in Figure 4. The resulting volume, calculated from a series of overlapping cones drawn in the area influenced by the section, will then be displayed. This spoil placement will also be shown on the plan view. The volume can then be adjusted by moving the cursor around the screen until the required volume is achieved (spoil volume = swelled waste volume), the spoil pile location accepted, and the section updated.

At the beginning of spoiling for each successive cut the top of the accumulated spoil piles will be shown as a heavy line on the section to indicate the starting topography for the next cut (see Figure 5).

Using a similar method, the coal can be digitized and the resulting swelled volume calculated. The coal can then be lifted out of the cut and placed onto a predetermined coal pad. Rehandling of waste material will be simulated by digitizing the area to be moved and by going through the same spoiling procedure as was previously described (see Figure 5). If the mine becomes spoil bound it will be possible to backtrack to any previous cut in the plan in order to begin a new plan from this point. A log of the steps that have been taken will be kept in order to facilitate this procedure.

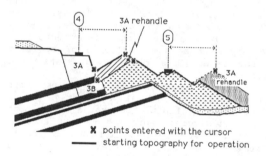

X points entered with the cursor
— starting topography for operation

Fig. 5 Rehandling spoil material

7 VOLUME CALCULATIONS

In-situ coal and waste volumes will be calculated by dividing each 3D shape in the model into a series of tetrahedra and totalling their volumes. Spoil volumes will be calculated from a 3D model incorporating a series of overlapping cones. Although the placement of the spoil will not be identical to that resulting from the actual mining operation, the total spoil volume placed within the area of influence of one section will be a close approximation.

Along with calculating cut and spoil volumes by section, cumulative cut volumes and total waste and coal volumes will be tabulated concurrently with the planning process. These will be available for viewing at any time in tabular form at the terminal or by generating a printout.

Because this program tabulates the waste and coal volumes for each section, each cut, and for the total mine, it will be possible to use this data along with production rate information for the dragline to generate a production schedule for the mining operation. This data could also be incorporated with cost data in order to generate economic feasibility studies for the property.

It would also be possible to later 'automate' this planning process for the development of plans and schedules in areas with a more regular geology such as in a plains environment. Here several adjacent geologic sections would be identical. Therefore, once planning has being completed on one section, the process can be duplicated on the other identical sections and the total volumes accumulated without further input from the user. Upon encountering a change in the geology the user will be prompted for further input.

Because a full set of sections is used it should also be possible to adapt this process to the planning of oil sands operations and to other draglines applications.

Computer Applications in the Mineral Industry, Fytas, Collins & Singhal (eds)
© 1988 Balkema, Rotterdam. ISBN 90 6191 760 3

The use of solid modelling in underground mine design

S.Henley & A.J.Wheeler
Mineral Industries Computing Ltd, London, UK

ABSTRACT: The block modelling capability of the DATAMINE software system allows the splitting of any part of a model into cuboid sub-cells. This cell division allows the incorporation locally of any amount of detail into any region of a model. A computerised model of a three-dimensional surface can be generated by linking perimeters on adjacent sections. In DATAMINE , this procedure allows triangulated wireframe models to be developed for geological boundaries or mined excavations. Such models can be converted into block models, or intersected with existing block models. This combination of modelling techniques has many applications in underground mine planning.

1 INTRODUCTION

During the past few years, there has been rapid development of computing techniques in general, and in the application of new techniques to the fields of geological modelling and mine planning. In the earliest years of such developments, the sixties and early seventies, three-dimensional geological models consisted of complete lattices of cuboid cells. For modelling of layered deposits, two-dimensional grids were quite often used. Many mining software systems still use these methods, but even in the mid-seventies their shortcomings in terms of flexibility were becoming apparent, particularly in the storage inefficiencies produced by attempting to model detailed geology (such as of narrow vein deposits) with larger numbers of ever smaller blocks. A development of the technique, exemplified by the Control Data SEAMSYS system, provided for variable block dimensions in one direction, allowing seam thicknesses to be modelled accurately. In the DATAMINE system of Mineral Industries Computing Limited (MICL), variable block dimensions are allowed in all three dimensions as sub-cells within a regular lattice of cells, thus giving detailed modelling capabilities for any type of geological feature.

A completely different development path has been followed by systems that have evolved independently from survey data handling, from finite-element analysis, and from computer-aided design (CAD). In these systems, rather than modelling the contents of three-dimensional volumes of space, it is the boundaries that are modelled. Known points are joined into an irregular lattice consisting usually of polygonal (usually triangular or rectangular) patches which represent faces on a polyhedral approximation to the desired three-dimensional boundary. In the case of survey systems, this was originally the topographic surface; in CAD it is the surface of a three-dimensional solid body; in finite-element systems, it may sometimes be a space-filling lattice designed to represent the mechanical properties of the volume being studied.

In the new generation of mining software currently being developed by MICL and others, both volume (cuboid) and boundary (wireframe) modelling are provided, with facilities to combine information from the two types of model. The reason for this is that the two techniques, though radically different, are complementary in that cuboid-based models allow the handling of data relating to volumes of space but are not so good for surface description; wireframe models are basically models of surfaces and have no easy provision for detailed modelling of variations within enclosed volumes.

2 WIREFRAME MODELLING

There have been many independent attempts at developing wireframe modelling methods for specific applications and for general use. However, although there is a great deal of commonality in principle between a three-dimensional CAD model and a triangulated model of a topographic surface, these are rarely, if ever, encountered in a single software package. Both, however, require identical data structures, and there is really no difference in principle between a topographic surface (which converges in the limit to a spheroid closed at the antipodes), and a three-dimensional model.

It is standard practice in surface modelling to triangulate the surface, and commonly the Delaunay (or Thiessen, Dirichlet, or Voronoi) triangulation is now used for this purpose as it provides the unique set of most equi-angular triangles for any set of points. Sometimes controls are superimposed such as external or interior boundary constraints. In three-dimensional wireframe modelling, however, it is very commonly found that no such discipline is followed, and commonly CAD and other software packages allow polygons with more than three sides. It is usually presumed in such packages that the polygon is flat, but much computational effort must be expended to ensure that this is so. By making all facets triangular, it is possible to simplify code (by simplifying the data structures to be handled as well as eliminating unnecessary checks) with a trade-off of slightly increased storage. In the wireframe modelling extensions to the DATAMINE mining software, all surfaces are modelled as sets of triangular facets, each necessarily guaranteed to be flat, and thus producing an unambiguous model.

Wireframe models with non-triangular facets will not be considered any further in this paper. There are many ways in which triangulated surface models may be defined; all surface geometries may be described by the same set of data structures, but the way in which the user – the geologist or engineer – wishes to define a surface depends on its nature and geometry. A topographic surface will often be modelled by an automatic or semi-automatic Delaunay triangulation algorithm, though there may be requirements for manual intervention to insert boundaries or other controls. For three-dimensional objects, such as orebody outlines, the geologist will often wish to specify the geometry slice-by-slice either in section or in plan, and then join up the shapes subsequently into three-dimensional

models. This joining up can be done in very many different ways. There is a wide choice of algorithms available, both constrained and unconstrained, owing much to film animation (automated in-betweening algorithms) and to the biological sciences (joining up slice views of complex organs and organisms into three-dimensional models). The available software packages for mining reflect this wide choice of methods.

An alternative approach to three-dimensional modelling is more relevant to the mining engineer than the geologist: in mine layout design, it is often more convenient to define a drift or shaft by a single line, representing an opening of given constant cross-section. This body itself can be triangulated automatically. Alternatively, a stope may be defined as a perimeter representing footwall or hanging wall which must be translated a fixed distance to define the opposite wall. Again the resulting triangulated solid can be generated automatically.

A prime constraint on three-dimensional wireframe modelling for geology and mine planning is that it must be possible to describe the contents of a solid body, localised within it. This is a function which block models fulfill very well. Intersection of wireframe models with block models must be convenient and efficient. The two ways in which this can be done are both important: a wireframe model can be converted to a block model (and in DATAMINE this allows subdivision of blocks at the boundaries to give detailed modelling of the enclosed volume): this may then be integrated with block models from other sources (e.g. kriged grades). Alternatively, the wireframe may be used to intersect the block model to provide an evaluation block-by-block or slice-by-slice - for example to determine the tonnage and classify the ore by grade within a stope or other extraction unit.

In DATAMINE the intersection/ classification program passes results into a results file (which can then be used by a general purpose reserve reporting program), and at the same time updates a 'mined-out' variable in the block model. This 'mined' flag marks the evaluated volume as unavailable for further evaluation, so avoiding the danger of double-counting of tonnages even if wireframe model volumes overlap, as they are allowed to do.

3 UNDERGROUND MINE PLANNING

The best way to demonstrate how the

wireframe and block modelling facilities of
DATAMINE can be used in underground mine
planning is to describe some different
applications.

3.1 Transverse Longhole Stope Design

An orebody wireframe model may initially be
generated, based on digitised ore zone
perimeters from diamond drillhole sections,
as shown in Figure 1. A block model may
then be generated, with sub-cell creation
at the edge of the ore zone. A section
through such a block model is shown in
Figure 2.

Wireframe models may then be generated
for existing or designed undercut and
overcut stope development. Plots may then
be produced against the orebody wireframe
model, as shown in Figure 3. Such plots,
which incorporate hidden line removal,
allow very good appreciation of underground
features.

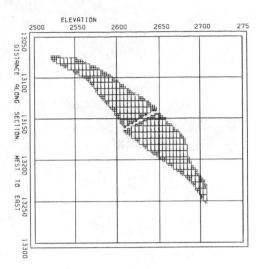

Fig.2 Section through orebody block model

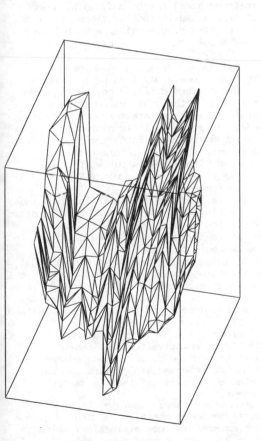

Fig.1 Orebody wireframe model

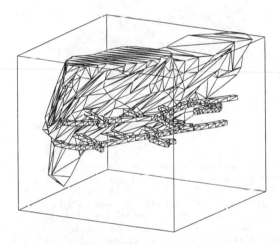

Fig 3. Stope development and orebody
wireframe models

From interactive examination of the block
model at the required elevations, undercut
and overcut perimeters may be defined for
each individual stope. Wireframe models may
then be generated for each stope, and the
overall layout may be viewed, from
different orientations, as shown in Figures
4. and 5.

Individual longhole rings may often be
oblique to the original diamond drillhole
sections. By producing oblique sections
through the block model along the same
plane as the longhole rings, the longhole

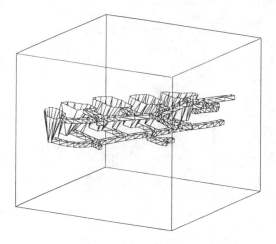

Fig.4 Stope layout, viewed from east side

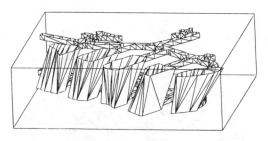

Fig.5 Stope layout, viewed from west side

design may optimised for maximum recovery and minimum dilution. An oblique section produced for this purpose is shown in Figure 6. Grade values are annotated on this section for each model cell. Very accurate wireframe models may then be developed for each individual stope, using the longhole layout perimeters, and evaluated against the block model.

As evaluation can leave a 'mined' flag in the block model, valuable information can be obtained on the residual tonnage left in pillars, which often becomes the ore for stoping in a secondary phase.

Many of the features just described can be equally well applied to many different stoping methods. These features include:

1. Accurate mining layout definition against a block model.
2. Three-dimensional visualisation of geological and mining features.
3. Production of oblique block model sections for drilling layouts.
4. Evaluation in stages which represent the different phases of mining.

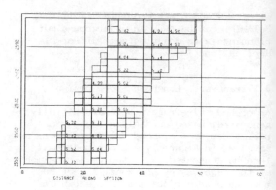

Fig.6 Oblique section through block model

3.2 Longitudinal Longhole Stope Design

A wireframe model, followed by a block model, may again be generated for the orebody, as described in section 3.1. Alternatively, perimeters for the orebody wireframe model may be defined by assay cut-offs along various sections, based on interpolated grades within a much larger block model, that has been generated previously.

Existing and designed development wireframe models may then be generated. Further diamond drilling may be required from some of the sublevel development before a final satisfactory orebody block model is produced. It is simple to continuously update both wireframe and block models as more information is produced. The stope layout will then finally be defined, producing a third type of wireframe model.

By selecting the cell size longitudinally to coincide with the longhole ring burden, evaluation of the stope design can automatically be sub-divided according to the longhole layout.

As mining progresses, a wireframe model representing the actual stoped excavation may be updated, as shown in Figure 7. If caving of the hanging wall was occurring, this procedure would enable a determination of the dilution resulting and its effect on the mined grade. Three-dimensional views of the stope, relative to other nearby workings, can be quickly produced and remaining intermediate pillars can be evaluated in terms of their geometry or content.

Upon stope completion, the final wireframe model can also be used to give a very accurate volume evaluation, which is often very important for determination of fill and cement quantities. All of these

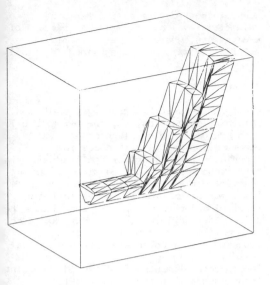

Fig.7 Wireframe model of stoped excavation

features could be equally well applied to vertical crater retreat stopes.

For narrow vein stopes, which are to be mined by a longhole method, an accurate wireframe model of the orebody would be of great value in determining the ore profile in the plane of each drilling ring.

3.3 Cut-and-Fill Stope Design

The mining width for narrow vein cut-and-fill stopes, which is often dependent on the machinery employed, has a major effect on the mined grade. Wireframe modelling of such deposits, followed by conversion to block models, as shown in Figure 8, allows very accurate representation. Stoping layouts employing different stoping widths may then be rapidly evaluated.

For large scale cut-and-fill stopes, the initial panel and pillar design can be optimised against a block model for maximum recovery, honouring existing geometric and geological constraints. This design process could take into account any existing excavations, any neighbouring orebodies (which have to be mined at a future date) and of course the orebody to be mined itself. The vertical cell size of the block model would be the same as the lift height for the stope. Thus single perimeters defined in plan at the mid-lift height could conveniently be used for evaluation. Along with the progressive updating of a stope wireframe model, as shown in Figure 9, distinct models of individual pillars

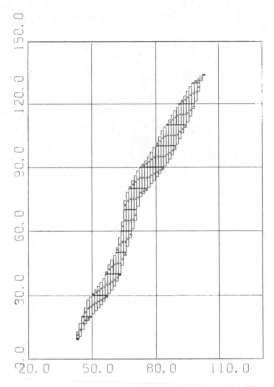

Fig.8 Section through narrow vein block model

could easily be built up. This would be valuable for pillar tonnage and grade evaluation, and would allow three-dimensional viewing of pillars, as shown in Figure 10.

The wireframing techniques would be particularly useful when cut-and-fill stopes are being mined through or around existing mined excavations. Three-dimensional views at any angle may be obtained, together with sections through remaining sill or crown pillars along any plane.

3.4 Backfilling of Mined Workings

For major backfilling operations of mined stopes or caved workings, the volume and exact shape of the excavations is of critical importance. A wireframe model may be generated for such excavations by using information from a number of different sources:

1. By examination of a block model which has been updated with the 'mined' flag, as

363

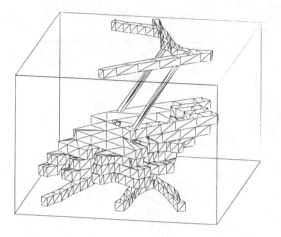

Fig.9 Wireframe model of cut-and-fill stope

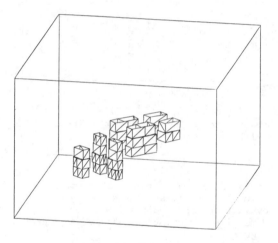

Fig.10 Wireframe models of the pillars in a cut-and-fill stope

accurate cost determination from the volumetric evaluation. Flatter regions of excavations, where high density fill may not flow easily, may be examined in detail.

4 CONCLUSIONS

CAD techniques are increasingly used in a number of industries, and the benefits of such systems in mine planning are now becoming apparent. Solid modelling techniques have important applications in both orebody modelling and mine design. However, mining applications require detailed information on the variation of physical and chemical properties within the solid model defining the orebody, and such information is very difficult if not impossible to incorporate within a general solid model. Thus there is a need to incorporate elements of CAD solid modelling techniques within the block models traditionally used to define orebodies.

The wireframe modelling techniques, described here, point the way forward to a synthesis of the two different approaches, meeting the particular needs of the mining industry. Furthermore, the approach described can be implemented on a desk-top micro, and does not need the specialised computing and graphics facilities required for many CAD systems. Developments which marry the two concepts are underway within DATAMINE, extending its underground mine planning capabilities to incorporate surveying, scheduling and interactive stope layout generation.

stoping has progressed.

2. From previously surveyed perimeters of mined-out stopes.

3. From surveyed points of the remaining surfaces, such as diamond drillhole intersections into a caved hanging wall.

4. From drift intersections with caved areas.

The wireframing approach will allow information from all these different sources to be correctly assimilated, producing one wireframe model for each excavation. Examination of these models will greatly assist the positioning of the backfill placement points and will allow

Computer Applications in the Mineral Industry, Fytas, Collins & Singhal (eds)
© *1988 Balkema, Rotterdam. ISBN 90 6191 760 3*

Les contours optimaux d'exploitations à ciel ouvert par un algorithme de flot avec SAS/OR

C.Plasse & J.Elbrond
Ecole Polytechnique, Montreal, Quebec, Canada

ABSTRACT: The conventional methods for the design of open pits, as for examples moving cone or Lerchs and Grossman, require considerable efforts of programming. However, similar operations research problems have been solved in other industries using program packages, which are commercially available. In the present paper, we describe a method for the optimization of open pit contours using a flow algorithm and the SAS program package's components SAS/OR and SAS/GRAPH, which are commercially available. The advantages of this method, other than low cost, include the ease of applying different angles at all points of the mine. A graphic representation of the optimal pit by SAS/GRAPH is given.

RÉSUMÉ: Les méthodes traditionnelles de design de fosses, comme par exemple cône mobile ou Lerchs et Grossman, impliquent un grand effort de programmation. Pourtant, des problèmes similaires de recherches opérationnelles ont été résolus dans d'autres industries à l'aide de progiciels qui sont commercialement disponibles. Dans le présent papier, nous décrivons une méthode de design de fosse utilisant un algorithme de flot SAS (avec ses composantes SAS/OR et SAS/GRAPH), qui est commercialement disponible. Les avantages de cette procédure, autre que le coût, comprennent aussi l'application facile d'angles variables en tout point dans la mine. Une présentation graphique de la fosse optimale par SAS/GRAPH est donnée.

1. INTRODUCTION

Dans la publication classique de Lerchs et Grossman "Optimum Design of Open-Pit Mines", plusieurs procédures pour trouver les contours optimaux d'une exploitation à ciel ouvert sont mentionnées.

1. Programmation dynamique.
2. La cloture maximale d'un graphe orienté.
3. Le flot maximal d'un réseau.
4. Une analogie hydrostatique.

Parmi celles-ci, la programmation dynamique est décrite à un certain niveau de détail tandis que le flot maximal dans un réseau et l'analogie hydrostatique ne sont que mentionnées. C'est la méthode par la clôture maximale d'un graphe qui est décrite en détail et qui est connue sous le nom de l'algorithme de Lerchs et Grossman. Cette méthode applique une approche directe qui possède certaines similarités avec une exploitation pratique. Au moment de la publication de cet article, la méthode par un algorithme de flot posait des problèmes d'un point de vue espace de mémoire d'ordinateur. Cet état de chose ne se présente plus.

2. DE LA THÉORIE DES GRAPHES AU FLOT DANS UN RÉSEAU

Picard a montré que la clôture maximale d'un graphe orienté est équivalente à la coupe minimale dans le réseau correspondant suppléé d'une source et d'un puits. Dans ce réseau les noeuds, qui représentent les vertex de valeurs positives dans le graphe orienté, sont liés avec la source par des arcs de capacités égales aux valeurs des vertex, et les noeuds, qui représentent les vertex de valeurs négatives, sont liés avec le puits par des arcs de capacités égales aux valeurs numériques des vertex. Les capacités des arcs internes sont infinies. Dans ce réseau, le flot maximal définit la coupe minimale par les arcs ayant un flot marginal 0-positif ou 0-négatif.

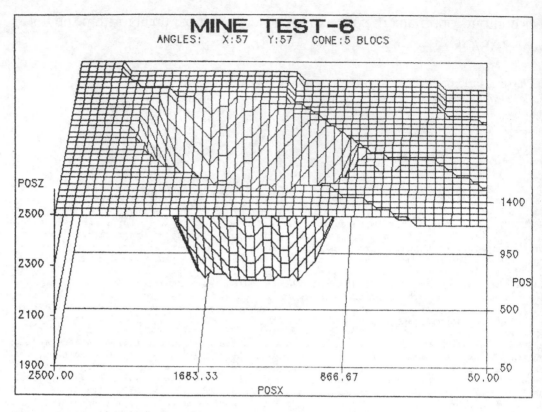

Figure 1: Vue isométrique de la fosse optimale par SAS/GRAPH.

Dans SAS/OR, la procédure NETFLOW trouve la coupe minimale par l'algorithme de Ford et Fulkerson. Elle est rapide mais elle perd sa performance si utilisée pour des grands réseaux et pour des arrangements complexes des arcs. Une autre procédure dans SAS/OR disponible pour le même problème est TNETFLOW. Elle résout le problème à l'aide de la méthode Simplex encore plus rapidement que NETFLOW. Les deux procédures sont plus rapides que la méthode de la clôture maximale dans un graphe orienté.

3. TRANSPOSITION À L'EXPLOITATION DES FOSSES

Ces procédures de SAS/OR produisent donc les contours optimaux d'une exploitation à ciel ouvert plus rapidement que la procédure de Lerchs et Grossman qui est celle de la clôture maximale dans un graphe orienté.

Le système de banque de données de SAS est très convenable car il permet de présenter la valeur des blocs à exploiter ou à laisser par plusieurs paramètres c'est-à-dire par une fonction de coûts et de prix. Il est donc facile de procéder aux études de sensibilité.

La définition des arcs, qui indiquent les priorités d'excavation, est basée sur l'angle maximal de la pente permise dans des endroits donnés dans la mine. Cet angle peut varier suivant la position du bloc dans la mine et l'orientation de l'angle. Le nombre d'arcs nécessaires afin d'écrire un angle peut devenir assez grand ce qui constitut la complexité d'arrangement d'arcs mentionnée plus haut. La définition des arcs est un développement original qui a été ajouté aux procédures de FLOW en langage SAS dans la présente recherche afin de pouvoir produire rapidement les assemblées d'arcs relatives aux contraintes d'exploitation reliées aux différents angles dans l'étude.

Avec les procédures de flot, il est

facile comme initiation d'utiliser des
résultats d'optimisations précédentes ou
heuristiques et ainsi encore sauver du
temps de calcul aussi bien dans l'optimisation subséquente que dans l'étude de
sensibilité. Contrairement à ce qui est
le cas de la méthode par la clôture
maximale dans un graphe, l'optimisation
subséquente à une première tentative
arrivera avec un algorithme de flot
toujours à l'optimum même si les contours
de départ auraient dépassé les contours
finaux. Tout le réseau reste intact lors
de l'optimisation tandis que celle-ci dans
la méthode de graphe commencerait avec les
contours de la tentative comme contours
initiaux.

4. GRAPHISME

Le système SAS possède aussi une
puissante procédure de graphisme
SAS/GRAPHE. La figure 1 a été dessinée à
l'aide de cette procédure.

5. CONCLUSION

L'existance de progiciels généraux et
commercialement disponibles pour l'établissement des contours optimaux des
exploitations à ciel ouvert permet à une
compagnie minière d'exécuter des études
d'optimisation et de sensibilité plus
souvent et d'une manière plus facile que
par des progiciels spécialement développé et probablement moins accessible
pour une utilisation fréquente.

6. RÉFÉRENCES

LERCHS, H. et GROSSMAN, I.F. 1965. Optimum
    Design of Open-Pit Mines. CIM Bulletin,
    Vol. 58, No. 633 (Jan. 1965).
PICARD, J.-C. 1976. Maximal Closure of a
    Graph and Application to Combinatorial
    Problems. Management Science, Vol. 22,
    No. 11 (July 1976).
SAS/OR and SAS/GRAPH. User's Guide. SAS
    Institute.

Computer Applications in the Mineral Industry, Fytas, Collins & Singhal (eds)
© 1988 Balkema, Rotterdam. ISBN 90 6191 760 3

Planning and scheduling by mine equipment and mine simulation programs for combined cross pit spreader (XPS) and dragline operations

Chris Niemann-Delius
Mannesmann Demag Baumaschinen, Geschäftsgruppe Lauchhammer-Bergwerktechnik, Düsseldorf, FR Germany

W.Thiels
Institut für Bergbaukunde III, Rheinisch-Westfalische Technische Hochschule, Aachen, FR Germany

ABSTRACT: Mine planning and operation management aim at obtaining maximum utilisation and output for individual machines and the mining system in its entirety. These objects can effectively be suppoted by a planning and scheduling program based on equipment simulation to be run on PC's. The XPS-PC program has a modular structure with sub-programs for BWE's(s) and Dragline simulation, mine geometry and simulation of mining.

1. INTRODUCTION

Todays processing capacities of a micro-computer /e.g. PC - AT) attain levels which can manage extensive computing programs within reasonable short periods. This permits to make simulations which some years ago were the sole domain of high-capacity computing systems. At the same time these important rates can be achieved at favorable prices for rapid amortization of the investment cost by savings in working time. PC programs are used to an ever increasing extent to assist in planning of open cast mines. An improved determination of target sizes with respect to quantity can be achieved in partial fields of planning, with sim-ultaneous optimization of the planning itself by advance checking of different planning variants through simulation.

Running operations are checked for weak points by comparison with simulation results and thus optimized. The degree of improvements depends on one side on the quality of the software used and on the other side on the reliability of the input data. Input data have to be con-stantly subject to critical checking to be more exact and to prevent erroneous results.

2. PRINCIPLE ADOPTED FOR SIMULATION OF MINING

The software presented here comprises a program package for simulation of an open pit with its main equipment of BWE,

XPS and Dragline. This simulation can both be made in a combined manner for the machines as a whole but also separately for each machine. For these simulations a read-in of deposit geometry data is made, with the possibility of read-in and variation from a file program. For the mining operation a single input of the Dragline and BWE data is made under use of the same scheme. The specific deposit data required for the individual simu-lations always make access to one single file, in order to avoid data confusion. This deposit file serves also as basis for a program which can show the mining geometry in form of a cross-section on the terminal or plotter.

The modular system of the programs is shown in figure 1.

The operation of the machine is followed by stepwise simulating the cinematics of mining. All time shares obtained from the partial activities of the mining operation are summed up. These time shares are determined by the geometry and the per-formance data of the machines, for instance acceleration and maximum speed.

The result of the simulation shows the optimum digging time for a defined deposit block.

To determine the time required to mine several successive blocks, several shut-downs are inserted into the calculations to cope with the real operating con-ditions. This gives an operational efficiency factor. The individual charac-teristics of the subprograms are described below.

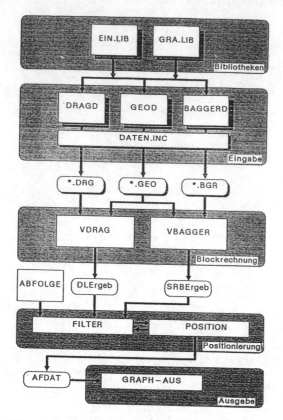

Figure 1: Modular Layout of the program
package

3. PROGRAM DESCRIPTION

3.1 DATA PROGRAM

The programs were developed in TURBO
PASCAL language, a product of Borland,
under MS-DOS. This compiler supports the
mathematical co-processor 80287 and is
necessary due to the high number of
mathematical functions to be processed by
the programs within short time. Principally, processing can, however, be made
without the use of a co-processor. The
simple user can work with the programs
directly from the operating system level,
since the compiler can also place the code
to the data carrier as directly executable
file.

3.2 FILE PROGRAMS

The files are produced and administered by
the relevant data programs. Data input is
made in dialogue mode, corrections can be
made at any time. In addition thereto the
possibility is offered to copy data sets
from the same file and to solely edit the
differing data.

To a limited extent the data programs
make logic checks in order to prevent
extreme erroneous inputs.

3.3 BWE SIMULATION PROGRAM

This program again is subdivided into the
following subprograms:
- Single version
  * simpe simulation
  * complex simulation
- combined version
  * indication after every block
  * indication of the results only

The single versions compute the capacity
of a bucket wheel excavator for one block,
with the complex version permitting the
user to change several parameters, e.g.
selection of slice height and advance. In
the single version, these two values are
automatically calculated, whilst the max.
value is assumed for advance.

The advance is always limited if more
than one slice are mined. Depending on the
block geometry and the machine concerned,
limitations are imposed by e. g. the
travelling mechanism, the boom etc. The
program uses the first limiting value for
further computation. For selection of
advance and slice height, the input data
is subject to a logic check, i. e. no
data input beyond the max. advance rate is
admitted.

The combined versions compute the capacities of one machine over practically as
many blocks as wanted, whilst only the
starting and ending block number has to be
entered. The results of the individual
block calculations are optionally read
out or firstly indicated in the consolidated result and placed in a results file.

After read-in of both the excavator and
geological file, the excavator travelling
path is determined with the max. slewing
angle at the excavator side. An optimization of the travelling path is
possible in the complex version. Subdivision of slices is done automatically,
but can also be fixed within the admissible
limits upon requirement. The following
tables determine the slewing, travelling,
switching and adjustment periods, with
consideration of the cos-phi regulation.
The major data are summarized in the
printout (fig. 2).

```
      OUTPUT DATA. HD_800      BLOCK NO 9010      PAGE  1

Continue (Y/N) ?   (<RETURN : Y>)

blockwidth                        (ft) :    94.34
block height                      (ft) :    50.00
sidewall slope                    (deg) :   50.00
front slope                       (deg) :   60.00
swell factor                      () :      1.35
max. bucket fill factor           () :      0.80
total advance in sub block        (ft) :    24.59
(limited by lifting cylinder)
number of cuts each terrace       () :      7

   Terrace height (ft)   slewing angles (deg) - open bench on left
Terrace                     left        right        vO
   1       17.95          18.86        80.00        20.99
   2       15.98          30.42        53.92        23.65
   3       16.06          43.52        42.90        23.56
```

```
      OUTPUT DATA HD_800      BLOCK NO 9010      PAGE  2

Continue (Y/N) ?   (<RETURN : Y>)

total sub block excavation       (bcy) :   4293.15

total slewing time               (min) :    108.71
adjustment time for cut          (min) :      2.83
adjustment time for slice        (min) :      3.22
adjustment time for sub block    (min) :      0.52
switch over time                 (min) :      1.60
total sub block cycle time       (min) :    116.88

production rate                  (bcy/h) :  2203.90
max. rate during cycle           (lcy/h) :  3858.82
max. theoretical output          (lcy/h) :  4823.53

efficiency                       () :        0.77
```

Figure 2: Output BWE Simulations

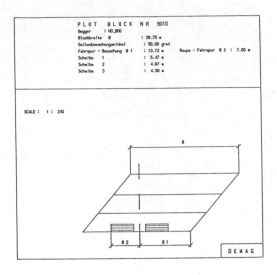

Figure 3: Plot cross-section BWE Block

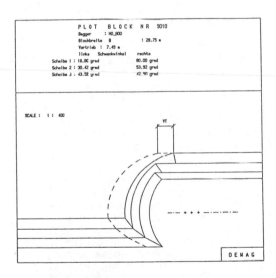

Figure 4: Plot plan view BWE Block

To demonstrate the block geometry, the single version shows front and top view on the terminal and if required also on the plotter. The output shows correctness of the angles and the paths, so that it can be directly used for further computations with the aid of the scale.

If inadmissible operational situations are determined during the computations, the simulation is interrupted with indication of the interrupting value. This is for instance the case if on the lowest slice either the free-cutting angle or the transport-sided slewing angle is surpassed.

3.4 DRAGLINE SIMULATION

Principally, the Dragline simulation developes similar to the BWE simulation. The Dragline simulation is limited to the operating mode ' direct casting '. It is, however, also possible to make a direct input of mining times or indirect input of empirical values, for which the mining times are determined by means of a stored table and placed in a results file as for the BWE simulation.

3.5 POSITIONING

When the mining times of the machines are available after the simulation, the position of the machines absolutely and to each other can be determined over a fixed mining sequence. The positions are determined and recorded in the 'results' file. This permits to recognize any critical operational situations in advance (e. g. inadmissible approaching of the machines).

371

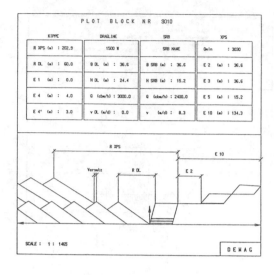

Figure 5: Plot Cross-Section

3.6 CROSS-SECTION-PROGRAM

The Cross-Section-Program is used for
representation of any wanted mining
block in cross-section, with representation of the mining and dumping side.
From the data of the mining side, a
cross-section of the dump is drawn under
consideration of the swell factor and the
reach of the XPS. This can give to the
user a clear view of the pit.

Inadmissible situations are notified
in advance, e. g. when the spoil room is
not sufficient. Since the dump layout is
also determined by the spoil of the
preceding strip, the computation enters
a fixed numer of strips into the planning
as determined by the program. The output
of the cross-section by the plotter is to
scale, the geometrical indications can
be directly used for further calculations
(fig. 5). The geometrical information is
additionally read out in the form of
coordinates.

4. LIMITS OF SIMULATION

Any simulation can make an output of its
computation results with almost every
wanted accuracy. Its reliability however,
can only be as exact as the input data.
Every previously fixed efficiency factor
represents an average value. It always
means a loss of accuracy and information,
comparable to the coarse screen resolution
of a terminal. Factors such as efficiency
of an individual operator, bad weather
influences and not scheduled shutdowns
cannot be determined in advance.

Even scheduled varations in capacity
(e.g. shutdowns for reason of maintenance)
can only be integrated into a simulation
system with corresponding expenditure.

Another weak point of all simulations is
the prevention of erroneous inputs. An
erroneous input can in most cases be
propped by program technique, but logic
erroneous inputs cannot be avoided.
Naturally, it is possible to partly eliminate such erroneous inputs by a type
of cross referencing, but still they can
not be finally eridicated.

In principle this is no specific
problem of computers, errors can also be
made with pocket calculators of slide
rules. The user of complex mining programs,
however, must not neglect his experience in
mining for a critical check of the
simulation results.

5. SUMMARY

The present lecture gives an example for
the reasonable use of PC´s for open pit
planning and shows both possibilities and
limits. The presented program package
permits to demonstrate nearly all desired
mining situations of the XPS by cross-
sections and to early recognize critical
mining positions.

The BWE simulation permits previous
selection of the best working mode and
ongoing operations can be constantly
checked for weak points.

The modular system permits further
extension to improve mining planning and
control. The calculated and effective
mining times offer further possibilities
to analyze the effectiveness of the
selected operating mode. The addition of
cost aspects is conceivable, too, to make
rentability checks.

9 Mine transport

Computer Applications in the Mineral Industry, Fytas, Collins & Singhal (eds)
© 1988 Balkema, Rotterdam. ISBN 90 6191 760 3

Computerized truck dispatching at Quintette Coal Limited

T.R.Farrell
Quintette Coal Limited, Tumbler Ridge, BC, Canada

ABSTRACT: Computer based truck dispatching in open pit truck/shovel operations is now an accepted, proven technology. Primary benefits of computerized haul truck dispatching include substantial production increases with an existing fleet of equipment, or a significant reduction in equipment working time required to meet production targets. Additional benefits such as achieving an optimum and continuous blend of run of mine ore or coal while minimizing rehandle, or collection of detailed production and equipment availability data are equally important. A computerized truck dispatch system purchased by Quintette Coal Limited was commissioned in early 1984. An overview of the system installed is presented, including descriptions of start up problems and solutions. A comparison of mine productivity before and after system application is discussed.

INTRODUCTION

Quintette Coal Limited (QCL) operates a large open pit coal mine located in north-eastern British Columbia on the eastern edge of the Rocky Mountains. The mine currently produces 4.75 million tonnes of metallurgical coal annually using conventional truck/shovel mining methods. Mining commenced in October 1982, and a heavy-media/froth flotation coal washplant was commissioned in November 1983.

The active mining areas are comprised of the Mesa Pit, Wolverine Pit, and the newly opened Shikano Pit, (Figure 1). Raw coal is exposed by stripping the overlying waste and is then mined and hauled to in-pit stockpiles or to a Bradford Breaker. Post-breaker product from Mesa and Wolverine must be transported 13.7 kilometers via an overland conveyor system to the plant. The Shikano breaker is located at the plantsite, to which raw coal is transported 2 kilometers from the pit by bellydump coal haulers.

The current load and haul equipment fleet consists of:

4 P&H 2800 cable shovels
2 P&H 2300 cable shovels
7 Demag H241 hydraulic excavators
5 Letourneau L1000 front end loaders
4 Letourneau L800 front end loaders
52 Wabco 155 tonne haulage trucks

3 Belly dump trucks, 127 coal tonnes;
   Cat-776 tractor with Atlas trailer
4 Terex 77 tonne haulage trucks, 311
   series

TRUCK DISPATCH SYSTEM JUSTIFICATION

The Quintette project is characterized by geographic remoteness, extreme climate conditions, and thin, complex, multi-seam coal deposits. The mining operation at QCL is large in scale, and fast moving. From the onset of operations QCL management realized that in order to maximize the effectiveness of the decision making process, and to maximize the use of equipment and resources, advanced computer-based technology would be required. QCL created project teams to investigate the latest mining technology in the areas of geological modeling, mine planning, surveying, process plant control, and haul truck dispatching. Since a significant portion of mine operating costs involve haul trucks, the potential savings are relatively high. Therefore the acquisition of a computer truck dispatching system was one of the first undertakings.

After careful review of operating costs and equipment capital acquisition costs, it was soon realized that an improvement of 10% in truck fleet production and the corresponding reduction in truck fleet requirement would generate an early pay back for the cost of any available computer truck dispatching system purchased.

SYSTEM SELECTION

In late 1982 a QCL project team consisting of one mine operations person, who served as project leader, and two Management Information Services personnel were assigned to justify, research, select, acquire and install a truck dispatch system which would:

1. improve truck production and reduce truck fleet requirements by minimizing truck idle time, and to minimize truck travel time. (Figure 2)

2. provide central control over haul truck operations by continuously monitoring the location, status and performance of haul trucks.

3. respond to changes in the pit and/or blending requirements, and provide full reporting capabilities.

Initially manual truck dispatching was considered an alternative, but was quickly dismissed. Because of frequent inclement weather conditions such as fog and snow, and the geographical locations of the pits, current truck locations would only be confirmed by voice radio. Additionally, a great deal of record keeping was required, and using manual methods, a dispatcher would not be able to adjust rapidly enough to change pit and blending conditions.

The use of a computer assisted manual dispatching system was considered, and although it offered the advantages of tracking trucks and recording important data, it would not dispatch trucks according to changing pit conditions or blending requirements. Computer based truck dispatching fulfilled all requirements.

The QCL project team investigated several computer based truck dispatching systems available to the mining industry. Extensive research from articles published in mining journals and exposure to various vendors of computer based systems indicated that these systems were not only viable, but were in fact in use at various mines around the world. After visits to mines with installed systems, and discussions of the success achieved, it was concluded that a computer truck dispatching system would be beneficial to Quintette.

Following careful investigation and review of the systems offered to the mining industry at the time, the DISPATCH system developed by Modular Mining Systems of Tucson, Arizona was selected because it most clearly met all of the selection criteria.

DISPATCH CONTROL LOGIC

The basic aim of DISPATCH is to maximize mine production with available equipment, or alternatively to achieve a desired production with minimum equipment. Dispatching trucks to meet either of these objectives is a dynamic operation which requires continuous monitoring of equipment status and location, route selection, and load/haul cycle timing parameters.

The first consideration under DISPATCH is to ensure optimal route selection whenever opportunities for exploiting proximity of shovels to dumping points exists. The haul routes (Figure 3) selected by DISPATCH require the minimum haulage capacity to adequately cover all operating shovels. In order to determine the optimum haulage model, DISPATCH uses the simplex method of Linear Programming. This model is regenerated whenever significant events, such as shovel or crusher downs occur, or by default, once every twenty minutes. Using Linear Programming, a steady state is assumed to exist, which in any mining operation is untrue. Therefore DISPATCH uses secondary dynamic assignment logic to ensure that any deviation from the steady state is addressed.

Once route selections and mine material flows are optimized, DISPATCH assigns trucks to minimize queueing of trucks at shovels, and to minimize shovel idle times. Current truck locations, speed factors and status, shovel digging rates, locations and status, are all considered when determining truck assignments. Dynamic Programming is used to assign trucks based on shovel need, shovel priority, blending considerations, rehandle costs, equipment cycle times, available equipment, other trucks expected to request an assignment in the future, and any operating constraints imposed by operations.

Since DISPATCH maintains running current time averages of all mining operations, the system is able to automatically adjust to changing pit conditions such as altered digging rates, fog or snow storms, or upsets in allowable dumping rates. DISPATCH tracks the location of trucks using data gathered from location beacons and from information entered into field control units by truck drivers. Beacons located at shovels, dumps, and along haul routes verify that each truck is traveling along the assigned route and in the expected amount of time. Shovel and dump beacons record truck arrival times at end points, while designated beacons along haul routes cause the assignment of trucks to be re-assessed to take into account any last minute changes in cycle time parameters which would alter shovel requirements. Trucks are only reassigned to shovels which are ahead of the current truck position.

A full description of the principles and theory upon which DISPATCH assignment logic is based is given by White and Olson (1983).

HARDWARE

A detailed description of the hardware is beyond the scope of this paper. For further reference Baker, Coburn, and White (1983) have fully described the operation of the system and the functionality of the associated hardware.

Central Computer: The central computer system used is a Digital Equipment Corporation VAX 11/780 with 12 megabytes (MB) of internal memory, 456 MB of storage, and is VAX clustered with another VAX 11/780 with 8 MB of internal memory, used to run mine planning software. The two VAXES share 7 RA81 disk drives. If the main dispatching VAX goes down, the other VAX takes control. (Figure 4)

Communications Hardware: Communications between the VAX and the control consoles in each truck and shovel is achieved through the use of a two channel UHF FM two way digital data radio system. Two repeaters, both with a 100% working cycle are located on a prominent hill overlooking all three pits. (Figure 5)

An interface panel was developed specifically to support a communications protocol for efficient error free communication on each radio channel.

The communication system between the VAX and the DISPATCH terminals in the dispatching office is designed to utilize a fully redundant system that produces uninterruptable communication. (Figure 6)

The DISPATCH control center consists of 6 terminals and a interface panel running at 9600 bits per second (BPS). An active switching system allows the simultaneous switching of all the terminals and interface panel between two communicator devices and data lines. Each side of the switching units is connected to its own communicating device, the Gandalf SMUX or the Motorola Codex 6050.

The Gandalf SMUX is a statistical multi-plexer capable of efficiently multiplexing up to 48 channels of input data on to one or two synchronous communication links. The input data may be either asynchronous or synchronous and have a variety of operating characteristics. Each link is capable of transmitting up to 64 kilobytes per second (KBS).

The Motorola Codex 6050 is a 40 channel Distributed Communication Processor or statistical multiplexer that employs dynamic data compression and adaptive call routing with redundant links capable of speeds up to 64 KBS. The link one's and two's from the two multiplexers are then connected to two Gandalf GLM504 units. These two units concentrate the data from the two multiplexers and transmit it down two 56 KBS lines to an exact mirror system located at the data centre 17 kilometers away. The communications system are then routed through a Gandalf PACX 2000 to all computer resource systems used by Quintette.

With the addition of the PACX 2000 it became possible, with just the throwing of one switch in the dispatch office, to request the back-up computer. Therefore there exists the complete redundancy in the host units, the local communication equipment, the remote communication equipment, the communication lines between the dispatch office and the data center while also supplying a total management of, security and control of all computer resources.

Field Control Units: Field control units are mounted in all trucks and shovels. These consoles are used to enable equipment operators to communicate with DISPATCH and vice-versa. The field control units consist of a vehicle interface panel, an operator interface panel, a radio and a power supply. The vehicle interface panel is of rugged construction and contains three cards which control communications, receipt of beacon transmissions, and vital signs monitoring. The operator interface panel features a 32 character display and data entry via the keypad. All field control units include three 6809 microprocessors and function in a multiprocessing environment. (Figures 7 and 8)

Location Beacons: Location beacons are small low power VHF transmitters located at strategic points in the haul route network. Each beacon is compact, robust and mounted in a stainless steel tube for maximum protection against the elements. As a result of their usually remote location, they are typically powered by solar panels. The range of transmission of the beacons is approximately 50 m which enables DISPATCH to determine the position of any truck which should detect a specific beacon. Beacons serve many purposes, and each beacon has a unique identification so that it can be defined as a call point, crusher reassignment, shovel reassignment, or auto-arrive location.

SOFTWARE

DISPATCH Software is written in "C" for maximum system efficiency. The Software installed in the field hardware is written in Pascal and Assembler.

When DISPATCH was first commissioned at QCL, maintenance on the DISPATCH system was easy; one mine, one set of source code. However, as more systems were sold, Modular experienced difficulty in maintaining a number of mine sites, each with its own unique operating problems and versions of software, while developing software for others. The DISPATCH software has therefore gone through many radical changes in the initial years of operation, many of them designed to turn the system into a set package which can be easily configured to suit any specific mining operation. This is primarily true of the newest version of the DISPATCH system, which allows Users to create custom reports, change control logic, and modify all User interface displays using a Forms Command Language which can be easily manipulated by mine personnel. A report that previously would take a system programmer 15 days to write can now be written in as little as one day by Users trained in the Forms Command Language.

OPERATOR INPUT

Truck drivers, shovel operators, dispatchers and the dispatch general foreman form key links in the smooth and effective operation of the system at QCL.

Truck Drivers: The basic function of each truck driver using the system is to inform DISPATCH when the truck arrives at a shovel, when it is being loaded, when a dump assignment is required, to confirm arrival at the dump, and to request an assignment back to a shovel. (Figure 9) Additionally, truck drivers must relay a change of status of the truck, such as a down, or delays experienced, the quantity of fuel put in the truck, and to properly login to the truck at the start of shift. Communication between DISPATCH and the operator is via 32 character digital

display on the operator interface panel.

Shovel Operators: Shovel operators use the same equipment as truck drivers, except that the shovel operator has a different set of actions which he is required to perform. Shovel operators are only required to specify the material type which is being loaded, and when the loading of a truck is complete. Change of status is also relayed via the operator interface panel.

Dispatchers: The control centre of the entire load/haul operation is the DISPATCH office. In the DISPATCH office there are six computer terminals in use, each with a different function in assisting the dispatcher to retain maximum control over the mining operation.

Dispatch General Foreman: At QCL, the upkeep of the DISPATCH system, particularly with respect to primary pit configuration data such as haul routes, blast information, and production equipment information are all the responsibility of the DISPATCH general foreman. Since DISPATCH has many configurable system parameters which affect the efficiency with which it operates, it is also the responsibility of the DISPATCH general foreman to ensure that these parameters are set to optimize the efficiency of the dispatching operation at all times.

PROBLEMS

Although the computer truck DISPATCH system now runs relatively trouble free, from system commissioning time to present there have been certain problems which had to be addressed.

Hardware: Various problems were encountered during the initial startup and installation of the DISPATCH system. There were many instances where radio communications on the data channel suffered periods where the panels could not be accessed through the communications interface panel. After intensive investigation into ther problem, it was diagnosed to be radio related. Temperature inversion layers in the atmosphere caused the radio signals from the repeater site located at the highest point in the mine to radiate upwards and away from the mine, rather than out and down into the mine. The antenna supplied with the original radio system was an omnidirectional type designed to radiate radio waves toward the horizon for communications as far away as possible. Unfortunately, a temperature inversion layer often formed between the repeater site and the mine, thus the radio waves would "skip" across the inversion layer and never be received by equipment working below. To solve the problem, a "down-tilt" antenna, which is designed to radiate most of the energy downwards at a sufficient angle into the mine to penetrate the inversion layer, was installed and proved to be effective.

At system installation the original repeater site was correctly located to adequately cover all areas of the mine. However, the topography has now been vastly altered by three years of mining operations. Many new and permanent high walls have developed, and the new Shikano pit has been started approximately 14 km away from the repeater site. Many areas of the operation have become "shadow areas" where radio signals cannot be received. The solution of this problem is to install a new repeater station at a location on a mountain across the valley from the major pits. This site should also cover any future changes in pit design or the startup of other pits considered in the twenty year mine plan.

The existing repeater site will be left as a standby unit in case of any failures of the remote system. This project is currently underway and should be completed by the end of 1987.

Haul truck electrical systems consist of a 24 volt battery charged from the truck's alternator, a 24 volt to 12 volt converter changes the truck's operating voltage to one which is used to power the data radio and field control units. The alternator initially installed on the haul trucks in use at QCL provided 74 amperes to charge the battery. Since the trucks purchased included many electrical options, the drain on the electrical system even without the DISPATCH panel was between 80 and 90 amperes. The DISPATCH control panel and the data radio combined draw 4 amperes. The battery supplied with the truck was a standard 180 amp-hour battery. The result was that the truck batteries drained after several hours of use. This proved to be an even bigger problem during winter months due to severe temperature extremes, which have an adverse impact on battery performance.

As the battery lost its charge, the field control panels began to power up and down because the threshold voltage required to power the panels was not always available. This created havoc with the operators of the trucks with low battery power conditions. The frequency of power up messages from trucks with low batteries became a

good indicator of pending truck electrical problems.

When selecting equipment for DISPATCH, QCL elected to purchase power converters external to the contract. These supplies were much cheaper in price, but were also lower quality converters which could not track wide fluctuations in input power. The resultant output voltage caused problems with the DISPATCH field control units and the voice grade data radios. The data radios were designed so that should the power begin to drop, the radio keyed up for transmission, bypassing the built-in time-out switch. When this occurred, the radio channel was effectively jammed preventing any radio traffic. Finding the truck causing the problem proved to be no easy task. Since the mine continued to run during the periods that the radio waves were jammed, the radio shop personnel had problems locating the truck.

After the problem was identified, Quintette began an intensive program to upgrade the electrical systems on the entire truck fleet. The 75 ampere alternators were replaced with 100 ampere models and alternator pulleys were reduced to give an increased output at lower engine RPMs. The standard truck batteries were replaced with 240 amp-hour batteries. Additionally the batteries were enclosed to insulate them from the severe climate. The DC converters were replaced with the units originally specified by Modular. The radios were modified to eliminate the keyup problem when the power dropped.

Central Computer: Radio communications and power were not the only problems encountered during the initial installation. Since the original QCL VAX 11/780 was located in the Management Information Services' initial office in Vancouver, QCL MIS provided a much smaller VAX 11/730 for use with the dispatching system until the VAX 11/780 moved to the QCL mine site. Since DISPATCH was designed to run on the more powerful VAX 11/780, there were problems with user response time when real-time DISPATCH was running on the VAX 11/730. Aside from the CPU resource problems, the VAX 11/730 was installed in the DISPATCH office, and the VAX 11/730, which is designed to operate in a controlled environment, soon experienced a number of system crashes due to coal dust, which had penetrated the machine. The QCL VAX 11/780 was soon moved up to the mine site and DISPATCH commenced operation on this machine by mid-1984.

Other computer related problems included poor land communications between the data centre (administration building) which houses the VAX 11/780 running DISPATCH and the DISPATCH centre in the mine office and dry, a distance of 14 km, and problems with VAX resource management.

In 1985 when the second VAX was purchased a VAX cluster environment was acquired with it. At this time no one could foresee the impending problems that the cluster would bring with it to real time dispatching. Two VAXES in a cluster, when either are being rebooted, impacts processes on the other VAX while VMS on the rebooting VAX reconfigures. In such cases processes on the impacted VAX halt for a couple of minutes while high priority VMS jobs reconfigure the VAX cluster. Although this has little effect on batch jobs or interactive users, real-time DISPATCH operation is halted and field transactions are lost causing operators in the field to wait for assignments or arrives etc. reducing the efficiency of the overall system. This to date still has not been resolved either by Digital nor QCL.

The other major problem was the sharing of data bases among VAXES in a cluster. For fast real-time access to shared data, DISPATCH DBMS (DataBase Management System) routines must utilize global sections in shared random access memory (RAM). The DBMS routines keep frequently accessed data in these global sections and do not update files onto disk until the file is closed. Because VAXES in a cluster can share disk files but not global memory sections, DBMS file sharing routines do not function properly when multiple VAXES in a cluster attempt to share the same data base. If the DISPATCH system was running on its primary machine, another process on the backup machine could modify the same data base. Both processes would see inconsistent sets of information which then resulted in data base corruption and many DISPATCH system crashes. VMS does not lock data bases which are accessed on one VAX from a process on the other VAXES in the cluster; extremely complex system software would be required to prevent such access. The solution to this by QCL and MMS was to have all user logins to look for the defined DISPATCH batch job (dbupdate) which is submitted by the DISPATCH processes at its login and if it is not found automatically disable the login to that VAX.

Training: At system installation time, operators were not available for adequate training on the use of field control units. This caused many operator misunderstandings and management misconceptions about the dispatching system. Many errors which

were caused by operator mistakes were interpreted as system mistakes: many problems that were caused by system mistakes were intepreted as operator mistakes. Since the system is not a totally automatic "computer box", but a tool to be used by any mine, Users must learn to use it correctly in order for it to be a success. This starts with a firm commitment from management, and in turn translates into a firm commitment from operators and line supervisors. After the system was initially commissioned, QCL began an extensive program to properly train all operators, dispatchers, supervision and management, which eliminated many of the supposed problems with the dispatching system.

SYSTEM BENEFITS

Since QCL installed the DISPATCH system during the mine startup phase, it was difficult to compare QCL using the truck DISPATCH system versus QCL not using the truck DISPATCH system. The truck and shovel fleet buildup was rapid; the operators inexperienced. Early DISPATCH sessions were said to be like turning on the autopilot while the plane is in a tailspin.

Whereas the mining operation using manual truck allocation methods with a stable fleet and workforce has no problem comparing productivity prior to system commissioning and after system commissioning, this was not the case at QCL. This made it extremely difficult to define system benefits.

Unconvinced of the system's effectiveness, QCL operations supervisors proposed that a test be performed in an attempt to define the productivity improvement through the use of the truck DISPATCH system. Therefore, Quintette ran several tests in the fall of 1984, ten months after the system was commissioned.

To form the basis of comparison, Quintette ran the DISPATCH system with DISPATCH making truck assignments for four days. The system was then run "locked" for the next four days, followed by four more days of DISPATCH truck allocation. When running "locked", foremen eager to "beat the system" made batch truck allocations. The foremen were distributed around the pit at strategic lookout points to watch the operation and inform the dispatcher of any change or reassignment they deemed necessary. Dispatchers were not allowed to assist the foreman in any way by using displays and data gathered by the DISPATCH system; dispatchers could only assign and relock trucks based on instructions from the pit supervisors.

Mine engineering compared production for the shifts run with the system running "unlocked" versus "locked". When DISPATCH made truck assignments, mine productivity increased by 10% based on excavator BCM/work hour and truck BCM/work hour over the batch allocation methods used by the dispatchers and foremen.

Quintette engineering then reviewed productivity rates for equipment in 1983 prior to installation of the DISPATCH system versus 1984 productivity rates after system installation and the 10% improvement in productivity rates was confirmed.

Besides increases in productivity, the two dynamic functions of dump management and coal blending predominant are in the day to day operation of Quintette.

Dump Management: Because of the mountainous location of the pits and geological nature of the coal reserves there existed a need within the DISPATCH system to establish control logic to limit not only the amount of material dumped on any existing dump but also to be able to regulate the amount dumped over a period of time (Tassie, 1987). Using software control logic it is possible for a Dispatcher to regulate dumping of material by using one, all, or any combination of the following:
1. dumping by time
2. dumping by material
3. dumping by feed rate
4. dumping by the blending of material
5. dumping by maximum number trucks arrived at location.

By using any combination of the above it is possible for Quintette to regulate the amount of material dumped at any of the dumping points while still maintaining a safe and efficient operation.

Blending of Coal: To maintain pit production to it's maximum while also providing a consistent coal blend to the plant; the use of the DISPATCH system in this area alone has proved paramount. By inputting control parameters the dispatcher can regulate the proportions of coal types trucked to the conveyor. The system also regulates the amount of coal to be sent to inpit stockpiles, thereby reducing rehandle costs. Keeping a proper coal blend is essential to efficient wash plant operations and yield. (Morash, 1987)

The DISPATCH system includes a comprehensive reporting system to help management identify potential problem areas before they occur, track production and availability trends throughout the year, and examine mine efficiency rates. The mine management system carries information on coal seams, blending, blasts and loading statistics, drills, auxiliary equipment, haul routes and grades, dumping, and truck tires. The information flow from this system is depicted on Figure 10.

CONCLUSIONS

The truck dispatching system has gainfully operated at QCL since early 1984, demonstrating anticipated productivity increases and providing a new era of management control over the mining operation. In financial terms the pay back upon which the system was originally justified has been realized. The system has now received a high degree of operator acceptance. The possible uses of data generated by the DISPATCH system are still being explored; the expandability is limited only by the imagination.

ACKNOWLEDGEMENT

The author wishes to thank the management of Quintette Coal Limited for permission to present this paper.

BIBLIOGRAPHY

Arnold, M.J. and White, J.W. October 1982. Benefits of computer-based dispatching in open pit mines. Presented at American Mining Cong., Las Vegas, NV.

Baker, M.R., Coburn, J.W. and White, J.W. 1983. Hardware, software and system considerations in computer-based open pit mine truck dispatching. ISA Trans.

Byles, R.D. May 1986. The installation of an automatic truck dispatch system at Bougainville Copper Limited. Presented 13th Congress of Council of Mining and Metallurgical Institutions, Canberra, Aust.

Chironis, N.P. March 1985. Computer monitors and controls all truck-shovel operations. Coal Age, pp. 50-55.

Clevenger, J.G. Sept. 1983. DISPATCH reduces mining equipment requirements. Mining Engr., pp. 1277-1280.

Morash, B.J. May 1987. Matching mining methods to complex geology. Manager of Mine Production: Quintette Coal Ltd. CIM Annual General Meeting. pp. 11-12.

Pelley, M.H. Apr. 1985. Quintette Coal Limited's Experiences to date with its computerized truck dispatch system. Presented at CIM 87th Annual Meeting,

Vancouver, B.C.

Quintette Coal Limited. Assorted memoranda
and unpublished reports.

Tassie, W.P. Sept. 1987. Waste dump management at Quintette Coal Lilmited. Geotechnical Engineer. CIM District 5
Annual General Meeting. pp.5.

White, J.W. and Olson, J.P. 1983. Computer
based dispatching in mines and concurrent operating objectives. Min. Engr.

White, J.W., Arnold, M.J. and Clevenger,
J.G. 1982. Automated open pit truck
dispatching at Tyrone. E.M.J.

White, J.W. and Olson, J.P. Oct. 1983.
Recent developments in computer based
dispatching. AIME-SME Preprint 83-413.
Presented at AIME-SME Fall Meeting.

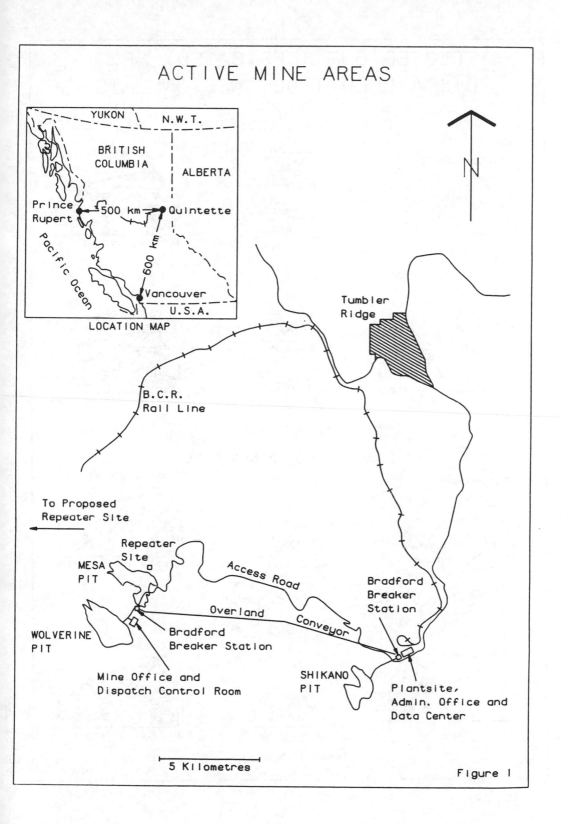

ACTIVE MINE AREAS

Figure 1

CLOSED OUT (DEDICATED) VS. DISPATCHED TRUCK ASSIGNMENTS

DUAL SHOVEL CLOSED OUT

Waste Dump

1 Minute to dump

8 Minutes

12 Minutes

Waste Shovel

3 Minute loading time

Coal Dump

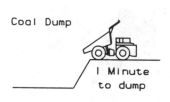

1 Minute to dump

8 Minutes

12 Minutes

Coal Shovel

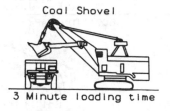

3 Minute loading time

Total = 16 trucks

DUAL SHOVEL DISPATCHED

Waste Dump

1 Minute to dump

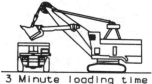

12 Minutes

Waste Shovel

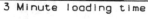

3 Minute loading time

3 Minute Travel

4 Minute Travel

Coal Dump

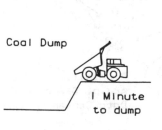

1 Minute to dump

12 Minutes

Coal Shovel

3 Minute loading time

Total = 13 trucks

Figure 2

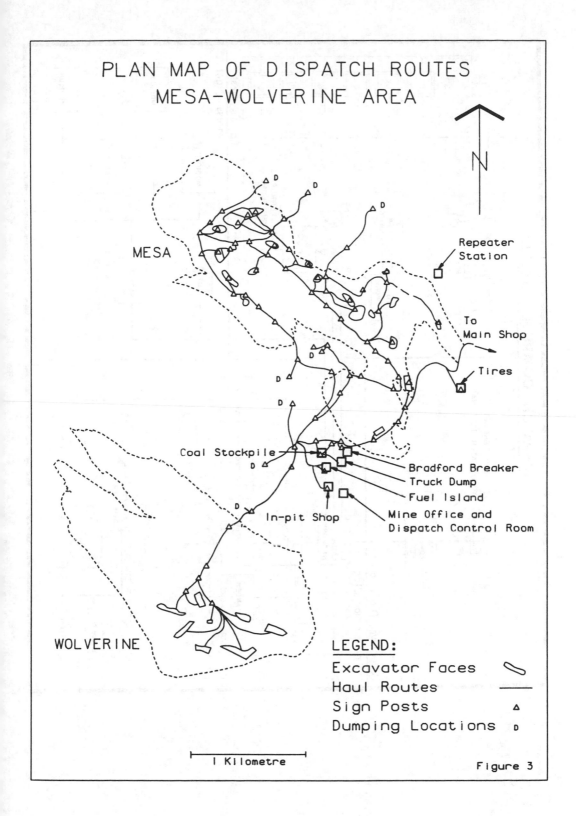

PLAN MAP OF DISPATCH ROUTES
MESA-WOLVERINE AREA

N

MESA

Repeater
Station

To
Main Shop

Tires

Coal Stockpile

Bradford Breaker
Truck Dump
Fuel Island

In-pit Shop

Mine Office and
Dispatch Control Room

WOLVERINE

LEGEND:
Excavator Faces
Haul Routes
Sign Posts   △
Dumping Locations   D

1 Kilometre

Figure 3

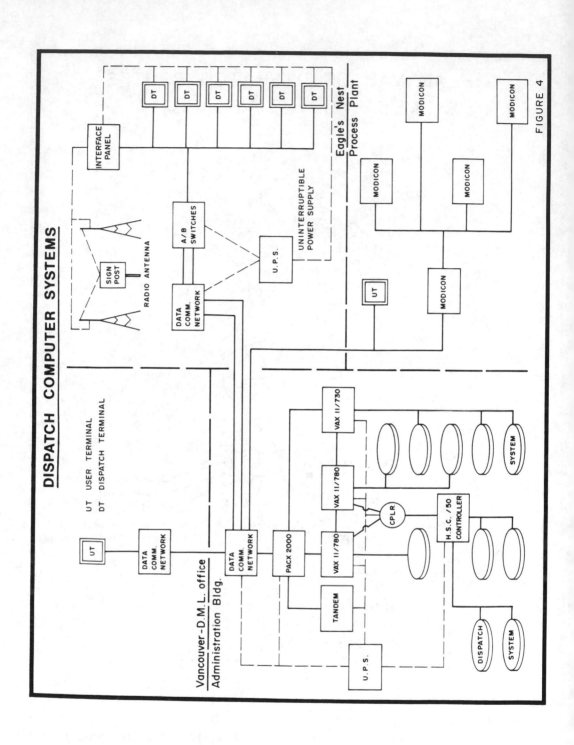

DISPATCH COMPUTER SYSTEMS

UT USER TERMINAL
DT DISPATCH TERMINAL

Vancouver-D.M.L. office
Administration Bldg.

Eagle's Nest
Process Plant

FIGURE 4

386

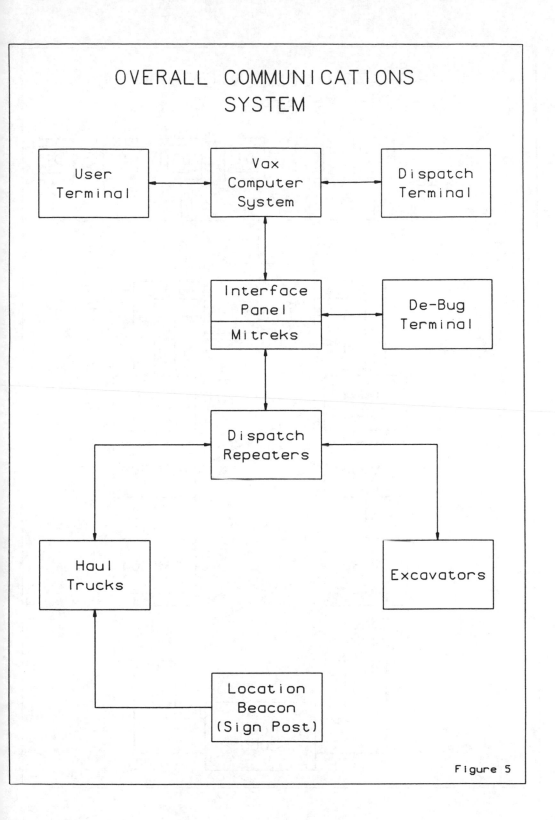

OVERALL COMMUNICATIONS SYSTEM

| User Terminal | ←→ | Vax Computer System | ←→ | Dispatch Terminal |

Interface Panel / Mitreks ←→ De-Bug Terminal

Dispatch Repeaters

Haul Trucks

Excavators

Location Beacon (Sign Post)

Figure 5

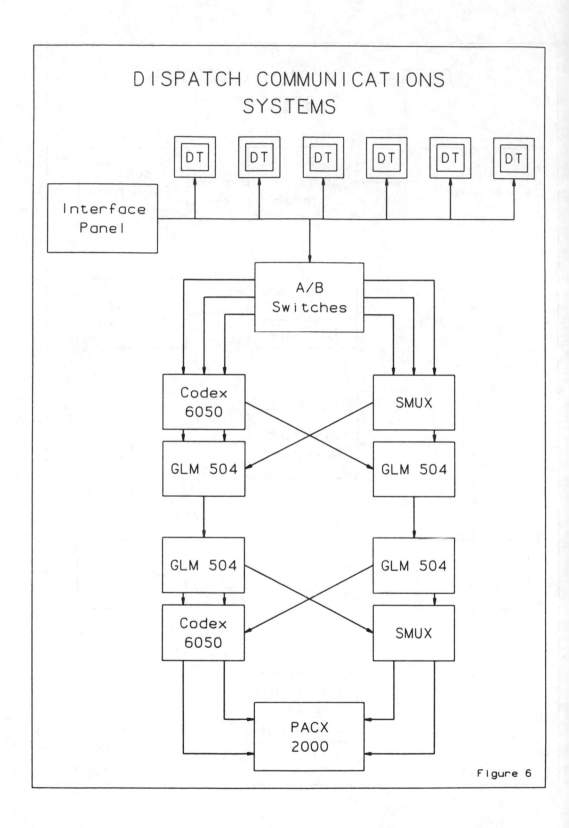

DISPATCH COMMUNICATIONS SYSTEMS

Figure 6

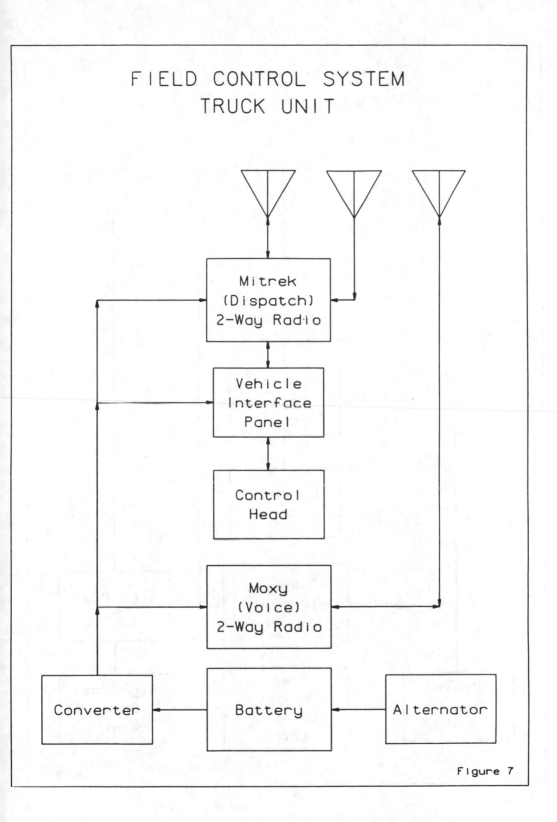

FIELD CONTROL SYSTEM
TRUCK UNIT

Mitrek
(Dispatch)
2-Way Radio

Vehicle
Interface
Panel

Control
Head

Moxy
(Voice)
2-Way Radio

Converter

Battery

Alternator

Figure 7

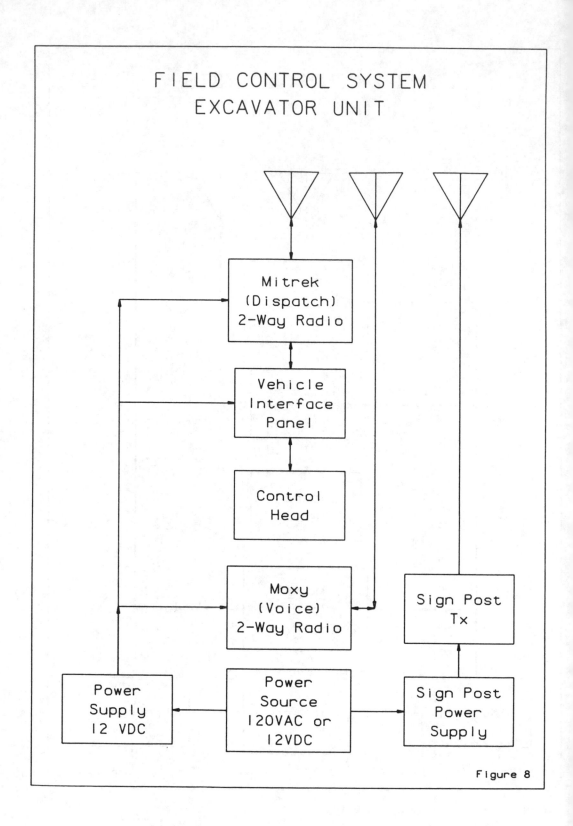

FIELD CONTROL SYSTEM
EXCAVATOR UNIT

Mitrek
(Dispatch)
2-Way Radio

Vehicle
Interface
Panel

Control
Head

Moxy
(Voice)
2-Way Radio

Sign Post
Tx

Power
Supply
12 VDC

Power
Source
120VAC or
12VDC

Sign Post
Power
Supply

Figure 8

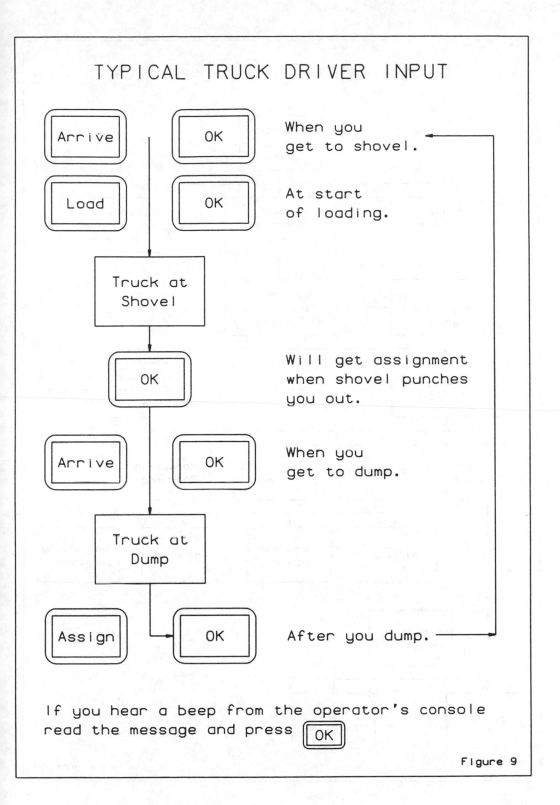

TYPICAL TRUCK DRIVER INPUT

| Arrive | OK | When you get to shovel. |

| Load | OK | At start of loading. |

Truck at Shovel

| | OK | Will get assignment when shovel punches you out. |

| Arrive | OK | When you get to dump. |

Truck at Dump

| Assign | OK | After you dump. |

If you hear a beep from the operator's console read the message and press OK

Figure 9

DISPATCH INFORMATION FLOW

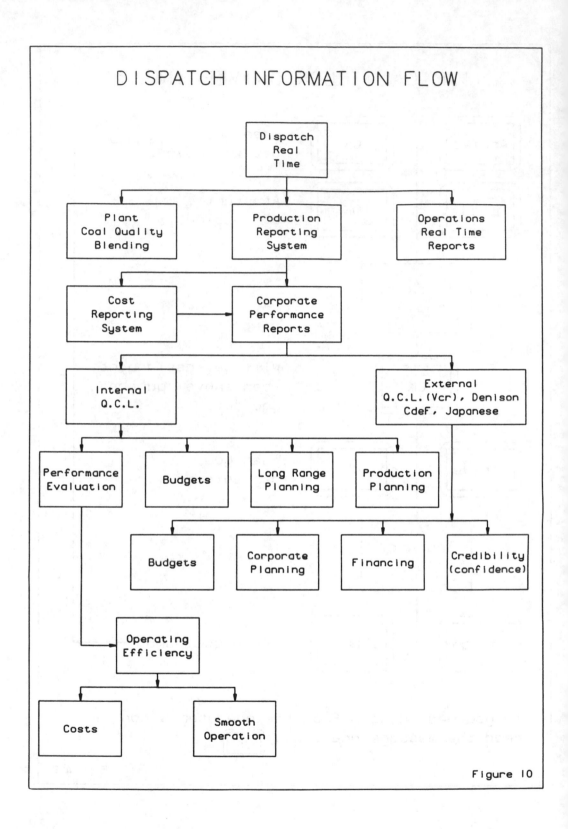

Figure 10

392

Computer Applications in the Mineral Industry, Fytas, Collins & Singhal (eds)
© *1988 Balkema, Rotterdam. ISBN 90 6191 760 3*

Algorithmes d'optimisation pour la gestion d'une flotte de camions dans une mine à ciel ouvert

François Soumis, Jean Ethier, Daniel McInnis & Jorgen Elbrond
Ecole Polytechnique de Montreal, Quebec, Canada

RÉSUMÉ: Nous proposons un système avant-gardiste d'assignation en temps réel des camions dans une mine à ciel ouvert utilisant des techniques d'optimisation mathématiques: programmation en nombre entier, réseaux à coûts non linéaires, problèmes d'affectation, etc. Ce système assure une utilisation de l'équipement disponible meilleure que celles obtenues avec les algorithmes existants. De plus, il fournit des outils de planification pour le choix et la localisation des équipements et pour trouver un équilibre économique entre les objectifs de quantité et de qualité de la production.

1. INTRODUCTION

La gestion des camions dans les mines à ciel ouvert a beaucoup évolué au cours des 15-20 dernières années. Trois générations de systèmes d'affectation des camions peuvent être identifiés.

Avant l'avènement des systèmes d'affection des camions dans le sens propre du terme, la pratique étant d'assigner chaque camion à une même pelle pour toute la durée du quart de travail. Le nombre de camions par pelle était déterminé suivant le plan de production. L'utilisation de "circuits fixes" qui prévalait avant l'avènement des systèmes d'affectation des camions est identifié comme étant la génération zéro à la figure 1 qui présente le gain en productivité en fonction de l'évolution de ces systèmes. La productivité des trois générations de système est comparée à la productivité avec "circuits fixes".

L'affectation des camions commença modestement avec un opérateur observant les mouvements de camions à portir d'une cabine située à un point stratégique. Il pouvait à partir de son information sur la situation dans la mine, changer la destination d'un camion vide de façon à réduire son attente avant le changement. La destination étant transmise au camion en affichant le numéro de la pelle choisit sur un tableau lumineux situé sur la route du camion vide. Il est toutefois difficile pour l'opérateur

sans moyen de calcul de réaliser simultanément les deux objectifs de l'affectation des camions; réduire l'attente des camions et suivre le plan de production. Ceci a conduit à l'introduction de la table d'affectation (Meller, 1977), qui est un modèle analogique de la mine, où des blocs remplacent les camions. Par la suite ce type de modèle de la la mine fut supporté par micro-ordinateur (Young, 1984). L'ordinateur représente la position des camions et les attentes prévues, calcule la production réalisée et sa composition, propose l'affectation correspondant le mieux aux deux objectifs. Toutefois, la décision finale appartient à l'opérateur qui transmet la prochaine destination à chaque camion. La communication radio entre l'opérateur et les camions a aussi été introduite pour cette première génération de système d'affectation des camions. Les gains de cette génération représentés à la figure 1 à "opérateur" sont dues à la réduction des temps d'attente et au suivi des opérations qui augmente les performances des conducteurs de camions.

La seconde génération de système a précédé certains développements de la première génération. Elle est apparue avant l'avènement du micro-ordinateur. Ce type de sytème est carctériser principalement par la transmission automatique des données. La position des camions est détecter par des capteurs et transmise automatiquement à l'ordinateur

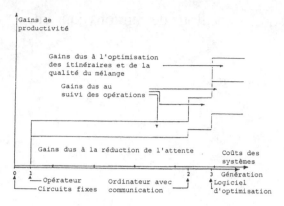

Figure 1: Gain en productivité suivant
l'évolution des systèmes.

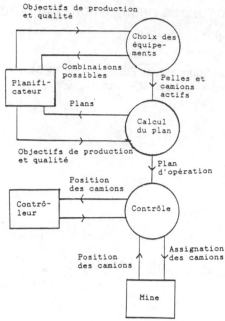

Figure 2: Schéma du système

par un réseau de communication. L'ordinateur prend les décisions d'affectation
avec des procédures semblables à celles
utilisées par les systèmes de première
génération. Les affectations sont com-
muniquées aux camions directement et
apparaissent sur un écran dans la cabine
du camion. Les gains de cette génération représentés à la figure 1 à "ordinateur avec communication" comportent
une réduction supplémentaire du temps
d'attente et une meilleure performance
des chauffeurs dues aux informations
plus précises sur la position des
camions. De plus, ces systèmes produisent automatiquement les rapports de
production. (Crosson, 1977; Charlemagne, 1979).

A la troisième génération les décisions d'affectation des camions et de
planification sont prises à l'aide
d'algorithmes d'optimisation mathématique. L'équipement utilisé est semblable à celui de la génération précédente.
A ce niveau l'ordinateur devient l'instrument central de la décision et intè-
gre certaines fonctions de la planification et de l'opération. Le changement
et le transport sont effectués en accord
avec un plan de production, ce plan
déterminé par l'ordinateur est réajusté
au besoin suivant l'état de l'équipement
et de la mine. Cette planification
effectuée par l'ordinateur considère les
objectifs de mélange de minerais en plus
du volume de production. Cette troisième génération de système est appelée
"logiciel d'optimisation" à la figure 1.
Les gains de productivité augmentent au
niveau de la réduction des temps d'attente à cause des procédures d'optimisation, des gains additionnels sont dues à
l'optimisation des itinéraires et à la

qualité du mélange (White, 1981; White,
1983; Clemenger, 1983; McInnis, 1985).
Cet article présente un nouveau système
de troisième génération avec une meilleure intégration des décisions de
planification et des algorithmes d'optimisation plus puissants.

2. DESCRIPTION GÉNÉRALE DU SYSTÈME

La figure 2 illustre le schéma général
du système et son interaction avec
l'environnement minier. Le système
informatique permet de réaliser les
trois étapes dans le processus de
planification-production: le choix des
équipements, le calcul du plan opération, et le contrôle en temps-réel.

Le planificateur est responsable des
deux premières étapes. Il choisira
d'abord quels sont les équipements à
opérer pour atteindre ses objectifs de
production et de qualité. Le choix se
fait en interaction avec l'ordinateur
par la fonction choix des équipements.
Lorsqu'un choix d'équipement est fait,
il est transféré à la fonction calcul du
plan opération où est calculé l'estimé
de la production. Le planificateur peut
alors réviser ses objectifs, réviser son
choix d'équipement et recommencer le
calcul ou accepter ce plan. Un plan

accepté sera transféré à la fonction contrôle en temps-réel où il servira de critère à l'assignation des camions.

La fonction contrôle est liée directement à la production. Elle interagit d'abord avec la mine, en détectant la position des camions et en transmettant les assignations aux camions via un canal de télécommunication. Il y a aussi interaction avec le contrôleur qui peut observer sur un écran le déplacement des camions et signaler à l'ordinateur les bris d'équipement.

2.1 Choix des équipements

La première étape du processus planification-production consiste à déterminer l'équipement à utiliser pour atteindre les objectifs de quantité et de qualité: Le nombre de camions, le nombre de pelles et les zones de la mine à exploiter sont déterminés à cet étape.

Du côté camions, il y a peu de marge de décision; les chauffeurs devront être prévenus à l'avance de leurs horaires de travail. Le paramètre d'entrée est le nombre de camions prévus. Exceptionnellement, si le nombre de camions n'est pas suffisant pour atteindre les objectifs, on pourra ajouter quelques conducteurs à la dernière minute s'il y a des véhicules en état de rouler. De même, si le nombre de camions excède le besoin, on le diminue en immobilisant temporairement quelques-uns. Dans de tels cas on recommence en modifiant le paramètre d'entrée.

Du côté des pelles, on connaît l'ensemble des pelles en état d'opérer et leurs localisations dans la mine. Chaque pelle a accès à quelques bancs de matériaux aux propriétés différentes (1 à 4 bancs). Notons qu'il n'est pas facile de déplacer une pelle sur une longue distance car il s'agit de pelles électriques alimentées par câbles. Les horaires de travail fournissent un nombre d'opérateurs de pelle disponibles et on voudra en général qu'ils travaillent tous. Donc, le problème consiste à déterminer les pelles actives et les bancs associés de façon à atteindre au mieux les objectifs de quantité et de qualité de minerai. C'est un problème difficile étant donné le grand nombre de combinaisons de bancs et de pelles possibles. Mais, en plus des facteurs quantité et qualité, on doit considérer la préférence pour les fonctions actuelles, les tâches nécessaires pour le déplacement vers un autre banc, l'ordre dans lequel il est plus facile d'extraire les bancs, la priorité de finir

un banc pour avoir accès au cours des prochains jours à une zone située derrière, etc.

Pour faire un choix judicieux, nous proposons une approche combinant le jugement de l'opérateur et la puissance de calcul de l'ordinateur. L'élément central est un modèle de programmation linéaire permettant à l'ordinateur d'évaluer la production maximale tout en respectant des contraintes de qualité pour une localisation des pelles actives. Ce modèle permet d'évaluer plus de 2 000 combinaisons, de localisations des pelles par minute sur un PDP-11.

Pour réduire le nombre de combinaisons à évaluer, l'opérateur fixe d'abord les principaux paramètres pour décrire le type de solutions souhaitées. Il détermine le nombre de pelles opérant dans le minerai, le nombre dans le stérile, le nombre dans chaque mine. De même, il limite les choix de bancs pour chaque pelle suivant les priorités et les préférences. L'opérateur demande alors à l'ordinateur combien de combinaisons sont possibles sous ces conditions et obtient un estimé du temps de calcul nécessaire pour toutes les évaluer. Si le temps de calcul est jugé raisonnable l'opérateur demande l'évaluation et obtient les meilleures combinaisons dans l'ordre décroissant de production. Si le temps de calcul estimé est trop grand l'opérateur peut préciser ses préférences pour restreindre les possibilités. Si les meilleures combinaisons de localisations évaluées ne sont pas satisfaisantes du point de vue de la production, l'opérateur peut modifier ses préférences pour explorer d'autres localisations. Par exemple une pelle peut être ajoutée si nécessaire pour atteindre les objectifs de production.

Cette étape de planification est réalisée au début des quarts si on modifie les objectifs, le nombre d'opérateurs de pelles ou le nombre de conducteurs de camions. Elle est aussi réalisée au besoin durant un quart de travail si un bris d'équipement ou une autre perturbation vient modifier les conditions de fonctionnement.

2.2 Plan d'opération

La seconde étape du processus planification-production consiste à déterminer un plan d'opération pour l'équipement choisi. Le nombre de camions, le nombre de pelles et leur localisation sont fixés. Le plan détermine le niveau d'activité de chaque pelle et les itinéraires des camions réalisant le mieux les objectifs de qualité et de quantité de minerai transporté.

Nous proposons à cette étape un modèle plus réaliste du problème de transport de minerai que les modèles de planification utilisant la programmation linéaire. Les modèles linéaires tel celui proposé par White et Olson (1983) ne considère pas l'augmentation de l'attente quand on approche la limite de capacité d'une pelle ou du concasseur. De plus, ils traitent les exigences de qualité en imposant des limites inférieures et supérieures sur la teneur et les autres paramètres du mélange. De tels modèles produisent des solutions extrémistes: certaines pelles au maximum de leur capacité et d'autres peut utilisées. Dans certains cas, deux pelles dans le même type de minerai peuvent avoir des niveaux d'activité différents. De plus, les solutions sont généralement aux limites des intervalles acceptés pour les paramètres de qualité. De telles solutions sont acceptables en première approximation pour le choix de localisation des pelles, mais ne peuvent servir de plan pour les opérations en temps-réel.

Pour obtenir un plan plus réaliste, nous proposons un modèle de programmation non linéaire intégrant le croissance des temps d'attente avec la congestion des pelles et du concasseur et réalisant un équilibre entre la quantité et la qualité. Le plan d'extraction est déterminé en résolvant un problème de réseau à coûts non-linéaires. Un exemple de réseau est présenté à la figure 3. Les variables sont le nombre de camions chargés à chaque pelle et voyageant sur chaque route pendant la période considérée. Les seules contraintes explicites sont les équations de conservation de flot; le nombre de camions arrivant à un point est égal au nombre en partant. Tous les autres facteurs sont considérés dans l'objectif à minimiser.

Un premier groupe de termes de la fonction économique assure la maximisation de la production en mesurant le carré de l'écart entre la capacité de chaque pelle et le nombre de camions chargés. Un second groupe de termes représente les objectifs de qualité. Par exemple, pour approcher la teneur moyenne désirée, nous utilisons la fonction de pénalité présentée à la figure 4.

La branche de droite mesure la valeur du fer perdu dans les rejets du concentrateur quand la teneur du minerai dépasse la valeur pour laquelle il est

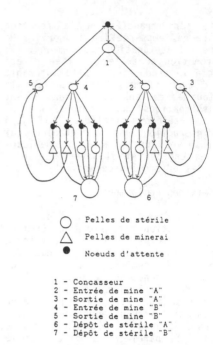

Pelles de stérile ○
Pelles de minerai △
Noeuds d'attente ●

1 - Concasseur
2 - Entrée de mine "A"
3 - Sortie de mine "A"
4 - Entrée de mine "B"
5 - Sortie de mine "B"
6 - Dépôt de stérile "A"
7 - Dépôt de stérile "B"

Figure 3: Structure du réseau de la mine du Mont-Wright.

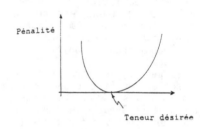

Figure 4: Pénalité associée aux écarts de teneur.

ajusté. La branche de gauche mesure la perte de qualité du concentré et la réduction du prix de vente quand la teneur du minerai décroît. De même, les autres objectifs de qualité sont assurés par des fonctions de pénalité mesurant le coût de s'écarter des valeurs cibles. Ainsi le modèle réalise l'équilibre économique entre le tonnage et la qualité. On diminue le tonnage pour augmenter la qualité seulement si c'est rentable.

Le dernier terme de la fonction économique assure que le nombre de camions utilisés correspond au nombre disponible. Ce terme est égal à K fois le carré de la différence entre le temps de camions disponible et le temps de camions utilisé. Le paramètre K est ajusté pendant la résolution pour obtenir la précision désirée sur le nombre de camions. Le temps de camions disponible est le nombre de camions fois la durée de la période de planification. Le temps utilisé comprend les temps de chargement et déchargement, les temps de déplacement et les temps d'attente. Ces temps sont estimés à partir des variables de décision: nombre de camions chargés à chaque pelle et nombre voyageant sur chaque route. Les temps de chargement sont calculés comme la somme du nombre de voyages à chaque pelle fois la durée moyenne du chargement. Les temps de déchargement sont la somme du nombre de voyages destinés à chaque point fois le temps moyen de déchargement. Les temps de déplacement sont calculés comme la somme du nombre de passages sur chaque route fois le temps moyen de déplacement. Les temps d'attente en chaque point sont donnés par une fonction non-linéaire croissante du nombre de camions servis. Cette fonction a été développée en utilisant la théorie des files d'attente.

Ce problème de réseau à coût non-linéaire se résoud assez facilement. Pour une mine avec 15 pelles et 5 points de déchargement le réseau comprend 40 noeuds et moins de 100 arcs. Un réseau de cette taille demande moins de 4K bytes de mémoire et se traite même sur un petit ordinateur. Nous avons développé pour sa résolution une implantation de l'algorithme du gradient réduit spécialisé pour ce type de problème. Notre algorithme réalise l'ajustement des différents paramètres pendant la résolution pour obtenir une solution ayant la précision désirée. Une solution donnant les nombres optimaux de voyages sur chaque routes a 3 chiffres significatifs est obtenu en 10 secondes sur PDP-11. Ce modèle à coût non-linéaire a l'avantage de produire des plans d'opération beaucoup plus stables que les modèles linéaires par rapport aux modifications des paramètres. La figure 5 présente un exemple illustrant cette qualité du module; les nombres de voyages par pelle varient très régulièrement en fonction du nombre total de camions utilisés pour calculer le plan. Cette qualité facilite la tâche d'un opérateur qui ajuste les paramètres de l'exploitation en vue d'obtenir un plan permettant d'atteindre certains objectifs de production.

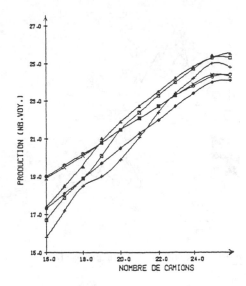

Figure 5: Nombre de voyages par pelle en fonction du nombre total de camions.

2.3 Contrôle en temps-réel

La dernière étape du processus de décision détermine en temps-réel l'affectation des camions. Après chaque déchargement, le système communique au camion son prochain point de chargement. Après son chargement, le camion se rend au point de déchargement qui est déjà fixé pour chaque pelle. Les décisions d'affectation sont prises de façon à réaliser au mieux le plan d'opération et ainsi obtenir les meilleures performances pour les objectifs de quantité et de qualité.

Le système de contrôle assure d'abord une fonction de communication. Cette fonction comprend la localisation des camions, l'évaluation des temps de parcours et de chargement et l'estimation des heures où chaque camion et chaque pelle deviendront disponibles; elle assure aussi la communication des assignations aux camions. La figure 6 illustre le matériel utilisé.

D'un côté, il y a des bornes émettrices radio 1 placées à des endroits stratégiques dans la mine (intersection de routes, pelles, entrée des haldes, etc.). Chaque borne émet un signal unique qui est enregistré dans la boîte noire du camion 2 lorsque celui-ci passe près de la borne. La transmission radio 3 permet au camion d'informer l'ordinateur 4 du contenu de la boîte noire.

397

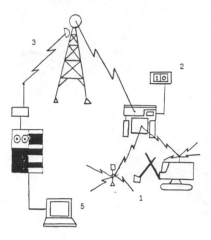

Figure 6: Le système de communication.

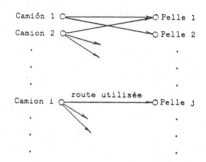

Figure 7: L'affectation des camions.

Cette information sera traitée dans l'ordinateur pour établir la position des camions et enfin afficher toute l'information nécessaire au terminal du contrôleur 5.

Puisque l'ordinateur est en mesure de connaître où sont les camions il sait aussi à quel moment ils requièrent une nouvelle assignation. L'ordinateur 4 peut décider de ces assignations et les transmettre au camion via la communication radio 3 pour afficher sur un écran numérique 2 le code associé à la nouvelle destination du camion.

La méthode de décision développée pour cette étape considère, en plus du camion à assigner, les prochains camions qui seront disponibles. Elle réalise alors l'affectation optimale d'une dizaine de camions aux prochaines pelles qui seront disponibles. C'est une amélioration importante par rapport à Charlemagne,

actuellement en opération à la mine du Mont-Wright et plusieurs autres systèmes qui ne considèrent que le premier camion. Ces décisions à courte vue peuvent être assez loin de l'optimum. La figure 7 présente ce problème d'affectation traité par notre système.

L'arc du camion i vers la pelle j est présent si la route allant du point de déchargement du camion i vers la pelle j est utilisée dans le plan d'opération. Les arc représentent alors toutes les affectations possibles de chaque camion et toutes les provenances possibles des camions pour chaque pelle suivant le plan. Le problème est alors de choisir un et un seul arc pour chaque camion et, au plus, un arc arrivant à chaque pelle.

Les coûts à minimiser lors de cette affectation mesurent les écarts par rapport au plan. Plusieurs mesures peuvent être définies. White et Olson (1983) proposent un coût basé sur l'écart entre l'estimé de l'heure d'arrivée des camions à la pelle et l'heure à laquelle il est attendu suivant le plan. L'heure attendue est donnée par les fréquences prévues par le plan en considérant des camions également espacés. Cette méthode à l'inconvénient de ne pas profiter des chargements plus rapides pour envoyer plus de camions. D'autre part, elle continue à envoyer les camions à la même fréquence même si le temps de chargement s'allonge et que les camions attendent. Une telle mesure produira des problèmes de fonctionnement pour les pelles très occupées présentant une variation importante du temps de chargement. Nous proposons donc une mesure différente qui évite ce problème. La figure 8 présente l'idée de cette mesure.

La mesure proposée s'évalue de la manière suivante: on calcule les probabilités de présence à la pelle du camion à assigner et du dernier camion déjà assigné, ensuite on estime l'espérance de l'attente du camion et de la pelle pour l'assignation considérée. Notre mesure est alors donnée par l'écart entre ces attentes et celles prévues par le plan. Avec cette mesure, les affectations ont tendance à reproduire les mêmes surfaces hachurées entre les camions indépendemment de la longueur des chargements. Ainsi, quand les temps de chargement augmentent, les camions sont plus espacés et plus rapprochés quand les temps de chargement diminuent. En moyenne, les camions arrivent à la fréquence prévue par le plan mais, en plus, leur espacement est synchronisé

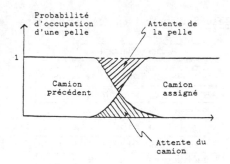

Figure 8: Évaluation des probabilités d'attente pour une affectation.

Tableau 1: Plan d'opération vs Mont-Wright.

	PRODUCTION (#Voy.)			ATTENTE CAMIONS (Min.)	ATTENTE PELLES (Min.)	TENEUR DU MÉLANGE
	Minerai	Stérile	Tot.			
PLAN	65.4	64.7	130.1	73.3	274.1	.3107
MONT-WRIGHT	65.0	51.0	116.0	267.5	338.7	.317

avec les fluctuations de la vitesse des pelles.

Le problème d'affectation d'une dizaine de camions est résolu de façon optimale en moins d'un quart de seconde sur PDP-11. Ce problème est résolu à chaque fois qu'un camion est disponible pour affectation. Le problème d'affectation est résolu à l'intérieur de la procédure suivante selon une méthode d'horizon fuyant.

Procédure d'affectation des camions:
1- Interrogation des camions et estimation des heures où chaque camion et chaque pelle seront disponibles;
2- Résolution du problème d'affectation pour une dizaine de camions;
3- Communiquer l'affectation au(x) camion(s) disponible(s) pour affectation, oublier les affectations des camions encore occupés, retourner en 1.

3. EXPÉRIMENTATION

Nous verrons par la suite que l'estimé de la production donné par le modèle de planification peut être réalisé en utilisant le modèle de contrôle en temps-réel. Premièrement nous évaluons le gain potentiel de production à la mine de Mont-Wright en utilisant les nouveaux modèles de prise de décision. Le test est réalisé en comparant la production prévue par le modèle de planification avec la production obtenue en utilisant Charlemagne.

En considérant les rapports d'opération de la mine, nous avons choisi une période passée de 2h:30 où il n'y avait pas de fluctuation du nombre de pelles ni du nombre de camions et où les teneurs en fer à chacune des pelles sont demeurées constantes. En utilisant les moyennes pour cette période des temps de chargements, déchargements et de parcours un plan d'opération a été calculé et les résultats ont été comparés à la production observée à Mont-Wright. Le tableau 1 présente les nombres de voyages effectués par les camions, les temps d'attente des camions et des pelles et les teneurs de mélange obtenues avec les deux systèmes. Durant cette période, 6 pelles et 15 camions étaient utilisés et l'objectif de teneur était de 31.0% (±.5%).

Le plan d'opération donne 14 voyages de plus que la production effectuée à Mont-Wright pour la même période. Ce qui fait une différence de 11% du nombre de voyages. D'autre part, on remarque que la teneur du mélange produit à Mont-Wright était de 0,317 alors que celle du plan d'opération était beaucoup mieux avec 0,3107.

Il est évident que le gain réalisé sur la production est attribuable à la diminution du temps d'attente des camions. En effet, on remarque une diminution de 72,5% du temps d'attente des camions dans le calcul du plan d'opération par rapport à ce qui a été réalisé à Mont-Wright.

Il est difficile de comparer les résultats du contrôle en temps-réel avec la production obtenue par un autre système. On ne peut pas faire fonctionner en même temps deux systèmes de contrôle et comparer leurs performances car un seul système doit communiquer les affectations aux camions. De même, on ne peut pas les faire fonctionner un après l'autre dans les mêmes conditions; une fois que le minerai a été transporté, il n'est plus là pour évaluer le deuxième système.

Pour obtenir une situation permettant de comparer deux systèmes de contrôle, nous avons développer une simulation de la mine. Le simulateur génère aléatoirement les temps de parcours, de chargements et de déchargements des camions et met à jour la banque de données du système. Il remplace l'interrogation

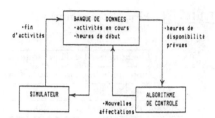

<u>Figure 9</u>: Contrôle des opérations dans une mine simulée.

des camions qui permet de connaître leurs heures de passages près des bornes émettrices dans la mine. Le simulateur génère des temps d'activité suivant des lois de Erlang avec des moyennes et des écarts-types correspondant aux valeurs observées. La figure 9 représente l'échange d'information entre l'algorithme de contrôle et le simulateur via la banque de données.

Une comparaison a été effectuée entre Charlemagne et le nouveau système de contrôle des opérations, en utilisant le simulateur pour reproduire un fonctionnement semblable à 15 heures d'opération. Cette durée se divise en 12 périodes avec des nombres de pelles et de camions et des temps moyens d'activité distincts. Un plan d'opération a été calculé pour chaque période et le simulateur a été utilisé avec les distributions appropriées. Les résultats sont présentés dans les tableaux 2 et 3.

On remarque d'abord une très bonne correspondance pour chaque période entre les nombres de voyages prévus par le plan et ceux réalisés par notre système de contrôle, autant pour le minerai que pour le stérile. La correspondance est encore plus forte au total. On peut donc affirmer:
1) Le contrôle des opérations suit le plan.
2) Les plans sont donc réalistes.

On remarque ensuite un gain de production de 3,16% par rapport à Charlemagne. La période de 15 heures avec 750 voyages est assez longue pour que ces résultats soient significatifs malgré la variance des durées d'activités simulées. On remarque encore une fois une bonne correspondance entre les résultats du plan et ceux réalisés par notre système de contrôle. On remarque de plus la réduction de l'attente des camions de 35% par rapport à Charlemagne.

Dix scénarios préposés par Québec-Cartier, avec des objectifs de qualité portant sur deux ou trois paramètres

<u>Tableau 2</u>: Comparaison des productions.

PERIODE	PLANIFICATION MIN.	STE.	SUR.	CONTROLE MIN.	STE.	SUR.	CHARLEMAGNE MIN.	STE.	SUR.
23H30-00H30	23.6	19.2	9.4	25	20	9	23	18	7
00H30-2H30	50.4	19.8	23.4	55	17	25	54	17	18
2H30-3H30	20.9	7.45	8.65	20	9	9	21	7	8
3H30-4H30	20.9	7.45	8.65	19	8	10	24	5	5
4H30-5H30	24.7	18.4	8.7	26	20	8	24	20	9
5H30-6H30	20.7	29.7	8.1	20	27	9	21	32	5
6H30-7H30	20.7	29.7	8.1	20	28	7	22	32	5
7H30-8H30	21.5	23.0	8.8	20	24	10	23	25	7
13H30-15H30	34.2	63.8	10.4	34	61	10	37	58	5
16H30-17H30	24.4	10.3	8.5	25	11	7	24	8	7
17H30-18H30	19.5	19.7	11.0	21	19	8	21	20	11
21H30-23H30	37.4	48.6	20.8	35	54	20	39	49	16
TOTAL	318.9	297.1	134.5	320	298	132	333	291	103
	750.5			750			727		

<u>Tableau 3</u>: Comparaisons des temps d'attente.

PERIODE	PLANIFICATION CAMIONS	PELLES	CONTROLE CAMIONS	PELLES	CHARLEMAGNE CAMIONS	PELLES
23h30-00h30	27.95	172.62	30.7	109.1	65.30	99.40
00h30-2h30	117.08	146.96	119.8	144.0	157.0	96.80
2h30-3h30	14.95	74.89	29.4	74.84	32.3	85.30
3h30-4h30	14.9	74.89	19.4	68.11	41.4	90.7
4h30-5h30	17.5	130.48	9.7	141.1	15.0	137.3
5h30-6h30	29.4	104.8	47.7	107.5	52.3	95.8
6h30-7h30	29.4	97.4	29.9	109.8	54.3	97.4
7h30-8h30	33.08	77.17	42.1	79.8	43.2	73.0
13h30-15h30	35.88	279.2	20.1	272.7	30.0	302.7
16h30-17h30	12.59	107.4	34.9	100.2	28.2	89.5
17h30-18h30	46.4	51.54	38.5	64.3	60.4	68.8
21h30-23h30	96.8	116.8	96.8	116.9	125.8	103.3
TOTAL	475.93	1434.2	521.0	1388.35	705.2	1340.0

ont été études en simulation. Les résultats ont démontré une fois de plus que le système de contrôle en temps-réel suit le plan de très près à la fois pour les nombres de voyages par pelle et la qualité du mélange. Nous avons obtenus un gain de production de 2.0% par rapport au système Charlemagne.

Notons que le gain pour le plan sans contrôle de qualité pour ces scénarios était de 3.2% par rapport au système Charlemagne. Ceci démontre que la perte de production pour atteindre l'objectif de qualité est faible quand les types de minerai disponibles sont compatibles avec le mélange désiré. Le module de choix d'équipements assure cette compatibilité et permet d'obtenir de bons résultats à la fois pour la qualité et la quantité.

Le tableau 4 présente à titre d'exemple deux des dix scénarios. Pour le premier scénario le teneur en fer désiré était de 31.5% (±0.5%). Quatre pelles étaient disponibles avec des teneurs en fer de 27.0%, 29.0%, 34.0% et 40.0% pour une moyenne de 32.5%. Le second scénario a le même objectif de teneur en magnétite inférieure à 0.4%. Quatre pelles étaient disponibles avec des teneurs en fer du 27.0%, 30.0%, 32.0% et 40.0% pour une moyenne de 32.3 et 40.0% et des teneurs en magnétite de 1.0%, 0.4%, 0.1% et 0.7% pour une moyenne de 0.55%.

3. CONCLUSION

Le système proposé permet d'obtenir de très bons résultats sur la quantité et la qualité du minerai extrait et ce, dans des temps de calcul très faibles grâce à l'utilisation d'algorithmes d'optimisation très efficaces. En plus des gains au niveau de l'opération illustrés dans ce texte, ce système permet des gains au niveau de la planification grâce à nos deux premiers modules. Les modules de choix des équipements et de plan d'opération permettent de réaliser des conditions où il est possible d'atteindre en même temps les objectifs de qualité et de quantité. En effet, plutôt que de constater en fin de journée que les conditions étaient malheureusement peu favorables et que les objectifs n'ont pas été atteints, ces modules permettent d'estimer la production prévue et de changer les conditions s'il n'est pas possible de rencontrer les objectifs. De plus, ces modules permettent d'ajuster à la hausse les objectifs si les essais montrent qu'il est possible de faire mieux.

4 BIBLIOGRAPHIE

Charlemagne, 1979, "Manuel de répartition des camions", La compagnie minière Québec Cartier, 1979.

Dallaire, R., Laplante, A.R. and Elbrond, J., 1978, "Humphrey's spiral tolerance to feed variations". The Bulletin of the canadian Institute of Mining and Metallurgy 71, 796, 128-134.

Clevenger, J.G., 1983, "Dispatch Reduces Mining Equipment Requirements", Mining Engineering, Sept.

Crosson, C.C., Tonkino, M.J.H. and Moffat, W.G., 1977, "Palabora System of Truck Control" Mining Magazine, February.

McInnis, D., 1985, "Planification du transport du minerai par programmation non-linéaire", mémoire de maîtrise, Département d'informatique et de recherche opérationnelle, Université de Montréal, Decembre.

Mueller, E.R., 1977, "Simplified Dispatching Board Boosts Truck Productivity at Cyprus Pima", Mining Engineering, August.

Orisek, I., and Diantonio, K.K., 1986, "Productivity Improvment with Automatic Truck Dispatching, a Case Study", preprint, Society of Mining Engineering of AIME, March.

White, J.W., Arnold, M.J. and Clevenger, J.G., 1981, "Automated Open Pit Mine Dispatching at Tyrone", Preprint, Mining Information Systems Technology Exchange Seminar, November.

White, J.W. and Olson, J.P., 1983, "Recent Developments in Computer Based Dispatching", preprint, Society of Mining Engineering of AIME, October.

Young, C.K., Rakshit, A. and Kolb, W.E., 1984, "Automatic Truck Dispatching Using a Micro-Computer", preprint, Society of Mining Engineering of AIME, October.

Computer Applications in the Mineral Industry, Fytas, Collins & Singhal (eds)
© 1988 Balkema, Rotterdam. ISBN 90 6191 760 3

A combined approach to solve truck dispatching problems

Eduardo Bonates & Yves Lizotte
Department of Mining and Metallurgical Engineering, McGill University, Montreal, Quebec, Canada

Abstract: Truck dispatching systems offer the potential for significant improvements in productivity and these systems will be extensively used in the future by mine operators as a means of optimizing equipment utilization.

This paper describes a model designed to study the feasibility of continuously dispatching trucks in open-pit mines based on different operating procedures. The system developed is semi-automated and is composed of two stages. The first stage deals with a mathematical programming module that can indicate average desirable shovel coverage while the second stage is concerned with the simulation of the system to assess operating dispatching policies.

The mining industry is frequently faced with an undercapacity of the operating shovels. Literature concerning the use of mathematical programming to solve the scheduling problem, taking into account grade requirement, is abundant. However, these techniques have been proved inefficient to solve the entire dispatching problem in order to allow the dispatcher to take the necessary decision at the right time given the computing time involved. Computer simulation models are designed to study the performance of mining operations and they can be conveniently structured to assess sets of rules relating the interactions between the components of the system.

The proposed system performs an active role by suggesting to the dispatcher the optimal truck allocation based on heuristic techniques, usually used to obtain solutions at the right time.

1 INTRODUCTION

The advances in computer technology and the associated increasing demand for mineral resources has created many problems for mine management. Fortunately, many problems in the mining environment can be solved by **Operations Research** techniques. These techniques, particularly simulation and linear programming, have been used extensively by the mining industry due to their inherent effectivenesss in the process of decision-making. Usually the models developed are complex requiring analytical solutions, which can be performed fast and accurately by computers. Computing costs have greatly decreased in recent years. Inexpensive hardware has also been developed in the last five years which can withstand the inclement mining environment while possessing autonomy in terms of processing capabilities.

The introduction of computers in the mining industry was very slow. The main reasons being the high capital cost for computer installations, the lack of adequate training of mining engineers to manipulate the computer efficiently, and rapid obsolescence. However, this has rapidly changed and today most mining engineers have sufficient knowledge of computer languages and systems. Computers are now cheaper, smaller, more reliable and easier to program; most mining companies have now realized the potential benefits of their use.

Mineral markets are now very competitive. Mine management must promote the improvement of operating techniques. One of the key requirements in such developments is the need for speed in decision-making to keep up with the growth in the face of stiff competition.

Computer applications in the mineral industry are principally for geological modelling, ore reserve estimation, pit design and mathematical programming for production scheduling. Computer applications for monitoring and controlling manufacturing type processes in the industry in general are increasing steadily. The special nature of the mining industry, i.e, environmental, technical and economical, has delayed the entry of computers for controlling mining equipment. In the last 5 years, however, equipment manufacturers and several mining companies have funded projects for the development of computerized systems for mining equipment. A great deal of attention is being given to equipment and procedures to optimize productivity with several mines worldwide using computerized truck dispatching systems to increase productivity and reduce costs. Fully automated computerized truck dispatching systems are commercially available but too expensive for small and medium size open-pit operations.

The search for ways to raise the efficiency of mining operations is based on the truck dispatch method and the analysis of the operational data. Traditional manual methods of truck dispatching generally only deal with the mechanical availability of the shovels and the truck fleet. Unexpected changes in the ore grade and specific production objectives also require truck reallocations. Data gathering through shift reports produced by the operators is restricted in volume and is susceptible to human error. This can now be done be done very efficiently through computerized truck dispatching systems. It is highly likely that most open-pit mining operations will use one form or another of truck dispatching in the near future.

2 TRUCK/SHOVEL SYSTEM

The truck/shovel system is well accepted by the mining industry because of the large operational flexibility it offers. Although mining trucks represent the most widely used method to transport the material from the pit, the cost of materials handling can exceed 50 percent of the total operating costs. This can also be affected by spiraling energy costs, increasing haulage distance from large and deeper pits and the need for higher material ton-

nage outputs. Since the first mining trucks were built, they have grown to a stage where trucks with a wide range of capacities up to 350 tons are in use. This is almost 20 times the size of the earlier trucks. During the 50's and 60's developments in truck technology and increased capacities allowed several low-grade, low-profit ore deposits to be brought into production using mobile and versatile equipment. However, it is unlikely that the scale of trucks or any other surface mining equipment can continue to grow at the same rate as desired productivity increases in the near future. To improve productivity, surface mines have relied on large machines. This trend, however, has decreased according to most equipment manufacturers. Mine operators are now turning to more sophisticated mine planning techniques to get the most out of equipment on hand.

The planning of a materials handling system for removing the material from the pit is not an easy task. Although truck/ shovel systems are complex to organize and not the most cost effective system in all cases, they are still presently the most widely used in open-pit mine operations worldwide. Truck fleet productivity has the lowest improvement rate among the three major unit operations: drilling, blasting and hauling. Thus, techniques for improving truck/shovel system efficiency need to be developed for continued improvement in productivity.

3 SIMULATION

The use of simulation models is one of the most powerful and versatile method for predicting the performance of mining operations. Its use has become routine as computers become more powerful and inexpensive. It can also handle the models virtually at the same time as the data is being collected. This is important to management due to the need for fast and accurate results. Also the real system can usually best be imitated by a simulation model.

Benefits from simulation use are potentially greatest when the real operation cannot be studied directly because it does not already exist, or it is not available, or to work with it directly is very expensive. Simulation can be used to test several alternatives before actual decisions are implemented. Simulation can be defined as the process of designing a

404

model of a real system and conducting experiments with this model for the purpose of either understanding its behavior or evaluating various strategies for its operation. The key components must be identified and the interactions quantified. The model can then be constructed in order to make the final recommendations.

Simulation models for truck/shovel systems are deterministic or stochastic. Deterministic simulation is used when the components of the system are represented by a discrete value. In this case, the results provided by the model also give a discrete value as long as the values of the components remain the same. Stochastic or Monte Carlo simulation must be applied when the components of the system can be predicted by their probability of occurrence in each event. In this case the results obtained from the model must be expressed as a distribution of possible outcomes because of the stochastic nature of each component. It is a technique for selecting discrete values randomly from probability distributions for use in each execution of a simulation program. The objective is to reproduce as realistically as possible the variability of the random variables and the interrelationships between components in the system being studied.

In dealing with a stochastic simulation it is necessary to obtain the frequency distributions for each system component. Time study data, interpretation of results and statistical analysis are required. After the probability distribution for each event is determined, then random numbers between zero and one are generated. In this way a series of several random variables representing each component are obtained each time from a series of random numbers generated. These random variables when used in the specified interaction between the different components of the system will produce different outcomes for each simulation program execution.

Simulation programs can be used to produce a great deal of information regarding the expected operation. Simulation results can indicate trends and are useful for predictions to help improve and optimize mining operations.

4 LINEAR PROGRAMMING

Among the several mathematical techniques available in Operations Research, Linear Programming (LP) is one of the most widely used. It is a powerful method for ensuring the efficient distribution of limited resources between competing activities with the final objective of optimizing the operations.

The common denominators that determine if linear programming will help arrive at a decision are the presence of several available alternatives which the mining engineer must choose from, and the presence of limiting factors that prevent him from choosing all alternatives simultaneously. The requirements of a decision to be made between alternative solutions, coupled with the presence of limiting factors is common to several mining problems and serve to explain the increasing use of linear programming. LP problems are always characterized by a large number of possible solutions which satisfy the conditions of the problem. The best solution depends on the aim or overall objective that is implied in the statement of the problem.

Mining engineers generally work within the confines of limited resources to attempt to take decisions. Many of these decisions are based on personal judgement or, if available, the experience of other operations and thus, the results from prior decisions greatly affect subsequent decisions. As the choices to be made become more complicated, engineers have tended to study methods to aid the decision process and have found LP to be useful, especially in solving problems related to production scheduling and grade control. One of the major problems in determining the optimum production schedules is ensuring that the required production standards can be achieved not only during the scheduled period but also in the future. This is because of the variability of the material and the difficulty in estimating and predicting variability in the future. The major difficulties in using this technique is in creating a valid model, and solving large scale problems (variables and equations) when confined by hardware and/or time. The formulation may have to be less elaborate in order to be feasible. A truck/shovel operation is a dynamic process, which is stochastic in nature. In this case, full optimization of the decision is not possible without previous knowledge of all future events. In general, mathematical models do not describe a realistic situation with complete accuracy and are only an approximation of the true situation. It must be pointed out that the LP algorithm does not solve the entire dispatching problem. Since the system is not really in steady-state conditions, the solution must be adjusted using heuristic rules in an attempt to reinforce it. These heuristic rules do not optimize the decision, but must produce acceptable results for the majority of the situations envisaged.

5 A TRUCK DISPATCHING SYSTEM

The purpose of this project is to develop a semi-automated computerized truck dispatching system for open-pit mines. It will assist mine operators in the dynamic assignment of trucks. The computer is programmed to display information recorded earlier and/or suggest the most appropriate truck allocation to meet a specified production objective. The system is semi-automated in the sense that the computer is not in direct contact with the operators, that is, the intervention of the dispatcher is necessary to communicate all assignments. The dispatcher correlates this information with the actual position of the equipment in the pit and takes an independent decision which may or may not be in agreement with the computer generated assignment. The dispatcher and the equipment operators are linked by two-way radio communications.

5.1 Functional Structure

The general structure of the system is shown in figure 1. The program is divided into three different components with different attributes: dynamic, operational and analytical. The first two components and the simulation programs used in the analytical component are nearly completed.

The dynamic component is used primarily for setting up the screen display and perform input/output operations. A simplified version of the main screen display which appears throughout the shift is illustrated in figure 2. The display uses colour capabilities of the computer to add information. The status of all equipment is displayed in the screen allowing the dispatcher to visually estimate the position of each unit. The equipment is represented on the screen by a number inside a colour coded block, such as; red for ore, blue for waste, yellow for overburden, white for trucks travelling empty and black for any additional equipment. The blocks move

on the screen which is updated approximately every minute. Blocks move at various speeds up and down the screen proportionately to cycle time durations which are constantly updated with current information. The dispatcher must be familiar with the current operation to correlate the position of the trucks on the screen with their actual position in the pit.

The operational component generates the usual daily reports, such as; general summary of truck and shovel productions, operating hours of equipment, sequence of trucks loaded by each shovel, scheduled and un-scheduled breaks, oil, fuel and water reports, etc... and eventually statistical analysis of equipment performance.

The full development of the dynamic and operational components is primarily based on the generation of the dispatching system presently in use at LAB Chrysotile's Black Lake operations, Quebec.

5.2 Linear Programming Model

The analytical component functions in two stages. In the first stage a mathematical programming model considers long-term production objectives. The second stage deals with the stochastic simulation program for testing and improving the performance of specific operating dispatching policies. The individual shovel production levels are obtained with the LP model and the simulation program assesses the usefulness of the dispatching policies for achieving the desired production objectives.

A variety of different objectives could be associated to short to long-term production plans, such as unconstrained productivity maximization, constrained cost minimization, profit/ore grade objectives, ore grade or hardness constrained productivity maximization, maximization of total shovel utilization, etc.. Each mining operation considers one or several of these objectives, implicity or explicity, to guide the execution of production plans. Based on prior work with dispatching systems, association with an operating mine and, primordially, the fact that the LP would be used in a semi-automated system (not on-line decision-making), the **first** LP formulation is based on the objective of constrained productivity maximization.

Three specific problems often limit the applicability of LP in mining production scheduling:

a) optimization of a linear objective function with linear constraints for real problems not purely linear,

b) variability and uncertainty associated to model input parameters,

c) non-stationarity of the real problem.

Because the LP will be used on a regular basis (once a shift) and the results are used to guide dynamic assignment policies, it is believed that LP use will be helpful for this restricted application.

The first LP formulation associated to the production planning objective is as follows:

$$\text{Maximize } Z = \sum_{i=1}^{n} X_i + \sum_{j=1}^{m} X_j \quad (1)$$

Subject to:

$$\sum_{i=1}^{n} X_i \text{ .LE. } CC \quad (2)$$

$$\sum_{i=1}^{n} (G_i - G) X_i \text{ .GE. } 0 \quad (3)$$

$$X_k \text{ .LE. } MAXR_k \quad k = 1,\ldots,n+m \quad (4)$$

$$X_k \text{ .GE. } MINR_k \quad k = 1,\ldots,n+m \quad (5)$$

$$\sum_{i=1}^{n} (X_k / A_k) \text{ .LE. } TT \quad (6)$$

$$X_k \text{ .GE. } 0 \quad k = 1,\ldots, n+m \quad (7)$$

where:

i = shovel in ore, index
j = shovel in waste, index
n = total number of shovels in ore
m = total number of shovels in waste
X_i = production of shovel i
    (tons of ore/production period)

X_j = production of shovel j
CC = crusher capacity
    (tons of ore/production period)
G_i = ore grade, shovel i
G = ore grade objective for production period
k = general shovel index
$MAXR_k$ = maximum possible (or required) shovel k production
    (tons/production period)
$MINR_k$ = minimum required shovel k production
    (tons/production period)
A_k = unit increase in shovel k production for a truck allocated to the shovel k for the production period (tons/trucks)
TT = total number of trucks available during the production period

A number of possible refinements to the basic model are apparent. A relative priority (P) could be associated to ore removal, the objective function then being:

$$\text{Maximize } Z = P \sum_{i=1}^{n} X_i + \sum_{j=1}^{m} X_j \quad (1')$$

Conversely a waste/ore ratio (R) requirement could be added:

$$R \sum_{i=1}^{n} X_i - \sum_{j=1}^{m} X_j = 0 \quad (8)$$

Limits to dump points can be considered by adding constraints similar to equation (2).

A major concession is made to model the problem as a LP. This is the assumption that shovel production increases linearly with the number of trucks. Figure 3 illustrates how the A_k values can be determined for each shovel. A greater precision can be obtained in the second application of the LP, where $MINR_k$ values are specified and b_k (figure 3) such that:

$$\sum_{k=1}^{n+m} X_k/b_k \text{ .LE. TT - MINTT} \qquad (6')$$

where b_k is the unit increase in shovel production for the portion of the shovel production function considered and MINTT is the total number of trucks already allocated to satisfy $MINR_k$'s specified.

The LP solution indicates the specific production level of shovels which must be sought in order to maximize production. Shovel coverage in terms of average trucks assigned during the production period can be derived by refering back to figure 3. The fact that the solution does not correspond to integer number of trucks to shovels does not matter at this point since the actual assignments will be done dynamically with a heuristic dispatching procedure.

5.3 Simulation program

The simulation program was designed to assess four dispatching policies; locked-in, maximize truck use, maximize shovel use and match factor. Most of the dispatching systems described in published literature attempt to maximize mine production with available equipment, either by trying to minimize truck waiting time or trying to minimize shovel idle time. Recently the concept of match factor was also introduced as a dispatching policy. The underlying objective of any dispatching system is to maximize production fleet efficiency.

Stochastic simulation was selected as the most appropriate technique due to the variability of the interdependent components of truck/shovel operation. This variability appears to be the major cause of lost potential production.

To elaborate a computer simulation model it is necessary first of all to establish the objectives of the study which will dictate the precision required. In the execution of a dispatching model the most important section is the simulation of the system. When simulating a truck/shovel operation it is essential that the method used be precise and reliable. Modelling of truck movement in conjunction with shovel productivity is the most critical aspect of the simulation. The entire decision-making process is influenced by the expected equipment performance.

The stochastic simulation program attempts to simulate the actual operation as realistically as possible. Reasonable, actual restrictions have been imposed whenever possible. The program is written in FORTRAN 77 and is based on an advance clock approach rather than direct event generation. In programming a model using this approach, the concept of a real clock plays a vital role because the events in a truck/shovel system occur at different points in the simulated time. The sequence of events is established by the advance of the clock. One of the major concerns related to the clock is the time increment. The guiding principle being that none of the events should be missed, but also that judicious use is made of computer time. This approach results in greater computation accuracy and increased computational efficiency. A precise comprehension of this technique makes the programming process very straightforward.

The first step was to design a simulation program to estimate fleet productivity using fixed dispatching. Figure 4 is a schematic representation of the basic program structure.

The general program consists of a main control program with five subroutines: CHECK, LOAD, DUMP, EMPTY and FULL. Besides these subroutines, a function RANDOM is used to generate random numbers. The first subroutine checks the status of the trucks at different points in the haulage network. The others correspond to events that change the appropriate characteristics. They are used to generate the respective event times for loading, dumping, travelling empty and travelling loaded. Because of the complex nature of the truck/shovel operation, several types of probability distributions can be used to fit the collected data. For modelling the loading event either log-normal or split normal distributions could be used because this is basically a repetitive process. However, the Weibull distribution was selected due to the wide range of experimental data it can fit. It is possible to ensure that this distribution does not yield values below some minimum imposed by the distribution parameters because of the physical limitations of the equipment. The dump event was modelled considering the same characteristics as those used for the load event.

The normal distribution was selected for travel times, since it appears to be most commonly used. In fact, the normal distribution models most physical phenomena as well as the distributions generated by physical measurements.

An important aspect of the simulation of a truck/shovel operation is to obtain a detailed layout of the mine. Figure 5 illustrates how the main haulage network was translated into a form acceptable to the computer. This network was also used to determine the haulage segments for each assignment in order to estimate particular travel time components with a deterministic simulator. Because the program does not consider any interference of trucks at intersection points, the network was further reduced to an array of travel times between the loading and dumping points. Figure 6 shows a flow diagram of the deterministic simulation program.

In every mine basic differences do exist, by way of truck fleet and shovel size and type, number of dumps, types of haulage network, etc... Several mines operate with more than one type of equipment. Moreover, subtle differences in the operational policies make the task of structuring a general mine simulator almost impossible.

The parameters used to develop the program were gathered according to the conditions prevailing at the mine site. A reasonable effort has been made to simplify the model by making appropriate assumptions. However, the results obtained from the simulation model can be extrapolated to suit other situations. As all mines are unique, definition of the parameters for them must be customized for specific needs.

A major point of interest in truck/shovel operations is the wide range of fleet productivity which may be achieved. This variability is due to the inherent variability of the human operator's reactions, changing conditions during normal shift operations and, particularly, due to the use of mixed equipment fleet. In an attempt to assess this variability the program allows as many different equipment types as needed to be specified.

5.4 LP - Simulation Analysis Procedure

A critical aspect of the combined approach to truck dispatching is establishing the correct link between the LP results and the heuristic dispatching procedures. Linear programming assumes system stationarity while dispatching involves dynamic decision-making in a system constantly changing. The methodology of conciliation (coordination) of these two elements proposed herein constitutes the novelty of this combined approach.

Figure 7 schematizes the procedure for linking LP results to improving and testing of the heuristic dispatching rules considered for the first LP model. Two heuristic dispatching rules are appropriate to be associated to a production maximization LP model: an adjusted maximize truck rule (AMTR) and a heuristic dynamic priority rule (HDPR). The choice of these rules is partly based on heuristic dispatching rules described by other authors but involves a formal method of testing and improving the empirical parameters selected. The maximize truck rule is applied in the AMTR with an added correction factor, continuously updated during the production period, to consider objectives and their relative achievement at any point in the simulated time. Similarly the HDPR requires constantly updating the relative priority number which considers individual shovel production objectives, current truck coverage and expected truck waiting. The stochastic simulation program is used to test the dispatching procedures with the selected parameters; statistical analysis is done to measure the deviations achieved from the planned objectives, enabling heuristic adjustments to the AMTR and HDPR parameters.

It is impossible to guarantee the validity of the parameters determined for all possible truck/shovel configurations. A variety of configurations will be tested, however, to achieve guidelines for mine management of any mining operation, to determine appropriate dispatching procedure parameters. The restricted objective of the semi-automated active dispatching system is to **suggest** to dispatchers appropriate truck assignments.

6 CONCLUSIONS

The structure of the simulation program to test automated truck dispatching should reflect the complexity of the operation. However, any unnecessary and superfluous details should be avoided to make the program efficient and not too lengthy. This will ensure rapid program execution and facilitate analysis.

Semi-automated and automated dispatching systems require a procedure for logi-cal assignments of trucks, be it for sug-gesting to the dispatcher or direct in-structions to the truck operator. Dis-patching policies will seek different ob-jectives, using varying degrees of sophis-tication. The policy selected is the most important component of a computerized truck dispatching system. There is a con-siderable range of complexity between the variety of dispatching procedures a mining operation can consider. Each situation is unique and no mining operation should con-ceive of implementing an off-the-shelf system without some adjustments to fulfill its particular needs.

The semi-automated dispatching system requires a dispatcher to take the final decisions, assisted by the computer. In fact, the dispatcher is in an ideal posi-tion, particularly when he/she can see a large part of the mining operation during the shift. A computer can be programmed to take the decisions according to a set of rules. Frequently it also takes a better immediate decision than the dispatcher. However, the dispatcher has a more comple-te control of the operation, besides his intuitive ability to predict unforeseen events that may occur and react according-ly. This is impossible to program into a computer. Furthermore, implementation of procedures to resolve any problems are easier to undertake with a dispatcher than by fully automated computer procedures for all possible situations.

The cost of development and implemen-tation is one of the major factors in the choice of an automated and semi-automated dispatching system. A semi-automated sys-tem is far cheaper. Digital signal commu-nications between equipment operators are not necessary and the system can be based on an enhanced micro-computer. The primary limitation of a semi-automated system is the ability of the dispatcher to handle many transactions in a short period of time. An active system is helpful in ra-pidly suggesting destinations but semi-au-tomated systems are limited to medium size open-pit operations.

Acknowledgments

The authors wish to acknowledge the UFPb (Universidade Federal da Paraiba), the CNPq (Conselho Nacional de Desenvolvi-mento Cientifico e Tecnologico), Brazil, and NSERC, Canada for their financial sup-port. Also, Mr. A. Leclerc, Chief Mine Engineer of LAB Chrysotile Inc. is grate-fully acknowledged.

References

Bonates, E., Analysis of Truck/Shovel Dis-patching Policies Using Computer Simula-tion, M.Sc. Thesis, McGill University, Dept. of Mining & Metallurgical Eng., August 1986, 209 pp, Canada.

Bonates, E., Lizotte, Y., A Computer Simu-lation Model to Evaluate the Effect of Dispatching, SME-AIME Annual Meeting, Phoenix, Arizona, January 1988, paper No 88-17.

Lizotte, Y., Bonates, E., Truck and Shovel Dispatching Rules Assessment Using Simu-lation, Mining Science & Technology, El-sevier Science Publishers, Amsterdam, Netherlands, June 1987, Vol. 5, No 1, pp. 45-58.

Lizotte, Y., Bonates, E., Leclerc, A., De-sign and Implementation of a Semi-Auto-mated Truck/Shovel Dispatching System, 20th APCOM, October 1987, South Africa.

Lizotte, Y., Scoble, M., Bonates, E., Ap-plication of Simulation to Assess Truck/ Shovel Dispatching Policies, Mining E-quipment Selection Symposium, The Uni-versity of Calgary & CANMET, Calgary, November 1985, Paper No. 24.

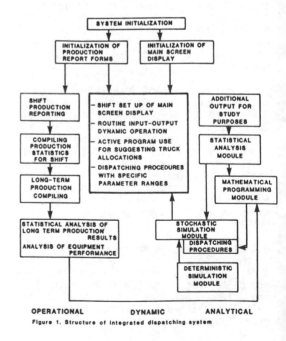

Figure 1. Structure of integrated dispatching system

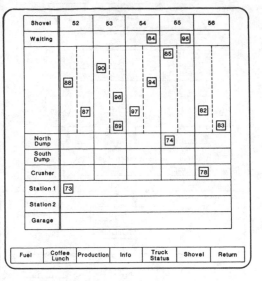

Shovel	52	53	54	55	56
Waiting				84	95
					85
			90		
	88			94	
			96		
		87		97	82
			89		83
North Dump				74	
South Dump					
Crusher					78
Station 1	73				
Station 2					
Garage					

Fuel	Coffee Lunch	Production	Info	Truck Status	Shovel	Return

Figure 2. Main Screen Display (simplified) on Dispatch System

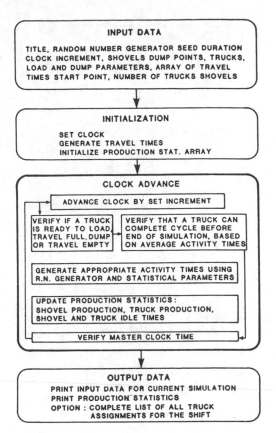

INPUT DATA

TITLE, RANDOM NUMBER GENERATOR SEED DURATION
CLOCK INCREMENT, SHOVELS DUMP POINTS, TRUCKS,
LOAD AND DUMP PARAMETERS, ARRAY OF TRAVEL
TIMES START POINT, NUMBER OF TRUCKS SHOVELS

INITIALIZATION

SET CLOCK
GENERATE TRAVEL TIMES
INITIALIZE PRODUCTION STAT. ARRAY

CLOCK ADVANCE

ADVANCE CLOCK BY SET INCREMENT

| VERIFY IF A TRUCK IS READY TO LOAD, TRAVEL FULL, DUMP OR TRAVEL EMPTY | VERIFY THAT A TRUCK CAN COMPLETE CYCLE BEFORE END OF SIMULATION, BASED ON AVERAGE ACTIVITY TIMES |

GENERATE APPROPRIATE ACTIVITY TIMES USING
R.N. GENERATOR AND STATISTICAL PARAMETERS

UPDATE PRODUCTION STATISTICS:
SHOVEL PRODUCTION, TRUCK PRODUCTION,
SHOVEL AND TRUCK IDLE TIMES

VERIFY MASTER CLOCK TIME

OUTPUT DATA

PRINT INPUT DATA FOR CURRENT SIMULATION
PRINT PRODUCTION STATISTICS
OPTION : COMPLETE LIST OF ALL TRUCK
ASSIGNMENTS FOR THE SHIFT

Figure 4. General structure of the simulation program

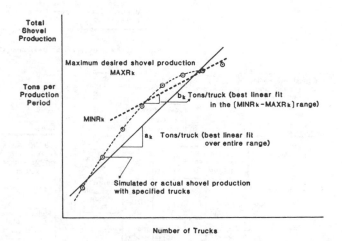

Figure 3. Assessment of A_k and b_k parameters for shovel k

411

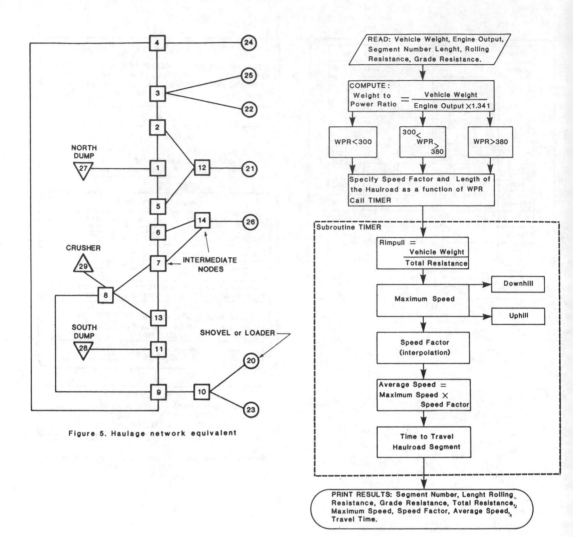

Figure 5. Haulage network equivalent

Figure 6. Flow diagram of deterministic simulation program

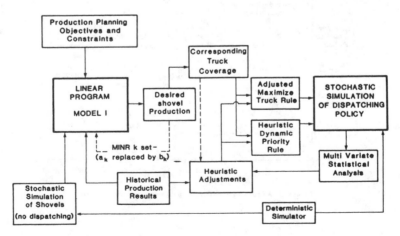

Figure 7. LP–Simulation links in the combined approach

Computer Applications in the Mineral Industry, Fytas, Collins & Singhal (eds)
© 1988 Balkema, Rotterdam. ISBN 90 6191 760 3

PROFITIS: A CAD simulator for automated guided LHD in underground mining

Nikos Vagenas
Division of Mining Equipment Engineering, Luleå University of Technology, Sweden

ABSTRACT: Automation of LHD machines has recently become an attractive possibility, due to, among other reasons, the rapid development and use of microprocessors.

This paper presents the basic features and principles of a computer program, called PROFITIS, which is used as a scientific tool for analyzing and evaluating the potential and the limitations of an automated guided LHD in mining operations.

Specifically, reasons for creating PROFITIS, modelling principles and a demonstration example of utilizing the program are presented and described.

The outcome of this paper will indicate that the utilization of automation and robotics in mining is strongly related to the computer science as well as to the development and use of appropriate software.

1 INTRODUCTION

In a worldwide view Load-Haul-Dump vehicles have continuously affected mining operations. Designed initially for fast tramming of ore from stopes to dumping points (ore passes), their use has extended to make them the workhorse of underground transport systems in many mines (Fig. 1).

Automation of these machines has become an attractive development step, but so far very little has been achieved.

An approach to the development of a LHD system with a high degree of automation can be as follows (Fig. 2):

* The tramming and the dumping are automated by a well known and proven navigation technique
* An operator, placed in the safe and comfortable environment of a control room, operates the filling of the bucket by remote control via a TV-camera system.

The reason for not automating the loading operation is mainly due to the uneven fragmentation in hard rock which demands the presence of operators.

A remote-controlled/automatic LHD system as previously described can offer the following advantages:

* Better utilization of working hours
* Less interruption of production at shift changes
* Better utilization of the vehicles

Fig. 1a. A diesel LHD.

Fig. 1b. An electric LHD.

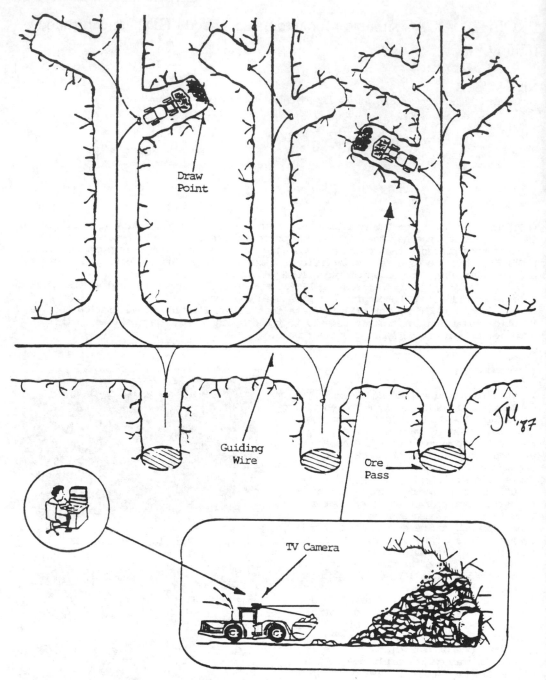

Figure 2. Concept of a Remote-Controlled/Automated LHD Loading System.

* Easier adoption to variations of
  production demands
* Smoother operation of the LHDs,
  resulting in lower maintenance costs
* One operator might be able to operate
  more than one unit simultaneously
* Improved safety
  In order to investigate, determine and
evaluate the potential for applying a
remote-controlled/automatic LHD, the
computer program PROFITIS has been
developed.
  PROFITIS is a graphical
animator/simulator for IBM-compatible
XTs/ATs with CGA or EGA graphics adapter.
The compiled version utilizes about 75 kb
of memory. The program's source code has
been written in TURBO PASCAL.

2 REASONS FOR CREATING PROFITIS

PROFITIS, which is a typical simulation
program, provides the tools for evaluating
various factors that play vital roles in
the successful operation of an Automated
Guided LHD (AGL) system.
  The main reasons for developing PROFITIS
are the following:
1. Simulation enables the study of and
   experimentation with the internal
   interactions in a system, or of a
   subsystem within an integrated system,
   without having to interfere with the
   actual process
2. Informational, organizational and
   environmental changes can be simulated
   for the AGL transport system and the
   effect of these alterations on the
   model's behavior can be observed
3. The knowledge gained in designing
   PROFITIS can be of great value toward
   suggesting improvements in the LHD
   system
4. By changing simulation inputs and
   observing the resulting outputs,
   valuable insight can be obtained into
   which variables are most important and
   how variables interact
5. PROFITIS can be used prior to
   implementation of the AGL system so as
   to prepare for what may happen
6. Once PROFITIS is developed, it can be
   used repeatedly to analyze proposed
   designs on the AGL system.

3 PROGRAM DESCRIPTION

The model structure is composed of five
major submodels:
* The Computer Aided Design (CAD) submodel
* The Route submodel
* The Loading points submodel
* The Ore-passes submodel
* The Vehicle submodel

3.1 CAD submodel

In the CAD submodel the user creates the
transport layout which is to be simulated.
  The transport layout is modelled as a
network. The nodes of the network are
points in the system, where two or more
segments meet. The segments are the
portion of the roads that the automated
guided LHD can follow between two nodes.
Curves are modelled as a sequence of
straight segments.
  Positions within the network are
identified using the cartesian-coordinate
system. For every node, a point location
is provided in terms of (x,y) coordinates.
The point location is useful for purposes
of measuring distances between points and
identifying node locations. All loading
points and dumping points are also
modelled as nodes.
  The program can be utilized for
transport layouts with a maximum number of
15 loading points, 8 ore passes and with a
total number of 99 nodes. Fig. 3
illustrates a layout designed by using the
CAD routines of PROFITIS.
  Rules that have been considered in
designing the CAD submodel are:
1. **Simplicity:** The graphics should be
   easy to use
2. **Consistency:** The package should
   operate in a consistent and predictable
   way to the user
3. **Completeness:** The set of graphics
   functions should be complete, with no
   inconvenient omissions
4. **Robustness:** The graphics system
   should be tolerant of minor instances
   of misuse by the operator
5. **Performance:** Graphics routines
   should be efficient and speed of
   response should be fast and
   consistent.
  The following is an outline of the two
dimensional graphics-package
specifications:
1. Menu driven: menu on the screen
2. Tasks supported
   a. Points (Method: Pointing to the
      location on the screen by means
      of cursor control).
   b. Lines (Method: Using two previously
      defined points).
   c. Uppercase letters (Method: Pointing
      to the location on the screen by
      means of cursor control).
3. Modes supported
   a. Add
   b. Delete

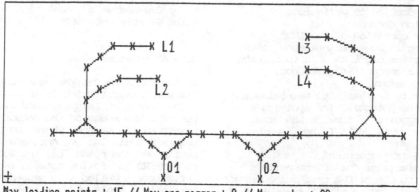

Max loading points : 15 // Max ore-passes : 8 // Max nodes : 99
Exit CAD screen : ESC // Create a node : F1 // Delete a node : F2
Create a line between two nodes (A and B) : Press F3 on A and F4 on B
Erase a line between two nodes (A and B) : Press F3 on A and F5 on B
Label an existed node as a loading point : F6 on the node // Delete label : F8
Label an existed node as an ore-pass : F7 on the node // Delete label : F8
Save design to disk-files : F9 // Load a saved design from a disk : F10

Fig. 3. A layout designed by using the CAD routines.
L = Loading point/node
O = Orepass point/node

4. Input/Output
   a. Disc store and retrieve image
      (picture) files
   b. Printer
5. Cursor control
   a. Arrow keys on keyboard
6. Menu selection
   a. Primary drawing tasks
      1. Points
      2. Lines
   b. Other tasks
      1. Uppercase letters
      2. Save drawing
      3. Retrieve drawing
      4. Exit

3.2 Route submodel

In the route submodel the user defines the routes that the vehicle is supposed to follow during the simulation. The routes are determined based on the network that has been created at the CAD submodel. Each route consists of a sequence of nodes of the network.

The time required to travel from one node to another depends on the distance between the nodes and the vehicle's travel speed.

To calculate the time in a segment the following variables are defined:
a = node of the start of the segment
b = node of the end of the segment
U_a = speed when the vehicle enters the segment (speed at node a)
U_b = speed when the vehicle leaves the segment (speed at node b)
D = length of the segment
t = time through the segment

The time in the segment is calculated by the formula: $t = 2 \cdot D/(U_a+U_b)$.

For defining the routes, special command-routines have been developed which direct the vehicle from a loading point node to an ore-pass one and vice versa. These command-routines have user-friendliness input formats. For example, the input format of the command which directs the vehicle from a loading point to an ore-pass is [L → O] and the input format of the command which dispatches the vehicle from an ore-pass node to a loading point node is [O → L].

The user also, for each segment, enters the node of the start of the segment, the node of the end of the segment, the length of the segment, the speed at the start node, the speed at the end node and the direction of the vehicle's movement (backwards or forward).

416

*** ROUTE MENU ***

<------------------------ COLUMN ------------------------->
	1	2	3	4	5	6	7
				<-- Section Definition -->			
				(m)	Speed (km/h) at		Move
Row	Route	EN	LN	Length	EnterNode	ExitNode	Backwards
1	O --> L	1	2	5.00	0.00	8.00	Y
2		2	19	5.00	8.00	11.00	Y
3		19	20	5.00	11.00	8.00	Y
4		20	41	3.00	8.00	0.00	Y
5		41	20	3.00	0.00	8.00	
6		20	4	20.00	8.00	20.00	
7		4	5	30.00	20.00	20.00	
8		5	6	8.00	20.00	15.00	
9		6	9	10.00	15.00	15.00	
10		9	10	5.00	15.00	15.00	
11		10	11	5.00	15.00	8.00	
12		11	12	6.00	8.00	8.00	
13		12	18	5.00	8.00	0.00	
14	L --> O	18	12	5.00	0.00	8.00	Y
15		12	11	6.00	8.00	12.00	Y

Fig. 4. An example of a route input menu.
Y = Yes
EN = Entering Node
LN = Leaving Node

The Priority Rule for the sequence of the input routes is First In First Out, (FIFO), which means that the first entered route, is first executed. Fig. 4 shows a part of a route input menu.

3.3 Loading points submodel

In the loading points submodel, the loading points nodes are defined by program's specific routines that can read both the created network at the CAD submodel and the input routes and thus the nodes that correspond to the loading points can be identified and determined. The user enters the tons to be loaded per loading point as well as the loading times per loading point. Four probability distributions (Normal, Poisson, Exponential and Uniform) can be used in order to fit the loading times. Fixed values for these times can also be entered. Fig. 5 shows a loading points input menu.

3.4 Ore passes submodel

In the ore passes submodel, like in the loading points submodel, the ore passes nodes are defined by program's specific routines which can read the created

network at the CAD submodel and the input routes and then these routines can identify the nodes that correspond to the ore passes points. The user enters the dumping times per ore pass from a selection of probability distributions (Normal, Poisson, Exponential, Uniform). Moreover, as in the case of the loading points submodel, fixed values for dumping times can also be entered. Fig. 6 shows an ore-passes input menu.

3.5 Vehicle submodel

In the vehicle submodel, maintenance characteristics for the LHD as Mean Time Between Failures, Mean Time To Repair, Delay for replacing a broken down vehicle and number of stand by vehicles are defined and entered by the user. These variables are used for estimating vehicle's availability and number of breakdowns. The actual bucket load of the LHD (the bucket that is really been picked up at the loading points) is entered by the user too. All these parameters, except the number of stand by vehicles that is an integer variable, are determined by the set of probability distributions which have been mentioned previously in the loading points and ore passes submodels. Fig. 7 shows a vehicle's input menu.

417

```
Press F1 for Help                          Press F2 to see the animation screen

                        ** LOADING POINTS MENU **
                                              (sec)
  Node Number   Qual   %        Tons     Load Time   Probability        Names :
                                          Val1  Val2                   (N)ormal
     18          0    0.0       550.0     20    40          u         (U)niform
     17          0    0.0       550.0     20    40          u         (P)oisson
     32          0    0.0       550.0     20    40          u      (E)xponential
     36          0    0.0       550.0     20    40          u      (F)ixed Value

  Commands
      -Position Cursor // Ins -Type New Value // Esc -Exit Menu
```

Fig. 5. An example of a loading points input menu.

```
Press F1 for Help                          Press F2 to see the animation screen

                          *** ORE-PASSES MENU ***
                        (sec)
    Node Number  Unloading Time   Probability  -----------------> Names :
                   Val1    Val2                                   (N)ormal
                                                                  (U)niform
       1           20      5           n                          (P)oisson
      24           20      5           n                       (E)xponential
                                                               (F)ixed Value
    Commands
        -Position Cursor // Ins -Type New Value // Esc -Exit Menu
```

Fig. 6. An example of an ore-passes input menu.

```
                                                           Press F1 for Help

                          *** VEHICLE MENU ***


                                  Val1       Val2    Probability  --> Names :
                                                                     (N)ormal
Bucket size         (tons) :     11.00      13.00        u          (U)niform
                                                                     (P)oisson
Mean Time Between                                                 (E)xponential
Failures - MTBF (hours) :         0.0        0.0         f        (F)ixed Value

Mean Time To
Repair - MTTR (hours)    :        0.0        0.0         f

Delay in order to replace a
broken-down vehicle (hours) :     0.0        0.0         f


Number of Stand-by Vehicles :       0     (Max 2)

Commands
    -Position Cursor // Ins -Type New Value // Esc -Exit Menu
```

Fig. 7. An example of a vehicle's input menu.

3.6 Model's activities

The processes and activities which are simulated are:
a. Loading activity
b. Dumping activity
c. Vehicle's movement
d. Break downs

To undertake a loading operation, a loading time for the specified loading point is generated by a predefined probability distribution. On completion of the loading activity, the loaded vehicle travels to the dumping point. The vehicle moves in the network in the following ways:
* with constant speed in a segment
* decelerating in a segment
* accelerating in a segment

When the vehicle arrives at a dumping point, the process is identical to the loading activity. A dumping time for the specified point is generated by a predefined probability distribution.

The model is based on the principles of discrete event system simulation. The event is the movement of the vehicle from node to node. The logical conditions are:
* If the node is a loading point then the loading activity is executed and the next event is the movement of the vehicle to the successive node which is defined by the currently executed route.
* If the node is an ore pass then the dumping activity is executed and the next event is the movement of the vehicle to the successive node which is defined by the currently executed route.
* In any other case the next event is the movement of the vehicle to the succesive node which is defined by the currently executed route.

If during the execution of the above logical conditions, a break down occurs, then the program checks if there is available a stand by LHD, in order to replace the broken one, otherwise the simulation is terminated.

The user can select a grafical simulation or a text oriented simulation.

In the grafical simulation, computer animation algorithms are incorporated for visualizing the movement of the vehicle. The main algorithm for the animation belongs to the family of Digital Differential Analyzer (DDA) computer algorithms and it has been modified for the purposes of PROFITIS. Animation is incorporated mainly for demonstration and communication reasons. However, based on the author's experience during the development of PROFITIS, it can also serve as a powerful debugging tool for correcting errors in the program's logic.

It has been estimated that up to 20% of developing time has been saved by utilizing the animation as a mean for identifying errors and uncomplete logic in the program's source code.

The program contains Help-menus and built in error messages for each input menu. Moreover, the menus appear as windows on the screen and they can easily be printed on a printer.

4 PROGRAM DEMONSTRATION RUN

One of the main purposes of developing PROFITIS, among others, is to compare a remote-controlled/automatic LHD, as it has been described in the introduction of the paper, with a manually driven LHD. In order to demonstrate this, let assume the following scenario:

In an underground mine, the utilization of a remote-controlled/automatic LHD is considered for the transport layout of Fig. 3. The mine engineers are interested in comparing the current manually driven LHD with the automatic one on the base of tons/shift.

Due to the mining method's characteristics, the LHD must follow certain priorities for accomplishing the loading process. These priorities (P) for the routes are (see Fig. 3):
P1: From loading point L2, the LHD transports loads to ore pass O1.
P2: From loading point L1, the LHD transports loads to ore pass O1.
P3: From loading point L4, the LHD transports loads to ore pass O2.
P4: From loading point L3, the LHD transports loads to ore pass O2.

Assumptions for the performance characteristics of the ACL:

Loading time: 20 - 40 seconds, Uniform probability.

Dumping time: Mean value 20 seconds, standard deviation 5 seconds, Normal probability. The dumping time is not considered as fixed value, even if the activity is automated. This is due to the assumptions that the bucket of an AGL can be equipped with a sensor system which will react if the load has not been completely dumped. In this case, the bucket automatic control system will shake the bucket again, until the sensor system stops reacting.

Speed characteristics: It is assumed that the AGL will have the same speed limits in the straight segments of the layout compared to the manually driven LHD, but it will be about 30% slower at curves due to the limitations of the automatic steering control.

419

Assumptions for the performance characteristics of the manually driven LHD:

Loading time: 15 - 30 seconds, Uniform probability. Based on (Int. Mining, Oct. 1985), it is assumed that the loading time of the manually driven LHD can be 25% less than the loading time for the AGL. The same assumption is also considered for the dumping time.

Dumping time: Mean value 15 sec, standard deviation 5 sec, Normal distribution.

For both cases the bucket load is: 11 - 13 tons, Uniform probability. Tons per loading point: 550 tons. Total tons: 2200 which can be transported during a shift period. Shift duration for both cases: 8 hours. Effective operating hours per shift for manually driven LHD: 6.0 to 6.5 hours.

Since the simulation will run only for one shift period, maintenance characteristics of the vehicle have not been considered in this scenario.

The above assumptions do not represent a real situation in a specific mine, since a remote-controlled/automatic LHD is not yet available and therefore they concern a hypothetical case. However, an attempt has been made to simulate an actual situation in a typical underground mine with no great loading and dumping problems due to the fragmentation of the rock.

Fig. 3 shows output from the CAD screen, where the layout has been created.

Fig. 8 illustrates the same layout, with node labelling that the program defines automatically.

Fig. 4 shows a part of the route screen menu output for the case of the AGL.

Fig. 5 and Fig. 6 illustrate the loading points and ore-passes menus for the case of AGL.

Fig. 7 shows the vehicle menu, which is identical for both cases.

For 10 repeated runs using different random number sequences and 95% confidence limits, the output of each case is shown in Table 1. A demonstration text-output of the program for the case of AGL is shown in Fig. 9, where cycle times per route, information about the loading and dumping times per loading/dumping point, number of arrivals and the average tons per arrival to each loading point, are displayed. The difference for 2200 tons/shift with the previous scenario's assumptions is the following:

In average, the time for transporting the 2200 tons is 0.86 hours more for the case of AGL and the average production (in tons/hour) is 43.8 tons/hour less for the case of the AGL.

However the utilization of a remote controlled/automatic LHD will decrease the delays per shift due to transportation of the LHD driver to and from the loading area for coffee, lunch and shift changes and due to the discussions between the foreman and the driver since both of them they will be in a control room away from the loading area.

Thus, it is anticipated that for the assumptions of this scenario, the production per shift for the AGL can be very close to the production which is achieved by the manually driven LHD. If the performance characteristics of the AGL (speed, loading/dumping time) are similar to the manually driven LHD, then the production per shift will be higher for the AGL, assuming that the reliability and availability of the automatic LHD are adequate for the demands of the mining environment.

In such cases the additional advantages of utilizing remote-controlled/automatic LHD as they have been mentioned in the introduction of this paper will increase further the handicap between a manually driven LHD and an AGL.

Table 1.

Confidence interval 95%	Simulation time (hours)	Production (tons/hour)
Manually LHD	6.146 to 6.214 Average: 6.180	354.012 to 357.916 Average: 355.964
Automatic LHD	7.017 to 7.079 Average: 7.048	310.644 to 313.618 Average: 312.131

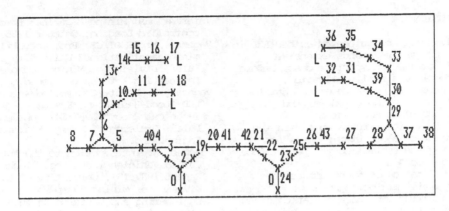

Fig. 8. Illustration of the layout of Fig. 3 with node labelling.

*** TEXT OUTPUT ***

Total Simulation time (hours):   7.03   Total loaded Tons :   2200.00
Vehicle's travelling kilometres :   43.3   Production (tons/h) :   312.75

	<--- Time in minutes --->			<--- Metres --->	
Route	Unloaded	Loaded	Total	Unloaded	Loaded
1-18- 1	0.6	0.7	1.3	110.0	106.0
1-17- 1	0.7	0.7	1.4	119.0	115.0
1-32-24	0.9	0.8	1.7	167.0	115.0
24-32-24	0.7	0.8	1.5	118.0	115.0
24-36-24	0.8	0.8	1.6	128.0	125.0

ress F1 to exit

				<-------------- Time -------------->	
oad node	Tons	Aver Tons/Arr	Arrival	Average (min)	Total (hour)
18	550.0	12.02	46	0.485	0.372
17	550.0	11.95	47	0.511	0.401
32	550.0	11.99	46	0.510	0.391
36	550.0	12.07	46	0.505	0.387

Dump node   :   1   24
Tot-Time-Hour: 0.51   0.50

Fig. 9. An example of a text-output.

Various operational conditions can be tested using PROFITIS for analyzing and estimating the capacity of a remote-controlled/automatic LHD and therefore to judge the potential or the limitations of utilizing such advanced LHD in underground mines. Vehicle's speed, loading/dumping times, production demands, vehicle's maintenance characteristics, transport layout's designs, vehicle's routes can be altered by using PROFITIS.

Finally it should be mentioned that the capacity of an AGL can be estimated not only with simulation models like PROFITIS, but also with other mathematical techniques. Simulation was selected mainly for the reasons which have been described in the beginning of this paper and also due to the intention that the next step is to analyze the performance of an automated LHD transport system consisting of a fleet of vehicles operating simultaneously and interacting to each other.

5 CONCLUSION

The utilization of automated guided LHD in underground mines demands extensive analysis of the transport process, before any attempt of a real application.

Simulation can be a proper scientific tool to indicate the potential and the limitations of an AGL system.

PROFITIS is the first step toward the development of a computer package for simulating and estimating the performance of a fleet of AGLs.

The utilization of such software can greatly assist mining engineers in decision making for the application of an AGL system in a mine. Moreover, PROFITIS can be used as a consultation tool by the manufacturer who considers the possibility to develop and sell an AGL to a specific mine.

PROFITIS indicates that the integration of CAD features and animation in simulation models for microcomputers extends and enhances the capabilities and the flexibility of modelling with simulation.

A successful application of automation and robotization in mining transport systems is based upon the development and utilization of reliable electronics and computer technology. From this point of view it is important that mining engineers are fully aware of the implications and the impact of this technology into the mining process and moreover it is indispensable that in the development and design of such advanced and complicated transport systems, mining engineers will play a major role.

6 ACKNOWLEDGEMENTS

I would like to thank prof. Sven Granholm for his valuable advise for preparing and completing this paper.

7 REFERENCES

Banks J. and Carson J.: Discrete-event system simulation. Prentice Hall Inc, 1984.

Granholm, S., Vagenas, N., Morris, J.: The Robotized Mine -A Utopia ? Proceedings of the 2nd International Conference on Innovative Mining Systems, Pennsylvania State University, U.S.A, October 1986.

Groover, M., Zimmers, E.: CAD/CAM - Computer Aided Design and Manufacturing, Prentice Hall International editions, 1984.

International Mining Magazine: Remote-controlled Loading, October 1985.

Vagenas, N: ROBOTICS: From industrial to mining applications? University of Luleå, Division of Mining Equipment Engineering - to be published.

Vagenas, N: Remote-Controlled/Automatic LHD (Load-Haul-Dump) System, Technical Report 1987:14T, University of Luleå, Division of Mining Equipment Engineering.

Vagenas, N: Introduction to the Automated Guided Vehicles Technology, Technical Report 1987:13T, University of Luleå, Division of Mining Equipment Engineering and Division of Industrial Electronics.

Vagenas, N., Morris, J, Granholm, S.: A concept for Remote-Controlled Automated LHD Systems, Proceedings of the 3rd International Conference on Innovative Mining Systems, University of Missouri-Rolla, U.S.A, November 1987.

World Mining Equipment Magazine: LHD Technology, January 1986.

Computer Applications in the Mineral Industry, Fytas, Collins & Singhal (eds)
© 1988 Balkema, Rotterdam. ISBN 90 6191 760 3

An application of queuing analysis to the production of aggregates in quarry operations

D.B.Ogbonlowo
Federal University of Technology, Akure, Nigeria

Y.J.Wang
West Virginia University, Morgantown, W.Va., USA

ABSTRACT: Simulation is the most desirable of all the analytical tools because it can be used to analyze the operating characteristics of a mine with varying number of service points and conditions in much less time than would be required by other methods. Thus simulation is used to evaluate several alternatives, test assumptions and decisions, and measure the impact of changing conditions on the various facets of mining. Analytical tools such as the manual and graphical methods have been used, and are still being used, successfully, but they have become too cumbersome with regard to time, accuracy, and economics. In this paper, the queuing theory is used to analyze production of aggregates in quarry operations. Specifically, the model is constructed for use on microcomputers. Results obtained from the application of the model to a real quarry show that the method is fast, accurate, and reliable.

1 INTRODUCTION

Many real life problems which can be solved by simulation technique involve many objectives such as control of production cost and time. The multiple objective nature of these problems usually poses difficulties of solution when other methods are used. Queuing analysis minimizes the difficulties of computation and representation of the preferences of the decision maker in a form amenable to solution.

The objectives of the queuing analysis being presented in this paper are to investigate the operating characteristics of a quarry with two service channels, namely loading and crushing, and to ascertain that the quarry functions satisfactorily with acceptable queue lengths. They are also to minimize the total cost of operating the quarry by considering such changes as an increase or decrease in the number of service channels. Evidently, quarry costs arise from two sources, namely the cost of capital equipment at the service channels and the cost of waiting in the system. Most invariably, the cost of waiting is due to loss of production when equipment breakdowns occur. Since costs due to loss of production are difficult to estimate, it becomes necessary to aim for satisfactory rather than optimum performance of the quarry.

2 THEORETICAL CONSIDERATION FOR MODEL FORMULATION

Most queuing systems are specified according to certain characteristics which govern such bases of comparison as the average queue length or the average time spent in queue waiting for service, and the operating cost (Ellison and Wilson, 1984). Consequently, the model is formulated according to the truck arrival pattern, loading and crushing patterns, quantity of service facilities, queue discipline, ultimate queue size, arrival population size, and costs of queuing (Whitaker, 1984).

2.1 Truck arrival pattern

The arrival pattern of dump trucks to the loader and the crusher can be used to determine the statistical distribution of arrivals. Arrivals are normally classified according to the number that arrive in a given time period. For example, the probability distribution of 172 observations of dump trucks joining a queue at the loader in a quarry are classified in Table 1. The relative frequency plot of the probability distribution of the data in Table 1 is presented in Fig. 1 which gives an instant picture of the characteristics of the arrival pattern.

The average arrival rate of 3.14 trucks per hour is a single measure that conveys a brief idea about the statistical distribution of arrivals. Figure 1 is similar to the pattern to be expected from the theoretical Poisson distribution. This assertion has been confirmed by comparing the graph to known probability curves (Shannon, 1975; Ogbonlowo and Wang, in press) and by using the Chi-square and/or Kolgomarov-Smirnov test to prove it (Clelland et al., 1973).

An equivalent of the Poisson distribution is the Negative Exponential distribution. If the number of arrivals per time period is Poisson distributed then it follows that the time for servicing a batch of arrivals is necessarily negatively exponentially distributed (Ogbonlowo, 1983). This fact holds true for the problem encountered in the quarry being analyzed.

2.2 Loading and crushing patterns

The pattern of service emanating from a loader or crusher can be described in analogous terms as those used for the truck arrival pattern where observed and theoretical distributions refer to service rates. For example, the loading or crushing rate may be Poisson distributed with a determined average loading or crushing rate per hour. The loading or crushing pattern is normally viewed in terms of the time taken to provide the service. Thus the loading and crushing rates that are Poisson distributed must necessarily have their equivalent service rates follow a Negative Exponential distribution. The loading or crushing time is always the reciprocal of the loading or crushing rate and vice versa (Ogbonlowo, 1983).

The empirical distributions of the loading and crushing patterns are shown in Tables 2 and 3 while their histograms are conveyed by Figs. 2 and 3, respectively.

It is observed from Figs. 2 and 3 that the histograms exhibit the characteristics of a Negative Exponential distribution. A necessary constraint is to formulate arrival and service patterns in the same units of time. The hour is used in this paper.

2.3 Quantity of service facilities

A service channel (loading or crushing) may comprise single or multiple server queues. Single server queues have one service channel while multiple server queues have varying number of service channels which may be in parallel or in series. Parallel loading and crushing facilities are treated in this analysis.

2.4 Queue discipline

Although several queue disciplines such as first come, first served (FCFS) or first in, first out (FIFO); RANDOM selection; and last come, first served (LCFS) or last in, first out (LIFO) exist in this formulation, the FCFS discipline is stressed since it is the most commonly applied discipline to queues awaiting service.

2.5 Ultimate queue size

The size of the queue is limited by the number of trucks in the quarry and possibly by the space available for queuing. Available space is inconsequential in this analysis because provision is made for finite and infinite arrival population.

2.6 Arrival population size

The size of population from which arrivals are generated influences the pattern of arrival. A large arrival population necessitates that the arrival pattern be specified in terms at which customers join the queue. On the other hand, if arrivals are from a small finite population then the arrival rate as observed at the queue is affected by the number of the population already in the queue and being served (Whitaker, 1984). A small quarry operation is an example of this where a finite number of trucks, loaders and crushers exist. For example, the quarry operation being analyzed has only one crusher, one loader and two dump trucks. When the crusher is being repaired all other operations cease. Thus the truck arrival (to the loader) rate is zero. The same is true when all the trucks break down or the loader is out of action. One way to circumspect this complication in the analysis is to use average arrival rate to specify the

Table 1. Frequency distribution of trucks arriving at the loader per hour.

Frequency of arrivals per hour	Observed frequency	Relative frequency	Cumulative frequency	Expected frequency/ 250 arrivals
0	0	0.00	0.00	0
1	44	0.26	0.26	34
2	63	0.37	0.63	53
3	54	0.31	0.94	55
4	11	0.06	1.00	43
> 4	0	0.00	1.00	0

Average truck arrival rate = 3.14/hour

Table 2. Frequency distribution of loading time (hours).

Loading Time Intervals (hrs.)	Observed Frequency	Relative Frequency	Cumulative Frequency
0.00–0.09	5	0.03	0.03
0.10–0.19	40	0.23	0.26
0.20–0.29	66	0.38	0.64
0.30–0.39	48	0.28	0.92
0.40–0.49	8	0.05	0.97
0.50–0.59	5	0.03	1.00
> 0.60	0	0.00	1.00

Average truck loading time = 0.64 hours

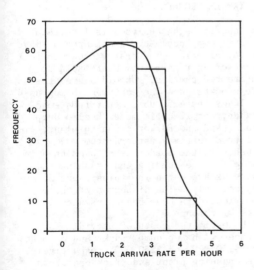

Fig. 1. Histogram and frequency plot of truck arrival to the loader.

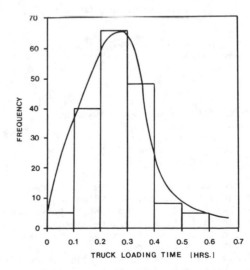

Fig. 2. Histogram and frequency plot of truck loading time.

Table 3. Frequency distribution of crushing time (hours).

Crushing time intervals (hrs)	Observed frequency	Relative frequency	Cumulative frequency
0.00–0.19	1	0.01	0.01
0.20–0.39	65	0.40	0.41
0.40–0.59	53	0.33	0.74
0.60–0.79	25	0.15	0.89
0.80–0.99	13	0.08	0.97
1.00–1.19	4	0.02	0.99
> 1.20	2	0.01	1.00

Average crushing time = 0.65 hours

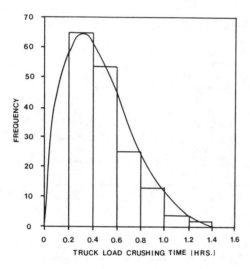

Fig. 3. Histogram and frequency plot of crushing time.

arrival pattern when all equipment are working.

2.7 Costs of queuing

The two most important costs associated with queuing systems, according to Whitaker (1984), are the cost of providing service facilities (crushers, loaders, etc.) and the cost of waiting. The former consists of a fixed cost of installation and a variable cost--cost per unit time if the server is paid by the hour (human labor), or cost per service where the server is paid piece rate for the service (mechanical servers).

Since it is often difficult to determine waiting cost, a cautious approach needs to be adopted. Generally, loss of future income plays an important role in estimating the cost of waiting. The difficulty of estimating the cost of waiting emphasizes a satisfactory rather than an optimum solution. In order to find an optimum solution, the accumulated waiting costs must be balanced against rise in the costs of facilities as more of these service facilities are added while waiting costs fall and vice versa. Thus the total cost of operating the system is the sum of the costs of the total service facilities and the accumulated costs of waiting.

3 EXAMPLE PROBLEM

The quarry being analyzed is located at Ifon in Owo Local Government area of Ondo State, Nigeria. The quarry produces four sizes of aggregates which are sold to interested parties. Unsold aggregates are stockpiled. The equipment being used includes one 4.5-cubic meter front-end loader, one 12-cubic meter Euclid dump truck, and one Parker crushing plant. The truck arrival pattern from the crusher to the loader is as presented in Table 1. The loading and crushing patterns are shown in Tables 2 and 3, respectively. The histogram of these empirical data have been presented in Figs. 1 through 3.

The estimated costs of providing a crusher with two dump trucks and a loader are $350,000 and $250,000, respectively. The crushing plant and truck operators are currently paid $15.87 per hour while the waiting cost due to facility breakdown is estimated at $3.82. The data are input as answers to questions

426

are currently paid $15.87 per hour while the waiting cost due to facility breakdown is estimated at $3.82. The data are input as answers to questions asked by the computer. The objective of the analysis is to examine the queue pattern with a view to minimizing time losses which can result in huge economic losses because of unsatisfactory performance of the quarry. By varying the number of loaders and crushers and balancing the total equipment costs against lost production costs, it is possible to ascertain the optimum system.

4 COMPUTATION AND PRELIMINARY RESULTS

In order to obtain an effective analysis, theoretical and simulation runs were performed on the microcomputer. Tables 4 and 5 (Whitaker, 1984) summarize the comparisons between the computer results relating to loading and crushing, respectively. They show the measures and costs of the operation of the queuing system at equilibrium. Equilibrium is reached when arrivals equal loading or crushing capabilities. The run length of each simulation was 1,200 hours for loading and 1,400 hours for crushing because different run length was required for system equilibrium to be achieved. The results are recorded under queue, system, utilization, and cost for ease of interpretation. A cursory examination of both tables reveals that no difference exists between theoretical and simulated values. This is a clear indication that the model is performing according to expectation.

The application of queuing theory to the analysis of the problem in question produces results which are obtained theoretically for single loader or crusher and multiple loader or crusher queues that have Poisson arrivals and Negative Exponential loading and crushing times. The maximum queue length represents the longest queue observed during the simulation of the system while the theoretical analysis uses probabilistic terms to produce the same measure which is presented as a 95 percent probability of finding up to a specified number in the queue. At equilibrium, queue length equals the derived average length while the average wait is the length of time that arrivals are expected to wait having joined the queue. As expected, all the queue and system parameters decrease with increases in the number of service channels while total costs increase. In non-equilibrium systems, queue length and average wait increase without limit upon extending the simulation time.

One loader and one crusher are required for the system to give a satisfactory performance since they are adequate for the arrival population. Output for the same queue was obtained with increasing number of loaders and crushers until a maximum number of five was reached for each of them. Results also show that both the queue and the system waiting times are highest for one loader and one crusher but become constant thereafter. This result was expected because of the very low finite arrival population. The overall analysis shows that increasing the number of loaders and crushers without corresponding increases in the arrival population does not result in an optimum system.

Similar interpretations are given to the system as those described for the queue above. The only difference is that the system comprises the loading and crushing facilities. The average loading or crushing time is an input for the problem and, when simulation is performed, confirms that the randomly generated times have the same average loading or crushing time as that input. Utilization refers to the proportion of time that the loader or crusher is busy doing useful work. For instance, a utilization factor of 0.96 for a loader working eight hours per day indicates that the loader is busy for 7.68 hours per day. Similarly, a utilization factor of 0.57 for two crushers working eight hours per day indicates that each of the crushers is busy for 4.56 hours per day. It has been observed that utilization decreases with increases in the number of loaders and crushers. The total cost which is the sum of the costs of waiting, fixed service and variable time is used in this analysis.

5 CONCLUSIONS

This paper has presented the application of queuing analysis to the production of aggregates in quarry operations. The comparison of theoretical and simulated results indicates that the queuing analysis carried out in this work is sound. The microcomputer program is particularly useful as a decision-support tool for quarry and other mine operators. It is encouraging to note that this operations research approach provides a fast, accurate and economic analytical tool.

Table 4. Comparison between theoretical and simulated values of truck loading operation.

No. of loaders	Queue			System			Utiliza-tion	Total cost
	Max	Length	Wait	Max	Length	Wait		
Simulation*								
1	1	0.57	0.38	2	1.52	0.00	0.96	250,018.22
2	0	0.00	0.00	2	1.34	0.00	0.67	500,029.92
3	0	0.00	0.00	2	1.32	0.00	0.44	750,042.25
4	0	0.00	0.00	2	1.33	0.00	0.33	1,000,054.67
5	0	0.00	0.00	2	1.34	0.00	0.27	1,250,067.11
Theory+								
1	1	0.62	0.43	2	1.54	1.06	0.92	250,018.28
2	0	0.00	0.00	2	1.33	0.64	0.64	500,029.90
3	0	0.00	0.00	2	1.33	0.64	0.44	750,042.30
4	0	0.00	0.00	2	1.33	0.64	0.33	1,000,054.70
5	0	0.00	0.00	2	1.33	0.64	0.27	1,250,067.10

*Poisson average truck arrival rate = 3.14 trucks/hour
 Histogram average truck loading time = 0.64
+Negative Exponential loading with same mean

Table 5. Comparison between theoretical and simulated values of stone crushing operation.

No. of loaders	Queue			System			Utiliza-tion	Total cost
	Max	Length	Wait	Max	Length	Wait		
Simulation*								
1	1	0.48	0.37	2	1.34	0.00	0.86	350,029.66
2	0	0.00	0.00	2	1.14	0.00	0.57	700,043.49
3	0	0.00	0.00	2	1.13	0.00	0.38	1,050,059.29
4	0	0.00	0.00	2	1.14	0.00	0.28	1,400,075.21
5	0	0.00	0.00	2	1.13	0.00	0.23	1,750,091.14
Theory+								
1	1	0.49	0.37	2	1.34	1.03	0.86	350,029.72
2	0	0.00	0.00	2	1.13	0.65	0.57	700,043.42
3	0	0.00	0.00	2	1.13	0.65	0.38	1,050,059.29
4	0	0.00	0.00	2	1.13	0.65	0.28	1,400,075.16
5	0	0.00	0.00	2	1.13	0.65	0.23	1,750,091.03

*Poisson average truck delivery rate = 2/hour
 Histogram average truck loading time = 0.64
+Negative Exponential loading with same mean

It is observed that the real case study quarry is operating efficiently within the range investigated. Thus it is concluded that the queuing model is fast, accurate and reliable.

REFERENCES

Clelland, R.C., J.S. deCani & F.E. Brown 1973. Basic statistics with business applications. New York: Wiley.
Ellison, E. & J.C.T. Wilson 1984. How to write simulations using microcomputers. London: McGraw-Hill.
Ogbonlowo, D.B. 1983. Analysis of surface coal mining using computer simulation, PhD thesis, West Virginia University, Morgantown.
Ogbonlowo, D.B. & Y.J. Wang (in press). A case study in surface mining simulation with special reference to the problem of model evaluation. International Journal of Mining and Geological Engineering.
Shannon, R.E. 1975. Systems simulation --the art and science. Englewood Cliffs: Prentice-Hall.
Whitaker, D. 1984. OR on the micro. Chichester: Wiley.

Computer Applications in the Mineral Industry, Fytas, Collins & Singhal (eds)
© 1988 Balkema, Rotterdam. ISBN 90 6191 760 3

Software for the numerical evaluation of conveyor belt lengths as a function of the installed power

J.K.Szymanski & N.J.Stuart
CANMET, Coal Research Laboratories, Energy Mines & Resources Canada, Devon, Alberta

ABSTRACT: This paper presents a public domain program to calculate the maximum permissible length of a single flight conveyor. The authors derive an analytical relationship between the coefficient of terminal friction and the length of a conveyor. This, with a standard method of deriving the conveyors resistance to motion, forms the basis of the program.

At present this kind of evaluation tends to be performed by the vendor with their own proprietary software. Although the assistance of a conveyor manufacturer is very valuable, users should be in a position to make their own estimation of their requirements. This paper is designed to fulfill the need of the engineer to quickly calculate the permissible length of a belt conveyor as a function of installed power.

A thoroughly documented FORTRAN 77 program is given to evaluate equations and produce tables or results and plots of relationships on Calcomp plotters. The use of the program is described in a "user manual" section. The practical significance and use of the results is discussed.

INTRODUCTION

When designing conveying systems consideration must be given to the maximum possible length of a single flight.

At present this evaluation is often left to the vendor, because the user or purchaser does not have sufficient background knowledge on this subject. Although the assistance of a conveyor manufacturer is very valuable, the user should basically make his own estimation based on this system requirements. The need, therefore, arose for a quick method to calculate the permissible length of the belt conveyor as a function of the power installed. In order to estimate the length of an inclined conveyor, tedious calculations are required. In this case the length is calculated using a method of subsequent approximation because the necessary coefficients are a function of the length itself.

To assist engineers and operators in the mining industry in the selection of conveyor belts a computer program has been developed which will enable a rapid calculation of conveyor belt lengths. The results of this analysis are presented in both tabular and graphical forms.

ANALYTICAL APPROACHES TO CONVEYOR FRICTION AND LENGTH

Vierling (1) and Zur (2) give basic theory and experimental results. Vierling proposes the relationship:

$$a_2 LC + a_1 L + a_0 = 0$$

Where L is the conveyor length, a_0, a, and a_2 are coefficients and C is the coefficient of terminal friction. Of more use perhaps is a table of values given in (1) from which we have plotted figure 1 and derived the following equation:

$$CL = 69.52 + 1.011L$$

This equation is the straight line plotted on figure 1 with a very good fit of better than 0.99.

TOTAL EFFECTIVE FORCE REQUIRED TO MOVE A LOADED BELT AROUND A CONVEYOR

As is normal in such problems the derivation of power requirement formulae is done by listing the power consuming

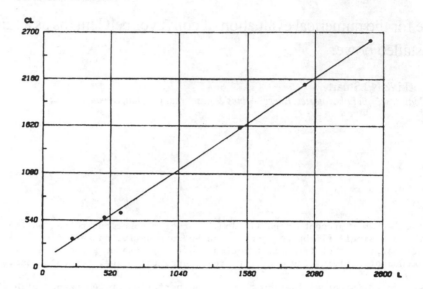

Fig. 1 - Relationship CL = f(L).

elements and then expressing each of these elements in terms that are meaningful to the engineer in the field. Thus,

Total effective = general friction force
force
+ force to raise the load

+ force to accelerate the load

+ frictional force along the skirtboards   (1)

The force necessary to move loaded belt along the conveyor is dependent upon the length of the conveyor, the weight of the moving parts, the rate of loading and the coefficient of rolling friction between the belt and the idlers. In order to allow for the terminal friction effects of pulleys etc. the coefficient C is introduced. The general friction force expands to:

$$Cfg \; (2M_t + M_k + M_d + M_n) \cos (alpha)$$

where:

C - coefficient of terminal friction
f - coefficient of friction
g - gravitational acceleration
M_t - mass of belt, kg
M_k - mass of rotating elements of return idlers, kg,
M_d - mass of rotating elements of return idlers, kg

M_n - mass of carried material, kg
alpha - angle of inclination, rad

We can thus expand our equation (1) to

$$P_u = Cfg \; (2M_t + M_k + M_d + M_n) \cos (alpha)$$
$$+ W_h + W_z + W_o \qquad (2)$$

where:

P_u - the total effective force, N
W_h - the force necessary to raise the load, N
W_z - the force necessary to accelerate the load, N
W_o - frictional force along the skirtboards, N.

M_t the mass of the belt can be expressed in terms of the unit mass as this is obtained from the suppliers specs

$$M_t = L \; B \; m_t$$

where:

L - length of conveyor, m
B - width of the belt, m
m_t - unit mass of the belt, kg/m^2

The mass of the troughing rotating idlers can be expressed as

$$M_k = \frac{L}{l_{g_s}} \; M_g \; Z_g \qquad (4)$$

432

where:

L - conveyor length, m

l_{g_s} - average idlers set spacing in upperstrand, m

Z_g - number of idlers per troughing set

The mass of the rotating idlers in the return strand can be expressed as

$$M_d = \frac{L}{l_{d_s}} M_d \, Z_d \qquad (5)$$

where:

L - conveyor length, m

l_{d_s} - average idlers set spacing in lower strand, m

Z_d - number of idlers per lower set

The mass of the material carried is

$$M_n = \frac{L k_n \, k_z \, Q_m}{V} \qquad (6)$$

where:

L - conveyor length, m

k_n - a value (close to 1) called the "load cross-section correction coefficient" that takes into account the different cross-section shape of the carried load on inclined conveyors

k_z - capacity coefficient

Q_m - mass carring capacity, kg/s

V - belt speed, m/s

Note that this also serves to define the capacity coefficient k_z.

W_o is the resistance to friction between material on the conveyor and (stationary) skirtboards.

$$W_o = \frac{\mu_o \, Q_m^2 \, g \, L_o \, k_z^2 \, k_n^2}{g \, V^2 \, B_o} \qquad (7)$$

where:

μ_o = coefficient of friction between load and the skirtboards

g = bulk density, m^3/kg

B_o = skirtboards width, m

L_o = skirtboards length, m

The effect of the acceleration of the material at the loading point is

$$W_z = k_z \, k_n \, Q_m \, V \qquad (8)$$

where:

k_z = capacity coefficient

k_n = as in equation (6)

Q_m = mass carring capacity, kg/s

The force necessary to lift the load can be expressed as

$$W_h = L \, g \, k_z \, k_n \, \frac{Q_m}{V} \, \sin (alpha) \qquad (9)$$

This gives us the following expression, which represents the total effective force

$$P_u = LC \, f \, g \, (2Bm_t + \frac{a \, M_g}{l_{g_s}} g + \frac{z \, M_d}{l_{ds}}$$

$$+ \frac{k_n \, k_z \, Q_m}{V}) \cos \alpha +$$

$$\frac{\mu_o Q_m^2 \, g \, L_o \, k_n^2 \, k_z^2}{g V^2 B^2 B_o^2} + k_z \, k_n \, Q_m \, V +$$

$$L \, g \, k_z \, k_n \, \frac{Q_m}{V} \, \sin \alpha \qquad (10)$$

The general formula for the total power required by the driving pulley is:

$$N_c = \frac{V P_u}{1000 n_m} \quad [kW] \qquad (11)$$

where:

V = belt speed, m/s

P_u = total resistance to motion of the conveyor belt (N)

n_m = mechanical efficiency of the power train

Substituting (10) into (11) and solving for L gives the final expression for conveyor length.

$$L = \frac{A^* - bC^* \cos \alpha - k_n \, k_z \, (bD^* \cos \alpha + E^*)}{a \, C^* \cos \alpha + k_n - k_z \frac{D^*}{f} (afn \cos \alpha + \sin \alpha)}$$

$$- \frac{k_n^2 \, k_z^2 \, F^*}{a \, C^* \cos \alpha + k_n - k_z \frac{D^*}{f} (afn \cos \alpha + \sin \alpha)} \qquad (12)$$

where:

$$A^* = \frac{100\ N_c\ n_m}{V},\ [N] \qquad (13)$$

$$C^* = 9.81f\ \left(2Bm_t + \frac{z_d\ M_d}{1g_s} + \frac{z_d\ M_d}{1d_s}\right),\ \left[\frac{N}{m}\right] \qquad (14)$$

$$D^* = \frac{9.81\ \frac{Q_j fp}{3600}},\ \left[\frac{N}{m}\right] \qquad (15)$$

$$E^* = \frac{Q_j\ v^2 p}{3600},\ [N] \qquad (16)$$

$$F^* = \frac{\mu_o\ Q_j\ 2p - 9.81\ L_o}{6.48\ \ 10^6\ \ B_o^{\ 2}},\ [N] \qquad (17)$$

Evaluation of this formulae may be tedious but is straight forward. However, as well as wishing to know the power required for a given length one often wishes to determine the length possible with a given motor.

This equation is the basis for the program described here (Figure 2).

Title: A numerical method for the calculation of belt conveyor lengths as a function of the installed power.

Author: J. K. Szymanski and N. J. Stuart, Energy Mines and Resources Canada. Surface Mining Laboratory, Devon, Alberta.

Description:

This program calculates the conveyor lengths as a function of angle of inclination, capacity coefficient and installed power.The results can be presented in both tabular and graphical form.

Description of data:

The data are kept in the external file TABLE.DAT in the following form:

η - mechanical efficiency
g- acceleration of gravity, m/s^2
p - mass density of conveyed material, kg/m^3
μ - coeffcient of friction between conveyed material and skirtboards
l- skirtboards length, m
f- friction coefficient (composite friction factor)
m- mass of belt per unit length, kg/m^t
z_r- number of idlers per troughing set
z_r- number of idlers per lower set
l_a- average idler spacing in upper strand, m
l_d- average idler spacing in lower strand, m
B- width of the belt, m
α- angle of inclination, rad
m- mass of the revolving parts of troughing idlers, kg
m- mass of the revolving parts of return idlers, kg
Q- unit carring capacity of conveyor belt with the belt speed v=1m/s
N- total belt horspower at the drive shaft, kw
v- belt speed, m/s
k- capacity coefficient
B- skirtboards width, m

Usege:

$ Run TABLE.FOR

Subprograms:
Data: This suroutine reads the data from file imput, unit 2 and performs the required computations.

Grapf: this subroutine produses the following graphs: L=f( ) for v=constant and k= constant; L=f( ) for n=constant k= constant, and L=f( ) for n=constant and v= constant.

```
      real nm,g,s,uo,lo,f,mt,zg,zd,lds,lgs
      real b,alfa,mg,md,qj,n,v,kz,bo
      character*20 input

      print*,'Enter the name of the data file:'
      read 99,input
99    format (a)
      call data(input,nm,g,s,uo,lo,f,mt,zg,zd,lds,lgs,
     $          b,alfa,mg,md,qj,n,v,kz,bo)

      stop
      end


      subroutine data(input,nm,g,s,uo,lo,f,mt,zg,zd,lds,lgs
     $          b,alfa,mg,md,qj,n,v,kz,bo)

      character*20 input
      real nm,g,s,uo,lo,f,mt,zg,zd,lds,lgs
      real b,alfa,mg,md,qj,n,v,kz,bo
      real l(20)
      integer nl

      open(unit=2,file=input,status='unknown')
      read(2,*,end=999)nm,g,s,uo,lo,f,mt,zg,
     $                 zd,lds,lgs,b,alfa,mg,
     $                 md,qj,n,v,kz,bo

      an=1.11
      bn=69.52

      type*,nm,g,s,uo,lo,f,mt,zg,zd,lds,lgs,
     $      b,alfa,mg,md,qj,n,v,kz,bo
      type*,' '
      type*,'            B =',b
      type*,'            Qj =',qj
      type*,'            N =',n
      type*,'            V =',v

      as=1000*n*nm/v
      cs=9.81*f*(2*b*mt+zg*mg/lgs+zd*md/lds)
      ds=9.81*qj*f*s/3600
      es=qj*v*v*s/3600
      fs=uo*qj*qj*s*9.81*lo/(6.48*(10**6)*b*b)
      type*,'       0.3      0.4      0.5     0.6
     $    0.7      0.8      0.9     1.0'

      do 20 i=0,20
      j=1
      ai=3.1416/180*i
        do 10 t=0.3,1.00,0.1
        t1=bn*cs*cos(ai)
        t2=t*kz*(bn*ds*cos(ai)+es)
        t3=t*t*kz*kz*fs
        b1=an*cs*cos(ai)
        b2=t*kz*ds/f*(an*f*cos(ai)+sin(ai))
        l(j)=(as-t1-t2-t3)/(b1+b2)
        j=j+1
10      continue
      print 200,i,(l(j),j=1,8)
200   format(' ',i4,3x,8f8.2)
20    continue

999   close(unit=2)
      call graph(nm,g,s,uo,lo,f,mt,zg,zd,lds,lgs,b,
     $ mg,md,qj,n,v,kz,bo)
      return
      end

*****************************************

      subroutine graph(nm,g,s,uo,lo,f,mt,zg,zd,
     3 lds,lgs,b,mg, md,qj,n,v,kz,bo)
      parameter(xlen=18.0,ylen=10.,max=10)

      character title*20,prompt*1,num*22
      character*26 linfmt,nfmt,kfmt,vfmt
      integer nl,pen
      real nm,g,s,uo,lo,f,mt,zg,zd,lds,lgs
      real b,alfa,mg,md,qj,n,v,kz,bo
      real l(max,int(xlen*2.+3.)),lin(int(xlen*2.+3.)),
     $ gr(int(xlen*2.+3.))
      real vp(max),lmax,x
1013  write(*,*) ' Do you wish to produce a graph (Y/N)?'
      read(*,'(a1)') prompt
      if(prompt.eq.'N'.or.prompt.eq.'n') return
      write(*,*) ' Number of lines to be drawn on graph?'
      read(*,*) nl
      write(*,*) ' Position paper on plotter '
      an=1.11
      bn=69.52
```

434

```
1       write(*,*) ' Enter correction coefficient'          730     format('md  =',f8.3,' kg       ')
      $ 'of conveyor incline'                                       call dnum(xpos,ypos,csize,num)
        read(*,*) t                                                 write(num,740) t
        if(t.lt.0.3.or.t.gt.1.0) goto 1                     740     format('kn  =',f8.3,'          ')
2       write(*,*) ' Vary N, Kz or V ?'                             call dnum(xpos,ypos,csize,num)
        read(*,'(a1)') prompt                                       kfmt='(''kz  ='',f8.3,''           '')'
        if(prompt.eq.'N'.or.prompt.eq.'n') then                     vfmt='(''V   ='',f8.3,'' m/s       '')'
          title='V,kz'                                              nfmt='(''N   ='',f8.3,'' kW        '')'
          goto 5                                                    if(prompt.eq.'N'.or.prompt.eq.'n') then
        elseif(prompt.eq.'K'.or.prompt.eq.'k') then                   linfmt=nfmt
          title='N,V '                                                write(num,kfmt) kz
          goto 5                                                      call dnum(xpos,ypos,csize,num)
        elseif(prompt.eq.'V'.or.prompt.eq.'v') then                   write(num,vfmt) v
          title='N,kz'                                                call dnum(xpos,ypos,csize,num)
          goto 5                                                   endif
        endif                                                       if(prompt.eq.'K'.or.prompt.eq.'k') then
        goto 2                                                        linfmt=kfmt
5       write(*,4) nl                                                 write(num,nfmt) n
4       format( ' Enter ',i2,' values of varied parameter')          call dnum(xpos,ypos,csize,num)
        read(*,*) (vp(k),k=1,nl)                                     write(num,vfmt) v
        do 100 k=1,nl                                                call dnum(xpos,ypos,csize,num)
          if(prompt.eq.'N'.or.prompt.eq.'n') n=vp(k)               endif
          if(prompt.eq.'K'.or.prompt.eq.'k') kz=vp(k)              if(prompt.eq.'V'.or.prompt.eq.'v') then
          if(prompt.eq.'V'.or.prompt.eq.'v') V=vp(k)                 linfmt=vfmt
                                                                     write(num,nfmt) n
        as=1000*n*nm/v                                               call dnum(xpos,ypos,csize,num)
        cs=9.81*f*(2*b*mt+zg*mg/lgs+zd*md/lds)                       write(num,kfmt) kz
        ds=9.81*qj*f*s/3600                                          call dnum(xpos,ypos,csize,num)
        es=qj*v*v*s/3600                                          endif
        fs=uo*qj*qj*s*9.81*lo/(6.48*(10**6)*b*b)
        j=1                                                        gr(int(xlen*2.+2.))=0.0
        do 20 x=0.,xlen,.5                                         gr(int(xlen*2.+3.))=1.0
          ai=3.1416/180.*x                                         pen=2
          t1=bn*cs*cos(ai)                                         ypos=ylen-1.
          t2=t*t*kz*(bn*ds*cos(ai)+es)                             xpos=xlen-4.
          t3=t*t*kz*kz*fs                                          do 150 k=1,nl
          b1=an*cs*cos(ai)                                           do 200 i=1,int(xlen*2.+1.)
          b2=t*kz*ds/f*(an*f*cos(ai)+sin(ai))                         lin(i)=l(k,i)
          l(k,j)=(as-t1-t2-t3)/(b1+b2)                     200      continue
          gr(j)=x                                                    call newpen(pen)
          j=j+1                                                      lin(int(xlen*2.+2.))=0.0
20        continue                                                   lin(int(xlen*2.+3.))=yfact
100     continue                                                     call line(gr,lin,int(xlen*2.+1.),1,2,k-1)
                                                                     call symbol(xpos-.4,ypos+.1,.2,' ',k-1,0.,-1)
        lmax=-99999.                                                 write(num,linfmt) vp(k)
        do 120 i=1,nl                                                call dnum(xpos,ypos,csize,num)
          if(l(i,1).gt.lmax) lmax=l(i,1)                             pen=pen+1
120     continue                                            150      if(pen.gt.4) pen=1
        call plots(0,0,6)                                   150     continue
        call plot(3.5,5.5,-3)
        yfact=aint(lmax/1000.+1.)*100                               call plot(0.,0.,999)
        call axis(0.,0.,'GRADE (%)',-9,xlen,0.0,0.,1.)              goto 1013
        call axis(0.,0.,'LENGTH (m)',10,ylen,90.,0.,yfact)          end
        call axis2(0.,ylen,-xlen,0.)
        call axis2(xlen,0.,ylen,90.)                        ******************************
        call symbol(5.5,ylen+.2,.3,
      $ 'CONVEYOR LENGTH VS GRADE',14,0.0,24)                       subroutine dnum(xpos,ypos,csize,num)
c draw input parameters                                            character*22 num
        ypos=ylen-1.                                               real xpos,ypos,csize
        xpos=xlen+.3                                                call symbol(xpos,ypos,csize,num(1:22),14,0.0,22)
        call symbol(xpos,ypos,.25,                                 ypos=ypos-csize*1.5
      $ 'CONSTANT PARAMETERS',14,0.0,19)                            return
        ypos=ypos-.7                                               end
        csize=.2
        write(num,600) f                                   **********************************
600     format('f   =',f8.3,'          ')
        call dnum(xpos,ypos,csize,num)                             subroutine axis2(xo,yo,axlen,angle)
        write(num,610) b                                           real xo,yo,len,angle
610     format('B   =',f8.3,' m        ')                          len=axlen
        call dnum(xpos,ypos,csize,num)                             ang=angle*0.0174533
        write(num,620) mt                                          ticlen=.2
620     format('mt  =',f8.3,' kg/sq.m ')                           if(len.lt.0.) then
        call dnum(xpos,ypos,csize,num)                               ticlen=-ticlen
        write(num,630) zg                                            len=-axlen
630     format('Zg  =',f8.3,'          ')                          endif
        call dnum(xpos,ypos,csize,num)                             cosang=cos(ang)
        write(num,640) zd                                          sinang=sin(ang)
640     format('Zd  =',f8.3,'          ')                          ntic=int(len)
        call dnum(xpos,ypos,csize,num)                             xtic=-ticlen*sinang
        write(num,650) lgs                                         ytic= ticlen*cosang
650     format('lgs =',f8.3,' m        ')                          xpos=xo
        call dnum(xpos,ypos,csize,num)                             ypos=yo
        write(num,660) lds                                         call plot(xpos,ypos,3)
660     format('lds =',f8.3,' m        ')                          do 100 i=1,ntic
        call dnum(xpos,ypos,csize,num)                               xpos=xpos+cosang
        write(num,670) qj                                            ypos=ypos+sinang
670     format('Qj  =',f8.3,' kg/s*s/m')                             call plot(xpos,ypos,2)
        call dnum(xpos,ypos,csize,num)                               call plot(xpos+xtic,ypos+ytic,2)
        write(num,680) uo                                           call plot(xpos,ypos,2)
680     format('uo  =',f8.3,'          ')                   100     continue
        call dnum(xpos,ypos,csize,num)                             call plot(xo+len*cosang,yo+len*sinang,2)
        write(num,690) lo                                          return
690     format('Lo  =',f8.3,' m        ')                          end
        call dnum(xpos,ypos,csize,num)
        write(num,700) bo
700     format('Bo  =',f8.3,' m        ')
        call dnum(xpos,ypos,csize,num)
        write(num,710) s                                   Fig. 2 - FORTRAN program design for a
710     format('p   =',f8.3,' kg/cu.m ')                            calculation of belt conveyor
        call dnum(xpos,ypos,csize,num)                              length as a function of installed
        write(num,720) mg                                           power.
720     format('mg  =',f8.3,' kg       ')
        call dnum(xpos,ypos,csize,num)
        write(num,730) md
```

435

```
0.9000000        9.810000      1800.000      0.3500000     2.000000
2.8000001E-02    7.100000      3.000000      2.000000      5.000000
1.250000         1.000000      5.000000      10.00000      21.70000
398.0000         200.0000      2.000000      1.000000      0.8000000
```

```
              B =    1.000000
              QJ =  398.0000
              N =   200.0000
              V =    2.000000
        0.3      0.4      0.5      0.6      0.7      0.8      0.9      1.0
 0   2698.78  2261.70  1943.32  1701.02  1510.42  1356.53  1229.66  1123.24
 1   2053.27  1671.32  1406.64  1212.37  1063.69   946.22   851.05   772.36
 2   1657.37  1325.69  1102.51   942.05   821.11   726.66   650.86   588.66
 3   1389.83  1098.81   906.76   770.50   668.80   589.97   527.06   475.69
 4   1196.99   938.49   770.25   651.99   564.31   496.70   442.96   399.21
 5   1051.44   819.23   669.65   565.24   488.20   429.02   382.11   344.02
 6    937.72   727.07   592.47   499.00   430.31   377.68   336.06   302.32
 7    846.44   653.74   531.39   446.80   384.81   337.42   300.01   269.73
 8    771.58   594.03   481.88   404.60   348.11   305.01   271.03   243.55
 9    709.09   544.47   440.94   369.80   317.90   278.36   247.23   222.08
10    656.17   502.69   406.53   340.62   292.61   256.08   227.35   204.15
11    610.77   467.01   377.22   315.80   271.13   237.18   210.49   188.78
12    571.43   436.19   351.96   294.44   252.67   220.94   196.02   175.93
13    537.01   409.32   329.97   275.88   236.63   206.85   183.48   164.64
14    506.85   385.68   310.67   259.60   222.58   194.52   172.50   154.78
15    479.69   364.73   293.59   245.21   210.18   183.63   162.81   146.04
16    455.59   346.05   278.38   232.41   199.14   173.95   154.20   138.31
17    433.94   329.29   264.75   220.95   189.27   165.29   146.51   131.39
18    414.37   314.18   252.47   210.63   180.39   157.51   139.59   125.17
19    396.62   300.49   241.35   201.29   172.36   150.47   133.34   119.56
20    380.44   288.03   231.25   192.81   165.07   144.09   127.67   114.48
```

Fig. 3 - The program output

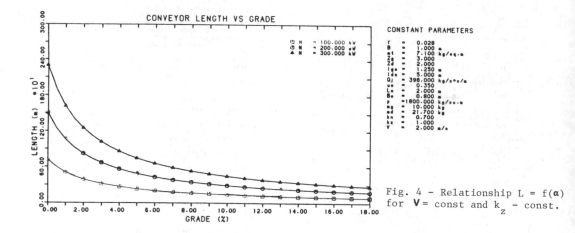

Fig. 4 - Relationship L = f(α) for V = const and k_z - const.

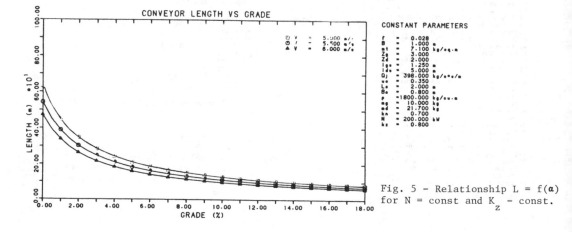

Fig. 5 - Relationship L = f(α) for N = const and K_z - const.

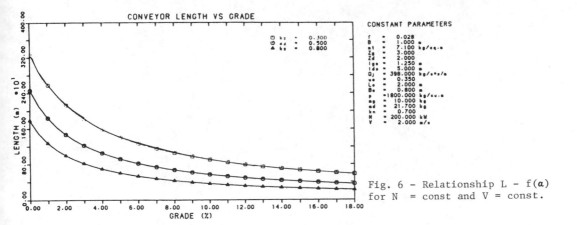

Fig. 6 – Relationship L – f(α) for N = const and V = const.

The program requires the following imput data in the following order.

n_m, g, p, μ_o, L_o, f, M_t, z_g, z_d, l_{ds}, l_{dg},

B, a, M_g, M_d, Q_j, N, V, k_z, b_o

Initially the program was written to give tables of solutions. Figure 3 shows such tables, generated for the case of 200kw of installed power. The first 5 lines repeat the input constants in order. Below is the table conveyor lengths for varying capacity coefficients and angles.

The program is now linked to calcomp (R) software to provide graphical output. All the figures have angular inclination as an X axis and length as a Y axis. Figure 4 shows relationships for a variety of installed powers. Figure 5 shows the effect of differing speeds and Figure 6 shows the effect of capacity coefficients.

CONCLUSIONS

1. This program is a public domain tool to enable the on site mining engineer to evaluate conveyor configurations, expecially with reference to install power and belt length.
2. Graphical output enables the creation of useful nomograms.
3. This analysis is limited to the steady state case where the angle of inclination is constant for the length of the conveyor. However, the analysis given would clearly be a useful foundation should a more complex approach be desired.

REFERENCES

Vierling, A. "Ergebnisse Weiterer messnngen on fordenbandanlagin", Braunkohle Warme und Energie, No. 6, 297-302, 1970.
Zur, T., "Mining Conveyor Belts", Siask, Katowice, 1979.

Computer Applications in the Mineral Industry, Fytas, Collins & Singhal (eds)
© 1988 Balkema, Rotterdam. ISBN 90 6191 760 3

Using the personal computer to simulate mining operations

John R.Sturgul
South Australian Institute of Technology, Adelaide, Australia

Raj K.Singhal
Coal Research Laboratories (CANMET), Devon, Alberta, Canada

ABSTRACT: GPSS is one of the most popular computer simulation languages which has found application to solving a wide range of problems. These problems have been in areas such as manufacturing, queueing theory, inventory control, etc. As yet, it has not found a wide acceptance by the mining engineer. Two possible reasons for this are:

1. It is a language (not a package) and so must be learned. Few mining engineers have been taught to program with GPSS in their studies. Being a language, a different program must be written for each problem.

2. It has not been widely available and some versions for mainframe computers are expensive to install. Until recently it was not available for the personal computer. (PC)

With the widespread availability of PCs, it is now possible to run a very inexpensive, but powerful version of GPSS on any MS-DOS PC. One such version of GPSS is described here. This is GPSS/PC which has numerous features such as excellent graphical capabilities, microwindows for isolating on different parts of the system, and the ability to view the simulation in progress.

Several samples of the simulation of working mines in Australia are given to indicate the power, speed and versatility of GPSS/PC. These include a surface coal mine which is considering the installation of a dispatching system, a power plant that is going to expand and wishes to know what size storage bunkers to build and how a mine might use simulation to determine the optimum number of spare parts to have on hand.

INTRODUCTION

The GPSS (General Purpose Simulation System) computer programming language for discrete system simulation had its origins in a language known as GPS in 1961. This was a very simple, and almost crude version with limited usage. However, the language has managed to survive through the years and every few years has been modified and improved. Previous versions have included GPS K, GPSS II, GPSS III, GPSS/360 and GPSS V. At the present, GPSS does not appear to be widely used in the mining industry, although the language is one of the more popular simulation languages for a wide variety of engineering and manufacturing problems. Current popular versions for the main frame computers are GPSS V/S from IBM and GPSS/H from Wolverine Software.

However, there are now several versions available for the personal computer (PCs). One of these is GPSS/PC from Minuteman Software and this has been available for use on MS or PC-DOS computers for only a few years. There are several reasons why the mining engineer might be interested in using the PC version. These are as follows:

1. The GPSS language in itself is excellent for discrete system simulation for mining situations as shown by Sturgul and Ren (1987) and Sturgul (1987). Greenberg summarizes the history of the language up the mid 1970s and Henrikson (1985) discusses the latest evolution of GPSS as well as GPSS of the future.

2. The PC versions are inexpensive, with a cost comparable to software for, say, database analysis or spread sheets.

3. The new features of GPSS/PC allow for

excellent animation and colour graphics.

4. Complete mining systems can be simulated in a matter of days or even hours.
This contrasts with using languages such
as Fortran or Pascal that may take months
or even years to write the programs.

5. The programs can be run quickly -
often the execution speed of a GPSS program written in Fortran. Functions such
as EXP, INT, LOG, SIN, SQR, TAN and others
now exist. (Their meansings are as in
Fortran.)

6. The memory limit can be increased
beyond a million bytes of memory, although
this is rarely needed for mining applications.

7. For a typical PC, such as an IBM AT,
there can be around a million block entries
in 20 minutes.

Most of the full power of GPSS versions
found on main frame versions exists in
GPSS/PC. For those familiar with GPSS,
features such as dynamic allocation, indirect addressing, transferring to a Fortran
program via the HELP block, etc are still
available. There are, in fact, a few
features that are unique to GPSS/PC which
make it easy to use (assigning commands to
function keys, using a STEP command to de-
bug a program, etc). To illustrate just
a few of the many applications of GPSS/PC
to mining problems, several examples from
actual Australian mines are next presented.
In some cases, the data has been altered
to protect the proprietary nature of the
data.

Note: Company references are for identification purposes only. The address of
Wolverine Software is 7630 Little River
Turnpike, Suite 208, Annadale, VA 22003-
2653. Minuteman Software is at P.O. Box
171, Stow, MA 01775. IBM needs no further
introduction. Other versions of GPSS are
also available.

SHOULD A DISPATCHING SYSTEM BE USED IN A
MINE?

A small coal mine in the state of Queensland was considering the purchase and
installation of a simple dispatching system. Figure 1 illustrates the schematic
of the mine.

There were only two shovels in the mine
and they haul their loads to 1 of 3 places.
The following statistical data was known:

1. Time to spot and load each truck.
2. Time to haul along each path.
3. Time to dump at each of the 3 dump
areas.
4. Time to return.
It should be emphasised that the various

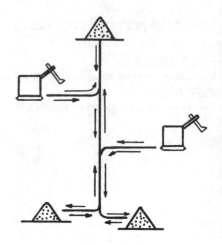

Figure 1. Schematic of Coal Mine

times were given by the exact statistical
distribution and were not deterministic.
The mine was operating with no dispatching,
ie trucks that were originally assigned to
shovel 1 returned there as did trucks assigned to shovel 2.

The chief engineer wanted to install a
dispatcher to route the trucks to possibly
different shovels when they finished dumping. The dispatching criteria would send
an empty truck to the shovel where the expected time to be loaded was a minimum.

The GPSS program to simulate the mine as
it presently was working was constructed
in less than 1/2 hour. It first was used
to determine the production as additional
trucks were added to each shovel. This in
itself was quite useful, as it showed that
the program could be used later to assist
In making production forecasts as the mine
expanded as travel paths increased. It
could also be used to study changes in
production levels as different equipment
was considered for purchase. One of the
main features of GPSS is that it is so easy
to make changes to the original program.
Often changes can be made by changing only
one or two program lines.

Once the program was accurately simul-
lating the actual performance of the mine,
it was very easy to change it to see if the
effect of a dispatcher would indeed justify
its purchase for any combination of trucks.
With the mine as it presently is, it was
found that the production would increase
by around 12% and, so, the dispatching system was justified. The program consists of
less than 100 lines of code. Moficiations
and changes take a matter of minutes, after
only seconds.

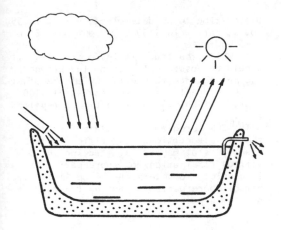

Figure 2.   Schematic of tailings pond

WILL THE TAILINGS POND OVERFLOW?

A medium sized surface mine in the Northern
Territory of Australia was expanding their
operations and so were going to build larger dams. The hydrologist in charge wanted
to simulate the conditions that would affect the water level in the ponds and see
how many times over the next 20 years
there might be spillage for different size
ponds. Figure 2 gives a schematic of a
pond and sources and sinks for the water
level.
   The possible sources of water were:
   1. Normal tailings from the mill.
   2. Rainfall, especially during the
rainy and cyclone season.
   Water was removed from the tailings as
follows:
   1. Natural evaporation.
   2. A pump was installed that would be
turned on when the level reached a certain
level.
   3. Miscellaneous losses.
   Statistical data was (somehow) obtained
for all of the above. The greatest difficulty in obtaining data was for the rainfall as this was primarily available only
as monthly summaries. Thus, the statistical data was not as accurate as could
have been. Even so, the GPSS program was
written and run to consider a wide range
of changes in the system parameters. As
more accurate data is obtained, the program can be correspondingly changed. This
was easy to add to the program as the rainfall statistics are given by a single function and as more data is added and/or
refined, only a few lines of the program
need be changed.

The model will be used to predict the
correct depth of the walls of the dam and
the size of the overflow pump. The whole
program took less than 100 lines of code.
As with other GPSS models, the hardest
part is in ensuring that the data used is
correct.

SIZE BUNKERS FOR COAL STORAGE

A major utility company in Australia was
considering the increase in its power
plant capacity and wanted to know what
size bunkers for storage of coal would be
needed. It was essential that over the
next 30 years there would be no power
failures due to lack of coal. The coal
is delivered to the power plant by trains
that arrive at random times. The statistical arrival of the trains and their loads
was known. There were several restrictions
on the trains, such as, if one train was
unloading and 3 were waiting to unload,
the facility was closed to any other trains.
If a loaded train should arrive when this
condition took place, it left. If a train
was to be unloaded and it was closer than
20 minutes to a shift change, the train
was delayed until a new unloading crew
came on duty. Coal was unloaded from the
trains and sent by conveyor belt to the
bunker. The conveyor was subject to random
failures. There were two FELs available
to pile surplus coal outside of the bunker,
although this was considered undesirable
due to the fact that it tended to rain frequently in this area. If the bunker level
was below a certain point, the FELs could
add coal to it from the outside stockpile
(if any coal was there). The availability
of these loaders was known via statistical
distributions of the failure time and repair times. The fuel rate for the power
station was also known. Other variables,
such as the frequency and severity of the
train and power plant workers going on
strike were known (in Australia this is a
significant factor in nearly all simulation
studies). There were, all told, something
like 54 conditions imposed in the model.
The GPSS/PC version was around 400 lines of
code (a comparable Fortran program would
probably have taken 40,000 lines of code -
assuming one could have been constructed,
that is). To simulate for 30 years took
several minutes to run on a main frame
and the PC version took over an hour. This
contrasts to comparable (but nowhere nearly
as complicated) Fortran models, which take
over four hours of CPU time on large main
frames. It is doubtful if such Fortran
based programs could be loaded onto a PC,

much less run in times that would be useful. Obviously, the operation conditions at the power plant will change over the next 30 years. However, the GPSS/PC model can be easily changed to accommodate them.

THE SPARE PARTS PROBLEM

The problem of determining the optimum number of spare parts to have for a mining vehicle is an ideal one for a GPSS simulation. Suppose a truck has n-parts that periodically fail. It is desired to study what the optimum inventory controls for spare parts of each should be. If there are too few parts, this causes loss of production. Having too many results in capital being needlessly tied up. To consider how a failure of a key piece of equipment can be simulated, consider a single truck with only 5 parts that can fail. Call them A, B, C, D and E. Suppose each of these parts is due to fail in the following times:

Part	Time Before Failure
A	1200 hours
B	850 hours
C	1400 hours
D	2000 hours
E	1600 hours

Thus, the equipment will work for 850 hours and then part B will fail. Once this happens, the following needs to take place:
1. The failed part must be recovered.
2. It is sent to a repair shop.
3. If a spare is available, it is used to replace the failed part.
4. If no spare is available, another must be ordered or a failed one repaired.
5. The repair shop may be used to repair other equipment. The failed part B may have to wait until other equipment is fixed (low priority situation), may be fixed as soon as the repair shop is finished working on whatever it may be fixing (high priority), or perhaps the failed part will be fixed immediately upon arrival at the repair shop (preempt situation).
6. The repair shop and the mine may not work the same hours. For example, the repair shop may work an 8 hour day while the trucks 24 hours a day.
7. Once a repaired part is installed, new time for it to fail must be determined. As a part gets older, the failure rate may change. For the other parts, the new time for failure has to be calculated as they have been running for 850 hours up to the time when part B failed. Thus, A will fail in 350 hours, new part B in some sim-

ulated time to be determined, part C in 550 hours, part D in 1150 hours and part E in 1050 hours.
8. While the truck is in for repairs, it should be possible to do routine preventive maintenance to see if any of the other parts is due for failure, say within 100 hours. If so, the part should be repaired also.

The GPSS program to simulate for all of the above conditions was written in less than 100 statements. By studying the various combinations of spare parts, it is possible to determine the optimum number of spare parts to have by running the program successive times.

CONCLUSIONS

GPSS/PC is a powerful, inexpensive tool for the mining engineer to consider for a wide range of mining problems. The programs tend to be quite sort and easy to write. Thus, a great deal of time is not spent on this aspect of a simulation study. Instead the engineer can devote time to ensuring that the data used is correct and modifying and implementing the results.

REFERENCES

Greenberg, S. 1972 GPSS Primer, Wiley-Interscience, New York.
Henriksen, J. O. State of the Art GPSS. This is available from Wolverine Software of from J. R. Sturgul. This paper presents a history of the recent developments of GPSS and dispels several myths about the language that seem to have grown up with GPSS. These include it being a slow language (it isn't); hard to construct formatted output, non-interactive, etc. It indicates what changes are planned for the GPSS of the future.
Sturgul, J. R. 1987 Simulating mining engineering problems using the GPSS computer language. Bull. Proc. Aust. Inst. Min. Met., pp. 75-78. Vol 292, No. 4, June.
Sturgul, J. R. and Ren Yi. 1987 Building simulation models of surface coal mines using the GPSS computer language. The Coal Jour., no. 15 pp. 11-17.

Computer Applications in the Mineral Industry, Fytas, Collins & Singhal (eds)
© *1988 Balkema, Rotterdam. ISBN 90 6191 760 3*

A new approach to determine the transportation system in underground mines

Q.X.Yun & D.B.Zhai
Xian Institute of Metallurgy and Construction Engineering, Shaanxi Province, People's Republic of China

ABSTRACT: Transportation system of underground mines is of importance in the planning and operations. The traditional method to determine the transportation system
is based on the multi-alternative comparison with timeconsuming manual calculation,
which might igrore the optimal solution. This paper presents a new approach --- the
minimum-cost flow algorithm --- to identify the optimal transportation system. By
means of operations research and computer techniques, the new method has been successfully used in QIANDOGSHAN lead-zinc mine of China.

1 INTRODUCTION

Underground mining activities can be
successfully carried out only with the
help of an optimum access from underground to surface. Therefore, a series of
openings need to be driven to constitute
a transportation system, which is the
focus in the planning and design of mine.
Traditionally, a transportation system
is determined by multi-alternative comparison, in which the best alternative
is sought by a detailed manual calculation from a set of feasible alternatives.
Obviously, the number of possible candidates is strictly limited on account of
the timeconsuming manual operations.
This paper, based on the operations research and computer techniques, presents
a new approach --- the minimum-cost algorithm --- to identify the optimum
transportation system among a great number of alternatives.

Fig.1 illustrates a cross-section of
orebody. It is reasonable that shaft,
inclined shaft, blind shaft, blind inclined shaft as well as tunnel-chute can
be combined to form many alternatives
for transportation system, such as the
single shaft (AB), the single inclined
shaft (CD), the tunnel-chute with blind
shaft (EFG & JK), the shaft with blind
shaft (AIFG), the shaft with blind inclined shaft (AIHG) as well as the inclined shaft with blind shaft (CHFG) and
so on. If the traditional method mentioned above is employed, a large number of

alternatives have to be omitted in advance. Only 2 or 3 of the remainders are
compared with each other to find the
best one. Therefore, it is inevitable
that the real optimum candidate might
be ignored with this course.

From the view of operations research,
an underground transportation system is,
in essence, a network flows which takes
the ore transportation as the focus and
includes the displacement of waste, materials, staff and equipment. Therefore,
the transportation system illustrated in
Fig.1 can be transformed into a network
flow system shown in Fig.2, in which

O_i (i=1,2,3,4,5) express the source, T_i
(i=1,2,3,4,5) the sink, A,B,C,... the
nodes and AB,CD,EF,... the arcs on which
the arrows represent the flow direction
and the digits the transportation amount
that passes through an arc(opening). By
calculation of the minimum-cost flow an
optimum alternative, which has the mimi-
mum transportation cost and capital investment, can be identified. If an arc
(opening) has a flow to pass through,
this means that it has to be kept for
the sake of transportation; otherwise,
it should be omitted from the network.
Thus, the remaining system is an optimum
transportation system that can be adopted to exploit the deposit. Furthermore,
the identified system can help to organize the haulage operations according to
the flow amount and its origin. For instance, the arc (opening) which carrys a

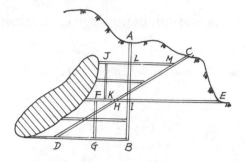

Fig.1   A cross-section of orebody

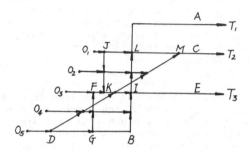

Fig.2   Network flow system

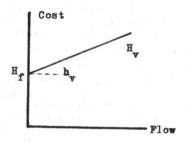

Fig.3   Cost function

great amount of flow has to be well
maintained or expanded its section. It
may even be necessary to open a new
parallal opening to mitigate the transportation amount.

This method has been applied to seek
the optimum transportation system in
Qiandonshan lead-zinc mine, northwest of
China. The results show that it is effective and can be employed in other mines.

2 NETWORK FLOW ALGORITHM

In network flow theory there are two
main parameters: cost and capacity. The
former is defined as the expenses when
flow passes through an arc (opening).
The cost is composed of two parts. The
first one is the capital investment (H_f)
that is paid for the establishment of
the arc (opening). This cost is consumed
only onec. The second part is the handling cost (H_v) that is expended on the
material transportation and is continuously being consumed in mining production. For the given arc the cost H_f is
a constant whereas the cost H_v is a
variable that varies with flow amount
and can be considered as a linear relationship with flow amount(Fig.3). Obviously, this is a concave fuction which
is very dificult to handle due to the
formidable computational problem.

In order to simplify the calculation
of concave function, only the handling
cost (H_v) will be considered for those
arcs that are indispensible for operations while their capital investment
(H_f) is ignored, hence the concave function of these arcs becomes a oblique
line passing through the original point.
However, for those arcs whose existence
has yet to be determined their cost H_v
and H_f are wholly taken into account.

Therefore, it is noticed that the actual
total cost is the flow expenses plus the
capital investment (H_f) which has been
previously ignored.

Another consideration about the cost
is that the capital investment and handling cost will be handled dynamically
with time. For the sake of convenience
to compare the alternatives, the cost of
each item is discounted into the first
year to obtain the present value using
the discount rate.

The second parameter of the minimum-
cost flow problem is capacity, which
expresses the maximum flow amount an arc
can carry. Strictly speaking, the capacity described in this paper has no

bound because the section of a opening
and the transportation facilities can be
expanded if necessary. However, to alleviate the burden of computation for concave function, the total flow amount that
can be carried over an arc is defined as
the capacity of this arc.

Based on these parameters described
above, the following algorithm can be
derived.

(1) The minimum-cost flow algorithm

The basic primal algorithm that stems
from the simplex theory in linear programming problem is employed here as the
minimum-cost flow algorithm.

In operations research the minimum-cost
flow can turn out to be a linear programming problem as the following form:

$$\text{Min} \sum_{ij} h_{ij} * f_{ij} \qquad (1)$$

$$\text{s.t.:} \sum_{j} f_{ji} - \sum_{j} f_{ij} = 0 \quad \text{for all i's}$$
$$\text{except the}$$
$$\text{source \& sink}$$

$$f_{ij} \geq 0 \qquad \text{for all i,j}$$

$$f_{ij} \leq C_{ij} \qquad \text{for all i,j}$$

where:

f_{ij} --- the flow amount passing through
 the arc i-j;
h_{ij} --- the unit flow cost of the arc
 i-j;
C_{ij} --- the capacity of the arc i-j;
i,j --- the nodes.

Based on the simplex algorithm the
minimum-cost flow algorithm is as follow-
s:

1) Use the artificial arcs, which connect each node to the slack node, to
establish an initial basic tree that
satisfies the primal feasible solutions.

2) Select an arc that violates the
complementary slackness conditions most
seriously from nonbasic arcs, and use it
as the entering arc K_E.

3) Determine the cycle composed of
basic arcs and the entering arc K_E, and
pick out the leaving arc K_L that has
to be dropped from basis.

4) Modify the basis tree, use K_E to
replace K_L.

5) Modify the value of dual variables
so as to satisfy the following condition:

$$T_i + h_K = T_j \quad \text{for arc } K(i,j) \qquad (2)$$

where:

T_i --- the value of the dual variable
 at the node i;
h_K --- the unit flow cost of the arc
 K(i,j).

Then, go to step 2.

This process repeats until the complementary slackness conditions are satisfied for all arcs.

(2) The minimum-cost flow algorithm in
 the case of concave cost function

The algorithm mentioned above is based
on the network flows in which the cost
function is linear. If the cost function
is of concave form in which the capital
investment is contained, the implicit
enumeration algorithm needs to be used.

Generally, a switching parameter X_i is
attached to the arc with concave cost
function. If $X_i=1$, this means that the
i-th arc is switched on to allow flow to
pass through; while if $X_i=0$, then the
i-th arc is swithed off to prevent any
flow. Thus, the minimum-cost flow problem in the case of concave function is
transformed into a 0-1 mixed integer
programming as:

$$Z = \text{Min} \sum_{i=1}^{m_c} H_f(i) * X_i + \sum_{i=1}^{m} h_v(i) * f_i \qquad (3)$$

$$\text{s.t.:} \sum_{j} f_{ji} - \sum_{j} f_{ij} = 0 \quad \text{for all i's}$$
$$\text{except the}$$
$$\text{source \&}$$
$$\text{sink}$$

$$0 \leq f_i \leq X_i * C_i \quad \text{for i=1,2,..}$$
$$m_c$$

$$0 \leq f_i \leq C_i \quad \text{for i=1,2,..}$$
$$m$$

$$X_i = 0 \text{ or } 1$$

where:

m --- the total number of arcs;
m_c --- the number of arcs whose existence has yet to be determined;
$H_f(i)$ --- the capital investment of the
 i-th arc;
f_i --- the flow amount passing through
 the i-th arc;
f_{ij} --- the flow amount passing through
 the arc i-j;
C_i --- the capacity of the i-th arc.

445

In order to find the optimum solution for this integer programming the branch-and-bound method is employed. At an intermediate stage of the enumeration process, some of X_i are set to 1, some of them are set to 0, and the remainder of the variables are free. For convenience, define W^+ as the variables with value 1, W^- as the set with 0, and W^o as the set of the free variables. The mixed integer programming described previously becomes as:

$$\text{Min} \sum_{i \in W^+} H_f(i) + \sum_{i \in W^o} H_f(i)*X_i$$

$$+ \sum_{i=1}^{m} h_v(i)*f_i \qquad (4)$$

$$\text{s.t.:} \sum_j f_{ji} - \sum_j f_{ij} = 0 \quad \begin{array}{l} \text{for all i's} \\ \text{except the} \\ \text{source and} \\ \text{sink} \end{array}$$

$$0 \le f_i \le C_i \qquad \text{for } i=1,2,...m$$

$$f_i \le X_i*C_i \qquad \text{for } i \in W^o$$

$$f_i = 0 \qquad \text{for } i \in W^-$$

$$X_i = 0 \text{ or } 1 \qquad \text{for } i \in W^o$$

To obtain a lower bound for all solutions associated with W^o, it is necessary to drop the condition that X_i must be an integer and replace it with

$$X_i = f_i/C_i \qquad \text{for } i \in W^o$$

Thus, the later two items of the formula(4) may be written as:

$$h_i * f_i = \left[H_f(i)/C_i + h_v(i) \right] * f_i$$

Using this result, the lower bound of the mixed integer programming(3) is found to be:

$$Z_{LB} = \text{Min} \sum_{i \in W^+} H_f(i) + \sum_{i=1}^{m} h_i*f_i \qquad (5)$$

$$\text{s.t.:} \sum_j f_{ji} - \sum_j f_{ij} = 0 \quad \begin{array}{l} \text{for all i's} \\ \text{except the} \\ \text{source \&} \\ \text{sink} \end{array}$$

$$0 \le f_i \le C_i \qquad \text{for } i=1,2...m$$

$$f_i = 0 \qquad i \in W^-$$

where:

$$h_i = H_f(i)/C_i + h_v(i) \qquad \text{for } i \in W^o$$

$$h_i = h_v(i) \qquad \text{for } i \in W^- \text{ and } W^+$$

W^+--- the variables set with $X_i=1$;

W^---- the variables set with $X_i=0$;

W^o--- the set of the free variables.
the remainders are the same as those in the formula (3).

This is a linear network flow problem that can easily be solved using the minimum-cost flow algorithm. Then, the algorithm of this method can be described as follows:

1) Let $Z_o = \infty$, and incorporates all arcs characteristic of concave cost function into the set W^o.
2) Use the minimum-cost flow algorithm to calculate Z_{LB} of formula (5).
3) If $Z_{LB} = Z_o$, go to step 6; otherwise calculate Z of formula (3).
4) If $Z = Z_o$, go to step 5; otherwise let $Z_o = Z$, then if $Z = Z_{LB}$, go to step 6.
5) Select the arc J on which the maximum difference exists between the actual concave function cost and the linearized cost. If $f_i/C_i = 0.5$, incorporate the J arc into the set W^+; otherwise into W^-, then go to step 2.
6) Backtrace one level and transformed the arc J from W^- into W^+ (or from W^+ into W^-). If the arc has been transformed, continue to backtrace until an arc that has not been transformed is found. Then, go to step 2.
Repeat this process until all arcs are searched.

3 PROCEDURE AND CASE STUDY

Associated with the actual study on Qiandongshan lead-zinc mine, the procedure to seek the optimal transportation system is described below.

(1) Determine the range of subsidence area

To derive the possible alternatives for transportation system, it is necessary to identify the range of subsidence area that appears after the orebody has been mined out.

At the beginning, the intersection points between the subsidence and surface must be identified on the cross sections of orebody according to the subsidence angles of the hanging wall, foot wall and side wall. Then, by means of the third-order natual spline function, the approximations of the identified points are smoothly connected into a curve that expresses the range of subsidence area on the topographic map.

The interpolating formula of the third order natual spline function within the interval $\left[X_i, X_{i+1}\right]$ is:

$$S(X)=S(X_i)+(X-X_i)*S(X_i,X_{i+1})$$
$$+(X-X_i)*(X-X_{i+1})*S(X,X_i,X_{i+1})$$

where:

X_i, Y_i --- the given coordinate;

$S(X)$ --- the third order natual spline function;

$S(X_i) = Y_i$;

$$S(X_i,X_{i+1}) = \frac{S(X_{i+1})-S(X_i)}{X_{i+1} - X_i}$$

$$S(X,X_i,X_{i+1})= \frac{1}{6}\left[S''(X)+S''(X_i) +S''(X_{i+1})\right]$$

$$S''(X)=3''(X_i)+(X-X_i)*S''(X_i,X_{i+1})$$

Using the No.2 to No.13 cross-sections of orebody 1# and the No.48 to No.58 cross-section of orebody 2# in Qiandongshan lead-zinc mine, the range of subsidence area of the whole deposit is determined.

(2) Calculate the location of the reference point with the minimum transportation work

The location of the reference point with minimum transportation work implies the optimum location for the main transportation opening. If an ore block is considered as a point, the location of the shaft with the minimum transportation work can be expressed with the following approximate formula:

$$\text{Min } f(x,y) = \sum_{i=1}^{m}\left[Q_i * \sqrt{(X-a_i)^2+(Y-b_i)^2}\right]$$

where:

X,Y--- the coordinate of reference point with minimum transportation work;

$f(X,Y)$--- the total work in mine with the location of shaft on (X,Y);

Q_i --- the reserve of the i-th ore block;

a_i, b_i --- the coordinate of the central point of the i-th ore block.

According to the first order partial derivative of $f(X,Y)$ with regard to X and Y respectively, the location of reference point with minimum transportation work can be defined.

In the case of Qiandongshan lead-zinc mine, the coordinate of the reference point is:

X = 3749754;

Y = 375374.4.

(3) Combine the possible alternatives of transportation system

Based on the range of subsidence area and the location of reference point as well as the surface feature and occurrence of deposit, all the possible alternatives that might be adopted to develop the deposit can be produced. Because the algorithm is computer-oriented, there is no limit to the number of candidates. As for Qiandongshan lead-zinc mine, 20 possible alternatives have been found by combination.

(4) Plot network system diagram

On the base of the possible alternatives of the transportation system, it becomes easy to plot the whole transportation network which illustrates the flow amount of materials from each source node, the capital investment and handling cost on each arc (opening) as well as the nodes of each arc.

It has been deduced that there are 379 nodes and 606 arcs, in which 60 integer arcs are contained, in the transportation network of Qiandongshan mine (Fig.4).

(5) Seek the optimum alternative of transportation system

Using the minimum-cost flow algorithm described previously, the optimum candidate can easily be determined. It is necessary to point out that the most

447

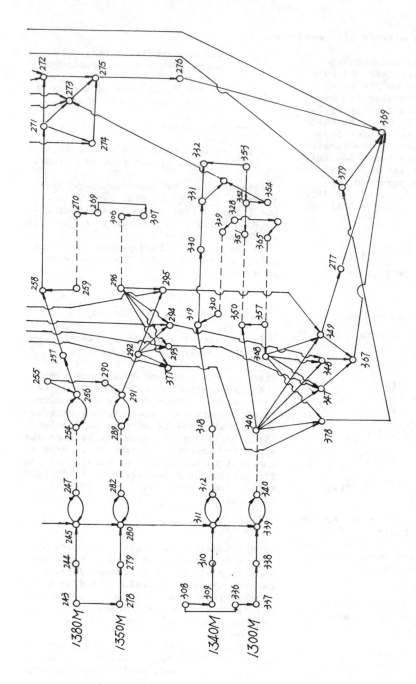

Fig.4 Part of the network flow diagram for Qiandongshan mine

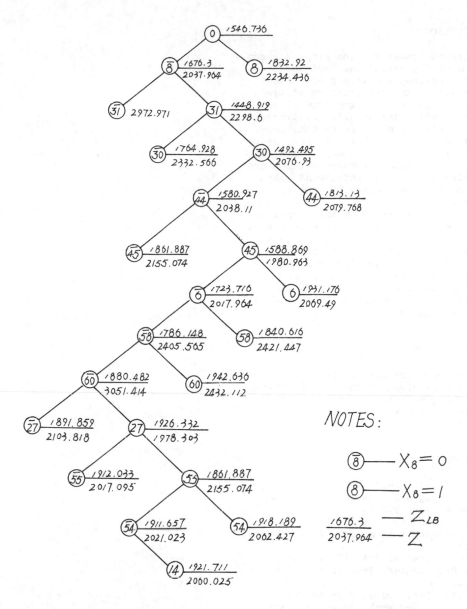

Fig.5 Part of the result from branch-and-bound algorithm

profitable one derived may be different from any original alternatives since the process of optimization is based on a synthetic combination among all possibilites. Therefore, this is another outstanding advantage this paper holds.

By calculation, the tunnel-chute with a single shaft arrangement is identified as the optimal transportation system for the development of Qiandongshan lead-zinc mine. Fig.5 illustrates parts of the result from the branch-and-bound algorithm.

The procedure described above has been written in computer language FORTRAN-77 and successfully run on micro-computer IBM-PC/XT.

4 CONCLUSION

Underground transportation plays a great
role in mining operations, and the production cost depends to a great extent
on the design and organization of transportation system in underground mines.
Mathematically, the underground transportation system in mine can be considered
as a network flow problem. Making use of
the minimum cost algorithm, especially,
employing the branch-and-bound technique
to handle the capital investment, the
best decision can be easily made by
computer.

   The method described in this paper has
been successfully applied to seek a best
design decision for Qiandongshan lead-
zinc mine, and has proved to be a new
optimization technique for planning and
design in underground mines. In fact,
this approach can be employed in the
arrangement of openings for underground
mining method.

ACKNOWLEDGEMENT

   The authors are greatly indebted to
Lanzhou Non-ferrous Metallurgy Institute
of Planning & Design, Lanzhou, China,
for its cooperation and financial
support.

REFERENCES

Bazara,M.S.;Jarris,J. 1977. Linear
   programming and network flows. John
   Wiley & Sons. U.S.A.
Garfinkel,R.S.;Newhauser,G.L. 1972.
   Integer Programming. John Wiley & Sons.
   U.S.A.
Jensen,P.A.;Barnes,J.W. 1980. Network
   Flow Programming. John Wiley & Sons.
   U.S.A.
Thesen,A. 1978. Computer Method in
   operations Research. Academic Press.
   U.S.A.
Yun,Q.X.; Yegulalp,T.M. 1983. Optimiza-
   tion of Underground Transportation
   System Using a Network Model. Inter-
   national Journal of Mining Engineer-
   ing. U.K.

10 Monitoring

Computer Applications in the Mineral Industry, Fytas, Collins & Singhal (eds)
© 1988 Balkema, Rotterdam. ISBN 90 6191 760 3

Load-Haul-Dump vehicle monitoring system

G.R.Baiden
Inco Limited, Mines Research Department, Copper Cliff, Ontario, Canada

ABSTRACT: A CANMET sponsored, NRC financed project was let to Inco Limited to develop
an underground Load-Haul-Dump Vehicle monitoring system. Inco with subcontracting to
Advanced Electronic Systems Limited and the Ontario Research Foundation has been
working on this microprocessor based monitoring system since December 1986. The system
uses an on-board microprocessor and sensors which are linked to a base station computer
(IBM PC AT) by radio. The main objectives of the project are to increase component
life, improve safety, reduce fleet size, lower maintenance costs and aid in the
diagnosing of potential problems. This paper describes the monitoring system and
preliminary results.

1.0 INTRODUCTION

LOAD-HAUL-DUMP (LHD) machine maintenance
costs have been steadily climbing while
nickel prices have remained relatively
weak. This fact has necessitated the
development of a system which would
increase LHD availability and decrease
maintenance costs while improving overall
machine life. By developing a mobile
equipment monitoring system all these
goals should be attainable.

One such electronic monitoring system
(mounted on LHD Figure 1) is in the
process of development at Inco Limited in
conjunction with Advanced Electronic
Systems Limited and the Ontario Research
Foundation. The system will monitor a
variety of sensor points on the LHD.
This information will be fed to an
on-board computer for data compaction and
limited operator diagnostics followed by
transmission to a base station computer
for further analysis. The base station
computer will allow the maintenance
professional to diagnose potential
problems with the aid of an "Expert
System". As well, a permanent historical
log can be developed for each machine and
operator. This paper discusses the
different components of the monitoring
system and the current status of the
project.

2.0 SENSORS

A total of 72 sensors are included on the
prototype monitoring system. These will
monitor pressures, fluid levels,
temperatures, gear position, voltages,
currents and CO concentration at the
operator for the seven systems on an LHD.
These systems consist of:

1) Engine
2) Braking
3) Hydraulics
4) Transmission and Drivetrain
5) Electrical
6) Fuel
7) Exhaust

The engine will have a total of 31
sensors. Engine oil will be measured for
pressure, temperature and level. The
rotation of the flywheel and timing mark
will be measured. As well, the Deutz 12
cylinder engine will have temperatures
determined at each cylinder head and
exhaust location. The pressure drop for
both filters will be measured. These
parameters should give an outline of the
engine's current performance
characteristics.

Several points are monitored in the air
over hydraulic dry disc braking system.
These include the main air pressure,
hydraulic pressures and fluid levels.

453

FIGURE 1: MONITORING SYSTEM
MOUNTED ON LHD

FIGURE 3: SENSOR INTERFACE UNIT

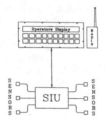

FIGURE 2: ON-BOARD
MONITORING EQUIPMENT

FIGURE 4: OPERATOR DISPLAY

Hydraulic pressures and levels are to be measured at three locations; the Service brakes (front and rear axles) and the Emergency/Parking brake.

Hydraulic system monitoring will consist of pump pressure, oil temperature and level, left and right steering cylinder pressure and the lift and dump cylinder pressures. This will allow the monitoring of all components in the hydraulic system.

To evaluate the transmission and drivetrain system 11 sensors were used. These consisted of oil temperature and level in the transmission, pressure in the torque converter, gear shift position, driveshaft speed, grease pressure of the auto lubrication system and vehicle attitude.

The electrical system is monitored using voltage readings across the battery.

The fuel system is measured at both tanks for high and low level. Fuel consumption is measured by the fuel pump rack position.

Exhaust system monitoring uses seven

sensors which measure inlet and outlet temperatures and back pressures on both filters along with carbon monoxide levels in the operator compartment.

3.0 ON-BOARD EQUIPMENT

Two units make up the electronic hardware located on the vehicle, the sensor interface and the operator display. These units to be powered by the vehicle's battery are shown in figure 2.

3.1 On-Board Hardware

The sensor interface unit shown in figure 3 contains three components, the termination panel, input board and the Z80 processor board. The on-board display consists of a 24 character LED, keypad and radio interface as shown in figure 4.

The termination panel accepts inputs from digital, analog and pulse sensors. These feed to terminal blocks which are connected to overvoltage protection, signal conditioning and filtering. From the termination panel the signals are fed via three bus connectors to the input interface card. The input interface card consists of an analog multiplexing stage, a 10KHz 10 bit analog to digital (A/D)

454

converter, a digital input stage and a pulse counter timer stage. All the conditioned information is then fed to the Z80 Board which runs a Z80C processor. This processor controls all input functions and communication with the master processor located on the Display Unit.

The Z80 board sends signals to the Display Unit via SAE bus communication which is controlled by an 8088 microprocessor. The 8088 also controls the operator display and keyboard. The Display Unit contains 96Kb of nonvolatile (battery backed up) RAM, 64Kb of ROM, a radio communication circuit and a Real Time Clock (RTC) circuit.

3.2 On-Board Software

The on-board software is controlled by the AMX 86 real time multitasking operating system. Software is written in Assembly, Pascal and "C" languages. All software is burned into EPROM's so reloading of routines is not required if power is cutoff to the unit. The majority of software was written in high level languages to allow easy understanding future development and expansion.

The software has three modes; power-up, operate and maintenance. Initialization occurs upon powering up the on-board computer. In this mode ROM, RAM, communications, sensor I/O and battery checks are performed. If the diagnostics routine is completed successfully, the Operate Mode will be entered otherwise the operator will be informed of the problem.

After the operating mode is entered, two checks are automatically performed the Initial Fluid Level check and the Brake test. This is done by starting a thirty minute timer. If the timer expires before the operator performs the tests, then an alarm is generated. Otherwise the timer is stopped upon test completion and normal sensing operation is continued. The sensors are read in real time and compared with threshold values. If the reading is above or below a threshold value, the operator will be warned of the condition. The operator can then inform the mechanic of the problem so maintenance can be performed. As the machine is operating, data is compacted to ready for RF transmission to the Base Station computer for mechanical diagnosis.

Maintenance mode is entered using a password. This mode is intended for the

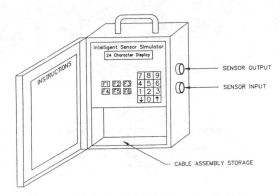

FIGURE 5: INTELLIGENT
SENSOR SIMULATOR

technicians who will be involved in system configuration, servicing and trouble shooting of the LHD computer.

The software has several features which include initial level checks, initial brake checks, operator alarm messages, and Harmonic Engine Analysis.

3.3 Production Monitoring

Present production monitoring at Inco consists of the operator counting buckets. The counts are then passed on to the engineering department where daily tonnage is calculated using a bucket fill factor. This fill factor is established for surveyed tonnages. With the LHD monitoring system each bucket will be weighed and recorded for retrieval during the shift. The weights will only be recorded upon completion of a loading, tramming and dumping cycle. By weighing the buckets, accurate daily production reports can be generated for mine management.

4.0 INTELLIGENT SENSOR SIMULATOR

A test set has been designed to enable underground mechanics with little or no electronics background to troubleshoot the on-board equipment.

The portable test set is computerized and automatically performs sensor tests and on-board electronics checkout.

This equipment, called the Intelligent Sensor Simulator (ISS), has a keypad for entry of auto-test or single step tests, a scrolling four line by 40 character alpha-numeric display, and built-in connection and operational instructions.

Components of the ISS are a termination panel, an input stage, an output stage, a

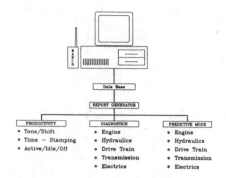

FIGURE 6: BASE STATION FACILITIES

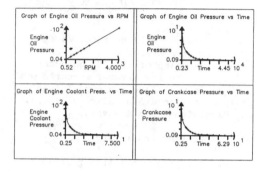

FIGURE 7: TYPICAL GRAPHIC OUTPUT

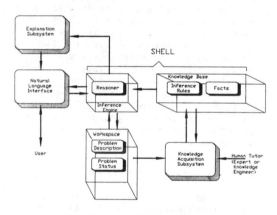

FIGURE 8: EXPERT SYSTEM

CPU, the keypad/display, and a communications interface as shown in figure 5.

The ISS connects between the Sensor Interface Unit and the sensors themselves by means of MIL style circular connectors.

The CPU stage is a Z80 based unit similar to the monitoring unit with diagnostics software written into the system ROM.

The ISS communicates with the system

using the SAE bus interface. This allows communication without disconnecting any other stage.

The unit is battery operated with a built in charger for 115 VAC.

5.0 BASE STATION FACILITIES

The Base Station facilities, shown in figure 6, consist of a computer running a radio modem, software, and an Expert System for diagnosing potential maintenance problems.

The computer used at the base station is an IBM Industrial AT which has 640 Kilobytes of memory, a 40 megabyte hard disk drive, 60 megabyte tape backup and a radio modem. The computer runs the PC DOS operating system. The Industrial AT was selected for its protective features like dust filtering, surge protection and vibration resistance.

The radio modem is controlled by the Data Collection Software which runs in the background as the Presentation Software runs in the foreground of the computer system. A typical plot of graphs showing real time data is illustrated in figure 7. Several combinations of sensor output vs. another sensor output or time can be plotted to allow the maintenance mechanic to review vehicle performance.

6.0 Expert Maintenance System

The Maintenance Expert System (MES) was developed to aid the LHD mechanic in diagnosing problems on the Scooptram. The expert system shell is Artificial Intelligence (A.I.) software which captures the knowledge of maintenance experts to determine fault conditions on an LHD. The shell has been written for an IBM PC environment using the LISP (List Processing) language.

An Expert System (figure 8) is composed of five parts:

A knowledge base.
An inference engine.
An explanation subsystem.
A user interface.
A knowledge acquisition subsystem.

The knowledge base consists of facts which are coded for use by the inference engine in determining a maintenance problem. There are three principle mechanisms used to store knowledge:

Production Rules
Semantic Networks
Frames.

456

A Production rule is an IF THEN rule linked by AND's or OR's. For example:

IF the starting air pressure is too low,
OR, the valves in the starting air lines are not sufficiently open,
OR out of fuel,
OR gear is not neutral,
OR Governor not in starting position,
THEN engine fails to start.

Semantic Networking is a method of relating abstract relationships between objects using links and modes. Objects are represented by modes and relationships by links.

Objects are physical objects that can be seen or touched. As well, they can also be events, acts, abstracts, categories or descriptors. Descriptors represent additional information about objects.

Links represent relationships between objects. Typical links are:

IS-A: Indicates an object is a member of a larger class, i.e., Cylinder is part of an engine.
HAS-A: Indicates the object is a property of another mode. i.e., Engine Cylinder has a piston.
CAUSED BY: Used to represent casual relationships. i.e., Flywheel slowing caused by misfiring of cylinder.

Frames are a knowledge structure, a framework, within which new data is interpreted in terms of concepts acquired through previous experience.

The maintenance professional's knowledge is represented in the knowledge base using a combination of these three types of mechanisms. This will allow maximum flexibility and ease of representation.

The inference engine is the reasoning mechanism which uses the rules of the knowledge base. Each rule that is applied, changes the status of the system and either activates or deactivates the appropriate rules leading to the correct conclusion.

There are two methods of reasoning either forward (data-driven) or backward (goal-driven) inferencing. Forward inferencing reaches its goal by collecting information. This is similar to solving a jigsaw puzzle if each piece is data and the picture is the goal. Backward inferencing is similar to proving a mathematical hypothesis by finding evidence to support it.

The explanation subsystem is a very useful feature of the expert system in that it allows the user through the user interface to question conclusions. Therefore, the mechanic can ask why a conclusion was reached if he does not agree or if he does not know it is correct. This interface is the essence of the usefulness of the expert system for aiding and teaching the mechanic.

Since understanding of LHD maintenance will continually grow a knowledge acquisition subsystem is required. This will allow easy updating and addition of the knowledge to be used in the expert system.

Figure 9 shows the architecture of the Expert Trouble-Shooting system to be developed for use by the underground maintenance mechanic. The system gathers data on the machine and operator from historical records. These records are reviewed along with the current symptom and relevant information is input into the Expert System. Processing is performed using the knowledge in the knowledge base and additional information as requested by the computer. This information will give a diagnosis and a repair description.

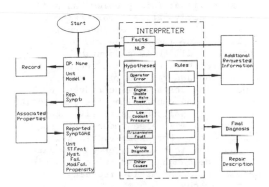

FIGURE 9:   EXPERT TROUBLE-SHOOTING
SYSTEM ARCHITECTURE

7.0 COMMUNICATIONS

Communications between the LHD and the Base Station can occur in two ways. The first is a data cartridge which collects data during the shift. This cartridge is removed and plugged into the Base Station computer for downloading. The second system uses a continuous communications network in which data is constantly downloaded to the Base Station computer in real time. The data cartridge system is the backup system for the continuous communication network which will be the primary method of communication.

8.0 BENEFITS OF MONITORING

Equipment monitoring with predictive maintenance techniques has one main purpose, that of increased production from each item of machinery.

An on-board computer is therefore, an item of production equipment with its own substantial return on investment.

Increased production is achieved by the following:
- accurate load measuring in order to provide for the setting of realistic incentive rates,
- advance warning of problems enabling early corrective action,
- reduced quantity of breakdowns,
- reduced severity of breakdowns,
- lower equipment life costs,
- reduced diagnostic, repair, and repair verification time.

Breakdown reduction will occur as a result of improved LHD maintenance practices. For example, if the monitoring system can determine a misfire in one cylinder this problem can be repaired early before a more catastrophic failure occurs i.e. premature engine failure. The result of early problem diagnosis will be less downtime. Therefore there will be an increase in production and reduced fleet size.

Operating costs will be reduced by enhancing and controlling the consumption of fuel and parts respectively. Better diagnostics will also reduce maintenance labor costs. For example, if fuel consumption is becoming excessive then the engine can be tuned up to maintain proper fuel consumption reducing operating cost.

Information from the LHD can be reviewed by mine management to aid in operating decisions. Both maintenance and production reports can be generated for statistics on availability, utilization and tonnage trammed by the LHD. These reports can allow determination of optimum fleet requirements at a particular mine.

9.0 CONCLUSIONS

This paper has described the LHD monitoring system under development at Inco Limited. Upon completion of this system, both mine operators and maintenance professionals should receive

a tool which will aid in diagnosing maintenance problems and reducing LHD downtime.

ACKNOWLEDGEMENTS

The author would like to thank CANMET for sponsoring and the National Research Council for funding the project. As well, Mr. A. Bowman of Advanced Electronic Systems for his contribution to the paper.

REFERENCES

Barr, A., Feigenbaum, E. A. 1981. The Handbook of Artificial Intelligence Volume 1. Los Altos. William Kaufmann Inc.

Barr, A. and Feigenbaum, E. A. 1981. The Handbook of Artificial Intelligence Volume 2. Los Altos. William Kaufmann Inc.

Barr, A. and Feigenbaum, E. A. 1981. The Handbook of Artificial Ingelligence Volume 3. Los Altos. William Kaufman Inc.

Considine, D. M. 1985. Process Instruments and Controls Handbook Third Edition. New York. McGraw-Hill Book Company.

Levine, R. I., Drang, D. E., and Edelson, B. 1986. A Comprehensive Guide to AI and Expert Systems. New York. McGraw-Hill Book Company.

Olaf, J.H.E. 1979. Automation and Remote Monitoring and Control in Mines. Essen. Verlag Gluckauf GMBH.

Computer Applications in the Mineral Industry, Fytas, Collins & Singhal (eds)
© *1988 Balkema, Rotterdam. ISBN 90 6191 760 3*

SENTRIE MINOS – Mine monitoring and control systems

Stewart Charles Gibbins
Huwood Electric Limited, Newcastle upon Tyne, UK

Douglas Bernard Owen
Coventry Polytechnic, UK

ABSTRACT

The data communication strategy adopted by British Coal, ICON (Integrated Communications Network), provides the mining industry with a framework within which to develop and introduce new technology. SENTRIE MINOS, a mine monitoring and control system, designed using the above standards, enables a colliery to install a small or medium sized system using a low speed transmission, and later as the system is extended, to add the high speed system simply and at relatively low cost. This means that there can be a low initial cost without restricting future developments either in size or choice of manufacturer.

INTRODUCTION

The aim of this paper is to create a greater awareness of the importance of standards to provide an industry with a framework within which to develop and introduce new technology. In particular, in the field of information technology,without such standards the automation of the mining industry would stagnate due to the multiplicity of subsystems being unable to communicate.

Successful exploitation of technology within an industry depends on a thorough understanding of the problems of the industry and an awareness of the opportunities and limitation offered by the technology in solving them. Neither the problems nor the technology are static, a previous paper described various strategies adopted to match technology with the perceived needs of the British Coal Mining Industry since the 1960s [1].

The focus of this paper are the standards for data communications infrastructure adopted in the British Mining Industry. An objective of the standards is to provide a framework so that new developments to the standard will fit into an overall plan.

We give an example of how this strategy has enabled a manufacturer to design systems which will not suffer from premature obsolescence and which can be expanded to meet future needs with flexibility for the customer.

PIONEERING PROJECTS OF THE 1960S

A number of projects undertaken during the 1960s emphasised a "systems strategy". There are described in [1]. The majority of these projects were not regarded as 'successful' at the latter end of the 1960s. Tregelles gives three reasons for this perceived lack of success [2].

First, the technologies were new and the electronic equipment was not reliable. Secondly, the information could not be digested, and so its value, in the form of action taken was never established. Thirdly, the state of automation in any industry was not advanced. Tregelles also reminds us that this was a decade of mechanisation, and management were unwilling to take on board automation when they were trying to float mechanisation. He summarises by stating that, with hindsight, the UK mining industry had the right ideas too soon.

MINICOMPUTERS PROCESS CONTROL AND MINOS

In 1972 the NCB installed a minicomputer, Archurus [3]. It formed the heart of a monitoring and control system for the coal transport of Bagworth Colliery. The trials were successful. The problem was how to achieve a rapid, rational and safe introduction of computer technology throughout the industry, without incurring large development costs.

At this time there was a multiplicity of remote monitoring and control systems, using a variety of data transmission methods. There had been many failures of suppliers leaving the collieries with unsupported systems. The manufacturers of Arcturus went bankrupt. A contract for rewriting the software for another machine made it clear that software development could be very expensive.

The NCB decided to replace the Arcturus with a supplier whom they judged to have stability, good field maintenance support, compatibility between each generation of computers and good software support.

Digital Equipment (DEC) was chosen as the supplier and their PDP 11/34 replaced the Arcturus.

The NCB introduced a standard software package for analysing and presenting data called MINOS (Mine Operating System). The full system consists of transducers connected to transmission outstations from which the information collected is passed regularly to a surface computer. Four suppliers were selected to supply these systems, and they were to develop the skills necessary for installing, commissioning and maintaining MINOS. Each supplier was to make use of its existing and approved data transmission equipment, thus retaining an element of competition.

Today over 160 such computer systems are operated at collieries for coal-transport control, environmental monitoring and general plant monitoring and control. Full descriptions of MINOS are given in [2,3,4].

A DATA COMMUNICATIONS STRATEGY FOR COLLIERIES

In the 1970s when MINOS was being introduced, the monitoring and control of such areas of activities as conveying had its control station located centrally on the surface. This was because, at that time the computer power available had to be provided at the surface for technical and economic reasons. With the availability of improved microprocessors, which can now be safely and economically used underground, there is a choice to be made as to where the computing should take place.

It is recognised by the NCB development engineers that such a choice necessitated a new standard for data transmission, not only because there would be a dramatic increase in information to be transmitted between items of underground plant, from surface to underground and from underground to surface; but also because electrical noise levels would increase, distances would increase, response time would become more critical, and the need for high integrity data would increase.

The low speed system (600 bits/sec) was defined as a British standard, ie BS6556:1985 Low Speed Digital Signals for use in Coal Mines. The system is of low cost and uses standard NCB telephone cable (NCB specification 492). BS6556 is published in three parts. Part 1 defines an optically coupled current loop transmission standard with a capacity for eight slave stations and a range of 2 km. This standard, because of its simplicity and hence low cost, is particularly useful for the transmission of the data between transducers and underground outstations. It is also ideal for interfacing between flameproof equipment (FLP) and intrinsically safe (IS) equipment since it provides high electrical isolation. Part 2 defines a transmission system which employes transformer coupling, a technique providing a maximum segregation of 1KV, making it more suitable for transmission between items of IS equipment. 15 slave stations can be supported on one ring, with a range of 10 km. For many collieries BS6556 Pt. 2 will meet the initial requirements for underground to surface data transmission.

BS6556 Pt. 3 Simple Asynchronous Protocol (SAP) is used with either BS6556 Pt. 1 or Pt. 2 and provides the required data integrity and outstation addressing facility. SAP may contain between 1 and 128 bytes of application data in each message. The British Coal Applications Protocol is used to define the format, quantity and the criteria for data transmitted using SAP. Only data which has changed since the outstation was previously addressed is transmitted. This means that response times will normally be faster then earlier systems which transmitted all data from an outstation being addressed whether it had changed or not. A medium speed, 64 kb/s system for pit-wide application, has been certified for use underground (HSE[M] certificate number is IS 11186). This system uses the same standard cable as the low speed system and has a range of 10 Km which may be extended up to 30 Km by the inclusion of up to 2 series repeaters.

There are over 160 NCB approved data transmission systems in operation throughout the UK coal fields and a framework was needed which would allow the coal industry to introduce the benefits of the newer standards while not completely abandoning existing investment and at the same time stimulate traditional equipment suppliers to develop commercial equipment incorporating the new standards. The strategy which resulted from this need became known as ICON (Integrated Communications Network), based on the International Standards Organisation (ISO) model for Open Systems Interconnect (OSI). In order to protect the existing investment, manufacturers of the existing equipment have been encouraged to develop 'upgrade kits' to convert, in situ, some of the early equipment to BS6556 standard.

MINOS is being developed so that it can accept data from both BS6556 data transmission systems and any one of the NCB approved systems into the same computer. This means that collieries with existing transmission systems can introduce rings of BS6556 equipment as they expand the MINOS system, progressively phasing out the earlier transmission equipment. New systems were specified to conform to BS6556 from early 1986.

ICON recognises that many collieries will outgrow the capacity of BS6556 transmission and a second stage of the strategy involves the introduction of medium speed trunk transmission at 64 Kb/s using the high level data link control protocol (HDLC), ISO 3309, 1979. A low speed to medium speed conversion unit is called a waystation. A typical waystation may have up to four BS6556 links feeding into it, the data being concentrated into a single 64Kb/s link to the surface. To upgrade a system to include a medium speed trunk transmission involves cutting the underground cable, inserting a waystation, and installing the appropriate software and high speed modems on the surface. The low speed modems previously fitted on the surface may now be used in the waystation.

The third stage of the strategy involves the development of the waystation to broaden its facilities e.g. provide an underground colour terminal, control console and programmable logical controller (PLC) facilities.

The fourth stage is concerned with networks. A single 64 Kb/s trunk transmission link has both capacity limitations and leaves the colliery vulnerable to a complete shutdown should the link become damaged.

It is proposed that a number of physically separate 64 Kb/s links should provide alternative routes in and out of the colliery. Their links would be interfaced with the surface computer via 'gateways'. The individual 64 Kb/s links would be connected underground via special waystation which would act as 'routing nodes' providing alternative data paths should a 64 Kb/s link become damaged or overloaded. At the surface, to integrate the monitoring and control processes with the management data bases, the surface computers need to communicate with each other. As the surface computers are of DEC origin, initial development trails have used DEC's implementation of Ethernet forming an high speed local area network (LAN).

SENTRIE MINOS - A SYSTEM WITHIN THE ICON FRAMEWORK

Huwood Electric Limited has grasped the challenge offered by the ICON strategy and supplies SENTRIE MINOS, which marries together British Coal's MINOS software with Huwood's range of sophisticated monitoring and control equipment.

The surface equipment comprises a DEC Micro PDP 11/73 Computer, two colour monitors with inbuilt character generators and a touch membrane keyboard with 128 user defined plant keys, plus QWERTY keyboard. Production and mine status reports are produced by a printer. The Huwood Electric BS6556 central station containing the data transmission modems is installed in the computer cabinet. The surface station is connected to the mine transmission cables through intrinsically safe line barriers providing up to 6 transmission circuits per system. The system uses a standard operating system, DEC micro RSX-11M and the MINOS applications software developed by British Coal. The system is upwardly compatible for future MINOS software versions.

Underground there are four types of outstations available called by the generic name SENTRIE.

SENTRIE-1, -2, -3 and -4 are all intrinsically safe (IS) to BS1259:1958. All types are configurable underground. (This means that authorised personnel can, for instance, assign to each input channel, local alarm values and trip values). Should the data transmission system or the surface computer fail, the outstation will still continue to provide complete plant protection and environmental monitoring, thus ensuring the safety of men and mine.

SENTRIE-1 was primarily designed to monitor and control conveyors, pumps, fans etc as a local stand alone system. However, BS6556 transmission can be fitted allowing it to be linked to the mine monitoring and control system; in such a case the inputs and outputs of the unit can be monitored from the surface while maintaining the option of local supervisory control.

SENTRIE-2 has all the facilities of SENTRIE-1, but with a greater power, flexibility and capacity. BS6556 is fitted as standard, with real time control functions available both locally and from the surface station.

SENTRIE-3 conforms with British Coal's waystation philosophy described above. One function is to act as a protocol converter, converting BS6556 low speed transmission to 64 Kb/sec, HDLC transmission and vice versa.

The unit can also be equipped with monitoring and control facilities similar to SENTRIE-2 and is capable of providing local programmable logic control (PLC) functions. For example, the logic will enable such activities on a coal-face as the haulage speed of a power loader and the load on the amoured face conveyor (AFC) to be integrated.

SENTRIE-4 is a dedicated low cost protocol converter allowing up to 30 outstations on up to 4 transmission circuits, either BS6556 Pt 1 or 2 or a combination of both, to be converted to 64 Kb/sec, HDLC transmission and vice versa.

CONCLUSIONS

The standards and strategy defined by British Coal have stimulated Huwood Electric to provide customers with a powerful and flexible system which will not suffer from premature obsolescence.

Customers are able to install small and medium sized installations using BS6556 low speed data transmission. As the installation is extended, the customer can add the HDLC (medium speed system) simply and at relatively low cost. This means that customers have a low initial cost without restricting future developments either in size or in choice of manufacturer.

**Since its launch in mid 1986 Huwood Electrics SENTRIE equipment has been installed in British Coal's, Scottish, North East, North Yorkshire, South Yorkshire North Derbyshire (now Central Area) Nottinghamshire and South Wales Areas

REFERENCES

[1] Douglas Owen. A Review of the Development of the application of Micro-Electronics in Underground Mining Systems in the U.K. Mining Science and Technology, Elsevier Science Publishers B.V., (Amsterdam), 4 (1987) 133-153.

[2] P.G. Tregelles. MRDE's contribution towards improving machine running time. Min. Eng. (London), 140 (236) (1981) : 807-814.

[3] D. Hartley. The development of remote monitoring and control for mining systems. Min. Eng. (London), 140(227) (1980) : 89.106.

[4] S. Gibbins, and D. Owen. Fan control and continuous monitoring, In: Proc. Mine Ventilation, Third Int. Mine Vent Congr., Harrogate. Institution of Mining and Metallurgy, London, (1984).

Computer Applications in the Mineral Industry, Fytas, Collins & Singhal (eds)
© 1988 Balkema, Rotterdam. ISBN 90 6191 760 3

Equipment performance monitoring

G.B.Gould & B.McConechy
Gregg River Resources Ltd, Hinton, Alberta, Canada
V.Srajer
CANMET Surface Mining Laboratory, Devon, Alberta, Canada

ABSTRACT: This paper describes Gregg River's (as a cost shared project with the Department of Energy, Mines and Resources-EMR, Canada Centre for Mineral and Energy Technology-CANMET and Coal Research Laboratories) evaluation of OBDAS; a real time equipment performance monitoring system developed by Philippi-Hagenbuch Inc.

For the evaluation OBDAS was installed on two-154 tonne haul trucks and operated for a six month trial period starting in January, 1987. During this time a series of studies were carried out to test the reliability, accurracy, and repeatability of the system. The presentation will describe the system and its installation, the findings and conclusions of the studies performed during its evaluation and its potential role as a building block in developing an equipment management system.

1 INTRODUCTION

Since 1984 the CANMET, Surface Mining Laboratory, presently located in Devon, Alberta investigated truck/shovel management systems applicable to surface coal mines in Western Canada. During the investigation it became apparent that a comprehensive truck management system is needed to address the needs of the mining industry.

Several mining companies were interested in various data collection and truck management systems that may be used to collect reliable field data and use this information for optimization studies. It was expected that optimization studies of the mining operation will identify ways to improve mine production, reduce operating cost and improve efficiency. To achieve these objectives, the mine operator must fully utilize the existing fleet of mine equipment, eliminate non-productive times, have control over operation and above all must know how the equipment is performing at any given time. There are several highly sophisticated and reliable systems, such as electronic and computerized control of machinery, sensing of unit operations or totally integrated data collection and management systems.

During the course of the investigation, information on such data collection and management systems were collected and evaluated. The search was limited to systems that could be directly applied to the trucking operation at surface coal mines in Canada and restricted to on-board systems.

Industry interest in this review of on-board truck management systems led to a co-operative demonstration and evaluation project of an integrated control system on two Euclid R-170 trucks at Gregg River mine in Hinton, Alberta. After evaluation of all existing data acquisition systems, the On-Board Data Acquisition System (OBDAS) developed by Philippi-Hagenbuch, Inc. of Peoria, Ill., was selected for field trial.

The OBDAS system was tested at Gregg River mine from January to November 1987. The evaluation of system installation, performance, and reliability is discussed in the presented paper. A study is also under way to extend this system to fully integrated data management system at the mine.

2 DESCRIPTION OF SYSTEM COMPONENTS

Central to the OBDAS system is an on-board micro processor, used to manipulate input signals and provide output essential for equipment performance monitoring. The processing unit used by the system is a Motorolla 68HC11, capable of receiving both digital and analogue input signals and time relating these signals using a built-in clock. Presently eight digital and eight analogue signals are being sent to the processor, with capacity to handle 24 more of each type of signal.

OBDAS's primary on-board sensing function is weight or payload monitoring. It is through this function that the most direct benefit, through cost savings, can be realized by a mining operation. The OBDAS weighing system (described for Euclid R-170 trucks) features four load sensors installed on the truck frame and forming the contact area for the box rails. Each load sensor consists of a 9" wide by 79" (front) or 83" (back) long urethane jacketed flexible hose that is partially filled with oil. Each hose sits in a base plate channel assembly that has been mounted directly onto the truck frame. The ends of the hose are sealed shut with a clamp, with one end having a fitting for a $\frac{1}{4}$" diameter nylon hose that leads to a pressure transducer. The pressure transducers, one for each load sensor, are mounted outside the cab of the truck and provide a voltage signal (varying with pressure) to the micro processor. A floating cap channel with an inside contact plate having a specific area sits on top of the hose assembly. By applying the area of the contact plate to the pressure created in the hose by the truck's load, the weight being carried is produced.

Second to knowing what weight is being carried by the truck is the information required to complete a time study of each haul cycle; including times for loading, travel full, dumping, travel empty, and waiting. This is achieved by combining the built-in clock feature of the processor with input signals received from additional OBDAS hardware.

Travel/non-travel time and direction of travel are gained using a transmission directional (forward-neutral-reverse) switch. The reverse signal is picked up by using the truck back-up alarm power signal, the neutral signal from the truck neutral safety switch power signal and forward when there is no reverse or neutral signal.

The start/end point of the haul cycle is picked up from the dump switch; a mercury activated switch mounted near the pivot point of the dump body. An input signal is received from this switch when the truck box is raised for dumping, indicating completion of a haul cycle.

Travel distance is determined by a proximity switch which serves as an odometer by sending a signal to the processor at each complete tire revolution. The proximity switch is mounted on the left front wheel of the truck and consists of a sensing oscillator, a solid state amplifier and a switching device. Proximity switching is achieved when a metallic object (target)

passes through the sensing field, representing one tire revolution. This changes the internal impedance of the oscillator, giving a signal to the micro processor. By applying tire circumference to the number of revolutions (sum of signals received from the proximity switch), the distance travelled in each gear direction is determined.

To have the information required for a complete analysis of a haul segment the hardware installation includes an inclinometer for developing haul road profiles. A varying voltage signal, based on the slope the truck is on, is sent to the micro processor from the level indicator. The inclinometer must be activated through the processor to receive output (slope indication is not monitored continuously); with output provided at every one degree change in road grade.

The remaining input to the micro processor is entered by the truck operator through the processor key pad. This input includes an operator code number (corresponding to the employee number) and a combination code for the loading unit and dump destination that the truck is assigned to. The operator code (3-digit number) is entered at the start of the shift or when there is a change of operator on the truck. The loader/destination code is changed whenever there is a change to these assignments; and is replaced with an operating delay or maintenance code when the truck goes out of production. Through this coding the ability to report productivity on a hourly basis and equipment utilization is achieved.

Manipulation of the incoming signals and information output by the OBDAS micro processor has been developed to provide real time monitoring on-board the truck and to remote locations.

Output signals for on-board viewing include a LED display on the front of the processing unit that provides a reading to the truck operator of actual load weight (tons). Through the key pad the date, time and a check for coding (operator, loader/destination) can also be displayed.

To give the loader operator an indication of load status a series of loading lights and imbalance lights are mounted on the truck (both sides). The loading lights consist of five lights: a green light indicating one or more bucket passes required, three amber lights indicating portions of a bucket required, and a red light indicating that the rated payload has been reached. The fore-to-aft imbalance lights give an indication to the loader operator if the next bucket should be placed towards the front or back of the truck to give a balanced load.

Additional on-board output includes a printer that is used as backup for data collection, gathering road profile information when required, and running diagnostic checks on the system.

Real time remote monitoring is achieved through a radio frequency (RF) data link (built into the processor) to the VAX 11/750 computer located in the engineering/administration office: 300m to 3000m away from the truck operating location.

Input signals to the micro processor are compiled and output, either to internal storage or to RF transmission, whenever there is a change in the truck transmission position; as signalled by the forward-neutral-reverse directional switch. The line of data compiled at this point, as illustrated in Figure 1, includes truck number, clock time, load weight in tons, transmission position, length of time spent in this position, and distance travelled in this position. The dump signal indicates the end of the haul cycle, with the raw data output at this time indicating cumulative time and travel distance since the last dump signal was received.

```
TRUCK #26   OBDAS 6811 VER 787 PAD SQ.IN.    2712
TRUCK #26  TRUCK LAST RUN 10/01/87 11:10:05
TRUCK #26  TRUCK STARTED 10/01/87 11:10:11
TRUCK #26  TRUCK NUMBER    0026
TRUCK #26  OPERATOR NUMBER 000259
TRUCK #26  SHOVEL NUMBER 000305
TRUCK #26   OBDAS 6811 VER 787 PAD SQ.IN.    2712
TRUCK #26  TRUCK LAST RUN 10/01/87 11:10:17
TRUCK #26  TRUCK STARTED 10/01/87 11:16:46
TRUCK #26  TRUCK NUMBER    0026
TRUCK #26  OPERATOR NUMBER 000259
TRUCK #26  SHOVEL NUMBER 000305
TRUCK #26  11:17 000 TON NEUT 00 MIN 33 SEC 00.0M
TRUCK #26  11:21 000 TON FORW 04 MIN 39 SEC 01.5M
TRUCK #26  11:22 000 TON NEUT 00 MIN 20 SEC 00.0M
TRUCK #26  11:22 001 TON FORW 00 MIN 27 SEC 00.1M
TRUCK #26  11:23 034 TON REVR 00 MIN 25 SEC 00.0M
TRUCK #26  11:24 181 TON NEUT 01 MIN 35 SEC 00.0M
TRUCK #26  11:34 174 TON FORW 09 MIN 52 SEC 02.5M
TRUCK #26  11:35 141 TON REVR 00 MIN 25 SEC 00.0M
TRUCK #26  11:35 176 TON DUMP 18 MIN 44 SEC 04.1M
TRUCK #26  11:35 000 TON NEUT 00 MIN 36 SEC 00.0M
TRUCK #26  11:39 000 TON FORW 03 MIN 44 SEC 01.2M
TRUCK #26  11:39 000 TON REVR 00 MIN 05 SEC 00.0M
TRUCK #26  11:39 000 TON FORW 00 MIN 11 SEC 00.0M
TRUCK #26  11:40 000 TON NEUT 00 MIN 42 SEC 00.0M
TRUCK #26  11:40 000 TON REVR 00 MIN 22 SEC 00.0M
TRUCK #26  11:44 000 TON FORW 03 MIN 50 SEC 01.5M
TRUCK #26  11:44 033 TON REVR 00 MIN 26 SEC 00.0M
TRUCK #26  11:46 191 TON NEUT 01 MIN 24 SEC 00.0M
TRUCK #26  11:57 189 TON FORW 11 MIN 05 SEC 02.6M
TRUCK #26  11:57 162 TON REVR 00 MIN 19 SEC 00.0M
TRUCK #26  11:58 187 TON DUMP 22 MIN 47 SEC 05.4M
TRUCK #26  11:58 000 TON NEUT 00 MIN 34 SEC 00.0M
TRUCK #26  12:05 000 TON FORW 07 MIN 11 SEC 02.5M
```

Figure 1   OBDAS Raw Data

3 INSTALLATION AND PERFORMANCE

OBDAS was installed on two, from a present fleet of eight, Euclid R-170 haul trucks being used primarily for waste rock haulage at the Gregg River Mine near Hinton, Alberta, Canada.

Initial installation of the system hardware was done by Philippi-Hagenbuch field personnel and minesite maintenance employees in January, 1987. Approximately four days per truck were required for the installation with the major delay being modifications required to the load sensor base plate assembly. These modifications required changes to the base plate-to-truck frame bolt positions to get around the fuel and hydraulic oil tanks that are also mounted to the truck frame.

Concurrent with the installation; orientation discussions were held with the truck drivers and pit foreman to familiarize them with the project and operation of the system and to start receiving their feedback on the system. From this feedback minor changes were made, aimed at making the system less distractive during night time operation. These included dimming the LED display and setting the loading lights so they only come on when the truck is in neutral; eliminating glare and mirror reflection when backing up to spot at the loader and dump.

As part of the evaluation process the mechanical reliability of the system with time and any subsequent effect on truck mechanical availability and operator acceptance was closely monitored.

The main source of OBDAS downtime during the evaluation period was on the load sensor assemblies. During the latter part of January the load sensor hoses started leaking oil at the end clamps on both trucks. Further investigations also found that the hoses were being cut by the cap channel contact plate when oil level became low. A review of these problems by Philippi-Hagenbuch concluded that steel rather than aluminum clamps were required to form a seal at the hose ends and that the contact plate would have to be machined to a narrower width (1/8") to prevent the cutting problem. Because of the time required to manufacture new clamps, the load sensor assemblies were not repaired until mid-March. Approximately three days per truck were required to implement these repairs.

Oil leaking at the load sensor hose clamps was evident within three weeks of the March repairs. At this time it was noticed that torque on the hose clamp bolts could not be maintained; indicating that the hose was stretching or creeping under the pressure exerted by the clamps. Temperature readings were also taken on the load sensor hoses to determine if nearby exhaust ducting might play a role in the hose stretching. Although temperature was not considered a factor it was determined that the material properties of the hose were at fault. As a result of production interruptions at the mine combined with the time required to manufacture and test new hose, these corrections were not completed until the end of August. Truck down time to make these repairs was reduced to less than 24 hours per unit. At the time of writing of this paper no further problems have been experienced with the load sensors, representing a two month trouble free period.

Although other minor problems have occurred with the system hardware, their effect on system reliability was that of losing output data versus losing truck production time. These include repairs/replacements to dump and proximity switches and problems related to radio communications.

4 DATA COLLECTION AND PROCESSING

The data collection software was programmed in FORTRAN on the VAX 11/750 minicomputer and acts as an interface between the OBDAS radio download system and the data storage and reporting system on the VAX.

The OBDAS radio download system consists of a radio frequency (RF) modem within the processor box, a GE portable FM (UHF) radio connected to a low profile colinear antenna (5 Db gain) on the truck cab, a UHF yaggie antenna (10 Db gain) on top of the 60 meter high loadout silos, another identical radio, a Telxon 202 RF modem, two lines drivers connected to each end of a four pair telephone circuit, and finally the RS232 connection to the terminal multiplexor on the VAX. The base station (antenna, radio, RF modem, and one line driver) is situated on top of the silos in a NEMA enclosure using a simple thermostat and light bulb for heat. The line drivers amplify the receive and transmit signals only. The other lines (signal ground, data terminal ready - DTR, and carrier detect) are connected directly, via the telephone cable, from RF modem to VAX multiplexor. Although DTR is mentioned as one of the operational lines, this acually controls the request to send (RTS) signal within the RF modem by jumpering. This was due simply to the lack of software control of RTS within the VAX. Therefore, in that which follows the references to RTS actually means DTR on the VAX which is jumpered to RTS on the modem.

The simplified sequence of steps for data collection is as follows:
- bring up RTS
- wait 40 msec.
- send out a poll sequence containing the truck number
- wait 200 msec
- turn off RTS
- get ready to accept data
- if no data is received in 2 seconds, go back to beginning and poll the next truck
- if data is received, verify it using a checksum and if not okay, go back and repoll the same truck
- if data is okay, store it, while still receiving
- if data is not received again within 2 seconds then send an okay to the truck and go back to poll the next truck.

The data sent from the OBDAS processor (duplicated on the printer when turned on) is a maximum of 40 characters per records, with a checksum and truck number. The latter information was added to the software after it was found that errors were occurring in the data especially during high winds. Errors were reduced substantially during calm weather and snow conditions, but as no errors were acceptable, the checksum method of error-checking was implemented.

The data collection software is submitted in batch with a higher priority than other users and is automatically initiated on reboot of the system. The various options (such as baudrate, parity, etc.) on the VAX multiplexor port are software selectable. The system is currently operating at 1200 baud. Initially there was a problem with RTS hanging constantly high after about eight hours of operation. The port settings were changed to disable the use of XOFF/XON and to increase the typeahead buffer and no problems have been encountered since.

As well as polling and repolling for data, and sending an okay, the same method can be used to: change OBDAS date and time, check to see if a unit is operational, and to send a four digit message to the remote porcessor's display. The latter could be useful to indicate which shovel and/or dump the truck operator should go to, assuming a dispatching system.

To avoid loss of data in the event of a VAX disk failure, the OBDAS data is stored on two different disks at the same time. If the computer goes down for maintenance, power outages, etc., the OBDAS processors can store up to four hours of data which will be sent in a stream once the computer comes back up and another poll sent out.

As seen in Figure 1, the raw OBDAS data

is quite wordy and descriptive. This is useful on the printer output but very inefficient for storage and retrieval of information for analysis and reporting. Therefore, a data base was created consisting of 6 files:

1. Gear Changes – key info for cycles, times, and tonnages
2. Shovel Changes – not only shovel info but also dump locations, down time, and maintenance reasons
3. Operator Changes – times of operator changes, average loads, and tonnages
4. Truck Off – tells when truck or OBDAS is turned off
5. Inclinometer Readings – when this option is turned on the data goes here
6. Test Data – records all test info when the test key is pushed – about 80 lines of data

These files contain all of the data in a much more efficient format. For example, the seven lines (200 characters) of a shovel change are condensed into a single 38 character records. This reduces both disk space and CPU and I/O in accessing the information. All data is stored in indexed sequential files with the real time date and time used as the key. Raw data is run through an editing routine before being stored in the data base. This allows data which does not conform to standard formats, or seems out of sequence, to be edited on the CRT screen, along with the most likely format for comparison. Before the use of checksums the editing procedure was very time consuming. Currently, errors average about one per shift per truck.

The first step in analysing the OBDAS data is to resolve the haul cycles. Since signposts are not used to locate the trucks, the haul full and haul empty portions of the cycle can be determined by the load tonnage. Thus when the load sensor system was inoperable for extended periods, cycles could not be resolved. All of the portions of a cycle, and the time spent, can be determined by the current, previous, and following gear, time, and tonnage. The DUMP record is also important in resolving cycles, and total time (dump to dump) on the record is used as a check on the total gear change times. However, this record does not give the most accurate tonnage for the haul cycle. Also, because the trucks are being loaded as they are spotting (in reverse) to load, and starting to haul (in forward), the tonnage in neutral (loading) is also not the most accurate. It has been verified using truck scales that the most accurate tonnage is the average generated during the haul, loaded.

5 OUTPUT EVALUATION

As part of the OBDAS evaluation a series of studies were carried out to confirm the accuracy of the systems output and repeatability of the output overtime. These studies included a series of independent weigh studies and time studies and a review of historic truck performance versus performance after the OBDAS installation.

The weigh studies were carried out to check the accuracy of the load sensors; advertised at $\pm$ 2%, and the overall impact on average fleet payload. Figure 2 summarizes the results (in tons) of the three weigh studies carried out in conjunction with the OBDAS evaluation. The initial study was done prior to installation of the system (November, 1986) followed by a study in January, shortly after initial installation (prior to fixing the load sensors) and a study in September after completing load sensor repairs.

Although it is evident that the scaled figures reflect errors as a result of scale accuracy and/or calibration; as indicated by the variance in NVW from each study, the trends indicated by the results are valid. From the January and September studies the scaled results indicate that the trucks equipped with OBDAS were hauling 6.1 and 5.9 ton more than the trucks without OBDAS. Comparing the separate weights obtained for the two trucks equipped with OBDAS (OBDAS vs. average of scales + OBDAS) variances of +2.4% and +1.9% are calculated.

	November 1986	January 1987	September 1987
Number of Scaled Loads	128	139	82
Average NVW	120.9	127.3	135.8
Scaled Weights:			
Fleet Average	132.6	140.1	169.1
Trucks without OBDAS	132.6	137.8	166.5
Trucks with OBDAS	-	143.9	172.4
OBDAS Weights:			
2 truck average	-	150.9	179.1

Figure 2 Summary of Weigh Studies (tons)

	Truck Cycle Analysis #1			Truck Cycle Analysis #2		
	OBDAS	Manual Study	Computer Simulation	OBDAS	Manual Study	Computer Simulation
Travel Empty	4:22	3:59	3:50	4:26	4:23	4:51
Wait at Shovel	0:03	0:04		1:10	1:37	
Spot at Shovel	0:27	0:27		0:33	0:21	
Loading Time	1:22	1:47		1:40	1:50	
Travel Full	6:27	6:31	6:25	6:30	6:30	7:07
Wait at Dump	1:42	1:44		2:03	2:10	
Spot at Dump	0:44	0:42		0:22	0:19	
Dump Time	0:34	0:53		0:44	0:43	
TOTAL CYCLE TIME	15:41	16:07		17:28	17:53	

Figure 3 Haul Cycle Time Studies (min:sec)

OBDAS MATERIAL MOVEMENT REPORT:
==================================

DATE/TIME: 87/10/01 07:00 TO 87/10/01 19:00
CREW: B SHIFT: Day   FOREMAN:J. Vinkle
TRUCK NUMBER: 6126
        ======

OPERATORS: 259

Loader Assignment	6003
Location	PQ1 1668 006
Dump	D5
Material: Type	Waste
Tonnes	4140
# Loads	24
Ave. Load	172.5

Truck Cycle Analysis:	# Resolved Cycles = 24	
(Decimal Minutes)	Average	Total
Travel Empty	7.74	185.67
Wait@ Shovel	1.93	46.27
Spot@ Shovel	0.60	14.47
Loading Time	1.75	42.03
Travel Full	10.62	254.87
Wait @ Dump	0.17	4.12
Spot @ Dump	0.49	11.65
Dump Time	0.61	14.58

Average Total Cycle Time	23.90
Average Cycle Distance(K)	5.2
Average Travel Speed	16.94
Average Speed Loaded	14.64
Average Speed Empty	20.10
Average Cycle Speed (Kph)	13.01
Fuel Consumption	N/A

Figure 4   OBDAS Material Movement Report

OBDAS PRODUCTION REPORT:
===========================

DATE/TIME: 87/10/01 07:00 TO 87/10/01 19:00
CREW: B SHIFT: Day   FOREMAN:J. Vinkle

TRUCK:	6126			6127		
LOADER	WASTE	COAL	TOPSOIL	WASTE	COAL	TOPSOIL
6001:Tonne	0	0	0	0	0	0
#Loads	0	0	0	0	0	0
Ave.Load	0.0	0.0	0.0	0.0	0.0	0.0
6002:Tonne	0	0	0	0	0	0
#Loads	0	0	0	0	0	0
Ave.Load	0.0	0.0	0.0	0.0	0.0	0.0
6003:Tonne	4140	0	0	4693	0	0
#Loads	24	0	0	28	0	0
Ave.Load	172.5	0.0	0.0	167.6	0.0	0.0
6021:Tonne	0	0	0	0	0	0
#Loads	0	0	0	0	0	0
Ave.Load	0.0	0.0	0.0	0.0	0.0	0.0
6022:Tonne	0	0	0	0	0	0
#Loads	0	0	0	0	0	0
Ave.Load	0.0	0.0	0.0	0.0	0.0	0.0
6031:Tonne	0	0	0	0	0	0
#Loads	0	0	0	0	0	0
Ave.Load	0.0	0.0	0.0	0.0	0.0	0.0
TOTAL:Tonne	4140	0	0	4693	0	0
#Loads	24	0	0	28	0	0
Ave.Load	172.5	0.0	0.0	167.6	0.0	0.0
	BCM	Tonnes	BCM	BCM	Tonnes	BCM
TOTAL:Payl.	1630	0	0	1848	0	0
#Loads	24	0	0	28	0	0
Ave.Load	67.9	0.0	0.0	66.0	0.0	0.0

Figure 5   OBDAS Production Report

To confirm the payload increases indicated by the weigh studies a review of actual truck production for the year prior to installing OBDAS and since installing OBDAS was completed. Using surveyed waste production figures, measured in bank cubic metres (BCM) and a count of truck loads from daily time card data, a figure for BCM per load or trip is obtained. In 1986 the fleet of R-170 haul trucks averaged 55.3 BCM per trip, or 154.8 ton using a specific gravity of 2.54. In the nine month period following the installation of OBDAS on two trucks, the fleet (8-trucks) has averaged 57.5 BCM per trip, or 160.4 ton.

To evaluate the time dependent information that OBDAS produces, and the breakdown of this information into haul cycle components (load travel full, dump, travel empty, etc.) a series of coincidental time studies were performed for comparison.

Figure 3 tabulates two examples from the results of these comparisons along with the truck travel times arrived at using the minesite simulation program (based on rimpull curves). These results indicate a close correlation between OBDAS and the manual time studies for all components of the haul cycle. The variances that do occur are largely the result of when the truck operator decides to change gears (OBDAS transmission/direction signal) when waiting, spotting and pulling away from the loader or dump.

To demonstrate the use of OBDAS as a means to compile truck production statistics a group of three reports were designed for output at any given time interval (shift, daily, weekly, etc.). Examples of these reports are shown as Figures 4, 5 and 6. These reports utilize the output

```
OBDAS UTILIZATION/PRODUCTIVITY REPORT:
=========================================

DATE/TIME: 87/10/01 07:00 TO 87/10/01 19:00
CREW: B SHIFT: Day   FOREMAN:J. Vinkle

             TRUCK: |  6126  |  6127  |
    All Times In Hours |--------|--------|
                       |        |        |
OPERATING DELAYS:      |        |        |
     Weather    (801)| 0.00 |  0.00 |
     Blasting   (802)| 0.00 |  0.00 |
     Eq. Moves  (803)| 0.00 |  0.00 |
     Waiting    (804)| 0.00 |  0.00 |
     Power      (805)| 0.00 |  0.00 |
                       |        |        |
         Sub-Total:| 0.00 |  0.00 |
                       |        |        |
FIXED DELAYS:          |        |        |
       Shift Start | 0.00 |  0.00 |
  Lunch/Rest Breaks | 1.24 |  1.23 |
   Servicing (806)| 0.00 |  0.00 |
         Shift End | 0.00 |  0.00 |
                       |        |        |
         Sub-Total:| 1.24 |  1.23 |
                       |        |        |
MAINTENANCE DOWN TIME: |        |        |
   Down-Travl  (807)| 0.00 |  0.00 |
   Down-Wait   (808)| 0.00 |  0.00 |
                       |        |        |
         Sub-Total:| 0.00 |  0.00 |
                       |        |        |
NET OPERATING HOURS:   |        |        |
           Waste | 10.76 | 10.77 |
            Coal | 0.00 |  0.00 |
         Topsoil | 0.00 |  0.00 |
                       |        |        |
         Sub-Total:| 10.76 | 10.77 |
                       |        |        |
TOTAL SHIFT HOURS:     | 12.00 | 12.00 |
                       |        |        |
PRODUCTIVITIES:        |        |        |
  Waste  (BCM/NOH) | 151.47 | 171.62 |
  Coal   (Tonne/NOH)| 0.00 |  0.00 |
  Topsoil (BCM/NOH) | 0.00 |  0.00 |
                       |        |        |
Mechanical Availability | 100.00 | 100.00 |
   Utilization Factor | 89.68 | 89.73 |
Effective Utilization | 89.68 | 89.73 |
                       |        |        |
```

Figure 6   OBDAS Utilization/Productivity
           Report

generated by OBDAS along with the data
presently compiled at the mine to track
daily equipment performance and production.
   The OBDAS material movement report,
Figure 4, provides a summary by truck for
each loader/destination assignment of total
quantity moved and average load. Based on
the number of truck cycles that can be
resolved (computer picks up all the proper,
transmission/direction signals to put a
haul cycle together) a cycle time analysis
for each loader/destination assignment is
also reported.

   The OBDAS production report, Figure 5,
summarizes truck production for each loader
assignment by material type. If OBDAS were
expanded to include all haulage units a
report of this nature would provide total
mine production with the possibility of
replacing surveys to generate production
volumes.
   The OBDAS utilization/productivity report
uses the codes designating "operating delay
and maintenance downtime (entered through
the processor key pad) to generate a complete breakdown of time utilization through
a shift (or other designated interval).
Mechanical availability and utilization
factor (measures effective use of mechan-
ically available time) is also reported
along with the trucks production on an
hourly basis. Extended over a period of
time this report could be used by maintenance planning to track specific downtime and repair time.

6 CONCLUSIONS

Applying the truck productivity increases
realized with the installation of OBDAS;
as shown by the increase in average load
per cycle from 55.3 to 57.3 BCM, to Gregg
River's present waste haulage cost indicates that a savings of $0.02 per BCM has
been achieved. At the present annual waste
capacity of 12 million BCM this represents
savings of $240,000; easily justifying the
capital expenditure of $60,000 for system
hardware (to fit two trucks plus radio
communications). Because the weigh studies
indicate that the two trucks having OBDAS
are carrying a greater payload than trucks
without OBDAS, expanding the system to cover
the entire fleet should further increase
savings. Providing the system proves to
be dependable, in particular the load sensor
assemblies, OBDAS can provide productivity
and cost improvements to the mines haulage
function.
   In addition to providing measurable production improvements, OBDAS also demonstrated
the ability to develop the information base
and real time monitoring needed to develop
a dispatching system. The microprocessor
is also able to accept additional input,
that could increase its use as a maintenance monitoring and planning tool.

7 ACKNOWLEDGEMENTS

The authors wish to thank the management
of Gregg River Resources Ltd. and all those
from Gregg River's Pit Operation and Mine/
Pit Maintenance Departments whose cooperation and assistance made this project possible.

REFERENCES

Hagenbuch, L.G. 1986. Equipment Management
and Vehicle Management Information Systems.
Presented at NSA Second Annual Operations
Division Meeting.
Hagenbuch, L.G. 1985. An Integrated Truck
Management Information System. Presented
at CIM Second Distruck Five Meeting.
Srajer, V. 1985. Truck Management Systems
Review.

Computer Applications in the Mineral Industry, Fytas, Collins & Singhal (eds)
© *1988 Balkema, Rotterdam. ISBN 90 6191 760 3*

An automatic vehicle status, location, and allocation system for surface mines

E.J.Krakiwsky & H.A.Karimi
University of Calgary, Alberta, Canada

A.B.Kempe & V.Srajer
Energy Mines and Resources Canada, SML/CANMET, Devon, Alberta

T.Lockhart & R.Wade
AVL Automatic Vehicle Location Systems Ltd, Calgary, Alberta, Canada

ABSTRACT: One of the main problems presently facing the efficient operation of a surface mine is the fact that the real-time status and position of vehicles in the mine are not known to the dispatcher. Further, the positions of vehicles are not being related to the existing route networks within the mine area to enable effective vehicle allocation. An Automatic Vehicle Location (AVL) system is recommended as a solution.

The system is composed of hardware (computer, input-output devices, positioning devices, etc.) and software (database management, best-route selection, etc.). The positioning system proposed in this application is a combination of, dead reckoning devices, and GPS satellite receivers. The combination of these positioning devices is to yield real-time 3-D vehicle coordinates accurate to a few meters. These vehicle coordinates are then related to an electronic map of the area and to a 3-D Digital Terrain Model (DTM). Consequently the link is secured between the real-time vehicle location, topography, route networks, and the dispatcher, enabling effective real-time route selection and other allocation decisions. A digital route map is an integral part of the AVL system. This map appears on a CRT visible to the dispatcher, illustrating all route networks in the mine domain, all vehicles linked to the system, and "best route" selections.

1. INTRODUCTION

1.1 OVERVIEW OF SURFACE MINING OPERATIONS

Surface mining involves the basic procedures of overburden removal, drilling and blasting, ore and waste loading, hauling and dumping and various auxiliary operations.

Loading of ore and waste is carried out simultaneously at several different locations in the pit and often in several different pits. Shovels and front-end loaders of various sizes are used to load material onto trucks. Loading times depend on shovel or loader capacity, digging conditions and truck capacity. Often trucks of different capacity are used at individual shovels, and since loading times will vary considerably with truck capacity, queues will form at the shovels. An effective real-time dispatching system can reduce time spent in queues.

Hauling material from the shovels to the dump sites must be accomplished through a network of road links of various lengths and grades. In many Western Canadian coal mines, the topography is very rugged.

Road links can become extremely complex, cover vast areas, and pass through extreme elevation changes. Often, segments are shared by more than one shovel-dump link, and because passing is not permitted, travel times are strongly influenced by the slower trucks. Many mines incorporate both on-road and off-road vehicles which haul specific materials. On-road vehicles are often used to haul coal from the pit along public road links to the stockpile or crusher, while off-road vehicles are restricted to the mine area. Thus, allocation of trucks to haul specific material from a specific pit or shovel becomes a complex problem.

Waste dumping and coal stockpiling is relatively straight forward. The queueing of trucks will also occur at these sites since only a limited number of trucks can dump at a given time. Dump sites must be continually relocated as they reach design volume limits.

Obviously efficient mining operations are strongly dependent on the proper allocation of trucks to shovels and the respective allocation of trucks along the appropriate road links and to dump sites. Due to equipment breakdowns, varying digging conditions and truck capacity, and even coal blending considerations, frequent truck re-allocations are required to maintain an efficient operation. Presently, the shift foreman, in order to maintain an up-to-date awareness of the truck-shovel operations in the mine, must circulate continually around the mine site observing queue lengths, digging conditions, etc., at each shovel and re-allocate trucks as required. A computerized dispatching system incorporating vehicle status and location information can be used to perform this task more effectively.

The often harsh Canadian climate creates many problems in conducting an efficient surface mining operation. Fog, driving rain, sleet and snow cause visibility problems for all vehicles in the mine. Visibility is often restricted to only a few meters making vehicle driving and manoeuvering extremely dangerous. Trucks in particular must reduce their speed considerably, and in areas of restricted manoeuverability trucks are forced to stop for fear of collision with other vehicles. For this reason, a vehicle location-dispatch system must yield accurate real-time vehicle position. In the event of excessive moisture some haul roads may contain grades too steep for trucks to climb or safely descend. An electronic map and dispatch system would enable the dispatcher to identify routes with problem segments, and detour vehicles around these trouble spots if possible.

Auxiliary equipment used in the mine includes bulldozers, graders, water and sand trucks, front-end loaders, service trucks, garage facilities, etc. All of this auxiliary equipment contributes extensively to the mining operation but their importance is sometimes overlooked. Locating and subsequently allocation such equipment throughout the mine is another very important function of an efficient dispatch system.

1.2 PROBLEM OF VEHICLE STATUS, LOCATION AND ALLOCATION

It is generally accepted that the principal objectives of any computerized dispatching system are to maximize shovel loader utilization (i.e. minimize shovel idel time) and maximize truck utilization (i.e. minimize truck queue time)[White et. al., 1982]. In order for the system to accomplish these objectives three basic functions: vehicle status, location and allocation, must be incorporated into the system. Dispatching is a dynamic operation that requires continuous monitoring of vehicle status and location to determine the optimal vehicle allocation, including best route selection, shovel and dump site selection.

To effectively position equipment throughout the mine, an instantaneous status report for every vehicle in the mine must be available to the dispatcher. This status report would include such conditions as queueing, loading, on route to dump, on route to shovel, and breakdown. This status report will be used by the dispatcher to make decisions concerning allocation of personnel and equipment. For example, if a shovel breaks down, trucks can quickly be redirected to a different shovel, distributing the numbers of trucks to suit the different materials, routes, etc. The implementation of a status report transmitted to the dispatcher is certainly a necessity of an efficient dispatch system.

In addition to vehicle status, an effective dispatch system must have knowledge of the real-time position of vehicles in the mine and at facilities outside the mine such as a remote stockpile and loading system. One of the main problems presently facing the efficient operation of a surface mine is the fact that the real-time position of vehicles is not known to the dispatcher. To make truly effective decisions concerning allocation, the dispatcher needs to know the location of all operating vehicles within the mine domain. For example, trucks originally designed to a shovel where a queue is presently forming, could, while "on route to shovel", be efficiently redirected to another shovel via the "best route" only if the real-time location of the trucks is known to the dispatcher.

For effective use of dispatch software, such as a "best-route" algorithm, the real-time position of vehicles must be related to the route networks within the mine domain. Through analysis of the position of vehicles, the dispatch software can conduct real-time allocation decisions such as the re-routing of an individual truck from its present shovel destination to reduce queueing at that shovel. To accomplish this, however, the real time position of vehicles must be related to the existing route networks, including all the vehicle's route alternatives before it becomes committed to a particular route.

A compounding problem is that the topography in the mine area is not related to the existing route networks and to the real-time position of vehicles in the mine.

In order to accomplish "best-route" selection (especially in mountainous regions), topography of the routes must be considered and the obvious effect is has on vehicle travel time.

Clearly, there is a need to obtain real-time vehicle status and location information and relate this to route networks and topography within the mine. Many dispatching systems yield real-time vehicle status information, however, few yield real-time vehicle location. Systems using "sign post" positioning yield position updates at only selected locations. This results in position "outages" between posts making these systems inadequate. The automatic vehicle location system (AVL Mining ^{TM} System) described in the following sections is a unique system. This system yields real-time vehicle status and location information and further relates this information to an electronic map and 3-D Digital Terrain Model (DTM) of the area. Together, the electronic map and DTM contain information on all route net-works including route topography. This ability to relate real-time vehicle status and location information to the route net-works and topography makes this a very effective dispatching and allocation system.

2. THE AUTOMATIC VEHICLE LOCATION (AVL) MINING ^{TM} SYSTEM

2.1 Overview

The AVL Mining ^{TM} System is comprised of a central dispatch unit and many mobile units (Figure 1). Each mobile unit receives position sensor input, processes data, and prepares output for transmission. In addition, each mobile unit is equipped with a communication unit for transferring information to and from the dispatch unit.

The dispatch unit also receives input, processes data, and prepares output for transmission or other purposes, and has a communication unit for the transfer of information to and from mobile units.

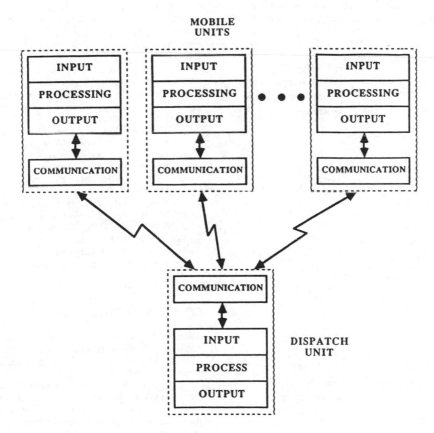

Fig. 1 AVL Mining ^{TM} System overview.

2.2 System Functions

Figure 2 depicts the AVL Mining ^{TM} System specifically designed for application in surface mining. Illustrated in detail is the dispatch unit, a mobile unit and the communication link.

Data input from each mobile unit is as follows:
a) vehicle status report, entered by means of a keypad; and
b) real-time vehicle position, resulting from a combination of one or more positioning sensors.

Data from the positioning sensors is input to the microprocessor, while the status report data is input directly to the communication unit, ready for transmission to the dispatch unit.

The operations performed in each mobile unit include operation initialization, calibration of sensors, computation of real-time positions, map matching, and transferring information to and from the dispatch unit.

A display device is used in each mobile unit to enable checking of information sent from the mobile unit and enable con-

firmation of instructions received from the dispatch unit. Voice synthesis (output) will be used primarily for transferring urgent information to the operator of the mobile unit to ensure that information is not missed, and to retain the safe efficient operation of the vehicle. For example, when two vehicles are approaching one another and visibility is limited to only a few meters, a voice synthesizer would first acknowledge the approaching vehicle to each operator, then simply "call-out" the relative tangential separation distance between the vehicles (at a predetermined interval) until they have safely passed one another. The operator would refer to the display device only if the distance was not heard.

The dispatch unit is responsible for overall information management, allocation and dispatching. The only manual input to the unit is through operator interaction by means of:
a) a keyboard; and
b) a mouse and/or light pen.

The tasks performed by a dispatch unit include operation initialization, simulation, best-route selection, allocation dispatching, Digital Terrain Model (DTM) data management, and communication with mobile units.

The only output device necessary for the dispatch unit is a Cathode Ray Tube (CRT) display for viewing the 2-D projected DTM map of the entire mining area, including remote facilities. The real-time position of all vehicles with mobile units is superimposed on the DTM along with pertinent shovel and vehicle allocation information. Note that a 3-D DTM map exists and is used throughout, especially for "best-route" selection, however, only the 2-D projected DTM map is used for display on the CRT to reduce the complexity of the 3-D image. The dispatcher does not need to visually observe the third dimension, its contribution is accounted for through system software.

3. HARDWARE

3.1 System Integration

Hardware components of the designed AVL Mining ^{TM} System are shown in Figure 3. The keypad necessary for each mobile unit is a customized product based on specific functions the vehicle operator requires for efficient interaction with the dispatch unit. Obviously, the necessary keys for a shovel will differ from those of a truck, grader, etc.

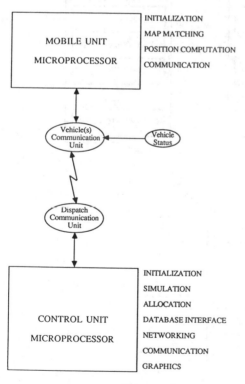

Fig. 2 AVL Mining ^{TM} System Components

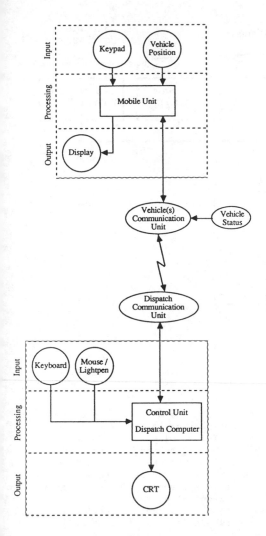

Fig. 3 AVL Mining^{TM} System hardware
        components.

The mobile unit positioning sensors are described in detail in the next section.

The mobile unit processor must be as compact as possible for obvious reasons. This dictates the employment of a microprocessor, either a custom design or simply a microcomputer. A small microcomputer with appropriate speed and capacity is proposed here. The criteria for the choice of a microcomputer should include the ability for present and future expansion. Therefore, a PC compatible system, such as the Little Board^{TM}/186, manufactured by AMPRO, is recommended. Some of its characteristics include: 8 MHz, 80186 CPU with DMA and counter/timers, 512K RAM

and 2 RS 232C serial ports. The position computation cycle time is approximately 1 second, therefore, this system is sufficient.

The display device in each mobile unit is either a Liquid Crystal Display (LCD) or a Light Emitting Display (LED), which is mainly used for checking sent text information. The voice synthesizer presently supports only a limited number of words and phrases which must be built into the system. Thus, voice synthesis would only be used for relaying urgent messages and could not replace standard radio communication. The voice synthesizer is an off-the-shelf hardware-software system. In the future, a flat panel display will also be installed in mobile units to display map data as well as text.

The PC/AT compatible (equipped with a keyboard, mouse, and/or light pen), with a high capacity hard disk drive and floppy disk drives, is proposed for the dispatch unit computer.

The CRT display unit, along with an appropriate graphics board at the dispatch unit, must include the following character-istics: high resolution, color graphics (at least 16 colors), and high speed graphics.

The data communication unit in each mobile unit is an off-the-shelf modem with a minimum data transfer rate of 1200 baud. The data communication unit in the dis-patch unit, having the same data transfer rating as that for the mobile unit, is interfaced to a separate microprocessor. The required microprocessor is a PC (XT/AT) compatible with a hard disk and a floppy disk drive without a keyboard and a CRT display. The Westar Production Monitor and Control System (PMCS) is an implement-ation of this system [Westar Engineering].

3.2 Positioning Sensors

The two basic categories of positioning$_{TM}$ sensors incorporated in the AVL Mining positioning system are Dead Reckoning (DR), and satellite positioning. (Figure 4).

The DR sensors include height sensing, speed sensing, and direction sensing.

Height sensing is accomplished using an AIR Intellisensor precision altimeter. This precision aneroid device is fully temperature compensated and has a resol-ution of 0.1 mb. This means the sensor can reliably measure vertical movement to 0.8 m. The device is microprocessor controlled and provides digital output in units of elevation (metres) or pressure (millibars). It can be calibrated against

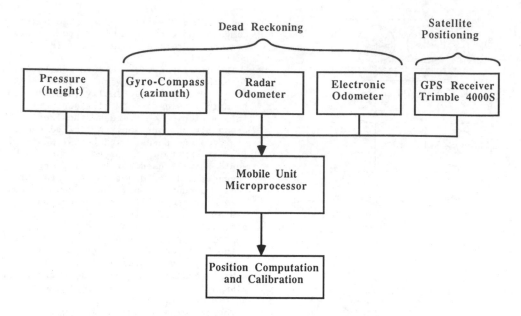

Fig. 4   Positioning Sensors

any known altimeter or a known elevation
to provide true elevation. Extensive
survey experience has proven a sensor
accuracy of about 2m in the majority of
cases.

All dead reckoning systems require position initialization. Initialization can be
accomplished simply by means of a button,
pressed by the operator when the vehicle
is located at a known position. Obviously,
the position initialization would include
a height initialization as well. In
addition, an Intellisensor would be installed at the dispatch centre at a known
elevation, enabling altimeter re-calibration
by vehicles at the centre. Experience has
shown that re-calibration every hour of
vehicle use is adequate to maintain the
desired accuracy.

Speed sensing is provided by two independent devices. One device, a Nu-Metric
precision odometer, counts revolutions of
a non-driving wheel. A micro-processor
converts this into speed using a simple
calibration routine. Nu-Metrics have been
used for electronic survey chaining. On
hard surfaces, they provide accurate speed
(or distance) measurement to one part in
one thousand.

Because the accuracy and reliability of
a precision odometer in a mining environ-

ment is unknown, a Doppler speed log will
also be incorporated to measure vehicle
speed. This device is also an off-the-
shelf product and retails for less than
$1000. They are used extensively on farm
machinery, and display speed in units of
kilometers and tenths of kilometers per
hour. Eventually, substantial field
evaluation will enable selection of one of
these two devices as the standard speed
sensor.

Similar to the height sensor, the speed
sensor is calibrated against known distances travelled between initialization
and check points. It is also proposed to
use dual Doppler speed logs to sense
change in direction.

Heading sensing is provided by a slaved
directional gyro. Slaving is provided by
a flux valve. The AIM 400 system, an off-
the-shelf avionics product utilized in
aircraft auto-pilot systems, is recommended.
The gyro is electric, requiring a 12 VDC
power supply. Flux valves are magnetic
compasses which sense the direction of the
earth's magnetic field by inductance
through their windings. They do not align
or point in the manner of ordinary
compasses and are considered accurate to
approximately 1.5 degrees. They are,
however, affected by magnetic variations

as is any compass, and they also must be "swung" after installation to account for the magnetic effect of metal in the vehicle itself. The long term stability of the flux gate compass is excellent, however, in the short term, it is still affected by proximity to other vehicles and steel structures in general.

The directional gyro drifts over time and thus has to be reset at regular intervals. The short term stability of the directional gyro is excellent, yielding an accuracy in the order of 0.5 degrees. The use of dual Doppler speed logs for detecting change in aximuth could eliminate the requirement for a gyro.

There are true north seeking gyros available which are used extensively for marine applications, but they are large and very expensive.

In certain mine environments the flux gate compass may be adequate without a directional gyro. The validity of this statement will be established in the first year of operation.

Similar to the speed and height sensors, heading sensors must also be calibrated during initialization and at check points.

The accuracy of a dead reckoning system is affected by two factors. First, sensor limitations and sensor drift contribute to position error (increasing with time from calibration) and second, map matching reduces or eliminates position error by comparing computed position with "observed" position.

For example, over a 10 km distance, an along track error of 10 meters would be expected from a system providing 1:1000 accuracy. An across track error of 262 metres would be expected (1.5 degree accuracy) and a height error of 2 meters would be likely. These figures assume the sensors are properly calibrated and operating within specifications. They also assume that the errors are systematic and not random. Map matching reduces these errors by comparing a computed position with the reality of the digital map. For example, if the vehicle is "constrained" to be on a road (at least in certain areas), heading errors are reduced to the width of the road. If the heading sensor observes a change of direction corresponding to a known change of direction for the section of road, this enables a position update and essentially all cumulative position error is eliminated.

The height sensor is used similarly, but its usefulness depends largely on the topographical relief found in the mine area.

The satellite positioning sensors incorporated into the positioning system are Trimble 4000S Global Positioning System (GPS) satellite receivers. In order to fully utilize the GPS in real-time, a receiver configuration is incorporated which uses kinematic relative positioning [Schwarz et. al., 1987]. With a stationary receiver located at the dispatch centre, all mobile receiver positions are determined relative to the stationary receiver. Using this method a real-time position accuracy of 0.5 - 1.5 m is attainable. Because of the present limited satellite constellation, GPS positioning technology must be used in conjunction with DR devices, which constitute the main positioning system. GPS could play a more important role in the future once the full constellation of satellites is available (early 1990's).

Measurements from all positioning sensors are input into the microprocessor resident in each mobile unit. The dispatch unit, via the data link, receives the result of calibrated and computed real-time positions which are determined by each mobile unit

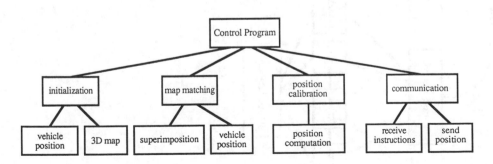

Fig. 5 AVL Mining ^{TM} System Mobile Unit Software.

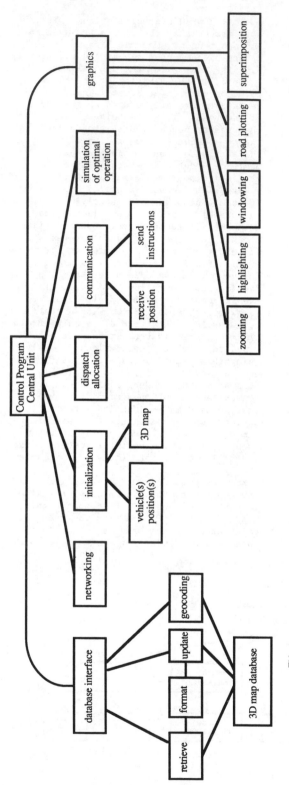

Fig. 6 AVL Mining ^{TM} System Dispatch Unit Software.

microprocessor.

4. SOFTWARE

4.1 Mobile Unit Software

The software required for each mobile unit microprocessor is composed of initialization, position calibration and computation, map matching and communication modules (Figure 5).

Initialization is the process of "setting up" operations. Ideally this would be done at the start of each shift. It essentially transmits the mobile units initial position to the dispatch unit. Consequently the mobile unit will start each shift by transmitting its calibrated real-time position, status report, etc. This transmission of information will be conducted at a predetermined time interval or when a request is made by the dispatch unit.

The map matching module tracks the mobile unit's real-time positions. Through this technique, real-time positions of the mobile unit are correlated to the DTM map of the mine area. At certain nodes in the mobile units tracks, coordinates from the DTM map are used to accurately define the units position, and enable a position update.

A communication module is used to transmit information (e.g. position, instructions, etc), between the dispatch unit and mobile unit.

4.2 Dispatch Unit Software

The software required for the dispatch unit of the AVL Mining ^{TM} System is composed of initialization, simulation, allocation-dispatching, communication, DTM data management and graphics modules (Figure 6).

The simulation module receives information on each mobile unit's real-time positions and status and uses this information to simulate an optimal mining operation based on status and location only.

The allocation-dispatching module takes each mobile unit's real-time position, status, and simulation results and converts this information into real-time vehicle allocation decisions. The module accomplishes this through application of a "best route" algorithm which simply determines the shortest distance between the vehicles present location and the desired destination. Several algorithms have been developed for the standard node to node shortest distance problem. Once appropriate vehicle allocation decisions

have been made, they must then be converted to routing instructions to be transmitted to the appropriate mobile units. This is one of the more important roles of the AVL Mining ^{TM} System and thus the effective performance of this module is crucial.

The communication module, similar to that of the mobile unit's transmits information between itself and the mobile units.

The DTM data management module is an interface between the control program and the physical storage of the DTM database. Its functions include retrieval, formatting, updating and geocoding of the DTM database.

The graphics module is a collection of subroutines to handle graphics output for display. Operational functions include zooming, high-lighting, windowing, road plotting and superimposition of each mobile unit's position on the display.

5. CONCLUSIONS AND RECOMMENDATIONS

In this paper, an overview of operations in a surface mine is presented in an attempt to recognize inefficiencies and problems that could be eliminated through the implementation of an effective automatic vehicle dispatch system. The two main problems recognized are the formation of queues at shovels and alternatively shovels sitting idle waiting for trucks. Effective vehicle allocation through an automatic vehicle dispatch system is seen as a means of minimizing truck queue time and maximizing shovel utilization. In addition, an automatic dispatch system would yield other benefits as follows:

1) reduction of distance travelled through optimal route selection,
2) improved response to emergencies and delays, such as vehicle breakdowns, shovel repositioning, etc.,
3) reduction of production differences, common with manual dispatching techniques,
4) improved safety through accurate vehicle location,
5) improved scheduling resulting from accurate vehicle location,
6) instantaneous status reports available on demand to management.

The AVL Mining ^{TM} System designed by AVL Automatic Vehicle Location Systems Ltd. of Calgary, Alberta, described in this paper is intended to reap the above mentioned benefits. This system is recognized for its ability to achieve the high standard of accuracy necessary, and because it satisfies all the functional requirements demanded of the system, while remaining very cost competitive with other systems.

Development, implementation, and component testing must be undertaken to eliminate unnecessary sensors, improve graphics capabilities, and optimize software components. The implementation of such a system in Canadian mines would substantially improve the operating efficiency of our mines and further improve Canada's competitiveness in world coal markets.

6. ACKNOWLEDGEMENT

This research is supported by an NSERC grant, Ottawa, CANMET, Edmonton, and AVL Automatic Vehicle Location Systems Ltd., Calgary.

REFERENCES

Automatic Truck Dispatching Increasing Productivity"; World Coal, Vol. 8, No. 2, March/April, 1982.

Chatterjee, P.K., Brake, D.J., "Truck Dispatching and Simulation Methods in Open-Pit Operations"; CIM Bulletin, November, 1981.

Chironis, N.P., "Automated Truck Dispatching .-. Study in Europe Finds it Pays"; Coal Age, Vol. 91, No. 5, May, 1986.

Dijkstra, E.W. (1959). A note on Two Problems on Corresiom with graphics, Numerische Mathematik 1, 269-271.

Karimi, H.A., Krakiwsky, E.J., Harris, C., Craig, G., and Goss, R. (1987). A Relational Database Model for an AVL System and an Expert System for Optimal Route Selection. Eighth International Symposium on Automation in Cartography, AUTO_CARTO 8, Baltimore, Maryland, March 30 - April 2.

Krakiwsky, E.J., Karimi, H.A., Harris, C., and George J. (1987) Research into Electronic Maps and Automatic Vehicle Location. Eighth International Symposium on Automation in Cartography, AUTO-CARTO 8, Baltimore, Maryland, March 30 - April 2.

Nenonen, L.K., Graefe, P.W.U., Chan, A.W., "Computer-Aided Study of Truck Dispatching in an Open Pit Mine"; National Research Council Report (Computers), Vol. 10, No. 1, Ottawa, Canada. 1981.

Schwarz, K.P., Cannon, M.E., Wong, R.V.C., "The Use of GPS in Exploration Geophysics: A Comparison of Models"' Presented at the XIX General Meeting of the International Union of Geodesy and Geophysics, Vancouver, British Columbia, August 9-22, 1987.

Westar Engineering, PMCS User Brief"; Brochure; Westar Engineering Ltd., Vancouver, British Columbia, 1986.

White, J.W., Arnold, M.J., Clevenger, J.G., Automated Open-Pit Truck Dispatching at Tyrone"; Open Pit Mining, E&MJ, June, 1982.

Computer Applications in the Mineral Industry, Fytas, Collins & Singhal (eds)
© *1988 Balkema, Rotterdam. ISBN 90 6191 760 3*

On-line on time: Electronic payroll from front line supervisor to bank deposit

R.C.Aelick & M.S.Chomiak
Inco Limited, Ontario Division, Copper Cliff, Canada

ABSTRACT: The development, implementation and highlights of a major payroll administration system at Inco Limited's Ontario Division is described. The network of nearly 500 terminals assists front line supervisors to input raw data and retrieve appropriate data base information. An organization of more than 6500 hourly paid personnel, distributed throughout 10 mines and 15 geographically separate surface plants is serviced on line. Manual auditing of input documents is all but obsolete. The result is a lower cost payroll process and multi-level access to current information. To date more than $100 million dollars of payroll have been dispersed.

1 BACKGROUND

Inco is a large, fully integrated mining, milling, smelting & refining organization. Approximately 360 million lbs. of finished nickel and 250 million pounds of finished copper as well as precious metal bi-products are produced annually. Revenue of about $1.5 billion per year gives an indication of the size of the company. The Ontario Division contributes more than 60% of the annual nickel production with a payroll in excess of 8000 people. The Division had, over time, built up a large, manually driven, infrastructure to handle the administration functions for the Company. Difficult market conditions and a low margin on our primary product brought 13 consecutive financial quarters showing a loss. The organization was faced with frightening prospects. In particular, the nickel business was faced with a three year decline that saw the price of our primary product slide from above $3.00/lb. to a low of $1.50/lb. nickel.

It was in this environment that the management team faced a precipitous situation. The managerial response was one of aggression. This meant major initiatives to drastically slash unit operating costs. The enterprise moved into a very basic survival mode. "Soft Spots" were attacked with an eye to making major improvements with only limited infusion of capital. Every system came under very close scrutiny. The topic of this presentation, computerized timekeeping is only one of the active and dynamic tactical maneuvers that contribued to the struggle to survive.

The development of a fully computerized system of payroll administration evolved. More than $100 million of payroll has been dispersed by the electronic system at time of writing. This is accompanied by substantial reductions in the administrative manpower requirements.

2 THE ANALYSIS - Problem Definition

The zero base approach was used in questioning every process that was at work within our organization. This approach was spear headed by our Mines Research Group. A small task force approach was taken to analytically detail our mining methods, systems and organization. The administrative system that impacted the front line supervisor was thoroughly investigated with the intent of stream lining. In this process the man hours of effort required on a daily basis to ensure accurate information for payment of employees seemed disproportionate and excessive.

2.1 Technique

The resultant analysis of paper flow and accounting systems led to the development of a typical "information (paper) flow" chart for the daily routine of paying our people (see figure 1). It became obvious that there was a great deal of manual involvement in our payroll and accounting system even though there was a mainframe computer "in charge" of the process. There were untold interations of checks and balances to ensure that every employee at work would be paid every day. The pay rate, the account number, the incentive time and numerous other variables were all audited daily by clerical staff to ensure accurate input into the computer. The intricacy of the manual process itself suggested the need for detailed analysis.

2.2 Analysis Results

The analysis of the existing paper flow led to the following conclusions about the traditional system:

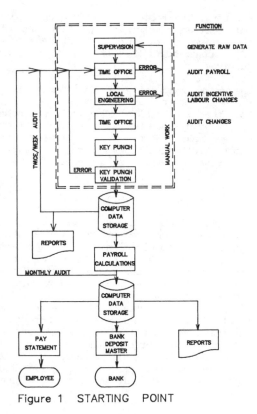

Figure 1   STARTING   POINT

* The existing computer based system did not include the total process.

* The manual input part of the system was supported by a substantial infrastructure of staff.

* The manual intervention was, by its very nature, a system to correct, support and question the quality of the initial input.

* This was a highly interactive process that continued on a daily basis. It was relentless, repetitive and demanded deployment of substantial human resource.

* The need to accomplish the objectives of the process were certainly long term. (We must account for and pay every employee with the appropriate pay)

* There was substantial opportunity to streamline the process through automated Management Information System techniques and to custom tailor it to our needs.

The analysis results mandated a critical review of several items involved in the process of paying our people.

2.3 Conclusions of Analysis

It was recognized that there are basic parameters of the process which must remain constant. The raw input data (i.e. who worked, when, on what job, etc.) must be generated by the front line supervisor. He is close to the action and in control of actually initiating the process. At the other end of the process, each employee expects and deserves accurate, punctual deposit of his earnings to the bank of his choice.

2.4 Opportunity For Change

The process in-between the aforementioned input and ouput is at the discretion of the system. "The System" was put in place by management and could well be massaged, revamped, discarded or replaced by management without impact on employees. The decisions and implementation of the process to revitalize the step between "input" and "output" are the subject material for the remainder of the paper.

482

3 DON'T AUTOMATE A POOR PROCESS

3.1 Side Stepping Historical Problems

It is a known fact that automating a "poor" process will provide less than satisfactory results. In an attempt to avoid such pitfalls, the total system was re-organized and planned to accommodate computer application. To accomplish this, it was necessary to critique the exsisting system. This process, as is usual, took us back to the basic philosophy. It became apparent that the original system had been conceptualized on a fully manual process of historic initiation. Subsequently the original process had been amended on numerous occasions to accommodate change through time. The administrative amendments were paralleled by "quick fix" responses to abuses of the system. Problems with quality of input were "repaired" by implementing inspection and audit procedures. This was an opportunity to start fresh.

3.2 Clearing Up The System Logic (Design)

Problems associated with the basic process not involving computerization were addressed first. An example of this process was the application of the All Mines Incentive Plan. Traditionally it was possible for an individual employee to work an 8 hour shift and be on incentive for only a portion of the shift. The decision of how long the employee worked on incentive was at the discretion, control and input of the immediate supervisor. When you consider the geographical remoteness of many of the underground workplaces, you can appreciate the problems that the supervisor faced each day. One supervisor may be responsible for 10 or more geographically distinct workplaces. He may have the opportunity to visit each workplace only once or twice a day for relatively short periods of time. Yet, he is expected to make a judgement call to the nearest ½ hour, as to how long each crew worked on incentive on a daily basis. Simply on the sheer number of judgement calls made each day it is easy to recognize the potential for error. This in turn resulted in a major staff auditing effort upstream to ensure consistency across all supervisors on both shifts throughout the individual mine as well as throughout the division. The auditing process was treating a symptom rather than the root cause in this particular case. The result was the building of substantial infrastructure.

3.3 Challenging Root Causes Not Symptoms Of the Problem

When the true cause of the problem was addressed, the system became much easier to manage. The incentive administration was revised to incorporate appropriate "allowances" such that there was no need to split time. Every incentive worker would be charged 8 hours of incentive work every day through appropriate adjustments to process allowances in the incentive rates. There would be no need for judgement calls on the part of supervision with regard to split shifts. This in turn negated the need for the incentive audit function of every man every day. There became no need to audit time changes. Every employee on incentive would be charged 8 hours consistently, every work day.

This basic change to the management of the incentive system had major dramatic impact on the "time keeping" function. It was this type of review of the basic philosophical reasoning for the existance of infrastructure and sub-systems that yielded results. The basic question of "WHY" applied at every stage of the process. Tracing the "WHY" back to its very roots. This system introspection took us back to the very basics. Doing this set the stage for the birth of a new system. We were trying not to automate a "poor" process! The streamlining would make automation simpler and easier.

4 LAYING THE FOUNDATION

4.1 Setting Objectives

Having clearly defined the bare essentials of the process (input & output), we embarked on dissecting the internals of the system to gain insight, critically assess the individual need and project consequences if each incremental step in the process were discarded.

The proposed computerized reporting system was intended to address time reporting, supplies ordering, job postings, absentee records, misconduct records and safety reporting. The system was to be accessible from key points located in the mine and surface plants (i.e. lunchrooms, surface offices, warehouses, etc.), all connected to a central computer.

The proposed system would have several advantages over the existing system. It would: -

* minimize manual work
* eliminate duplication
* improve utilization of the foremans time
* increase accuracy
* increase timeliness
* provide cross reference

Based on the above, a preliminary set of the basic objectives were established.

Objectives:

1. To permit the front line supervisor to report employee's time via computer terminals.

2. To minimize the amount of reporting done by supervision, by automating the repetitive data transactions.

3. To minimize or eliminate error checking and corrections at subsequent steps in the process by providing an editing and correction capability at the data source.

4.2 Development Strategy

The administrative housekeeping of the total system was now complete. It was time to develop a strategy for system development and pilot a new system to test the applicability. The process was put into motion. An Industrial Engineer was assigned to the project. His mandate was to develop a working system. He was temporarily assigned to computer services to learn Natural programming language and develop a prototype time reporting system.

4.3 System Mechanics (Management.)

The system that was designed was intended for limited application to test the concept. The basics of the system are highlighted in the attached flowsheet (Figure 2).

Input includes individual employee identification, record of attendance, the account number for the cost accounting system, the appropriate designation for the incentive system, work classification of the individual employee, as well as other basic data. The system was developed to minimize the time required for input by the supervisor each day. This was achieved by allowing the supervisor to recall any one day within the last 2 weeks to use as a base template. Once this was established it was only necessary to input

the hours worked and incentive code for each individual employee "if" everything else remained constant.

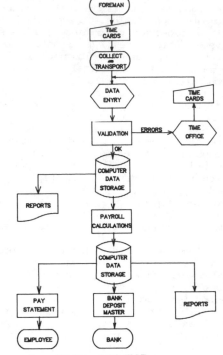

Figure 2   PILOT   PROJECT

The system would also generate a weekly summary on hardcopy at which time each supervisor would be required to review his weekly input to ensure its accuracy. He would then sign the hardcopy and submit it to the accounting people. This was deemed necessary to meet accounting and auditing purposes on the short term.

5 IMPLEMENTATION

5.1 Pilot Project

The system was essentially developed and co-ordinated by an industrial engineer. When it came time for the initial implementation this same industrial engineer was charged with developing the implementation process. The start up involved a trial at only one mine site.

This made it necessary for all of the front line supervisors at the mine to be trained on the system. Computer training was initially viewed as a frightening experience for the senior mine foremen who had never previously been exposed to actual input on a computer terminal. Surprisingly, these traditional, pragmatic people adjusted very quickly. Soon they became proud of their newly developed skills. When the system went live there were many growing pains. The introduction of the new system, combined with newly trained input personnel and the new concept provided many minor collision courses. All of these were resolved by having the development engineer readily available to respond to problems immediately.

5.2 Skeleton Up and Running:

The pilot system was up and running actively within a time frame of only man-months of work. This prototype was left in operation to assist in accruing some preliminary experience while the main company-wide system was under development. The trial period proved the basics of the operating system. There were no longer any doubts about the following:

The application of the concept, (i.e. to minimize manual intervention).

The reliability of the hardware in a hostile environment.

The ability to develop applicable software.

The recognition that the system was a skeleton.

5.3 Hardware

The appropriate hardware was installed, initially on surface and later underground in the lunchroom for ease of access for the supervisor. The first terminal was initially installed and covered with a plexiglass enclosure when not in use. There were very few problems with the hardware. The hostile environment for a computer terminal, i.e. dust, humidity, etc. did not present any major problems.

5.4 Software:

The pilot project dealt only with the basics of the system. It made few attempts to deal with the exceptions such as the numerous anomalies built into individual

callout provisions, hoisting delays and the accounting systems for miscellaneous training assignments such as first aid, fire brigade, mine rescue etc. The initial design approach was hardcoded, for a relatively small population in an attempt to test the operation of the system.

5.5 Human Intervention:

The developmental program covered all of the simplest of the input information. However, when the unusual circumstances started to occur, as they always do, then it was necessary to handle all of the exceptions manually. This was accepted in the initial process but was not deemed acceptable on the long term with the permanent system. Therefore as exception circumstances occurred, each was logged for reference to be included in the permanent system.

5.6 Meshing Objectives From Various Prespectives:

Once the pilot system was operating reliably, the total process was reviewed with an eye towards establishing an enhanced version. It would all but eliminate manual auditing and yet comply with all of the "organizational needs" such as assuring accuracy and control, providing the appropriate management reports and ensuring the existance of anticipated audit trails. This led to a clear definition of objectives for the division-wide system.

6 SYSTEM EVOLUTION, THE BIG PICTURE

6.1 Division-Wide Objectives:

The main objective was to reduce the cost of collecting, processing, distributing and reporting time and attendance data for unit work force.

To increase the efficiency and productivity of involved personnel through optimization of information flow, elimination of duplication and redundancy, automation of calculations and data retrieval, and validation of input data wherever possible.

To produce accurate, meaningful and timely data, along with reports (hardcopy or screen) for various purposes:
- Detail data for payroll
- Summary & status reports for management
- Summary data for accounting purposes

To make information available from data bases to all plants where applicable.

To provide corporate management with necessary current and historical information as required.

To store and make available reasonable amounts of permanent and temporary historical data for purposes of evaluation, forecasting, statistical analysis or reference. To provide security and reliability of data. To provide reasonable system flexibility and expansion capability.

6.2 Up-Dating The Development Plan:

The test results from the pilot project forced some re-evaluation of the mandate and provided detailed information for design of an all encompassing major system. A renewed development plan was established as outlined below:

(1) Define the objectives of the system and prepare a report.
(2) Review the reports with user groups.
(3) Format all output screens and end reports to be produced.
(4) Outline menues to lead users through application.
(5) Take reports back to users for explanation, review, and approval.
(6) Design the database.
(7) Select the software.
(8) Determine the type of computer technology to employ;
  - network of micros,
  - stand alone multi-user
  - terminals linked to central computer
  - medium for inter-machine communications: - twisted pair, co-axial cable, or fibre optics.
(9) Hardware selection
(10) Logical documentation of software to be developed.
(11) Write and debug programs.
(12) Prepare operating & training manuals
(13) Install system, debug hardware problems and open files
(14) Enter historic data.
(15) Train foremen and other users
(16) Further development

Highlights of the Completed System

The follow through of the above development plan led to the installation of 480 terminals, miles of co-axial cable and the need to upgrade existing mainframe capability (See Figure 3 for flow sheet).

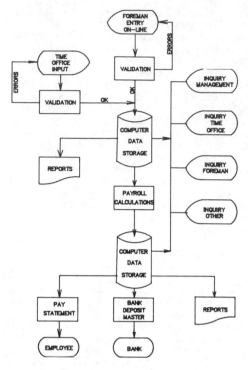

Figure 3   COMPLETED SYSTEM

Currently, one-line time reporting consumes between 30 to 35% of the main frame resources. In return for this substantial consumption of computing resources, the system meets all of the objectives outlined. The functional facilities available through the system include:

1. Appropriate payment and documentation for each individual employee.
2. The automatic development and maintenance of a work history for each employee.
3. Multi-level force networks allow an overview of the workforce by individual, or supervisory level i.e. foreman, general foremen, superintendency, etc.
4. Permanently documents time away form normal work activities such as:
  - vacation
  - injury or illness
  - bereavement
  - jury duty
  - union business
  - penalty, etc.
5. Maintains appropriate medical surveillance records such as miner's chest x-rays.

6. Document overtime records by individual employee and monitors overtime worked in accordance with company guidelines and government regulations.

7. Provides a major base of input data into cost accounting systems as it applies to labour charges.

8. Documents Employee Relations activities such as job postings and job transfers.

9. Provides basic reminder mechanisms as appropriate for two separate collective bargaining agreements, and all wage items. Some are as remote as wage maintenance due to impact of technological change.

10. Opportunity for any or every management employee to interrogate the system to gain information about his particular part of the organization.

11. Automatic generation of all appropriate management reports.

12. An excellent level of security which effectively prevents inadvertent access by unauthorized personnel.

7 CONCLUSIONS

The successful development and implementation of the timekeeping system was heavily dependent upon senior management commitment, an appropriate strategy and a willingness to accept the necessary change from the traditional method.

The management team recognized that the impetus for change had to be driven top down and that implementation must come from the bottom up. The technology and know how to implement this system had to be coupled with the people involved to establish a mind set. That meant creating a positive environment by involving people in the system development, communicating and surveying for feedback as the system was being developed.

The critical analysis of the "traditional" system led to recognition of the need to change long established practices. The system had to be re-designed and made simpler "before" automating the process. The removal of the root causes of problems which mandated manual intervention required both commitment and willingness to risk possible ramifications.

The inclusion of a pilot project in the strategy initially allowed confirmation of the concept and the opportunity to learn from mistakes on a small scale.

The resultant division-wide system makes a major contribution to improving supervisor and administrative staff productivity and efficiency. There are also a number of spin-off effects which are now available from "doing it". The terminals in the workplace set the stage for immediate future enhancements. The management team is on-line, on time!

Computer Applications in the Mineral Industry, Fytas, Collins & Singhal (eds)
© 1988 Balkema, Rotterdam. ISBN 90 6191 760 3

Computer applications in blast vibration monitoring

David Sprott
Noranda Research Centre, Pointe Claire, Quebec, Canada

ABSTRACT: The effects of underground blasting on the stability of hangingwalls,
pillars and other rock structures is an inherent part of the Noranda Geomech-
anics/Mine Design Program to improve mine stability, productivity and reduce ore
dilution. The blast vibration monitoring program at Noranda Research aims to
develop controlled blasting techniques. This is done through the measurement
and analysis of peak particle velocities and accelerations, actual delay firing
times and vibration frequencies. This paper describes the various computer
technologies used at Noranda to monitor and assess underground blasts. In
particular, the use of readily available software packages for waveform analysis
are presented, as are recent advancements in the use of digital recording and
personal computer based systems.

1 INTRODUCTION

Blasting is the first operation of the
many rock comminution processes within
a mining operation. The efficiency of
blasting relates to the economics of
producing a suitably fragmented product
while creating an excavation that must
be stable to a certain level of risk.
Unfortunately, rock breakage and rock
"damage" have the same source and only
the degree of damage differentiates the
two (i.e. good fragmentation or
dilution from overbreak).

The design of mining excavations must
be based on the quality of rock after
blasting. The effect that poor blasting can have on this quality can be
significant. Furthermore, a lack of
detailed monitoring of the blasting
operation often results in uncontrolled
blasting practices and thus stability
problems, explosive wastage, dilution
and overbreak.

This paper outlines the development
at Noranda of suitable equipment for
the monitoring of underground production blasts. In particular, the use
of readily available computer analysis
techniques are discussed as is the
employment of computer hardware in the
design of next generation instrumentation packages. Noranda is actively
pursuing the development of complete

mine monitoring technologies of which
blast vibration measuring plays an
important role.

2 BLAST MONITORING

To employ controlled blasting techniques in routine production blasting,
performance assessments must be performed. This is done using blast vibration monitoring equipment that consists
of multi-channel recorders, vibration
sensors, digitizing equipment and
analysis techniques. The level of
measured vibration, generally stated in
terms of peak particle velocity, can
depend on many variables including the
type of blast, explosive, confinement,
timing and rock mass quality. The
amount of "damage" experienced will
depend largely on the quality of the
rock mass and excavation shape or size.
Finally, the acceptable amount of
"damage" will depend on the purpose for
the final excavation - a stope with a
one year life or an access drift with a
10 year life.

Three primary purposes for monitoring
blast vibrations exist and are as
follow:

1. to provide a means of verifying
correct delay firing times and
sequence;

2. to allow effective control and assessment of delay interval, quantity of explosive charge and overall blast design (pattern), and;

3. to permit a quantitative analysis of blast induced vibrations as they relate to rock strength and damage to underground rock structures (pillars, hangingwàlls, drawpoints etc.).

Detailed blast vibration monitoring, and subsequent controlled blasting practices can thus provide a practical means of reducing damage, dilution of ore grades, secondary blasting of boulders and explosive consumption.

3 DESCRIPTION OF INSTRUMENTATION

3.1 Sensor Selection

A critical component of any vibration monitoring device is the selection of a suitable transducer. These may be either velocity measuring devices or acceleration measuring devices. Both are capable of providing the effects of passing strain waves in rock but practical limitations exist for both.

Ideally, an inexpensive transducer with response characteristics covering the range induced by underground blasting is desired. Practically speaking, an insensitive yet responsive, shock resistant and disposable unit is required. To meet these specifications a very low cost acceleration transducer was developed at Noranda Research by the Instrumentation Department. The unit consists of a   piezoelectric cell and simple voltage following circuitry, is responsive from 20 Hz to 3000 Hz at $\pm$ 3 dB and can withstand up to $\pm$ 150 g forces of acceleration. Of particular importance is that the unit can be permanently grouted in place and abandoned upon completion of testing. Commercial accelerometers would not permit this due to their high purchase costs.

3.2 Recording Equipment

To meet the criteria of being a portable, rugged and inexpensive monitor, tape recording equipment was selected for initial monitoring at Noranda operations. A Fostex 4-channel tape recorder, capable of recording frequencies from 20 Hz to 14000 Hz with a signal-to-noise ratio of 60 dB was therefore chosen. The unit employs compact cassette tapes and tape travel is 3 3/4 in/s, or 95.3 mm/s (double normal speed). Overall dimesions are 80 (h) x 295 (w) x 390 (d) mm and total weight is only 3.5 kilograms.

The recording unit allows the use of one well-located tri-axial sensor (three orthoganol transducers) for each blast recorded. It is realized that additional sensors will allow a more detailed analysis of a blast, however, with this system it is felt that about eighty percent of the analysis can be done. This data can go a long way to improving blasting practices with minimum cost and complexity.

As will be discussed in a later section, Noranda Research is presently active in the employment of digital recording systems capable of 8 to 16 channel monitoring.

3.3 Fibre Optic Blast Timing

A fibre optic blast timer was developed for two reasons. Firstly, it establishes the actual zero base time of a blast to permit evaluation of delay accuracy. Secondly, it allows unmanned operation with full event capture by automatically turning-on and off the recorder using a fibre optic connection to the blast. This system comprises a length of 50 micron fibre optic cable, a fibre optic reciever and a high frequency signal generator.

The light sensing end of the fibre optic cable is fastened to the surface primacord or to the first charge to fire. Once the recorder is turned on by a separate timer, an 8 kHz signal is recorded onto each of the four recording channels. At blast initiation the flash of light from the surface cord, or explosive, is transmitted to the blast timer and terminates the signal thus providing a very pronounced cut-off point as shown in Figure 1. This point then acts as a zero time base point for determining actual delay firing times. (The point can be adjusted to allow for primacord burning speeds).

This same system can also be used to determine accurate wave propagation velocities in rock and has the advantage of using an optical system so that no current is leading directly to the blast initiation system.

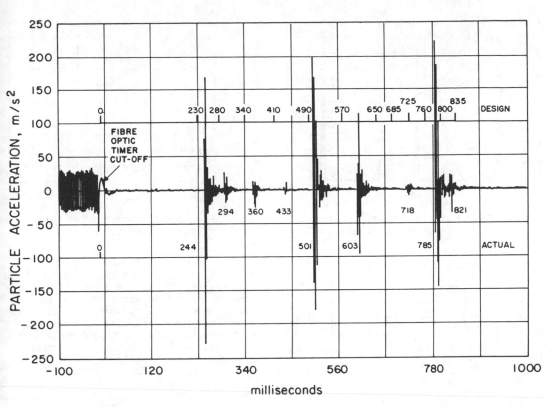

Fig.1 Vibrations produced by a large diameter underground blast showing fibre optic cut-off

3.4 A/D Conversion and Computer Equipment

An important aspect to the entire system is the final analysis of recorded information. The analysis must be within an acceptable level of accuracy, must be easy to use and should allow full waveform processing.

For the initial Noranda system utilizing the tape recorder, a two channel Nicolet 3091 Digital Oscilloscope is employed for analog signal digitization. The unit is relatively inexpensive, can sample up to 1 microsecond per point, employs 12 bit accuracy and is relatively portable. Furthermore, a waveform analysis software package (Waveform Basic) is readily available from Nicolet, allowing downloading of records to personal computer and full waveform analysis. Practically any IBM compatible personal computer with 640 kbytes of RAM and a graphics screen can be used with this packaged software. A disadvantage of this A/D conversion device for blast signal digitization is

the small record length of only 4 Kbytes. This limits the length of blast which can be digitized to an acceptable level of accuracy.

4 COMPUTER PROCESSING OF BLAST VIBRATIONS

The utilization of computer analysis of waveforms is certainly not a new development and similar analyses have been widely used in other scientific applications. Of interest, however, is the fact that many user-friendly software packages readily exist off- the-shelf for use with typical blast monitoring equipment, and for minimal cost. These packages also allow the operator to program simple equations and subprograms with ease, allowing quite flexible data manipulation.

Transfer from the A/D converter, or oscilloscope, to PC is simply done through either RS-232 or IEEE connection. Once sent into the computer, hard storage of data is possible and

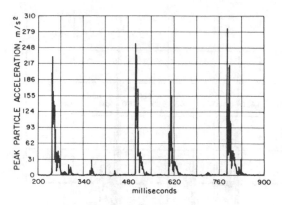

Fig.2 Peak particle accelerations determined from three channels

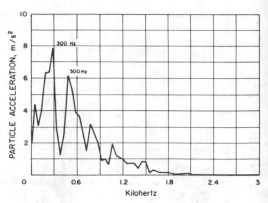

Fig.4 Frequency spectra of single charge vibration record

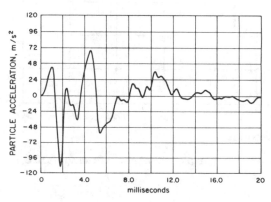

Fig.3 Vibration from a single charge in massive ore

detailed analysis can be done. Generally, the initial step is to calculate peak particle sums from the three channels of data. Due to the presence of the fibre optic cut-off point each channel can be adjusted to the same time domain with ease and peak particle acceleration can be obtained through the simple input of a formula on the command line. Figure 2 shows the results for the large diameter drop raise blast shown in Figure 1. Cursor control permits precise determinations of each peak donating particle acceleration in this case.

The determination of actual delay firing times can then be performed using the built-in cursor control and data expansion in both x and y directions. Figure 3 illustrates the waveform recorded from a single charge in massive ore expanded 55 times in the horizontal direction.

For full waveform analysis the built-in Fast Fourier Transform function can be used to determine dominant frequencies within specific wave packets. This is shown in Figure 4, which is the frequency spectra for the waveform in Figure 3. This information is useful as particle acceleration, velocity, displacement and frequency are interdependent variables.

As with most software packages, alterations to axis scales, labeling and plotting to printer or plotter are readily available. Hard copy storage and filing of data is then done to the floppy disk or hard disk of the PC.

5   FUTURE DIRECTIONS AT NORANDA

As previously mentioned in Section 3, the initial choice of equipment for the monitoring of production blasts was based on availability, ease of use, accuracy and ruggedness. With the recent miniaturization of computer systems, such as laptop computers, and advances in digital recording technology the future for blast monitoring at Noranda is both exciting and promising.

The next generation of portable underground monitoring equipment will consist of a laptop personal computer which is used to program an internal data aquisition card, built into a portable ruggedized carrying case. Such a system was recently developed by the Julius Kruttschnitt Mineral Research Centres' Commercial Division in Queensland, Australia, through a research program of which Noranda is an active sponsor. This commercially available Blastronics BMX Blast

Table 1. Summary of three blast monitoring systems

Parameter	1 Analog Self-Trigger Portable	2 Digital Self-Trigger Portable	3 Digital Self-Trigger Stationary
No. Channels	4	8	16
Transducer			
Frequency (Hz)	20-3000 at 3 dB	Same	Same
Axes	3	3	3
Recorder			
Type	Tape	Digital	Digital
Speed (mm/s)	95	Variable	Variable
Resolution	60 dB	8 Bit	12 Bit
Physical			
Weight	10 kg	15 kg	N/A
Dimensions (mm)	670 x 460 x 240	440 x 385 x 212	PC XT
Power	NiCad Recharge	Recharge Gel	120V DC
Digitizing			
Rate (max.) Samples/s	5 000	25 000	62 500
Cost (Cdn $)	~ 25 000.	30 000.	30 000.

Monitor is a light-weight, fully digital, self-powered, 8-channel recorder for collection and analysis of blast vibration waveforms. Noranda's present sensor technology, as well as the blast timer, can be easily connected to this equipment. The system allows the tri-axial sensors to be connected directly to an 8-channel, 8 bit, 200 kHz acquisition board with storage of blast events to CMOS memory where they can be held for up to 12 hours. Record length is also expandable to 64 Kbytes. The primary advantages of such a system are; user friendly operation by blasting operators; the use of additional sensors allowing triangulation of vibration records and, therefore, increased information on the performance of the blast, and; increased memory to allow greater accuracy (higher sampling rates) and length of individual records (longer blasts).

Due to the wide variety of mining operations within the Noranda group, some monitoring systems are more applicable at specific locations than at others. Of particular interest is a system to be implimented at both Brunswick Mining and Smelting (BMS) and the Matagami Division (Norita Mine and the new Ilse Dieu Project) which will work in conjuction with a mine-wide microseismic monitoring system. The system employs either a portable JKMRC-type unit (BMX monitor) or a 16-channel, 12 bit, 1 MHz A/D converter board situated in a PC. Both units can be located in the surface mine office and may function alongside a present microseismic system (in the case of the portable BMX unit), or as a stand-alone microseismic/blast monitoring system. This is made possible by simply connecting the cabling network for the blast monitoring sensor array to the microseismic monitoring cable network already in place, and connected to surface. The additional array of blasting sensors (located close to the blasting operation at less than 20 m) can then be switched into the microseismic monitor circuit prior to blasting, allowing whole waveform processing of blast vibrations. The switching process can be automated through the use of the fibre optic blast timer at the instant of blast initiation. The length of time the system would be monitoring the blast would only be as long as the blast itself - usually no longer than a few seconds.

This type of system would allow considerable ease of operation from surface once the connections are made underground. Normally, the array of blast sensors would be located so as to permit recording throughout the mining life of a particular stope. Blasting records can be easily accessed from the hard storage medium for analysis and then transferred onto either VHS video tape, floppy disk or transmitted by modem to the Research Centre for further analysis or consultation. Table 1 is a summary of the three monitoring systems available for blast diagnosis at Noranda.

6   CONCLUSIONS

As Noranda develops new technologies for the monitoring of all aspects of mining operations the continued use of computer techniques will unquestionably be strong. This is evident in the monitoring of mine equipment, ventilation systems, ore handling systems, underground rock stability and microsiesmic activity, and blasting operations. Only through the monitoring of these activities can improvements be made and assessed in a quantitative fashion.

The use of blast vibration monitoring has proven invaluable to the development of controlled blasting techniques and increasing efficiency. By employing readily available components and off-the-shelf computer hardware and software great gains can be realized in a short time at low cost. This trend will continue as new technologies continue to advance and be implimented in a similar fashion.

The next generation of portable multi-channel blast monitors and systems operated almost exclusively from surface will become routinely used as this technolgy is developed and transferred to the operations of the Noranda Group. This will then increase the level of engineering expertise in the design of blasts and underground openings. Improved blasting will mean improved recovery, better fragmentation and reduced stability problems leading to greater efficiency and productivity.

Computer Applications in the Mineral Industry, Fytas, Collins & Singhal (eds)
© 1988 Balkema, Rotterdam. ISBN 90 6191 760 3

Practical truck dispatch – A micro computer based approach

Warren M. Sadler
Wenco International Mining Systems Ltd, Vancouver, BC, Canada

ABSTRACT: Recent research into use of computers to optimize surface mine operations has produced some esoteric results. Current actual and planned applications include onboard condition monitoring, payload control and record keeping, automatic truck dispatching, even a potential for inertially guided driverless trucks and for satellite based location monitoring. To a large extent, these efforts focus on making the maximum use of computers at the expense of complexity. Typical of early attempts to fit the problem to the solution in the world of computers, many of these systems will fail, or succeed only after long and painful implementation periods.

This paper introduces a real time surface mine operations management system designed for ease of use, reliability and results. Designed by mine operators rather than by technical or research staff, the practical results and cost effectiveness should be of interest to all who are in the business of moving excavated materials. PMCS is a system that monitors the surface mine in real time, dispatches trucks to reduce or eliminate waiting time and provides minute by minute decision support for the first line supervisor. It also offers same day performance information for the mine manager and detailed past performance data for mine planners.

1 INTRODUCTION

Efforts to optimize truck and shovel mining operations, through the use of computers, have taken many forms during recent years. The most ambitious approaches have applied the power of computers to eliminate the fallible human element from the decision making process and provide fully automated truck dispatching functions without human intervention. Other systems have concentrated on gathering minute data from equipment at work for later analysis, with a view to optimization through a better understanding of the equipment, its operators and factors affecting cycle times. Some claim to be all things to all people; fully integrated systems involving truck dispatch, load control, vehicle condition monitoring and management information.

Mine operators who have explored the field have often found that all is not as wonderful as we would like to believe in this very complex area. Both initial costs and operating costs have been high, and in respect of operating costs, they have often been unpredictable with any degree of confidence. Reliability has proven disap-

pointing. Linear programming decision algorithms are well understood, but using them effectively for truck dispatching depends on the the correct interpretation of operating constraints or requirements and the maintenance of a current image of the mine, its roads, locations, materials and qualities in the computer. Not easy, when the mine changes every day.

It is probable that all of the present problems will be dealt with in the next decade. Many are being resolved now, at the expense of implementation periods during which trial and error processes drive out a workable solution. But meanwhile, how can we take advantage of the power of computers and data communications to improve productivity and control today? Without risking a great deal on success!

At Westar Mining, Canada's largest and perhaps most productive metallurgical coal producer, truck dispatch systems were examined with great interest. While real productivity gains could be forecast, the risks were so broad that justification was not possible. The greatest concern surrounds the time, effort and accuracy required to keep the computer's image of the

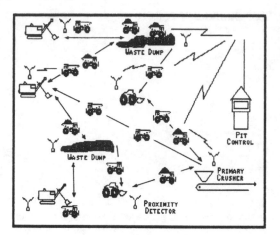

Figure 1. PMCS interaction with mining equipment

mine and mining requirements current considering that things change fast when you are moving 50 million cubic meters annually, on a mountaintop. A critical factor if the computer is to deliver sound dispatching decisions. Several such exercises ultimately caused the question to be raised, "Are we trying to do the right things with the computer?" The answer here was "Perhaps not." Through linear programming, a computer can find an optimum dispatching solution that will be beyond the capability of the man, as long as the operating constraints and requirements are correctly interpreted and kept in a current condition. In an operating mine, conditions are constant only at the moment of observation, so this may not be easy.

An approach was developed that had as its basic precept that experienced mine operators can make good decisions if they have good information. As mine roads are built and others abandoned, quality, market and mining needs vary, or shovels and dumps are moved, the people receive this knowledge in passing. They don't need to have the information formally translated and laboriously entered on a periodic basis. The unusual is simply another case, within our capacity to make a sound decision, even though we have not anticipated this particular event or problem. If the human dispatcher knows where his equipment is, what it's doing and can measure relative efficiency at all times, he can make very good decisions indeed.

At every stage of the development of the Production Monitor & Control System, or PMCS, every function was questioned. Not as to whether it could be done by computer, but whether it was better to do it by com-

puter. Simplicity and ease of use were the criteria. The objective was to produce a real time mine management system that provides an automated dispatching capability, effective decision support for the first line supervisor, same day performance information for the guidance of the mine manager and detailed records for the use of the mine planner.

The result was the development by Wenco International Mining Systems Ltd, then a subsidiary of the mining company, of the system that will be described here.

Acquisition costs for a PMCS system are a fraction of those for earlier automatic truck dispatch systems. Implementation is quick and sure. Maintenance costs are both low and predictable. Yet PMCS delivers, easily and flexibly, most of the promised benefits of the largest and most severely structured systems.

2 SYSTEM DESCRIPTION

PMCS is a real time approach to operations management that was designed by mine operators with the needs of mine operators firmly in mind. It uses powerful communication and information tools to improve productivity and control in the surface mining environment.

This is an integrated system comprised of custom designed data gathering and encoding hardware with the ability to receive and transmit data mounted in production equipment, radio repeaters when needed, a radio communications system interface, a twin microcomputer based dispatcher and pit monitor display, and a similar microcomputer that provides current and historic management information.

3 DATA CAPTURE

Equipment operators are provided with a dash mounted dataset with several clearly identified function keys that they will touch at certain points in each production cycle. This dataset replaces the usual mechanical load counter and adds memory functions to ensure accuracy. Operation is simple and foolproof. Each time the operator touches the data transmit key, a high speed burst of data is sent to the communications handling equipment in Pit Control. The dataset has a frequency lockout and an automatic retry function to handle bad radio reception or simultaneous transmit conditions.

While it would have been relatively easy to cut the driver out of the loop by sensing on board conditions, this is an area where we chose not to do so. Our reasons were twofold. The elimination of maintenance demands to serve the sensors and circuitry and the positive value of recognizing the driver as an important part of the production process.

The truck driver sends a data signal each time he sets the spotting brake; at the dump, when he stops in a shovel queue, under the shovel to begin loading and finally when he releases the spotting brake to depart loaded.

The signals generated are received by a custom designed modem that translates messages between the languages of the base radio and the computer. The primary Monitor computer continuously runs a software set called Acknowledger whose function is to determine that the message received is logical and meaningful in terms of this mine and its equipment. Acknowledger accepts and acknowledges valid messages, then time stamps and passes them to the Monitor program, running simultaneously on the same computer.

Several error correction routines are included in the validation process to ensure that erroneous entries are corrected rather than being lost to the system. Acknowledger's error traps are designed to be helpful and easy for the equipment operator to deal with, commonly requiring only a retransmission or confirmation to accept a message.

Since truck drivers are human and fallible, it is assumed an occasional signal will be forgotten. When an out of sequence signal is received, Acknowledger will seek confirmation. If the driver confirms that the message he sent is correct, by retransmitting without changing the data, the signal is accepted and the unusual sequence will be correctly interpreted by the Monitor computer.

Where there is a need, optional fixed point location monitoring functions are used to validate or extend the information available to the system. To automatically trigger a shovel assignment is a typical use of this function. Using very low power radio transmitters at extremely high frequencies, these devices cause the dataset to transmit identification and location codes or take other predefined action when they pass within a defined area nearby.

At the request of the mine operator, this function can replace manually entered data. For example, a proximity device at the primary crusher could eliminate the need for an operator entered "dump" signal, or other such functions that are critical to accurate production reporting. In this area, as in all others, practicality is maintained in preference to automation for the sake of automation.

The system is designed to receive and process production information from loading equipment that works alone, not using trucks. Tramming loaders or draglines, for example. In a dragline, data transmission is triggered by the operation of the dipper trip, providing a very accurate record of cycle times in this area where minor location, reach and swing changes can have an enormous impact upon productivity. At the option of the mine operator, PMCS can accommodate production, function and time recording from other equipment. Drills, dozers and graders can be tracked with equal efficiency and accuracy.

4 TRUCK DISPATCH

The system includes an optional automatic truck dispatching function that may be used at the discretion of the pit control supervisor. By linking a group of shovels that are working in reasonable proximity to one another as a set, trucks assigned to the set are directed to a particular shovel, through the mobile dataset, based on the first of the linked shovels that will be ready to serve this truck.

The dispatch decision is made by the computer on the basis of presently assigned units, priority, status, and observed loading and cycle times among the shovels and trucks of the set. In an uncomplicated way, it avoids trucks waiting for shovels while nearby shovels wait for trucks. It minimizes truck waiting times while keeping the shovels as fully occupied as available trucks permit. And it's simple.

The capability is also provided for the dispatcher to assign a truck through the computer keyboard. This assignment looks no different than the computer generated assignment to the driver. He hears an alerting tone, and a shovel number and destination appear on his dataset LED display. In keeping with the princple of doing things the easiest and most practical way, most such reassignments will be made by voice radio. In this case, PMCS will recognize the event upon reciept of the first signal from the equipment indicating a new status.

At the option of the mine operator, PMCS can give the driver direct feedback during the course of the shift. Each time he tips a load, for example, Monitor can return his or the mine's production this shift, unit productivity, load count, production forecast, or any specific piece of information that is most meaningful. Again, the operator is recognized as an important link in the production chain.

5 FIRST LINE OPERATION

Decision support for the first line supervisor is provided through the Monitor program. This program receives its data by drawing messages from the Acknowledger program whenever there are any to be read. It presents a display, continuously updated, that reflects events in the mine as they occur. A buffer is provided so that the Monitor program can be interrupted for short periods to perform other tasks. Acknowledger continues its work and when Monitor is returned to its primary function, it promptly rolls forward all activities that have occurred while it was away, again presenting a current and accurate picture of mine status. Linked together and running in parallel with this computer is the second microcomputer in the system. This computer serves to provide immediate backup in the event of a device failure and to allow two views of events in the mine to be observed at the same time.

The pit control supervisor, or dispatcher, is given a tool to use equipment much more productively through such things as accurate knowledge of truck and shovel waiting times in each setup, calculated optimum trucking levels and both the planned and actual blend produced during the present operating period.

The status of each unit is known, as well as the cycle segment in which each unit is now operating or, if it is delayed, the nature of that delay. Cycle times and parts of cycle times are known by unit. Production and forecast production are presented. Since this information reflects what is happening now, very finely tuned responses are made possible.

Monitor provides additional error trapping and correction capability to ensure that end of shift results are accurate before they are passed to Viewer, the Management Information module of PMCS, or to external Accounting or Management Information systems.

The Monitor computer also runs a set of system maintenance programs that provide the tools to make and modify assignments, blend requirements and an assortment of assignment and relationship tables used by Monitor, as well as to enter certain operating data to complete the system's knowledge of production events and results achieved during the shift.

Since all significant production inform-ation has been electronically recorded during the course of the shift, these results can be displayed in a comprehensive shift report. The shift report aids in the process of handing the mine over from one crew to the next and provides preliminary shift performance details. Equipment in use, or available and present assignments are set out in a form to help the oncoming shift supervisor assign his crew most effectively. In addition, the shift report contains the results obtained this shift in the areas of production, productivity and availability of equipment. The shift report is produced automatically, by the Monitor computer in the dispatch office, at the end of each shift.

6 MANAGEMENT INFORMATION

Detailed performance information is made available to mine planning staff without the time consuming effort of time and motion studies. Finally, these results can be fed directly to external computers providing Accounting or Management Information services. This eliminates the expense and delay involved in gathering, collating and entering such data after the fact. More importantly, we are provided the opportunity to report operating results for the guidance of staff responsible for production as a whole. These results are made available on the third microcomputer in the system.

This third microcomputer, identical to the Monitor micro and optionally linked to the computers in the dispatch office through a networking facility, is used to provide current operating results and budget or baseline comparators for the guidance of mine management. It runs a program set called Reporter that adds shift results to historic information files and updates a large number of report files. These report files contain all pertinent operating information, current to the end of the last shift, to stand ready for access by mine management personnel. The purpose is to give staff information on how they are performing during the course of the current accounting period, while there is time to affect results.

Those carefully balanced reports that we will see on the fifth of next month may be marginally more accurate, but it's a little late to help in today's decision making processes. Whether PMCS production information is accurate to accounting standards is always in the control of the mine operator. Corrections and adjustments to the record of events can be made during the course of the shift, at shift end, or at any time before the information is passed to the MIS computer or to external management information systems. Once in the MIS computer, final adjustments can still be made by authorized staff, if total accuracy is required in this primary reporting system.

Access to management information is achieved through a Viewer program, running on the same computer. Mine Management has the capability, through a terminal network, to see a comprehensive series of performance indicators on request. Nearly paperless, PMCS provides a wide range of management information, in graphic or tabular form, in response to simple, menu selectable, commands. These displays are presented on a high resolution color monitor. Of course, if the user wants to send a paper copy of a report or graph to others, or take it back to his desk for further study, the touch of a key sends the current display to the attached printer.

All facets of production information are included. Superintendents and Mine Managers are provided with results that are current to the end of the last shift. Each display contains budget, baseline and historic information to provide a frame of reference for evaluation. The range of operating results available is flexible and mine specific. Menus and report files are tailored to the users' requirements, using common site terminology.

A unique feature of this system is its ability to measure and report on the results it has achieved. A baseline, or actual results achieved in key performance areas during a normal pre-implementation period, is established. Viewer displays these baselines along with current operating results in its graphs and reports. This feature continues to be of value long after mine operating efficiency stabilizes at new levels. At any time in the future when changes in operating procedures are planned, baseline data are entered before implementation. Again, actual results attributable to the revised plan are accurately measured and reported.

7 HARDWARE

Remarkably low in hardware cost for such a powerful and complex system, PMCS requires three IBM PC/AT's, PS/2's or equivalent, in addition to the custom designed data communication devices. A maximum configuration would also require proximity devices when automatic location monitoring is specified.

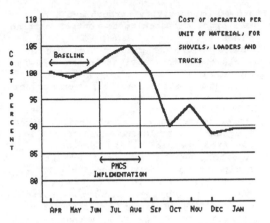

Figure 2. Shovel and truck operating cost per unit of material moved

8 RESULTS ACHIEVED

To obtain an objective measure of the results achieved through the implementation of PMCS required a detailed approach to cost accounting. Examination of the cost of operation of trucks and loading units in terms of the cost per unit of production provides a basis least affected by extraneous factors. We used average costs over a three month period immediately preceding implementation as the baseline. The ten month period under consideration contained relatively normal weather, including a fall break-up, and stable haul distances.

During the two month installation and training period, unit costs rose between four and five percent. Factors affecting productivity during implementation were observable and expected, as operators and supervisors learned how to apply their new tools effectively. In the third month, costs began to show a significant decline. In the fourth and subsequent months, costs between 88 and 90 percent of pre-implementation values became the norm. Similar relative results were observed in productivity figures, measuring product per gross operating hour for both shovels and trucks at a second mine when PMCS was introduced.

8.1 Significant sources of benefit

Benefits stem from a number of sources, some less tangible than others, and some impossible to quantify. Users comments indicate benefits in the following areas:

Improved productivity through the provision of real time decision support for first line supervisors.

Improved productivity through cycle by cycle truck dispatching to control waiting times.

Accurate control of product blend improves yield and offers savings in the processing plant.

Mine planning efforts are based on current, accurate performance data, for better results at lower cost.

Improved performance through prompt, direct feedback on results achieved at all levels from operator to mine manager.

More effective utilization of available equipment to achieve planned results on a shift by shift basis.

Improved productivity through the provision of detailed operating time information assisting mine engineers in the analysis of trends, and in optimization to reduce non-productive time.

As an effect of improved productivity and utilization, lower equipment and manpower requirements resulting in substantially lower costs over the long term.

Improved control for Superintendents and Mine Managers through the provision of production information while there is still time to affect this period's results.

Reduced cost through the elimination of effort in the acquisition and handling of data.

More effective use of time through reduced paper handling by operating staff.

9 DEVELOPMENT DIRECTIONS

Current development work centers around the expansion of the opportunity to optimize surface mining operations through a number of extended options. Some of these are: The application of simulation techniques and linear programming solutions to the dispatch function, while retaining the solid stability of the system; Integration of data from any of the several emerging shovel or truck based load weighing modules; Integration of data from equipment condition monitoring devices; Time and function recording for operating crews.

10 SUMMARY

In summary, the approach taken involves the use of computers for what they do best-- handling data, and leaves it to the humans involved to do what they do best--thinking and acting. The result is a reliable and easy to use surface mine management system that pays for itself in under six months and produces solid productivity gains sim-

ply through more effective handling of equipment by the first line of supervision and the operators themselves. Additional operating efficiencies that can be expected to result from the timeliness of good information at the superintendent level, and from reduced clerical effort, are bonuses.

SPECIFICATIONS

Frequency Range - Optional VHF or UHF, simplex or duplex
Computers - Three IBM PC/AT, PS2 or equivalent
Software - MS/DOS, Borland Turbo Pascal
Capacity - 70 trucks, 20 loading units at work

11 Information management systems

Computer Applications in the Mineral Industry, Fytas, Collins & Singhal (eds)
© 1988 Balkema, Rotterdam. ISBN 90 6191 760 3

A fully integrated production information system at Newmont Gold

Gary S.Flye
Newmont Mining Corporation, Danbury, Conn., USA

ABSTRACT: An essential factor in cost competitiveness for today's mining operations is
the efficient use of information to aid in decision making. This paper describes the development of a fully integrated production information system utilizing distributed database and fourth generation language technology. Areas of application include mine production, ore movement, mill production, laboratory automation, warehouse inventory,
equipment maintenance, exploration drill hole logging and finished metal accounting. The
techniques employed in designing this multi-database system are described with special
emphasis upon the integration of each database system with all others. Each component
database system is evaluated in terms of realized or anticipated benefits, followed by
a concluding evaluation of the entire project.

1 INTRODUCTION

Newmont Gold Company, a subsidiary of Newmont Mining Corporation, operates five
mines, two mills and three heap leaching
operations in the state of Nevada in the
United States. With a potential annual output of 346,000 ounces of gold, Newmont Gold
Company is one of the largest gold producers in the western world.

Ore mined from the various pits is evaluated against a cutoff grade. Ore below
the cutoff for mill feed is treated with
cyanide on leach pads. This pregnant solution is purified by carbon adsorption and
gold is recovered by electrowinning. Mill
grade ore is processed in one of two mills,
depending upon the source of the ore and
its carbon content. The older No. 1 Mill
uses the historic Merrill-Crowe zinc precipitation process whereas the newer No. 2
Mill uses the latest carbon adsorption
technology.

With the expansion of Newmont Gold in recent years, the need became evident for a
plant-wide computer-based system for data
capture, reporting and analysis. Newmont
Mining Corporation's Information Systems
(IS) group proposed the concept of a number
of functionally separate Data Base Management Systems (DBMS) which would be fully
integrated by sharing data with one another
on a real-time basis. The steps which were
taken to analyze Newmont Gold's production

environment are described below. After
assessing the needs of Newmont Gold and the
available software tools (in conjunction
with the existing hardware), the decision
was made to use a Fourth Generation Language (4GL) to create the required systems.

2 SYSTEM DESIGN

The first step taken in designing the integrated system was to analyze the production environment. Applying General Systems
Theory as described by Schoderbek (1985),
the Newmont Gold Company (NGC) was viewed
as a closed system in which the resources
of money, equipment, technology, etc. are
internally available for accomplishing organizational goals (Figure 1). The diagram
illustrates the system comprising NGC and
identifies its major resources. The resources input of gold ore is transformed by
the mining process into the output product
of gold bullion. All of these resources
and activities are internal to the system.
Note that 'gold extraction technology' re-
sides partially in the system and partially
in the environment. This division of technological input signifies a combination of
internal expertise by NGC employees and external expertise by outside consultants.
Environmental factors include corporate
policies, governmental regulations and the
exportation of gold to refineries, resul-

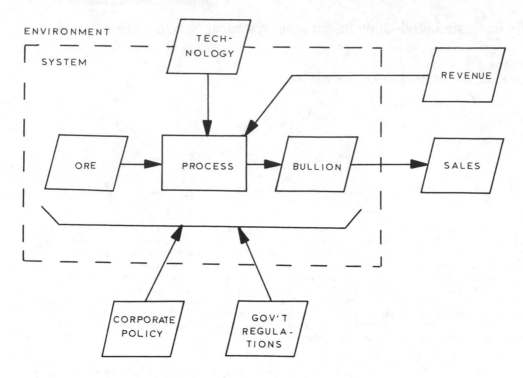

Figure 1. The Newmont Gold system and environment.

ting in revenue input.

A hierarchical approach in systems design was adopted because mining and metallurgical plant processes are best defined, and thus modelled from a hierarchical 'unit process' approach. Using the technique of Functional Decomposition as described by Fairley (1985), the system represented in Figure 1 was decomposed into functional modules as shown in Figure 2. Three major classifications resulted - production systems, production support systems and general support systems. Further decomposition of these modules was performed as indicated. This diagram provides the global view of functional areas in which the DBMSs were constructed. The category of general support is not within the scope of this paper, and thus will not be given further consideration.

An effort was made to avoid too little or too much decomposition. Insufficient decomposition yeilds a few excessively large and complex DBMSs with may require centralized hardware upgrades, are generally difficult to code and maintain, and tend to inhibit geographic distribution. On the other hand, excessive decomposition yields numerous simplistic DBMSs requiring complex integration, not to mention the unnecessary drain on computer resources due to the overhead of so many separate DBMSs.

Using the system decomposition of Figure 2, a more detailed analysis of the basic systems and their interrelationships is shown in Figure 3. Data flows are represented by lines with arrows pointing in the direction of data flow. These flows identify the required communication links between systems needed to produce a truly 'integrated' network of DBMSs.

3 SYSTEM DEVELOPMENT

To accomplish the goal of a fully integrated production information system as quickly and efficiently as possible, DB and 4GL tools were selected.

Data Base (DB) technology was selected over conventional file management because of the following advantages as identified by Date (1982):

1. Centralized control of operational

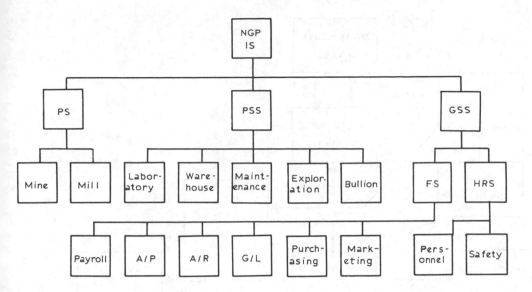

Figure 2. System hierarchy as a result of functional decomposition.

data.

2. Reduction of redundancy by eliminating duplications of private data throughout the company.

3. Increased data integrity by centralized control of data access and update.

4. Ability to share data among several
concurrent users.

5. Application of security restrictions.

6. Data independence (physical storage
independent of application programs, allowing changes to DB without affecting applications).

Database design followed the approach of
the Entity-Relationship Model described by
Chen (1976) in identifying the various entities in each system and the relationships
between them.

A Fourth Generation Language (4GL) was
used to create the required systems. In
addition to creating the database itself,
the 4GL produces the entire environment in
which the DB is to function, and all this
in a very high level language which reduces development time. The capabilities of
this 4GL of interest to Newmont were:

1. Network control and management for
support of the data communications network.

2. Creation and utilization of a DB.

3. Provision of an on-line environment
with transaction routing, message queueing
and audit control.

4. Interactive screen painting.

5. Report writing utility.

6. Security facilities.

7. Restart/Recovery feature to ensure
complete data and transaction integrity.

8. Support of distributed processing.

Newmont's mainframe hardware is exclusively Unisys (Burroughs). Newmont Gold has
two Unisys A3s with 48 bit word sizes and
the same operating system (MCP) as Newmont
Mining Corporation's large B7900 and A15
mainframes in Danbury, Connecticut and Tucson, Arizona, respectively. The smaller
DBMSs are able to reside locally on-site in
Nevada to provide rapid response and local
control of job execution. Newmont Gold has
no provision for high-speed backup and off-
site archiving for disaster recovery, so
all critical files are backed up nightly
over the network to the data center in Connecticut. A very reliable network using
conditioned analog lines at 19.2 K bits per
second provides response times within five
seconds per transmission at the Nevada
site.

Two development philosophies used with
good success have been prototyping and user
developed reports. Prototyping is helpful
when either the user does not know how to
describe what is wanted, or there is some
question whether the user's ideas will even
work. An example was the Laboratory Automation DBMS. Even after extensive discus-

505

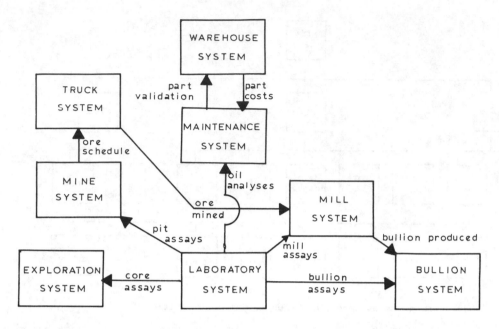

Figure 3. Integration of component systems.

sion, it was unclear how the computer might best accomplish sample logging and lot creation. In three weeks a prototype was developed which not only answered these questions, but even served as the basis for much of the final design. Another useful concept was to convince users that they should write their own reports. Reporting was simplified through the use of a 4GL reporting package which enabled non-programmers to write most of their own reports, freeing the IS staff to other activities, but most importantly, involving the users intimately in their own system.

4 SYSTEM DESCRIPTION

The entire Newmont Gold Production Information System (NGPIS) is an integrated network of individual DBMSs. Before discussing the integration issue, each component DBMS will now be described.

   The Mine System is not currently a compatable DBMS but data generated by this system is shared with other systems. Two primary functions are performed, daily ore control and mine planning. Daily ore control models the grade of blocks on a bench currently being mined from blast hole data, producing an annotated ore control map. A scheduling

module then uses the ore body model along with an outline of the proposed area to be mined and reports the tons and grade by cutoff of the material contained. Mine planning calculates block reserves within digital polygonal outlines on the basis of either kriged blocks derived from blast holes or reserves calculated from exploration drilling.

   The Mill System is an interactive on-line system for data acquisition, routine reporting and statistical analysis. Control room operators enter production data at various time intervals and are able to rapidly generate shift and daily reports. Metallurgists are able to generate daily, weekly and monthly material balance reports on demand. In addition, raw data from the DB may be downloaded to microcomputers for statistical and graphical analysis.

   The Truck System is an interactive DBMS for monitoring the movement of ore from pits and/or stockpiles to crushers and for tracking equipment availability for each shift. Clerks enter data on a daily basis from equipment usage tickets. Ore grades are correlated to material moved to aid in reconciling tons mined with tons milled.

   The Laboratory System is a real-time data acquisition DBMS which provides the functions of sample logging, preparation, assaying and reporting on a real-time basis. Electronic devices such as balances, pH meters, atomic absorption (AAS) and atomic flourescence (AFS) spectrophotometers are interfaced for direct data entry. Data is entered at key points throughout the lab and at any time, the progress of sample assaying is observable on inquiry screens. Verification of assay results is performed automatically for most samples through online table definitions.

The Warehouse System is a material management DBMS used by many Newmont properties. Modules within this system include purchasing, inventory control and maintenance/work orders. All aspects of equipment or material procurement from vendors is handled by this system. All items contained in the various warehouses is stored in the DB. Any issue of items from the warehouse is entered and the DB instantly updated, providing real-time monitoring and control of inventory levels. The maintenance/work order module concerns materials procured from the warehouse for repair or maintenance of plant equipment and is directly interfaced to the maintenance system.

The Maintenance System is a on-line system which provides comprehensive monitoring of equipment usage and maintenance history with full preventative maintenance and breakdown maintenance scheduling. Components of the maintenance system are as follows:

1. Equipment Maintenance module: defines all the various items of equipment on which maintenance is to be performed.
2. Procedure module: describes the procedures to be followed in the performance of maintenance.
3. Scheduling module: performs scheduling based on item usage and other factors.
4. Data Entry module: facility for entering data on equipment usage.
5. Work Order module: creation and processing of maintenance work orders.

The Exploration System is a data entry and editing DBMS for the maintenance of exploration drill hole logging. Within one central DB, all drill holes for all exploration projects at Newmont Gold are stored (approximately 8,000 logs containing more than 1,000,000 assays). Data is stored according to project, each project containing both geologic and assay information. In addition to full data entry and editing capability, an interactive error checking facility allows logs to be checked for inconsistencies such as overlapping intervals. Once a project (ore deposit) has been fully checked for errors, a FORTRAN77 extraction utility is used to extract log

data into a standard binary log file which is used for contour mapping, statistical analyses and other similar techniques.

The Bullion System is a DBMS which manages the gold bullion inventory, schedules shipments to refiners and provides real-time status tracking of gold bullion. There are three main modules - shipments, settlements and royalties. The shipments module contains a record of buyers, organizes the bullion into lots and schedules bullion shipments. The settlements module reconciles the final agreements as to bullion weight, fineness and price. Finally, the royalties module computes royalties for each lot based on agreements.

These, then, are the individual systems which comprise NGPIS. Individually, they are greatly useful to their respective application, however, none of these systems operates in true isolation. We have already seen the interaction between systems in Figure 3. The physical implementation of this integration will now be described.

5 SYSTEM INTEGRATION

Since the individual DBMSs are distributed on various machines at various geographic locations, inter-database communication is supported locally on the same host computer as well as remotely between two different hosts in the network. Several options for integration are possible depending upon whether the application requires real-time or batch response. Each DBMS itself operates as a real-time system and thus it is physically possible to make all data transfer real-time. In practice it is very expensive to support the overhead of numerous real-time linkages when the application only requires daily batch updating. Consequently, we have sought to install the most appropriate data transfer mechanism for each inter-DB linkage. Referring to Figure 3, the following integration mechanisms were selected:

Table 1. Data transfer mechanisms.

Data flow	Batch/real-time	Volume
Part valid.	Real-time	100/day
Part costs	Batch (monthly)	1000/mo.
Core assays	Batch (8 hour)	1000/day
Bul. assays	Batch (daily)	5/week
Mill assays	Batch (daily)	500/day
Bul. prod.	Batch (weekly)	5/week
Pit assays	Batch (daily)	1000/day
Oil analy.	Batch (weekly)	50/week
Ore sched.	Batch (daily)	10/day
Ore mined	Batch (daily)	10/day

Because of the geographic distribution of these DBMSs and the ability for any user on any host computer to access these systems, the Newmont Gold Production Information System (NGPIS) may be properly considered a simplified distributed database system as defined by Date (1983). The principal challenges of this distributed DBMS network have been the provision of a reliable network to support real-time inter-host data communications and also 'host transparency' (the response between remote hosts approximating local response so users are unaware of the physical location of the database).

6 SYSTEM EVALUATION

The Newmont Gold Production Information System was initiated in September 1985 with the development of a prototype for the Laboratory System. Anywhere from two to six people have been assigned to this project since its inception. A total of seven DBMSs are in production, including most of the inter-DB linkages. Although cost benefits are not yet available at this time, there have been many tangible benefits observed thus far.

The Truck System, with its functional and easy-to-use data entry screens reduces data entry time by a factor of five over the old system. This productivity improvement allows clerks to be more efficiently utilized.

Benefits from using the Mill System have been the ability to perform statistical analyses on production data, allowing a better understanding of the process and time and effort saved through automatic generation of reports.

With recent expansion of the laboratory, it is estimated that the Laboratory System will be processing over 1,000,000 analyses per year. This transaction volume could not have been achieved under the old manual system. In addition, accuracy has been increased through elimination of transcription errors. Finally, the real-time status tracking of samples has allowed lab supervisors to monitor and improve laboratory performance.

One of the principal benefits of the Warehouse System has been the money saved through minimization of inventory levels and elimination of emergency orders for items already existing in the warehouse. Finally, a variety of users have benefitted from the ability to inquire on the status of purchase orders, work orders and warehouse stock.

It is anticipated that the Maintenance System will prove in time to directly influence equipment longevity and reduce equipment downtime through more thorough and consistent preventative maintenance. Faster response to equipment breakdowns by instant access to maintenance procedures and more rapid procurement of repair parts will contribute to this end.

These preliminary observations indicate tangible benefits attributable to NGPIS. In addition to the obvious improvements in rapidity and accuracy of data manipulation, it appears that the organizational and procedural changes resulting from the installation of these systems have been very beneficial.

7 CONCLUSIONS

A project of the magnitude and complexity of a fully integrated production information system has been a new experience for the Newmont Mining Corporation IS staff. Not only has the scope of the project itself been ambitious, but NGPIS was the first major application of DBMS and 4GL technology within Newmont. In some respects, the development method evolved with the system, and was refined as we learned from our mistakes. Some of the lessons learned include the following:

1. Don't yield to the temptation of using the 4GL to build large and complex systems. It is always preferable to keep the systems of moderate size and complexity.

2. Consider report formats prior to DB design to ensure that the design will easily accomodate the various data access paths. Neglect of this point often makes DB reporting very difficult.

3. Real-time systems are very powerful and desirable, however, they do consume hardware resources and are not always warranted for every application.

4. 4GL reporting packages are not always the perfect solution for every reporting need. They are ideal for simple to moderately difficult applications, but some complex reports may need to be coded in a conventional programming language.

5. DBMS efficiency must be continually monitored because performance tends to degrade over time, which if uncorrected, could jeapardize user acceptance of the system.

Because NGPIS is still a new and evolving system, it has not yet been fully evaluated as to its benefits. This will undoubtedly be accomplished in the near future, but will be complicated by the very nature of the integrated system. Cougar (1982) has observed that there is a synergistic effect (the whole is more than the sum of the parts) from the integration of systems which is difficult to

quantify, and yet it is precisely this
synergy which makes a fully integrated
production information system so desirable
for the mining industry.

REFERENCES

Chen, P.P. 1976. The entity-relationship
    model -- toward a unified view of data.
    ACM trans. db. sys. 1:1:9-36.
Cougar, D.J. 1982. Cost/effectiveness an-
    alysis techniques. Advanced system de-
    velopment/feasibility techniques. New
    York, NY : John Wiley & Sons.
Date, C.J. 1982. An introduction to data-
    base systems. Reading, Mass. : Addison-
    Wesley Publishing Company.
Date, C.J. 1983. An introduction to data-
    base systems volume II. Reading, Mass.
    : Addison-Wesley Publishing Company.
Fairley, R.E. 1985. Software engineering
    concepts. New York, NY : McGraw-Hill
    Book Company.
Schoderbek, P.P. & Schoderbek, C.G. & Ke-
    falas, A.G. 1985. Management systems -
    conceptual considerations. Plano, Tex.
    : Business Publications, Inc.

Computer Applications in the Mineral Industry, Fytas, Collins & Singhal (eds)
© 1988 Balkema, Rotterdam. ISBN 90 6191 760 3

An advanced geo-information system for mining applications

Marinos Kavouras, Salem E.Masry & John R.Smart
Department of Surveying Engineering, University of New Brunswick, Fredericton, Canada

Edmund C.Reeler
Universal Systems Ltd, Fredericton, NB, Canada

ABSTRACT: Resource evaluation and mine design involve large amounts and differing types of spatial information. Efficient integration and handling of such information requires advanced geometric modeling techniques and data organizations. General purpose computer-aided design, and computer graphics systems are not designed to deal with such complexity. This paper presents efficient approaches to handle geo-information in order to facilitate mine design and planning. Particular emphasis is given to spatial information. The octree scheme is introduced as it is used to represent highly irregular ore bodies. A prototype system, which has been developed as a result of this research, is also introduced. The system facilitates geometric operations, and also spatial and attribute queries which are essential to mine design. An example of the practical utility of the system is also presented.

1 INTRODUCTION

Resource evaluation and Mining involve large amounts and differing types of spatial information. This information is used to solve the complex problems of mine planning and design. Efficient mine design requires good access to the available data in order to optimize the exploitation. This is particularly crucial in the case of paramarginal ores combined with high fluctuation of market values. These requirements cannot be practically satisfied without the use of digital procedures.

There is therefore a need for an information system which efficiently handles the complex geometry and geology of mining objects so that analysis and updates can be comfortably performed. Such a need cannot be satisfied by general purpose computerized drafting systems, CAD, or computer graphics systems some of which have recently entered the mining environment. Other smaller computer systems specialized in mining, also prove to be very limited in their capabilities. They are designed to automate mining procedures which were performed manually in the past. Their simplistic modeling and data structure makes handling of different operations very awkward, and future expansions difficult or impossible.

Four years of our research have been devoted to the analysis of mining needs, and the development of an information system which integrates all spatial mining data, has 3-D modeling capabilities, and offers analysis and design operations. The result has been a prototype system which is referred to here as Daedalus. This prototype system relates to CARIS -- a Computer Aided Resource Information System (Masry, 1982; Lee, 1983), which is used in many geo-applications.

The most important problems in designing a mining information system are:
1. The modeling of mining spatial objects.
2. The organization of the models in an integrated database.
3. The retrieval and manipulation of the existing models.

This paper presents the modeling of mining space as it is takes place in Daedalus, as well as, how these models are handled by the system. Our intention is not to give a full description of the system modules and implementation details, but to show the potential of handling mine design and feasibility studies in an information system.

2 MODELING OF MINING SPACE

Mining involves spatial and non-spatial information. Often these are known as geometric and attribute information. Geometry describes the location and shape of mining objects, such as the axis coordinates of a water line; whereas attributes contain textual information such as date of construction, last maintenance, and person responsible.

Spatial information is the hardest to handle. Mining objects can be classified into four different geometric types:
 • points (e.g., surveying marks, stations, single sample points),
 • lines (e.g., drill-holes and utilities such as power, water and ventilation lines),
 • surfaces (e.g., earth's surface, mineralogical boundaries),
 • solids: regular (most excavations), or irregular

511

(most ore bodies).

The representation of ore bodies and excavations, and the design of operations which involve both object types are major tasks in building a mine modeling system. An ore body representation scheme should possess the following:

1. Ability to represent both rigid and fragmented ores, and at variable resolutions.

2. Ability to represent the actual interior of the solid and spatial distribution of its quality.

3. It should be easy to generate it from, or relate it to, geostatistical block estimations.

4. It should facilitate efficient geometric operations such as projections, sections, booleans (union, subtraction, and difference), volume computations, and adjacency and interference analysis.

Given these requirements, evaluation of all available solid representation schemes (Requicha, 1980), shows that volume representations are more suitable than boundary representations. One volume representation in particular, that satisfies most of the above requirements, and has also other computational advantages, is the octree encoding. This scheme was used, in the system developed, to represent irregular geo-solids and particularly ore bodies. Octrees are hierarchical tree structures which have found many applications in computer graphics (Samet, 1984) and solid modeling (Meagher, 1982). Very recently, they have also been used to represent topography and geology (Kavouras, 1985; Mark & Cebrian, 1986).

In this scheme (Fig. 1), the areal extend of the application (such as the entire mine) is enclosed in a large cuboid called the "universe". The cuboid universe is subdivided into eight subcuboids of equal size which are indexed in a specific encoding scheme. These subcuboids are also known as octants or voxels (volume elements). Each voxel is attached a color depending on whether it lies inside (black), outside (white), or at the border (grey) of the geo-solid (e.g., the ore). The subdivision continues recursively only for the grey voxels, and terminates when either no grey voxels remain or when a preset resolution is reached. The smallest elements after the last partition are called resolution voxels.

Due to the spatial coherence of nature, neighboring voxels are likely to have the same properties. Octree aggregation can therefore achieve substantial storage compression, particularly in the case of homogeneous ores.

Octrees are attractive schemes for modeling ore bodies in underground or open-pit mines, water reservoirs, caverns, and other geo-solids. By storing both full and void space in the octree scheme, volume computations for ventilation analysis become trivial. Octree modeling can also be used in the finite element method, being therefore useful in deformation analysis and rock mechanics (Chrzanowski et al., 1983).

Geostatistical block models can be converted into octree models using an aggregation/classification procedure (Kavouras, 1987), which takes into account not only the estimated ore grades, but also

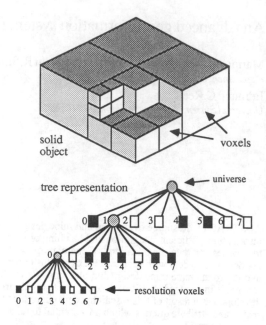

solid object   voxels

tree representation   universe

resolution voxels

Figure 1 : Octree encoding of a solid object.

their accuracy. Octree models of ore bodies can also be derived from geological sections. The procedure is the following:

-- The geologist retrieves all sample data of a specific section, and outlines zones of different ore quality.

-- The computer correlates digitally adjacent sections, and builds a boundary representation of the ore body (Smart, 1986).

-- Subsequent conversion builds the octree representation from the boundary representation (ibid.).

In the Daedalus system, excavations are represented implicitly by some geometric parameters. The actual solid needs to be constructed only locally when a boolean operation is to take place.

Surfaces are handled in two ways: either as thin solids, in which case they are represented by an octree; or as piecewise surfaces, in which case a simple polyhedral model with planar faces is used.

3 SYSTEM OVERVIEW

The modules of the Daedalus prototype system are organized in the following layers (levels):

-- The database stands at the lowest part of the system, and contains all definitional data (e.g., sample data, and surveying points), complete geometric descriptions (representations), object attributes, and object relations.

-- The next higher level — the data structure of the

512

system, contains low level facilities which access the database. The integrity of the object descriptions and the spatial searches are also part of this level.
-- The next level contains general modeling operations on a number of representation types. Boolean operations (Reeler, 1987) and also sections, projections and transformations are facilitated here.
-- At the top of the hierarchy, there exist a number of high level (application) modules, such as data editing, contouring, volume computations, geostatistical evaluations, mine design, interference and adjacency analysis, and so on.

The organization of the octree representation of the ore body stores voxel geometry and basic geology together, while detailed geology is stored in attribute files. This organization allows for direct accessing of single voxels and their attributes. The so far experience with mining applications shows that direct accessing to single pieces of information is always essential and demanded by the users.

An essential part of the system's data structure is the spatial indexing module which facilitates geometric search. The object descriptions are organized in a cell-based structure, so that objects are located without extensive searches in an extended database.

Textual information, such as utility maintenance records, detailed geological descriptions, and land ownerships are stored separately from the geometric information in attribute files. Attributes are accessed via a commercially available relational DBMS.

The software of the prototype system was written in Fortran 77 and runs on DEC's VAX series computers. A Tektronix 4115B device is used as a display device.

4 EXAMPLE

In order to assess the suitability of the system, both simulated and real test data were used. One case with a real gold deposit is presented below:

The block grades and their accuracy had been previously estimated using geostatistics. The block model consisted of 638,400 single (20x20x20) feet size blocks, with grades varying non-uniformly between 0.0 and 5.4 ounces of gold per ton. Classification of the deposit into ore and waste, and subsequent compaction of the ore into an octree model, resulted into 17,007 blocks of different size and grade.This ore body is shown generalized in Figure 2.

The surface topography was also encoded in the data base. Excavations were then designed in the form of shafts, ramps, drifts and stopes for different levels. The surface topography and portion of the underground utilities are shown in Figure 3. By intersecting designed excavations and the ore body, volumes of recoverable ore were computed and undesirable interference among was avoided. Figure 4 shows an embedded excavation in a slice of the ore body.

Combined attribute and geometric queries were also invoked, such as: "Retrieve all blocks with grade between Z1 and Z2", or "Find the shortest distance between a given drift and the richest block of ore", "Does this direction intersect any other utility?". Such queries generate attribute reports, while the geometric answers are also displayed on the graphics screen.

5 CONCLUSIONS

The overall assessment of the prototype system shows the following:
• The system uses suitable representations for point, line, and regular-solid type mining objects.
• The octree scheme is suitable for representing irregular geo-solids, and particularly ore bodies with variable distribution of ore quality.
• The octree scheme offers sufficient compaction for homogeneous and non-fragmented ore bodies.
• The data structure facilitates fast access to any piece of spatial information without extensive searches.
• The system offers a variety of geometric manipulations on block models which involve thousands of blocks. It is particularly efficient in boolean operations and sectioning, which are essential to mine design.
• It integrates all mining objects with surface topography.
• The system can answer geometric and attribute queries.

The assessment shows that the Daedalus design can efficiently serve the mining needs. The octree proved to be a very powerful scheme in representing and manipulating irregular ore bodies. Results have been very satisfactory and mining firms have expressed a strong interest in adopting a production system. A number of optimizations are required however in order to develop a production system. These relate to the peripheral utility programs and the user interface.

In summary, our experience shows that spatial information systems with 3-D modeling capabilities are excellent tools for geologists and mining engineers in defining ore bodies and performing mine design and feasibility studies. The prototype system presented here contributes to this goal. Our research will continue towards optimization of the design, incorporation of application programs, and extensive testing in order to satisfy many adverse applications.

ACKNOWLEDGEMENTS

This project has been funded by the Natural Sciences and Engineering Research Council of Canada; this support is greatly appreciated. We are also grateful to numerous organizations,companies and individuals who have contributed to the reported work with data and information.

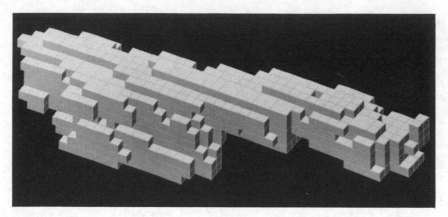

Figure 2: Generalized view of the ore body.

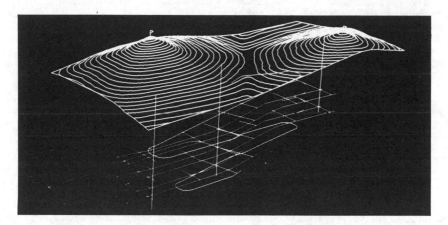

Figure 3: Surface and portion of underground utilities.

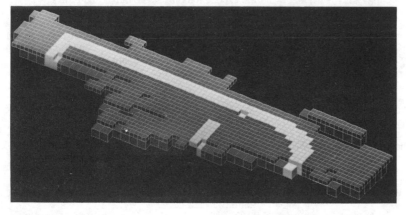

Figure 4: Embedded excavation in an ore body slice.

REFERENCES

Chrzanowski, A.; Y.Q. Chen; A.S. Chrzanowski
    1983. Use of the Finite Element Method in the
    Design and Analysis of Deformation
    Measurements. Proc. of XVII FIG Congress,
    Commission 6, 611.1, Sofia.
Kavouras, M. 1985. Design of a Geometry-System
    to Handle 3-D Mining Information. Proc. of VI
    Inter. Congress on Mining Surveying,
    Harrogate, England, September.
Kavouras, M. 1987. PhD Thesis in preparation.
    Dept. of Surveying Engineering, Univ. of New
    Brunswick, Fredericton, N.B., Canada.
Lee, Y.C. 1983. A Data Structure for Resource
    Mapping with CARIS. Proc. of AUTO-CARTO
    VI, Vol. I, Ottawa: 151-160.
Mark, D.M.; J.A. Cebrian 1986. Octtrees: A
    Useful Data Structure for the Processing of
    Topographic and Sub-surface Data. Proc. of
    ACSM-ASPRS Annual Convention, Vol. 1,
    Washington, D.C., March: 104-113
Masry, S.E. 1982. CARIS--A Computer Aided
    Resource Information System: An Overview.
    Paper pres. at the Institute for Modernization of
    Land data Systems, Georgetown University Law
    Centre, Washington, D.C., January (rev. Sept.
    1982).
Meagher, D.J.R. 1982. The Octree Encoding
    Method for Efficient Solid Modeling. IPL-TR-
    032, Image Processing Lab., Rensselaer
    Polytechnic Institute, Troy, N.Y.
Reeler, E.C. 1986. The Manipulation of Linear
    Octrees in a Three-Dimensional Digital Mapping
    System. MScE Thesis, Dept. of Surveying
    Engineering, Univ. of New Brunswick,
    Fredericton, N.B., Canada
Requicha, A.A.G. 1980. Representation of rigid
    solids: theory, methods and systems. ACM
    Computing Surveys 12: 437-464.
Samet, H. 1984. The Quadtree and Related
    Hierarchical Data Structures. ACM Computing
    Surveys 16: 187-260.
Smart, J.R. 1986. Three-Dimensional Modelling of
    Irregular Natural Objects. MScE Thesis, Dept. of
    Surveying Engineering, Univ. of New
    Brunswick, Fredericton, N.B.,Canada.

Computer Applications in the Mineral Industry, Fytas, Collins & Singhal (eds)
© 1988 Balkema, Rotterdam. ISBN 90 6191 760 3

The modernization of underground surveying at Kidd Creek Mines Ltd

Alfred Yetter & W.George Hughes
Kidd Creek Mines Ltd, Timmins, Ontario, Canada

ABSTRACT: This communication describes the evolution and modernization of survey methodology and practise at Kidd Creek Mines Limited, Timmins Ontario. The Kidd Creek implementation of the **Total Survey Station** concept is very simple and relatively inexpensive: Employing conventional survey instruments (eg.: **Wild T-1A**), the technician keys field observations and notes into a programme operated data-collector, at the point of underground survey; The data-collector (eg.: The **Psion Pocket Organiser II**), verifies or indicates discrepancy within the observations and provides the technician with feedback of simple survey computations; based on the results, the operator may then choose appropriate action (eg.: re-survey); At a later time, the collected observations (within the data-collector) are uploaded to a host computing system (eg.: the **Apple Macintosh** microcomputer), manipulated, checked, archived, printed in field sheet form, plotted to scale and optionally, transmitted (as data), to other mainframe computer-based software systems (eg.: **Mineval** on the **CDC CYBER 180/830**). The data-collector paradigm is also described as adapted for use with borehole survey instrumentation

1.0 Introduction

Traditional methods employed in the discipline of surveying have undergone a marked transformation over the past decade. The laborious technique of repeated transcription of theodolite data followed by manual computation, unit transformation and representation of results, can now be replaced with computerized tools providing automatic capture and manipulation of survey data as a singular field process, followed by reporting and plotting. Thus, survey observations can instantaneously be massaged, presented and archived, in the field, at their point of origin. Within the industry, this modernized form of surveying has come to be known as the **Total Survey Station** concept.

The technological economies afforded through the application of microprocessor instrumentation provide the modern surveyor with electronic theodolites which interface directly to computerized data collection units. However, such advancement is not without detractors as concerns the application to underground mining operations: - the instruments are generally expensive, have more bulk and are less robust in comparison to conventional theodolites; available software for the manipulation of data are poorly adapted for underground mining applications.

2.0 Description of mining operations

Since the discovery of the Kidd Creek orebody in 1964, the company and its operations have shown sustained growth. Basemetals mining commenced operation in 1965, by open pit methods. The orebody, characterized by widths of up to 600 feet, strike length of 1970 feet, and dip of 70° to 80° towards the east, appears open to depth. Owing to size and orientation, the deposit was considered amenable to sub-level, blasthole stoping.

In 1969, the underground development phase began with the collaring of Number One Shaft and an access ramp from surface. The shaft was subsequently sunk to 3000 feet. Operational Levels exist at 400 foot intervals, with ramp access sub-levels every

FIGURE 1: Aireal photograph of Kidd Creek Timmins Basemetals operation showing the relationship of the Number 1 and 2 mine headframes to the open pit.

FIGURE 2: A scale model (note the CN tower to the left), of the the Kidd Creek underground workings from surface to the 2800 level. At present, access to the workings can also be obtained via a ramp from surface to the 4600 level.

100 feet (termed the Number One Mine). Since that time, a second shaft has been sunk to the 5000 foot level (termed the Number Two Mine, operational levels at 200 foot intervals, ramp access sub-levels every 50 feet, commencing at 2600 feet), and the pit has ceased production. Three two-man crews are responsible for the surveying of the operations. The Number One Mine operations are surveyed in Imperial measure, while the Number Two Mine workings are surveyed metric. Within the next two years, the two mines will be merged into a single entity, employing metric measure.

Two smaller gold mines were brought on-line during the past seven years and are some seventeen miles south of the main basemetals mines. The primary form of production is through open pit at the Owl Creek Mine while the Hoyle Pond operation employs a number of small vein underground mining methods.

During 1984, Kidd Creek underwent an extensive and successful modernization of computer technology, throughout its operations. Purchase of several centralized super-minicomputer hosts and a large number of distributed **Apple Macintosh**

microcomputers contribute significantly to maximizing operating profits. The impetus underlying this modernization effort are based on the fact that reducing the labour and temporal components of any numerically intensive operation will necessarily enhance the precision and control over the tasks performed. Therefore more resource time is available, at reduced unit cost.

3.0 Rational for survey methodology modernization

While the current basemetal operations are not expanding in tonnage hoisted, the mines are (as within any mature workings), ever increasing in logical and physical complexity. For example: In 1979, Number Two Mine had no production below 2800 level but during 1989 production is planned above the 4600 level while still proceeding on the 2800 level. This complexity is illustrated in Figures 1 and 2.

3.1 The distance logistics problem

With this increase in mine physical dimensions, the distances that surveyors are required to frequent and maintain have become excessive. At Kidd Creek, as in most underground mining operations, survey crews travel to the work place via the cage; - walk/descend some combination of ramps, drifts, cross-cuts etc. before reaching the place of survey; effect transcription of instrument measurements to field data sheets and subsequently repeat some reverse order of their steps to move to the next survey area or ultimately regaining the surface office. It was determined that in order to increase survey crew effectiveness, modernization of underground survey practice was necessary. Since the majority of the survey crew's time was spent in travel from work place to work place, some form of rapid mobilization was required.

3.2 The calculation and mine engineering record maintenance problem

The second most time consuming category of survey work involves data manipulation on surface. Having accomplished the field work, the survey team must return to their office, transcribe field observations to a number of calculation sheets, effect translation by trigonometric and logarithm tables (possibly aided by a simple arithmetic calculator), check and double check results and then subsequently record entries on survey record sheets, a station record book and a station index book. Subsequently, all records are manually plotted on various sections/plans and grade chain lengths are optionally calculated. Such methodology requires numerous transcriptions of primary survey data into various engineering formats. Given that the data must only undergo format transformation, it is fair to conclude that automation of survey data entry and representation by use of computers would yield considerable savings in manpower resources expended, as well as reduce the possibility of data transcription errors.

Considering the arguments presented above, a programme of survey practice modernization at Kidd Creek was implemented consisting of improved field data collection, computerization of surface work and providing transportation for expediting underground mobility.

4.0 Expediting underground mobility

The costs of maintaining a skilled survey technician must be weighted against the actual time that these individuals can be employed in carrying out their function. After studying the amount of time that each survey crew spent in travelling from work place to work place on their beat, it became obvious that providing these crews with small diesel tractors would net an immediate 40% gain in the amount of actual surveying performed per day.

While this form of transportation is common in most LHD trackless mines for the purposes of supporting the mine production and maintenance functions it is usually not applied to survey or geology field technicians. The movement of men and materials on mechanized carriers is usually restricted to specific locations and times during a typical shift. As survey technicians generally try not to interfere with production functions in the underground, efficient use of this equipment in support of their more or less random logistical movements was not feasible at Kidd Creek.

Further study of the economics involved showed that purchase of two tractors dedicated to the survey function would show a pay back of less than three years. Figure 3 describes the units selected at Kidd Creek, as well as the retrofit adaptation of the units for the explicit purpose of supporting the survey function.

5.0 Modernization of survey practise

Since survey information is legally required as well as crucial to the efficient and safe operation of any mine, a concerted effort was expended in reviewing whether or not surface calculations and representations could be economically and effectively computerised. The historical development of a perspective on the problem and its solution is provided as a series of case studies.

FIGURE 3: Photograph of one of the two **FORD** model 1710 diesel powered, four wheel drive tractor units as adapted for the survey function. Each unit has been retrofitted to accomodate a three member survey crew. A 4' x 4' x 1' steel locker has been fitted to the rear of each unit to accomodate all equipment and supplies necessary to support the day to day survey function. Platforms over the hood and fenders are provided to facilitate reaching back heights up to twelve feet without the use of ladders.

5.1 CASE 1: Understanding new technology

In 1979, a computerized mine planning function was well in place within the Kidd Creek Mine planning department. Geological and mine engineering information could be readily modeled and grade calculation on planned excavations determined with relative ease. As development information was not central to the planning and design of excavations, very little effort was expended on the incorporation of survey data. Mine planners simply required boundary points to effect their designs and these were scaled from prints produced by the surveyors. While modernization of the mine engineering function was proceeding all about the survey department, there were few computer or temporal resources available to apply toward the modernization of surveying. After all, even the manufacturers of survey instruments payed little attention to the computer

revolution that was happening concurrently yet outside of their industry. However, the pocket and programmable calculator industry was maturing at this point in time. Recent survey graduates of the day, were familiar with these products and justified the purchase of two **Texas Instruments TI59** programmable calculators equipped with plug-in modules adapted for survey calculation. However, this first attempt to modernize underground surveying at Kidd Creek failed largely due to non-conformity and a lack of a single driving force to implement the system. At that time, there were some five survey parties but only two calculators to support their calculations. Indeed, most surveyors questioned the accuracy and fidelity of the results computed on the machines, and therefore status-quo prevailed with most crews returning to the use of trigonometric and logarithm tables.

5.2 CASE 2: Gaining acceptance of new technology

By early 1983, the survey instrument industry had progressed to the point that some new technology was available. The "total station" could now be purchased at some expense but these machines proved rather cumbersome and poorly adapted for use in underground applications. A number of vendor demonstrations were conducted onsite, with delegates of the survey section in attendance. Full input of the survey group was solicited for each demonstration. Employing this input, Kidd Creek subsequently purchased one **Geodometer** total station for application within pit survey operations. A complete assessment of vendor available offerings of the day again showed that no appropriate or cost justifiable total station systems were available for underground application. Survey parties were receptive of the concept of using a data collector if for nothing more than to reduce the number of times that they transcribed the same numbers into different record sheets.

In 1984, the mine maintenance department had purchased a small inexpensive data collector (the **Videx TimeWand**), in an attempt to replace the shift foreman's written timekeeping records . This same data collector appeared to be suitable to collect mine survey data as well. Preliminary testing showed that the data collector did not stand up to underground working conditions but the idea of using a generalized , industry-available data collector persisted and so the search for a suitable unit remained .

At the same time, the **Apple MacIntosh** microcomputer was introduced at Kidd Creek. Within six months, record keeping was transformed to spreadsheet ledgers in almost all areas of the company except within the survey department. The survey department agreed that the development of a computer survey database would be useful to them if only to perform audits of which points required check surveying. While it would be possible to construct such a database on the Macintosh, the backlog of station points exceeded then available capacity of the hardware and software tools.

Subsequently, in late 1984, the coordinates and elevation some five to six thousand previously surveyed stations were entered into a **Digital PDP 11/70** minicomputer and a program written that would calculate coordinates and elevation of all new stations from data contained on field sheets . These new survey points, as well as re surveyed points would then be added to a survey point data base . The intended use of this data base was to combine it with an existing mine planning system (three dimensional graphic stope design). This system was not accepted by the survey section due to lack of available terminal data entry equipment and poor computer system response time. Little could be done to rectify the matter at that time as the **PDP 11/70** was in the process of being replaced by a **CDC CYBER 180/810** with all existing software requiring conversion on a two year timetabled priority basis (surveying being one of the lowest priority tasks).

Although personnel kept in touch with what was purveyed in the market place and one **Geodometer** total station was purchased for the Owl Creek open pit, none of the available instruments appeared suitable to Kidd Creek's perceived needs for underground surveying therefore no further attempt to modernize was made until 1985 .

Progress having been made with the implementation of microcomputing applications in other engineering areas warranted the development of a spreadsheet document utilizing **Microsoft Multiplan** on an **Apple Macintosh** Microcomputer. The spreadsheet was designed to perform all survey calculations, but was only partially accepted by survey technologists. The reason for its partial acceptance was the fact that within the mine planning and surveying sections there existed only one **Macintosh** computer with some ten planning people already attempting its use. The addition of a further six users would have overloaded an already highly utilized but productive system. At that time no software was available to expand the application beyond a record keeping system since programmes to graphically display and plot survey data were not available in the marketplace. Most survey parties returned to the traditional method of computation with the aid of trigonometric and logarithm tables.

5.3   CASE 3: Back to square one, with a fresh approach

In late 1986, we again evaluated our options for yet another attempt at modernizing the survey function. We empirically arrived at the following:

1. The survey technicians were willing to accept technological upgrading if the changes met a certain set of criteria:

a) New instruments must not be bulkier or heavier than their present **Wild T1A** instruments.
b) Accuracy and fidelity of computation were assured.
c) Their jobs remained secure by adopting change.
d) There must be a reasonable chance of successful implementation of change.
e) Change facilitated and upgraded the quality of the work performed.

2. The Survey technicians recognized that computers were not going to go away and due to the numerically intensive nature of their work, computers would soon be an integral tool of their craft (whether by choice or evolution). As a database of all existing survey information had already been created, full utilization of this data could only be achieved through the use of a computerized retrieval system.

3. A complete vendor supplied solution amenable to the Kidd Creek method of underground surveying would not likely be purchasable as a packaged entity.

Therefore, what ever modernization course was to be applied to surveying methodology, the required software would have to be built utilizing in-house programming resources.

4. Microcomputer based spreadsheet applications were well suited to the implementation of survey calculation and database archiving of survey reference information.

5. Microcomputer based applications for the two dimensional rendering of scaled points in three space were now available, at reasonable cost.

6. Intelligent portable hardware was available for the application of data collection in the field. Operationally dissimilar computing equipment of varying sizes and capabilities could easily be made to communicate via standardized ASCII text files and the RS-232c protocol.

From these observations we concluded that the goal of modernization of survey practise could be achieved only by assembling components into a customized "Total Survey Station".

6.0 The Kidd Creek customized total survey station

The Kidd Creek implementation of the total survey station concept is very simple and relatively inexpensive: Employing conventional survey instruments, the technician keys field observations and notes into a programme operated data-collector, at the point of underground survey; The data-collector verifies or indicates discrepancy within the observations and provides the technician with feedback of survey computations; based on the results, the operator may then choose appropriate action (eg.: re-survey); At an appropriate time, the collected observations (within the data-collector) are uploaded to the Macintosh, manipulated, checked, archived, printed in field sheet form, plotted to scale and optionally, transmitted (as data), to other computer-based software systems.

6.1 Components of the Kidd Creek system

6.1.1 Survey instrument

The survey instruments would remain the highly reliable **WILD T1-A** theodolites

favored by the survey technicians. While a field data collection unit is not available for direct electronic interface to the instrument, keystroking of pertinent data via a structured yet flexible entry protocol is possible given a portable, programmable and self contained microcomputer. This method of data capture, while not as automated as the newer electronic theodolites, would prove almost as effective even though manual data entry is required. The main time consuming procedure in the underground was after all not the time at the transit, but rather, the time required to move from place to place within the mine workings.

6.1.2 Electronic data collector

Three types of portable data collectors were tested during the span of the project. These included the **VIDEX Time Wand, Hand Held Products MICROWAND II** and the **Hand Held Products Advanced Pocket Computer** (also marketed as the **Psion Pocket Organiser II**). Both the **Time Wand** and **MICROWAND II** have desirable features as concerns data collectors but both failed endurance tests in the underground environment due to condensation dew point. The **Psion Pocket Organiser II** hand held computer was chosen as a suitable electronic field data collection unit. The **Psion** unit can be configured with various memory sizes (from 8 Kb through 288 Kb), has a multi-purpose peripheral port capable of interfacing to RS-232c compatible devices or to a bar code reader wand, two line by sixteen character liquid crystal display and a full ASCII keyboard. The unit is housed in a self contained plastic case of about three by five by one inch, weighing eight ounces and is powered by a nine volt replaceable battery.

6.1.3 Data acquisition programme

Programming of the data acquisition procedure was accomplished through the BASIC like computer language provided with the Psion unit. Thus field data is assembled into electronic records, analogous to entry lines on the paper type records formally employed by the survey technicians. These electronic records may then be transferred over the RS-232c asynchronous communications link to any type of computing equipment, for further processing.

The development of the data acquisition programme was achieved in a series of phases, by a "divide and conquer" approach to the analysis of the problem. The surveyors were asked to specify the exact sequence of entry used to complete a survey field observation. A prototype programme was written in the **Psion Pocket Computer Language** (PCL), and field tested for a number of weeks. The programme was constructed about a series of menu options which were selectable by the use of cursor arrow keys. The main parts of the programme included: sequential data acquisition of observed data; a data store facility; a data edit facility; a data delete facility and a data transmit facility. Thirty-three fields of information comprise a full survey observation record (see Figure 4). Not all fields are required however and as few as eight fields may serve to make a complete survey observation record.

During the field testing, surveyors were asked to provide any changes they felt were necessary to the prototype programme. Most of the changes centered about the inflexibility of entry. In some instances, sequential entry of information served adequately (eg.: shooting in a new station), while in others, sequential entry slowed down the progress of the crew (eg.: tope shots in a noisy drift). Examination of the suggestions implied that some form of modicum could only be achieved through the use of a bar code menu of commands, followed by keyboard or wand entry of data. Thus, each survey team can enter data in any sequence they desire. The PCL driver programme checks to ensure that all data required to complete a survey record is present before storing the data. Should an essential piece of data be missing, the surveyor is prompted for the missing item(s). Only after all required data is assembled can the record be stored.

While check survey and traverse station doubled angles are calculated, no trigonometric computations are performed. Thus all data is stored in its raw, as entered form. A full editing capability of any stored survey record is provided to aid in those situations where is required.

Training in the utilization of the PCL programme has been facilitated through the use of menu driven programmes integrated with "help" tutorial texts on procedural functions. A miniature "flip-book" of bar codes and instructions on the use of programmes has also been provided to each surveyor. A sample of one flip chart

Survey Field Record Data Sheet		
Field Name	Raw Data	Field #
Record Number	7	
Field Party	AY,AY	1
Check Survey	N	2
Mine Number	2	3
Level Name	28-1	4
Location Name	28-1 CONZWAY	5
Back Site	6312	6
Setup Station	6381	7
Front Site	6503	8
Instrument Height	-3.075	9
1st Vertical Angle Degrees	91	10
Minutes	45	11
Seconds	0	12
1st Horizontal Angle Degrees	193	13
Minutes	22	14
Seconds	49	15
2nd Vertical Angle Degrees	0	16
Minutes	0	17
Seconds	0	18
2nd Horizontal Angle Degrees	0	19
Minutes	0	20
Seconds	0	21
1st Measured Distance	8.54	22
2nd Measured Distance	0	23
Height of Point	1.304	24
Tope from Point	Y	25
Horizontal Angle	14.2	26
Horizontal Distance	31.23	27
Offset Distances Up	2.25	28
Down	1.79	29
Left	2.42	30
Right	1.97	31
Vertical Angle	81.2	32
Remarks	CHAINS REQUIRED	33
Date Entered	4-May-87 7:46:32	

FIGURE 4: An example of the output produced by the **Excel** Field Sheet Macro Report System.

for the purpose of tope surveying is provided in figure 5.

6.1.4 Host computer system and data transfer

While almost any computer system could be used to receive data from the **Psion** data collector, for reasons already described, the choice of the Apple Macintosh was the most appropriate at Kidd Creek. Data within the **Psion Pocket Organizer II** is transferred by direct or modem link to the **Macintosh** operating under a general communications package (eg.: **Palantir InTalk**), as an ASCII text type file. Each record in the file consists of thirty-three variable length character fields, separated by tabs. Each record represents one complete set of survey observations relative to a single setup.

6.1.5 Host computer software applications

Three off-the-shelf vendor supplied applications form the nucleus of the host computer data manipulation and display system. These include generalized packages for data communications (**Palantir InTalk**), spreadsheet calculation and database archiving (**Microsoft Excel**), and computer assisted drafting (**Diehl Graphsoft MiniCAD**). Employing the **Apple Computer** supplied **SWITCHER** application, all three applications can be loaded onto the host **Macintosh** system, at the same time. While only one application can be active at any time, data conversion facilities within **SWITCHER** allow the user to rapidly transfer data between any permutation sequence of the three applications. In this way the three main functional components of the nucleus software can be integrated into an apparently contiguous and singular system.

While the excellencies and deficiencies can be argued amongst a host of other products, it had been determined that these three primary packages were the best fit to accomplish the goal of automating surface survey office operations.

6.1.5.1 Microsoft Excel database

Through the use of integrated user-friendly macro programmes operating on a spreadsheet, a database of survey point information has been constructed. The salient features of the **Excel** Macro programme design include the facility to:

1. Read field observation data transferred from the data collector, or manually entered information.
2. Retrieve coordinates and elevations of previously stored stations necessary for the trigonometric conversion calculation of observations into results.
3. Automatically store the results as a simple sequential list of database records, by station name.
4. Provide printed reports of field observations and calculation results as per excepted engineering standards as well as by ad hoc database query.
5. Automatically update, demand edit or query the database by any data criterion.

FIGURE 5: Sample of bar code command entry chart used in conjunction with the **Psion** data collector for the purposes of tope data field entry. Bar codes are implemented within the industry standard CODE 39 stripe system and are printed as required using an **Apple LaserWriter Plus** laser printer.

6. Calculate coordinates of all survey points (including topes), and provide output in a format compatible with the **MiniCAD** graphics system (an ASCII text file of x, y, z coordinate line segments).

The surveyors employing this system have had little difficulty in assimilating the technology. Through the use of pull-down menus and question windows, this application appears only slightly different then the type of interface employed within the data collector. In this way a consistent user interface has been achieved.

6.1.5.2 Diehl Graphsoft MiniCAD

The tope and station information calculated and stored within the survey database is converted to an ASCII text file of x, y, z coordinate line segments. These line segments are then read and converted (via an **Excel** macro programme), to graphical representation by the **MiniCAD** application. Standard features of **MiniCAD** which provide enhanced graphic rendering ability particularly amenable to survey drafting include:

1. Generation of a text file of coordinates abstracted from the line segments and/or polygons of either two dimensional (2D) or three dimensional (3D) graphic objects.
2. Generation of a scaled 2D or 3D graphic plot from a text file of line segment or polygon primitive coordinates.
3. Ability to calculate the area of any polygon.
4. Ability to switch from 2D to 3D graphic representation of an object on demand.
5. Ability to merge a number of 2D layers into a single 3D plot, at any orientation or projection.
6. Ability to render any graphic object according to any definable scale.

While **MiniCAD** provides the surveyor with many extended CAD functions, we have experienced one minor problem. Due to the size of our mine coordinates (eg.: Northing 215,000, Easting 215,000, Elevation 1020) concerted with the extent of the workings relative to the precision of our rendering scale (1 foot = 250 feet ± .001 foot), it is often difficult to locate a 12 by 12 drift round on the screen. **MiniCAD** presents a screen grid relative to Northing = 0, Easting = 0, Elevation = 0. Thus a drift round would appear as a single black dot in the extreme upper right quadrant of the screen. To overcome this problem, we have employed a coordinate transformation offset so that user will not have to perform extensive screen zooming in order to locate a relatively insignificant drift heading against a background of the whole Number 1 or Number 2 mines. Employing this method, graphical results are output in various scaled plan and sectional views to a 300 dot per inch laserwriter. The output is then manually traced to survey and mine engineering master maps.

6.2 Commissioning the Kidd Survey system

Since the complete survey modernization system is composed of many parts, commissioning was carried out through a phased approach. A single survey crew was selected to proceed through the stepped phases over a one month period. At the end of each phase, the crew would instruct the remaining technicians through the commissioned portion of the system. The phases of commissioning were:

1. Data collector field entry system.
2. Field data transmission to host computing system.
3. **InTalk** Host computer communication procedure programme.
4. **Excel** Database and Calculation procedure programme.
5. **MiniCAD** graphics presentation of survey data.
6. Report generation, printing and plotting of selected data under **SWITCHER** control.
7. **Macintosh** host computer system data maintenance procedures.

During the commissioning period all bugs were documented and required changes effected. As a safeguard, manual records were maintained according to traditional methods until such a time as the survey crews determined the old practise became redundant. In this way, training remained primarily an on the job exercise with a minimum of disruption in daily work routine.

7.0 Description of the blasthole survey project

7.1 The blasthole deviation problem

During 1986 a study was conducted to ascertain the extent of blast fragmentation and its relationship to blasthole placement and orientation. One of the basic findings was that hole deviation had a significant effect on fragmentation. Indeed, no matter how elaborate the instrumentation employed to insure that a hole was aimed in the proper direction and regardless of what equipment was used to maintain direction, blastholes deviated from what the mine planner had envisioned in layout. Unless a blasthole becomes a surveyable break-through, their is no certainty of what orientation it has assumed. The associated additional mining costs and grade dilution attributed to poor fragmentation can mean a rather sizable difference in operating costs.

Considering the time savings that are afforded through the modernization of conventional underground surveying, the purchase of an instrument capable of surveying blastholes would simply become another tool for use by the survey technician, at no additional cost in manpower.

7.2 New technology

Recognizing the blasthole deviation problem was in many ways similar to problems encountered in the oil exploration industry, a search for new survey instrumentation was carried out. A suitable vendor was identified (**Owl Technical Associates Inc.**), and invited to demonstrate their **OTA Borehole Directional Survey Tool**. The instrument was reportedly capable of providing an exact geodetic survey of a borehole and thus appeared amenable to the blasthole deviation problem.

The survey system consisted of :

1. A probe containing instruments capable of sensing inclination and direction of inclination.
2. A winch for lowering the probe and its electrical umbilical (complete with a metering distance device), safely and securely.
3. A console containing the circuitry necessary for:
    a)   operator controls
    b)   visual data and control display
    c)   thermal printer output
    d)   serial output RS-232c port

The console provides for the calculation of results such as probe direction , true vertical depth , distance north , distance east , net horizontal distance and net bearing as well as the capability of compensating for declination.

7.3 The customized Kidd Creek blasthole survey tool

Upon completion of a successful demonstration the **OTA Borehole Directional Survey Tool** was purchased subject to a few custom modifications:

1.   Provision of an environmental protected case to isolate the console electronics from the rigors of the underground.
2.   Provision for electronically capturing data output by the instrument as an ASCII text file.
3.   Conversion of the winch to electrical power and of a more robust design.
4.   Provision for the system to be completely powered by 12V DC and mountable on an underground service tractor.

The data collector field observation information provided by the **OTA Borehole Directional Survey Tool** is transferred to the Apple Macintosh host computer system as described for conventional surveying in section 6.1 above. Excel manipulated data may be reported or alternatively transformed for plotting by MiniCAD.

7.4 Commissioning the Kidd blasthole survey system

The modified instrument system was received in mid 1987. After a number of initial problems were worked out on surface, the instrument was mounted on a small **Ford** tractor and sent underground for initial testing on a hung-up orepass . In order to blast the hang-up down it was necessary to drill a 270 foot hole that was to intersect the orepass at the location of the hang-up . The hole was drilled with a **Robbins model R11D** but did not break through . The hole was surveyed using the borehole directional survey tool and it was determined that the hole had deviated some 20 feet from design . The next hole was layed out allowing for the deviation previously determined . The hole broke through into the orepass , was blasted and the hang-up knocked down .

8.0 Conclusions

While numerous products are purported to provide a total survey station facility for mining, it has been our experience that few of these offerings are applicable to the underground operations at Kidd Creek Mines Ltd. Many vendors claim increased throughput of survey work performed as justification for the purchase of their products. Indeed, in some cases this may be true. However, our research indicates that the cost of modern electronic instruments is better applied in providing transportation facilities to technicians, allowing more time at the conventional transit in performing survey work. While improvements in office practise are afforded by packaged vendor offerings, these are so difficult to learn and precarious in their operation as to almost negate the temporal improvements which they promise. For a much smaller investment in capital, we believe the Kidd Creek System provides facilities equal to the single vendor product for underground surveying, in a more easily learned and user friendly environment.

Utilizing the temporal savings incurred through modernization of underground survey practise, it is now possible to provide the mine engineering and production departments with valuable information concerning blasthole deviation. This blasthole survey system will allow Kidd Creek to determine whether or not a trend in hole deviation in various rock formations exists and if so, factor the information within future blast hole layouts. The blasthole survey system will also enable determination of the hole location at any pre-determined elevation. In this way mine management can make informed decisions as to the probable result of blasting a particular hole; sequencing a blast or determining the exact location of a hidden raise borer pilot hole.

Computer Applications in the Mineral Industry, Fytas, Collins & Singhal (eds)
© 1988 Balkema, Rotterdam. ISBN 90 6191 760 3

Practical aspects of technology transfer and systems integration in the area of micro-based computer applications at Noranda Minerals (Abstract)*

K.Urbanek & L.Merrick
Information Services, Noranda Inc., Toronto, Ontario, Canada

ABSTRACT: The paper discusses how systems dealing with various areas of a mining enterprise should be designed and implemented. Designing a well integrated system and transferring that system to the operators in the field, require a high-level of cooperation between operators, the Research personnel and Information Services staff. An important pre-requisite for a successful implementation is user-readiness, both in the area of computer literacy and up-to-date knowledge of various types of numerical analysis. One major obstacle in the process of developing and marketing a new system, is previous bad experience with micro technology. Quite often a system which allows more flexibility and much more advanced capabilities is questioned because of an existing system which is not performing as expected. The "art" of defining system requirements and the need to establish realistic expectations will be discussed. A case study of site implementations involving micro-based mine surveying, geological mapping and orebody modelling will be used to illustrate the concepts of technology transfer, and integrating these into the mine production environment.

*Paper not received on time for publication.

Computer Applications in the Mineral Industry, Fytas, Collins & Singhal (eds)
© *1988 Balkema, Rotterdam. ISBN 90 6191 760 3*

The use of integrated software systems in many Australian mines

G.K.Brew
Surpac Mining Systems Pty Ltd, Perth, Western Australia

T.Lee
Moriah Mining Consultants Pty Ltd, Sydney, New South Wales, Australia

ABSTRACT: Initially, there was in international trend to develop software for mining activities which tended to be site dependent as each activity was perceived to have unique conditions. The result of this approach was a collection of specialised software highlighting the difference between mines and the techniques used. Later, systems emerged which provided a general solution to a specific problem such as geological mapping, ore reserve estimating and mine planning.

More recently, particularly in Australia, the approach to software design for mining activities has been to provide a general solution to a very wide range of mining related applications. This approach has required the identification of common features of all mining activities which, in turn, has made it possible to develop a suite of software options appropriate to mainframes through to personal computers which integrate the mining activities of surveying, geology, planning and production aspects of mining operations (both surface and underground).

This paper provides an overview of the software systems in use around Australia in the mining industry, followed by an outline of an integrated system approach and the flexibility and benefits it offers.

1 INTRODUCTION

In the mainstream of software needs, there is a common need, by an extremely large number of computer users for spreadsheets, wordprocessing, communications, databases and statistical graphics. To meet this need, a variety of software packages exist, particularly in the PC domain, like Symphony, Open Access and Framework which offer these opportunities within an integrated system. The concept has been commercially very successful. While the integrated components may not be as powerful as specialised alternatives, the fact that they are offered as integrated packages is more attractive to a larger number of users.

In the mining industry, there is an increasing movement away from software which is mine specific, to systems which provide an integrated approach and provide modules that relates to all phases of a mine's development, from the initial exploration, such as geophysical and geochemical surveys, through to mine closure.

In Australia, a variety of internationally marketed mining software systems are used. The dominant ones are SURPAC, Datamine, PC-XPLOR/PC-MINE, MicroMODEL/MicroMINER, MEDSYSTEM, GEOMIN and MINEX. SURPAC and MINEX are systems which have been developed locally. In the case of MINEX, the initial success of the company developing mining software was associated with surface coal mining in the eastern states. SURPAC commenced by developing systems for metalliferous operations in Western Australia. The growth in the use of these packages has been dramatic in Australia with the diminishing cost of PC's, as well as the increasing reliability of PC's together with their increased processing power. This growth also complements the local expansion of mining operations; in particular, surface low grade gold mines with annual capacities less than 300,000 tonnes treated.

To create an integrated software system for mining, it is essential that it can provide computing support in the following areas:

- data management (all types of regional details associated with mineral prospects and properties),
- modelling from data bases,
- mine design,
- mine surveying,
- model updating due to exploitation to date,
- production scheduling,
- exploration and operational reporting,
- cartographic support.

These requirements are basically independent of the activities being on the surface or below. Until relatively recently, most broadly based mining systems tended to be most effective in the pre-production phase of a mining operation. There was a tendency to have limited flexibility to cope with the continual updating which is essential with an operating mine. Most of these packages were developed by consulting groups, project managers and software houses who were more orientated to the exploration and feasibility phases of mine development. Consequently, they frequently ignored mine surveying or gave it minimal attention. In ignoring this aspect of a mining operation the systems had restricted flexibility as operating tools.

2 COMMON SYSTEM FEATURES OF MINING SYSTEMS

2.1 Drillhole Data

The minimum data base of the vast majority of mineral properties is a set of drillholes. Consequently, there is a module within mining systems to provide some degree of manipulating, plotting and graphical capacity of the assay and geological logs associated with the drilling data. The capabilities of these modules vary and are normally compared on the basis of how friendly the systems are, and what is the relative programming efficiency, as reflected in the relative response time to achieve similar tasks.

The scope for interactive editing of data within systems can vary from nothing to extensive. Some systems orientate themselves around a database flexibility and easy access to specific data. These systems are often more capable of handling other relevant sample data, e.g. costeans, channel samples, grabs, face samples etc. without presenting the data as pseudo drill holes.

The flexibility of manipulating the data is another area where differences can occur. Most provide opportunities for statistical reviews, downhole and bench compositing.

The extent and ease to which the user can produce selective compositing does vary.

More recent systems are incorporating the capability to complete 2 and 3 dimensional variography on the drilling data with varying degrees of sophistication at evaluating trends within specific search windows. With these modules, the system's friendliness to users, and the programming efficiency to achieve similar results commence to become important. To enable rapid assimilation by users, modules are commonly menu driven and/or have the capacity to be run as a batch process. While menus are popular, they can significantly increase the storage size of modules, which can be a limitation with some personal computers.

2.2 Model Generation

The scope of modelling and variation between systems can be quite significant, from digitised section outlines to three dimensional trend surfaces. The extent interaction in modelling will often depend on the mathematical basis of the option.

Most systems will offer predominantly mathematically based modelling options on which to create models. The flexibility to create models will vary from simple grade distribution within a mineralisation, to models reflecting structures and rock types, to zones to weathering and groupings for metallurgical characteristics, like coal washability, grindability, operating cost and specific gravity.

The mathematically derived modelling options will normally include:

- polygonal,
- inverse distance to a specified power,
- 2 dimensional block kriging,
- 3 dimensional block kriging,

The systems which find international favour incorporate a geostatistical strength in the modelling options. The scope to develop realistic models is frequently a function of how friendly the module is to constrain the limits of the model and/or the data being used to derive the model.

In generating models, there is a tendency to prefer creating it with regular blocks. The dimensions of these blocks will be primarily dependent on the mineralisation geometry and the primary reasons for creating the model. Some systems can cope with sub-blocks too. Each block is defined by its centre, and is ultimately assigned a series of attributes such as grade, number of observations used to create estimate, rock type, specific gravity, state of weathering, etc. However, the majority of modellings packages have some attribute limit e.g. six that can be allocated to a block centre. The ease of generating these various attributes will vary between systems, as will the ability to merge the attributes to their associated block centre.

The ability to revise models in light of operational experiences and additional data e.g. grade control details has improved rapidly in more recent years.

2.3 Mine Design

The scope to create mining outlines varies significantly from package to package and often reflects the initial situations for which the system was created. In these cases, the packages have a bias, for example, to surface strip mining or underground. The majority have a bias to large scale open pit operations, where the need has been greatest to utilise computers to achieve some degree of operational efficiency.

Obviously, all mine design software systems need to draw on the geological models for their success. The way that this is achieved varies from system to system, with the degree of friendliness between model generation and mine design being similarly variable.

These modules have varying scope to incorporate economic, physical and operational constraints on the mine design process. The preferred international systems have the flexibility to create short and long term production schedules with the associated plans, sections and reports. Further, they include optimisation algorithms to identify the mining outlines which achieve particular operating strategies.

The ability to revise and update mine designs, due to recent exploitation has attracted significantly more attention in the past few years.

2.4 Cartography

Available systems vary widely in their cartographic capabilities. Many are linking with CAD systems to acheive extensive flexibility in producing plots appropriate to the varying needs of geol-

ogists, mining engineers, mine supervision and management. The scope of this module is limited often by the resources of the software designers and the system's structure of the associated modules. This structure may make it inconvenient to produce a desired plot which requires information from several modules to complete the plot.

With many systems orientated around interactive operating approaches, the cartography is increasingly being incorporated into the screen graphics too. In many instances, editing can be introduced to the system through changes to the screen graphics. The cartographic options are continually being extended as more systems provide some three dimensional capability.

3 INTEGRATED MINING SOFTWARE

3.1 Introduction

The key ingredient in an integrated system is to utilise an approach, which in its simplicity, enables the user to move freely between modules, with a structure that provides him with extensive flexibility because the system utilises versatile tools.

To achieve this approach, it is necessary for the various modules to adopt a common 'data structure' which will allow them to interact in the same language. One data structure approach, which is being used, is the 'STRING'. It was introduced in the UK with the MOSS system for more efficient highway design, which used the concept of String Digital Ground Models, which was derived in the late seventies.

A STRING is a sequence of three dimensional co-ordinates, which delineate some characteristic. A string could also identify pit crest, or mineralisation, or building. STRINGS can be described as open or closed. Any outline, or classification, which might be produced on drawing, can be described as STRING. Further, the sequence direction can also be used to segregate STRINGS, as well as the obvious being to identify the STRINGS by different numbers. These STRINGS could in turn be incorporated in the same file or different files to distinguish their character.

The STRING concept has been utilised in surveying software for sometime, because it provided economy in data storage and enabled increased speed and efficiency in computing volumes and other surveying

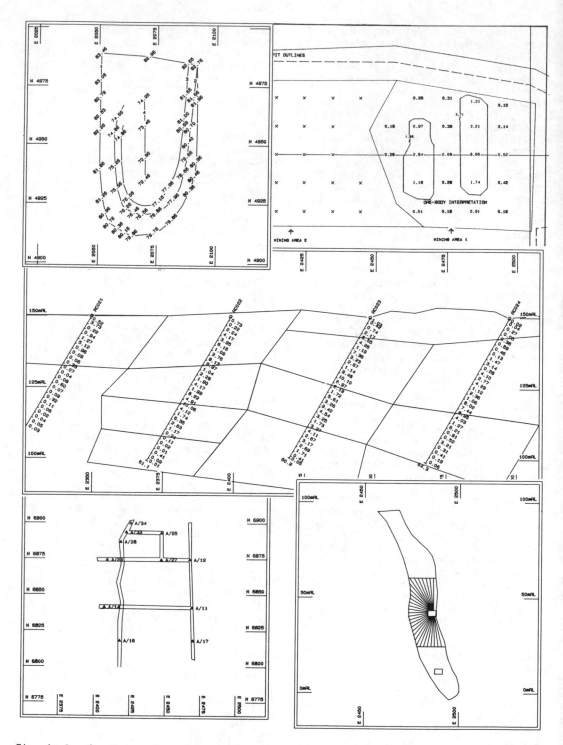

Fig. 1 Sample string plots - pit pickup, grade control plot and interpretation, interpreted drill hole section, underground development, ring drilling design.

activities.

3.2 The Power of Strings

The most commonly used models in mining are line drawings, which are themselves examples of string models. Some common mining examples of such drawings are shown in Fig 1. Such models are applied to an almost unlimited variety of modelling situations, and are universally understood. Software based on the string model provides the same flexibility in application to mining operations, and presents a 'familiar' interface to users. But more than provide 'static' models, computer string models are easily displayed, edited and made to interact with each other.

Strings can be used to model points, lines, surfaces and volumes. 3D surfaces are modelled as sets of strings; either contours or strings defining the features/discontinuities of a surface. 3D volumes are modelled as sets of layers of strings. Each layer can include many strings to define different regions within the layer. The number of points used to define a region, and the layer thickness may be selected to make the model as realistic as required. Layers may be in either plan or section, and it is a simple matter to convert between the two forms.

Thus strings may be used to accurately and realistically model surface topography, ore bodies, open pits, stopes and underground development, coal seams, leach pads, waste dumps, etc.

A model may be constructed for any features of interest, and then have it 'interact' with other models to create modified models. or to allow evaluation of models, Sample model interactions can be

a) topography with pit design
b) survey pickups with ore body
c) mineralisation with stoping
d) ore zone projections with structures
e) weathering with mineralisation
f) existing underground workings with pit design
g) grindability with mineralisation, etc.

(The interaction takes the form of an 'intersection' operation between string domains of the two models involved.) All string models are in actual co-ordinates, allowing them to be displayed and manipulated together, and strings may be set out directly in the field from the models. Thus string models effectively integrate the surveying, geology and mine planning functions within a mine.

3.3 Strings in Mine Design

Fundamental to mine design is the creation of an ore body model. This layered model is derived either from interpreted sections/plans, or the layers of a kriged block model.

An open pit model is a set of toe and crest strings generated subject to specific pit slope and ramp parameters. The pit is generated by the software but the designer can graphically edit the computed strings to accommodate non programmed geological or geotechnical constraints. The final pit model may be the combination of a number of design stages generated subject to different design parameters. The model is created for layers up to the maximum surface elevation.

The surface topography is modelled as a set of contour strings. Corresponding layers of the surface and pit models are intersected to derive the final pit outlines, as illustrated in Fig 2.

The pit model is then intersected with the ore body model to create a 'mineable ore model' from which the pit design may be evaluated, as illustrated in Fig 3.

Such is the flexibility of the string model, the pit generator may be used to design waste dumps and leach pads as 'inverted' pits. In strip mining operations, the strip and spoil pile models are generated in the same way. Dragline operations are simulated by imposing constraints on the relative positions of the strip and spoil pile determined by the dragline characteristics.

Strings are equally able to model underground workings. Stopes may be modelled in plan or section, and designs may be evaluated by intersection of stope layers with the orebody model. In addition, string stope outlines are interpolated between design sections to assist with ring drilling design.

3.4 Strings in Mine Production

Scheduling can be performed by imposing strings on the mine design which reflect various production time targets (e.g. month, quarter, year). Grade control drilling/channel sampling can be incorporated into the orebody model to assist in day-to-day scheduling. Mine limits

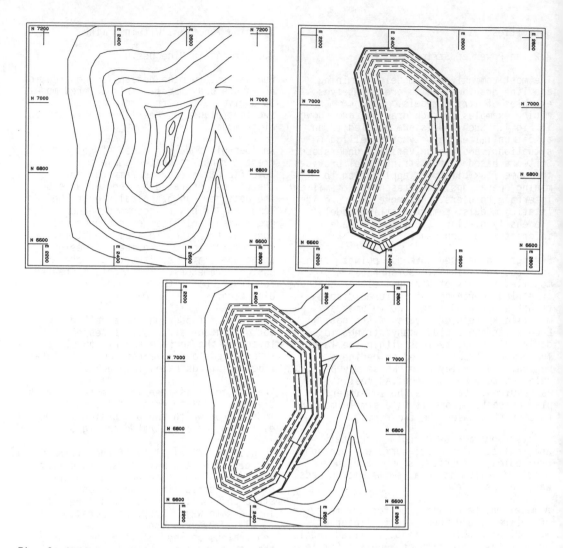

Fig. 2 Surface topography and open pit string models, and the result of the intersection of the two models to produce the actual pit design.

and ore zone boundaries are set out in the field directly from the respective models. This control provides for the possibility to undertake reconciliation between planned results and actual tonnes and grades achieved. This is an essential step in refining and improving the operational efficiency of the mine, to ensure that operational decisions are made with the most reliable appreciation of what remains to be extracted.

Strings surveyed during periodic mine surveys allow creation and maintenance of an 'actual mine model'. This model is used directly to compute volumes of material mined during each period.

3.5 Cartography with Strings

Strings used in modelling represent features, or entities, which are usually drafted with well defined attributes (e.g. pen thickness, colour, text size, etc.). These attributes are usually detailed in a legend on a map. A generalised plotting and display system for string data only needs a library of these entities (e.g. drill hole, ore block, crest, control station, etc.) in which their displayed attributes are defined, and a facility to define a map as a set of

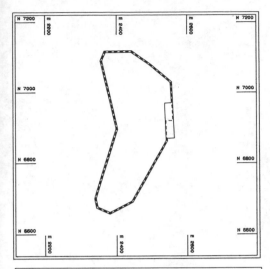

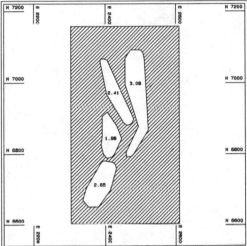

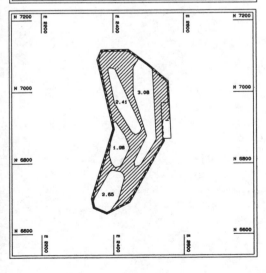

strings, each of which represent particular entities.

In this way, any string data, from any source (surveying, geology, mine planning), which lays in the same co-ordinate space can be combined onto a map. The string model, combined with such a general purpose plotting facility, provides for the ultimate integration of all disciplines and data involved in the mining process.

4. CONCLUSION

Mining system software has evolved from minesites which were biased to local conditions and operating perceptions. Rather than focus on the 'special cases' the approach described in this paper is to identify what is common to a wide variety of mining environments, and provide software tools of practical use in all those environments. This development is not unlike the experience associated with the spreadsheet packages on computers. Spreadsheets are a simple concept, easy to use, and whose uses are limited more by the imagination of the user, than limitations of the packages.

In Australia, the software system SURPAC is finding that its use of string models provides users with the flexibility to use the system in more diverse ways than current international alternatives. The system is currently used at over 130 sites, both surface and underground. It is being utilised in operating conditions as varied as iron ore, base metals, gold, coal and diamonds.

It is being used throughout the full spectrum of mine development; from exploration to day-to-day mining monitoring. As well, it is used by consultants providing services to the industry, particularly in the area of technical feasibility studies. It has also been incorporated into the curriculum of a number of mining colleges both in Australia and overseas.

The simplicity and integrated approach provided by the SURPAC system has aroused interest throughout the international mining industry, and is attracting users to the system from beyond Australia.

Fig. 3 Bench from pit model and corresponding layer from ore and waste model, and the results of intersection of the two layers to determine the materials within the bench.

12 Expert systems

Computer Applications in the Mineral Industry, Fytas, Collins & Singhal (eds)
© 1988 Balkema, Rotterdam. ISBN 90 6191 760 3

Multiple regression analysis vs expert system for the study of the relationship between mine productivity and other significant variables such as bonus

Marcel Laflamme & Jean-Luc Collins
Laval University, Quebec, Canada

ABSTRACT: The paper describes briefly the steps taken to analyze the relationship between underground mine productivity and other significant variables such as bonus. Two different approaches are tested with the data collected for eight room and pillar stopes over a two year period. The first one uses multiple regression analysis whereas the second one, after recognizing patterns, extracts a set of rules (knowledge) from the data which are then used by an expert system shell.

The computer results are presented along with a brief discussion about the relative merits of each technique.

1 INTRODUCTION

Over the past decade, many people have studied the relationship between bonus and accident rates in underground mines (1, 2, 3, 4, 5, 6, 7).

These studies were carried out after two commissions in both Québec and Ontario recommended the abolition of the bonus system since they assumed a direct link between bonus and accidents.

A most recent study (6) supported the above recommendations for activities related to production only, but not for development and services activities.

The same study also concluded that the term "productivity" was ill understood and needed a clearer definition and understanding of the relevant factors affecting it.

2 STATEMENT OF THE PROBLEM

A preliminary study was carried out in one underground mine in Québec based on productivity data collected from 1975 through 1984. The results have shown a net decrease in the number of short tons per man-shift over time irrespective of the mining method used (see figure 1). Indeed, all the regression lines (broken lines) show a negative slope and it is certain that the trend is in the right direction as indicated by the probability figure appearing between parentheses.

Figure 2 shows that the number of short tons increases when the number of man-shifts increases which is a normal situation. Each dot on the plot represents the production over a 6 month period. However, the same figure also shows clearly that some unidentified factors can play a significant role and can influence the productivity particularly for the room and pillar mining method.

Finally, figure 3 shows that in terms of current dollars the bonus paid per short ton has increased over time. Similar results have been obtained in terms of constant dollars with a less steeper slope. Figure 3 also shows that for the room and pillar and the cut and fill mining methods, the bonus paid per short ton is similar but the bonus paid per short ton for the shrinkage stope mining method was much smaller with a less steeper slope.

These preliminary results have indicated that it was essential to properly establish the relationship between productivity and other variables such as bonus, mining method, total number of man-shifts, etc.

3 IMPACT ON THE ECONOMY

Mine managers forecast a decrease in

productivity of about 30% if the bonus system is abolished. A decrease in productivity of that magnitude could certainly prove disastrous for most mining districts.

4 METHODOLOGY

The relationship between productivity and other significant variables can be established by means of two different techniques namely:
- Multiple regression analysis
- Inductive inference (expert systems)

Multiple regression analysis simply expresses a relationship in a mathematical form by determining an equation connecting some variables together. The equation connecting the variables is derived from a set of data showing corresponding values of the variables under consideration.

Inductive inference derives from a set of data an inductive assertion that explains this data.

Each technique will be described in more details in the following paragraphs but first the variables selected for this analysis are given.

5 LIST OF VARIABLES SELECTED

A series of twenty-eight quantitative variables have been collected from seven participating underground mines in Québec. Among the variables collected, fourteen have been selected for this preliminary analysis. The list is as follows:
- Production level (broken ore expressed in short tons)
- Bonus paid in dollars
- Rock quality (0 = hard rock, 1 = soft rock)
- Number of 5 foot rock bolts installed
- Number of 8 foot rock bolts installed
- Number of posts installed
- Total number of man-shifts in a stope (including all activities)
- Total number of man-shifts of drilling only
- Drilling footage
- Number of drillers in a stope
- Age of the stope
- Footage of sub-levels
- Footage of manways
- Number of square feet of bulkheads installed.

The data has been collected on a monthly basis over a two-year period. In the above list, the production level is considered the dependent variable (Y) whereas the other variables are considered the independent variables (X_i).

6 MULTIPLE REGRESSION ANALYSIS

The model studied was of the form:

$$Y = b_0 + a_1 X_1 + a_2 X_2 + \ldots + a_i X_i + \ldots + a_{13} X_{13}$$

The data collected for eight room and pillar stopes has been used for this test and the calculations carried out by a statistical package (8,9) called SPSS[x] (Statistical Package for the Social Sciences) available at Laval University on an IBM 4381.

The package allows three different procedures for the selection of the independent variables (X_i). These are:
- Forward selection
- Backward elimination
- Stepwise selection

In the forward selection procedure, the first independent variable considered for entry into the equation is the one with the largest correlation (positive or negative) with the dependent variable. To determine whether this variable (and each succeeding one) is entered, a F value is compared to an established value that can be either a minimum value for the F statistic or a minimum probability of F-to-enter.

While forward selection starts with no independent variable in the equation and sequentially enters them, backward elimination starts with all the variables in the equation and sequentially removes them based on a removal criterion which can be either a minimum F value statistic or a maximum probability of F-to-enter.

The stepwise selection is a combination of the above two procedures. Variable selection for entry into the equation terminates when no more variable meet entry or removal criteria (threshold values).

After several trials with the above three procedures, the stepwise selection has been finally selected for this analysis. It must be pointed out that several other models involving different combinations (e.g. multiplication, addition, etc.) and/or manipulations (e.g. log, inverse, etc.) of the variables have also been investigated.

However, the printouts obtained have convinced us that a simple linear combination of the independent variables was the best model capable of explaining the behavior of the dependent variable (i.e. the productivity level obtained).

7 RESULTS OF THE MULTIPLE REGRESSION ANALYSIS

The analysis has shown that among the thirteen independent variables selected, five of them were capable of explaining in a significant way the productivity level obtained in terms of short tons on a monthly basis. These independent variables are:
- Footage drilled in feet (X_1)
- Bonus paid in dollars (X_2)
- Total number of man-shifts of drilling (X_3)
- Total number of man-shifts in the stope (X_4)
- Number of 5 feet rock bolts installed (X_5)

Some of the computer results are shown in Table 1 (many more statistics are given by the package SPSS* but they are not reproduced here).

The value of R^2 (the coefficient of determination) was equal to .8531 meaning that more than 85% of the variations of the dependent variable (i.e. the production level obtained) could be explained by a linear model of the form:

$$\hat{Y} = \hat{b}_o + \hat{a}_1 X_1 + \hat{a}_2 X_2 + \hat{a}_3 X_3 + \hat{a}_4 X_4 + \hat{a}_5 X_5$$

where the values of the estimated coefficients ($\hat{b}_o$, $\hat{a}_1$, ... $\hat{a}_5$) can be found in the second column of Table 1.

Each of the independent variable (X_i) may vary between the two bounds (minimum and maximum) shown in the last two columns of Table 1.

The comparison between the one-hundred and sixteen observations and the corresponding expected values calculated from the model is summarized in Table 2.

The fact that both means are exactly equal is purely coincidental. Among the predicted values, two of them were considered as outliers i.e. suspiciously different from the others (see Table 3).

For observation number sixteen, the difference between the observed and the predicted values can be explained by the fact that most of the economic ore had already been extracted from the stope and very few rock bolts had been installed in that period. For observation number sixty, it is felt that unique ground conditions are responsible for the difference between the observed and the predicted values.

Finally, figure 4 shows the cumulative plot of the observed distribution of the residuals to that expected under the assumption of normality. Initially, the observed residuals are below the straight line (dotted line), since there is a smaller number of large negative residuals than expected. Once the greatest concentration is reached, the observed points are above the line, since the cumulative proportion exceeds the expected.

8 EXPERT SYSTEM

An expert system is a programming tool that captures expertise and then imitates the human thought process. The implementation of an expert-system always integrates a knowledge base with an inference engine. The knowledge base acquisition constitutes the first phase of the implementation of an expert-system.

9 EXTRACTING KNOWLEDGE FROM DATA

In our case, we are looking for a tool capable of extracting rules (knowledge) directly from the observations. The ID3 (Inductive Decision Trees) algorithm (10) provides a way of building a classification rule in the form of a decision tree in a recursive manner. The algorithm reduces redundancy in the knowledge base, organizes the data to recognize patterns and then produces a set of rules.

We have experimented with a computer program called INDUCE (11) on an IBM PC/AT which implements the ID3 algorithm for the generation of rules from a data set. The program is written in Pascal and works with text files created with a text editor.

10 LIST OF ATTRIBUTES SELECTED

First of all, it must be pointed out that in the expert system jargon, an independent variable (X_i) should be translated as a class attribute whereas the dependent variable (Y) should be translated as a class name. Furthermore, due to the limited size and capacity of the computer program INDUCE, it has been impossible to work directly with all the thirteen attributes listed in section 5.

Instead, it has been decided to consider only the five significant independent variables obtained from the multiple regression analysis (see section 7).

The corresponding translation between the independent variables and the class attributes is given in Table 4.

The dependent variable (i.e. productivity level obtained) is translated into the word BROKEN which represents the class name.

Here again, the limited size and capacity of the computer program INDUCE, did not allow us to work directly with the numeric data (i.e. the 116 observations). Instead, it has been decided to qualitatively segregate the variables based on their respective mean and standard deviation.

The following segregation process has been used:
- If (quantitative value) is greater than (variable mean + variable standard deviation) then substitute the quantitative value by the qualitative value HIGH.
- If (quantitative value) is less than (variable mean − variable standard deviation) then substitute the quantitative value by the qualitative value LOW.
- Otherwise substitute the quantitative value by the qualitative value MEAN.

This type of coarse segregation created some duplication for the observations. The duplicated lines had to be removed from the text file for proper execution by the computer program INDUCE.

The final result is shown in Appendix I. The resulting sorted text file has thirty-six observations instead of the original one-hundred and sixteen observations. The first line in the file starts with the class name followed by the five class attributes all separated by commas. The other lines contain the qualitative values (separated by commas) obtained from the segregation process explained above.

It must be pointed out that other types of segregation process have been tried in order to obtain more qualitative values but without success due to the limited capacity of the program used.

11 RESULTS FROM THE COMPUTER PROGRAM INDUCE

The use of the program INDUCE in connection with the text file shown in Appendix I, has produced a set of eighteen numbered rules and their associated prompts which are listed in Appendix II.

Each if-then rule is a series of conditions consisting of some class attributes and their corresponding qualitative values, followed by a single conclusion that contains the class name and its corresponding qualitative value. All the rules obtained seem quite logical except for rule #2 which states that if the bonus paid is high and the footage drilled is low then the production level is high. It must be pointed out that this rule is based on one of the outliers discussed earlier in section 7.

The rules may seem terse to read but short attributes were necessary in order to meet computer memory requirements for INDUCE.

The close examination of the set of eighteen rules shows that the bonus paid to the miners is greater than or equal to the production level obtained for fourteen of them. This means that for a reasonable amount of work, underground mine workers can expect an interesting bonus.

12 RULE CHECKING

The qualitative values shown in Appendix I have been replaced by their corresponding quantitative values shown in Table 5.

The quantitative values obtained for the independent variables (or class attributes) have then been inserted into the regression equation given in section 7 and the value for each dependent variable estimated. Then it has been possible to check if the estimated values were in accordance with those shown in Table 6.

It has been found that twelve rules were completely verified and five others were more or less verified. Only rule number 2 could not be verified.

It must be pointed out that the values shown in Table 5 and 6 are the averages of the values inside the ranges defined in section 10.

The set of rules and prompts found in Appendix II have then been inserted into a file that has become the knowledge base for the MicroExpert system which is next described.

13 MICROEXPERT SYSTEM

The expert system known as MicroExpert (12) has been selected for the following reasons:
- The file containing the rules and the prompts is directly accessible by MicroExpert providing an extension ".KB" is appended to the filename produced by INDUCE.
- The system provides an inexpensive way of learning what an expert system is, what it can do for us and how it can be built.

The MicroExpert system also provides some commands that can be used during a consultation. The summary of the commands can be found in Appendix III.

The following example shows how the system prompts the user for the value of each class attribute and then tries to reach a conclusion based on the information supplied at the computer terminal (Note: Q. means a question asked by the system and appearing on the screen, A. means an answer supplied by the user through the keyboard (underlined words) and appearing on the screen, a. means an answer supplied by the system and appearing on the screen).

Q.   What is the final goal for this consultation?
A.   BROKEN
Q.   What is the value of BONUS?
A.   why
a.   The system is trying to find a value for BROKEN using rule no. 1.
Rule 1
IF
1)   BONUS   IS HIGH
2)   DRILLING   IS HIGH
     THEN
3)   BROKEN   IS HIGH
Q.   What is the value of BONUS?
A.   high
Q.   What is the value of DRILLING?
A.   whatif mean
a.   The following conclusions are possible:
Rule 3
1)   BROKEN   IS MEAN
Rule 4
2)   BROKEN   IS HIGH
Rule 5
3)   BROKEN   IS MEAN
Rule 6

4)   BROKEN   IS HIGH
Rule 7
5)   BROKEN   IS HIGH
Rule 11
6)   BROKEN   IS MEAN
Rule 14
7)   BROKEN   IS HIGH
Rule 15
8)   BROKEN   IS MEAN
Rule 18
9)   BROKEN   IS MEAN
Q.   What is the value of DRILLING?
A.   mean
Q.   What is the value of BOLTS?
A.   what
a.   The following facts have been determined:
1)   DRILLING   IS MEAN
2)   BONUS   IS HIGH
Q.   What is the value of BOLTS?
A.   whatif mean
a.   The following conclusions are possible:
Rule 3
1)   BROKEN   IS MEAN
Rule 4
2)   BROKEN   IS HIGH
Rule 5
3)   BROKEN   IS MEAN
Rule 8
4)   BROKEN   IS MEAN
Rule 9
5)   BROKEN   IS HIGH
Rule 10
6)   BROKEN   IS MEAN
Rule 11
7)   BROKEN   IS MEAN
Q.   What is the value of BOLTS?
A.   mean
Q.   What is the value of BORING?
A.   whatif mean
a.   The following conclusions are possible:
Rule 4
1)   BROKEN   IS HIGH
Rule 5
2)   BROKEN   IS MEAN
Rule 9
3)   BROKEN   IS HIGH
Q.   What is the value of BORING?
A.   whatif low
a.   No conclusion can be determined
Q.   What is the value of BORING?
A.   mean
Q.   What is the value of SHIFTS?
A.   whatif mean
a.   The following conclusions are possible:
Rule 5
1)   BROKEN   IS MEAN
Rule 15
2)   BROKEN   IS MEAN

Q. What is the value of SHIFTS?
A. <u>whatif high</u>
a. The following conclusions are possible:
Rule 4
1) BROKEN   IS HIGH
Rule 14
2) BROKEN   IS HIGH
Q. What is the value of SHIFTS?
A. <u>mean</u>
a. Conclusion:
BROKEN   IS MEAN
The following facts have been determined:
1) BROKEN   IS MEAN
2) SHIFTS   IS MEAN
3) BORING   IS MEAN
4) BOLTS   IS MEAN
5) DRILLING   IS MEAN
6) BONUS   IS HIGH

This example shows that an expert system could be a useful tool on a stope basis for underground mine planning.

It should be pointed out that MicroExpert would also allow other things like:
- Adding or deleting rules and prompts into an existing knowledge base (see Appendix II).
- Making rules less terse to read by means of translations (i.e. a string of characters used to reword each attribute)
- Adding mathematical functions and numerical comparisons (for bonus calculation for instance)
- Listing the knowledge base, etc.

14 MULTIPLE REGRESSION ANALYSIS VS EXPERT SYSTEM

Experimentation with these two techniques has allowed us to find out the advantages and the drawbacks of each method as far as the analysis of the relationship between underground mine productivity and other significant variables is concerned. The advantages and the drawbacks of each method are summarized in Tables 7 and 8.

15 CONCLUSION AND RECOMMENDATIONS

The results of this preliminary study have clearly shown that underground mine productivity is not only affected by the total number of man-shifts but also by some other variables like the bonus paid to the mine workers.

However, it is recommended to enlarge the scope of this research and to study also the effects of some other independent variables on productivity.

A partial list of these variables is: drill hole diameter, equipment type and size, equipment clearance, drawpoint spacing, experience and age of the mine workers, etc.

This study has also shown that for a reasonable amount of work, the mine workers can expect an interesting bonus.

Finally, the usage of the everyday notion of rules should have a considerable intuitive appeal for underground mine planners as well as mine managers. However, a more powerful expert system shell such as RULEMASTER would be required for implementing such a planning tool on a personal computer for each participating mine.

16 ACKNOWLEDGMENTS

This project could not have been carried out successfully without the financial support of the Association des Mines de Métaux du Québec (AMMQ), the Centre des Recherches Minérales du Québec (CRM) and the Ministry of Mines, Energy and Resources of Canada (EMR). We are also thankful to the seven participating underground mines for providing us with easy and complete access to their historical records. Finally, we are thankful to our sponsors for granting their permission to publish this paper.

REFERENCES

1. Burkett, K. et al. (1980): Towards safe production. Rapport de la Commission conjointe d'enquête fédérale-provinciale sur la sécurité dans les mines et les ateliers miniers en Ontario. Avril.
2. Beaudry, R. et al. (1981-1982): La sécurité dans les mines souterraines, 3 volumes. Aussi connu comme la Commission Belmoral.
3. Fisher, J.H. (1983): The relationship between small crew incentives and other factors and accident rates in Ontario mines, final report. Juin. Peter Moon and Associates.
4. Chouinard, J.L. et Billette, N. (1984): Bonus-accidents: is there a link? 86ième congrès annuel de l'Institut Canadien des Mines et de la Métallurgie. Ottawa. Avril.
5. Billette, N. et Laflamme, M. (1986): Influence of bonus, age, and experience on Québec underground accidents. 88ième congrès annuel de l'Institut Canadien des Mines et de la Métallurgie. Montréal, Mai.

6. Laflamme, M. (1986): Etudes sur les liens présumés entre prime au rendement et productivité ainsi qu'entre prime au rendement et accidents du travail dans les mines souterraines du Québec. Rapport présenté à Energie, mines et ressources Canada, au Centre de Recherches minérales du Québec et à l'Association des mines de métaux du Québec. Juillet.

7. Billette, N.R. et Laflamme, M. (1987): Influence of various parameters on underground accidents: a multifactorial approach in Quebec mines. 89ième congrès annuel de l'Institut Canadien des Mines et de la Métallurgie. Toronto. Mai.

8. Norusis, M.J. (1985): SPSS^x: Advanced Statistics Guide. McGraw-Hill Book Company, SPSS Inc., Chicago, U.S.A., pp. 10-71.

9. Norusis, M.J. (1985): SPSS^x: User's Guide. McGraw-Hill Book Company, SPSS INC., Chicago, U.S.A., pp. 662-686.

10. Quinlan R.: Learning Efficient Classification Procedures and their Application to Chess End Games, in Machine Learning: An Artificial Intelligence Approach, R.S. Michalski et al. Palo Alto, California, Tioga Publishing Co, 1983..

11. Thompson, B. & Thompson, W.: Finding Rules in Data, BYTE, November 86, pp 149-158.

12. Thompson, B. & Thompson, W.: MicroExpert-IBM PC Version 1.0, McGraw-Hill Book Company 1985.

7. Galibois, A. Etude des Systèmes Experts pour la Planification Intelligente des Opérations de Soudage à l'Arc Electrique, rapport interne, Université Laval, Département de mines et métallurgie, 1986.

BIBLIOGRAPHY

1. Ryval, M. (1987): L'intelligence artificielle. Revue d'Air Canada (pp. 25-124), Mars.

2. Nilson, D.S. (1987): Underground mining productivity. Mining Magazine (pp. 539-547), June.

3. Harman, P. & King, D.: Expert Systems, John Wiley & Sons, 1985.

4. Hayes-Roth F., Waterman D.A. Lenat, D.B.: Building Expert Systems, Edison-Wesley Publishing Company, 1983.

5. Fytas, K., Collins, J.L., Flament, F., Galibois, A., Singhal, R.: Potential Applications of Knowledge Based Systems in Mining, 89 AGM of CIM, Toronto, May 1987.

6. Townsend C. & Feucht, D.: Designing and Programming Personal Expert Systems, Tab Books Inc, 1986.

BROKEN,	DRILLING,	BONUS,	BORING,	SHIFTS,	BOLTS
HIGH,	HIGH,	HIGH,	HIGH,	HIGH,	MEAN
HIGH,	HIGH,	HIGH,	MEAN,	HIGH,	HIGH
HIGH,	HIGH,	HIGH,	MEAN,	HIGH,	MEAN
HIGH,	HIGH,	HIGH,	MEAN,	MEAN,	MEAN
HIGH,	HIGH,	MEAN,	MEAN,	MEAN,	HIGH
HIGH,	HIGH,	MEAN,	MEAN,	MEAN,	MEAN
HIGH,	LOW,	HIGH,	MEAN,	MEAN,	MEAN
HIGH,	MEAN,	HIGH,	MEAN,	HIGH,	LOW
HIGH,	MEAN,	HIGH,	MEAN,	HIGH,	MEAN
HIGH,	MEAN,	HIGH,	MEAN,	MEAN,	HIGH
HIGH,	MEAN,	MEAN,	MEAN,	HIGH,	HIGH
LOW,	LOW,	LOW,	LOW,	LOW,	LOW
LOW,	LOW,	LOW,	LOW,	LOW,	MEAN
LOW,	LOW,	LOW,	LOW,	MEAN,	LOW
LOW,	LOW,	LOW,	LOW,	MEAN,	MEAN
LOW,	LOW,	LOW,	MEAN,	LOW,	LOW
LOW,	LOW,	LOW,	MEAN,	MEAN,	LOW
LOW,	LOW,	LOW,	MEAN,	MEAN,	MEAN
LOW,	LOW,	MEAN,	LOW,	MEAN,	HIGH
MEAN,	HIGH,	MEAN,	HIGH,	HIGH,	LOW
MEAN,	HIGH,	MEAN,	HIGH,	HIGH,	MEAN
MEAN,	LOW,	MEAN,	LOW,	MEAN,	LOW
MEAN,	LOW,	MEAN,	LOW,	MEAN,	MEAN
MEAN,	LOW,	MEAN,	MEAN,	MEAN,	MEAN
MEAN,	MEAN,	HIGH,	HIGH,	HIGH,	MEAN
MEAN,	MEAN,	HIGH,	MEAN,	MEAN,	MEAN
MEAN,	MEAN,	LOW,	MEAN,	LOW,	LOW
MEAN,	MEAN,	LOW,	MEAN,	LOW,	MEAN
MEAN,	MEAN,	LOW,	MEAN,	MEAN,	MEAN
MEAN,	MEAN,	MEAN,	HIGH,	HIGH,	MEAN
MEAN,	MEAN,	MEAN,	HIGH,	MEAN,	LOW
MEAN,	MEAN,	MEAN,	HIGH,	MEAN,	MEAN
MEAN,	MEAN,	MEAN,	MEAN,	LOW,	LOW
MEAN,	MEAN,	MEAN,	MEAN,	MEAN,	HIGH
MEAN,	MEAN,	MEAN,	MEAN,	MEAN,	LOW
MEAN,	MEAN,	MEAN,	MEAN,	MEAN,	MEAN

1
If   BONUS is HIGH
and   DRILLING is HIGH
then BROKEN is HIGH
•

2
If   BONUS is HIGH
and   DRILLING is LOW
then BROKEN is HIGH
•

3
If   BONUS is HIGH
and   DRILLING is MEAN
and   BOLTS is MEAN
and   BORING is HIGH
then BROKEN is MEAN
•

4
If   BONUS is HIGH
and   DRILLING is MEAN
and   BOLTS is MEAN
and   BORING is MEAN
and   SHIFTS is HIGH
then BROKEN is HIGH
•

5
If   BONUS is HIGH
and   DRILLING is MEAN
and   BOLTS is MEAN
and   BORING is MEAN
and   SHIFTS is MEAN
then BROKEN is MEAN
•

6
If   BONUS is HIGH
and   DRILLING is MEAN
and   BOLTS is HIGH
then BROKEN is HIGH
•

7
If   BONUS is HIGH
and   DRILLING is MEAN
and   BOLTS is LOW
then BROKEN is HIGH
•

8
If   BONUS is MEAN
and   BOLTS is MEAN
and   DRILLING is HIGH
and   BORING is HIGH
then BROKEN is MEAN
•

9
If   BONUS is MEAN
and   BOLTS is MEAN
and   DRILLING is HIGH
and   BORING is MEAN
then BROKEN is HIGH
•

10
If   BONUS is MEAN
and   BOLTS is MEAN
and   DRILLING is LOW
then BROKEN is MEAN
•

11
If   BONUS is MEAN
and   BOLTS is MEAN
and   DRILLING is MEAN
then BROKEN is MEAN
•

12
If   BONUS is MEAN
and   BOLTS is HIGH
and   DRILLING is HIGH
then BROKEN is HIGH
•

13
If   BONUS is MEAN
and   BOLTS is HIGH
and   DRILLING is LOW
then BROKEN is LOW
•

14
If   BONUS is MEAN
and   BOLTS is HIGH
and   DRILLING is MEAN
and   SHIFTS is HIGH
then BROKEN is HIGH
•

15
If   BONUS is MEAN
and   BOLTS is HIGH
and   DRILLING is MEAN
and   SHIFTS is MEAN
then BROKEN is MEAN
•

16
If   BONUS is MEAN
and   BOLTS is LOW
then BROKEN is MEAN
•

17
If   BONUS is LOW
and   DRILLING is LOW
then BROKEN is LOW
•

18
If   BONUS is LOW
and   DRILLING is MEAN
then BROKEN is MEAN
•

APPENDIX II CONTINUED

Prompt SHIFTS
What is the value of SHIFTS ?
●

Prompt BORING
What is the value of BORING ?
●

Prompt BOLTS
What is the value of BOLTS ?
●

Prompt DRILLING
What is the value of DRILLING ?
●

Prompt BONUS
What is the value of BONUS ?
●

APPENDIX III

SUMMARY OF MICROEXPERT COMMANDS

WHY — When we ask WHY a question is asked, we see the goal that is being searched for and the rule that is being used to try to find that goal. Each successive time WHY is typed, we see the previous goal along with the rule being used to find its value until we reach the main goal of the consultation.

HOW(n) — The lines displayed by each Microexpert command are numbered. When we enter the command HOW followed by a line number, we see a description of how that clause was proved true or false or, if the value is unknown, how it can be proved. Clauses can be proved either by a rule in the knowledge base or by information supplied by the user.

WHATIF — WHATIF lets us see what conclusions can be drawn if we give the specified response to the current prompt.

RULE (n) — RULE followed by a rule number displays the specified rule.

WHAT — WHAT displays all the facts derived during the consultation.

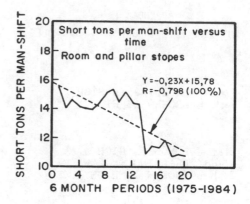

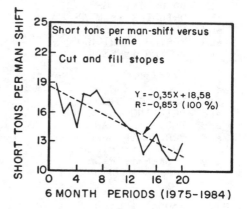

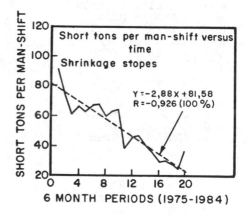

Figure 1

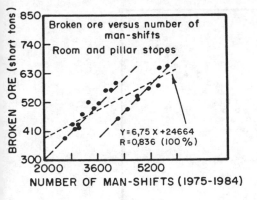

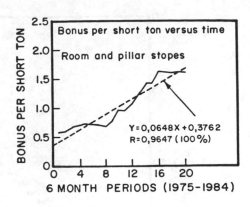

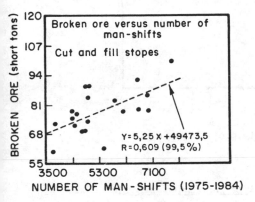

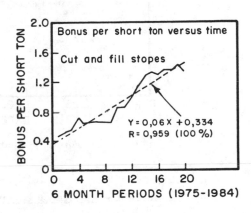

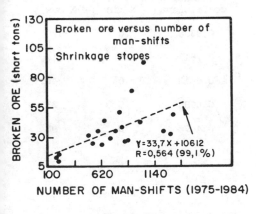

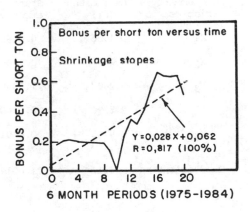

Figure 2

Figure 3

551

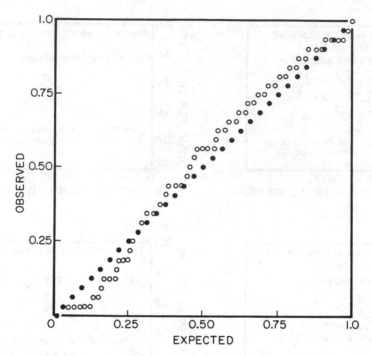

Figure 4. Normal probability plot for residuals.

Table 1 - <u>Results of the multiple regression analysis</u>

Independant variables	Coefficient $(\hat{a}_i)$	Standard error	Correlation coefficient	X min	X max
X_1	0.1956	0.0186	0.8336	0	3470
X_2	0.2647	0.0500	0.7817	55	1514
X_3	-8.2482	1.9998	0.5610	0	46.9
X_4	2.5090	0.8440	0.6988	5.5	85
X_5	0.9783	0.4101	0.4476	0	98
Coefficient $\hat{b}_o$	1.4211	21.6314	-	-	-

Table 2 - <u>General statistics for the productivity level</u>
<u>(broken ore expressed in short tons)</u>

Broken ore (short tons)	Minimum	Maximum	Mean	Standard Deviation
Expected from the model $(\hat{Y})$	29.78	1028.74	489.12	215.57
Observed from data	0	1056	489.12	233.40

Table 3. <u>Outliers for the productivity</u>
<u>level (short tons)</u>

Observation number	Observed value	Predicted value
16	890	491.80
60	954	683.44

Table 4. <u>Translation of the independent variables into class attributes</u>

Independent variables (X_i)	Class attributes
Footage drilled	DRILLING
Bonus paid	BONUS
Number of man-shifts of drilling	BORING
Number of man-shifts in the stope	SHIFTS
Number of 5 feet rock bolts ins-talled	BOLTS

Table 5. <u>Equivalence between qualitative and quantitative values for the</u>
<u>class attributes.</u>

	Qualitative values		
Class attributes	<u>LOW</u>	<u>MEAN</u>	<u>HIGH</u>
DRILLING	579	1765.1	3097.2
BONUS	153.6	463.2	978.4
BORING	3.3	14	32.8
SHIFTS	12.9	35.2	78.2
BOLTS	0	27.8	70

Table 6. <u>Equivalence between the qualitative and the quantitative values</u>
<u>for the class name.</u>

Class name	<u>LOW</u>	<u>MEAN</u>	<u>HIGH</u>
BROKEN	< 255.7	≥ 255.7, $722.5 \leq$	> 722.5

Table 7. Advantages and drawbacks of multiple regression analysis

Advantages

- Can detect easily the most significant variables

- Can detect outliers

- Allows a direct comparison between an estimate and an observed value on a quantitative basis

- Works with quantitative data of equal or different weights.

Drawbacks

- Most tests assume normality

- Data can be transformed and arranged in a multitude number of different ways

- Special cases are eliminated

- Requires adequate correlation between the dependent and the independent variables to produce meaningful results

- Past must repeat itself into the future for the equation to hold.

Table 8. Advantages and drawbacks of an expert system

Advantages

- Rules can be created and validated by an expert

- Special cases can be considered

- Can explain the line of reasoning it uses

- Very easy to add new rules or delete existing rules

- Provides an easy way of preserving knowledge and getting advice on the spot

Drawbacks

- All the rules have the same weight in the knowledge base

- Works best with qualitative data

- All possibilities must appear in the knowledge base

Computer Applications in the Mineral Industry, Fytas, Collins & Singhal (eds)
© *1988 Balkema, Rotterdam. ISBN 90 6191 760 3*

Using GENEX to develop expert systems

John A. Meech
Department of Mining Engineering, Queen's University, Kingston, Ontario, Canada

ABSTRACT: The use of GENEX, a microcomputer-based expert system software package developed in Canada, is described. Expert Systems are computer programming tools used in the field of Artificial Intelligence to capture expertise and imitate the human thought process. Expert problem-solving involves the analysis of large amounts of specialized knowledge. Much of this knowledge is based on rules-of-thumb or heuristics learned and refined over years of experience in a particular field. The knowledge gained from an experienced specialist is programmed to permit a novice to perform problem-solving tasks. The implementation of an Expert System integrates a knowledge base with an inference engine. GENEX contains a sophisticated inference engine to operate on rule-based knowledge. The package also contains tools to assist in the creation of a knowledge base. Rules can be written in plain English allowing non-technical people to participate in the use and creation of the Expert System.

INTRODUCTION

This paper presents an overview of GENEX, a GENeric EXpert System developed by Comdale Technologies Limited of Toronto. This micro-computer software package has extensive features that allow inexperienced users to create rule-based systems. These systems are the tools for use in the field of Artificial Intelligence to help train personnel (Tucker and Meech,1984), control processes (Harris and Meech,1987), diagnose problems (Buchanan and Shortliffe,1984), design new plants (Meech and Baiden,1987) or simply store information.(Boose,1986)

An expert system is software which uses specific knowledge to carry out specific tasks. The process of creating such a system begins with organizing the knowledge into a form that can be understood by a computer. With GENEX, this is accomplished by defining a series of rules that relate certain facts with other facts.

All expert systems consist of an inference engine and a knowledge base. An inference engine is the thinking part of the system; an algorithm which organizes the order in which the knowledge is to be considered and decides the sequence of activities.

A knowledge base is a file that contains the rules concerning facts of importance to the problem to be solved. These rules not only relate the facts to one another, but further, they influence the degree of certainty about the information.

The GENEX software package consists of two modules: an application module to execute the system during use and a rule compiler to convert a knowledge base into a compiled module to be used with the execution module. In order to build a knowledge base that can be compiled properly, there are certain techniques that must be mastered in developing the rules and other functions to be incorporated into the file.

WHAT IS A GENEX KNOWLEDGE BASE?

Keyword Triplets

In GENEX, all facts are stored in the form of word triplets that consist of three elements: an **Object**, an **Attribute**, and a **Value**. For example:

OBJECT	ATTRIBUTE	VALUE
Reagent	name	Sodium_ethyl_xanthate
Reagent	type	Collector
Reagent	function	Recover_copper_minerals
Reagent	addition	50_cc/minute

Note that more than one word can be used in each element of the triplet simply by linking words together into a continuous string by using a _ or a . symbol.

To each triplet, a **certainty factor** (CF) is assigned which determines whether the triplet is significant or not. The value of the certainty factor can lie between 0 and 100 or it can be unknown. Obviously, the higher the value of the certainty factor, the greater the likelyhood that the Object's Attribute has the Value of the triplet in question. Similarly, the lower the CF value, the greater is the belief that the Object's Attribute does **not** have the Value of the triplet in question.

When knowledge about a triplet is needed but the CF of the triplet is unknown, the system is forced to search through the knowledge base to find out if the certainty factor can be established. In this way, triplets and their associated CF values are at the very heart of the system in that they form the basis for data known by the system and provide the driving force for the system in its search for knowledge.

Certainty Factors

At the start of each session, the CF values of all triplets are unknown. The GENEX inference engine begins to search for knowledge by attempting to **"instantiate"** certain triplets.

A triplet is said to be "instantiated" when its certainty factor has been established. Triplets can be instantiated in several ways. One way involves the user, in that an answer is provided to a question posed by the system. For example, the system may ask:

"Does the mineral specimen show distinct euhedral crystals?"

By responding YES or NO to this query, the user causes the inference engine to instantiate the triplet

    mineral
        specimen
                distinct_euhedral_crystals

with a CF value of 100 or 0 respectively.

A second method by which a triplet can be instantiated is to read a data file that contains the triplet and its CF value. In this way, triplets that are associated with one another can be grouped in a file and instantiated together at one time. This may be considered equivalent to the grouping of knowledge in **frames** as is used in other expert systems. (Waterman, 1986)

For example:

mineral name	**galena**	-1.0	1.0
mineral luster	**metallic**	-1.0	1.0
mineral colour	**dark_gray**	-1.0	1.0
mineral streak	**black**	-1.0	1.0
mineral hardness	**#**	2.5	1.0
mineral SG	**#**	7.5	1.0

The -1.0 value informs GENEX that this triplet has a string value. Those triplets that have a # symbol as their value are numerical triplets with the numerical value being the next number shown. The final number listed for each triplet defines the CF value and so in this case all triplets are instantiated with 100 percent certainty

This method of storing data allows GENEX to distinguish between numerical and string variables and hence to conduct mathematical operations on the data.

A third and perhaps, the most important way to determine the CF value of a triplet is to use a rule.

Rules

In GENEX, the search for knowledge is carried out using rules to link one or more triplets to another triplet or group of triplets. A rule is a statement composed of a premise and a conclusion. Each condition of a premise or conclusion consists of a triplet in which the attribute and value are separated by a **predicate function**.

For example:

```
              RULE   42
 IF feed grade          IS  increasing
AND tailing grade       IS  high
AND concentrate grade   IS  OK

THEN reagent addition   IS  increased_alot
CF = 90
THEN tonnage feedrate   IS  decreased
CF = 100 #
```

When GENEX attempts to fire this rule, the inference engine examines each of the conditions of the premise in turn to establish the degree of certainty about each statement. After examining all of the conditions that make up the premise, an overall CF value is determined. This overall value is then multiplied by the CF value assigned to the triplet contained in each statement of the rule conclusion.

So, in the example above, if the system is 80 percent certain that feed grade is increasing, 90 percent sure that tailing grade is high but only 70 percent certain that concentrate grade is OK then the triplet

reagent addition increased_alot

would be instantiated with a CF value of 63
percent (0.70 x 90) and the triplet

tonnage feedrate decreased

would be instantiated with a CF value of 70.

In determining the overall CF value of
the premise, GENEX must also consider the
logical connectives between each condition.
The first condition must be preceded by an
IF to denote the start of a new rule. All
subsequent conditions commence with an **AND**
or an **OR**.

When conditions are connected by an **AND**,
the combined CF value is the **minimum** CF
value of the conditions as was shown above.
When triplets are connected by an **OR**, the
combined CF value is the **maximum** CF value
of the conditions.

Forward and Backward Chaining

A **START** 1 command tells the system which
rule to try first - in this case, rule 1.
Rules then, are used to control the flow of
information through the knowledge base.
The last triplet to be instantiated within
the conclusion of each rule establishes
which rule is to be tried next.

The system searches from top to bottom
through the list of rules until it finds
this triplet contained in the **premise** of a
rule. This rule is then selected for
firing. This process, known as **FORWARD
CHAINING**, is continued until another
triplet in the premise of a selected rule
is encountered with an unknown certainty
factor.

At this point, the system begins **BACKWARD
CHAINING** to search for this triplet
contained in the **conclusion** of a rule.
When this triplet has been instantiated,
the system returns to the rule that
initiated back-chaining activity. Forward
chaining is then resumed.

The control of triplet instantiation by
rules can be influenced further by defining
a confidence level at the start of each
session. This level, which is set between 0
and 100, establishes the degree of
certainty required of a triplet before it
can be instantiated by the conclusion of a
rule and then used to affect the flow of
knowledge.

So, in the example given above, if the
defined confidence level had been set at
65, the first triplet contained in the rule
conclusion would not be instantiated with a
CF value of 63. Rather, the degree of
certainty of the triplet would remain
unknown.

Predicate Functions

The connective between the value and
attribute of a triplet is called a
predicate function. The predicate used
within a triplet can be used to affect the
CF value of a condition of a premise.
These include:

IS
IS_NOT

which transfers the CF value of the triplet
without change.

IS_DEFINITELY
IS_DEFINITELY_NOT

which requires that the CF value of the
triplet be 100 or 0 respectively.

IS_ALMOST
IS_ALMOST_NOT

which requires that the CF value of the
triplet be between 75 - 100 or 0 - 25
respectively.

MIGHT_BE
MIGHT_NOT_BE

which requires that the CF value of the
triplet be between 50 - 100 or 49 - 0
respectively.

IS_MORE_OR_LESS
IS_NOT_MORE_OR_LESS

which requires that the CF value of the
triplet be between 25 - 75 or outside this
range respectively.

IS_VERY
IS_NOT_VERY

which squares the CF value of the triplet
before assigning it. So 90 becomes 81 in
the case of IS_VERY and 20 becomes 96 in
the case of IS_NOT_VERY. This process is
known as **Amplification** or **Concentration**.
(see Negoita, 1985 for more details)

Variable predicate functions can also be
defined within the knowledge base for use
when **displaying** certain keyword triplets.
For example, a predicate function COULD_BE
can be defined in the knowledge base to be
used when displaying triplets with CF
values above 50 percent. Then, should a
triplet such as

mineral - name - chrysocolla

be instantiated with a CF value of 52, the
displayed message reads

"I am <u>100</u> percent sure that the mineral name <u>could be</u> chrysocolla"

instead of

"I am <u>52</u> percent sure that the mineral name <u>is</u> chrysocolla".

Mutually Exclusive Sets

In certain cases it is useful to include data in the knowledge base which describes common states of an Object's Attribute. For example, the triplets:

  mineral hardness very_soft
  mineral hardness soft
  mineral hardness semi_hard
  mineral hardness hard
  mineral hardness very_hard

all represent possible terms for describing the hardness of a mineral. When one value has been instantiated with definite certainty, there clearly is no need to search for the CF value of the other possiblities. This can be prevented by defining these triplets as **mutually exclusive sets** in which case the remaining triplets are given CF values of 0.

Fuzzy Sets

A further refinement of the above concept is available to allow the user to define multiple sets with varying CF values depending upon a numerical value assigned to the triplet. For example, a number of triplets could be set up to characterize the SG of a mineral as follows:

mineral SG very_light
mineral SG light
mineral SG intermediate
mineral SG heavy
mineral SG very_heavy

Each numerical SG value can then be assigned a fractional membership number in each of these sets which can be interpreted as the CF value of the respective triplet. So, **fuzzy sets** can be created to represent the triplets

    mineral - SG - very_light
and
    mineral - SG - light

as follows:

Fuzzy 1 mineral SG very light 5
        1.0 1.5 2.0 2.5 2.6 2.7
        1.0 1.0 1.0 0.5 0.2 0.0

Fuzzy 2 mineral SG light 9
2.0 2.5 2.6 2.7 3.0 3.2 3.5 3.6 3.7
0.0 0.5 0.2 1.0 1.0 0.8 0.5 0.2 0.0

The number beside each triplet describes the number of elements contained in each set defined in the line below. The second line contains the elements of the set while the third line contains the respective membership number of each element immediately above.

Other elements that fall outside the range of the described set take on the membership number of the element on the extremity of the set. By defining sets that correspond with the other SG triplets, all of these triplets are instantiated when an SG value is identified for the mineral.

Fuzzy Sets can be extremely useful for storing relational data sets. For example, minerals can be found in nature in a wide variety of colours. As well, different people may describe the actual colour observed using a variety of words.

A mineral may occur as **olive-green** and yet the term you use to characterize the colour may range from **"green"** to **"light-green"** to **"pale-green"** to **"yellow-green"**; all of which may be appropriate, albeit to a different degree of correctness.

By assigning a reference number to each mineral, fuzzy sets consisting of these reference numbers can be created to describe each of the colours above and possibly others. Membership numbers can be assigned to each reference number in each colour's fuzzy set and hence, the CF value of each colour can be determined.

Library Functions

Instead of using a triplet in each statement of a rule's conclusion there are a number of library functions available to control the order of knowledge acquisition and/or the CF value of a specific triplet. Examples of these functions are listed below.

ACTIVATE
    to execute external processes or DOS commands.

APPLYRULE
    to fire every rule that contains a specific triplet in its premise. This is equivalent to conducting a Breadth-Search. (Waterman, 1986)

ASNCERTAINTY
    to assign the CF value of one triplet to another triplet.

DISPLAY

to display the contents of a GENEX complied file.

EXPORT

to write data to a file.

FIND

to fire every rule with a specific triplet in its conclusion until the triplet has been instantiated. This is the same as backward chaining except the backchained triplet does not affect the CF value of the rule's conclusion.

FORGET

to forget that a specific triplet has been instantiated.

FORGET CHORES

to forget that the system is presently backward or forward chaining.

FREERULE

to make already evaluated rules available for further firing.

GOTO

to dictate the next rule to be fired.

HALT

to terminate the consultation.

IMPORT

to read data from a file.

LOAD

to load a new knowledge module.

MACRO

to dictate the next rule to fire then return to present position.

REDO

to start a new consultation.

TELL

to show the current CF value of a specific triplet or set of triplets.

TEXT

to show textual material.

HOW IS A CONSULTATION WITH GENEX CONDUCTED?

Questions

Genex conducts an interview in a highly interactive manner. When information about a triplet cannot be obtained from within the knowledge base itself, the system asks

the user for data about the triplet.

There are several types of **questions** that GENEX is programmed to ask that depend on the type of information required.

If GENEX simply needs to know the CF value of a triplet, it asks:

"Is \<Object>\<Attribute>\<Value>?"

A typical question might be:

"Is reagent addition high?"

If the triplet in question is a numerical one, then the question asked is of the form:

"Tell me the value of \<Object>\<Attribute>?"

Finally, if GENEX needs to identify the correct value of an Object's Attribute when more than one possibility exists, the question reads:

"Can you describe \<Object> \<Attribute>?"

In this case the user is provided with a list of options to choose from.

Any of these questions may not be in proper English so GENEX has the facility to design customized questions that express the request for knowledge in a more friendly and perhaps, understandable form.

Explanations

Even so the user, particularly a novice, may not fully comprehend the question or why the information is needed. To assist in this situation, GENEX can provide an **explanation** about the data requested.

These customized explanations can give additional information about the triplet to help in understanding meaning or to tell the user about useful methods for obtaining the measurements that are needed.

Rule Translation into English

If the user simply wants to examine the reason GENEX needs this data, the current rule can be displayed with the flick of a key. Often the format of the rule may be difficult to decifer so GENEX has the capability of providing a customized English translation of the rule in question.

Control Strategy

GENEX begins its search through the knowledge base at any rule defined by the START command placed within the module. The system then begins to move from rule to rule in a forward or backward chaining fashion depending upon the triplets encountered in the rules or depending on any library functions located in a successful rule.

If a triplet is instantiated in the conclusion to a rule that does not appear in any premise of another rule then GENEX knows it has arrived at a **final conclusion** and ends the session by displaying the CF value of this triplet and any others that are also final conclusions.

To ensure that the overall task is accomplished successfully, a statement defining the **GOAL** of the consultation can be placed in the knowledge base. This forces GENEX to continue searching for a specific triplet even after some other final conclusion has been reached.

In GENEX, once a triplet has been instantiated, its CF value cannot be diminished unless the FORGET function is invoked. However, it is possible that additional information can contribute to an increase in the CF value of a triplet that has already been evaluated.

Query Facility

After arriving at a successful conclusion to a consultation, GENEX can permit the user to probe its knowledge base and discover how the final conclusions were arrived at. This is done by consulting the **Query Facility**. This unit allows the user to chain back through the knowledge base and view the rules that were used to arrive at all final or intermediate conclusions. This unit also permits the user to ask for specific information about any triplet and its relationship with other triplets in the knowledge base.

HOW IS A KNOWLEDGE BASE CREATED WITH GENEX?

Initial Steps

The creation of a knowledge base requires considerable preparation before writing the rules and other features. The initial step is to decide on the purpose and scope of the system to be created. Then a basic stategy should be evolved to establish a method to arrive at a final conclusion.

This is extremely important to insure that a sensible approach is taken in organizing the thought process along a clear line or procedure and avoiding the development of many isolated and unique rules for each area of the knowledge.

To illustrate these points, the development of MINEX - A Mineral Identification Knowledge Base will be discussed.

A Mineral Identification Knowledge Base

MINEX was developed to assist in the instruction of basic mineralogy to university students. It was considered that the data base available to such a system would provide a good first test of the features of a tool such as GENEX.

The initial approach taken attempted to create rules for each different mineral. It was quickly realized that this would mean building an extremely large knowledge base with perhaps, 2000 to 3000 rules - clearly outside the capacity of most PC computers.

As well, it became apparent that some technique was required to determine when sufficient information was available to make a final conclusion about a particular mineral. There are many different properties to consider about a mineral and many different combinations of knowledge available depending upon the situation.

Thus, an alternate approach was taken to resolve this problem. First, it was decided that for the system to be a truly useful teaching tool, the students should be forced to learn about the knowledge. Simply having the user input data and wait for the system to reach a conclusion was not a good way to learn.

Accordingly, the system was designed to provide the user with a list of possible mineral names after some preliminary data had been entered. The user is then required to select one name from this list that is felt to be the correct mineral. The system then requests additional information which is used to verify the suggested name.

Strategy for Concluding

In order to make a final conclusion (in this case; the mineral name is correct) it was necessary to develop a rule stategy in which the CF value of this belief would increase as more data was found to be acceptable.

This was accomplished by setting up a series of rules related to seven main properties - crystal characteristic, luster, hardness, specific gravity, colour, streak and one of several unique properties. These latter factors include radioactivity, fluorescence, magnetism, taste, swelling, special twinning characteristics, greasiness and reactivity with acid.

The system concludes with 100 percent certainty when all 6 or 7 properties of the specimen match the corresponding ones of the suggested mineral name. If only 5 properties are matched and the remaining ones are unknown then the CF value is 95 percent. For 4 properties the CF drops to 85 while for 3 properties the conclusion is made with 75 percent certainty. If any property is not matched then the system will conclude with 100 percent certainty that the mineral is not the one suggested.

Whenever a unique property is matched, the system informs the user that a final conclusion can be made if desired, however additional information may help to increase the certainty of the conclusion.

If the user's **educated guess** is judged to be incorrect by the system, a list of alternative mineral names can be provided on request that do have a match on the mis-matched property.

In this way, a novice to the field of mineralogy can learn quickly about the properties of the minerals of interest to him both directly through successful consults and indirectly through failure.

Areas of Code

The rules in MINEX are divided into several main areas of code:

1. Variable Initialization
2. Mineral Groups (90 groups)
3. Reference Numbers (240 minerals)
4. Property Matching
5. List Creation
6. Final Conclusion
7. Display Files

Data initialization occurs in two ways: first by IMPORTing a file of triplets and second by setting DEFAULT values. This is done to ensure the proper sequencing of rules and to allow minerals to have multiple reference property characteristics. For example - to allow a mineral to have both an adamantine and resinous luster.

In all cases the initialized triplets are given CF values of 0. This prevents certain redundant questions from being asked when the CF value of one of these is instantiated at a higher value by a subsequent rule.

The mineral group rules establish the list of minerals to be displayed on the screen from which the user is to select. These lists are devised according to similar properties such as crystal characteristics - twinning, shape and cross section, aggregate structure, cleavage and luster. Each of these main groups are further subdivided based on hardness and colour. Some groups have as few as one mineral while others are as large as 55.

Within each of these group rules, a list

of mineral reference numbers is IMPORTed from an external file and one of these numbers is identified from the selected mineral name in the section of code dealing with reference numbers.

This reference number is cross-referenced with a series of fuzzy sets that define the reference mineral properties. These sets include a great many descriptors for such attributes as colour, streak and luster which permit the system to establish matches for a variety of property values.

Two other methods for storing the reference properties were tried. One technique IMPORTed all appropriate triplets from a separate file for each mineral. This had the advantage of externalizing all of the data for easy update and reduction of RAM requirements but proved to be a slow process particularly when using floppy disks.

A second method involved creating what we termed "frame rules", in which the reference properties were stored in the conclusion of a rule in which the premise was unique for each mineral reference number. Again, while this was advantageous from the viewpoint of updating, memory requirements were significantly higher. The process of locating the reference properties was slow as the system was forced to search through all these rules.

In matching properties, a perfect match is not often required. The use of Fuzzy Sets provides considerable leeway in allowing matches to occur at less than 100 percent certainty. For example, although the mineral reference hardness may be between 2.5 and 3.5 on the Mohs scale, this can be matched through the fuzzy sets **very_soft** and **soft** with various membership values.

When the identification process has failed, the system can produce, on request, a list of alternate names from the mineral group list IMPORTed earlier in the session. This alternate list contains all minerals that pass the failed property test.

However, this does not necessarily mean that the correct name is on the list. If information has been improperly obtained or if measurements have been conducted without care then the wrong reference list may have been used. The user must be careful when requesting an alternative list that the data provided to the system are reliable.

Finally when a successful conclusion has been reached, the user is given the right to view a special display file that contains a one or two page description of the mineral identified. Thus, all other facts about the mineral can be examined and verified.

CONCLUSION

Our approach has been to make these systems extremely user friendly and to ensure that the user feels that she or he is in charge - not the computer program.

GENEX provides us with a tool that has tremendous flexibility and options. It allows us to develop systems for use on small PC computers - only 256K of RAM is needed to create a knowledge base. Larger bases, of course, require more memory.

Future versions of GENEX will allow string manipulation of triplets - a feature that can reduce the rule requirements and memory usage significantly. The current system allows values to be transferred from one triplet to another only when the value is numeric.

This paper has focussed on only some of the many useful aspects of GENEX. For those interested in applying this package to particular problem areas, you will not be disappointed - GENEX will prove to be the problem-solver you are searching for.

ACKNOWLEDGEMENT

Funding for this study was provided by the Microcomputer Working Group of the Faculty of Applied Science at Queen's University and by the Natural Sciences and Engineering Research Council of Canada.

The assistance of the staff of Comdale Technologies for providing and modifying many of the features described in this paper is gratefully acknowledged.

REFERENCES

Baiden, G.R. and Meech, J.A., 1987. Simulating the Mine/Mill Interface. Proc. 19th Annual CMP Meeting, Ottawa, No. 17: 443-478.

Boose, J.H., 1986. Expertise Transfer for Expert System Design, Elsevier Science Publishers, N.Y.

Buchanan, B.G. and Shortliffe, E.H., 1984. Rule Based Expert Systems:The MYCIN Experiments of the Stanford Heuristic Programming Project. Addison-Wesley, Mass.

Negoita, C.V., 1985. Expert Systems and Fuzzy Systems. Benjamin/Cummings Publ. Co., Calif.

Harris, C.A. and Meech, J.A., 1987. Fuzzy Logic - A Potential Control Technique for Mineral Processes. CIM Bull. Vol.80, No.905: 52-59.

Tucker, R.J. and Meech, J.A., 1984. A Dynamic Computer Model of Ore Crushing Plants for Teaching Purposes and Operator Training. Proc. Summer Simul. Conf. - Soc. Comp. Simul., Boston, Vol 2: 1107-1112.

Waterman, D.A., 1986. A Guide to Expert Systems. Addison-Wesley, Mass.

Computer Applications in the Mineral Industry, Fytas, Collins & Singhal (eds)
© *1988 Balkema, Rotterdam. ISBN 90 6191 760 3*

Expert systems: Application for process control (Abstract)*

Compton A. Harris & Eugene Woo
Comdale Technologies Inc., Toronto, Ontario, Canada

ABSTRACT: Expert systems technology has opened new doors to process control applications by providing new capability and flexibility for controlling processes. It offers the opportunity to capture, to retain and to utilize the invaluable process expertise that has accumulated over years of plant engineering and process operation. Modern plants possess high degrees of instrumentation which return vast amounts of data to the operator control room. In controlling the plant the operator is usually assisted by a computer which makes decisions by applying available data to rigidly constructed algorithms. Often situations arise which require the use of knowledge gained through experience with the process; this involves the extraction of data which is pertinent to the situations at hand, the filtering of enormous amounts of signals and the use of qualitative information in decision making. Traditional control systems offer only limited means of addressing these situations and are inflexible in the application of process 'know-how'. Above all they are incapable of explaining how their control decisions were made and are difficult to modify.

*Paper not received on time for publication.

Computer Applications in the Mineral Industry, Fytas, Collins & Singhal (eds)
© *1988 Balkema, Rotterdam. ISBN 90 6191 760 3*

Tool assessment in expert system prototyping: The KEE expert system shell

F.Flament & D.Hodouin
GRAIIM, Université Laval, Quebec, Canada

J.Vanderstichelen
Ciments Canada Lafarge, Montreal, Canada

D.Laguitton
CANMET, Energie, Mines et Ressources, Ottawa, Canada

ABSTRACT: KEE (Knowledge Engineering Environment) is an expert system shell distributed by Intellicorp for prototyping expert systems. Based on both a frame concept and a rule based system, KEE features user facilities such as active images, a Lisp interface, a rule tracing system and an advanced man-machine interface involving a window managment tool and on-line menus accessible through a mouse activated cursor. Although it is not necessary to learn a new language or hundreds of coded instructions, a basic KEE training is required, followed by in-depth practice in order to explore the extensive capabilities of the system. The paper reviews the basic KEE features and discusses an industrial application for simulating expertise in a cement grinding environment and should provide guidance for those contemplating the acquisition of expert system shells.

1 INTRODUCTION

Since the 50's, artificial intelligence has been rapidly moving from research to application. Expert systems are viewed as the branch of artificial intelligence which shows the highest level of activity as indicated by the long list of tools proposed on the market to help in developping expert systems. Such tools include, of course, software shells but also workshops, training and apprenticeship programs supplied by specialized companies and universities. Therefore, the task of developing an expert system has become much easier than it was a few years ago.

The development of traditional computer programs has always required a thorough phase of system analysis in which the requirements as well as the functional and the design specifications of the program must be defined. On the opposite, in expert system development, a prototype is started around a kernel of information and then continuously refined with new data provided by the customer. From one prototype to the other, the structure of the system can be thoroughly changed if the initial design can no longer comply with the supplied data. Expert systems are shaped by the knowledge that is incrementally coded while traditional software is structured principally around a chosen algorithm.

The expert system shell used to build a prototype should therefore be flexible enough to provide all the necessary attributes to structure a knowledge base while it is growing and all the facilities to manipulate it easily from both the knowledge engineer and the customer point of views. That means a versatile built-in data structure, a powerfull editor, a complete inference engine system, a good debugger and an attractive user interface for inputs and outputs. If most of the shells claim to include the necessary items, their quality varies greatly from one shell to the other.

This paper presents KEE (Knowledge Engineering Environment) from Intellicorp. The shell was used to develop a prototype of an expert system for the trouble-shooting of a clinker grinding circuit. Our intention is only to report on a tool which played a key role in the success of a first project for our team in the domain of artificial intelligence.

2 DESCRIPTION OF THE PROJECT

The project was initiated during the fall of 1986 by CANMET and Canada Cement

Lafarge. The target was to build a practical prototype of an expert system for the trouble-shooting of clinker grinding circuits. A four month period was allocated to the project and three persons were implicated:

1- A project leader from CANMET;
2- A cement expert from Lafarge;
3- A research engineer from Laval University to play the role of a knowledge engineer.

The objectives of the project were the following ones:

1- To evaluate the possibilities of using the expert system technology in the cement industry and in a more general way in the mineral processing industry;

2- To study a sophisticated expert system workstation and eventually to compare with less expensive tools;
3- To get a practical experience in knowledge engineering.

To fulfill these objectives a fourth partner was involved: UNISYS Corporation through its Apprenticeship Program. The program consists of 20 days of consultation with a knowledge engineer, used over a 4 month period. The program supposes the use of the KEE shell installed on an Explorer. The Explorer is a Lisp machine manufactured by Texas Instruments. The program warranties a practical prototype at the end of the 4 month period. KEE and Lisp trainings are two prerequisites to the program and courses are provided by Unisys. Two weeks were therefore devoted to KEE and Lisp by the three participants before the project started.

```
-------------------------------------------------
Unit CEMENT in knowledge base FINISHMILL
Created by   LAFARGE on  3-dec-1986  16:06:44
Modified by  LAFARGE on 23-Sep_1987  20:54:49
Member of:  (CLASSES in kb GENERICUNITS)
Members: FINAL.PRODUCT
-------------------------------------------------

Member slot:  READ.SO3 from CEMENT
  Inheritance: OVERRIDE
  ValueClass: (REAL in kb KEEDATATYPES)
  Cardinality.Min: 1
  Cardinality.Max: 1
  Values: Unknown

Member slot:  READ.C3S from CEMENT
  Inheritance: OVERRIDE
  ValueClass: (REAL in kb KEEDATATYPES)
  Cardinality.Min: 1
  Cardinality.Max: 1
  Values: Unknown

  ...

Member.slot:  TYPE from CEMENT
  Inheritance: OVERRIDE
  ValueClass: (ONE.OF T10 T20 T30)
  Cardinality.Min: 1
  Cardinality.Max: 1
  Values: T10

  ...
```

Fig. 1 A frame example

3 PRESENTATION OF KEE

KEE, a Knowledge Engineering Environment, is an hybrid artificial intelligence development tool. It integrates frames for representation of static knowledge, as well as rules, images and active values. It provides therefore one of the richest environments for development of knowledge based expert systems.

3.1 KEE Frames

A frame allows a user to describe any object, concrete or abstract. Each frame is given a name and slots by the designer. Slots are used to describe the object. They are not limited in number. Each slot itself is made of several attributes which in turn are used to describe the slot. An example of frame is given in Fig. 1. The object, CEMENT, is described in terms of its chemical composition and physical properties. In addition, the slot TYPE

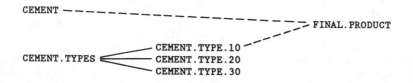

Fig.2 Classes, sub-classes and members

describes the technical names used to characterize the particular cement corresponding to the chosen operating conditions. Its attributes, as supplied by KEE, are, except for the Inheritance which will be described later:

1- the Value.Class which limits the possible values the slot can take to a given class of values such as a numerical interval or a list of characters or strings; in the present case, the value can be one of T10, T20 or T30.

2- Cardinality.Min is the minimum number of values the slot can take; it is set here to 1.

3- Cardinality.Max is the maximum number of values the slot can take; it is set her to 1 again.

4- Values is of course the slot value or list of values; it would be set to Unknown if no value had been given to the slot yet.

Other types of attributes can be defined by the user such as a Comment, the value Units...

With the combination of slots and attributes, frames can achieve the description of very complex objects.

3.2 Classes and Inheritance Laws

The frame based representation in KEE provides the options to describe individuals as well as classes of individuals or classes of sub-classes of individuals. Each individual or class is represented by a frame. Frames can be organized into groups of frames by creating links between them. KEE supplies two kinds of links to represent the type of relationship to be used: the member-link represents a class-membership, while the subclass-link represents a class subdivision for purposes of specialization for instance. In Fig. 2, CEMENT.TYPES is a class, while CEMENT.TYPE.10, CEMENT.TYPE.20 and CEMENT.TYPE.30 are subclasses and FINAL.PRODUCT is a member of class CEMENT.TYPE.10. That means the final product has all the slots of the T10 type of cement, and can also inherit their values.

The slots composing the frame CEMENT.TYPES describe the characteristics of the class as a whole as well as of the class as an individual. To do so, two types of slots are provided: the "member slot" which is a characteristic of the class as a whole and the "own slot" which is a characteristic of the individual only. Therefore, member slots are inherited by all individuals within the class while own

slots are not. Furthermore, a member slot stays a member slot when it is inherited through a subclass-link but becomes an own slot through a member-link (Fig. 3). By analogy, class, subclass and member frames can be seen as parents and children.

Remembering that each slot has an inheritance attribute (Fig. 1), the slot value itself is now inherited by a child from its parents according to an inheritance law contained in the Inheritance attribute. There are several types of inheritance laws to cover all possibilities when a child has several parents and/or has a Cardinality.Max of 1 or more.

Inheritance laws, slot attributes and subclass/member links implicitly contain rules that would have to be explicitly written in other environments than KEE. Therefore, KEE frames are very appropriate to depict objects and their relationships one with the others and to alleviate the rule base of many definition rules. This greatly accelerates the prototyping stage and improves the inference engine efficiency at time of reasoning since inheritance laws are wired in and do not involve any rule parser and interpreter.

```
------------------------------------------------
Unit FINAL.PRODUCT in knowledge base FINISHMILL
Created by   LAFARGE  on  3-Dec-1986  16:07:29
Modified by  LAFARGE  on 23-Sep-1987  19:10:02
Member of:  CEMENT, CEMENT.TYPE.10
------------------------------------------------

Own slot: READ.SO3 from CEMENT
  Inheritance: OVERRIDE
  ValueClass: (REAL in kb KEEDATATYPES)
  Cardinality.Min: 1
  Cardinality.Max: 1
  Values: Unknown

...

Own.slot: READ.C3S from CEMENT
  Inheritance: OVERRIDE
  ValueClass: (REAL in kb KEEDATATYPES)
  Cardinality.Min: 1
  Cardinality.Max: 1
  Values: Unknown

...

Own.slot THEO.SO3 from CEMENT.TYPE.10
  Inheritance: OVERRIDE
  ValueClass: (REAL in kb KEEDATATYPES)
  Cardinality.Min: 1
  Cardinality.Max: 1
  Values: 4.5

...
```

Fig.3 Slot inheritance

```
(
  IF      (  ( premise 1 )
             ( premise 2 )
             (   ...    ) )
  THEN    (  ( assertion 1 )
             ( assertion 2 )
             (   ...    ) )
  DO         ( action 1 )
             ( action 2 )
             (   ...    )
)
```

Fig.4 General rule format

3.3 Rules and Reasoning in KEE

KEE frames provide a powerful way of describing the objects on which the reasoning will take place. In addition, they provide a way of representing the rules themselves.

3.3.1 Rule Syntax

A rule in KEE is made of three parts (Fig. 4):
    1- a set of premises or conditions,
    2- a set of assertions or conclusions,
    3- a set of actions.

As shown in Table 1, each part of a rule may follow a different syntax. One will notice that assertions and actions are complementary. They are both optional but at least one assertion or action must be specified per rule.

TellAndAsk is a simple language following either an english-like or a predicate-like syntax called Well-Formed Formula. The english-like syntax is very attractive and practical although limited under some circumstances. A few examples and their meaning are given in Fig. 5. They show how easy it is to interact within a rule with the frame base and prove that it is not necessary to have a global overview of the data base before starting writing rules since frames and slots are naturally suggested and created as rules are being written. Under some circumstances however, TellAndAsk is limited in that there might be no provided formula to represent the english syntax. As an example, all mathematical operators for comparisons between values cannot be translated into TellAndAsk. In the same way, the negation operator although it exists, had very strange implications which, we understand

Table 1. Syntax of rules.

	TellAndAsk	Lisp	free syntax
Premises	Y	Y	Y
Assertions	Y	N	Y
Actions	N	Y	N

have been modified in the new release of KEE. However, in puzzling situations, Lisp palliative expressions can always be used.

Since TellAndAsk is implemented in Lisp, Lisp expressions can always replace TellAndAsk. It is even possible to intermix Lisp and TellAndAsk expressions as shown in Fig. 5. However, for various reasons including readability and comprehensive prompts issued by the system, we think the usage of Lisp should be avoided in rules whenever possible.

To overcome TellAndAsk limitations and Lisp drawbacks, KEE permits the usage of arbitrary text. An arbitrary text is a text which does not follow any of the recognized syntaxes in KEE. Arbitrary texts that have been evaluated as true, are stored in a list of unstructured facts. The Unstructured.Facts list is scanned over whenever an unstructured fact is encountered during a reasoning process.

3.3.2 Rules Chaining

The reasoning process consists of the selection of all rules pertinent to a situation, the resolution of all conflicts in choosing the next rule to be applied and the execution of the selected rule. There are two different reasoning processes: the forward chaining and the backward chaining.

In a forward chaining mode, the infernce engine is data driven: i.e. it starts from facts to derive conclusions. The rules that are selected are those the conditions of which match the facts. After conflicts are resolved, execution of rules takes place and conclusions are derived. These conclusions become eventually new facts that may drive to new conclusions and so on until all pertinent rules have been fired.

In a backward chaining mode, the inference engine is goal driven, i.e. it tries to prove a goal. The rules that are selected are those the conclusions of which match the goal. After conflicts are

```
------------------------------------------------------------------
TellAndAsk english-like syntax          Meaning
------------------------------------------------------------------
Final.Product IS IN CLASS Cement        Member link

THE Type OF Final.Product IS T10        Own slot value

Cement HAS AT MOST 1 Type               Own slot max.cardinality

ALL Cement.Type.10 ARE Cement.Types     Subclass link

--------------------------------------------------------------
TellAndAsk/Lisp intermixed syntax       Meaning
--------------------------------------------------------------
( >  (THE Theo.SO3 OF Final.Product)    Comparison between two values
     (THE Read.SO3 OF Final.Product) )
```

Fig.5 Syntax examples

resolved, execution of rules takes place,
All the premises of selected rules become
new goals to be proven and so on. The
process stops when all goals have been
proven. To prove a goal, the inference
engine first searches the facts base, then
the unstructured facts list. If the search
fails, only then the rule base is scanned.
If this also fails, the user will
eventually be prompted to evaluate the
goal.

Chaining is a crucial process during
reasoning. A reasoning could be stopped
prematurely because of an unproper coding
preventing the chaining to occur. In KEE,
fortunately, the usage of TellAndAsk forces
the usage of slots, frames and available
links between frames. As a consequence,
TellAndAsk helps in keeping a coherent
syntax between rules and therefore
facilitates establishing and maintaining
chainings.

3.3.3 Inference engines

KEE contains two different inference
engines for forward and backward
reasonings. If it is not possible to run
both of them simultaneously, it is however
possible to temporarily stop the inference
engine in use, start the other one and, on
completion, to resume with the original
one. The forward chaining mode is
activated with an ASSERT command against a
class of rules, while the backward chaining
mode is fired with a QUERY command. With
an ASSERT, the engine tries to see the
implications of an assertion, while the

QUERY command activates the evaluation of
whether the assertion is true.

The role of inference engines is to
organize the search of a solution through
the reasoning tree of chained rules.
Furthermore, if a premise cannot be derived
from a chaining to another rule, the
inference engine prompts the user for an
evaluation.

Both inference engines have many
qualities that are also found in other
shells such as switches for controlling and
flags for tracing the reasoning process.
Furthermore, since actions can be activated
when a rule is true (Fig. 4), there does
not seem to be any real limit to the power
of the inference engines.

3.4 Active Values and Methods

Active values, or demons, are Lisp methods
which provide a triggering mechanism.
Supplied with KEE as an integrated tool,
they can be attached to a given slot value.
They contain the address of a user
developed Lisp method which will be
activated whenever the slot value is
referenced or changed. Examples of
utilization of active values are the
updating of the relative composition of a
mixture following a change in the flowrate
of one of the additives, or the deletion of
the membership link between FINAL.PRODUCT
and CEMENT.TYPE.10 frames (Fig. 2) and
creation of a new link with frame
CEMENT.TYPE.20 as soon as the value of the
slot TYPE of frame CEMENT (Fig. 1) is
changed to T20.

Lisp methods, other than active values, can also be defined. They are stored in a frame slot as all other sources of knowledge. Methods are initiated by running the KEE function UNITMSG and specifying the unit and slot names of the method.

Methods can be integrated to rules as actions to be taken when rules are found true. They are mainly used to perform calculations or define actions or behaviours. They are also suggested in simulation type problems to create object-oriented programming.

3.5 Images and Icons

Unquestionably the most visible feature of KEE, images are of two types:

1- images attached to the slot of a frame, and

2- stand alone icons.

Images are supplied with KEE, they range from bar graphs, levelometers, alarms, dials, to simple numerical and text displays. They are user modifiable. Most of them can be defined as active images and as such can be used as actuators activated by the mouse cursor to change the slot value they are attached to.

Icons on an other hand are graphical representations of objects. A rather limited bitmap editor is available for their definition. An icon is not connected to any slot.

Images and icons are addressable through the mouse system to initiate Lisp methods or to give access to hiden menus for further actions. They provide many facilities to create a user friendly environment in an expert system.

4 PRESENTATION OF THE PROTOTYPE

4.1 The Grinding of clinker

ALthough it involves several mineral processing units, the process of Portland cement is a chemical process. It is generally divided into three main steps:

1- Extraction from quarries of raw materials: an argillaceous material such as clay and a calcareous material such as limestone, followed by crushing and mixing. It is essential that the mixture leaving the plant and entering the kiln be finely ground and thoroughly intermixed. Otherwise, the final cement quality will be affected.

2- Burning in a rotary kiln as long as 200 meters. After removal of water and CO_2, chemical reactions take place and new compounds are formed. The kiln new product is known as clinker.

3- Finish grinding of clinker and mixing with additives. Cement quality may be affected during this final step.

The grinding of clinker is the part of the process investigated in our study. A detailed flowsheet is depicted in Fig. 6. Four feeders supply the mill with clinker and additives: gypsum, anhydrite and filler. The mill itself is a two-compartment ball mill equipped with a fan for dust removal and a water injection system for cooling the product. The mill discharge is collected by an elevator and feeds an air separator. Coarse particles are recycled to the mill while fines are sent to the final product. Dust collected from the mill is processed in an electrostatic filter and eventually recirculated to the mill or sent to the final product.

4.2 The trouble-shooting

Three basic problems are investigated by the prototype:

- A decrease in the production rate,
- A bad workability of the product and
- A loss of strength of the cement.

Other concurrent problems are also identified in the prototype such as bad separation, bad milling and SO_3 changes in cement (Fig. 6).

A low feed rate problem is generaly related to a loss of efficiency of one of the two main units of the plant: the mill or the air separator. Bad workability is the result of an alteration of the cement chemical composition, while a low strength can be associated to a change in its physical characteristics. However, the three problems are not independant since they may have common causes. Trouble-shooting in such a system usually requires an expert.

4.3 The prototype screen

The prototype screen is shown on Fig. 6. Several icons, one per process unit, have been used to draw the plant flowsheet. In addition, all the data generally available from the plant control panels are displayed on the prototype main window. The

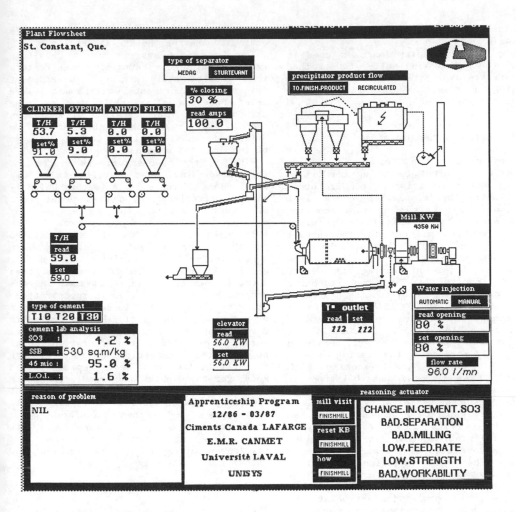

Plant Flowsheet

St. Constant, Que.

type of separator
| WEDAG | STURTEVANT |

% closing
30 %

read amps
100.0

precipitator product flow
| TO.FINISH.PRODUCT | RECIRCULATED |

CLINKER	GYPSUM	ANHYD	FILLER
T/H	T/H	T/H	T/H
53.7	5.3	0.0	0.0
set%	set%	set%	set%
91.0	9.0	0.0	0.0

Mill KW
4350 KW

T/H
read
59.0
set
59.0

type of cement
| T10 | T20 | **T30** |

cement lab analysis
SO3	:	4.2 %
SSB	:	530 sq.m/kg
45 mic	:	95.0 %
L.O.I.	:	1.6 %

elevator
read
56.0 KW
set
56.0 KW

T° outlet
| read | set |
| 112 | 112 |

Water injection
| AUTOMATIC | MANUAL |
read opening
80 %
set opening
80 %
flow rate
96.0 l/mn

reason of problem
NIL

Apprenticeship Program
12/86 - 03/87
Ciments Canada LAFARGE
E.M.R. CANMET
Université LAVAL
UNISYS

mill visit
FINISHMILL
reset KB
FINISHMILL
how
FINISHMILL

reasoning actuator
CHANGE.IN.CEMENT.SO3
BAD.SEPARATION
BAD.MILLING
LOW.FEED.RATE
LOW.STRENGTH
BAD.WORKABILITY

Fig.6   The prototype control panel

corresponding images are active images, i.e. actuators. Since the workstation is not on-line with any real plant, the active images allow the simulation of real situations. In addition, hiden menus, like behind the gypsum feeder image for instance, give access to data usually available only by a visit to the plant. A method actuator permits to fill in a mill visit report available to operators. Since extra data, such as extensive laboratory analyses, can also be requested by process engineers, the expert system may as well prompt the user for such data during the reasoning process.

Several operating conditions can also be selected from the main window:

1- The type of air separator in use is reflected by the icon displayed and, within the rule base, by different classes of rules;

2- The type of cement being produced can either be T10, T20 or T30. Following a change in production objectives, automaticaly all theoretical conditions of operation and cement characteristics are changed without the intervention of the user;

3- The type of control of the water injection to the mill may as well influence the issued diagnosis; it is user selectable.

A simple text display image is used to report the diagnosis while two method actuators are available to reset the system following a reasoning or to trace the reasoning a posteriori.

4.4 Operation of the prototype

For convenience, the system is always initialized to usual operating conditions (Fig. 6). The operation of the prototype starts with the simulation of a real situation. This is performed by changing on the screen the values that the plant operator reports as out of the normal operating level. For instance the mill temperature could be read at 115 C while the water injection system is under automatic control. Then the reasoning process can be started by posting a problem symptom such to BAD.WORKABILITY. After a few seconds, the diagnosis is displayed in the corresponding window: "Error in the Control Loop of the Water Injection System". By activating the tracing method, one can visualize the reasoning path followed. Among many possibilities, a temperature problem within the mill combined to the usage of gypsum as an additive may generate a product with a bad workability. From the many causes of a temperature problem, there is a full class of problems related to the water injection system. The water injection is used to control the temperature. In the present case, the valve is under automatic control and since it is open at its set point, the problem is not with the valve. Obviously with a temperature that high, the valve should be fully open. Since the set point is not 100%, one concludes that the control loop is not operating properly and that it should be verified.

5 TOOL EVALUATION

Following a one week course on KEE and four months of in depth practice on a real industrial application, we think KEE had a key role in the success of the project. Although KEE is a huge system, it can be mastered to a practical level in a few days. The KEE user interface is so user friendly that after the first three days of the course, the purpose of which is to cover the philosophy and major features, it is possible to start building a practical system. Such an easiness of interaction

from the first exposure and after is obviously one the the KEE assets.

The hybrid system for knowledge representation is certainly the feature that permits early prototype development followed by incremental refinement. The chief benefits of hybridization are the pseudo-natural representation of knowledge and the ability of the system to provide efficiency and flexibility in solving most of the problems when coding.

Finally, the openness of the system and the Lisp interface permit in house development of the code, a feature which is always appreciated by system developpers and very practical in the engineering domain.

The main drawbacks are of course related to the price of KEE. However, when contemplating the acquisition of a shell, one should consider two factors: the time spent to develop an expert system and the net return when the system is in operation. KEE with its potential in developping quick prototypes is unquestionably a valuable development tool. Besides this, its versatile knowledge representation model suggests that KEE could be used as a mean to investigate the best representation model for the problem addressed. Therefore KEE should be seen as a corporate tool shared by several members developping their prototypes and then moving their application to cheaper and more appropiate shells. All this should permit the rapid development of expert systems efficient and rather inexpensive.

6 CONCLUSION

The KEE expert system shell uses an hybrid system for knowledge representation. Even though all features are based on a frame structure, KEE provides also rule-based reasoning, Lisp interface, graphics, images and active values. These are integrated and they interact with one another. With such a variety of representation and reasoning techniques, the sytem developer can focus on knowledge extraction rather than knowledge representation. Hybridization helps the designer to investigate the problem and specify the solutions by developing prototypes which demonstrate an incremental fitness in the reasoning process. Furthermore, the supplied environment relieves the designer of a painful apprenticeship, even though expertise may come only after a long and in depth practice. KEE provides efficient tools for developing attractive screens and a user friendly interfaces. Such facilities may play a key role in making a success of an attempt to develop an expert system.

The application of KEE to the trouble-shooting of a clinker grinding plant has shown the assets of the shell regarding potential applications of the expert system technology in the mineral industry to solve problems such as: diagnosis of system malfunctions, design of configuration of objects, planning of actions, monitoring of system behavior, education of operators and interpretation of industrial data.

REFERENCES

Kehler, T.P and Clemenson, G.D. 1984. KEE, the Knowledge Engineering Environment for industry. Systems and Software January 1984.

Kunz, J.C., Kehler, T.P. and Williams, M.D. 1984. Applications development using a hybrid AI development system. The AI Magazine, Fall 1984: 41-54.

Fikes, R. and Kehler, T.P. 1985. The role of frame-based representation in reasoning. Communications of the ACM, 28(9): 904-920.

Fytas, K., Collins, J.L., Flament, F., Galibois, A. and Singhal, R. 1987. Potential applications of knowledge based systems in mining. 89th Annual General Meeting of CIM, Toronto, paper 202.

Vanderstichelen, J., Flament, F., Laguitton, D. and Hodouin, D. 1987. Prototyping of an expert system for trouble-shooting of clinker grinding mills. Proceedings of the Conference on Artificial Intelligence in Minerals and Materials Technology, University of Alabama, Tuscaloosa.

Computer Applications in the Mineral Industry, Fytas, Collins & Singhal (eds)
© 1988 Balkema, Rotterdam. ISBN 90 6191 760 3

Pattern classification of continuous mining duty cycle data for an intelligent decision support system

D.R.Schricker, P.J.A.Lever, R.E.Cameron & R.H.King
Mining Engineering Department, Colorado School of Mines, Golden, USA

ABSTRACT: The microcomputer based monitoring and control system at Western Fuels Utah Incorporated Deserado Mine is collecting data on production equipment power usage that can input to an intelligent production and maintenance management decision support system. The problem is to develop techniques that will classify patterns of kilowatt versus time data in the variable mine environment.

This paper will focus on mathematical algorithms developed to date for production and maintenance information from a continuous miner section. These algorithms classify the kilowatt vs. time patterns into machine operating tasks.

Data convolution with Fourier, least squares, running average, and other functions have been tested to determine which if any enhances classification. Simple and complex single and multiple step classification techniques based on parsing algorithms and statistical modeling have also been studied. However, because the research is ongoing, more optimal pattern classification techniques may be found in the future.

1 INTRODUCTION

1.1 Overview of the project

Mine monitoring systems typically measure and evaluate environmental factors such as CO, CH4, and air velocity. Although this valuable information supports certain federal regulation variances it does not approach the ultimate capability and usefulness of monitoring systems. For example, we are proving that monitoring systems can provide key production and maintenance information to support management decisions.

Our ability to extract such information from this resource has improved drastically due to the capabilities of inexpensive yet powerful microcomputers and engineering workstations. In our instance they can implement pattern recognition algorithms and techniques on extremely large data sets to economically provide key information.

The pattern recognition algorithms are part of an intelligent decision support system presently under development. The recognition algorithms define the beginning and end of patterns and identify the patterns. In themselves, they can be used

for output similar to a typical industrial engineering time study. However, many mines have file drawers of time studies that for one reason or another are not very useful. Therefore, the ultimate objective of our system is to provide key information to managers, that is, a small set of information extracted from an extremely large data base which they need for solving today's and tomorrow's problems. To anticipate the key components we are planning to incorporate maintenance and operating managers knowledge into the system. But first, we must recognize data patterns which is the topic of this paper.

1.2 Mine description

We are implementing the system at the Western Fuels-Utah Inc. Deserado Mine in Rangely Colorado. This is an excellent mine for our project since mine management has established a unique atmosphere by optimizing mine production and creating safe working conditions with appropriate advanced technology.

The mine is presently being developed by two continuous miner sections operating under 700 ft. of overburden in the 7.5 - 8.5 ft. thick D coal seam. Main, submain, and panel entries have been developed for longwall production which began in January of 1987. The continuous miner sections average about 500 raw tons per shift. One section operates 8 shifts per week and the other four shifts per week. The longwall produces about 5,500 tons per shift average and operates four shifts per week.

At present, we are working with data from a continuous miner section and the equipment shown in Table 1.

Table 1. Machine Specifications

TYPE	MANUFACTURER	MODEL	HP	VOLTAGE
Continuous Miner	Joy	12CM11−10BX	535	950 AC
Shuttle car	Joy	10CS22−56BXHE−4	100	440 AC
Roof Bolter	Lee Norse	TD2−30	40	440 AC
Feeder Breaker	Stamler	BF−14B−2−7C	125	440 AC

A 54-inch slope belt conveyor hauls the coal from underground to two 10,000 ton raw coal storage silos, then to a 750 tph preparation plant. After cleaning, a 3.5 mile overland conveyor carries the coal to a clean coal slot storage and rail load-out facility. An electric locomotive then transports the coal to the Bonanza Power Plant in Utah.

1.3 Monitoring system description

The mine management emphasis on appropriate technology led to the purchase of a state-of-the-art monitoring system. It includes a central polling/processing station located on the surface fed by accessors connected with trunk lines. The trunk lines contain four electrical conductors: two provide a power source for the accessor cards, one communicates to the accessors and the other relays information back to the central processor. The central computer connects to four trunk lines, each with a capacity of 127 accessor cards. Figure 1 shows the system layout.

The basic polling procedure begins with the computer sending out an accessor address name requesting an update on present status. After responding by identifying itself, the accessor outputs its current status. Since accessor cards are equipped with an eight bit A/D convertor, the card returns an eight bit word transmitted in serial format. Hence, eight binary or one analog value can be obtained per accessor. We are using analog values input to accessors by power transducers mounted in the section load centers.

The mine presently polls some 700 surface and underground points. Since we study data from power transducers connected to a continuous miner, shuttle cars, roof bolter, and feeder breaker from one section, we analyze just a fraction of the mine's total data base. However, principles and concepts developed by our research are applicable to other areas in this and other mines.

1.4 Statement of the problem addressed by this paper

At the present time information for evaluating or supporting design and equipment procurement decisions is difficult to obtain without sending an engineer onto the section to conduct time studies. This technique is expensive, considering engineers' time costs, and is many times biased and restricted to short time periods (i.e., a few shifts). However, a computer monitoring system can provide data continuously over long periods without an engineer present in the section. With proper analysis, this data can provide decision support information automatically without tedious low level data collection, manipulation and analysis.

However, it is not an easy task to replace the eyes, ears, and knowledge of an experienced industrial engineer with a monitoring system. It is relatively easy for the engineer to identify the stop and end points of a cycle element by looking at the machines and glancing at the stop watch. But monitoring system hardware and software doesn't have these unique capabilities, let alone the knowledge to describe and evaluate the task performed by the machine during the cycle element. Therefore, the major objective of the work described in this paper is to develop algorithms that automatically define begin and end points of machine tasks and identify the tasks based on power vs time data. Some of the tasks we plan to identify are: cutting and loading, cutting without shuttle car, trimming, cleaning, maneuvering at the face, and place changing. By collecting, analyzing and summarizing this type of data, production managers will have the information for improving production, costs, and safety; for example, through training, equipment modification, and rescheduling. The data will also provide automatic, accurate motor and component usage for preventative maintenance schedules and comparison studies.

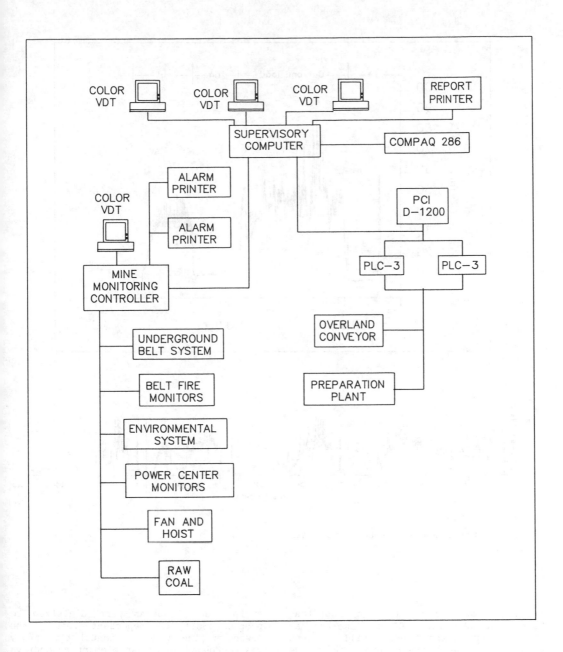

Figure 1. Deserado Mine Monitoring System

2 DATA COLLECTION TECHNIQUES

2.1 Strip chart recordings

Our ability to recognize and classify patterns produced from the kilowatt vs time measurements is dependent on the monitoring system sampling rate.
Initial tests using data taken at different sampling rates indicate that a rate of 10 samples/second produce sufficient detail for this task. However, the present mine monitoring system is not designed

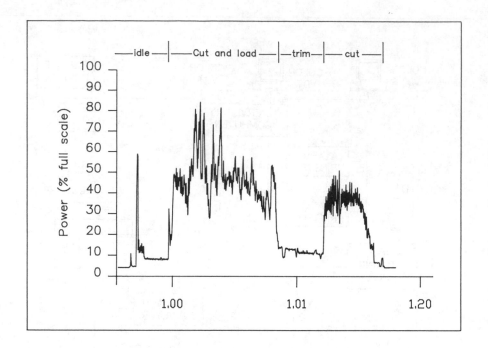

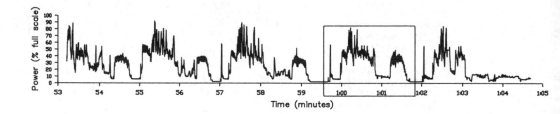

Figure 2. Continuous Miner Data Patterns

to sample at this rapid rate. Modifications in the near future include an "intelligent accessor" which will allow us to gather data at the required rate. We will connect a separate personal computer (located on the surface) with a dedicated cable (two twisted pair) to collect the intelligent accessor data before integrating it in the present mine system.

To facilitate algorithm development while the monitoring system is being modified, continuous miner, shuttle car, and roof bolter data were initially collected with a dual pen strip chart recorder placed at the load center. The strip chart tracings were digitized to produce ASCII file representations of the power vs time data. A manual time study was performed while strip chart recordings were made to produce data sets for testing algorithms. Examples of strip chart recordings of different machine operations identified from time study material are shown in Figure 2, eleven minutes of continuous miner cut and load recording taken from a typical shift.

3. DATA ANALYSIS METHODS AND RESULTS

3.1 Filtering

The first step in pattern recognition work was to identify cut and load cycles from the digitized data. Typical continuous miner data (see Figure 2) has high frequency short duration information not necessary for simple cut and load cycle classification. A common technique used to eliminate high or low frequency events is digital signal processing or filtering.

Filtering a digital wave form involves substituting a weighted average of surrounding data points for each datapoint. The number of surrounding points (n-1) in the filter is generally user specified. The first and last (n-1)/2 values of the wave form are not filtered since the filter requires (n-1)/2 points before and after each datapoint to calculate filtered values.

For this project we chose to use a triangular weighting function to smooth the data (Figure 3). A triangular filter was selected since the frequency characteristics of the filter tends to dampen the amplitude variation higher frequency events without totally eliminating them, thus better retaining the definition of the onset of an event than would be seen with a square filter.

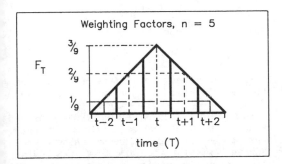

Figure 3. Five Point Triangular Filter

The specific triangular filter used in this work required input of the number of points in the filter (n) where n must be an odd number. The weighting numerator of the filter decrements from (n+1)/2 at the center point to one at both its extremities, and a value of zero elsewhere. The denominator is simply the sum of all numerators. The sum of the weighting factors is equal to one, thereby preserving the conditions required for unbiased filtering. The filtered value of each observation, P_t, is computed by multiplying each observed value by the filter function. The general formula is:

$$P_t = \sum_{i=1}^{M} (P_i * F_i) \tag{1}$$

Where

M = total number of observations

$$F_i = \begin{cases} \dfrac{K - (t - i)}{b}; & \text{for } t - (\dfrac{n-1}{2}) \le i \le t \\[2ex] \dfrac{K - (i - t)}{b}; & \text{for } t < i \le t + \dfrac{n-1}{2} \\[2ex] 0 & ; \text{ELSEWHERE} \end{cases} \tag{2}$$

and

$$K = \frac{n+1}{2} \tag{3}$$

$$b = \frac{n+1}{2} + \sum_{i=1}^{\frac{n-1}{2}} (2 * i) \tag{4}$$

The degree of smoothing varies directly with n, as shown in Figure 4; consequently, features can be filtered out or retained depending on analysis objectives. For example, if gross features such as significant waveform begin and end points are the objective features, high n values will smooth out many unnecessary features. The filter succeeded in doing this at n = 101.

Filtering was found to be computer time intensive, especially if it is employed numerous times at different n values to isolate different objective features. Because our procedure will be implemented at the mine on a micro computer, we searched for a faster means of accomplishing the same results.

3.2 Parsing

We developed a parsing algorithm to fit linear regression lines to the discrete data. The program produced line segment statistical summaries written to output files. Each statistical summary includes line segment begin and end times, slope, intercept, correlation coefficient, and line variance. The parsing technique divides a continuous data stream into data packets signifying association due to linearity. By visual analysis, machine operation change points correlate with

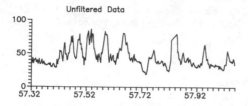

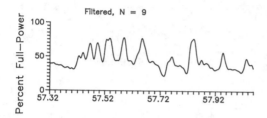

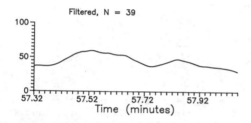

Figure 4. Filtered Continuous Miner Data Patterns

steeply sloping line segments. Machine idle and off conditions are flat low/no amplitude line segments. Parsing benefits pattern classification by allowing line segments to represent hundreds of individual data values. Searches that match certain criteria with line segment statistics can classify line segment machine operations.

Visual inspection of parsing segments showed a closer match when the program fitted low amplitude data tighter than high amplitude data. Therefore the algorithm allows greater error of fit for line segments through highly variable cut and load cycles than in lower power operations. This reflects the operating rule that changes in power at low amplitudes are more significant because there is a greater likelyhood that they identify a machine duty change.

The program begins by comparing three points for linearity. If the goodness-of-fit (GOF) tests pass, one more point is added and tested likewise for GOF. More points are added to the line segment until one or both GOF tests fail. The point that caused the fit error is deleted from the current line segment, and the line segment statistics are output. The whole cycle repeats, beginning with the rejected data point from the previous line segment.

The first GOF test is a point test:

$$\frac{|\ ERROR\ |}{AVE\ OF\ LINE + FIT\ PARAMETER} < TOLERANCE$$

$$(5)$$

Where:

TOLERANCE(T) can be specified by the user depending on feature distinction objectives.

ERROR(E) is the difference between the actual power level and the "best fit" power level for the last point in the segment.

AVE OF LINE(A) is the average of the best-fit line.

FIT PARAMETER (FP) allows the algorithm to either fit lines tightly at the low power scale while allowing a looser fit at the higher power levels (such as cutting and loading) or establish a fixed point variance. Larger FP values cause point variance to converge.

The second GOF test:

$$\frac{Sum\ Squared\ Errors}{n - 2} < Maxvariance \qquad (6)$$

This is a test which puts a ceiling on the overall maximum line variance for each fitted line.

For example, let us say we wish to allow a point error of plus or minus 10% when the miner is operating at 45% power. However, we wish to allow only a 2% deviation when the miner is operating at 5% power. This creates two equations with two unknowns.

From Equation (5), we form

$$\frac{10}{45 + FP} < T \qquad (7)$$

$$\frac{2}{5 + FP} < T \qquad (8)$$

In the general case, padding can be solved for, given two boundary conditions as follows:

$$FP = \frac{E_1 * A_2 - E_2 * A_1}{E_2 - E_1} \qquad (9)$$

Hence, in our example case

$$FP = \frac{10 * 5 - 2 * 45}{2 - 10} = 5.0 \qquad (10)$$

and

$$T = \frac{10}{45 + 5.0} = 0.2 \qquad (11)$$

Some parsing algorithms use fixed knots to determine line spacing. This is most often used with curve fitting packages. This did not fit our needs because our data varies between long zero periods to highly variable short-time events. The number of points in a line segment is variable, from two to infinity theoretically. In Figure 5, the ends of the line segments do not match up with each other. Therefore, the resulting curve is not "connected", because fixing the ends does not improve the classification algorithm.

This segmentation procedure identifies characteristic features in data (such as average sustained amplitudes, slopes, etc.) which are repeated patterns that will be incorporated into a database for automatic classification purposes. The data was reduced from 6,888 x,y data pairs to 211 line segments, while extracting information regarding slope trends, averages, minimums, and maximums, and correlation coefficients.

Parsed data can be used to distinguish special features. For example, to identify miner duty change points, where sharp increases or decreases in power exist, we want to distinguish line segments with steep slopes. Therefore, a simple program was written to extract steeply sloping line segments which fall below a certain power level, these results are given in Figure 6. Generally, this hypothesis is supported, although not in every case.

Statistical analysis performed on the line segment statistics output from the parsing routine showed the mean power level could distinguish general vs. cut and load line segment populations. Therefore, an algorithm was developed to identify cut and load patterns in the test data by searching for a line segment mean value to surpass a given threshold. The algorithm then traces to the cut and load cycle extremities, defining the endpoints by mean values falling below a certain threshold. Figure 7 shows the classification algorithm.

The identification output is imposed above Figure 5's parsed data. The time study data is added for comparison and the input is plotted below. The algorithm correctly identifies every cut and load cycle and excludes non-cut and load cycles; however, the end points are not precisely defined. This problem, however, will be solved when incorporating the shuttle car data into the algorithm.

Shuttle car data (in Figure 8) shows that when a shuttle car begins it's tram loaded cycle, the power increases to a maximum value and remains steady. This will be a relatively simple identification, enabling us to contrast miner data with the shuttle car data. Likewise, if shuttle car data indicates a cut and load sequence not supported by the miner classification algorithm, a diagnostics routine can prompt for user input or add new cut and load data into this pattern knowledge base.

After cut and load begin and end points are established and general features recognized, a more careful pattern distinction must be made. For example, amplitude distinguishes cut and load patterns, however this will not always find all the cut and load sequences. The

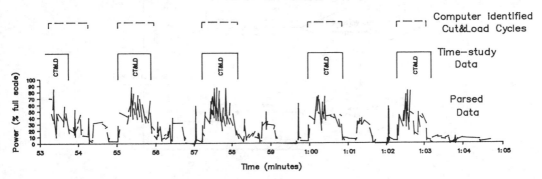

Figure 5. Parsed Continuous Miner Data Patterns

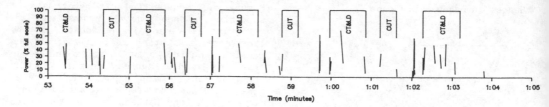

Figure 6.　Steep Slope Extraction from Parsed Continuous Miner Data Patterns

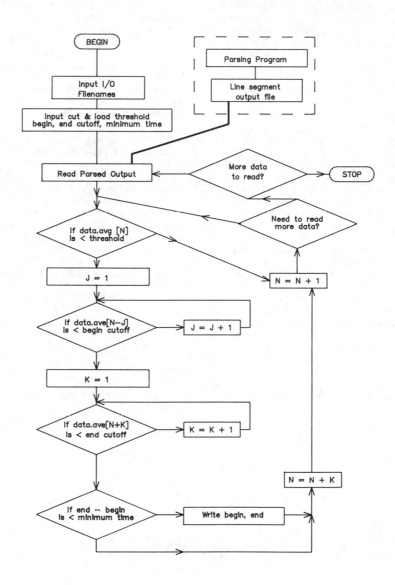

Figure 7.　Flow Chart for Cut and Load Pattern Recognition

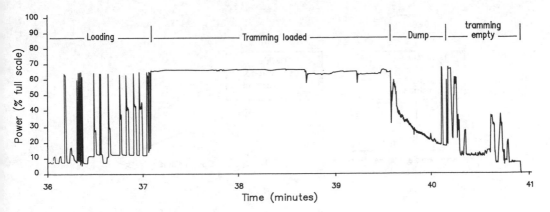

Figure 8. Shuttle Car Data Patterns

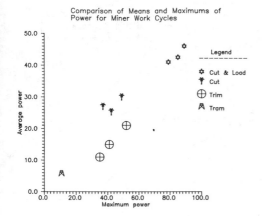

Comparison of Means and Maximums of
Power for Miner Work Cycles

Legend

☆ Cut & Load
† Cut
⊕ Trim
Ꭱ Tram

Figure 9. Scatter Diagram of Continuous
Miner Pattern Average and Maximum Power

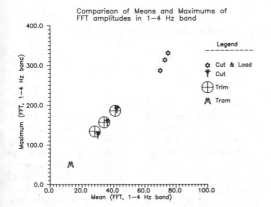

Comparison of Means and Maximums of
FFT amplitudes in 1–4 Hz band

Legend

☆ Cut & Load
† Cut
⊕ Trim
Ꭱ Tram

Figure 10. Scatter Diagram of Continuous
Miner Mean and Maximum FFT Amplitude

Table 2. Pattern Characteristics

Feature	C&Ld hi/lo/ave	Cut hi/lo/ave	Trim hi/lo/ave	Tram
Duration (sec)	56 51 54.1	23 19 21.7	38 14 27	96
Maximum	89 79 85	49 37 42.3	52 35 41	11
Minimum	10 7 8.7	9 4 7	14 2 6.3	0
Mean	46 41 42.5	30 27 25.3	21 11 15	5.8
Std. Dev.	16 11 14.3	6.9 5.5 6.2	7.0 5.4 6.0	2.8
Area	2542 2125 2313	690 541 639	407 300 361	558
FFT 1–4 Hz Max	331 287 314	189 125 157	186 134 157	50
FFT 1–4 Hz Mean	75 70 73	42 30 36	41 28 34	13
FFT 1–4 Hz St. Dev.	52 47 50	32 22 28	26 19 23	11

power differences between trimming and
tramming are indistinguishable in some
data, thus other features must be examined
in future classification refinement. Some
of the features we are investigating are
shown in Table 2 with values produced by a
standard signal processing program (DADisp
by DSP Development Corp.). Figures 9 and
10 are scattergrams showing some important
pattern recognition relationships.

Figure 9 show cut and load cycles all
clustered in the upper right corner of the
graph. One thing seen from looking at the
strip chart data is that no other miner
operation creates this high power demand

583

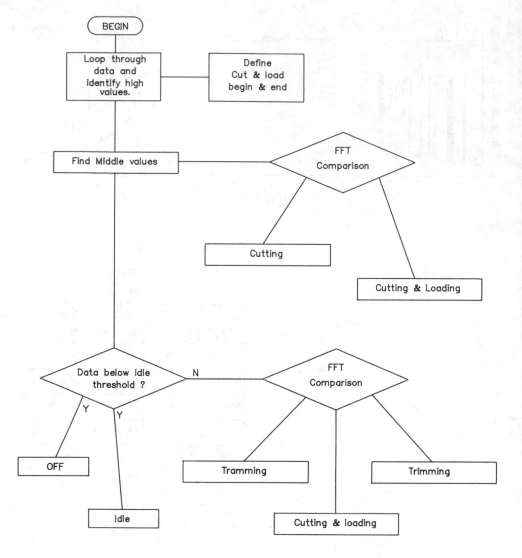

Figure 11. Emerging Power Amplitude-FFT Amplitude Classification Scheme

other than cutting and loading. But it is true that there are instances where the cut and load amplitudes would fall into the cut or trim category. Recognizing cut and load endpoints in shuttle car data will make our classification algorithm smarter. Characteristic steady high power use during tramming loaded can be easily identified. However, it would be beneficial to check with the shuttle car data, or have another method to fall back on if shuttle car data is not available.

Therefore, we chose to examine the frequency domain of each signal to determine if identifying characteristics exist. Figure 10 shows the frequency characteristics for four miner operations. The maximum and mean values were from the FFT (fast fourier transform) signal in the one to four Hertz bandwidth. This bandwidth was chosen in the one to four Hertz bandwidth. This bandwidth was chosen both by comparing plots of FFT's from different operations, which

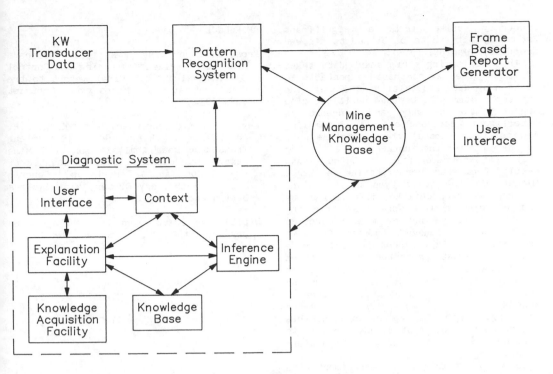

Figure 12. Emerging Intelligent Decision Support System Flowchart

showed significant differences clustered around two Hertz, and also by looking at the strip chart recordings. By eye, frequency differences in the data that look to be in the one to four Hertz band are apparent.

There exists a very definite clustering of values associated with miner operations. Cutting and loading cycles have a distinctly higher maximum frequency peak as opposed to other operations. However, frequency characteristics for cutting and trimming appear to be almost identical. This is expected, since the miner is essentially doing the same thing when cutting or trimming, except that "cutting" is used to identify hard and intense cutting whereas "trimming" is used to describe light, cleanup cutting. Mean FFT values can partition tramming and trimming data.

Our first pattern recognition algorithm proved successful on the test data, but much more obscure patterns have yet to be tested. By incorporating shuttle car data into a pattern recognition engine to identify miner operations, we will greatly enhance our recognition success and accur-acy. Adding frequency characteristics into the pattern classification for the miner will further enhance and refine pattern classification.

4 FUTURE OBJECTIVES

In the coming months, more sophisticated and robust pattern recognition algorithms will be tested. For example, work has begun on a new procedure for using FFT analysis of continuous miner data (Figure 11). The next step is to integrate shuttle car data.

The present monitoring system only saves data every seven seconds, which is not nearly enough for performing an FFT of the data. Assuming a lower bound for FFT frequency significance to be at 3 Hz, then a common rule for data acquisition is that the sampling rate should be ten times the required frequency. Therefore, we require 30 Hz sampling rates. Since the data used for present analyses was at 10 Hz, we will need to repeat the analysis at a higher sampling rate to investigate digital aliasing. Digital aliasing causes the frequency spike from a significant frequency pattern to be biased to a lower frequency (DaDISP, 1987).

Since digitizing strip chart data takes too much time, we purchased a portable PC containing data acquisition hardware. The lap-top system will be used until the mine upgrades the hardware and connects an intelligent accessor (capable of sampling 100 points a second) via a dataline to a dedicated PC on the surface.

The pattern recognition techniques that result from our work provide a large amount of time study type data. Since managers only require key information, we are building a knowledge base from the ideas and rules used by mine production and maintenance managers to identify "key" information. This expert knowledge base will analyze data presented by the pattern recognition system, detect production problems and suggest possible solutions such as revising preventative maintenance schedules, and recommending operator training. Figure 12 represents a possible structure for this system, however this is ongoing research and changes are anticipated.

We are also planning to experiment with symbolic representations of significant features within important waveforms. This type of representation will facilitate learning capabilities required to adapt to continually changing mining conditions. For example, correctly sequenced features of parsed line segments such as sharp peaks, vertical linear slopes and shallow valleys might distinguish one operator's pattern from another or identify geological conditions changes.

ACKNOWLEDGEMENT

Funding for this project was provided by the U.S. Bureau of Mines Generic Mineral Technology Center in Mine Systems Design and Ground Control which is directed by Dr. J. Richard Lucas in the Department of Mining and Minerals Engineering at the Virginia Polytechnic Institute and State University.

DISCLAIMER

The views and conclusions contained in this document are those of the authors and should not be interpreted as necessarily representing the official policies or recommendations of the Interior Department's Bureau of Mines or of the U.S. Government.

REFERENCES

Eros, L.A. and King, R.H. 1986. "Deserado Mine Computerized Monitoring and Control System Evaluation," Final Report to the U.S. Bureau of Mines on Contract J0348003.

Lever, P.J.A., Schricker, D.R., King, R.H. and Cameron, R.E. 1987. "Electrical Transducer Data Analysis for Coal Mine Management Reports," paper presented at the Fifth Annual Workshop, Generic Mineral Technology Center, Mine Systems Design and Ground Control.

DaDISP Worksheet Manual, 1987. DSp Development Corp., Cambridge, MA.

Computer Applications in the Mineral Industry, Fytas, Collins & Singhal (eds)
© 1988 Balkema, Rotterdam. ISBN 90 6191 760 3

Development of a knowledge-based decision support system for surface mining

Leo K.Nenonen
National Research Council of Canada, Ottawa, Ontario

Malcolm Scoble & John Hadjigeorgiou
McGill University, Montreal, Quebec, Canada

ABSTRACT : The application of computer tools to assist in tactical surface mine planning is investigated. The overall concept for a knowledge-based decision support system for proper utilisation of available tools is described. Furthermore, the paper reports on the status of current research work.

1 INTRODUCTION

Elbrond and Soumis (1987) described two levels of planning in industry, namely strategic planning which determines what to do and tactical planning which determines how to do it.

Mine planning encompasses several distinct systems which are not mutually exclusive - Production, Geological, Geotechnical, Environmental, Human and Financial Planning. The central and most complex system underlies Production Planning which necessitates a complex interaction between management, engineering, geological and administrative personnel. The volume and complexity of data flow in support of mine production planning activities present the obvious challenge to computer application, which intensifies with geological complexity. Production Planning is a multi-layered sequence of activities, Table 1, dependent on frequency and projection period. The layers span the levels of strategic and tactical planning, between which the dividing line is difficult to resolve.

It is in tactical planning, which is equivalent to Production Control, that new technology will likely exert the most immediate and obvious impact. Equipment monitoring and communications development will provide the data to facilitate tactical planning with speed and accuracy. Examples are becoming numerous, including excavator, drill and vehicle monitoring (International Mine Computing 1987b, Pomeroy 1982, Hagenbuch 1986, Lang et al 1984, Dawson 1987, Thompson 1986, USBM Technology News 1984). Support systems to exploit such technology will require intensive software development to facilitate accuracy, precision and timeliness in decision making. This will be based on the reconciliation of actual with planned production parameters such as tonnage, grade, equipment/manpower productivity, equipment availability and utilisation, and operating costs.

A recent Canadian survey of software for mining identified more than one thousand software modules (H.A. Simons 1985) of which 44% concerned exploration, mine design and development (i.e. strategic planning) and 19% mine production (i.e. tactical planning).

This paper describes research on new types of software-based tools and on more effective ways of using existing tools for decision making in the area of tactical planning in surface mining.

2 DEVELOPMENT OF A DECISION SUPPORT SYSTEM FOR SURFACE MINING

2.1 Background

While the term " Decision Support System " was first used in the context of strategic business or financial planning, it can be assigned a more general definition. The term is particularly appropriate since it reminds software developers of the ultimate purpose of the software modules they are developing, be they models, algorithms, etc. A more general definition of a decision support system (DSS) is a computer-based system which

TABLE 1 : Typical Layers and Activities in Mine Production Planning

	LAYER	ACTIVITIES
STRATEGIC	Life of Mine	Reserve development sequence, Production rates, Strip ratio, Tonnage/grade (per year)
	Annual	Manpower/equipment deployment, Production tonnage and grades, Bench development/blending (per week)
	Monthly	Drill/blast/load/haul sequences, Ramp/road access plans, Waste and ore tonnage/grade (per week)
	Weekly	Equipment moves, Maintenance schedules, Waste and ore tonnages/grades (per day)
TACTICAL	Daily	Operating instructions, Detailed bench plans (per shift)
	Real time	Equipment/manpower moves, Dispatching, Maintenance

aids specific types of decision-makers in a specific domain in carrying out specific types of tasks. It's scope and capabilities depends on the types of users, the types of decisions, the procedures they prefer to follow in making them and the information, data and criteria on which they base their decisions. In the domain of surface mining, the types of decision-makers include the mine manager, short-range planner, the geotechnical and mining engineers, and the operations supervisors. A DSS facilitates access to the most suitable software tools at each step in each decision-making task it encompasses. In effect, a DSS is a specialised tool kit with instructions on where and how to use the tools. However, as with use of any types of tools, the experienced user does not need to refer to nor follow the instructions and can use the tools, to a limited extent, in his own unique way.

Simulation models, knowledge-based systems and optimisation algorithms are being used for tactical planning (Fytas et al 1987). The proposed DSS will provide more " structured " ways of applying such tools which should increase the quality of the decisions made and, conversely, minimise the risks of making wrong decisions. Lane (1985) describes computer software as " dumb " tools requiring knowledge and judgement for proper applications by the mineral developer. Furthermore, he describes

software as providing " objective " support for decision making, a very important role for a decision support system. Objectivity is desirable by both the DSS user in interpreting the data and recommendations it produces and the DSS developer in explicitly acknowledging it's limitations.

Loy (1987) evaluated a number of public-domain underground computer models in terms of applications, scope and organisation, and computer aspects. While his comments were related to specific models, certain problem areas were identified that, in the authors' experience, apply to most modeling software :

1. Significant time and effort to enter data.

2. Need for highly detailed and precise input information.

3. Uncertainty re applicability of numerical techniques utilised.

4. Uncertainty re constraints on interpretation of outputs.

5. Large quantity of output generated.

6. Lack of interactive features.

7. Inadequate user guidelines.

8. Need for extensive theoretical background in mining or modeling.

This list encompasses a number of important problem areas which should be addressed by developers of decision support systems. In one particular case, Loy attributes the need for a significant amount of input data to the fact that the associated model was designed to be as flexible as possible for general applicability. This is a valid observation and reflects the desire of the model developers to maximise the cost-effectiveness of their efforts. However, by so doing, the effort required by the user may be increased to the point where use of the model is not cost-effective and that is the so-called " bottom line " on which developers of decision support systems should focus.

The topic of decision support systems is currently very popular and is the subject of many papers, conferences and books. Such systems are designed to solve so-called " ill-structured " problems which are defined as those which are not "well-structured ". Well-structured problems meet the following conditions (Bosman and Sol 1985) :

1. Set of solutions or actions is finite.

2. A good empirical model is available which provides consistent solutions.

3. Effectiveness of possible actions can be numerically evaluated.

By this definition, many tactical planning problems in mining can be described as ill-structured and needing a DSS-like problem solving aid. Bosman and Sol (1985) advocate construction of a "descriptive" model of the such problem situations and development of an "inquiry" system comprising various kinds of expertise with different knowledge representations. They cite a number of limitations of current technologies and approaches, including lack of adequate attention to the purpose of the DSS, to the information needs of it's users, to the gathering of good intelligence including empirical evidence to support it's hypotheses. They also distinguish between " prescriptive conceptual models " and " descriptive empirical models ". Conceptual models are said to be equation models which specify the problem and form the basis of algorithms for it's solution. Empirical models describe how a decision-maker solves a problem, possibly by rules of thumb. The generality of empirical models is low but their correspondence with reality is high while the reverse is true for conceptual models. Thus, both types of models have a role to play in a DSS for surface mining.

Bonczek et al (1981) advocate a DSS concept comprising a language system, a knowledge system and a problem processing system. The language system encompasses both those for data retrieval and for computation and is the sum total of the means by which the DSS user can enter information about his problem and his preferences re it's solution. The knowledge system is the knowledge bank of both heuristic and technically based knowledge about the problem domain, in our case, tactical planning of surface mining operations. Different representation methods will be required for the different types of knowledge to facilitate efficient storage, updating and accessing. Bonczek describes the problem processing system as guiding the decision maker in formulating his problem.

Keim and Jacobs (1986) discussed the merits of combining the features of " conventional " decision support systems with those of expert or knowledge-based systems.

Development of the proposed DSS constitutes a relatively large scale software development project. Parnas and Clements (1986) discussed the main problems encountered in software development and how some of these can be tackled by at least attempting to follow an "ideal" approach which starts with creation of a user requirements document. This document records the desired behaviour of the system as described by the users and perceived by the developers and can be reviewed by the users to ensure that the DSS developers have perceived their needs correctly. Parnas and Clements cite a number of compelling reasons for creation of such a document, including the desirability to avoid making requirements decisions accidentally while designing the program.

Since production planning and control procedures are expected to vary from mine site to mine site, the DSS should provide a variety of " standard " options and facilitate implementation of user-specified procedures. The system should also permit and facilitate stand-alone use of any of it's modules.

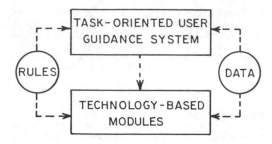

FIG. I STRUCTURE OF THE DECISION
SUPPORT SYSTEM

2.2 Description Of The Decision Support System

The system will include a task-oriented guidance system to assist the user in performing his selected task using the most appropriate tools (Fig. 1). The procedures for performing the decision-making tasks will be based on mining knowledge and the choice of tools will be based on modelling knowledge, user preferences and particular mine site characteristics. Since each DSS module will have it's own user guidance system to permit it's standalone use, the user guidance system referred to in Fig.1 will function at a high level.

Researchers in the field of Artificial Intelligence and knowledge-based systems are investigating development of " intelligent front ends ". These are knowledge-based user-friendly interfaces to relatively complex software packages, or in other words, task-oriented user guidance systems. Consequently, their work is expected to generate many ideas of value to our DSS development efforts. For example, referring to the work of Bonczek et al described earlier, the concept of a user guidance system replaces their concept of a problem processing system and, using their terms, involves use of the language system to not only define the user's problem but also to guide use of the knowledge system to solve it.

Some commercial software packages refer to concepts similar to that of a user guidance system. For example, the modules in MINEX, a mine planning package developed by Exploration Computer Services of Australia, are said to be organised to " task level processing " (International Mine Computing 1987a). From another perspective, Elbrond and Soumis (1987) refer to " the use of a complete simulation model for mine planning from start to finish, including optimisation procedures such as linear programming ". While this approach was suggested as a means to reduce computer costs associated with sole use of optimisation techniques, it's implementation could be facilitated by a user guidance system. In fact, the DSS described here may be the " complete simulation model " referred to. Karmarkar, a Bell Laboratory mathematician, has developed a much more efficient linear programming technique which is claimed to be ten to fifty times faster than conventional linear programming algorithms (Emmett 1985, Rockett and Stevenson 1987). If this algorithm proves to be so superior in practice, it's use in our DSS could reduce the need for applying modeling methods.

To illustrate the concept of the task-oriented user guidance system, consider the task of selecting excavating and haulage equipment for a new pit. This task involves a series of decisions concerning types, makes, models, unit and fleet sizes, and purchase schedules. The user may wish to select one set of equipment for mining out the entire pit. Before proceeeding with this task, the geological and geotechnical domains in the pit must be defined using core logging, geophysics and mapping. Commercial software packages exist which can process and store such data in the form of models which define the extent and quality of the ore body. Once this information on geological and geotechnical conditions has been prepared, the DSS guidance system for the equipment selection task would invite the user to:

1. Use a knowledge-based module to assist in selecting equipment types for each of the domains in the pit.

2. Use a deterministic-model-based equipment performance module to select makes and models of equipment for each domain and estimate their average expected performances.

3. Use a short-term event-based model module to determine the overall fleet size for the pit, taking into account equipment interactions, deployment and dispatching policies, and production quantity and quality targets.

4. Use a module for determining optimum equipment purchase schedules to determine a timetable for their acquisition, taking into account interest rates, possible production shortfalls, etc.

While the above description does not illustrate in detail the procedures to be implemented by the task-oriented guidance system, it does illustrate one way of using the various types of available tools to carry out a given task. Other ways of using the tools to accomplish the same task will also be investigated. For example, Bernold (1986) describes use of a deterministic model for calculating cycle times in an " intelligent simulator " which determines the optimal size of truck fleet for a series of cut and fill operations. The program uses two simple rules, namely, keep the constraining resource busy and minimise idle resources. For each phase of the cut and fill operation, the program uses a deterministic cycle time calculator to automatically determine the number of trucks needed to keep the loader busy while minimising truck idle time.

2.3 Status And Future Plans

To date, the following modules have been developed :

1. Database management

This is a general database management package (Trumbley et al 1986) that allows the user to develop tailor-made data bases using user-defined data prompting and data listing files. It has been used to create and manage databases containing specifications for mechanical and electric drive trucks and haulers, hydraulic excavators, cable shovels, front-end loaders, and draglines. It is also used to specify mine sites in terms of materials, layouts, production targets, equipment complements, shift schedules, etc. Menus facilitate access to and revisions or additions to database entries.

2. Deterministic model of load/haul operations.

This model was originally developed in collaboration with Canstar Oil Sands Ltd (Collins and Nenonen 1982, Nenonen 1986). It has been used by the CANMET Mining Research Laboratories in Edmonton (Srajer 1986) and versions have been made available for testing by two surface coal mining companies. The model accesses databases generated by the previously mentioned database package. The user can select data on specific mine sites and evaluate performance of different makes and models of load/haul equipment. The model calculates loading and travel times, fuel consumption and tire overheating indices. If the user chooses a mine layout, the model allows him to work with any load/dump site combination and uses a built-in routing algorithm to determine the connecting road segments. The user can also specify the route if he wishes. A variety of parameters are used to reflect weather, equipment and road conditions.

3. Detailed, short-term event-based models of conveyer/bin, dragline and bucketwheel excavator systems (Chan 1982, Graefe 1982a, Gibb 1982).

These models utilise a FORTRAN-based simulation package called ANEVENT developed by the Systems Laboratory (Graefe 1982b, Crate 1982). At present, they function on a stand-alone basis but will be adapted for access by the task-oriented DSS user guidance system. A simulation model of truck/shovel operations at the Lornex mine (now Highland Copper) was developed in 1980 (Nenonen et al 1981) for use by Glenayre Electronics (Pomeroy 1982) to design their MIDISS truck dispatching system. Specifications for a general version of this model have been developed which will form one of the DSS modules. This general model will be designed to facilitate evaluation of alternative equipment monitoring and dispatching systems. Other types of event-based models will also be developed to support short-range scheduling tasks.

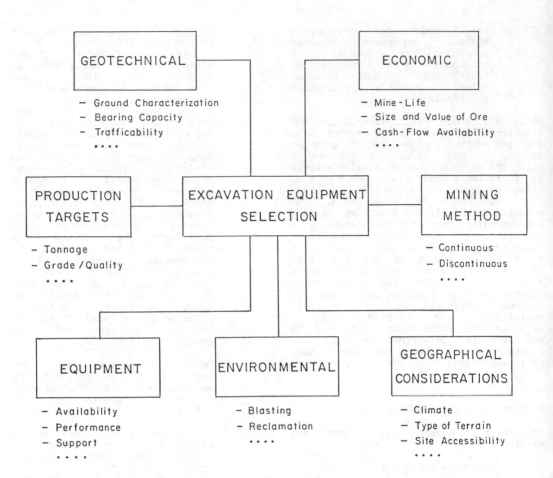

```
┌─────────────────┐                          ┌─────────────────┐
│  GEOTECHNICAL   │                          │    ECONOMIC     │
└─────────────────┘                          └─────────────────┘

 – Ground Characterization                    – Mine-Life
 – Bearing Capacity                           – Size and Value of Ore
 – Trafficability                             – Cash-Flow Availability
   ....                                         ....

┌────────────┐    ┌─────────────────────┐    ┌────────────┐
│ PRODUCTION │    │ EXCAVATION EQUIPMENT │    │   MINING   │
│  TARGETS   │    │      SELECTION      │    │   METHOD   │
└────────────┘    └─────────────────────┘    └────────────┘

 – Tonnage                                     – Continuous
 – Grade/Quality                               – Discontinuous
   ....                                          ....

┌────────────┐      ┌───────────────┐      ┌─────────────────┐
│ EQUIPMENT  │      │ ENVIRONMENTAL │      │  GEOGRAPHICAL   │
│            │      │               │      │  CONSIDERATIONS │
└────────────┘      └───────────────┘      └─────────────────┘

 – Availability        – Blasting              – Climate
 – Performance         – Reclamation           – Type of Terrain
 – Support               ....                   – Site Accessibility
   ....                                          ....
```

FIG. 2   FACTORS CONTRIBUTING TO EXCAVATION EQUIPMENT SELECTION

4.   Knowledge–based system to assist in selection of surface excavating equipment based on geotechnical parameters of the material to be excavated (Scoble et al 1987)

The task of equipment selection involves the interaction of several factors (Fig. 2). The existing prototype of the knowledge-based system (Hadjigeorgiou 1987) uses geotechnical and equipment characteristics, was developed using a generalised inference engine (Moran 1987) and is implemented on an IBM AT computer in Turbo PROLOG. The system uses rules which describe the feasibility and desirability of selecting various types of excavation equipment given values for geotechnical parameters such as block size, rock strength and degree of weathering. The rules were derived from documented case studies and from consultations with experts. Features of this knowledge-based system include a user-friendly interface, the ability to explain its line of reasoning and to justify asking the user specific questions. The knowledge base can be updated at any time by addition or deletion of rules. Confidence factors are used to represent any uncertainties inherent in the data provided by the user. Operations using these confidence factors arise both during data entry and during rule evaluation and firing stages.

Development and implementation of the described modules led to a number of observations. The most important, time-consuming and difficult task in development of " intelligent " simulators and knowledge-based systems is that of knowledge acquisition and representation. The methodology for collecting, analysing, interpreting and organising the conceptual and procedural skills constituting expert knowledge has not yet matured. Hart (1986) and Kidd (1987) describe the evolution and mechanics of several current knowledge acquisition techniques. As improved techniques are developed, their use can lead to improved simulators and decision support systems.

Communication between the various modules of the DSS can be complicated due to the incompatability of different computer languages. For example, while suppliers claim that knowledge-based modules written in Turbo PROLOG can be linked to model-based modules written in FORTRAN, the authors have not yet succeeded in doing so. The desire is to find a common basis for implementing the knowledge-based system features in the decision support system including the task-oriented user guidance system, the stand-alone modules · such as the geotechnical one discussed above, and knowledge-based or rule-based portions of the various model modules.

3 SUMMARY

A decision support system for tactical planning in surface mines has been proposed and some of it's modules are being developed. A task-oriented guidance system will be developed to assist the user in selecting the most suitable model and to highlight it's limitations in relation to the specific mining situation being considered. Use of such a system is expected to improve the quality of decisions.

4 REFERENCES

Bernold, L.E. 1986. Low level artificial intelligence and computer simulation to plan and control earthmoving operations. Earthmoving and heavy equipment. Oberlender, G.D., Editor, Published by the American Society of Civil Engineers, 1986, pp 156-165.

Bonczek, R.H., Holsapple, C.W., Whinston, A.B., 1981. Foundations of Decision Support Systems. Academic Press,   1981. pp. 393.

Bosman, A., Sol, H.G., 1985.   Knowledge representation and information systems design. Knowledge Representation for Decision Support Systems. L.B. Methlie and R.B. Sprague (Editors). Elsevier Science Publishers, B.V. (North Holland).   IFIP 1985. pp. 81-91.

Chan, A.W. 1982. Interactive computer modelling of conveyor systems. CIM Bulletin Nov. 1982, v.75, ⅰ.. 847.

Collins, J.L., Nenonen, L.K. 1982. Interactive computer aids for design of oil-sands mining operations.   1982 Annual General Meeting, Canadian Institute of Mining and Metallurgy.

Crate, G.F. 1982. ANEVENT: An interactive computer modeling package based on discrete event simulation: A user's guide. Laboratory Technical Report, Division of Mechanical Engineering, National Research Council of Canada, LTR-AN-53, Oct. 1982

Dawson, J.E. 1987. How on-board haulage truck monitoring systems are reducing maintenance and downtime costs.   1987 Annual General Meeting, Canadian Institute of Mining and Metallurgy, May 1987 Toronto.

Elbrond, J., Soumis, F. 1987.Towards integrated production planning and truck dispatching in open-pit mines. International Journal of Surface Mining, 1 (1987), pp 1-6, A.A. Balkema, P.O.Box 1675, 3000 BR Rotterdam, Netherlands.

Emmett, A. 1985. Karmarkar's algorithm : A threat to Simplex ?   IEEE Spectrum, December 1985, pp 54-55.

Fytas, K., Pelley, C., Calder, P.   1987. Optimization of open-pit short and long range production scheduling. CIM Bulletin, v.80, n.904, Aug. 1987,   pp 55-61

Gibb, G. 1982. Interactive computer modeling of bucketwheel excavators. 1982 Annual General Meeting, Canadian Institute of Mining and Metallurgy.

Graefe, P.W.U. 1982a.Interactive computer modelling of draglines. CIM Bulletin, Nov. 1982., vol. 75, no. 847.

Graefe, P.W.U. 1982b. ANEVENT: An interactive computer modeling package based on discrete event simulation. Laboratory Technical Report, Division of Mechanical Engineering, National Research Council of Canada, LTR-AN-48, Oct. 1982

Hadjigeorgiou, J. 1987. A knowledge-based system for selection of excavating equipment based on geotechnical parameters. Laboratory Technical Report, Division of Mechanical Engineering, National Research Council of Canada. (under preparation).

Hagenbuch, L.G. 1986. An integrated truck management information system concept. CIM Bulletin, Vol. 79, No. 892, August 1986, pp 62-68.

Hart, Anna. 1986. Knowledge acquisition for expert systems. Kogan Page. 1986.

H.A. Simons (International) Ltd., 1985. Investigation of computer software and hardware in surface and underground mining in Canada. DSS file no. 15SQ.234 40-3-9202; Contract serial no. OSQ83-00 304; April 1985.

International Mine Computing.1987a. MINEX Computerized geologic evaluation, mine planning and design.   May/June 1987, pp 4-15.

International Mine Computing. 1987b. WEN-CO's production monitor and control system. vol.2, no.4, July-august 1987, pp 4-11

Kidd, A.L. 1987. Knowledge acquisition for expert systems. (Editor). Plenum Press, New York, pp 194.

Keim, R.T., Jacobs, S. 1986. Expert Systems: The DSS of the future. Journal of Systems Management, Dec. 1986, pp 6-14.

Lane, A. 1985. Computer applications for the mineral development industry. In-industry. International Mine Computing, 1985 issue, pp 26-27.

Lang, A., Louw, P., Wyatt, N.P.G.   1984. Electronic monitoring of dragline performance in open pit mining operations. Proceedings, Electronics in mining symposium, Published by the South African Institute of Electrical Engineers.

Loy, M.D. 1987.   A comparison of public domain underground models.   Society of Mining Engineers, Preprint no. 87-59, Presented at SME Annual Meeting, Denver Colorado, Feb. 24-27, 1987.

Moran, S. 1987. General inference engine for expert systems : User's guide. Internal Laboratory Report, Division of Mechanical Engineering, National Research Council of Canada.

Nenonen, L.K., Graefe, P.W.U., Chan, A.W. 1981. Interactive computer modelling of truck-shovel operations in an open-pit Mine. Proceedings, 1981 Winter Simulation Conference, pp 133-139.

Nenonen, L.K. 1986. MINDET: Deterministic Model of load/haul operations. Internal Report, Division of Mechanical Engineering, National Research Council of Canada.

Parnas, D.L., Clements, P.C. 1986. A rational design process: How and why to fake it. IEEE Transactions on Software Engineering. V. SE-12, No. 2., Feb.1986

Pomeroy, R.J. 1982. Computer-aided dispatch systems for open-pit mining. Proceedings, 1982 Canadian Conference on Industrial Computer Systems.

Rockett, A.M., Stevenson, J.C. 1987. Karmarkar's algorithm.   Byte, Sept. 1987, pp 146-160.

Scoble, M.J., Hadjigeorgiou, J., Nenonen, L.K. 1987. Development of an excavating equipment selection expert system based on geotechnical considerations. Proceedings, 40th Canadian Geotechnical Conference, Regina, Saskatchewan,   19-21 Oct.1987.

Srajer, V. 1986. The development and validation of a truck and shovel computer model.   Proceedings, 19th Applications of Computers and Operations Research in the Mineral Industry. April 14-16, 1986 Pennsylvania State University, pp 377-393.

Thompson, R.A. 1986.   Computerized mine management systems in open pit mining. 1986 Annual General Meeting, Canadian Institute of Mining and Metallurgy, May 11-15, 1986, Montreal, Preprint No. 163.

Trumbley, K., Isnor, D., Nenonen, L.K., 1986. DATAENTRY: A database management system . Internal Report, Division of Mechanical Engineering, National Research Council of Canada.

USBM Technology News. 1984. Off-highway truck overload detection. Technology news from the Bureau of Mines, US Department of the Interior.   No. 199, April 1984.

594

13 Mine ventilation

Computer Applications in the Mineral Industry, Fytas, Collins & Singhal (eds)
© 1988 Balkema, Rotterdam. ISBN 90 6191 760 3

Computerized calculations of air temperatures increases in mine excavations

Julian Partyka
School of Engineering, Laurentian University, Sudbury, Ontario, Canada

ABSTRACT: Heat transferred from rock into mine excavations is considered. The complete system of equations is derived and used to predict air temperature increase and moisture evaporation change along typical mine excavations such as airway, raise, shaft and longwall stope. Measurements taken in bone-dry haulage are used to evaluate the applied formulae. In addition, different values for the wetness factor are used for estimating the amount of heat flow from wet surfaces. All computations are done in spreadsheet form for Lotus 1-2-3 version 2.01.

INTRODUCTION

The heat conducted through the strata into a mine excavation depends mainly upon the size and deposition of the excavation and upon the efficiency with which the heat passes from the rock surface to the airstream.

The results of the work done to develop a method for calculating heat flow from a wet surface in a non-circular excavation can not be considered satisfactory, and there is a very real need for further investigation of this problem. There is especially a need for evaluation of the wetness factor to achieve better approximation for the overall heat transfer coefficient at a wet surface, the temperature at which evaporation or condensation takes place, and for the change of moisture evaporation rates.

Most of the equations used to calculate the amount of heat rock can be applied if the surface temperature, t_s, is known. It is unlikely that there is any direct method of calculating the t_s; an iterative procedure has to be employed. A convenient method for taking the surface temperature is based on heat balance at a wet surface. An initial value of t_s is assumed and the value of heat transferred by conduction through rock is compared with the combined heat of convection and the latent heat of evaporation. This process is repeated iteratively until there is negligable error in the heat balance.

BASIC EQUATIONS

In non-horizontal airways, geothermal gradient and autocompression have to be taken into account. In such cases the solutions suggested by Koening (1952), Boldizsar (1960), and Greuer (1973) can be used. The change of moisture evaporation rates and the temperature of air along the horizontal tunnel can be evaluated from equations given by de Braaf (1951) or Starfield and Dickson (1967).

However, a practical solution is needed which can be applied to predict air temperature change for any typical mine excavation such as airway, raise, shaft, and longwall stope. It should consider the inclination slope and the wetness of the rock surface at the same time.

For inclined airways with slope angle α (negative for ascending airways) and when the part P_w of the perimeter P is covered with water the heat balance can be written as follows:

$$P*K*(t_{vr}-t_s)*dL-M*g*\sin(\alpha)*dL = M*C_p*d(t_d)+M*\lambda*dW \qquad (1)$$

$$M*g*dW=\beta*P_w*(e_s-e)*dL \qquad (2)$$

where
K   = heat transfer coefficient, $W/m^2 °C$
W   = absolute humidity, kg/kg
ß   = mass transfer coefficient, s^{-1}
e_s = saturated vapour pressure at t_s, kPa
e   = vapour pressure, kPa
λ = latent heat of evaporation, kJ/kg

t_{vr}=virgin rock temperature, °C
t_s =surface temperature, °C
t_d =dry-bulb temperature, °C
M =mass flow rate, kg/s
C_p =thermal capacity, kJ/kg°C
L =length of airway, m
g =gravitational acceleration, m/s²
P =perimeter of airway, m
P_w =perimeter covered with water, m

Equation (1) consists of heat flow from rock at surface temperature t_s and the heat due to autocompression. This complex heat is used for evaporation (or condensation) and to change the dry-bulb temperature of air. Equation (2) describes the increase (or decrease) in absolute humidity.

An approximate solution of the differential equations (1) and (2) satisfying practical purposes are:

$$t_{do}=t_{vri}-(t_{vri}-t_{di})*exp(-P*K*L/(M*C_p))-$$
$$[g*\beta*P_w*\lambda*(\delta*e_d-\sigma*W_i)/(\sigma*C_p*\beta*P_w-g*\delta*P*K)]$$
$$*[exp(-P*K*L/(M*C_p))-exp(-\beta*P_w*\sigma*L/$$
$$(g*M*\delta))]-g*sin(\beta)*L/C_p \qquad (3)$$
$$W_o=\delta*BP/\sigma-(\delta*BP/\sigma-W_i)*exp(-\beta*P_w*\sigma*L/$$
$$(g*M*\delta)) \qquad (4)$$

where
i,o =subscripts for inlet and outlet
e_i =saturated vapour pressure at t_{di}, kPa
BP =barometric pressure, kPa
δ =$R_a/(R_v-R_a)$=1.649
σ =$BP*R_v/(R_v-R_a)$, kPa
R_a=gas constant for air =0.287 kJ/kgK
R_v=gas con. for water vapour=0.461 kJ/kgK

Equations (3) and (4) have been derived under the assumptions that t_s=t_{di} and when BP, C_p, β, K, and e_d are constant. In this regard these equations can be applied only for a short distance, eg. less than 100 m, to achieve reasonable accuracy. However, there is still need to correct the results due to the fact that t_s is not equal to t_{di}.

ADDITIONAL CONSIDERATIONS

For non-circular excavations the equivalent radius, r_{eq}, should be used. The radius is calculated as below

$$r_{eq}=2*A/P \qquad (5)$$

where
A=cross sectional area of excavation, m²
P=perimeter, m

The Fourier number is determined by formula
$$F_o=\pi*\theta*\emptyset/A \qquad (6)$$

where
θ=age of excavation, s
$\emptyset$=thermal diffusivity of rock, m²/s

The overall heat transfer coefficient, h, is based on the equation given by McAdams (1954). This value, for a practical range of mine temperatures, is expressed by formula

$$h=3.283*M^{0.8}/D_{eq}^{0.2}+6 \qquad (7)$$

where
D_{eq}=equivalent diameter of excavation, m
h =overall heat transfer coefficient, W/m²°C

The first term in equation (7) represents convective heat transfer coefficient, h_c, and the value of 6 W/m²°C is assumed for 0.95*h_r, the radiative heat transfer coefficient.

The Biot number, B, is calculated from

$$B=h*r_{eq}/k \qquad (8)$$

where k=thermal conductivity of rock, W/m°C

The mass transfer coefficient, β, in equation (4) is derived from equation (2) and the equation given by Whillier (1982) for moisture evaporation rate, M_w,

$$M_w=f_{eq}*0.7*h_c*(e_s-e)*P/(BP*x) \qquad (9)$$

where
f_{eq}=$0.16*f^2+0.2*f$, the equivalent wetness factor
f=P_w/P, wetness factor
x=$0.92*f+1$

The value, M_w, is expressed in g/ms for h_c in W/m²°C, P in m, and the pressures e_s, e and BP are in kPa.
Following formula is derived

$$\beta=2.257*10^{-5}*(0.16*f^2+0.2*f)/(0.92*f+1)$$
$$*M^{0.8}/(BP*D_{eq}^{0.2}) \qquad (10)$$

where β in s⁻¹, M in kg/s, D_{eq} in m and BP in kPa.

In addition the following psychrometric equations are needed:

$$e_{sw}=0.6105*exp(17.27*t_w/(237.3+t_w))$$
$$e =e_{sw}-0.000644*BP*(t_d-t_w)$$
$$r =0.622*e/(BP-e_{sw})$$
$$S =1.005*t_d+r*(1.8*t_d-4.18*t_w+2501)$$
$$C_p =1005+1800*r$$
$$\lambda =2501-2.383*t_s$$
$$S_o =S_i+C_p*\Delta t_d$$
$$q =M*(S_o-S_i) \qquad (11)$$

where

e_{sw}=saturated vapour pressure at t_w, kPa
e =vapour pressure, kPa
r =moisture content, kg/kg (dry basis)
S =sigma heat, kJ/kg
λ =latent heat,kJ/kg
C_p =thermal capacity, kJ/kg°C
S_o,S_1 =sigma heat for outlet and inlet
Δt_d=increase in dry-bulb temperature, °C
q =heat flow, kW

The equations (11) are detailed by Hemp (1982) and Barenbrug (1974).
The surface heat transfer coefficient

$$K=2*\pi*k*BF(F_o,B)/P \qquad (12)$$

where the function $BF(F_o,B)$ represents the Carslaw and Jaeger (1959) solution of Fourier equation for radial heat conduction with varying in time surface temperature. The dimensionless temperature gradient $BF(F_o,B)$ can be approximated for computation by the following algorithm (Hemp 1985):

$$BF(F_o,B)=T*(b*y^2+c*y+d) \qquad (13)$$

where
$y=1/(1/B+a)$
$a=-.016218*x^3+.163336*x^2+.558742*x+.622702$
$b=.012454*x^3-.120683*x^2-.398410*x-0.255263$
$c=-.009933*x^3+.103367*x^2+.76269*x+1.041458$
$d=.000007*x^5-.000250*x^2+.001292*x-0.001661$
$x=\log_{10}(F_o)$
$T= (0.001073*x^5-0.004525*x^4-0.01572*x^3+$
$\quad 0.1459*x^2+0.7288*x+1.017)^1 \qquad (14)$

COMPUTER PROGRAM

An excavation is divided into a number of small sections, n_z. Starting at the first section, the inlet air conditions are used to compute values for moisture evaporation rates and dry-bulb temperature of the air leaving the section. These values are used then to calculate the outlet wet-bulb temperature and the heat flow. The procedure is repeated for the next section, and for all subsequent sections until the end of the excavation is reached.

This procedure is applied for a wet and dry excavation, and for horizontal and non-horizontal airway. In the case of an inclined airway, the values for the barometric pressure and the virgin rock temperature have to be recalculated before applying them as the input data for inlet to the subsequent section. These velues are assumed to be constant for one section.

The following flow chart is used

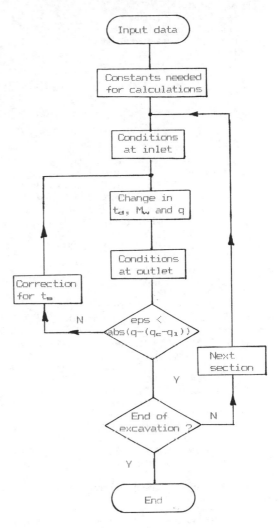

Figure 1. Flow chart for computing heat flow in mine excavation

The input data consist of following values:

Excavation
  − length, m
  − perimeter, m
  − area, m²
  − inclination slope, deg
  − depth, m
  − age, years
  − wetness factor

Rock properties
  − diffusivity, m²/s
  − conductivity, W/m°C

599

Air parameters
- air quantity, m³/s
- wet-bulb at inlet, °C
- dry-bulb at inlet, °C
- barometric pressure, kPa
Additional data
- virgin rock temperature (VRT), °C
- VRT gradient, °C/m
- pressure gradient, kPa/m

Part of the program is represented in Figure 2.

f	0.00		i	5.00	
nz	6.00		d1	1.65	
dL	71.12	m	d2	333.84	kPa
dLz	0.00	m	rsh	900.00	J/kg°C
PB_i	126.20	kPa	wr	2670.00	kg/m³
Pws_i	2.00	kPa	Rv	461.49	J/kg°C
Pw_i	1.42	kPa	Ra	287.03	J/kg°C
r_i	0.01	kg/kg	b_JP	0.00	1/hr
S_i	42.25	kJ/kg			
v_i	0.68	m³/kg			
Cp_i	1017.76	J/kg°C			
db_i	24.60	let °C			
db_o	24.83	°C	Pir_w	0.00	m
wb_i	17.50	let °C	r_oo	0.01	kg/kg
wb_o	17.60	°C		0.62	
tsc_i	23.67	°C	del_r	0.00	g_w/kg_da
VRT_i	41.30	°C	del_r	0.00	g_w/s
VRT_o	41.30	°C	exp_1	1.00	
S_o(r_o)	42.48	kJ/kg	K	0.83	
Pb_o	126.20	kPa	exp_2	0.99	
Pws_o	2.01	kPa	ed	3.09	
Pw_o	1.42	kPa	con1	0.00	
r_o	0.01	kg/kg	con2	-15.04	
S_o	42.50	kJ/kg	con3	0.00	
v_o	0.69	m³/kg	t_a	24.83	°C
Cp_o	1017.77	J/kg°C	q_tot	70.22	kW

Figure 2. Part of the presented program

Symbols used:

f   = wetness factor
nz = number of sections
dL = length of section
dLz = elevation change
Pws= saturated vapour pressure
Pw = vapour pressure
td = dry-bulb temperature
wb = wet-bulb temperature
r   = moisture content
r_oo= moisture content of outgoing air
S   = sigma heat
Cp = thermal capacity
b_JP= mass transfer coefficient
del_r=increase in moisture content
t_a =dry-bulb temperature of outgoing air
q_tot=total heat flow
-i = subscript for inlet
-o = subscript for outlet

Formulae used in the same ranges of spreadsheet are displayed in Figure 3.

```
D26:  'dL
E26:  (F2) +$LENG/$NZ
D26:  'dLz
E27:  (F2) +$DL*@SIN((@PI*$INCL/180)
E28:  'Pb_i
E28:  (F2) +$BP(IN)-0.01**$DLZ**$I
D29:  ^Pws_i
E29:  .6105*@EXP(17.27*$WB_I/(237.3+WB_I))
D30:  ^Pw_i
E30:  +$PWS_I-0.000644**$PB_I*($DB_I-$WB_I)
D31:  ^r_i
E31:  (F2) 0.622*PW_I/(($PB_I-PW_I)
G31:  'b_JP
H31:  6.867*0.001*3.6*($F^2*0.16+0.2*$F)/
      (0.92**$F+1)*$H(CON)/$PB_I
D32:  ^S_i
E32:  1.005**$DB_I+R_I*(1.8**$DB_I-4.18
      *$WB_I+2501)
D37:  'db_o
E37:  (F2) $VRT_I-($VRT_I-$DB_I)*$EXP_2-
      $CON3*($EXP_2-$EXP_1)+9.81*(1+R_OO)/
      (CP_I/1000)*(-DLZ/1000)
G38:  'r_oo
H38:  $H$39-($H$39-$R_I)*@EXP(-$B_JP/3600*
PIR_W*$D2*@ABS($DL)/(9.81*MASS_FLOW*$D1))
```

Figure 3. Cell formulae

The sequence of computations is controlled by macros. Some of the macros used are presented in Figure 4.

```
\a  {goto}E36~/rvE4~.E36~
    {windowsoff}~{paneloff}
    /rvE5~.E38~
    {let m,0}~{let i,0}~{calc}~
    {\b}~{calc}~

\b  {recalccol K50..K54}~
    {if
dif1<.01}~{beep}/rvK51~.E39~{SUBR5}
    {let m,m+1}~
    {\b}

SUBR5 {if (i+1)=nz}~{beep}{quit}
      {let i,i+1}~{let m,0}~
      /rvE37~.E36~
      /rvE39~.E38~
      {recalc E24..H51}~
      {RETURN}
```

Figure 4. Macros applied in program

Additional macros are used to find wet-bulb temperature from the calculated values of dry-bulb, moisture content, and barometric pressure, and the surface temperature. A method used for computing the corrected heat flow and the corrected moisture evaporation rates is not presented in this paper. A procedure similar to that described by McPherson (1986) has been adopted.

PROGRAM APPLICATIONS

Measurements have been taken at eight places for the drive 0090 of the 7200 level at Creighton Mine, Ontario, spaced 200 feet from each other. Following data were reported:

length   426.7 m
perimeter   11.0 m
area   11.2 m²
inclination   0 deg
wetness factor   0 (dry conditions)
wet-bulb (in/out)   17.0/17.6 °C
dry-bulb (in/out)   23.4/24.7 °C
barometric pres.   126.2 kPa
surface temp. (inlet) 23.7 °C
mass flow   45.9 kg/s

Additional specified values:
type of rock   – quartzite
age of the drive   – 15 years
thermal diffusivity   – 2.55*10⁻⁶ m²/s
thermal conductivity – 5.5 W/m°C

Calculated results (see the attached part of spreadsheet) are:
wet-bulb temp. at outlet, wb_o=17.60 °C
dry-bulb temp. at outlet, db_o=24.83 °C
total heat flow, q_tot=70.22 kW
difference between surface temperature and dry-bulb temperature of air at inlet Δ_1=tsc_i–db_i=0.27 °C.
The heat flow calculated from air side is
q_tot=mass flow rate*increase in Σ heat
= 45.9*(42.5–41.0)=68.85 kW

The above evaluation shows very good agreement between the presented physical model and the measurements (2%).

REMARKS

The wetness factor applied in the program can be selected in the range from 0 (bone-dry conditions) to 1 (totally wet conditions). It is unlikely that a wet excavation will be found in this region. However, some calculated results are given below (table 1). The following input data is assumed:

length   1000 m
perimeter   12.0 m
area   9.0 m²
age   4 years
quantity   80 m³/s
barometric pressure

at inlet   105.0 kPa
wet-bulb (inlet)   25.0 °C
dry-bulb (inlet)   30.0 °C
type of rock   quartzite

Results

factor f	slope deg	temperatures wb_o/db_o °C	heat flow q, kW
0.0	90	29.2/41.0	913.70
0.4	(downcast	29.3/36.6	967.70
1.0	shaft)	29.4/34.6	1009.20
0.0	45	27.3/36.2	511.50
0.4	(down)	27.4/32.6	555.10
1.0		27.4/31.1	589.30
0.0	0	25.3/31.3	129.48
0.4	(hor)	25.4/28.7	143.30
1.0		25.5/27.5	170.10
0.0	–45	23.4/26.5	–292.90
0.4	(up)	23.4/24.7	–267.80
1.0		23.5/23.9	–248.40
0.0	–90	21.4/21.6	–695.10
0.4	(upcast	25.5/25.5	–678.70
1.0	shaft)	25.5/25.5	–667.40

Table 1. Computation results of heat flow and airtemperature changes in different mine excavations.

CONCLUSIONS

The presented program allows prediction of a change in air temperature of air passing through typical mine excavations. It can be applied for inclined, non-circular, dry or wet excavations.

Evaluation of the derived equations is based on measurements in-site. However, there is still need for further investigation for determining the wetness factor and the equivalent surface temperature.

All calculations have been made on the IBM XT under LOTUS 1-2-3 release 2.01. LOTUS macro language gives the tremendous ability to automate functions, expand program capabilities, and develop applications in a short time.

A spreadsheet implementation allows interactive adjustment of existing formula. In addition each formula is tested immediately upon entering it step by step, without the necessity of having all the program done.

The size of the spreadsheet is 20KB. Minimum hardware requirement for the system development is: 256KB, and one floppy disk drive. Time of recalculation without mathematical co-processor is less than 30 s (for nz=6).

REFERENCES

de Braaf, W. 1951. Heating of air in shafts and intake airways.Geologie en Mijnbouw. pp. 117-154.

Koening, H. 1952. Mathematical investigation of the mine climate. Bergbau-Archiv 13, H 3/4, pp.1-14.

McAdams, W.H. 1954. Heat Transmission. McGraw-Hill. Third edition.

Carslaw, H.S. & Jaeger, J.C. 1959. Conduction of heat in solids. Second edition. Clarendon Press: Oxford.

Boldizsar, T. 1960. A numerical graphical method for the calculation of temperature changes of mine air. Bergbau-Archiv 21 ,H. 2, pp. 17-27.

Starfield, A.M. & A.J. Dickson. 1967. A study of heat transfer and moisture pick-up in mine airways. Journal of the South African Institute of Mining and Metallurgy. Vol. 68.

Greuer, R.E. 1973. Influence of mine fires on the ventilation of underground mines. USBM Contract Report No. S0122095.

Barenbrug, A.W.T. 1974. Psychrometry and Psychrometrics Charts. Third edition. Chamber of Mines of South Africa.

Hemp, R. 1982. Environmental Engineering in South African Mines. Psychrometry. Chapter 18. The Mine Ventilation Society of South Africa.

Whillier, A. 1982. Environmental Engineering in South Africa Mines. Chapter 19. The Mine Ventilation Society of South Africa.

Hemp, R. 1985. Air temperature increases in airways. Journal of the Mine Ventilation Society of South Africa. Vol. 38 No. 2.

McPherson, M.J. 1986. The analysis and simulation of heat flow into underground airways. International Journal of Mining and Geological Engineering. 4. pp. 165-196.

Computer Applications in the Mineral Industry, Fytas, Collins & Singhal (eds)
© 1988 Balkema, Rotterdam. ISBN 90 6191 760 3

A review of computer programs available for underground mine climate simulation (Abstract)*

Michael Heeley & Pierre Mousset-Jones
Mackay School of Mines, Reno, Nev., USA

ABSTRACT: A number of computer programs are now available which can be used to assist in the design of ventilation and air conditioning systems for underground mines experiencing hot working conditions. This paper will provide an overview of these programs discussing their differences and similarities, ease of use, limitations, and accuracy. A number of different case studies using measured data from operating hot underground mines, will illustrate input requirements, operation, and output for these programs. In conclusion, the paper will indicate the procedures a mine ventilation engineer should take to use one of these programs to assist effectively in the design of the ventilation system for a new or operating mine.

*Paper not received on time for publication.

Computer Applications in the Mineral Industry, Fytas, Collins & Singhal (eds)
© *1988 Balkema, Rotterdam. ISBN 90 6191 760 3*

Computer simulation of face ventilation to dilute high methane concentrations developed by blasting oil shale

C.E.Brechtel
J.P.T. Agapito & Associates, Inc., Grand Junction, Colo., USA
E.D.Thimons
US Bureau of Mines, Pittsburgh, Pa.

ABSTRACT: A one-dimensional finite element model was used to simulate turbulent mass transfer in face ventilation of oil shale mines. The computer model was applied to study dilution of methane released by rubblizing oil shale during face blasting. A methane release function was back analyzed from field measurements and the finite-element model was calibrated using tracer gas characterization of full-scale face ventilation systems.

1 INTRODUCTION

Methane saturation of the kerogen in Green River Formation oil shales of the Western United States impacts face ventilation design in room and pillar oil shale mining. Face blasting has the potential to release large quantities of methane gas in dead-end headings. Rates of release are high and in conjunction with the large tonnages excavated by each blast, have the potential to produce dangerously high concentrations of methane soon after the blast.

Ventilation in the face after blasting must be capable of safely diluting the methane to keep it below the explosive range as it is released from the muck pile. Data produced in a research and development project that tested large capacity face ventilation equipment in an oil shale mine suggested that the use of freestanding, jet fans would be an effective method of diluting blast released methane. The fan could be left on during the blast, thereby, preventing the buildup of methane in still air.

Analysis of this problem required a technique that allowed the coupling of the effects of turbulent mass transfer with the time rate of production of methane after the blast. An existing computer code, BASFEH, was adapted for the simulation of turbulent mass transfer and was back-calibrated using tracer gas measurements of face ventilation systems in an oil shale mine. A methane generator function for simulating the time rate of

methane production was developed using field measurements of methane emitted in oil shale mining operations.

The objective of this work was to study the effect of the time lag between blasting the face and activating the fan, on methane concentration. The results would provide design guidelines for flow quantities and procedures for a range of gas saturations and fan activation times. This paper discusses preliminary results of the work.

2 THE BASFEH COMPUTER CODE

BASFEH is a two-dimensional finite element model developed to analyze conductive heat transfer, by Nelson (1969), and modified to include convective heat transfer, by Hardy and Hocking (1978). The program allows the use of one-dimensional elements to solve a conduction-convection equation identical in form to mass transfer models presented by Thakur (1976) for use in mine ventilation and shown as equation 1

$$\frac{\partial c}{\partial t} = \frac{\partial}{\partial x}\left(E_x \frac{\partial c}{\partial x}\right) + u(x)\frac{\partial c}{\partial x} + f(x,t) \quad (1)$$

where: c = concentration (kg/m^3)
t = time (s)
E_x = turbulent dispersion coefficient (m^2/s)
$u(x)$ = airflow velocity in the x direction (m/s)
$f(x,t)$ = a source term (kg/s)

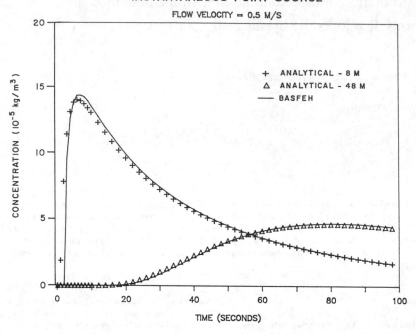

Figure 1   Comparison of BASFEH results to the analytical solution in equation (2)

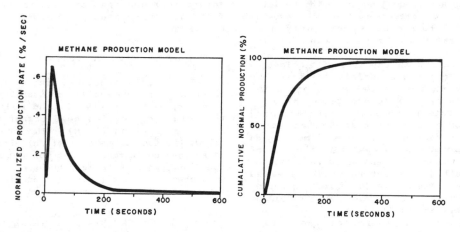

Figure 2   Normalized production rate and cumulative production curves illustrating the
methane production model

BASFEH solves equation (1) explicitly in time and is, therefore, well suited for application using microcomputers because storage requirements are small and execution is very fast. These analyses were performed using a PC-AT computer.

This work was oriented towards application of the modeling technique to mine ventilation problems, therefore, discussion of the mathematics and implementation of the programming is not presented. Other work in this area, presented by Bandopadhyay and Ramani (1984), discusses mathematical formulations for solution by finite difference schemes and may be of interest to the reader.

BASFEH was tested against a one-dimensional analytical solution for concentrations produced by an instantaneous stationary point source (Thakur, 1976) shown in equation 2

$$C(x,t) = \frac{Q}{2A(\pi E_x t)^{0.5}} \exp\left[-\frac{(x-ut)^2}{4E_x t}\right] \quad (2)$$

where: C = concentration
Q = mass introduced instantaneously
E_x = turbulent dispersion coefficient
u = air velocity
A = tunnel area
t = time

The results are compared in Figure 1 and confirmed that the formulation in BASFEH could successfully simulate turbulent mass transfer. The figure shows the concentrations at 8 and 48 m from the source for a velocity of 0.5 m/s (symbols) compared to the BASFEH output (lines) at the same location.

3 METHANE PRODUCTION MODEL

The methane production model was developed from field measurements of methane liberated during mining operations at the Horse Draw experimental mine in the Piceance Creek Basin of Colorado, USA and at the White River Shale Oil Mine in the Uinta Basin of Utah, USA. Experimental data presented by Sapko, et al. (1982) and Sapko, et al. (1986) indicated that methane release began immediately upon rubblization and reached peak rates within five minutes of the blast. The rate of production then began to fall, with the shape of the curve very similar to a log-normal statistical distribution.

A log-normal curve was chosen as the basis of the methane production model using equation 3

$$f(t) = \frac{1}{\sigma t \sqrt{2\pi}} \exp\left[-\frac{1}{2\sigma^2}(\log t - \mu)^2\right] \quad (3)$$

where: f(t) = normalized rate of production
σ, μ = shape factors
t = normalized time

This equation produces a normalized curve with an area of 1.0. Parameters are then used to scale the time axis and area to represent the quantity of methane produced as shown in equations 4

$$t = \frac{T}{T'}$$

and   (4)

$$F(t) = f(t) * Q$$

where: T = time (s)
T'= scaling factor (s)
Q = total quantity of methane liberated (kg)

Normalized curves of production rate and total production are shown in Figure 2 to illustrate the production model. The rate model is very similar to the shape of time-concentration data developed in the mining experiments. Integration of the production rate curve produces a cumulative production curve that is very similar to laboratory desorbtion behavior for methane in finely ground oil shale, presented by Matta, et al. (1977). This indicates that the use of the log-normal curve is consistent with the physical process of gas desorbing from the solid kerogen in the oil shale.

4 BACK-ANALYSIS OF METHANE PRODUCTION DATA

BASFEH was used to model experimental measurements from the Horse Draw data developed by Sapko, et al. (1982) in order to back analyze the correct shape parameters for the log-normal curve. Figure 3 shows the underground configuration in the mine and the finite element mesh used to simulate the field measurements. The blast modeled was in a face heading being ventilated by 3.25 m^3/s of fresh air and liberated 137.7 kg of methane (assuming an air density of 1.189 kg/m^3). A total of

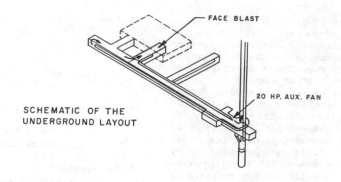

FACE BLAST

20 HP. AUX. FAN

SCHEMATIC OF THE
UNDERGROUND LAYOUT

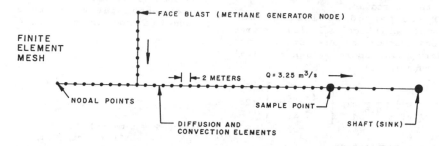

FACE BLAST (METHANE GENERATOR NODE)

FINITE
ELEMENT
MESH

2 METERS

$Q = 3.25 \text{ m}^3/\text{s}$

NODAL POINTS

SAMPLE POINT

DIFFUSION AND
CONVECTION ELEMENTS

SHAFT (SINK)

Figure 3 Schematic illustrating Horse Draw Mine configuration and the finite element mesh used for simulation

69.5 metric tons of oil shale was rubblized for an estimated methane saturation of 1.98 kg/ton. The drift was assumed to be 3.7 by 3.7 meters square. The log-normal curve was scaled to a 10 minute duration pulse with a total quantity of 137.7 kg, based upon the shape of the experimental data. A coefficient of turbulent dispersion of 20.0 was used in the simulation.

The results of the BASFEH simulations are shown in Figure 4, which presents the time-concentration curve at the experimental sample point and compares the BASFEH output to experimental measurements. A very good fit was obtained with both of the log-normal shape factors equal to 1.0.

5   BACK ANALYSIS OF THE TRACER GAS EXPERIMENTS

A model of the face heading configuration utilized in room and pillar oil shale mining was developed using the results of large-scale face ventilation experiments reported by Brechtel, et al. (1985). Those tests measured the efficiency of a 1.4 m diameter freestanding, jet fan in ventilating a dead-end heading 91 m long, with a cross section of 16.8 m by 9.1 m.

The face ventilation experiments were conducted in the Exxon Colony Shale Oil Mine at expected full-scale conditions. The dilution of SF_6 tracer gas was used to measure efficiency of the face ventilation system and the combined effects of full-scale turbulence and recirculation.

The experimental configuration used in the face ventilation tests is compared to the finite element mesh used for the simulation in Figure 5. Air leaves the fan as a turbulent jet of high velocity air directed towards the face area. The jet expands and loses velocity as it approaches the face, due to frictional entrainment of slower velocity air at the perimeter of the jet. The entrainment action produces great turbulence throughout the test room and very effective mixing of any pollutants. The rate of pollutant dilution is controlled by the net flow of fresh air in and out of the room.

Evaluation of the tracer gas data indicated that mixing throughout the test room was very uniform due to the great turbulence. For a steady state release of SF_6, the standard deviation of all measurements throughout plan and vertical area of the room averaged 12 percent of

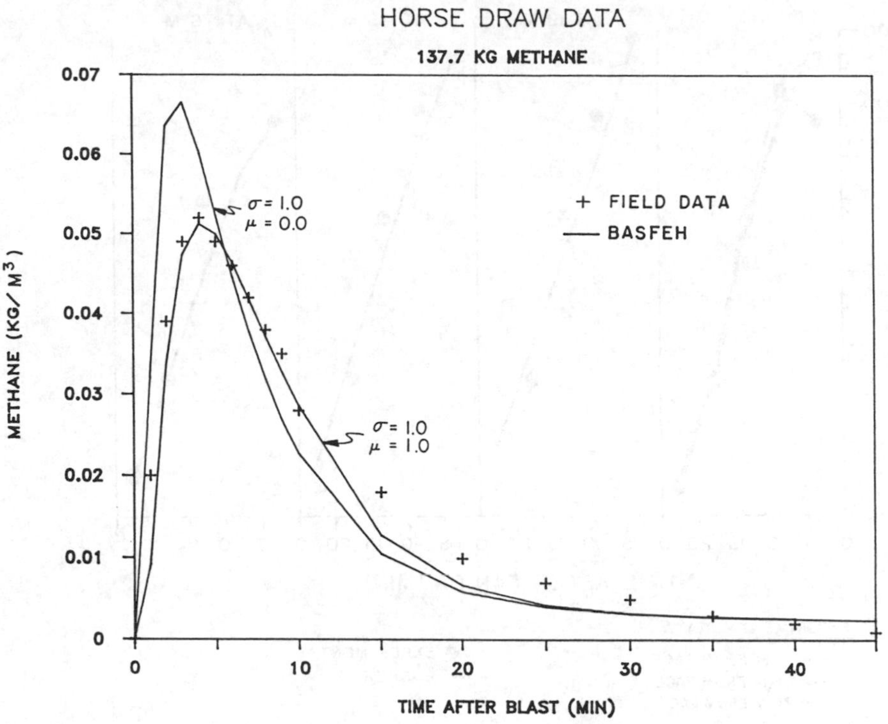

HORSE DRAW DATA

137.7 KG METHANE

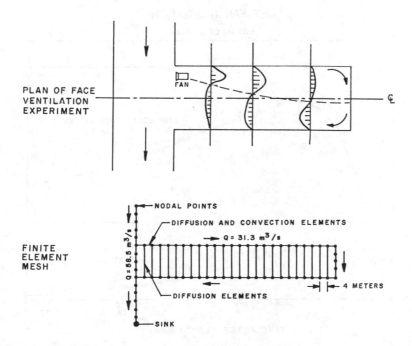

Figure 4   Time-concentration predicted by BASFEH versus experimental data

Figure 5   Schematic illustrating Jet fan experimental configuration and finite element mesh used for simulation

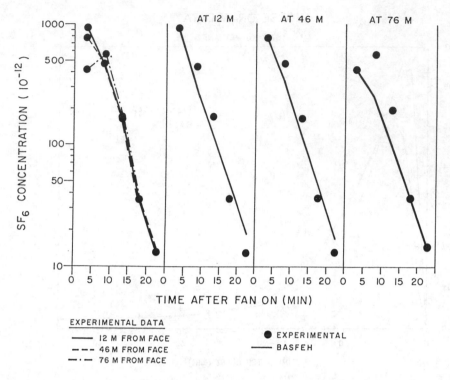

Figure 6   Comparison of tracer gas data with BASFEH simulation of full-scale face ventilation experiments

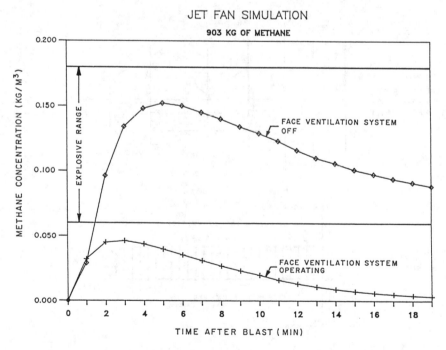

Figure 7   Methane concentration versus time for a simulated full-scale blast with and without ventilation

the mean value. This suggested that use of the one-dimensional approximation, which assumes that turbulent dispersion is the same in all directions, would produce acceptable results.

The finite element mesh was made up of one-dimensional diffusion-convection elements forming the average path of air flow past the room in the last open crosscut and within the room. Nodal points were cross connected by diffusion elements to simulate the large turbulence created by entrainment along the jet.

Figure 6 compares the experimental data to the results predicted by the BASFEH model at three distances from the face. The tracer gas had been released throughout the room producing a uniform concentration of 920 ppt (parts per trillion). The main ventilation system was then actuated and the air flow from the last open crosscut began to remove some SF_6 from the room producing the concentration gradient indicated by the first data points. After actuating the jet fan, the turbulence produced a uniform concentration throughout the room within 5 minutes and diluted the SF_6 to levels of minimum resolution within 20 minutes.

The BASFEH simulation was begun at the same gradient of concentration in an attempt to reproduce the experimental data. The coefficient of turbulent dispersion was raised to 120 m^2/s in an attempt to reproduce the uniform mixing throughout the room.

Simulation of the tracer gas test was not able to reproduce the degree of turbulence observed experimentally, however, the dilution of the tracer gas was duplicated within 20 to 30 percent. Behavior in the simulation was very similar in trend to the experimental data.

6 SIMULATION OF FULL-SCALE FACE BLASTING

The methane generator function and the jet fan model have been exercised to produce preliminary indications of the application of the work. Sapko, et al. (1986) reported average methane saturations of 0.4 m^3/ton of oil shale from blasting tests at the White River Shale Oil Mine. This value was used for an assumed face blast rubblizing 1900 tons of oil shale to produce 903 kg of methane. The time normalization for the release was identical to that used in the back analysis. Air ventilation rates were the same as used in the tracer gas simulation.

Figure 7 compares the time-concentration data for a point 12 m from the face for two cases:
- o Face ventilation system OFF
- o Face ventilation system ON during blast

Without face ventilation, the methane concentrations reach explosive levels and are diluted very slowly. The operation of the fan during the blast or immediately after the blast prevents the methane concentrations from reaching the explosive range.

7 DISCUSSION

These preliminary results illustrate the utility of this type of simulation in design of face ventilation systems. The one-dimensional approximation is currently unable to simulate the full turbulence of the system, however, it does produce a useful simulation of the time rate of production and dilution of the tracer gas and methane.

The results suggest that jet fans would be effective in preventing the large concentration of methane produced by blasting from reaching dangerous levels.

REFERENCES

Bandopadhyay, S. & R.V. Ramani. 1984. Convection-diffusion equations in mine ventilation planning. Proceedings of the Third International Congress for Mine Ventilation. Harrogale, England: Institution of Mining & Metallurgy.

Brechtel, C.E., M.E. Adam, J.F.T. Agapito & E.D. Thimons. 1985. Characterization of the performance of large capacity face ventilation systems for oil shale mining. Proceedings of the Second U.S. Mine Ventilation Symposium, Reno, Nevada: University of Nevada-Reno.

Hardy, M.P. & G. Hocking. 1978. Numerical modelling of rock stresses within a basaltic nuclear waste repository, phase II - parametric design studies, vol. II. A report to Rockwell Hanford Operations, Rockwell International under DOE prime contract EY-77-0-06.1030, University of Minnesota.

Matta, J.E., J.C. LaScola & F.W. Kissell. 1977. Methane absorption in oil shale and its potential mine hazard. Bureau of Mines RI 8243.

Nelson, C.R. 1969. Investigations of modes of thermal fracture of some brittle materials. Ph.D. Dissertation, Massachussettes Institute of Technology.

Sapko, M.J., J.K. Richmond & S.P. McDonnel. 1982. Continuous monitoring of methane in a deep oil shale mine. Proceedings of the Fifteenth Oil Shale Symposium, Golden, Colorado: Colorado School of Mines.

Sapko, M.J., E.S. Weiss & K. Cashdollar. 1986. Methane released during blasting at the White River Shale Project. Proceedings of the Nineteenth Oil Shale Symposium, Golden, Colorado: Colorado School of Mines.

Thakur, P.C. 1976. Turbulent mass transfer in mine ventilation systems. Proceedings International Mine Ventilation Congress, Johannesburg, South Africa: Mine Ventilation Society of South Africa.

Computer Applications in the Mineral Industry, Fytas, Collins & Singhal (eds)
© 1988 Balkema, Rotterdam. ISBN 90 6191 760 3

Air temperature calculation in hot airways: A critique and a new solution

G.Danko & P.Mousset-Jones
Mackay School of Mines, University of Nevada, Reno, USA

M.J.McPherson
University of California, Berkeley, USA

ABSTRACT: A general solution of the heat conduction in the infinite rock mantle is provided to allow air temperature change with time using the superposition technique, and the non-dimensional age function. Completely dry and wet surfaces are assumed, and the development of a computation algorithm is described to predict sensible and latent heat transfer in an airway developed in a homogenous infinite solid. The model can be used with arbitrary temperature and moisture variation of the inlet air with time, while the variations can continuously adjust for changes with distance. Therefore the computation algorithm correctly handles the prehistory of the heat and moisture transfer, and can be used when seasonal or periodic variations, or a rapid continuous temperature increase, e.g. during mine fire simulation, occur.

1 INTRODUCTION

Calculation of the temperature increase of air flowing through hot underground airways is the basic task of any mine climate simulation program. The non-dimensional age function first tabulated by Jaeger, et.al. (1966) has been used in the mining engineering practice e.g. by McPherson (1979) or Hemp (1982) to calculate heat inflow from the surrounding rock mass. This heat conduction model is correct only if the air temperature is changed according to a step function at the start of ventilation. Even if such an air temperature change is assumed at the entrance of an airway section, the air temperature will vary with time at a distance from the entrance, resulting automatically in a calculation error because of the lack of modeling this variance. The error in the wall heat flux calculation depends on the actual function of the temperature variation with time. Using the superposition solution described in the following section, it is possible to compare this correct solution with the erroneous wall heat flux which is calculated using the age function and disregards the shape of the temperature function. Such a comparison is shown in Table 1 for linear air temperature change with time, and for a few input parameters. The differences indicated in Table 1 are not negligible

and since a linear air temperature variation with time may not be far from reality at a further distance from the entrance of a mine airway, there seems to be no longer any reason for using the simple age function concept in mine climate simulations.

The paper considers with the derivation of a more correct wall heat flux model for a variable air temperature, and uses this model to predict the heat and moisture transfer along an airway. The model can be used for arbitrary temperature and moisture variation of the inlet air with time, and governs these variations along the airstream using the conservation and transport laws. Consequently, it provides an extremely flexible application field in addition to improved precision. The vector-matrix representations throughout the derivation result in a clear structure and fast computational time.

Firstly, the generalized wall heat flux model will be introduced for the case of a completely dry airway, followed by the more complex model needed for the wet airway.

2 DRY ROADWAY

2.1 Wall Heat Flux Density

A continuous air temperature variation

Table 1. Comparison of wall heat fluxes for linear inlet air temperature variation

		5	20	50
	$Q_a[m^3/s]$ v [m/s] h [W/Km^2]	5 0.71 3.73	20 2.83 11.3	50 7.07 23.52
1/4 year	q_w^* [W/m^2]	9.25	12.05	12.98
	q_w^{**}[W/m^2]	8.42	10.66	11.38
	error %	-8.9%	-11.5%	-12.3%
2 years	q_w^* [W/m^2]	6.50	7.79	8.18
	q_w^{**}[W/m^2]	6.04	7.14	7.46
	error %	-7.0%	-8.4%	-8.8%

* Calculated by the superposition solution of equation (1)
** Calculated directly by the age function

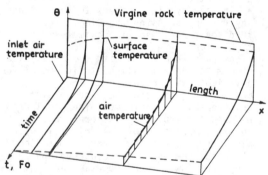

Figure 1. The two dimensional air temperature.

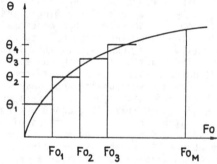

Figure 2. The step-change approximation at a location x.

with time is assumed at each airway cross section, which will be approximated by a number of steps taking place at time t_1, t_2, ... t_M. Figure 1 shows the assumed air temperature variation with time t and length x. Using $\theta = \theta_{vr} - \theta_a$ and the Fourier number Fo instead of time, the approximation of an arbitrary temperature function is shown in Figure 2.

The instantaneous wall heat flux density at a Fourier number Fo can be expressed by the following superposition equation using the original age function with the corresponding incremental time and temperature step:

$$q(Fo) = \frac{k}{r_s} \sum_{i=1}^{M} G(Bi, Fo-Fo_{i-1}) \Delta\theta_i \quad (1)$$

614

Applying equation (1) for $Fo=Fo_1$, Fo_2, ... and denoting the wall heat flux density with the vector $\bar{q}$ which consists of components $q_i=q(Fo_i)$, the following equation is obtained:

$$\bar{q} = \frac{k}{r_s} \underline{\underline{A}} \bar{\theta} \qquad (2)$$

In equation (2) vector $\bar{\theta}$ has components $\theta_i=\bar{\theta}(Fo_i)$, and A is a matrix with the following structure:

$$\underline{\underline{A}} = \{a_{ij}\}, \quad \text{where:} \qquad (3)$$

$a_{11}= G(Bi, Fo_1-Fo_0)$

$a_{21}= G(Bi, Fo_1-Fo_0)-G(Bi,Fo_2-Fo_1)$

$a_{22}= G(Bi, Fo_2-Fo_1)$

$a_{31}= G(Bi, Fo_3-Fo_0)-G(Bi,Fo_3-Fo_1)$

$a_{32}= G(Bi, Fo_3-Fo_1)-G(Bi,Fo_3-Fo_2)$

$a_{33}= G(Bi, Fo_3-Fo_2)$

.
.
.

2.2 Air temperature prediction

It is straightforward to describe the variation of the air temperature, i.e., the vector $\bar{\theta}$ along the roadway. The heat balance of the airflow gives:

$$- \frac{d\bar{\theta}}{dx} = \frac{2 \, k\pi}{Q_a \rho c_p} \underline{\underline{A}} \, \bar{\theta} \qquad (4)$$

where the negative sign on the left hand side comes from the expression $\bar{\theta}_a=\theta_{vr}-\bar{\theta}$, while the virgin rock temperature θ_{vr} is considered to be constant.

The solution of the linear differential equations (4) expressed in vector form can be obtained either analytically or numerically. The latter maybe quicker during computation and offers somewhat higher flexibility for further purposes. The Reverse Euler solution will be used in the following form, substituting

$d\bar{\theta} = \bar{\theta}_{n+1}-\bar{\theta}_n$ into equation (4).

$$- \frac{\bar{\theta}_{n+1} - \bar{\theta}_n}{\Delta x} = \frac{2 \, k\pi}{Q_a \rho c_p} \underline{\underline{A}} \, \bar{\theta}_{n+1} \qquad (5)$$

Rearrangement gives the following recursive formula to calculate temperature functions at subsequent locations n, n+1, along the airway:

$$\bar{\theta}_{n+1} = (\underline{\underline{E}} + B\underline{\underline{A}})^{-1} \bar{\theta}_n \qquad (6)$$

where B is the dimensionless product of Δx and the constant multiplicator of $\underline{\underline{A}}$ in equation (5) while $\underline{\underline{E}}$ is the unit matrix. Further simplification is made by introducing the following resolvent matrix, $\underline{\underline{R}}$:

$$\underline{\underline{R}} = (\underline{\underline{E}} + B\underline{\underline{A}})^{-1} \qquad (7)$$

The calculation of the temperature vector $\bar{\theta}_n=\theta_{vr}-\bar{\theta}a_n$ along the airway at subsequent distances of Δx means a repeated multiplication with matrix $\underline{\underline{R}}$:

$$\bar{\theta}_1 = \underline{\underline{R}}\bar{\theta}_0. \qquad (8)$$

$$\bar{\theta}_2 = \underline{\underline{R}}\bar{\theta}_1,$$

.
.
.

where $\bar{\theta}_0$ is either a constant (step-change) or an arbitrary temperature vector, describing the inlet air temperature variation with time.

3 WET ROADWAY

3.1 Wall heat flux density

Firstly, the instantaneous wall heat flux density, generated by both the sensible and latent heat transfer on the surface, is expressed. An apparent air temperature can be determined, which results in the same wall heat flux on an assumed dry surface as that on a wet surface, experiencing both sensible and latent heat

615

transfer. This apparent air temperature is called the pseudo-base temperature by McPherson (1986) and can be expressed by the following thermal boundary condition at the surface:

$$h(\theta - \theta_w) + q_L = h(\theta_b - \theta_w) \qquad (9)$$

where θ, θ_w and θ_b are the temperature difference between the virgin rock temperature θ_{vr} and the air, wet wall surface, and pseudo-base temperature, respectively, while q_L is the latent heat flux density. The pseudo-base temperature difference can be expressed from equation (9):

$$\theta_b = \theta + \frac{q_L}{h} \qquad (10)$$

Consequently, the wall heat flux vector can be expressed using equation (2) and the "dry" pseudo-base temperature in equation (10):

$$\bar{q} = \frac{k}{r_s} \underline{\underline{A}} (\bar{\theta} + \frac{\bar{q}_L}{h}) \qquad (11)$$

where $\bar{q}_L$ denotes the latent heat flux density vector representing the evaporation from the surface at time t_1, $t_2, \ldots t_M$.

3.2 Wet surface temperature

Although the surface temperature does not appear in equation (11), and is quite unimportant for the prediction of the heat flux coming from the rock, it is necessary to express it explicitly for both the dry heat and the moisture transfer prediction on the surface. The overall heat flux density q in equation (11) is the sum of the convective and latent heat flux density terms, i.e., equals the L.H.S. of equation (9), which is applied now for t_1, $t_2, \ldots t_M$:

$$h(\bar{\theta} - \bar{\theta}_w) + \bar{q}_L = \bar{q}_{convective} + \bar{q}_L = \bar{q} \qquad (12)$$

From equations (11) and (12) the surface temperature difference vector $\bar{\theta}_w$ can be expressed:

$$\bar{\theta}_w = (\underline{\underline{E}} - \frac{k}{hr_s} \underline{\underline{A}})(\bar{\theta} + \frac{\bar{q}_L}{h}) \qquad (13)$$

3.3 Air temperature prediction

When both sensible and latent heat transfer occur on the surface, the heat balance equation for the airflow involves only the dry convective heat transfer. Consequently the corresponding equation to equation (4) will be the following:

$$- \frac{d\bar{\theta}}{dx} = \frac{2\pi r_s h}{Q_a \varrho c_p} (\bar{\theta} - \bar{\theta}_w) \qquad (14)$$

The numerical solution of this differential equation system, using the Reverse Euler scheme gives an entirely different equation from equation (6), namely:

$$\bar{\theta}_{n+1} = \frac{1}{C+1} \bar{\theta}_n + \bar{\theta}w_{n+1} - \frac{1}{C+1} \bar{\theta}w_n \qquad (14)$$

$$\text{Where } C = \frac{2\pi r_s h}{Q_a \varrho c_p} \qquad (15)$$

In equation (14) the heat transfer from the surrounding rock is represented implicitly by the surface temperature vectors expressed by equation (13). A slight simplification is made by assuming that the temperature difference between θw_{n+1} and θw_n is small and can be estimated by the averaged wall temperature over distance Δx:

$$\bar{\theta}w_n \cong \bar{\theta}w_{n+1} \cong \bar{\theta}w_m \qquad (16)$$

Substituting equation (16) into equation (14) gives:

$$\bar{\theta}_{n+1} = \frac{1}{C+1} \bar{\theta}_n + \frac{C}{C+1} \bar{\theta}w_m \qquad (17)$$

Since the heat flux density is represented by the temperature vector $\bar{\theta}w_n$, equation (17) and (13) must be solved simultaneously. Applying equation (13) for the averaged wall surface temperature vector, gives:

$$\bar{\theta}w_m = (\underline{\underline{E}} - \frac{k}{hr_s} \underline{\underline{A}})(\bar{\theta}_m + \frac{\bar{q}_L}{h}) \qquad (18)$$

$$\text{Where } \bar{\theta}_m = \frac{\bar{\theta}_n + \bar{\theta}_{n+1}}{2}$$

There is no advantage to be gained by further substituting equation (18) into (17), since the surface temperature is needed explicitly for the calculation of the moisture transfer and the latent heat flux density, $\bar{q}_L$.

3.4 Moisture and latent heat transfer on the surface

The mass flux density q_m on a completely wet surface is expressed by the following equation used e.g. by McPherson (1986):

$$q_m = \beta \frac{e_{ws} - e}{P} \qquad (19)$$

where:

- e_{ws} is the saturated moisture partial pressure at the wet surface temperature and can be calculated by the following equation:

$$e_{ws} = 610.5 \exp \frac{17.27(\theta_{vr} - \theta_{wm})}{237.3 + \theta_{vr} - \theta_{wm}} \qquad (20)$$

- e is the vapor partial pressure in the main airstream, taken at the mean air temperature, $\theta_{vr} - \theta_a$.

- β is the moisture transfer coefficient, which is in this case:

$$\beta = 0.00073 \ h \qquad (21)$$

- P is the barometric pressure

Finally the latent heat flux density will be expressed with the specific heat of evaporation, L:

$$q_L = L q_m \qquad (22)$$

Comparing equations (22), (19), (20) and (18), it is clear that the sensible and latent heat transfer are coupled, and that these equations need to be solved simultaneously.

3.5 Prediction of the vapor partial pressure in the air

The mass balance equation for the moisture transfer in the airstream is the following:

$$\frac{dQ_m}{dx} = \frac{2\pi r_s \beta}{P} (e_{ws} - e) \qquad (23)$$

The moisture flux, Q_m can be expressed by the air flux, Q_a and the specific humidity of the air, ω:

$$Q_m = Q_a \omega \rho \qquad (24)$$

Using the well known psychrometric relationship between ω and the partial vapor pressure, the following equation is obtained:

$$Q_m = Q_a \rho 0.622 \frac{e}{P-e} \qquad (25)$$

The derivative of Q_m with respect to x gives:

$$\frac{dQ_m}{dx} = 0.622 Q_a \rho \left[\frac{1}{P-e} + \frac{e}{(P-e)^2}\right] \frac{de}{dx} \qquad (26)$$

Substituting equation (26) into equation (23), and rearranging, gives the following differential equation, which governs the vapor pressure change with distance at any time instant:

$$\frac{de}{dx} = \frac{2\pi r_s \beta}{\rho 0.622 P Q_a \left[\frac{1}{P-e} + \frac{e}{(P-e)^2}\right]} (e_{ws} - e) \qquad (27)$$

Once e_{ws} is determined from the solution of the coupled sensible and latent heat transfer problem at a specific location, the vapor pressure variation with distance can be further predicted by the numerical solution of equation (27). This solution is achieved by using a central difference scheme, at any time instant, as follows:

$$e_{n+1} - e_n = D_n (e_{wm} - \frac{e_n}{2} - \frac{e_{n+1}}{2}) \qquad (28)$$

where

$$D_n = \frac{2\pi r_s \beta \Delta x}{\rho 0.622 P Q_a [\frac{1}{P-e_n} + \frac{e_n}{(P-e_n)^2}]} \qquad (29)$$

In equation (28) e_{wm} stands for the mean value of the saturated partial moisture pressure over section Δx. Since equation (28) and (29) hold for any time instant, it is straightforward to turn into vector notations, where the component of vector $\bar{e}_n$, $\bar{e}_{n+1}$, $\bar{e}_{wm}$ and even $\bar{D}_n$ will be the corresponding values taken at time instants t_1, t_2, ...t_M. The partial air vapor pressure vector at the end of section Δx can be written in the following form:

$$\bar{e}_{n+1} = (\underline{\underline{E}} \ \underline{D}_{-n}^1)\bar{e}_n + (\underline{\underline{E}} \ \underline{D}_{-n}^2)\bar{e}_{wm} \qquad (30)$$

where the components of $\underline{D}_{-n}^1$ and $\underline{D}_{-n}^2$ are the following

$$\left.\begin{array}{c} D_n^1 = \dfrac{1 - D_n/2}{1 + D_n/2} \\[4mm] D_n^2 = \dfrac{D_n}{1 + D_n/2} \end{array}\right\} \qquad (31)$$

3.6 Solution of the coupled sensible and latent heat transfer

An iteration procedure is needed to solve equations (18)-(22) simultaneously. Firstly, vector equation (18) is solved, for an initial latent heat flux density vector q_L. Using the resulting surface temperature vector components, the saturated moisture partial pressure is calculated from equation (20). From equation (19) and (22) finally the latent heat flux density vector can be calculated for an initial air vapor partial pressure distribution with time, and compared with the first input vector.

A fast convergent iteration procedure is achieved using the following chord method, illustrated in Figure 3.

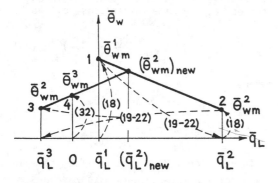

Figure 3. Scheme of a fast iteration procedure to calculate wet surface temperature vector, $\bar{\theta}_w$.

Firstly, θ_{wm}^1 is calculated for an initial vector $\bar{q}_L^1$ using equation (18). Secondly the components of $\bar{q}_L^2$ are calculated using equations (19)-(22). Another wet surface temperature vector $\bar{\theta}_{wm}^2$ is then calculated, and with this result the component for a new latent heat flux density vector $\bar{q}_L^3$ is obtained from equations (19)-(22). It is noted that the inverse solution of equations (19)-(22) will result in $\bar{\theta}_{wm}^2$, since $\bar{q}_L^3$ is calculated from this vector. Another point concerns the zero latent heat transfer when $e_{ws}=e$, and equation (20) gives directly a surface temperature vector component as follows:

$$\theta_{wm}^3 = \frac{237.3 \ Ln(e/610.5)}{17.27 - Ln(e/610.5)} \qquad (32)$$

From the geometrical relationship of the points described above and denoted by numbers 1-4 in Figure 3, coordinates of the components of the new vectors $\bar{q}_L^1$ and $\bar{\theta}_{wm}^1$ can be calculated. Continuing the iteration with these new input vectors, rapid convergence is achieved, in a few steps. In addition to the iterated mean surface temperature vector $\bar{\theta}_{wm}$ used for the air temperature prediction in equation (17), the components of the saturated partial moisture pressure vector $\bar{e}_{wm}$ is also available for the prediction of the vapor pressure of the flowing air using equation (30).

3.7 Coupled heat and moisture transfer prediction

The final objective of the calculations is to predict the simultaneous air temperature and moisture content, the latter being represented by the partial vapour pressure. The block diagram for the complete calculation is shown in Figure 4.

The solution of the coupled heat and moisture transfer prediction in an airway starts with the solution of the coupled sensible and latent heat transfer in section Δx.

The second and the third steps are the air temperature and vapor pressure predictions using equation (17) and (30), respectively. To be able to continue the

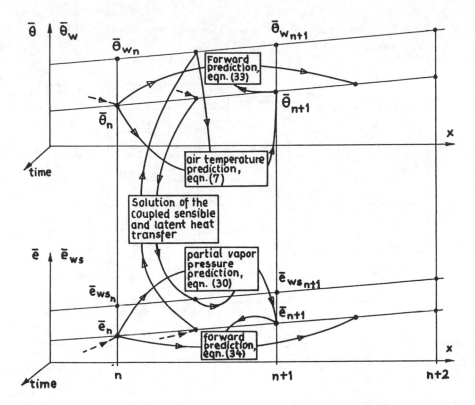

Figure 4. Scheme of the coupled heat and moisture transfer prediction.

prediction of vectors $\bar{\theta}_a$ and $\bar{e}$ for the next subsequent Δx interval, the forward prediction of $\bar{\theta}_m$ and $\bar{e}_{wm}$ for this new section needs to be estimated. This is carried out by linear extrapolations as follows:

$$\bar{\theta}_m \Big|_{n+1} = \frac{\bar{\theta}_{n+1} + \bar{\theta}_{n+2}}{2} = \frac{3}{2} \bar{\theta}_{n+1} - \frac{1}{2} \bar{\theta}_n \quad (33)$$

$$\bar{e}_{wm} \Big|_{n+1} = \frac{\bar{e}_{ws}\big|_{n+1} + \bar{e}_{ws}\big|_{n+2}}{2} =$$

$$\frac{3}{2} \bar{e}_{ws}\big|_{n+1} - \frac{1}{2} \bar{e}_{ws}\big|_n \quad (34)$$

In practice, a final loop of the iteration is advantageous to check and modify these forward predictions at the end of the calculation in each section. This iteration is necessary in the very first section, and may be discontinued during the subsequent steps.

3.8 Partially wet roadway surface

If the surface is uniformly damp but not completely wet, a wetness factor $W < 1$ can be used to multiply, i.e. decrease the moisture transfer coefficient. Other impacts such as the increase in the convective heat transfer coefficient as described by Luikov (1966) are neglected. The important practical case of a partially completely wet and partially entirely dry airway along the perimeter, is not dealt with here because of its extreme complexity.

619

Table 2. Comparison of the calculated dry (θ_a) and wet bulb (θ_{wb}) temperatures using different codes

		Dry Surface		Wet Surface W=0.15	
		3 months	3 years	3 months	3 years
CLIMSIM	θ_a	31.47	30.20	27.56	26.74
	θ_{wb}	25.99	25.62	26.11	25.68
TUNNEL	θ_a	31.64	30.41	30.52	29.38
	θ_{wb}	26.04	25.69	26.12	25.74
Proprie-tary Program	θ_a	31.55	30.33	30.93	29.78
	θ_{wb}	26.01	25.67	26.05	25.69
Present Work	θ_a	31.48	30.33	28.69	27.88
	θ_{wb}	26.03	25.69	26.61	26.11

The simple treatment of the uniformly damp surface using the wetness factor, allows the use of the coupled heat and moisture transfer prediction model even for completely dry roadways by substituting $W\beta = 0$ into the equations. The predicted air temperature along x, shows excellent agreement with the application of equations (8) to some test examples, even without reiterating equations (33) and (34). Consequently, the general model is used for either a dry or wet roadway.

3.9 Air wet bulb temperature calculation

Although the temperature and the partial vapor pressure are unique characteristics, using the dry bulb and wet bulb temperatures is more convenient. The relationships between the vectors of the air partial pressure $\bar{e}$, dry bulb temperature difference $\bar{\theta}$ and the wet bulb temperature $\bar{\theta}_{wb}$ are the following:

$$\bar{e} = \bar{e}_w - \frac{Pc_p}{0.622L}(\theta_{vr} - \bar{\theta} - \bar{\theta}_{wb}) \quad (35)$$

where $\bar{e}_w$ is the saturated vapor partial pressure at the air temperature:

$$\bar{e}_w = 610.5 \exp \frac{17.27(\theta_{vr} - \bar{\theta})}{237.3 + \theta_{vr} - \bar{\theta}} \quad (36)$$

4 COMPARISON AND AGREEMENT

Three mine climate simulation programs were used for comparison, namely CLIMSIM and TUNNEL developed by M.J. McPherson and R. Hemp respectively, and a proprietary program not yet publicly released. The following input data set was chosen for this test:

Perimeter of airway = 12 m
Cross section area = 9 m^2
length = 500 m
k = 6.0 W/mk
α = 2.5 X 10^{-6} m^2/s
Q_a = 35 m^3/s
θ_a = 28°C
θ_{wb} = 25°C
θ_{vr} = 45°C
P = 120KPa

The new computer code was written on an IBM PC, and has been slightly modified and further developed from the mathematical background described previously. The applicaton for a non-circluar cross section was made possible by using the hydraulical diameter, and modifying the surface area. The convective heat transfer coefficient was calculated automatically based on the recommended equation of Mousset-Jones et al. (1987), giving a value of 18.0 W/m^2K for this example.

The results are shown in Table 2 for completely dry and slightly wet (W=0.15) airway surfaces and for two different ages. It is noted that the reference simulations used only two district time intervals, i.e. 3 months and 3 years, while the new calculation followed the instantaneous heat and moisture transfer processes from the beginning to the end, using 20 steps to represent the complete time interval.

It can be seen in Table 2 that the agreement for a dry surface is excellent. For a partially wet surface, the reference values are scattered and the range of uncertainity is too wide to draw any significant conclusion. However, CLIMSIM produces lower estimates of both the dry and wet bulb temperatures, when compared to the others. The present model has lower dry bulb and higher wet bulb temperatures, which is expected because of the special surface wetness involved in the new model.

5 APPLICATIONS

The new algorithm was first applied to a sensitivity analysis, including the cumulative thermal effect on the input parameters for a climate simulation, and was published by Danko et al. (1988). The model can be used as a core procedure in any of the more complex mine climate, or fire simulation programs.

To further emphasize the advantage in increasing the precision by using the new model, two more examples using sinusoidal air temperature variation are presented and the results compared with those of the non-cumulative models.

The previous data set is used except for the inlet dry and wet bulb air temperature, which are now varied sinusoidally. A daily and a weekly periodical temperature change is shown in Figures 5 and 6, respectively. The outlet temperature changes at the end of the airway were calculated using the new method both with and without the cumulative thermal effect. The latter was simply simulated using the straightforward age function instead of equation (1) and, correspondingly, a diagonal matrix instead of equation (3) for matrix A. The differences in the amplitudes $\bar{\bar{a}}$re far too significant to be considered negligable, namely 40% and 25% in the air temperature amplitudes for the one and the seven days simulation, respectively.

6 CONCLUDING REMARKS

1. The new method using superposition solution and a continuous simulation throughout the time interval compares well for long periods of time with the other methods neglecting the cumulative thermal effect caused by the changing air temperature. The vector-matrix representation provides a simple computational structure.

2. It is easy to apply the new method for arbitrary input temperature and humidity distribution, e.g. to predict daily or weekly temperature variations. In these cases the new method provides much higher precision besides the convenient application.

8 ACKNOWLEDGEMENT

This research has been supported by the Department of the Interior's Mineral Institutes program administered by the Bureau of Mines through the Generic Mineral Technology Center, Mine Systems Design and Ground Control, under the allotment grant number G1125151.

NOTATIONS

t	time [s]
θ	air temperature difference = $\theta_{vr} - \theta_a$ [K]
θ_a	air temperature (dry bulb temperature) [K]
θ_{vr}	virgin rock temperature [K]
k	thermal conductivity of the rock [W/(mK)]
r_s	radius of the airway [m]
Bi	Biot number = $h\, r_s/k$
Fo	Fourier number = $\alpha t/r_s^2$
q	heat flux density [W/m^2]
Δx	length of an airway section [m]
ρ	density of air [kg/m^3]
c_p	specific heat of air [J/(KgK)]
$\bar{\theta}a_n$	air temperature vector at length division n [K]
θw	wall surface temperatue difference = $\theta_{vr}-\theta_s$ where θ_s denotes the surface temperture [K]

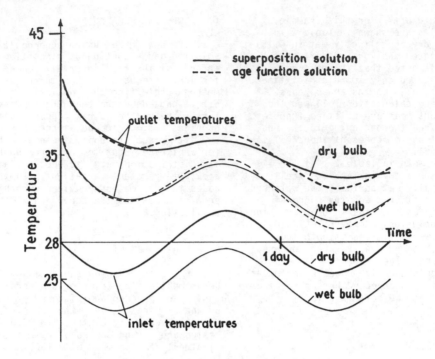

Figure 5. Daily temperature variation.

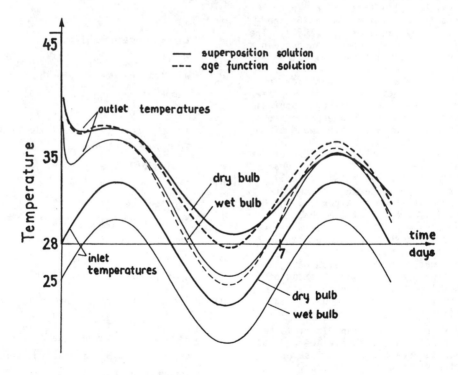

Figure 6. Weekly temperature variation.

θ_b apparent, or pseudo-base temperature [K]

$\bar{\theta}w_n$ wall temperature difference vector at length division n [K]

C constant, equation (15)

B constant, $B=2k\pi\Delta x/(Q_a\rho c_p)$

Q_a air flux [m³/s]

h heat transfer coefficient [W/(m²K)]

q_m mass flux density [kg/m²]

β mass transfer coefficient [kg/m²]

e moisture partial pressure [Pa]

ω specific humidity [kg/kg]

D_n, D_n^1, D_n^2 constants, equations (29,31)

$\bar{\theta}_{wb}$ air wet bulb temperature vector [K]

W wetness factor

α thermal diffusivity of the rock [m²/s]

Superscripts

$(\bar{\ })$ vector with components of ( ) at time intervals $t_1, t_2, \ldots, t_M$

$(\ )^{-1}$ inverse of matrix ( )

$=$ $=$

Subscripts

n division according to the length, x

L latent heat

w value at the wall

s saturated

REFERENCES

Luikov, A.V. 1966. Heat and Mass Transfer in Capillary-porous Bodies. Oxford: Pergamon Press.

Danko, G., P. Mousset-Jones & M.J. McPherson 1988. Sensitivity Analysis on Selected Input Parameters for Mine Climate Simulation Programs. Proceedings, Society of Mining Engineers, Annual Meeting (Accepted).

Hemp, R. 1982. Psychrometry. Environmental Engineering in South African Mines. Mine Ventilation Society of South Africa.

Jaeger, J.C. & T. Chamalaun 1966. Heat Flow in a Infinite Region Bounded Internally by a Circular Cylinder with Forced Convection at the Surface. Aust. Jour. Phys. 19:475-488.

McPherson, M.J. 1979. Psychrometry: the Measurement and Study of Moisture in Air. University of Nottingham, Mining Department Magazine. 31.

McPherson, M.J. 1986. The Analysis and Simulation of Heat Flow into Underground Airways. International Journal of Mining Engineering. 4.

Mousset-Jones, P., G. Danko & M.J. McPherson 1987. Heat Transfer in Mine Airways with Natural Roughness. Proceedings, 3rd Mine Ventilation Symposium. Pennsylvania State University.:42-52.

Computer Applications in the Mineral Industry, Fytas, Collins & Singhal (eds)
© 1988 Balkema, Rotterdam. ISBN 90 6191 760 3

Development of a rule-based program for the elaboration of typified ventilation projects for the Spanish mining industry

J.Samaniego, J.Ruiz Pellicer & D.Alarcon
AITEMIN, Spain

V.Luque
HUNOSA, Spain

ABSTRACT: AITEMIN is a non-profit mining research association whose members are mining companies and mining equipment manufacturers. For the last two years, AITEMIN has been involved in the development of computer programs to solve ventilation networks applied to the Spanish mining environment. That work has resulted in two computer programs that run in microcomputers, which include interactive graphics capabilities and that check that the resulting ventilation air flow complies with the Spanish Mining Law Regulations. Air flow resistance for each ventilation branch is calculated by the program, although the obtained values can be overriden by the user.
Both systems are designed for small mines with no more of 100 branches, as it covers over 90% of the Spanish mines and it seems the upper limit of the hardware graphics limitation (640x350 resolution). A brief description of the programs is presented emphasizing their appropiateness to the Spanish Mining Industry.
On the other hand, AITEMIN is one of the few companies in Spain entitled to audit mining projects for the Administration through the Directorate of Mines. Engineers of AITEMIN have been checking ventilation projects for some years as well. All these facts merged in the idea of developing a computer system which should be able to help the mining engineer at the mine site to do ventilation projects to be presented to the Directorate of Mines. The objective of the system is to facilitate the standardization of mining ventilation projects achieving good quality work. The system will direct the user to the important facts and calculations to be made according to the objective of the proposed project. Related Mining Law Regulations are taken into account and mentioned, as the user may not know which rule applies to the specific mining conditions of the case study. Rules derived from the experience from the engineers auditing ventilation projects are also included. The system is designed to automatically generate the reports in the appropiate format, as required by the Administration.
The paper reflects the experience of the development of the system, its structure and contents.
The system was initially developed using an expert system shell, written in LISP, incorporating various mathematical functions that provide the needed calculation capability. However, it was later rewritten in compiled BASIC, for speed and versatility sake.

1 INTRODUCTION

The Spanish mining industry is largely characterized by small companies. Particularly, in the coal mining industry, 80% of the 300 producing companies have less than 100 workers, averaging a production of 35,000 tons per year.
Most coal producing companies have limited technical and economical resources, and face difficult and narrow seams.

2 MINE SAFETY INSPECTIONS

Recently, both Central and Regional Governments have developed safety inspection programs aimed mainly at the coal industry, with the objective of checking the safety measures taken at the operating mines, and henceforth to increase the safety level at the mine working place.
AITEMIN was the institution in charge of developing the Mine Safety Integral Inspection Plan, that lasted two years in different autonomic regions of Spain. While that work was performed, it was realized that the knowledge of ventilation engineering and of the Spanish Mining Law was somewhat limited and sometimes confused around the smaller mines.

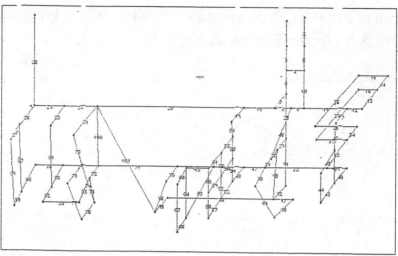

(A)=Añado rama (B)=Borro última (C)=Borro una rama (X)=Veo rama (F)=Fin

Figure 1. Computer display while editing a mine ventilation layout

3 MINE VENTILATION COMPUTER PROGRAMS

The department of Microcomputers at AITEMIN was contracted to develop a computer program to run on IBM-PC-type micros, that should help solve small ventilation networks, and with the objective of selecting the main ventilation fan.
The program developed was divided into two independent modules, which have the following characteristics:

Data entry module:
1.- Data entry is simplified, as the user does not need to go through the tedious task of numbering nodes nor branches
2.- Data entry is based on interactive graphics
3.- Mine ventilation layouts can be entered into the computer so that they resemble the original ones on paper
4.- Other information concerning oxigen consumption and methane releases is input as well, to estimate the minimum air fintake required by mine regulations
5.- The resistance of each drift (branch) is also estimated, whenever it is not available from in-situ measurements.
6.- Input information is stored in disk data files for later use and modification, allowing for simulation of a mine under various developing stages.

Calculation module:
1.- The working point of the main fan is calculated based on a trial input flow, specified by the user. The fan heading is calculated based on that trial air flow.

2.- The original version of the program could only simulate one fan.
3.- Air velocity, head loss, power consumption, and air flow and direction of flow are calculated for each branch.
4.- The problem can be recalculated modifying resistances in up to ten branches simultaneously, allowing simulation of ventilation doors, new roof support, or work progress.
5.- Results are shown both in tabular form and as a bar chart, so that the extreme values can be easily and rapidly detected.

Figure 1 shows a screen dump while data is being input for a coal mine, and Figures 2,3 and 4 show various screens of the results provided by the calculating program.

A completely new version of the calculation module has been recently released, incorporating capabilities for up to 16 fans located throughout
the mine, and represented by their characteristic curves.

4 MINE SAFETY REGULATIONS

One of the objectives of the work was to attempt to include the relevant Mine Safety Regulations into the program so that it would help checking poor ventilation conditions. The program calculates the minimum required air flow, and tells the user which branches do not comply with the law regarding air velocity and air flow, according with personnel and power of diesel engines

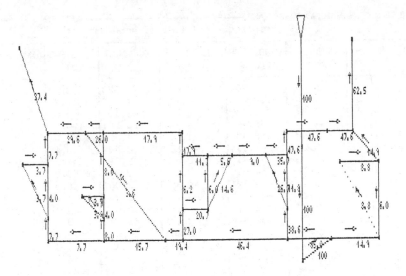

Figure 2. Output of the calculation program showing the mine layout, air flow and direction of flow for each branch

present in the branch, quantity of explosives used and air clearing time, and methane leaks.

However, many practical considerations as where to install the fan, how to account for secondary ventilation, or many other regulations about methane control were not included.

AITEMIN was contracted later to develop a workable system to develop ventilation projects that follow the Spanish Mining Regulations. This project is part of a wider program to develop models of technical projects (or "typified projects"), which are intended to serve as guidelines to mine engineers for calculating and presenting mine engineering projects in a standardized manner.

5 THE TYPIFIED MINE VENTILATION PROJECT

The typified projects, funded by the Spanish Ministry of Industry through the Spanish Geological and Mining Institute (IGME) were to be accompanied by a diskette containing computer programs that would help to perform the calculations using a standard personal computer.

The objective that we faced was the development of a system flexible enough to be able to provide calculation power, law consulting capabilities, and technical recommendations for solving ventilation problems, all packaged to implement standard calculation procedures and to generate standard reports.

Since this was the first project out of several that were developed on paper then, we

also faced the idea of building a standard "project generator", that should serve to develop future typified projects.

DESIGN OF THE PROJECT MODEL

The model that we had to build was divided into modules as shown in Figure 5

5.1 Calculation module

The structure allowed us to interface with existing ventilation programs, (the calculation module), hence taking advantage of previous work performed on this field. However, some calculating subroutines were extracted from the ventilation programs to be included within the law consulting module, such as drift resistance calculation routines.

By doing so, there is no need to call external programs that would slow down the process.

The system was designed to be able to interrupt the development of a given project for whatever reason, store on disk all the information entered, and retrieve it later to modify existing data, and add the rest of the information

5.2 Report generator

The program writes onto a temporary file all the information needed for the report, and the the report generator merges that information with the standard written project skeleton, maintaining a specific format.

627

resultados del fichero sabero2

TRAMO	NUDOS		CAUDAL (m3/s)	PERD.PRES (kp/m2)	VELOCIDAD (m/s)	POTENCIA (kw)	RESIST. (µ)	NOTAS
1	1	2	-100.00	24.00	10.00	23.520	2.4	V. ALTA
2	2	3	-7.73	278.24	0.77	21.077	4656.7	
3	3	4	-33.67	3.40	3.37	1.122	3.0	
4	4	5	-47.15	266.79	4.72	123.281	120.0	
5	4	6	13.48	-186.45	-1.35	24.632	1026.0	
6	6	7	8.00	-0.01	-0.80	0.001	0.1	
7	7	8	8.00	-34.61	-0.80	2.713	541.0	
8	6	8	5.48	-34.62	-0.55	1.860	1152.0	
9	8	9	13.48	-22.72	-1.35	3.001	125.0	
10	9	10	92.27	-35.76	-9.23	32.334	4.2	V. ALTA
11	2	10	-92.27	3.41	9.23	3.079	0.4	V. ALTA
12	9	11	-78.79	55.87	7.88	43.140	9.0	
13	11	12	-23.64	8.38	2.36	1.942	15.0	
14	12	13	-12.62	148.83	1.26	18.402	935.0	
15	13	14	-32.16	16.54	3.22	5.214	16.0	
16	13	15	19.54	-10.31	-1.95	1.974	27.0	
17	15	12	11.02	-138.05	-1.10	14.915	1136.0	

Figure 3. Table of results showing branches that do not comply with regulations

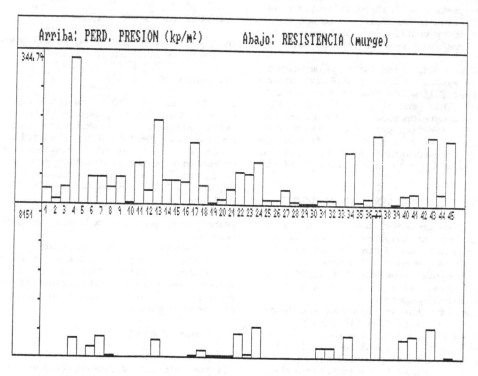

Figure 4. Example of a bar chart used to point out which branches are more important to the overall result

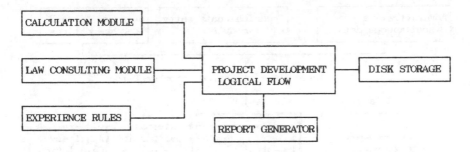

Figure 5. Project structure

5.3 Project development logical flow

The project development logical flow module acts as the control module of the project, guiding the user through the appropiate steps of the project. It is a rule-based subsystem, as new information redirects the program flow accordingly to the calculation and law requirements.

5.4 Law consulting module

The law consulting module is another rule base subsystem that calls the regulation number and parragraph recalled whenever required.

5.5 Experience rules

As it is reflected in the typified project, experience is incorporated into the project through recommendations, advises and rules of thumb. Examples of those are:

Context: ventilation doors
Rule:
IF
    THERE ARE SEVERAL DOORS IN PARA-
    LLEL IN THE MINE
AND
    THE VENTILATION CIRCUIT IS NOT
    OF DIAGONAL TYPE
THEN
    IT IS ADVISABLE TO USE A SAFETY
    FACTOR OF 1.5 OVER THE AIR FLOW
    REQUIRED.

Context: secondary ventilation
Rule:
IF
    THERE IS FORCED SECONDARY VENTI-
    LATION
AND
    SECONDARY FAN IS COMPRESSED-AIR
    POWERED
AND

COMPRESSED AIR ALSO FEEDS NEARBY
DRILLS
THEN
    DO NOT TRUST FAN
    CHARACTERISTICS: THERE IS A
    CHANCE THAT C.A. PRESSURE AND
    FLOW IS NOT ENOUGH TO PROVIDE
    FULL POWER TO FAN. TAKE IN-SITU
    MEASUREMENTS.

Context: old workings
Rule:
IF
    DRIFTS ARE DEVELOPED IN COAL AND
    WERE ARE OLD WORKINGS ABOVE
THEN
    ALLOW FOR SENSIBLE AIR LEAKS.
    TAKE IN-SITU MEASUREMENTS.

As it is shown here, the project seemed to be suitable for a solution as an expert system with capabilities to interface with other calculating programs.
Based on the experience provided by a research project on application of expert systems to mining, funded by the CAICYT, programming started to be carried out using an expert system shell written in LISP. Formulae and calculation functions had to be programmed directly in LISP, while the rules were input using the shell's own english-like language interface.
Soon it was realized that the system was becoming way too large and slow to be used in the field, so it was decided to reprogram all the previous work done in a more conventional language.
Besides speed, this decision brought more advantajes, namely full screen formatting, access to our own libraries, and graphics, as well as an easier interface with the rest of the calculating programs already available.

629

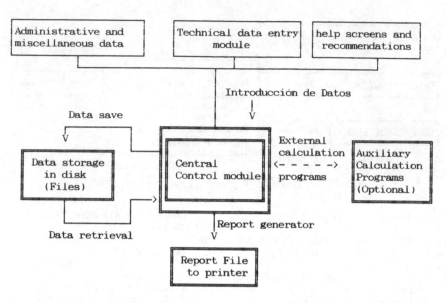

Figure 6. Final program structure

So, the whole system was reconverted to a BASIC compiler.

The resulting system has the structure shown in Figure 6.

The system once it was finished, was made of 11 programs and subprograms and over 4000 lines of code.

Input data is performed through formatted screens which can be forwarded or backwarded as pages of a book, and the program performs all the required recalculations as needed. Input data screens are divided by themes as follows:

- Miscellaneous and administrative data about the mine and the project.
- Objetives of the project.
- Initial situation of the ventilation before the changes proposed by the project and expected situation afterwards.
- Ventilation and resistance data for each branch. Mine alyout.
- Overall mine resistance. Optional connection with the calculation programs.
- Air flow requirements.
- Checking on air velocity and flow requirements for each branch, and other considerations regarding the use of explosives.
- Mine safety regulations.
- Recommendations on changing the ventilation direction. Fires.
- Ventilation fans.

6 CONCLUSIONS

A tool to help developing ventilation projects has been constructed. Certain speed and size limitations have been encountered while trying to use the available expert system shell tool available at the time the project started, and a more conventional programming approach has been taken.

The computer tool developed allows the use of existing programs to perform technical calculations of ventilation networks, and incorporates a guideline to fill out a standardized ventilation project report.

Common pitfalls concerning data estimation are highlited, and the user is awared of potential misscalculations.

During 1988, the program will be released to the Spanish mines, hance improvements and feedback from field experiences are expected.

REFERENCES

AITEMIN. 1987. Manual de ventilación. AITEMIN, Madrid.
AITEMIN, CCAA de Castilla y León. Plan Integral de Seguridad Minera. Internal Reports. Valladolid. 1986

Computer Applications in the Mineral Industry, Fytas, Collins & Singhal (eds)
© *1988 Balkema, Rotterdam. ISBN 90 6191 760 3*

Digital simulation of blast fumes, dust and diesel exhausts in sublevel caving

Wu Chao & Agne Rustan
Division of Mining and Rock Excavation, Luleå University of Technology, Sweden

ABSTRACT: In order to be able to monitor and control the contamination from blast fumes, diesel exhausts and dust, etc, it is necessary to know the releasing rates of pollutants, ventilation time, required air quantity, and geometric conditions of the drifts in different cases in networks, especially in sublevel caving. Based on the mathematical equation of diffusion mechanics and combined with network analysis theory, a computer program which can be used to simulate the concentration distribution of the pollutants in the working place was developed by MSc Wu Chao. The program is written in FORTRAN 77 and can be run on a PC-computer. According to the results of field measurement, the digital simulation is basically coincident with the field data.

1 INTRODUCTION

In order to realize the real time control of local fans at the working place and achieve the goal of safety and economy, it is necessary to know the releasing rate of pollutants, ventilation time, required air quantity and geometric conditions of the working place. Fortunately, several mathematical formulas to calculated the concentration and distribution of different pollution sources in a single drift have been set up by Thakur (1975), but we still do not know how precise the results coincide with the field. Furthermore, it seems necessary to combine these formulas with a network analysis program so that more obvious results can be gotten quickly and conveniently. Therefore, the following study has been done.

2 THE MATHEMATICAL MODELS

According to diffusion mechanics, the basic diffusion equation relevant to most air pollution problems can be written in the form

$$\frac{\partial C}{\partial t} + V\nabla C = \frac{\partial}{\partial x}(E_x \frac{\partial C}{\partial x}) + \frac{\partial}{\partial y}(E_y \frac{\partial C}{\partial y}) +$$

$$\frac{\partial}{\partial z}(E_z \frac{\partial C}{\partial z}) + source + sink \quad (1)$$

where C = the concentration of pollutant
    t = time
    V = the air velocity vector
    ∇ = the Laplace operator
    E_x, E_y, and E_z = the components of
                      diffusion in x, y
                      and z-direction

source = pollution source
sink   = pollution collector
           (filter, wet drift walls etc)

Equation (1) is a non-homogeneous partial differential equation and it can not be solved by analytical methods. However, with simplification in some particular cases, several solutions can be achieved. These solutions are still useful when they are used to describe the concentration distribution in a drift.

2.1 Dust source contamination at drilling or loading

Equation (1) is modified and solved in this case with the following assumptions:
  1. The dust source occupies the entire width of the drift. In fact, the respirable dust concentration is almost homogeneous in the entire cross section 10 meters away from the drilling or loading machine.

2. The problem is considered as a one dimensional case.

3. The decay coefficient for respirable dust is a constant. Generally it depends on air velocity, friction factor, wetness of the drift walls, etc. On the basis of the above assumptions and by the Laplace transform method, the solution of concentration in equation (1) is as follows:

$$C(x,t) = \frac{q_d}{2V_x A} \exp(\frac{V_x x}{2E_x}) \; (\exp(- x (\frac{\alpha}{E_x})^{1/2})$$

$$erfc(\frac{x}{2\sqrt{E_x t}} - \sqrt{\alpha t}) + \exp(x (\frac{\alpha}{E_x})^{1/2})$$

$$erfc(\frac{x}{2\sqrt{E_x t}} + \sqrt{\alpha t})) \qquad (2)$$

where q_d = the respirable dust quantity produced by pollution source
A = cross sectional area of the drift
V_x = average air velocity in the drift
x = length away from the pollution source
E_x = diffusion coefficient
t = time
α = $(V_x^2 + 4E_x\lambda_d)/(4E_x)$

erfc = the complementary error function
λ_d = decay coefficient for respirable dust along the drift

2.2 Blast fume contamination in the case without recirculation of blast gases

This case may be taken as an instantaneous stationary point source. Equation (1) can be modified and solved by using the following assumptions:

1. There is no sink or additional source in the drift.
2. The diffusion coefficient is a constant.
3. The problem is considered as a one dimensional case.

The concentration function can then be written as:

$$C(x,t) = \frac{F_1 C_i}{2(\pi E_x t)^{1/2}} \exp(- \frac{(x-V_x t)^2}{4E_x t})$$

where F_1 = the length of the drift which is filled with blast fumes after blasting
C_i = the initial concentration of blast fumes

The meaning of other parameters is as given above.

2.3 Blast fume contamination in the case with recirculation of blast gases after blasting

The following formula was derived from the diffusion mechanics equations and actual field results reported by Rustan et al 1986.

$$C(x,t) = \frac{F_1 A C_i}{t^n+1} \exp(- \frac{V_x}{E_x} x) \qquad (4)$$

where n = the factor which depends upon the real situation of recirculation of blast fume.

The meaning of other parameters is as given above.

2.4 Diesel exhaust contamination when the engine is fixed

This case is similar to (2.1) except the decay coefficient is equal to zero. Therefore, the one dimensional solution of the problem is as follows:

$$C(x,t) = \frac{q_e C_i}{2V_x A} \exp(-\frac{V_x x}{2E_x}) \; (\exp(- \frac{V_x x}{2E_x})$$

$$erfc(\frac{x}{2\sqrt{E_x t}} - \frac{V_x}{2} (\frac{t}{E_x})^{1/2}) + \exp$$

$$(\frac{V_x x}{2E_x}) \; erfc(\frac{x}{2\sqrt{E_x t}} + \frac{V_x}{2} (\frac{t}{E_x})^{1/2}))$$

$$(5)$$

where q_e = the diesel exhaust quantity produced by the engine
C_i = the concentration of species I in the diesel exhaust or the combined concentration of all gases

The meaning of other parameters is as given above.

2.5 Diesel exhaust contamination when the engine is moving

This model was set up by Stefanko et al (1974). The solution is as follows:

$$C(x,t) = \frac{q_e C_i}{2 V_r A} \left(erfc\left(\frac{x - V_r t}{2\sqrt{E_x t}}\right) - (1 + \frac{V_r x}{E_x} + \right.$$

$$\frac{V_r^2 t}{E_x}) \exp(\frac{V_r x}{E_x}) \ erfc(\frac{x + V_r t}{2\sqrt{E_x t}}) + 2 V_r$$

$$\left. (\frac{t}{E_x \pi})^{1/2} \exp(-\frac{(x - V_r t)^2}{4 E_x t}) \right) \qquad (6)$$

where C_i = the concentration of species I in the diesel exhaust or the combined concentration of all gases

q_e = the quantity of diesel exhaust emission

V_r = $V \pm U$ (the relative velocity)

V is the velocity of the moving engine, and U is the air velocity in the drift. When the engine and air travel in the same direction, the effective velocity is (V − U). When they travel in opposite directions, the effective velocity is (V + U).

A special case is when the diesel engine travels in the same direction as the air flow and with the same velocity. The solution is as follows:

$$C(x,t) = \frac{C_i}{2} \left(erfc\left(\frac{x - \frac{q_e t}{2A}}{2\sqrt{E_x t}}\right) - (1 + \frac{q_e x}{2 E_x A} + \right.$$

$$\frac{q_e^2 t}{4 A^2 E_x}) \exp(\frac{q_e x}{2 E_x A}) \ erfc(\frac{x + \frac{q_e t}{2A}}{2\sqrt{E_x t}}) +$$

$$\left. \frac{q_e}{A} (\frac{t}{E_x \pi})^{1/2} \exp(- (\frac{x - \frac{q_e t}{2A}}{2\sqrt{E_x t}})^2) \right) \qquad (7)$$

The meaning of all parameters in formula (7) is as given above.

2.6 Diffusion coefficient.

The formula for calculating the diffusion coefficient recommended by Stefanko (1974) is used in the program. It is as follows:

$$E_x = 28.8 \ V_x R \ (\frac{\lambda_s}{\lambda_r})^{1/2} \qquad (8)$$

where V_x = the average air velocity in the drift

R = the equivalent radius of the drift

λ_s = the coefficient of friction for a smooth pipe

λ_r = the coefficient of friction for the drift

3 EXPLANATION OF THE COMPUTER PROGRAM

According to the mathematical models described above, a computer program was written using the language FORTRAN 77. The program can be run in a PC-computer. The flow chart for the calculation system is shown in Fig 1. For the sake of the simplicity and practicallity, the program is used only to estimate the pollutant concentrations in a drift or stope. In the appendix, only the program which can be used to calculate pollutant concentration are listed. The part of the network analysis program is omitted because it was published by several authors many years ago, such as Wang et al (1967). The program listed in the appendix can be easy combined with a network analysis program based on the flow chart in Fig 1. The program given in Appendix 1 is written in an interactive form, and a complete discription of all the variables used is contained in the source code listings.

4 THE COMPARISION BETWEEN CALCULATING VALUES AND FIELD DATA

The following results of in situ measurement in a sublevel caving area were done by Luleå University of Technology at LKAB in Malmberget, Rustan et al (1986).

4.1 Respirable dust distribution

An example of calculated values is shown in Table 1. The parameters used for the calculation are based on the real conditions and are listed at the bottom of the table. The symbols used have been explained earlier in this paper.

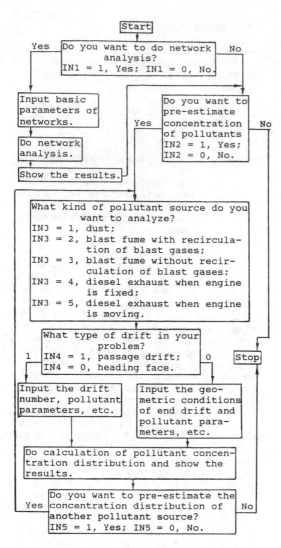

Fig. 1 The flow chart of the calculation system.

Start

Do you want to do network analysis?
IN1 = 1, Yes; IN1 = 0, No.
Yes   No

Input basic parameters of networks.

Do network analysis.

Show the results.

Do you want to pre-estimate concentration of pollutants
IN2 = 1, Yes;
IN2 = 0, No.
Yes   No

What kind of pollutant source do you want to analyze?
IN3 = 1, dust;
IN3 = 2, blast fume with recircula-tion of blast gases;
IN3 = 3, blast fume without recir-culation of blast gases;
IN3 = 4, diesel exhaust when engine is fixed;
IN3 = 5, diesel exhaust when engine is moving.

What type of drift in your problem?
IN4 = 1, passage drift;
IN4 = 0, heading face.
1   0   Stop

Input the drift number, pollutant parameters, etc.

Input the geo-metric conditions of end drift and pollutant para-meters, etc.

Do calculation of pollutant concen-tration distribution and show the results.

Do you want to pre-estimate the concentration distribution of another pollutant source?
IN5 = 1, Yes; IN5 = 0, No.
Yes   No

Table 1. The calculated values of dust concentration.

Time (min)	1	2	3	4	5	6	7	8	9	10	11	12	13	14
Conc (mg/m^3)	0.46	1.26	1.55	1.71	1.80	1.86	1.90							
		0.96	1.42	1.64	1.76	1.83	1.88	1.91						
Conditions	$q_d = 3.64$ (mg/m) $\lambda_d = 0.0$ $\lambda_S = 0.01$ $\lambda_\gamma = 0.08$ $V_x = 0.1$ (m/s) $x = 23$ (m) $R = 1.96$ (m)													

Table 2. The calculated values of blast fume concentration.

Time (min)	20	30	40	50	60	70	80	90	100	110	120	130
CO-conc (ppm)	114.4	47.1	28.1	19.4	14.6	11.6						
		68.1	35.4	23.0	16.7	12.9	10.4					
Conditions	$C_l = 1100$ (ppm) $\lambda_S = 0.01$ $\lambda_\gamma = 0.08$ $V_x = 0.25$ (m/s) $n = 1.28$ $x = 30$ (m) $R = 1.96$ (m) $F_l = 50$ (m) $A = 18.2$ (m^2)											

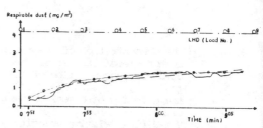

Fig.2 A comparison between real respir-able dust concentration and calculated as a function of time.
---------- calculated values
—·—·—·— average field data (Re-spirable dust concentration when loading ore in cross cut No. 551, Kapten 572 m, level at LKAB in Malmberget)

According to the values in Table 1 and field data, a comparison between the concentration of respirable dust and ventilation time is presented in Fig 2.

From Fig 2, it is clear that the cal-culated and measured values coincide well.

4.2 Blast fume distribution in the case with recirculation of blast gases

An example of calculated values is shown

in Table 2. The parameters used are listed at the bottom of the table.

According to the values in Table 2 and field data, a comparison between cal-culated and real concentration of blast fume versus ventilation time is shown in Fig 3.

From Fig 3 it is clear that the calculated and measured values coincide well.

4.3 Blast fume distribution in the case without recirculation of blast gases

It can be pre-estimated that the concen-tration decrease more quickly in the case without recirculation of blast gases than that with recirculation. An example is shown in Fig 4.

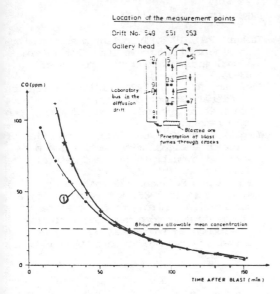

Fig. 3 A comparison between calcuated and real concentrations of CO versus time after a sublevel caving round has been blasted. LKAB in Malmberget, Captain ore-body 572 m level.

+———+———+ calculated values

•——•——• real measured values 30 m from the blast. (Measurepoint No. 1)

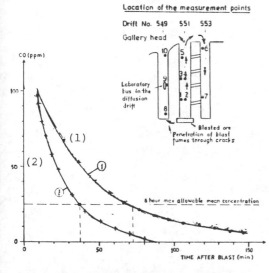

Fig. 4 A comparison between calculated CO-concentration with (1) or without (2) recirculation versus time after a sub-level caving round has been blasted.

+———+———+ calculating values without recirculation

‘lated values with
‘lation

Table 3. The calculated concentrations of diesel exhaust.

Time (min)	1	2	3	4	5	6	7	8
Conc (ppm)	162.2	198.0	214.8	224.3	230.4	234.6	237.7	240.0

Conditions $C_i = 10^4$ (ppm) $q_e = 0.0455$ (m³/s) $\lambda_S = 0.01$
$\lambda_r = 0.08$ $V_x = 0.1$ (m/s) $x = 10$ (m)
$R = 1.96$ (m)

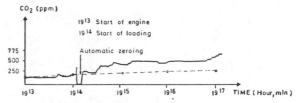

Fig. 5 A comparison of calculated and real concentration of diesel exhaust gas versus ventilation time.

--------- calculated values

————— field data (CO_2 concent-
ration in the drivers cabin,
TORO 500 D, during loading of
ore in a short cross cut,
Fabian 548 m level, LKAB in
Malmberget)

From Fig 4, it is clear that there is much difference in the required ventilation time in the two cases with or without re-circulation of blast gases. Therefore, it is very important to avoid the recircula-tion of pollutions in the working face.

4.4 Diesel exhaust contamination when the engine is not moving

An example of calculated values is shown in Table 3. The parameters used for cal-culation are listed at the bottom of the Table.

According to the values in Table 3 and field data, the relationship between the concentration of diesel exhaust gas and ventilation time is shown in Fig 5.

From Fig 5, it can be seen that the cal-culated values and field data are well coincident during the period of engine start. After the time when loading star-ted, the engine emits more exhaust, and there exist a large difference between the calculated values and the field data. Therefore, it is better to divide the ope-rating time into two periods when making the calculation. Otherwise the quantity of exhaust emission has to be represented by a period function. Another example of cal-culated values is shown in Table 4. The

635

Table 4. The calculated concentrations of diesel exhaust.

Time (min)	1	4	8	12	16	20
Conc (ppm)	1503	1857	1944	1972	1985	1991
Conditions	C_i = 10 (ppm) Λ_γ = 0.08 R = 1.96 (m)		q_e = 0.02 (m^3/s) V_x = 0.1 (m/s)		Λ_S= 0.01 x = 7.3 (m)	

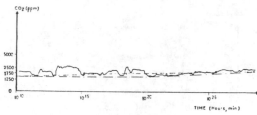

Fig. 6 A comparison between calculated and real CO-concentration of exhaust versus ventilation time.
----------- calculated values
—·—·—·— measured values data (Gas concentration measurement in cross cut No. 161, Fabian 548 m level at LKAB in Malmberget)

conditions are also listed at the bottom of the table.

According to the values in Table 4 and field data, a comparison between calculated and real CO-concentration of diesel exhaust versus ventilation time is shown in Fig 6.

From Fig 6, it can be seen that the calculated values and field data are well coincident. Unfortunatly, there is not enough field data to test all variables in the mathematical models described above.

5 CONCLUSIONS

Summarizing above analysis, the following conclusions may be made:
- Generally, the mathematical equations listed in this paper are basically coincident with the field data, but they have to be tested more thoroughtly.
- According to the comparision in (4.3), it is very important to avoid recirculation of pollutions.
- There are several parameters in the program which are difficult to choose correctly. In order to make the simulation program more accurate, much more field measurement work is required.

REFERENCES

Thakur, P.C. 1975. Turbulent mass transfer in mine ventilation system. 1st international mining ventilation congress. 93-96.
Stefanko, R. et al 1974. Digital simulation of diesel exhaust contamination of mine ventilation systems. Bureau of Mines open file report.
Wang, Y.J. et al 1967. Computer solution of three-dimensional mine ventilation networks with multiple fans and natural ventilation. Int. J. of Rock Mech. and Mining Science. 129-154.
Rustan, A. et al 1986. Automatic analogue ventilation control through measurement of gas or dust concentration including continuous environmental control. Luleå University research report TULEA 1985:20, May 1986.

APPENDIX 1: A PROGRAM FOR CALCULATION OF POLLUTANT CONCENTRATION DISTRIBUTION

The program listed in Appendix 1 can be used to calculate the pollution concentration distributions in a single drift independently. They are written in interactive language. The meaning of all variables is explained in the programs. They can also be combined with a network analysis program if a little improvement is done with the help of the flow chart of a calculation system given in Fig 1.

```
C     ****************************************************
C     ****************** MAIN PROGRAM ******************
C     ****************************************************
C
      DIMENSION X(50),T(50),CONC(50,50)
      DATA IYES/'YES'/
C
      OPEN (6,FILE='DATA',STATUS='NEW')
C
C     INPUT THE INDEX FOR CALLING DIFFERENT SUBROUTINES
5     WRITE(*,10)
10    FORMAT(10X,'PLEASE INPUT THE INDEX FOR CALLING SUBROUTINES'//
     *      10X,'IND = 1 -- DUST'/
     *      10X,'IND = 2 -- BLAST FUME WITH RECIRCULATION OF'/
     *      10X,'            POLLUTION GASES'/
     *      10X,'IND = 3 -- BLAST FUME WITHOUT RECIRCULATION'/
     *      10X,'            OF POLLUTION GASES'/
     *      10X,'IND = 4 -- DIESEL EXHAUST WHEN ENGINE IS FIXED'/
     *      10X,'IND = 5 -- DIESEL EXHAUST WHEN ENGINE IS MOVING'/
     *      10X,'WHICH ONE ?')
      READ(*,*) IND
      GO TO (20,30,40,50,60),IND
20    OPEN (1,FILE='DATA1',STATUS='NEW')
      CALL DUST
      GO TO 70
30    OPEN (2,FILE='DATA2',STATUS='NEW')
      CALL FUMERC
      GO TO 70
40    OPEN (3,FILE='DATA3',STATUS='NEW')
      CALL FUMENR
      GO TO 70
50    OPEN (4,FILE='DATA4',STATUS='NEW')
      CALL EXHAUF
      GO TO 70
60    OPEN (5,FILE='DATA5',STATUS='NEW')
      CALL EXHAUM
70    WRITE(*,80)
80    FORMAT(10X,'DO YOU WANT TO ANALYZE ANOTHER KIND OF '/
     *      10X,'POLLUTANT CONTAMINATION (YES/NO) = ?')
      READ(*,90) IN
90    FORMAT(A4)
      IF(IN.EQ.IYES) GO TO 5
      STOP
      END
C
C     ****************************************************
C     ** THIS PROGRAM IS USED TO PRE-ESTIMATE THE DUST **
C     ** CONCENTRATION DISTRIBUTION IN A SINGLE DRIFT  **
C     ****************************************************
      SUBROUTINE DUST
      COMMON X(50),T(50),CONC(50,50)
C
C     READ AND PRINT THE BASIC PARAMETERS IN THE PROBLEM
      WRITE(*,10)
10    FORMAT(5X,'** PLEASE INPUT BASIC PARAMETERS IN THE PROBLEM **'
     *      //)
      WRITE(*,15)
15    FORMAT(5X,'THE RESPIRABLE DUST QUANTITY (IN MG/S) PRODUCED'/
     *      5X,'BY POLLUTION SOURCE = ?')
      READ(*,*) QD
      WRITE(*,20)
20    FORMAT(5X,'DARCY FRICTION COEFFICIENT FOR SMOOTH PIPE = ?')
      READ(*,*) YLAM
      WRITE(*,25)
25    FORMAT(5X,'DARCY FRICTION COEFFICIENT FOR DRIFT = ?')
      READ(*,*) XLAM
      WRITE(*,30)
30    FORMAT(5X,'THE AVERAGE AIR VELOCITY (M/S) IN DRIFT = ?')
      READ(*,*) V
      WRITE(*,35)
35    FORMAT(5X,'THE HEIGHT (IN M) OF DRIFT = ?')
      READ(*,*) HEIGHT
      WRITE(*,40)
40    FORMAT(5X,'THE WIDTH (IN M) OF DRIFT = ?')
      READ(*,*) WIDTH
```

```
      WRITE(*,45)
45    FORMAT(5X,'THE DECAY COEFFICIENT FOR RESPIRABLE DUST ALONG'/
     *      5X,'DRIFT = ?')
      READ(*,*) RANDA
      WRITE(*,50)
50    FORMAT(5X,'THE TIME (IN MINUTE) AFTER POLLUTION SOURCE PLAYS'/
     *      5X,'FUNCTION = ?')
      READ(*,*) NMIN
      WRITE(*,55)
55    FORMAT(5X,'THE NUMBER OF POINTS AT WHICH CONCENTRATIONS ARE'/
     *      5X,'TO BE CALCULATED = ?')
      READ(*,*) NX
      WRITE(*,60)
60    FORMAT(5X,'THE THRESHOLD LIMIT VALUE (IN MG/M**3) = ?')
      READ(*,*) TLV
      WRITE(6,65)
65    FORMAT(5X,55('*')/
     *      10X,'THE VALUES OF INPUT PARAMETERS ARE AS FOLLLOWS'/
     *      5X,55('*')//)
      WRITE(6,70) QD,YLAM,XLAM,V,HEIGHT,WIDTH,RANDA,NMIN,NX,TLV
70    FORMAT(5X,'THE RESPIRABLE DUST QUANTITY PRODUCED BY POLLUTION'/
     *      5X,'SOURCE            QD = ',F10.3,'(MG/S)'/
     *      5X,'DARCY FRICTION COEFFICIENT FOR SMOOTH PIPE'/
     *      5X,'                  YLAM = ',F10.7/
     *      5X,'DARCY FRICTION COEFFICIENT FOR DRIFT'/
     *      5X,'                  XLAM = ',F10.7/
     *      5X,'THE AVERAGE AIR VELOCITY IN DRIFT'/
     *      5X,'                  V = ',F10.4,'(M/S)'/
     *      5X,'THE HEIGHT OF DRIFT'/
     *      5X,'                  HEIGHT = ',F10.4,'(M)'/
     *      5X,'THE WIDTH OF DRIFT'/
     *      5X,'                  WIDTH = ',F10.4,'(M)'/
     *      5X,'THE DECAY COEFFICIENT FOR RESPIRABLE DUST'/
     *      5X,'ALONG DRIFT       RANDA = ',F10.7/
     *      5X,'THE TIME AFTER POLLUTION SOURCE PLAYS FUNCTION'/
     *      5X,'                  NMIN = ',I5,'(MIN)'/
     *      5X,'THE NUMBER OF POINTS AT WHICH CONCENTRATIONS ARE'/
     *      5X,'TO BE CALCULATED  NX = ',I5/
     *      5X,'THE THRESHOLD LIMIT VALUE'/
     *      5X,'                  TLV = ',F10.4,'(MG/M**3)'//)
C
C     DO CALCULATION
      RH=(HEIGHT*WIDTH)/(HEIGHT+WIDTH)
      AREA=HEIGHT*WIDTH
      PI=3.1416
      T(1)=60.0
      X(1)=10.0
C
C     CALCULATE THE DIFFUSION COEFFICIENT BY FORMULA (8)
      EX=28.8*SQRT(YLAM/XLAM)*V*RH
      DO 75 I=1,NMIN
      T(I+1)=T(I)+60.0
      DO 75 J=1,NX
      X(J+1)=X(J)+20.0
      C1=V*X(J)/(2.0*EX)
      C2=(V*V+4.0*EX*RANDA)/(4.0*EX)
      C3=X(J)*SQRT(C2/EX)
      C4=X(J)/(2.0*SQRT(EX*T(I)))-SQRT(C2*T(I))
      C5=X(J)/(2.0*SQRT(EX*T(I)))+SQRT(C2*T(I))
C
C     CALL COMPLEMENTARY ERROR FUNCTION TO CALCULATE FORMULA (2)
      CONC(I,J)=QD/(2.0*V*AREA)*EXP(C1)*(EXP(-C3)*CER(C4)
     *          +EXP(C3)*CER(C5))
75    CONTINUE
C
C     PRINT THE RESULTS OF CALCULATION
      WRITE(6,80)
80    FORMAT(5X,48('*')/
     *      10X,'THE CALCULATION RESULTS ARE AS FOLLOWS'/
     *      5X,48('*')//)
      WRITE(6,85)
85    FORMAT(5X,'TIME IN SECOND',3X,'DISTANCE IN METER',3X,
     *      'CONCENTRATION IN PPM'//
     *      10X,'T(I)',12X,'X(J)',12X,'CONC(I,J)',8X,'TLV'//)
      DO 95 I=1,NMIN
      DO 95 J=1,NX
      WRITE(6,90) T(I),X(J),CONC(I,J),TLV
90    FORMAT(5X,4(2X,F15.7))
95    CONTINUE
C
C     STORE THE RESULTS IN FILE "DATA1"
      WRITE(1,100) NX,NMIN
100   FORMAT(10X,'NX',10X,'NMIN'/10X,I5,8X,I5/
     *      10X,'T(I)',10X,'X(J)',10X,'CONC(I,J)',10X,'TLV')
      DO 110 I=1,NMIN
      DO 110 J=1,NX
      WRITE(1,105) T(I),X(J),CONC(I,J),TLV
105   FORMAT(5X,4(2X,F15.7))
110   CONTINUE
      RETURN
      END
C
C     **********************************************************
C     ** THIS PROGRAM IS USED TO CALCULATE THE COMPLEMENTARY  **
C     ** ERROR FUNCTION BY NEWTON-COTES APPROXIMATE METHOD    **
C     **********************************************************
      FUNCTION CER(Z)
```

```
c
       IF(Z.LT.0.0) GO TO 10
       GO TO 15
    5  A0=17280.0
       A1=751.0/A0*EXP(-Z*Z)
       A2=3577.0/A0*EXP(-(Z+(4.0-Z)/7.0)**2)
       A3=1323.0/A0*EXP(-(Z+2.0*(4.0-Z)/7.0)**2)
       A4=2989.0/A0*EXP(-(Z+3.0*(4.0-Z)/7.0)**2)
       A5=2989.0/A0*EXP(-(Z+4.0*(4.0-Z)/7.0)**2)
       A6=1323.0/A0*EXP(-(Z+5.0*(4.0-Z)/7.0)**2)
       A7=3577.0/A0*EXP(-(Z+6.0*(4.0-Z)/7.0)**2)
       A8=751.0/A0*EXP(-(Z+7.0*(4.0-Z)/7.0)**2)
       C0=2.0/SQRT(3.1416)*(4.0-Z)*(A1+A2+A3+A4+A5+A6+A7+A8)
       CER=2.0-C0
       GO TO 30
   10  Z=ABS(Z)
       IF(Z.GT.4.0) GO TO 20
       GO TO 5
   15  IF(Z.GT.4.0) GO TO 25
       A0=17280.0
       A1=751.0/A0*EXP(-Z*Z)
       A2=3577.0/A0*EXP(-(Z+(4.0-Z)/7.0)**2)
       A3=1323.0/A0*EXP(-(Z+2.0*(4.0-Z)/7.0)**2)
       A4=2989.0/A0*EXP(-(Z+3.0*(4.0-Z)/7.0)**2)
       A5=2989.0/A0*EXP(-(Z+4.0*(4.0-Z)/7.0)**2)
       A6=1323.0/A0*EXP(-(Z+5.0*(4.0-Z)/7.0)**2)
       A7=3577.0/A0*EXP(-(Z+6.0*(4.0-Z)/7.0)**2)
       A8=751.0/A0*EXP(-(Z+7.0*(4.0-Z)/7.0)**2)
       CER=2.0/SQRT(3.1416)*(4.0-Z)*(A1+A2+A3+A4+A5+A6+A7+A8)
       GO TO 30
   20  CER=2.0
       GO TO 30
   25  CER=0.0
   30  RETURN
       END
c
c
c
c   ***********************************************************
c   ** THIS PROGRAM IS USED TO PRE-ESTIMATE THE CONCENTRATION **
c   **    DISTRIBUTION OF DIESEL EXHAUST IN A SINGLE DRIFT     **
c   **              WHEN DIESEL ENGINE IS FIXED               **
c   ***********************************************************
       SUBROUTINE EXHAUF
       COMMON X(50),T(50),CONC(50,50)
c
c      READ AND PRINT THE BASIC PARAMETERS IN THE PROBLEM
       WRITE(*,5)
    5  FORMAT(5X,'** PLEASE INPUT BASIC PARAMETERS IN THE PROBLEM **'
      *        //)
       WRITE(*,10)
   10  FORMAT(5X,'THE CONCENTRATION (IN PPM) OF SPECIES I IN THE'/
      *        5X,'DIESEL EXHAUST OR THE COMBINED CONCENTRATION OF'/
      *        5X,'ALL GASES = ?')
       READ(*,*) CI
       WRITE(*,15)
   15  FORMAT(5X,'THE DIESEL EXHAUST QUANTITY (IN M**3/S) PRODUCED'/
      *        5X,'BY DIESEL ENGINE = ?')
       READ(*,*) QE
       WRITE(*,20)
   20  FORMAT(5X,'DARCY FRICTION COEFFICIENT FOR SMOOTH PIPE = ?')
       READ(*,*) YLAM
       WRITE(*,25)
   25  FORMAT(5X,'DARCY FRICTION COEFFICIENT FOR DRIFT = ?')
       READ(*,*) XLAM
       WRITE(*,30)
   30  FORMAT(5X,'THE AVERAGE AIR VELOCITY (IN M/S) IN DRIFT = ?')
       READ(*,*) V
       WRITE(*,35)
   35  FORMAT(5X,'THE HEIGHT (IN M) OF DRIFT = ?')
       READ(*,*) HEIGHT
       WRITE(*,40)
   40  FORMAT(5X,'THE WIDTH (IN M) OF DRIFT = ?')
       READ(*,*) WIDTH
       WRITE(*,45)
   45  FORMAT(5X,'THE TIME (IN MINUTE) AFTER ENGINE OPERATES = ?')
       READ(*,*) NMIN
       WRITE(*,50)
   50  FORMAT(5X,'THE NUMBER OF POINTS AT WHICH CONCENTRATIONS ARE'/
      *        5X,'TO BE CALCULATED = ?')
       READ(*,*) NX
       WRITE(*,55)
   55  FORMAT(5X,'THE THRESHOLD LIMIT VALUE (IN PPM) = ?')
       READ(*,*) TLV
       WRITE(6,60)
   60  FORMAT(5X,55('*')/
      *        10X,'THE VALUES OF INPUT PARAMETERS ARE AS FOLLOWS'/
      *        5X,55('*')///)
       WRITE(6,65) CI,QE,YLAM,XLAM,V,HEIGHT,WIDTH,NMIN,NX,TLV
   65  FORMAT(5X,'THE CONCENTRATION OF DIESEL EXHAUST'/
      *        5X,'                       CI = ',F10.3,'(PPM)'/
      *        5X,'THE DIESEL EXHAUST QUANTITY PRODUCED BY ENGINE'/
      *        5X,'                       QE = ',F10.7,'(M**3/S)'/
      *        5X,'DARCY FRICTION COEFFICIENT FOR SMOOTH PIPE'/
      *        5X,'                     YLAM = ',F10.7/
      *        5X,'DARCY FRICTION COEFFICIENT FOR DRIFT'/
      *        5X,'                     XLAM = ',F10.7/
      *        5X,'THE AVERAGE AIR VELOCITY IN DRIFT'/
      *        5X,'                        V = ',F10.3,'(M/S)'/
```

```
      *        5X,'THE HEIGHT OF DIRFT'/
      *        5X,'                   HEIGHT = ',F10.3,'(M)'/
      *        5X,'THE WIDTH OF DRIFT'/
      *        5X,'                    WIDTH = ',F10.3,'(M)'/
      *        5X,'THE TIME AFTER ENGINE OPERATES'/
      *        5X,'                     NMIN = ',I5,'(MIN)'/
      *        5X,'THE NUMBER OF POINTS AT WHICH CONCENTRATIONS ARE'/
      *        5X,'TO BE CALCULATED       NX = ',I5/
      *        5X,'THE THRESHOLD LIMIT VALUE'/
      *        5X,'                      TLV = ',F10.5,'(PPM)'//)
c
c      DO CALCULATION
       RH=(HEIGHT*WIDTH)/(HEIGHT+WIDTH)
       AREA=HEIGHT*WIDTH
       PI=3.1416
       T(1)=60.0
       X(1)=10.0
c
c      CALCULATE THE DIFFUSION COEFFICIENT BY FORMULA (8)
       EX=28.8*SQRT(YLAM/XLAM)*V*RH
       DO 70 I=1,NMIN
       T(I+1)=T(I)+60.0
       DO 70 J=1,NX
       X(J+1)=X(J)+20.0
       C1=V*X(J)/(2.0*EX)
       C2=(V*V)/(4.0*EX)
       C3=X(J)*SQRT(C2/EX)
       C4=X(J)/(2.0*SQRT(EX*T(I)))-SQRT(C2*T(I))
       C5=X(J)/(2.0*SQRT(EX*T(I)))+SQRT(C2*T(I))
c
c      CALL COMPLEMENTARY ERROR FUNCTION TO CALCULATE FORMULA (5)
       CONC(I,J)=QE*CI/(2.0*V*AREA)*EXP(C1)*(EXP(-C3)*CER(C4)+
      *            EXP(C3)*CER(C5))
   70  CONTINUE
c
c      PRINT THE RESULTS OF CALCULATION
       WRITE(6,75)
   75  FORMAT(5X,48('*')/
      *        10X,'THE CALCULATION RESULTS ARE AS FOLLOWS'/
      *        5X,48('*')//)
       WRITE(6,80)
   80  FORMAT(5X,'TIME IN SECOND',3X,'DISTANCE IN METER',3X,
      *        'CONCENTRATION IN PPM'//
      *        10X,'T(I)',12X,'X(J)',12X,'CONC(I,J)',8X,'TLV'//)
       DO 90 I=1,NMIN
       DO 90 J=1,NX
       WRITE(6,85) T(I),X(J),CONC(I,J),TLV
   85  FORMAT(5X,4(2X,F15.7))
   90  CONTINUE
c
c      STORE THE RESULTS IN FILE "DATA4"
       WRITE(4,95) NX,NMIN
   95  FORMAT(10X,'NX',10X,'NMIN'/10X,I5,8X,I5/
      *        10X,'T(I)',10X,'X(J)',10X,'CONC(I,J)',10X,'TLV')
       DO 105 I=1,NMIN
       DO 105 J=1,NX
       WRITE(4,100) T(I),X(J),CONC(I,J),TLV
  100  FORMAT(5X,4(2X,F15.7))
  105  CONTINUE
       RETURN
       END
c
c
c   ***********************************************************
c   ** THIS PROGRAM IS USED TO CALCULATE THE CONCENTRATION   **
c   ** DISTRIBUTION OF BLAST FUME IN A DRIFT AFTER BLAST     **
c   **    IN THE CASE WITH RECIRCULATION OF BLAST GASES      **
c   ***********************************************************
       SUBROUTINE FUMERC
       COMMON X(50),T(50),CONC(50,50)
c
c      READ AND PRINT THE BASIC PARAMETERS IN THE PROBLEM
       WRITE(*,10)
   10  FORMAT(5X,'** PLEASE INPUT BASIC PARAMETERS IN THE PROBLEM **'
      *        //)
       WRITE(*,15)
   15  FORMAT(5X,'THE INITIAL BLAST FUME CONCENTRATION (IN PPM)'/
      *        5X,'AFTER BLASTING = ?')
       READ(*,*) CI
       WRITE(*,17)
   17  FORMAT(5X,'THE LENGTH (IN M) OF DRIFT WHICH IS FULL WITH'/
      *        5X,'BLAST FUME AFTER BLASTING = ?')
       READ(*,*) FL
       WRITE(*,20)
   20  FORMAT(5X,'DARCY FRICTION COEFFICIENT FOR SMOOTH PIPE = ?')
       READ(*,*) YLAM
       WRITE(*,25)
   25  FORMAT(5X,'DARCY FRICTION COEFFICIENT FOR DRIFT = ?')
       READ(*,*) XLAM
       WRITE(*,30)
   30  FORMAT(5X,'THE AVERAGE AIR VELOCITY (IN M/S) IN DRIFT = ?')
       READ(*,*) V
       WRITE(*,35)
   35  FORMAT(5X,'THE HEIGHT (IN M) OF DRIFT = ?')
       READ(*,*) HEIGHT
       WRITE(*,40)
   40  FORMAT(5X,'THE WIDTH (IN M) OF DRIFT = ?')
       READ(*,*) WIDTH
```

```
        WRITE(*,45)
45      FORMAT(5X,'THE FACTOR OF RECIRCULATION OF BLAST GASES = ?')
        READ(*,*) FN
        WRITE(*,50)
50      FORMAT(5X,'THE TIME (IN MINUTE) AFTER BLASTING = ?')
        READ(*,*) NMIN
        WRITE(*,55)
55      FORMAT(5X,'THE NUMBER OF POINTS AT WHICH CONCENTRATIONS ARE'/
     *         5X,'TO BE CALCULATED = ?')
        READ(*,*) NX
        WRITE(*,60)
60      FORMAT(5X,'THE THRESHOLD LIMIT VALUE (IN PPM) = ?')
        READ(*,*) TLV
        WRITE(6,65)
65      FORMAT(5X,55('*')/
     *         10X,'THE VALUES OF INPUT PARAMETERS ARE AS FOLLOWS'/
     *         5X,55('*')//)
        WRITE(6,70) YLAM,XLAM,V,HEIGHT,WIDTH,FN,NMIN
70      FORMAT(5X,'THE INITIAL BLAST FUME CONCENTRATION AFTER BLAST'/
     *         5X,'                                   CI = ',F10.2,'(PPM)'/
     *         5X,'THE LENGTH OF DRIFT WHICH IS FULL WITH BLAST'/
     *         5X,'FUME AFTER BLASTING       FL = ',F10.3,'(M)'/
     *         5X,'DARCY FRICTION COEFFICIENT FOR SMOOTH PIPE'/
     *         5X,'                             YLAM = ',F10.7/
     *         5X,'DARCY FRICTION COEFFICIENT FOR DRIFT'/
     *         5X,'                             XLAM = ',F10.7/
     *         5X,'THE AVERAGE AIR VELOCITY IN DRIFT'/
     *         5X,'                                V = ',F10.3,'(M/S)'/
     *         5X,'THE HEIGHT OF DRIFT'/
     *         5X,'                           HEIGHT = ',F10.3,'(M)'/
     *         5X,'THE WIDTH OF DRIFT'/
     *         5X,'                            WIDTH = ',F10.3,'(M)'/
     *         5X,'THE FACTOR OF RECIRCULATION OF BLAST GASES'/
     *         5X,'                               FN = ',F10.5/
     *         5X,'THE TIME AFTER BLASTING'/
     *         5X,'                             NMIN = ',I5,'(MIN)')
        WRITE(6,72) NX,TLV
72      FORMAT(5X,'THE NUMBER OF POINTS AT WHICH CONCENTRATIONS'/
     *         5X,'ARE TO BE CALCULATED     NX = ',I5/
     *         5X,'THE THRESHOLD LIMIT VALUE'/
     *         5X,'                            TLV = ',F10.3,'(PPM)'//)
C
C       DO CALCULATION
        RH=HEIGHT*WIDTH/(HEIGHT+WIDTH)
        AREA=HEIGHT*WIDTH
        PI=3.1416
        T(1)=60.0
        X(1)=0.0
C
C       CALCULATE THE DIFFUSION COEFFICIENT BY FORMULA (8)
        EX=28.8*SQRT(YLAM/XLAM)*V*RH
        DO 80 I=1,NMIN
        T(I+1)=T(I)+60.0
        DO 80 J=1,NX
        X(J+1)=X(J)+5.0
        C1=V*T(I)
        IF(C1.LT.X(J)) GO TO 75
C
C       CALCULATE THE BLAST FUME CONCENTRATION BY FORMULA (4)
        CONC(I,J)=CI*FL*AREA/(T(I)**FN+1.0)*EXP(-V/EX*X(J))
        GO TO 80
75      CONC(I,J)=0.0
80      CONTINUE
C
C       PRINT THE RESULTS OF CALCULATION
        WRITE(6,85)
85      FORMAT(5X,48('*')/
     *         10X,'THE CALCULATION RESULTS ARE AS FOLLOWS'/
     *         5X,48('*')//)
        WRITE(6,90)
90      FORMAT(5X,'TIME IN SECOND',3X,'DISTANCE IN METER',3X,
     *         'CONCENTRATION IN PPM'//
     *         10X,'T(I)',12X,'X(J)',12X,'CONC(I,J)',8X,'TLV'//)
        DO 100 I=1,NMIN
        DO 100 J=1,NX
        WRITE(6,95) T(I),X(J),CONC(I,J),TLV
95      FORMAT(5X,4(2X,F15.7))
100     CONTINUE
C
C       STORE THE RESULTS IN FILE "DATA2"
        WRITE(2,105) NX,NMIN
105     FORMAT(10X,'NX',10X,'NMIN'/10X,I5,8X,I5/
     *         10X,'T(I)',10X,'X(J)',10X,'CONC(I,J)',10X,'TLV')
        DO 115 I=1,NMIN
        DO 115 J=1,NX
        WRITE(2,110) T(I),X(J),CONC(I,J),TLV
110     FORMAT(5X,4(2X,F15.7))
115     CONTINUE
        RETURN
        END
C
C
C       *****************************************************
C       ** THIS PROGRAM IS USED TO CALCULATE THE CONCENTRATION **
C       ** DISTRIBUTION OF BLAST FUME IN A DRIFT AFTER BLAST **
C       ** IN THE CASE WITHOUT RECIRCULATION OF BLAST GAESE   **
C       *****************************************************
        SUBROUTINE FUMENRE
        COMMON X(50),T(50),CONC(50,50)
```

```
C
C       READ AND PRINT THE BASIC PARAMETERS IN THE PROBLEM
        WRITE(*,10)
10      FORMAT(5X,'** PLEASE INPUT BASIC PARAMETERS IN THE PROBLEM **'
     *         //)
        WRITE(*,15)
15      FORMAT(5X,'THE INITIAL BLAST FUME CONCENTRATION (IN PPM)'/
     *         5X,'AFTER BLASTING = ?')
        READ(*,*) CI
        WRITE(*,20)
20      FORMAT(5X,'THE LENGTH (IN M) OF DRIFT WHICH IS FULL WITH'/
     *         5X,'BLAST FUME AFTER BLASTING = ?')
        READ(*,*) FL
        WRITE(*,25)
25      FORMAT(5X,'DARCY FRICTION COEFFICIENT FOR SMOOTH PIPE = ?')
        READ(*,*) YLAM
        WRITE(*,30)
30      FORMAT(5X,'DARCY FRICTION COEFFICIENT FOR DRIFT = ?')
        READ(*,*) XLAM
        WRITE(*,35)
35      FORMAT(5X,'THE AVERAGE AIR VELOCITY (IN M/S) IN DRIFT = ?')
        READ(*,*) V
        WRITE(*,40)
40      FORMAT(5X,'THE HEIGHT (IN M) OF DRIFT = ?')
        READ(*,*) HEIGHT
        WRITE(*,45)
45      FORMAT(5X,'THE WIDTH (IN M) OF DRIFT = ?')
        READ(*,*) WIDTH
        WRITE(*,50)
50      FORMAT(5X,'THE TIME (IN MINUTE) AFTER BLASTING = ?')
        READ(*,*) NMIN
        WRITE(*,55)
55      FORMAT(5X,'THE NUMBER OF POINTS AT WHICH CONCENTRATIONS ARE'/
     *         5X,'TO BE CALCULATED = ?')
        READ(*,*) NX
        WRITE(*,60)
60      FORMAT(5X,'THE THRESHOLD LIMIT VALUE (IN PPM) = ?')
        READ(*,*) TLV
        WRITE(6,65)
65      FORMAT(5X,55('*')/
     *         10X,'THE VALUES OF INPUT PARAMETERS ARE AS FOLLOWS'/
     *         5X,55('*')//)
        WRITE(6,70) CI,FL,YLAM,XLAM,V,HEIGHT,WIDTH,NMIN,NX,TLV
70      FORMAT(5X,'THE INITIAL BLAST FUME CONCENTRATION AFTER'/
     *         5X,'BLASTING                  CI = ',F10.3,'(PPM)'/
     *         5X,'THE LENGTH OF DRIFT WHICH IS FULL WITH BLAST'/
     *         5X,'FUME AFTER BLASTING       FL = ',F10.3,'(M)'/
     *         5X,'DARCY FRICTION COEFFICIENT FOR SMOOTH PIPE'/
     *         5X,'                             YLAM = ',F10.7/
     *         5X,'DARCY FRICTION COEFFICIENT FOR DRIFT'/
     *         5X,'                             XLAM = ',F10.7/
     *         5X,'THE AVERAGE AIR VELOCITY IN DRIFT'/
     *         5X,'                                V = ',F10.3,'(M/S)'/
     *         5X,'THE HEIGHT OF DRIFT'/
     *         5X,'                           HEIGHT = ',F10.3,'(M)'/
     *         5X,'THE WIDTH OF DRIFT'/
     *         5X,'                            WIDTH = ',F10.3,'(M)'/
     *         5X,'THE TIME AFTER BLASTINF'/
     *         5X,'                             NMIN = ',I5,'(MTN)'/
     *         5X,'THE NUMBER OF POINTS AT WHICH CONCENTRATION ARE'/
     *         5X,'TO BE CALCULATED          NX = ',I5/
     *         5X,'THE THRESHOLD LIMIT VALUE'/
     *         5X,'                            TLV = ',F10.4,'(PPM)'//)
C
C       DO CALCULATION
        RH=WIDTH*HEIGHT/(HEIGHT+WIDTH)
        AREA=HEIGHT*WIDTH
        PI=3.1416
        T(1)=60.0
        X(1)=0.0
C
C       CALCULATE THE DIFFUSION COEFFICIENT BY FORMULA (8)
        EX=28.8*SQRT(YLAM/XLAM)*V*RH
        DO 75 I=1,NMIN
        T(I+1)=T(I)+60.0
        DO 75 J=1,NX
        X(J+1)=X(J)+5.0
        C1=FL*CI/(2.0*SQRT(PI*EX*T(I)))
        C2=-((X(J)-V*T(I))**2)/(4.0*EX*T(I))
C
C       CALCULATE THE BLAST FUME CONCENTRATION BY FORMULA (3)
        CONC(I,J)=C1*EXP(C2)
75      CONTINUE
C
C       PRINT THE RESULTS OF CALCULATION
        WRITE(6,80)
80      FORMAT(5X,48('*')/
     *         10X,'THE CALCULATION RESULTS ARE AS FOLLOWS'/
     *         5X,48('*')//)
        WRITE(6,85)
85      FORMAT(5X,'TIME IN SECOND',3X,'DISTANCE IN METER',3X,
     *         'CONCENTRATION IN PPM'//
     *         10X,'T(I)',12X,'X(J)',12X,'CONC(I,J)',8X,'TLV'//)
        DO 95 I=1,NMIN
        DO 95 J=1,NX
        WRITE(6,90) T(I),X(J),CONC(I,J),TLV
90      FORMAT(5X,4(2X,F15.7))
95      CONTINUE
C
```

```
C     STORE THE RESULTS IN FILE "DATA3"
      WRITE(3,100) NX,NMIN
100   FORMAT(10X,'NX',10X,'NMIN'/10X,I5,8X,I5/
     *       10X,'T(I)',10X,'X(J)',10X,'CONC(I,J)',10X,'TLV')
      DO 110 I=1,NMIN
      DO 110 J=1,NX
      WRITE(3,105) T(I),X(J),CONC(I,J),TLV
105   FORMAT(5X,4(2X,F15.7))
110   CONTINUE
      RETURN
      END
C
C
C
C     ************************************************************
C     ** THIS PROGRAM IS USED TO PRE-ESTIMATE THE CONCENTRATION **
C     ** DISTRIBUTION OF DIESEL EXHAUST IN A DRIFT WHEN DIESEL   **
C     **   ENGINE IS MOVING. IT CAME FROM STEFANKO, R. (1974).   **
C     **         SOME IMPROVEMENTS HAVE BEEN DONE                **
C     ************************************************************
      SUBROUTINE EXHAUMO
      COMMON X(50),T(50),CONC(50,50)
C
C     READ AND PRINT THE BASIC PARAMETERS IN THE PROBLEM
      WRITE(*,10)
10    FORMAT(5X,'**PLEASE INPUT BASIC PARAMETERS IN THE PROBLEM**'
     *       //)
      WRITE(*,15)
15    FORMAT(5X,'THE CONCENTRATION (IN PPM) OF SPECIES I IN DIESEL'/
     *       5X,'EXHAUST OR THE COMBINED CONCENTRATION OF ALL GASES'/
     *       5X,'IN THE EXHAUST = ?')
      READ(*,*) CI
      WRITE(*,20)
20    FORMAT(5X,'THE DIESEL EXHAUST QUANTITY (IN M**3/S) PRODUCED'/
     *       5X,'BY ENGINE = ?')
      READ(*,*) QE
      WRITE(*,25)
25    FORMAT(5X,'DARCY FRICTION COEFFICIENT FOR SMOOTH PIPE = ?')
      READ(*,*) YLAM
      WRITE(*,30)
30    FORMAT(5X,'DARCY FRICTION COEFFICIENT FOR DRIFT = ?')
      READ(*,*) XLAM
      WRITE(*,35)
35    FORMAT(5X,'THE AVERAGE AIR VELOCITY (IN M/S) IN DRIFT = ?')
      READ(*,*) U
      WRITE(*,40)
40    FORMAT(5X,'THE VELOCITY (IN M/S) OF DIESEL ENGINE = ?')
      READ(*,*) V
      WRITE(*,45)
45    FORMAT(5X,'THE HEIGHT (IN M) OF DRIFT = ?')
      READ(*,*) HEIGHT
      WRITE(*,50)
50    FORMAT(5X,'THE WIDTH (IN M) OF DRIFT = ?')
      READ(*,*) WIDTH
      WRITE(*,55)
55    FORMAT(5X,'THE TIME (IN MINUTE) AFTER ENGINE OPERATES = ?')
      READ(*,*) NMIN
      WRITE(*,60)
60    FORMAT(5X,'THE NUMBER OF POINTS AT WHICH CONCENTRATIONS ARE'/
     *       5X,'TO BE CALCULATED = ?')
      READ(*,*) NX
      WRITE(*,65)
65    FORMAT(5X,'THE THRESHOLD LIMIT VALUE (IN PPM) = ?')
      READ(*,*) TLV
      WRITE(6,70)
70    FORMAT(5X,55('*')/
     *       10X,'THE VALUES OF INPUT PARAMETERS ARE AS FOLLOWS'/
     *       5X,55('*')///)
      WRITE(6,75) CI,QE,YLAM,XLAM,U,V,HEIGHT,WIDTH
75    FORMAT(5X,'THE CONCENTRATION OF DIESEL EXHAUST'/
     *       5X,'                CI = ',F10.3,'(PPM)'/
     *       5X,'THE DIESEL EXHAUST QUANTITY PRODUCED BY ENGINE'/
     *       5X,'                QE = ',F10.7,'(M**3/S)'/
     *       5X,'DARCY FRICTION COEFFICIENT FOR SMOOTH PIPE'/
     *       5X,'              YLAM = ',F10.7/
     *       5X,'DARCY FRICTION COEFFICIENT FOR DRIFT'/
     *       5X,'              XLAM = ',F10.7/
     *       5X,'THE AVERAGE AIR VELOCITY IN DRIFT'/
     *       5X,'                 U = ',F10.3,'(M/S)'/
     *       5X,'THE VELOCITY OF DIESEL ENGINE'/
     *       5X,'                 V = ',F10.3,'(M/S)'/
     *       5X,'THE HEIGHT OF DRIFT'/
     *       5X,'            HEIGHT = ',F10.3,'(M)'/
     *       5X,'THE WIDTH OF DRIFT'/
     *       5X,'             WIDTH = ',F10.3,'(M)')
      WRITE(6,77) NMIN,NX,TLV
77    FORMAT(5X,'THE TIME AFTER ENGINE OPERATES'/
     *       5X,'              NMIN = ',I5,'(MIN)'/
     *       5X,'THE NUMBER OF POINTS AT WHICH CONCENTRATIONS ARE'/
     *       5X,'TO BE CALCULATED     NX = ',I5/
     *       5X,'THE THRESHOLD LIMIT VALUE'/
     *       5X,'               TLV = ',F10.3,'(PPM)'//)
C
C     DO CALCULATION
      RH=HEIGHT*WIDTH/(HEIGHT+WIDTH)
      AREA=HEIGHT*WIDTH
      PI=3.1416
      T(1)=60.0
      X(1)=0.0
      VR=V-U
C
C     CHECK WHETHER TO USE FORMULA (6) OR (7)
      IF(VR) 80,125,85
80    VR=-VR
C
C     CALCULATE THE DIFFUSION COEFFICIENT BY FORMULA (8)
      EX=28.8*SQRT(YLAM/XLAM)*U*RH
      GO TO 90
85    EX=28.8*SQRT(YLAM/XLAM)*V*RH
      GO TO 90
90    DO 95 I=1,NMIN
      T(I+1)=T(I)+60.0
      DO 95 J=1,NX
      X(J+1)=X(J)+10.0
      C1=(X(J)-VR*T(I))/(2.0*SQRT(EX*T(I)))
      C2=(X(J)+VR*T(I))/(2.0*SQRT(EX*T(I)))
      C3=VR*X(J)/EX
C
C     CALL COMPLEMENTARY ERROR FUNCTION AND CALCULATE FORMULA (6)
      A=CER(C1)
      B=(1.0+(VR*X(J)/EX)+(VR*VR*T(I)/EX))*EXP(C3)*CER(C2)
      C=2.0*VR*SQRT(T(I)/(EX*PI))
      D=EXP(-(C1*C1))
      F=C*D
      CONC(I,J)=(CI*QE/(2.0*AREA*VR))*(A-B+F)
95    CONTINUE
C
C     PRINT THE RESULTS OF CALCULATION
97    WRITE(6,100)
100   FORMAT(5X,48('*')/
     *       10X,'THE CALCULATION RESULTS ARE AS FOLLOWS'/
     *       5X,48('*')///)
      WRITE(6,105)
105   FORMAT(5X,'TIME IN SECOND',3X,'DISTANCE IN METER',3X,
     *       'CONCENTRATION IN PPM'//)
      WRITE(6,110)
110   FORMAT(10X,'T(I)',12X,'X(J)',12X,'CONC(I,J)',8X,'TLV'//)
      DO 115 I=1,NMIN
      DO 115 J=1,NX
115   WRITE(6,120) T(I),X(J),CONC(I,J),TLV
120   FORMAT(5X,4(2X,F10.7))
C
C     STORE THE RESULTS IN FILE "DATA5"
      WRITE(5,140) NX,NMIN
140   FORMAT(10X,'NX',10X,'NMIN'/10X,I5,8X,I5/
     *       10X,'T(I)',10X,'X(J)',10X,'CONC(I,J)',10X,'TLV')
      DO 150 I=1,NMIN
      DO 150 J=1,NX
      WRITE(5,145) T(I),X(J),CONC(I,J),TLV
145   FORMAT(5X,4(2X,F15.7))
150   CONTINUE
      GO TO 135
125   VR=QE/(2.0*AREA)
      CI=CI/2.0
      EX=28.8*U*RH*SQRT(YLAM/XLAM)
      DO 130 I=1,NMIN
      T(I+1)=T(I)+60.0
      DO 130 J=1,NX
      X(J+1)=X(J)+10.0
      C1=(X(J)-VR*T(I))/(2.0*SQRT(EX*T(I)))
      C2=(X(J)+VR*T(I))/(2.0*SQRT(EX*T(I)))
      C3=VR*X(J)/EX
C
C     CALL COMPLEMENTARY ERROR FUNCTION AND CALCULATE FORMULA (7)
      A=CER(C1)
      B=(1.0+C3+(VR*VR*T(I)/EX))*EXP(C3)*CER(C2)
      C=2.0*VR*SQRT(T(I)/(EX*PI))
      D=EXP(-C1*C1)
      F=C*D
      CONC(I,J)=CI*(A-B+F)
130   CONTINUE
      GO TO 97
135   RETURN
      END
```

Computer Applications in the Mineral Industry, Fytas, Collins & Singhal (eds)
© 1988 Balkema, Rotterdam. ISBN 90 6191 760 3

Study on the computational method for the optimization of ventilation network

Lu Songtang
Beijing Graduate School, China Institute of Mining and Technology

ABSTRACT: The law of power conservation of ventilation network is put forward in this paper. With this law the property and character of calculation for air-flow rate distribution depending on needs can be better understood. The computational method and program for the determination of the optimal location of resistance increase branch and resistance decrease branch, and optimal location of auxiliary fan in the air-flow regulation are studied. A computational example demonstrates that the optimal design should produce notable profit.

1 INTRODUCTION

For a long time, the ventilation network calculations have been carried out with the aim of the natural splitting of air-flow or air-flow distribution depending on needs by the resistance regulation in the fixed quantity branches. Only a set of input data is needed for such a calculation and then the corresponding result can be obtained. Such a calculation has a unique solution. However, the more important problem is the optimal design of the ventilation network, in which the air-flow rate in the fixed quantity branch (or working area) is given, and the needed resistance factor of the regulating branch and operating points of the main fan are unknown. Such a problem has infinite solution. Not all of the solutions are the best. The aim of optimal design of ventilation network is to find the most rational in technology and most economical solution. For a rational solution, it should be simple, safe and workable in mining. It is obvious that the various air-flow regulations should meet the above basic demands. To say that the solution is economical means the sum of expenses on all engineering facilities and the consumption of ventilation power is minimum.

The optimal design of ventilation network has attracted the attention of researchers. Scientists of the Soviet Union and China have studied the problem of the optimization of ventilation network by linear programming method. However, with this method air-flow rates of all branches is fixed, and so the application scope is strongly confined. On the other hand, some researchers attempted to solve the problem by the non-linear programming computational method, such as the steepest descent method and so on. But, due to the complexity and specialty of the optimization of the ventilation network, these methods are not effective. Therefore, the optimal calculation of ventilation network has not evolved a completed theory or effective algorithm. Further study in this area is necessary.

2 STANDARD STRUCTURE OF VENTILATION NETWORK WITH AIR-FLOW RATE DISTRIBUTION DEPENDING ON NEEDS

2.1 A cut set composed of all fixed quantity branches separates the network into two parts

The design of the ventilation system must obey the principle that the air-flow from entrance to exit must be used and used only once. So, in the real mine the fixed quantity branches should form a cut set and separate the network into two independent subnetworks. One is connected with fan nodes (called fan subnetwork). The other is connected to the nodes of atmosphere (called atmosphere subnetwork). The above mentioned network is shown in Figure 1. Air-flow rate through the fan in Figure 1 is the sum of all air-flow rates of fixed quantity branches and does not change when the fan pressure changes. The structure of the ventilation network shown in Figure 1 is the standard structure of ventilation networks with air-flow rate distribution depending on needs.

If the network of the ventilation system does not meet the above demands such as the air-flow rates of one or two branches are free, then we may close these branches, or treat them as branches with constant air-flow rates. Therefore the air-flow rate through the fan will reduce and the ventilation power consumed by the fan will decrease. The standard structure of ventilation networks with air-flow rate distribution depending on needs is not only reasonable, but also optimal. The first step of ventilation network optimization is the standard treatment of the ventilation network.

2.2 A complicated ventilation network consisting of a number of fans may be divided into several subnetworks with standard structures

Generally more than one fan is at work in the real mine ventilation system, such as Figure 2. No branches are allowed to connect subnetwork 1 with subnetwork 2 in Figure 2. If there is a branch connecting subnetwork 1 with subnetwork 2 in some special cases, the branch may be considered as a fixed quantity branch. Therefore, any complicated ventilation network with a number of fans can be divided into several subnetworks connecting with fans or atmosphere, as shown in Figure 1.

2.3 The calculation of ventilation network with standard structures must be carried out with a given total power.

Apart from the air-flow rates of all fixed quantity branches (the sum of which is total air-flow rate), the fan pressure is demanded in order to carry out the network calculation. In Figure 1, if only the air-flow rates of all fixed quantity branches are given and the fan pressures are not known, then the nodal pressure cannot be calculated. The needed resistance factors of fixed quantity branches (regulated branches) also cannot be calculated. Total fan pressure should be given in order to calculate the needed resistance factors of fixed quantity branches. The input data of some ventilation network computational programs need only total air-flow rate rather than total pressure. In the

calculation of standard structure ventilation networks by these programs, one fixed quantity branch must be free for allocation of air-flow rate. The reason is that in these programs the total pressure is equivalent to the sum of the pressure drop of the air-flow path from entrance to exit through the branch, in which the air-flow rate is free. Obviously these programs are not perfect.

3 THE LAW OF POWER CONSERVATION OF THE VENTILATION NETWORK WITH AIR-FLOW RATE DISTRIBUTION DEPENDING ON NEEDS

This law is as follows: In the standard ventilation network with air-flow rate distribution depending on needs, the absorbed power of the fan subnetwork and atmosphere subnetwork does not change if in the network only the main fan pressure changes.

This law can be described in the following form. In the standard ventilation network with air-flow rate distribution depending on needs, the air-flow rate of any branch in the fan subnetwork and atmosephre subnetwork does not change, despite the change of main fan pressure, if in the network only main fan pressure changes. Meanwhile the pressure of nodes in the fan subnetwork also changes with the same value ΔHf and the pressure of nodes in the atmosphere subnetwork does not change. The increase of the main fan power due to the increase ΔHf of the main fan pressure is absorbed completely by the fixed quantity branches in which the regulator is located.

In the following we give a brief proof of the law. Suppose there are N branches in the subnetwork which are connected with fans. Hn and Qn are given as:

$$Hn = (Hn_1, Hn_2, \ldots \ldots Hn_N)^T$$

$$Qn = (Qn_1, Qn_2, \ldots \ldots Qn_N)^T$$

Here Hnk is the pressure drop and Qnk is the air-flow rate through the n_k-th branch.

According to Kirshhoff's Law for electric current, one may obtain:

$$AQn = 0 \qquad (1)$$

$$BRnQnQn = 0 \qquad (2)$$

Here A is an augmented incidence matrix, B is an augmented mesh matrix and Rn is a resistance matrix. Then M simultaneous equations are obtained with solutions for the air-flow rate of N branches.

Due to the topological structure of the network, all resistance parameters do not change. Therefore, the matrices A, B and R do not change. The only changed quantity is the fan pressure. But the air-flow rate through the fan remains as a constant which is the sum of air-flow rates through all fixed quantity branches. In the equations AQn = 0 we move all of the given air-flow rate items to the right side and the constant items do not change. So the change of the fan pressure has no effect on equations (1) and (2), and the air-flow rates Qn of the branches do not change with the change of fan pressure. For the equation Hn = RnQnQn, since Qn and Rn do not change, then the pressure drop of every branch also does not change. In the condition that the fan pressure changes, each nodal pressure should also change to maintain the pressure drop of every branch unchanged.

The change of fan pressure has no effect on the atmosphere subnetwork. So the nodal pressure, air-flow rate of the branches and the pressure drop of the branches of this atmosphere subnetwork do not change with the change of fan pressure.

Tellegan's Law says that in a network the power from the active elements is absorbed completely by the passive elements in any instant. Applying this law to the ventilation network, one may explain that the power from the main fan is absorbed wholly by the resistances in the network. When the fan pressure changes, the air-flow rates and pressure drop of branches in the fan subnetwork and atmosphere subnetwork do not change. Therefore, the absorbed power of the two subnetworks also does not change. The increase in the main fan power arising from the increase in main fan pressure is absorbed completely by the fixed quantity branches, in which the artificial resistance is regulated.

4 A NEW ALGORITHM FOR PRESSURE DROP CALCULATION OF DIFFERENT ROUTES AND THE OPTIMAL AIM OF RESISTANCE INCREASE REGULATION

4.1 New algorithm for pressure drop calculation of different paths

The air-flow from the entrance passes through the fan subnetwork, fixed quantity branches, and atmosphere subnetwork, and finally flows out through the exit. The route through which the air-flow passes is called the path, in which all nodes are distinct and the final node of one branch is the initial node of the next branch.

The old method of calculating the total pressure drop (or resistance) starts by solving the air-flow rate and pressure drop of the branches in the network, and then sums up all of pressure drop of branches in the path. This method is trivial. The new algorithm for the total pressure drop calculation of the path introduced in the following is simpler and more effective than the previous ones.

Taking Figure 3 as an example we calculate the pressure drop of the path through the fixed quantity branch i-j. It is not necessary to know with which branches the route is made up in the path. The total pressure drop of the path should be the sum of the pressure drop between fan node and i-th node, the pressure drop between the i-th node and j-th node, and the pressure drop between the j-th node and atmosphere node. The formula for the calculation of the resistance (or total pressure drop) of the path is given as follows:

$$h_{ij} = Hf - Hi + R_{ij}Q_{ij}^2 + Hj \qquad (3)$$

Here h_{ij} is the total pressure drop of the path through the i-j branch, Hf is the fan pressure, Hi and Hj are the pressure on the nodes i and j, R_{ij} and Q_{ij} are the resistance and air-flow rate factors of fixed quantity branch i-j respectively.

Let:

$$\Delta h_{ij} = (R_{ij}' - R_{ij})Q_{ij}^2 \qquad (4)$$

In equation (4), Δh_{ij} is the non-equilibrium pressure of fixed quantity branch i-j, R_{ij}' is the necessary resistance which should make the air-flow rate through i-j branch be the value given, i.e. the resistance after regulation. R_{ij}' can be calculated by the following formula:

$$R_{ij}' = (Hi - Hj)/Q_{ij}^2 \qquad (5)$$

In equation (5), if the fan pressure given is too small or the air-flow rate of fixed quantity branch i-j is too large, then Hi may be smaller than Hj and R_{ij}' may be negative.

Now substitute equation (5) into equation (4):

$$\Delta h_{ij} = ((Hi-Hj)/Q_{ij}^2-R_{ij})Q_{ij}^2 \qquad (6)$$

Simplify to arrive at equation (7):

$$\Delta h_{ij} = Hi - Hj - R_{ij}Q_{ij}^2 \qquad (7)$$

Then substitute equation (7) into equation (3):

$$h_{ij} = Hf - \Delta h_{ij} \qquad (8)$$

In the calculation of ventilation network with air-flow rate distribution depending on needs, after the solving of non-linear simultaneous equations one can obtain the value of Rij'. By substituting Rij' into equation (4), Δh_{ij} can be obtained. From equation (8) one may calculate the total pressure drop of path Δh_{ij} if Hf is given. Obviously, it is not necessary here to sum up the pressure drop of all branches in the path.

Δh_{ij} represents the difference between the natural splitting of air-flow and air-flow distribution depending on needs in the i-j branch. If Δh_{ij} = 0 (Rij' = Rij), then the natural air-flow rate is equal to the given air-flow rate depending on needs. If Δh_{ij} > 0 (Rij' > Rij), i.e. Δh_{ij} is positive, then the natural flow rate is larger than the given air-flow rate depending on needs in the i-j branch and the resistance of the regulator in the i-j branch should be increased correspondingly. If Δh_{ij} < 0 (Rij' < Rij), i.e. Δh_{ij} is negative, then the natural air-flow rate is smaller than the given air-flow rate depending on needs in the i-j branch and the resistance of regulator in the i-j branch should be decreased correspondingly. In other words, h represents the degree of air-flow rate in the i-j branch and is presented in pressure units.

From equation (8), Δh_{ij} also represents the air-flow resistance of the path through the i-j branch. For the i-j branch, if Δh_{ij} is negative and its absolute value is maximum, then the path resistance through the i-j branch is maximum. If Δh_{ij} of all fixed quantity branches is positive, then the branch which has the minimum absolute value of Δh_{ij} has the maximum path resistance.

4.2 The aim of optimal regulation is to equalize and raise the non-equilibrium pressure of fixed quantity branches

The common method to equalize the non-equilibrium pressure Δh_{ij} or the equivalent path resistance is to choose the maximum resistance of a path as standard, and then increase the resistance of other paths to be maximum.

If the fan pressure Hf is given, the non-equilibrium pressure of each fixed quantity branch is shown in Figure 4. From Figure 4 one may see that the non-equilibrium pressure of branch 3 is zero. So the natural air-flow rate is equal to the given air-flow rate, and the resistance of the path through branch 3 is at

a maximum. The resistances of branches 1,2,4 and 5 should be increased. However, in the case of choosing the maximum path resistance as standard, the amount of ventilation power expended is too large. The energy consumed on the branches is proportional to $\Sigma \Delta h_{ij} Q_{ij}$.

If the increase regulation of resistance is carried out, not only in the fixed quantity branches, but also in the suitable branches of the whole network, then the non-equilibrium pressure of the branches can be equalized with the value of $\Delta h'$ greater than zero. This is shown in Figure 5. From Figure 5 one may see that the air-flow rate of all fixed quantity branches are rich and all of these richness are $\Delta h'$, which is the non-equilibrium pressure after equalization of each branch.

When the Δh of all fixed quantity branches is equalized and greater than zero, we may decrease the fan pressure to make the Δh_{ij} of all fixed quantity branches equal to zero, i.e. the natural air-flow rate equates the given air-flow rate. If the fan pressure decreases by $\Delta h'$, the Δh_{ij} of all fixed quantity branches will be zero. From equation (7) we have:

$$\Delta h_{ij} = Hi - Hj - RijQij^2$$

And from the law of power conservation of the ventilation network, if the decrease in fan pressure is ΔHf the pressure of the i node also decreases by ΔHf and the pressure of the j node does not change. Therefore, the non-equilibrium pressure Δh_{ij} also decreases by ΔHf. Accordingly, if the fan pressure drops by $\Delta h'$, Δh_{ij} of all fixed quantity branches will be zero. In other words, we can increase the resistance of the branches to carry out the predetermined distribution of air-flow rate depending on needs on the condition that the fan pressure is lower than the total pressure drop of the maximum resistance path. The optimal aim of increase of resistance is to equalize and raise the non-equilibrium pressure of all fixed quantity branches. Meanwhile, the greater the Δh_{ij}, the better the optimal effect.

5 OPTIMAL DESIGNATION FOR AIR-FLOW RATE REGULATION IN VENTILATION NETWORK

5.1 The optimal location of the resistance increase branch and optimal increase values

Suppose the air-flow path with the maximum resistance is shown in Figure 6. In Figure

6 the branch between node 3 and 4 is the
fixed quantity branch. The air-flow rate
and the original resistance factor are
given as $Q_{3,4}$ and $R_{3,4}$ respectively.
$\Delta h_{3,4}$ would increase in the positive direction if the resistance factor of outlet
branch 2,7 in the fan subnetwork and inlet
branch 8,5 of the atmosphere subnetwork
was increased.

Because $Q_{3,4}$ is fixed, the decrease in
$Q_{2,7}$ will cause the $Q_{1,2}$ to decrease, and
the pressure of node 2 will increase. The
pressure drop between nodes 2 and 3 is
fixed. The pressure on node 3 of the
fixed quantity branch will increase the
same as node 2. Meanwhile the pressure
of node 4 does not change. From the formula,

$$\Delta h_{ij} = Hi - Hj - RijQij^2$$

one may find that $\Delta h_{3,4}$ also increases.
Therefore Δh_{ij} will increase in the positive direction. For the same reason, increasing the resistance in the inlet
branch of the atmosphere subnetwork will
also cause the Δh_{ij} to increase in the
positive direction.

Since paths 1-6 have the maximum pressure resistance, then nodes 7 and 8 will
be located on the path with lower resistance. The opposite to these branches,
the branches 2,7 are inlet branches of
the lower resistance path in the fan subnetwork and branches 8,5 are the outlet
branches of the lower resistance path in
the atmosphere subnetwork. This means
that increase H_{ii} of the path with maximum
resistance results in decreasing Δh_{ij} of
the paths with lower resistance. Therefore
the increase Δh_{ij} of the high resistance
path is accompanied by the Δh_{ij} of the
relevant low resistance path. This is the
procedure of the equalization of Δh_{ij} in
different paths.

To equalize ΔH_{ij} of K paths with different resistances, it is necessary to have
K-1 regulated branches with certain conditions. These conditions are that along
the high resistance path there must be outlet branches in the fan subnetwork and inlet branches in the atmosphere subnetwork.
The air-flow rates of these branches may
change arbitrarily without any restriction.
The resistance value of the regulated
branch should be estimated comprehensively
on the basis of the change in the trend
of Δh_{ij} of all fixed quantity branches
after every resistance regulation. The
final aim is to equalize the Δh_{ij} of all
fixed quantity branches to the maximum
possible value.

Real ventilation network often does not
meet the condition that Δh_{ii} of all fixed
quantity branches be equalized completely.
However, if we can reduce the total pressure of the maximum resistance path, the
main aim would be achieved. So the best
location for resistance increase should be
on the inlet branches of the maximum resistance path in the atmosphere subnetwork
and outlet branches in the fan subnetwork.

Following we will analyse the optimal
location for resistance increase in the
region, in which the air-flow rate of
every branch is fixed. The region includes the fixed quantity branches (working area) and all the other branches connected with fixed quantity branches which
air-flow route is determined uniquely by
the air-flow rate of the fixed quantity
branches. We call this region as fixed
quantity region. To increase the resistance at the different branches of the
fixed quantity region, the air-flow rate
must remain constant regardless of any
extent of changes in resistance increase.
If the air-flow rate in this region is
fixed, then according to the law of power
conservation of the ventilation network,
the absorbed power of the whole network
except for the region with fixed quantity
is fixed. Since the power from the fan
is fixed, then the absorbed power of region with fixed quantity is also fixed and
does not change despite of the change with
location of regulator and the resistance
increase value. In other words, for different resistance increase locations in
the region with fixed quantity, the consumed energy of the region is the same.

5.2 The optimal location of the auxiliary
    fan

If the resistance difference between the
highest resistance path and the other low
resistance paths is large, and the air-
flow rate needed by the fixed quantity
branch in the highest resistance path is
relatively small, it is reasonable to install auxiliary fans in this fixed quantity branch in order to decrease the main
fan pressure.

Suppose the pressure drop of the highest
resistance path is h1, the pressure drop
of the second highest resistance path is
h2, and the difference between h1 and h2
is great. If choosing the maximum pressure drop as the standard for resistance
increase regulation, then the fan pressure
should be chosen as h1 and the non-equilibrium pressure of this branch is zero.

If choosing the second highest pressure drop as the standard for regulation, then the main fan pressure may drop down to h2 and the non-equilibrium pressure of the fixed quantity branches in the highest resistance path will drop from zero to -(h1-h2). In this case, to guarantee the air-flow rate needed through the fixed quantity branch, an auxiliary fan with pressure (h1-h2) should be installed in this branch. From this relation, the power saved after the installation of the auxiliary fan can be calculated by:

$$P = (h1Qf - h2Qf - (h1-h2)Aij)/102$$

$$= (h1-h2)(Qf-Qij)/102, \qquad (9)$$

in which Qf is the fan air-flow rate, which is the sum of air-flow rates in all fixed quantity branches, Qij is the air-flow rate needed through the fixed quantity branch i, j of the highest resistance path, and is the total efficiency of the ventilation system.

From equation (9), if h1 >> h2 and Qf >> Qij, the amount of the ventilation power saved due to the installation of the auxiliary fans i significant. The decrease of the main fan power is great and the costs on the main fans will drop subsequently with the drop of fan power. Meanwhile the power required by the newly installed auxiliary fans is relatively small and the cost is small. Suppose the amount of ventilation power saved within the service period of the auxiliary fan installation is B1, the cost saved on the main fans from the drop of the main fan power is B2, and the cost and installation expense of the auxiliary fans is B3. If the sum of B1 and B2 is greater than B3, the installation of auxiliary fans is worthwhile economically. Detailed computational formulas will not be repeated here.

5.3 The optimal location of the resistance decrease branch and optimal decrease values

The location of the optimal decrease resistance branch, which reduces obstacles and increases the area of the tunnel, must meet the following requirements: (1) be located at the branches in the highest resistance path and need to decrease resistance. (2) the branches for resistance decrease should have the maximum air-flow rate.

The significance of the above two requirements is obvious. It is only on the highest resistance path for resistance decrease that we can drop the main fan

pressure effectively. The pressure drop of the branch is proportional to the square of the air-flow rate, so the greater the air-flow rate of the branch, the more the pressure drops.

From the above and within the scope of real mining, it is appropriate to decrease the resistance of the chief tunnel and the other main tunnels with large air-flow rates. Since the air-flow rate through the chief tunnel is fixed, decreasing resistance in the chief tunnel by R is equivalent to increasing the fan pressure by ΔRQf. In other words, to guarantee the predetermined distribution of air-flow rate, the amount saved on the ventilation power can be calculated according to the amount of fan pressure drop. Suppose the ventilation power saved within the service period for the resistance decrease tunnel is B4 and engineering cost to decrease resistance by ΔR is B5. If B4 is larger than B5, then the resistance decrease is economically feasible. The above method cannot be used for branches in which air-flow rate changes with resistance regulation. However, we can calculate the drop in fan pressure by finding the increase of non-equilibrium pressure Δh_{ii} of all fixed quantity branches, if the resistance of the regulated branch is reduced by ΔR.

Besides, the pressure drop of the high resistance path can be dropped by decreasing the resistances of the outlet branches of the high resistance path in the fan subnetwork and the inlet branches in the atmosphere subnetwork. Under some suitable conditions one may also consider decreasing the resistance of these branches.

6 NOTES ON THE COMPUTATIONAL PROCEDURE AND PROGRAMS

The procedure of optimal design for the ventilation network is to solve the network again and again under conditions of different fan pressures and different resistance factors of the branches. According to the calculation result, one then regulates the fan pressures and resistance factors of the branches to design an optimal ventilation system.

The basic computational method for ventilation networks is the nodal method. The nodal method has strong functions, high computational speed, and need not to select mesh. This method is especially suitable for the optimal design.

The optimal computational program employes the block structure. The main blocks are: (1) the calculation of natural splitting of air-flow with the given fan, (2) the calculation of natural splitting of air-flow with the given fan pressure, (3) the calculation of resistance increase regulation, (4) the calculation of resistance decrease regulation, and (5) the calculation of auxiliary fan.

In the procedure of optimal computation, users may control the running of the program in the form of meny and chose the wanted block to calculate. Menu is shown in Chinese. Each block will return to the state of main menu after calculation. Moreover, some common subroutines for general purpose are designed in the main program for the use by blocks. The message exchanges between blocks through the data file.

Since the resistance increase regulation is simple and workable, so the first step of optimal designation is to calculate the resistance increasement regulation in order to drop the head pressure of the main fan. If the head pressure of the main fan has not dropped enough after increasing resistance, one may employ auxiliary fan or resistance decrease according to the non-equilibrium pressure Δh_{ij} of fixed quantity branches.

The original data for the program are stored in the input file. Table 1 is the file with N rows. N is the number of branches. Each row consists a complete set of parameters of a branch. These parameters are branch number, initial node number, final node number, resistance factor, and natural pressure value et al. In the table the given air-flow rate of each corresponding fixed quantity branch should be also input. The parameters of the fan character curves are stored in the fan parameter input file. The data of the data file are written in the free form. Different data file has different name. Input the file name wanted and then corresponding data file is read in. So the program is convenient in use.

The calculation result is saved in the output data file and can be shown in the form of table on the screen. The output results consist of nodal pressure, branch number, initial node number, final node number, air-flow rate of branch, resistance factor of branch, and branch pressure drop et al. For the fixed quantity branches, the result also consists of resistance factor after regulation and non-equilibrium pressure Δh et al. The calculation result saved in the data file can be read in for next calculation.

For the sake of convenience to modify the resistance factor of branch in the network, the program has the function to change resistance values of any branch. When the corresponding response row appears on the screen, type in the branch number and modified resistance value and resistance will be modified. For the sake of convenience, to calculate with different fan pressure, fan pressure is also input during the program running.

The optimal design program for the ventilation network is written in FORTRAN 77. This program has been run using IBM-PC, APPLE, CEOMENCO, BCM and other microcomputers, and also the VAX11/780.

7 EXAMPLE

The ventilation network with the resistance factors of each branch are shown in figure 7. The air-flow rates given for the fixed quantity branches are $Q_{3,43} = 35$, $Q_{8,10} = 30$, $Q_{8,9} = 30$, in units of m^3/s. The problem is to: (1) calculate the pressure drop of different air-flow paths. (2) calculate all of the nodal pressures and branch air-flow rates when the fan pressure is changing and (3) determine how to increase the resistance factors of the branches in order to make the fan pressure minimum.

Solve the problem by the nodal method. The total fan pressure is given firstly as Hf = 300 in units of millimeter water-bar. The output of the computational result after 6 iterations with the given error of E = .001 are as follows:

Nodal Pressure Values

300.0   299.0975 276.3814 31.2834 13.7420
25.6632 268.2369 264.6369 29.3230 26.1219
0.0

To save space, Table 1 presents only the computational results for a few typical branches.

With Hf and h, the pressure drop paths can be calculated by the formula:
$H_{ij} = Hf - h$:
Branch 3,4:   1 - 2 - 3 - 4 - 6 - 5 - 11

$$h_{3,4} = 300.0 - 24.5980 = 275.402$$

Branch 8,10: 1-2-3-7-8-10-6-5-11

$$h_{8.10} = 300.0 - 202.5151 = 97.4849$$

Branch 8,9:   1-2-7-8-9-10-6-5-11

$$h_{8,9} = 300.0 - 228.114 = 71.886$$

In the above results the pressure drop of the path with branch 3,4 is maximum, equal

to 275.402. If the fan pressure is given
as 275.402, then the Δh of branch 3,4 must
be zero and the Δh of the other branches
are positive. This is the increasing resistance regulation.

Now for the Hf = 275.4020, the results
are:

Nodal Pressure Values

275.4020 274.4995 251.7834 31.2834 13.7420
 25.6632 243.6390 240.0390 29.3230 26.1219
 0.0000

From Table 2, when Hf drops by 24.5980
(i.e. the fan pressure drops from 300.0 to
275.402), all of the nodal pressure in the
fan subnetwork drop by 24.5980 and the Δh
of all fixed quantity branches also drop
by 24.5980. The nodal pressure and air-
flow rate of each branch in the atmosphere
subnetwork remained unchanged. The computational results coincide completely with
the law of power conservation of the ventilation network. Therefore, it is not necessary to solve the non-linear simultaneous
equations for the whole network when the
main fan pressure changes only ΔH. The
nodal pressure, branch air-flow rate and

the non-equilibrium pressure can be calculated directly by the above method.

To equalize the Δh, we increase $R_{3,7}$,
$R_{10,9}$, $R_{10,6}$ to $R_{3,7} = 1.02$, $R = 0.22$,
$R_{10,6} = 0.0107$. Let the total fan pressure
$Hf = 275.402$. The new results are:

Nodal Pressure Values

275.4020 274.4995 265.3381 27.8198 11.6896
 21.5781 96.7087 93.1087 68.8155 39.9629
 0.0000

From the law of power conservation of the
ventilation network we know, only by the
resistance increase regulation the predetermined air-flow rate distribution will
completed if the total fan pressure drops
by 17.0183, i.e. Hf = 258.3837. In this
case, the fan pressure drops by 6.2%, main
fan power drops by about 21 KW, and about
200 thousands KWH may be saved in one year.

REFERENCES

(1) HARTMAN, H.L. 1961. Mine ventilation
     and air conditioning.

Table 1

No. of upper nodes	No. of lower nodes	regulated resistances	flow rates	pressure drop	original resistances	Non-equilibrium pressure
1	2	0.001	94.9998	0.9025	0.0001	0.0000
2	3	0.0040	75.3594	22.7162	0.0040	0.0000
2	7	0.0800	19.6407	30.8606	0.0800	0.0000
3	7	0.0050	40.3595	8.1444	0.0050	0.0000
3	4	0.2001	35.0000	245.0980	0.1800	24.5980
8	10	0.2650	30.0000	238.5150	0.0400	262.5151
8	9	0.2615	30.0000	235.3140	0.0080	228.1140

Table 2

No. of upper nodes	No. of lower nodes	regulated resistances	flow rates	pressure drop	original resistances	Non-equilibrium pressure
1	2	0.0001	94.9998	0.9025	0.0001	0.0000
2	3	0.0040	75.3593	22.7161	0.0040	0.0000
2	7	0.0800	19.6407	30.8605	0.0800	0.0000
3	7	0.0050	40.3594	8.1444	0.0050	0.0000
3	4	0.1800	35.0000	220.5000	0.1800	0.0000
8	10	0.2377	30.0000	213.9171	0.0400	177.9171
8	9	0.2341	30.0000	210.7160	0.0080	203.5160

Table 2

No. of upper nodes	No. of lower nodes	regulated resistances	flow rates	pressure drop	original resistances	Non-equilibrium pressure
1	2	0.0001	94.9998	0.9025	0.0001	0.0000
2	3	0.004	47.8577	9.1614	0.0040	0.0000
2	7	0.0800	47.1422	177.7908	0.0800	0.0000
3	7	1.0200	12.8578	168.6294	1.0200	0.0000
3	4	0.1939	35.0000	237.5183	0.1800	17.0183
8	10	0.0591	30.0000	53.1458	0.0400	17.1458
8	9	0.0270	30.0000	24.2972	0.0080	17.0972

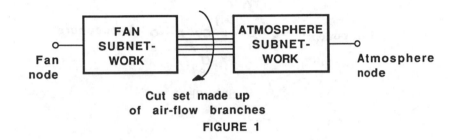

Cut set made up
of air-flow branches
FIGURE 1

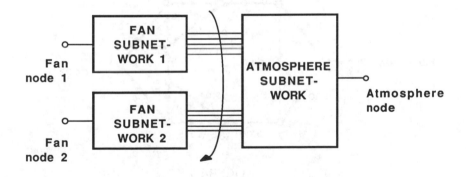

FIGURE 2

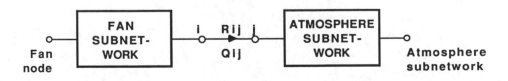

FIGURE 3

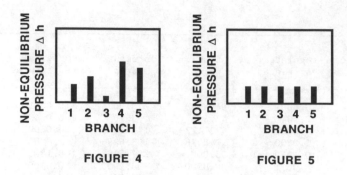

FIGURE 4

FIGURE 5

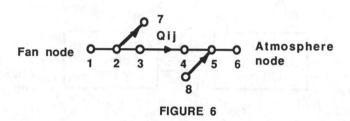

FIGURE 6

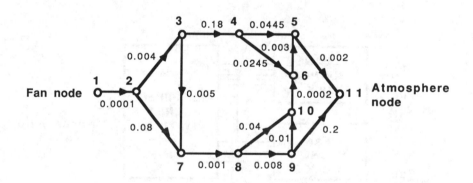

FIGURE 7

14 Exploration

Computer Applications in the Mineral Industry, Fytas, Collins & Singhal (eds)
© 1988 Balkema, Rotterdam. ISBN 90 6191 760 3

Computers in exploration in the 80's

William R.Green
Placer Dome Inc., Vancouver, BC, Canada

ABSTRACT

The past decade has seen a great increase in the use of computers in mineral exploration. The development of small inexpensive personal computers means that any exploration organization (including one person) can afford a computer to aid in data acquisition and analysis. In addition, the portability of these machines allows much of the computing work to be done in the field. The rapid spread of personal computers has also led to great improvements in software, particularly in allowing computers to be used without special training.

Placer Development (now part of Placer Dome Inc.) has a long history of computer applications, although the way they are used in exploration has changed greatly. In 1980, efforts were just starting to modify programs designed for computer specialists to allow geologists and geophysicists to use them effectively. All computer work was done on a large mainframe computer. Starting in 1981, microcomputers came into play for recording data in the field and doing initial data checks. By about 1984, this had progressed to the point where a microcomputer was considered standard equipment in the field, and essentially all exploration data (geochemical sampling, geophysical surveys, drill-logs, etc) were entered in the computer. At the same time, software improvements continued on the mainframe computer, with the result that almost all normal computer work was run directly by exploration staff, and not handed over to computer specialists.

Placer is currently undergoing a transition to the next phase of technical computing: the use of powerful individual workstations in place of a mainframe. Our new configuration is based on the UNIX operating system, which provides a much greater level of portability and adaptability than any other system. While conversion to UNIX is a major task, it enables the addition of more computing power when needed at minimal effort and expense. It also provides a wealth of software not normally considered part of an operating system. The strong emphasis on computer graphics in a UNIX workstation is another significant advantage. UNIX will undoubtedly be a standard computing environment for many years to come.

1. General trends in computers

The 1980's have seen great changes in the computer industry, and resultant changes in almost all aspects of business and society. In some areas of mineral exploration, the effect has been profound, since prior to this time there was almost no direct use of computers. In other areas, improved technology means that much more of the job is now being done with computers, and at a much lower cost.

- lower costs, greater power

The most obvious change is the price performance ratio of today's computers. A typical "workstation" has the processing speed of the top "superminis" of the mid 70's, at a price not much greater than the top "personal computers" (all of these terms are somewhat nebulous, since the definitions tend to change as each new model is announced). The main result is that almost any organization can now afford a powerful computer, while a decade ago only large companies were able to come up with the required funds.

The first mass-market computer was the Apple I (1977), which was applied mainly in education and home use, although many technical users found great value in these machines. The real breakthrough for business computing was the IBM PC, first released in 1981 (not so much for technical innovation, but because of the backing of the industry leader). For exploration projects, development of portable machines (such as the HP 85, Kaypro 2, and later the COMPAQ) was equally important, since it allowed a computer to be taken to a field camp for on-site data analysis. At current prices, a new computer for a field project is often considered standard equipment, since the cost is a small fraction of more traditional expenses such as mobilizing drill-rigs, helicopter support, and so on.

- "end user" replacing computer experts

A perhaps unexpected result of the rapid spread of personal computers is that many jobs formerly left to computer experts were now being done by people with limited computer knowledge. The success of computerization then depended on software that was easy to learn and use. This has now become a standard principle of software developers, although sometimes more evident in the sales literature than in the actual programs.

The "end user" also frequently must be involved in setting up hardware, for example attaching a printer or plotter. To make this chore manageable, the emphasis has been on standardization of components, or in the terms of the manufacturers, compatibility. This is especially true of devices for IBM PC's, and also applies to other computers that work just like an IBM PC (IBM "clones" or "compatibles").

2. Types of applications in exploration

With the arrival of affordable computers, many applications previously restricted to large main-frames could be undertaken by a single-user machine. In addition, the portability of some of the new machines lead to new uses for computers in exploration. Computers are now used in all phases of exploration. Reviews of specific applications are often included in the literature of each discipline (for example, see GEOPHYSICS, 1985 and Garrett 1987).

- data entry in the field

Perhaps the most visible change in exploration is the now common practice of entering data in the field at the time of acquisition. This developed originally in geophysical surveys, when increasing sophistication of the electronic instruments allowed data to be stored in the instrument (rather than be being manually copied onto a log-sheet). When general purpose computers are taken to the field, other types of data (e.g. geochemical sample description, drill-holes) may be typed directly into a computer file. Often existing written forms are retained, but the data are entered into the computer almost immediately (for example during evenings in a camp or field office). In other cases, a "lap-top" computer operating on batteries is used as an electronic notebook, and there is no manual writing involved.

- data checking and analysis

Once the data are in a computer format, even a small computer can provide extensive data checking and basic statistical analysis. This allows errors to be found (and corrected) much earlier, and generally results in cleaner data than the traditional approach of entering all data into a large computer after the field work is completed. The preliminary analysis may also show anomalies to be followed up immediately, or indicate a need for changes in survey procedure.

- computer graphics: mapping etc

Computer graphics and map drawing was well established in the 1970's: the major change has been a vast increase in usage. As with many other applications, this stems from the much lower costs involved in setting up a computer mapping facility. Sophisticated computer-aided design work can now be done on a standard personal computer. A decade ago, the simplest CAD systems were in the several hundred thousand dollar range.

The use of computer graphics in exploration is largely a replacement for manual drafting. Geophysical and geochemical surveys lead to sets of maps in posted or contoured form. With a computer, it is easy to go beyond simply plotting raw data, and produce various composite or derived quantities in the search for anomalies. Analysis of drill-logs similarly involves sets of cross-sections or plan views, drawn at different location and showing different categories of data.

- geophysical surveys - instruments

Geophysical instruments now are usually powerful computers, albeit specialized to the task at hand. The readings are stored in internal memory (which will have room for at least one full day's work). The attached computer allows entering additional information (such as time of day, line identifiers), and performs basic data verification. Standard output channels are provided to allow printing of results, and to transfer data to a general purpose computer (usually available in the field camp, for processing each day's work during the evening). As noted above, the development of geophysical systems has led to the wider use of computers in field exploration, through both shared hardware and the transfer of computer skills from geophysicists to other field staff.

- modelling

Another area where geophysics has led the way is computer modelling in interpretation. This was originally a job for mainframe computers, and for many types of problems is still beyond the capacity of a personal computer. Most of the development effort seems to be at the PC level, however, since so many mineral exploration teams do not have ready access to larger machines.

Effective modelling demands both a powerful processor and well designed software. The numerical algorithms may be complex and require lengthy iterations to converge on a solution. Software must be written both with efficiency and ease of use in mind, as it is common to try many variations on a model to come up with an acceptable solution.

Modelling of ore deposits is also traditionally a job for large computers, but is also rapidly becoming practical to do on smaller machines. Several powerful geostatistical packages are now commercially available, and in the past year or so have been supplemented by sophisticated mine design programs using interactive graphics.

- remote sensing

Processing of satellite images and other remote sensing data is still largely a specialist function. The vast quantities of data are difficult to effectively handle on a small computer, although it is practical to work with subsets of an image on a PC; provided that it is linked to a larger system for extraction of the desired data. The training and experience needed to work with multispectral images does not come quickly, so this area of application will likely never spread to the majority of exploration geologists and geophysicists. Still, the availability of remote sensing interpretations has been greatly improved. Along with the advances in the sensors (particularly the LANDSAT Thematic Mapper), the ability to map meaningful geology from space is becoming standard practice (at least in areas with sufficient exposure - i.e lack of vegetation).

3. Key developments

Several significant developments in the computer field have already been outlined. In terms of relevance to mineral exploration, the key points are described below.

- portability

Computers that can be easily taken to the field are becoming indispensable in exploration. This is not only a matter of size, but also of capacity and reliability. The latest portables are about the size of a briefcase, providing the power of an IBM PC-AT and at least 20 megabytes of hard-disk data storage. The general experience with portables is that they are rugged enough to tolerate frequent moving. Another important consideration is power source: battery operated units are acceptable in some cases, but most portable computers require AC power (or are more dependable when running off AC). This is not usually a restriction, since even in remote areas exploration camps will have a power generator (computers can generally be used directly off a generator, although a power filter is often interposed to protect against surges).

- inexpensive

The affordability of powerful computers is the main reason they have become so widespread (in all industries, not just exploration). A consulting geologist can now buy a system providing the computing power that was only available to major companies a decade ago. Another aspect of declining prices is that computers can be justified as a short-term expense, and incorporated into the budget of a single exploration program.

Price decline (or looking from a different point of view, more computer for the same price) is sometimes a problem in deciding on new equipment. Almost inevitably a new and more powerful computer will be introduced within only a few months of a purchase. This can lead to indecision - in the worst case, upgrades are never made for fear of spending too much today compared to what will be available tomorrow.

- easy to use software

Typical computer programs of the 1970's were not easy to learn and use. The rapid spread of personal computers was as much dependent on improved techniques for writing software as on the hardware. Programs that cannot easily be learned by people without a background in computers generally fail to gain wide acceptance. An associated problem is that people who try to use unacceptable software develop a negative attitude about computers in general. Fortunately the demands of the marketplace have largely taken care of this problem, although many packages may seem deceptively simple in demonstrations, and prove troublesome when first tried by the end user.

- greater computer literacy

Related to the above is the increasing level of computer knowledge in the general population. Most people in office or technical jobs now have at least some direct use of computers. The great majority of university and technical school programs include computer courses as a basic requirement.

4. Placer's history

Placer Development Ltd. (which became part of Placer Dome Inc. in 1987) had a long history of computer usage in mineral exploration and mine development. Following is a brief outline of recent developments to illustrate the general patterns discussed above.

- mainframe orientation

Starting in the late 1960's on timeshare systems, and after 1975 with its own computer, Placer used a UNIVAC 1100 series mainframe computer. Initial emphasis was on ore reserves and mine designs for large-scale open-pits. Use of the system quickly spread to other departments, and the UNIVAC became the corporate system.

Exploration use at first concentrated on geophysical applications (which were already a well-established technology). In 1979, work started on a more general exploration software system, designed to allow basic analysis and display of all types of exploration data. In 1980, a large electrostatic plotter was installed to allow routine drafting chores to be taken over by the computer. By 1983, a core package of programs were running in an interactive, self-prompting mode, and were used by several geologists (in addition to the geophysicists, who were already comfortable with using a computer).

A related development in exploration was a computerized logging system (GEOLOG), purchased in 1980. As software was developed to exploit the standardized log data, plotting of cross-sections and statistical analysis of drill-hole assays became a routine job in the exploration group.

- field data entry

The use of GEOLOG for drill-hole logs led to trials of a microcomputer for data entry (in 1981). The aim was to have the logs entered and checked as soon as possible after the coded log forms were filled out. In 1982, several Kaypro portable microcomputers were purchased, and used in field camps to enter GEOLOG, run basic data verification, write preliminary reports, and so on.

Software to enter and check other types of data was quickly developed, and the Kaypro became an indispensable part of most of our field work. This included entry of geophysical data (both keyed in manually and automatically unloaded from digital memory in the instruments), description and location of geochemical samples, and drill-logs. Some trials were made of direct entry of drill-logs into a lap-top computer, but in most cases data entry forms are written first, and entered into a computer in the field camp.

- communications

A natural extension of having a microcomputer in the field was to use it to communicate with the head office. This presented a problem, since in 1982 there was no commercially available software that would allow error-checking file transfer between a microcomputer and the UNIVAC mainframe. To provide this capability, we developed an extension to the public domain program MODEM7, which was then in wide use for communication between CPM based microcomputers.

With our local version of MODEM7, it became practical for field camps to transfer data to the mainframe, run more complex analysis programs there, and return the results to the field. In addition, it allowed smaller offices to use all of the computing facilities of the head office in Vancouver.

- data analysis packages

Throughout this period, the software package on the UNIVAC continued to grow. Enhancements were aimed both at providing new types of analysis, and making all programs easy to use. The result of these efforts was that nearly all computer work related to exploration was done by the people assigned to the projects (rather than relying on computer specialists). In general, this means a faster route to the desired maps, statistical summaries etc. More details on the evolution of the software system have been given elsewhere (Green and Barde 1984a, 1984b; Green 1985).

- computer plotting

As noted above, much of the computer work related to exploration involves graphics: i.e. the plotting of maps, sections, and other visual forms of data presentation. The key to successful use of computer graphics at Placer were twofold. First, the development of easy to use software allowed project staff to produce their own maps. Secondly, a fast reliable plotter meant there was no effective limit on the number of maps that could be produced. Until this year, our only plotter was a 36-inch Benson electrostatic plotter, which handled all the plotting demands of the head office (mostly exploration and mine evaluation), as well as the needs of the operating mines in B.C. The electrostatic plotter on occasion produced over 100 large plots in a single working day (i.e. less than 8 hours).

- current status

Placer has just completed a major conversion of computer systems. The UNIVAC mainframe has been replaced by separate systems for commercial applications (e.g. accounting) and technical applications (exploration and engineering). The technical system is a network of "workstations" running under the UNIX operating system. Each workstation is a powerful computer in its own right, although in our configuration a central "file-server" is used to store all datafiles. This means that the workstations cannot be used standalone (although each could have its own disks if desired).

The benefits of making this switch are numerous. First, the equivalent computing power of a mainframe is obtained at much lower cost. It is much easier to add more capacity when required: simply buy additional workstations. The technology is more advanced than most mainframes, providing each user with a high-resolution graphics terminal, a multi-tasking machine, ability to use multiple windows for concurrent processes, and many more state-of-the-art facilities.

A multi-tasking, window-based system is also a great boon to software developers (as opposed to end users). The ability to concurrently edit multiple files and edit while compiling and testing made the job of converting programs from the UNIVAC versions to UNIX relatively simple. The program debugging facilities of UNIX were also very useful.

The first phase of conversion was simply to use the programs in the same style as before. They were designed for running on character terminals, with graphics produced mainly on hard-copy devices. The next step is to take advantage of the workstation technology, and adapt the programs to make greater use of interactive graphics.

5. The future

There is no end in sight to the rapid advances in computer technology. The trend towards more extremely powerful desktop computers will continue. It also seems likely that the truly portable computers will take an increasing share of the market, as they now can provide all the features of the traditional PC

- rise of "workstations" and UNIX

The major thrust of small computer development is towards "workstations". These are generally full 32-bit computers, running at much greater clock speeds than a PC. While generally considered a single user machine, they are distinguished from a PC in being full multi-task machines, capable of running several independent processes concurrently. Normally they have a sophisticated window system so that user can easily monitor multiple tasks. The UNIX operating system (originally developed by AT&T in the early 1970's) is becoming a de facto standard for this class of machine.

- power of UNIX

UNIX has several unique features among popular operating systems. First, it runs on machines of almost all sizes, from PC's to super-computers. It has less machine dependence than other systems, since it is built on a relatively small number of native commands, and relies on the widely available C language to develop the more sophisticated functions.

Perhaps the most significant feature of UNIX is the large library of software that is considered part of the basic system. Unlike other operating systems which simply provide the tools for operating the hardware, many user functions are included as standard features in UNIX. These include text editing, document formatting, file searching routines, spelling checkers, and a host of others.

- UNIX and MS-DOS coming together

The continued spread of UNIX is partly a result of enhancements to MS-DOS. Many MS-DOS commands are identical (or nearly so) to their UNIX equivalents, which is not surprising since MS-DOS was originally a derivative of UNIX. The increasing power of PC's means that more functions can be built into MS-DOS than was formerly possible.

A commonly heard complaint about UNIX is that many of the commands are rather cryptic. Another problem is that command syntax is not always consistent. These problems should be alleviated with the continued evolution of MS-DOS towards UNIX, as the greater user base will demand an easier-to-learn syntax. Our experience with UNIX indicates that these perceived problems are not a severe problem in training people how to use UNIX, in any event.

The 32-bit PC's now coming out have all the power necessary to provide full multi-tasking: these machines are in fact a match for what are now called workstations. Within a very few years the basic PC will have all the features now associated only with engineering workstations.

- communications

As UNIX becomes more widely accepted, many current problems of communication between different systems should disappear. Remote file transfer, mail facilities, and information services are already incorporated into UNIX. UNIX user groups already run world-wide networks, so that messages can be automatically sent to anyone in the group.

In conclusion, the future of computers in mineral exploration looks bright. In the years to come, most of the manual chores in data collection and analysis will be taken over by computers. In addition, more complex analysis and modelling procedures will become practical to apply as standard practice. The result should be more effective exploration, and an increased ability to find elusive targets.

REFERENCES

Garrett, R.G., 1987, The Role of Computers in Exploration Geochemistry, "EXPLORATION 87" conference (proceedings in press)

GEOPHYSICS, vol. 50, no. 12, December 1985, Golden Anniversary Issue

Green, W.R., 1985, Computer Aided Data Analysis, John Wiley, New York

Green, W.R. and B.W. Barde, 1984a, A General Purpose Computer Program for Data Analysis in Exploration, CIM Bulletin, vol. 77, no. 870

Green, W.R., and B.W. Barde, 1984b, An Interactive Computer System for Data Analysis in Exploration, CIM 86th Annual General Meeting

Computer Applications in the Mineral Industry, Fytas, Collins & Singhal (eds)
© *1988 Balkema, Rotterdam. ISBN 90 6191 760 3*

Geological data flow and structure at BPM-Brazil

Monica Chueke & Miguel Armony
BP Mineração, Rio de Janeiro, Brazil

ABSTRACT: Decisions and calculations during the exploration and evaluation phases are based on information collected at an earlier stage involving areas such as topography, geophysics, geochemistry, petrography, rock mechanics, chemical analysis, drillhole surveying etc. Part of these data is collected directly in the field, part comes from later analyses, and part is conditioned to personal or subjective decisions. In these processes, data are accessed and manipulated, while graphics and calculations are produced. A healthy flow of data, an easy way for retrieval and a secure mode for their organization and archiving are essential to high quality work. In this paper the way in which this is done by BPM-Brazil is shown. Emphasis here is given to the conceptual model behind the Data Base Project. Field information flow and data organization are also presented and discussed. In this way, all the information compiled is integrated, and is used through the particular userviews, to help the decision process in the different phases of mineral exploration.

1 INTRODUCTION

BP Mineração is involved in several projects and produces many thousands of chemical or geochemical analysis, ground geophysics and geological information every week.

The increasing volume of data available and the complexity of the processing require advanced techniques for building an integrated information systems. At BPM-Brazil, the SINATEC (Sistema de Informações Integradas da Area Tecnica) system was developed. SINATEC is a system to integrate the technical information through Data Base Technology for the Exploration Department. SINATEC was built to support all exploration phases from initial topography and geology to final ore reserve calculations and mine planning. In the future, this project will be extended to other areas of the company.

Special emphasis in this paper will be given to the presentation of the the data base model. This model has to be able to capture the characteristics of the information in the mineral exploration area.

An efficient data capture system is fundamental for building any reliable system of information. General rules for data capture may be easy to list, but in practice is difficult to implement especially in the case of exploration. Instead of theoretical considerations, the procedures used by BPM for data capture and treatment will be described and applied to two examples: geochemistry and borehole data.

It is important to note that a set of graphical symbols connected by arrows representing complex operations may be impressive, but it hides the hard and difficult reality associated with them. So, lastly, besides the ideal scheme for data flow and organization, several problems associated with system implementation and with system analysis will be pointed out.

SINATEC's system was implemented under a VAX 11/785 mainframe with 12 MB memory and about 2GB disk storage. Microcomputers are Brazilian made IBM-PC compatible. Simple terminals are also Brazilian made. Graphics terminals are Calcomp and HP (plotters and digitizers), Tektroniks and Digital (screen).

2 DATABASE TECHNOLOGY

A database is a collection of integrated data stored in a structured way which can be shared by several different users. Each user may have access to only part of the database. The data are said to be integrated in the sense that the database integrates the user's different points of view. The implementation of a database requires a special software called Data Base Management System (DBMS). BPM uses a relational database managed by the INGRES software developed by Relational Technologies.

Database technology brings several benefits for the Information System. Among them are:

. data independence - application programs do not depend on the way data are stored physically or accessed. This significantly simplify its maintenance;

. data usage sharing - the system presents to each user his own view of the data according to the processing needs. Moreover, data may be used simultaneously in several different applications;

. central control - data centralization allows the adoption of homogeneous control policies. Among the principal control features are:

.. integrity - ensure data precision and validation procedures thus avoiding inconsistencies;

.. concurrence - ensure a synchronous access to the data by multiple users;

.. security - allocate for each type of user different permissions of data and procedures access;

.. recovery - ensure database restoration in the event of system failures.

Many of the DB management functions are achieved using direct DBMS facilities and utility programs helping the data base administrator (DBA). End users may have access to data without programming by using special DBMS oriented languages or menus and simple keyed commands. A good DBMS simplifies database implementation and administration. However, the main problem of reaching an efficient Information System is building an adequate project guided by a good methodology.

2.1 Database system project and development methodology

For SINATEC the project was split into four phases:

. preliminary system analysis
. conceptual database project
. implementation and physical project
. loading and operation project.

The first step was the delimitation of the system comprehension. The second step was an evaluation of the principal subsystems. These subsystems were chemical analyses, geochemistry, geophysics, geology, topography and petrography. Interviews were held and several questionnaires prepared and filled out at the management and operational levels. The purpose was to identify the information needs for the functioning of each subsystem. This collected information permitted the building of a preliminary version of the external views which corresponded to each isolated subsystem, thus completing this first phase.

In the second phase, the conceptual database project, a first integration of the different external views was produced to obtain the primary global conceptual model. By incorporating new information and subtracting some, according to users real needs and possibilities, the model was improved. This lead to the final version of the conceptual model which will be illustrated in the next chapter.

During the third phase, that of implementation and physical project, the normalized relational model was produced. It represents the general scheme of the database in terms of tables and restrictions, which clearly express the meaning of the data. The strategy for the DB implementation was defined from this scheme using INGRES features and languages.

The last project phase consists of the implementation of the real system. This was comprised of initial data loading, as well as the definition of standards for the database usage and security procedures.

3 THE SINATEC PROJECT

SINATEC's goal is to follow and support all the geological research process from the beginning up to the mine planning phase. Integration of the different phases allows better estimation of the deposits and gives substantial information for the first stage of mine planning and control.

There is a strong interaction between the different exploration areas during data processing. Each area supplies the SINATEC DB with information resulting from its own activities. At the same time, each area accesses and recovers from this DB information fed by the other areas. The success of

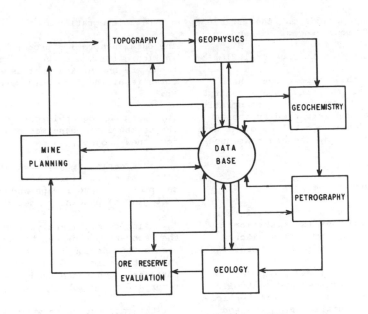

FIGURE I - DATA FLOW GENERAL VIEW

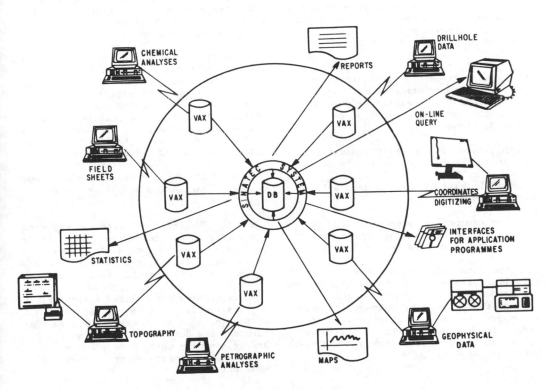

FIGURE II

the global system depends on the efficiency of this continuous process of feeding and recovering.

Figure I illustrates this general flow of information.

An example of the scheme above can be given by chemical analysis. Assay results are supplied to the DB by the Geochemistry sector (the lab) and are used by the Geology sector which also supplies other data to the DB. This is a cyclic and continuous process.

3.1 System inputs and outputs

The SINATEC database is fed by different devices in several ways (work stations, digitizers, etc.) and from many places (head office, field offices, laboratories, etc).

The system data model was projected to cater for, in an organized and controlled way, the development of different applications through DRMS utility programs and normal computer languages. The system proposes solutions to guarantee efficiency and integration of the data flow through the peripheral systems, and between the peripherals and the mainframe.

In Figure II the different entries can be seen, as well as the types of outputs produced by the system.

Types of entries:

. Work Stations: data entry by local systems via microcomputers (laboratory, chemical analysis, geochemical sheets in the field, petrographic analysis at head office, geophysical and geological data in the field, etc.).

. Digitizing: data entry via digitizers (maps in the head office or in the field).

. Local terminals: on-line data entry for general information.

. Other equipment: automatic entry by special devices (ICP in lab, cartography at head office, geophysics in the field).

All the data referred to above are automatically transferred to the mainframe, and then loaded into the SINATEC database.

The system produces as outputs: reports, online queries, maps, statistics, an interface for micros and other devices, as well as an interface for packages and application programs.

3.2 Data integration

The data integration process is illustrated in Figure III where the principal interactions are shown. The most important features resulting from this data integration are:

. using cartographic data one can draw the cartographic base map and line profiles for a new project;

. integrating geophysical and topographic data, while digitizing the coordinates, one can produce contour maps and cross sections to begin the process of defining anomalies;

. using chemical analysis information, the laboratory flow can be controlled. The backlog of the samples can be monitored, while statistics and quality control features aree generated producing also cost reports for the lab;

. the interaction between analytical results, digitized coordinates and geochemical field sheets can produce geochemical maps, statistical reports and graphics;

. petrographic results and digitized coordinates provide petrographic sample locations and allow statistical comparison with geochemical samples;

. the integration of chemical analysis, geophysical results, topography and digitized coordinates give the possibility of a more precise definition of the areas of interest. In a subsequent phase, after geophysical definition for possible interesting mineralized bodies, the analytical results integrated with the coordinates, will define targets and drillhole locations;

. basic drillhole data allow the production of location and elevation contour maps;

. basic drillhole data, analytical data and survey data provide the production of sections and contour maps for different variables. This allows the definition of the mineralization within each hole;

. integrating mineralized intervals, analytical results, lithology etc. allows the production of a first ore reserve estimation and the definition of optimum grid and/or a mining unit. Mine planning can begin in an interactive process with ore evaluation;

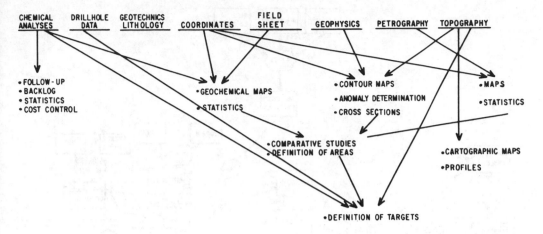

IIIa

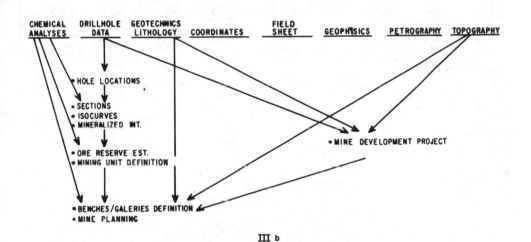

III b

FIGURE III DATA INTEGRATION PROCESS

a) Target definition

b) Ore reserve estimation and mine planning

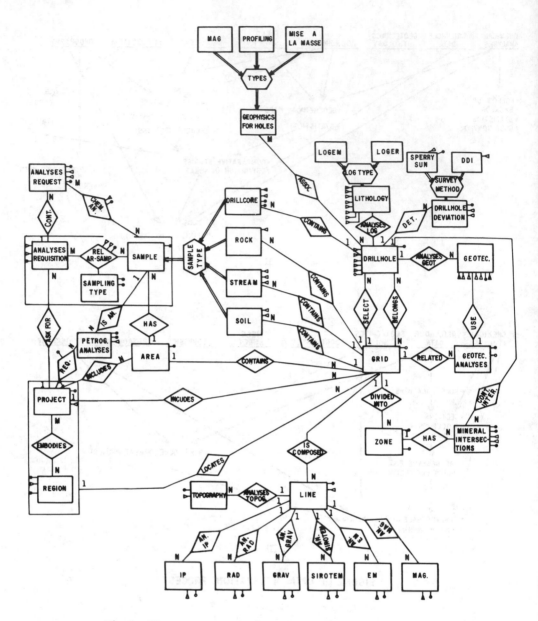

Figure IV-a – General view of the data relations

Atributes are named characteristics of an entity

————○ Simple (ex.: analytical results)

————● Simple identifier (ex.: sampling code: identifies the sample)

————▷ Multiple (ex.: mineralized intersection – for each cutoff a different
                intersection is defined)

————◁ Composed (ex.: drillhole general data)

664

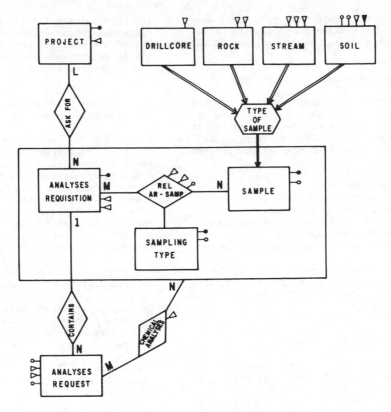

FIGURE Ⅳ-b THE LAB VIEW OF THE CONCEPTUAL MODEL

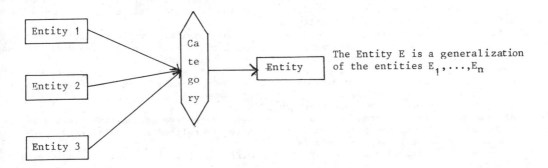

Entity – fundamental point of interest to the organization about which
information is to be collected

Relationship – is an association between one or more entities

The Entity E is a generalization
of the entities $E_1, \ldots, E_n$

FIGURE IV– THE CONCEPTUAL MODEL OF THE DB

. the first project of mine development done by integrating geotechnical, drillhole, lithological and topographic data.

3.3 Expected benefits

SINATEC's project implementation will provide important tools for a quick and efficient analysis to aid decision making at all company levels. Among the principal benefits expected from the SINATEC project are :

- quick updating
- increased data reliability
- better quality information
- crossed information possibilities
- improved data security
- easier data and system maintenance

The rapid development of new applications combined with the query facilities allow an increase in user productivity and overall efficiency.

4 THE CONCEPTUAL PROJECT

The principal goal of the conceptual project is to specify the Final Conceptual Model of the system and the different data userviews. This model must represent the information in an abstract way and be independent of the DB software or the available hardware.

The conceptual data modeling produces a description called the 'conceptual design'. This conceptual design represents both the perspective of the information structure and its intended use.

The information structure is represented graphically through a Relations and Entities Diagram (E-R). All the relations between the data, all the connections and the most important information are represented in this diagram. The meaning of each item of information is defined by 'integrity restrictions' that determine the value that each variable may assume, individually or in connection with other variables.

Figure IV shows the essentials of the conceptual model of the SINATEC data and one of its particular userviews.

As an example it can be seen that the relationship between drillhole and sample (drillcore) represents the fact that a hole is composed by several sample intervals and that one sample interval belongs only to one hole.

As an example of restriction of integrity one can say that: "a given sample interval may have more than one code". It means that

the physical sampling interval can have different denominations or values; this is due to the possible resampling procedures! All these controls must be foreseen in the system implementation.

4.1 Some characteristics of the SINATEC conceptual model

In SINATEC case the integration between the different subsystems is done via coordinates information. An exploration system is essentially geographic. The intrinsic nature of the stored data provides, a geographical/spatial database view. The database contains information at the levels: Region - Project - Grid - Line - Station. Spatiality is the dominant characteristic for the following points of view:

- topographic view: measured elevations in each station of the grid lines;

- geophysical view: radiometric, IP, gravimetric, sirotem, electromagnetic and magnetic measurements in the different stations;

- geological, geochemical and petrographical view: by drillholes and samples corresponding to spatial points, and so on.

It was observed that certain entities such as drillhole and sample are very important to the conceptual design, and can be focused under different view points as follows:

Different points of view of drillhole entity -

- spatial: by collar coordinates and angles, and drillhole deviation;

- geological: by geotechnical and lithological characteristics;

- geophysical: by the results obtained by geophysical measurements like profiling and "mise-à-la-masse";

- geochemical: by analytical results of its samples;

- ore reserve estimation: via the drillhole mineralized intervals, etc.

Different points of view of sample entity-

- laboratory: by physical flow, chemical analysis and quality control;

. petrographic: by petrographic analysis;

. geochemical: via analytical results and qualitative characteristics;

. ore reserve estimation: using drillcore characteristics, and so on.

5 PHYSICAL FLOW

Besides the ideal data scheme that was discussed above, the data production phase must be considered. In the first moment there exist only physical entities. These entities are to be taken, measured, analyzed, transported, split, transformed, etc., producing information at each one of these operations.

A healthy physical flow scheme, the design of appropriate data sheets and procedures and an efficient data capture system are prior conditions for the building of a reliable information system.

This chapter will describe the physical flow and the data capture scheme as done by BPM in two phases: geochemistry and borehole information.

5.1 An example from geochemistry

When a sample is collected in the field, it immediately produces two types of data:

. qualitative data: a field data sheet is filled out by the geologist; identification key may be the sample number or relative geographical position;

. geographical data: the geologist marks on the map or sheet the position where the sample was collected.

The field sheet has an appropriate size, the geologist pocket's size, and tries to be easy to fulfil. The sample is sent to BPM's laboratory together with a request for analysis (R.A.). In the lab the sample is prepared and analyzed according to the corresponding requisition. A set of micro-computers workstations perform quality control. Data concerning the analyses of the sample (elements, method, grain size, etc.) are stored into floppy disks and verification reports are produced. Once a day the data are transmitted by phone lines to the mainframe and stored in the relational database. Here, all information concerning the samples can be accessed through user's terminals. A complete follow-up can be reached for each sample. New requisitions

may be made for the same or new elements, using the same or different procedures for the analysis.

Maps with sample positions are digitized and their results are also stored in the database. If the grid is to be a regular one, then only the end points are needed, and coordinates can be calculated by program. Analytical information is now ready for application programs such as contouring or statistics. These can be produced in the head office or in the field office.

Field sheets remain in the field office, often very far from headquarters. Magnetic storage is done locally using microcomputers. Each Project is responsible for managing its own floppy disk library. After double keying, checking and confirmation, data are sent to the mainframe and stored in the database.

Some projects use phone lines, others send in the floppy disks. In some places the system is not yet implemented. Everything depends on cost and priority.

Field sheets also describe qualitative physical and geological properties of the samples and their environment. Some of these are objective observations and part depend on the geologist's interpretation. Application programs can be used for map some of these properties.

For the systems analyst, geochemistry is an endless accumulation of information that, except for some peaks, will seldom be consulted; keeping all of them online in the same database will be a space and time consuming procedure that will make the system running too much slow. For this reason it was decided to create one database per region keeping online only undergoing projects.

5.2 An example on borehole information.

As the input sheet is just a document for entering data into the system, it behaves like an information unit and is filled out in one place by the same person. If possible all its fields are filled out at the same time. Trials to mix information in the data entry phase generally produces a mess in the system.

Boreholes may have many data sheets associated to them. But let us concentrate only on these that have been the most important along the time:

Drillhole general data - contains general information about the drillhole: hole number, local and UTM collar coordinates, collar elevation, dip and azimuth, drillhole survey method, hole length, total re-

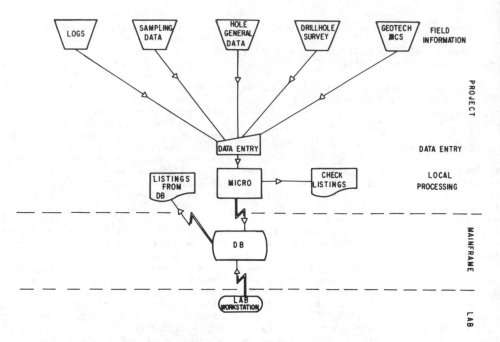

FIGURE Ⅴ- PROJECT INFORMATION ROUTINE

covery, drilling diameters, equipment, level of water etc.. Many fields are mandatory and some are optional and can be replaced. It is filled out in the project office and its set of information will be complete only when the hole will be ready. But data may be stored as they become available so reports can be always produced and updated.

Log - contains geological information; each data sheet refers to one borehole interval and is filled out immediately in the field. Intervals are not sampling intervals but follow the changes in geological properties.

Sampling - drillcore is divided into samples; after splitting, one of the halves is milled and is assigned a code number, and the procedure becomes similar to the geochemical samples: a requisition analysis is prepared and sent to the lab together with the milled sample. Sampling sheet gives us the correspondence between sample code, hole and interval and is filled out at the project office after setting codes to the samples.

Geotechnics - contains information about rock mechanics; each record describes one borehole interval. The intervals are not

the same as for log or sampling but depend on rocks' characteristics.

Drillhole Survey - up to now two different types of equipment were used for drillhole deviation measurements. Raw data produced by the equipment are preprocessed locally in microcomputers yielding the information needed to fill out the input sheets. This needed data are the angles end the local coordinates of the points inside the borehole.

From system's point of view, the Hole General Data sheet is mandatory. So are sampling sheets when samples are assayed, and drillhole survey data when hole deviation is measured. All data sheets have fixed, optional and replaceable fields.

Data flow can be seen in Figure V. A brief description is given below:

. after the data sheet is filled out and verified, it enters into local microcomputer via programs that require double keying. Listings are produced for visual check;

. after approval, a backup copy is made and data are sent to the mainframe via phone line or into floppy disks by plane, thus entering into the relational database;

. routine reports are produced locally
for each kind of data and/or merging different types of data, using microcomputers,
according to geologists' specifications;

. in the same way statistics, graphics
and other specific reports may be produced
locally;

. chemical analyses and other information, can be accessed by phone line to allow its usage in the field; microcomputers
can work as terminals when necessary, or as
local workstations.

6 GENERAL ASPECTS OF SYSTEM ANALYSIS

Exploration data concentrate most of the
nightmares a System Analyst can imagine:

. for each deposit there are different
data requirements, causing difficulties to
develop general data sheets;

. even for a known deposit, the system
must be built (and the data stored) before
it can be completely defined; this involves
crucial changes when system is already
running;

.it happens very often, in system analysis, that the user does not know how to
express what he wants or what he needs; in
the case of exploration it is worse: frequently the geologist really cannot know
what he needs and can not preview what will
be the evolution of the system;

. one cannot wait until all data will be
available to fulfill the data sheet: information must be stored as it arrives; the
sheet is to be typed even if vital information is missing;

. sometimes data to be altered belong to
key fields;

. there is not a real secure way for data
consistency: what can be done is to retype
data for comparison and make the best visual check as possible. It's not like accounting where cross sums are possible or
like payroll where each employee acts as an
inspector;

. data come from many places and in different ways and sometimes it is very difficult to confirm or to complete information.

6.1 The system at management level analysis

A top level manager generally wants an
easy, quick and immediate access to all
data produced by the company. If possible,
a terminal on the table so that at any moment he can be updated on exploration results.

These requirement brought together with
system analyst's difficulties pointed out
above, produce for the Data Processing management level people, their own nightmares.
They must know for each kind of data what
is possible and what is not. They must e-
valuate the cost of each system, the time
and personnel requirements. For each set of
data they must be able to evaluate if the
users necessities are clear enough to allow
system planning or development. It is up to
them to inform the top level managers about
the real possibilities and actual limitations, thus giving them the conditions for
decision making on data systems development
priorities. Then, according to these decisions, they have to produce robust data
models, to schedule and implement them.

6.2 Data sheets and reports

A geological data system must not start
without having both data sheet and reports
well analyzed and ready so the complete
procedure can be automated. Otherwise computerization will be for the geologist just
an endless process of feeding a machine
with data sheets.

When a decision is taken to build a system for certain type of data or procedure,
the system analyst must sit down with the
field geologist to plan the data entry
sheets.

The data processing systems can be
broadly divided into two different types.
The first comprises data capture, flow, and
organization. The second involves application programs on these data.

In the data capture phase care must be
taken not to mix concepts. A data sheet is
built together with the geologist who will
define his information needs and with the
system analyst who will put this information in a way that the data system can as-
similate it. The data sheet does not replace geologist's notes. It is not a report, but only a way of entering data into
the computer. It is a mistake to enter
data into computer without having implemented the programmes which produce the
reports required by the geologists. The
system must be understood as a whole: the
computerized procedure is being compiled

to replace the manual procedure. So, an
input sheet must be implemented simultaneously with the output reports. Otherwise,

. geologist will continue to use his
handmade notes dissociated from the data
system and we will not have a really computerized system but a hybrid one with
different data sources that can lead, in
the future, to data contradictions;

or,

. geologist will take the data-entry
sheet as a report and will tend to introduce into it output data or data of other
nature causing confusion into the system.
He will not understand the essence of computerization and will be unhappy with the
results. On the other hand system will
loose coherence and efficiency.

6.3 Data nature and data value

Ore reserve evaluation is the end point -
before entering in the mining stage - of a
long sequence of procedures that involved
several types of data capture and acquisition in many different ways. In each phase
there is the danger of loosing and mainly
corrupting information. The rupture of a
single link on the chain will make the information arrive at the end point completely adulterated. From data capture point
of view special care must be taken with
  . data transcription
  . data reading
  . data typing.
   Mistakes in these phases can cause ir-
reparable damage, if there does not exist a
parallel manual procedure. But attention
must be paid also to the application phase
where there are many types of manipulation
involving data transference, selection,
composition or usage for calculation. Undetected errors can easily change - among
other things - sample position, chemical
analysis or geophysical information risking
"transforming" ore into waste or waste into
ore.
   Each company must be conscious that this
simple final value which is to be used in
ore reserve estimation is the result of a
long sequence of difficult operations:
geochemical samples collection, geophysical
ground measurements, drilling, borehole
survey, geographical and topographical
measurements, drillcore split and mill, lab
preparation and chemical analysis etc.
Indeed, these values are capital goods, a
high-cost fixed asset that must be treated
as a precious stone.

BIBLIOGRAPHY

Date, C.J. 1986. Introducao a Sistema de
   Banco de Dados. Rio de Janeiro : Campus
Theorey, T.J. & Fry, J.P. 1982. Design of
   Database Structures. New Jersey.
De Antonellis, V. & Demo, B. & Bussolati, &
   Bertaina, P. 1983. Methodology and Tools
   for Database design. North Holland
   Publishing Company.
Sinatec's Project Documents 1986/1987. Rio
   de Janeiro : BPM.
Chueke M. & Mello R. & Collares A. & Sa'
   Maria , 1987. Sistema de Informação para
   Acompanhamento de Exploração Mineral
   (Internal Paper). Rio de Janeiro: BPM.

Computer Applications in the Mineral Industry, Fytas, Collins & Singhal (eds)
© 1988 Balkema, Rotterdam. ISBN 90 6191 760 3

MINE-FORECAST: Software for the financial evaluation of mineral exploration projects

John B. Gammon
M.S.S. Mining Software Services Ltd, Delta, BC, Canada

ABSTRACT: The MINE-FORECAST software package has been developed to assist the exploration geologist and junior mining company to evaluate the potential economics of projects at an early stage. Important consideration in developing the software has been to ensure a user-friendly format to encourage computer novices to carry out evaluations while retaining maximum flexibility to allow sophisticated sensitivity studies and for modifications to accommodate non-standard projects.

MINE-FORECAST operates within the environment of LOTUS 1-2-3 as a self-loading macro. The user is presented with a variety of menu choices and input prompts which allow for the rapid build-up of a model of the project in worksheet format. The resulting worksheet model can be saved for further modifications using the many tools packaged with LOTUS 1-2-3.

1.0 INTRODUCTION - REASON FOR DEVELOPING THE SOFTWARE

One of the major challenges that faces the exploration geologist is how to make the decision on whether a particular project should be further pursued. By spending additional exploration dollars, more information will be obtained. Conversely, by abandoning the project, in favour of one with better chances of success, losses will be cut in a timely fashion.

This critical decision is usually based on the geologist's perception of the economic viability of the target being investigated. Even at the earliest stages of project development there is likely to be sufficient information available on which

to attempt this assessment. The target commodities will be known, the location of the deposit will suggest what infrastructure will be required, the size of anomalies, alteration patterns, and mineralized zones in trenches and initial drill holes will suggest the likely tonnage potential of the target. Having arrived at this initial estimate of likely grades and tonnages the geometry of the target is assessed to decide which mining method would be most appropriate.

The critical decision on whether this perceptual target would be economically viable is usually strongly biased by the level of knowledge and experience of the explorationist concerned. It will likely be based on knowledge of capital and operating costs for similar deposits that the decision maker might be familiar with through the literature, or from personal experience.

Assistance in assessing the risk on a more rigorous basis is available from the literature. Landmark papers have been published by O'Hara (1980, 1982, 1987) which give rules of thumb for estimating most of the operating and capital costs associated with mining operations. On a more detailed basis both the CIM (Mular, 1982) and the U.S. Bureau of Mines (1975, 1987) have published data from which individual cost components can be estimated. Canmet (1986) has recently published a useful overview of such techniques for the assistance of the developers of small mining operations.

Few geologists have the time, or expertise, to fully utilize these available tools effectively. M.S.S. Ltd. realized that it would be both appropriate and useful to create a software package that would make them available, in a user-friendly fashion, within the environment of IBM-PC and compatible micro-computers.

The software was designed to have the following characteristics:
- Simple to operate
- Reliable
- Inexpensive
- Capable of estimating all but the most basic of parameters
- Well documented
- Capable of performing sensitivity analyses on key parameters

- Good graphics and reporting capabilities
- User modifiable to incorporate other data available

2.0 CONCEPTUAL DESIGN OF MINE-FORECAST

In order to allow the user to rapidly make "what-if" evaluations on the project it was essential that the data be presented in spreadsheet format. Provided that all the cells within the spreadsheet are suitably interlinked, it is then possible to carry out sensitivity studies by changing one or more of the input variables and observing the resulting changes as they ripple through the cells containing the dependent variables throughout the spreadsheet.

Of the available spreadsheet software, LOTUS 1-2-3 was selected because of its widespread user base and because of the built-in software features such as data table construction, graphic interaction and macro programming tools.

A user could learn to work with LOTUS 1-2-3 and then construct a suitable spreadsheet to model the deposit type under consideration. To do this would require selecting the appropriate estimating equations from those available in the literature. In order to facilitate this step it was decided to construct a library of such equations and store them within a LOTUS apreadsheet.

An "autoexec" macro was then designed which presents the user with choices as the model is built up.

Depending on the selections made, the appropriate estimating equations are taken out of the "library" and installed, with suitable explanatory text, within a spreadsheet framework.

3.0 THE MINE-FORECAST PROGRAM

The program has been designed around a series of modules, some of which require user input, and others which are completely under the control of the program. A brief outline of these modules is presented below.

3.1 Introduction

In this module the basic parameters are chosen to describe the property under evaluation. The user's name, the name of the deposit and the date of the run are all incorporated into this section of the spreadsheet.

The user is asked to give mineral inventory grades and tonnages for the deposit. Initial metal prices are also required at this time. To assist in the selection of appropriate prices a data-base is provided which gives the historical prices of the metal, in both constant and current dollars, over an eight-year period. The metals currently available for analysis by MINE-FORECAST are gold, silver, copper, lead, zinc, molybdenium and mixed ores of these metals.

3.2 Mining Parameters

The user is asked to choose between an open-pit or underground operation.

If an open-pit scenario is chosen then questions are asked about waste:ore ratios, pre-production stripping and prevailing topography. Given the ore tonnage and the stripping ratio, the total tonnage to be mined is derived. Using Taylor's Rule for an open-pit, an optimum operating rate is selected. This in turn gives the expected lifetime of the operation.

If underground operation is selected the user can choose between the common mining methods (cut and fill, shrinkage stoping, room and pillar, and blast hole stoping). In order to estimate mine dilution and recovery, the user is required to estimate the dip of the ore zone and the average stoping width expected. These dilution and recovery factors are applied to the mineral inventory tonnage and grade to provide revised figures which will more closely approximate the mill head grades to be expected. The revised tonnage is input into Taylor's Rule for underground operations to derive an optimum mining rate and mine lifetime.

3.3 Milling Parameters

In this section, the program uses the adjusted millhead grades and annual mining rates derived in the previous module. The metallurgical recovery of each element is estimated and the quantity of each element recovered per year is calculated. The number of tonnes of concentrate produced are calculated. The net smelter value of this concentrate, less transportation charges, gives an Annual Revenue at the Minesite. If several concentrates are produced, provision is made for sending them to different smelter destinations and the overall annual revenue is the sum of that derived from each concentrate. Provision is made for precious metal credits in the base metal concentrates.

Most of this proceeds automatically. The user sees graphs displayed illustrating the metallurgical recoveries to be expected for the different metals as a function of the millhead grades and the complexity of the ore type selected. The only user input required relates to estimation of transportation charges for concentrates. For each concentrate the transportation distances by road and/or rail are requested. Provision is made for ocean freight to Europe or Japan from either the eastern or western seaboards. The resulting concentrate transportation charge per tonne is subtracted from the Net Smelter Value per tonne of concentrate in estimating the Annual Revenue at the minesite.

The vertical axis of the spreadsheet required for this section varies greatly between a simple gold ore and a complex base metal ore with three different concentrate specifications. Thus, cell addresses for this, and subsequent sections, will not be constant. The use of range names within the controlling macro is therefore indispensible.

3.4 Operating Costs

No user input is required in this section. The program looks at the annual tonnage of ore produced from the mine and converts this to daily operating rates, assuming that mining will proceed for 5 days/week and that mill-

ing will be on a continuous 7 days/week basis.

The labour and supplies cost/tonne for the mining operation will vary depending on the mining method selected. The similar milling costs will vary depending on which ore type was selected.

A detailed estimation is then made of the number of employees that will be required to carry out the different functions in the mine, mill, and mine-services areas.

A similar estimation is also made for operating costs attributable to administration and general mine services.

The sum of all these costs are then totalled to give an overall Annual Operating Cost.

3.5 Capital Cost of Process Plant

The cost of constructing a mill can vary widely depending on local factors of climate and topography as well as ore type. The user is required to give information on the local topography, the medium on which foundations will be based, the temperature and snowfall conditions, and the likely hardness of the ore.

Given this information, appropriate estimating equations are provided for the costs of plant-site clearing, pouring foundations and construction of crushing plants, ore storage facilities, concentrator building, grinding section, flotation section, thickening and filtering sections, and concentrate storage and loading facilities.

The sum of these individual items gives the overall capital cost of the milling plant.

3.6 Mine Capital Cost Estimation

The format of the spreadsheet for this module will vary depending on whether an open-pit or underground operation is envisaged.

In the open-pit case, capital items considered include:

- The cost of pre-production overburden removal which is assumed to be done on a contract basis.

- Preproduction rock stripping - it is assumed that equipment used for this purpose will rarely be suitable for production mining, thus allowance is made here for equipment amortization costs over the pre-production period.

- Site preparation, plant area and pit roads - the cost of these items is a function of the local topography and tree cover.

- Mining equipment and maintenance facilities - capital cost estimates for shovels, haulage trucks, and other production equipment is based on the total daily pit tonnage of ore and waste. This in turn will be dependent on the waste:ore ratio provided earlier.

For an underground mining operation, provision is made for either shaft or adit access. If a shaft is required user input is requested on development depth and rock competence to decide which type of shaft is appropriate. The cross-sectional area of the shaft, be it circular and concrete or rectangular and timbered, depends on the daily tonnage to be hoisted.

The parameters for the Hoisting Plant sizing and cost are then estimated. The hoist drum diameter depends on the depth of the shaft and the number of tons to be hoisted daily. The hoisting speed is a function of the same parameters and is also estimated at this point. Given the diameter of the hoist drum, and the hoisting speed, we can estimate the motor horsepower required. The motor horsepower and drum diameter are the main controlling features on the capital cost of the hoisting equipment. The size of the hoistroom must be sufficient to accommodate the hoisting drum and mine compressors. This information is used to estimate the construction cost for the hoistroom and the cost of installing the hoist.

Headframe height must be sufficient to permit ore skips to dump into ore bins of sufficient size and allow for sufficient room for skip overtravel and braking. This height is estimated based on hoisting drum diameter and daily tonnage hoisted. The weight and cost of structural steel required to build the headframe will depend on the headframe height and safety requirements related to drum diameter.

The amount of necessary pre-production development in the form of drifts, ramps, and raises is based on the assumption that initial mine development will develop ore reserves sufficient for

675

40% of the planned mine life. A formula based on the daily milling rate and indicated stoping width is used to estimate this figure.

The size, capital cost, and installation cost of required compressor equipment depends on the daily mining rate. This rate also determines the cost and installation charges for the underground mining equipment used for development, loading, stoping, haulage, pumping, ventilation, and crushing. The final figure estimated in this section for an underground mine is the cost of maintenance facilities for repair of drilling, loading, and haulage equipment. The sum of all the above is added together on the final line of this module to give the Capital Cost of the underground mine.

3.7 Plant Utilities and General Services

This area of capital costs can be a surprisingly large proportion of the total pre-production capital expenditures. To help in their estimation, the user is asked to answer some questions about prevailing logistics at the site.

If a transmission line is available locally, provision has to be made for the cost of extending this to the mine site, as well as for the cost of a stepdown transformer substation at the mine and local distribution of power around the plant.

If it is necessary to install a diesel generator on site, the peak load figure is estimated to determine the capital cost requirement. The cost of low voltage local distribution of power to the mine and the mill will be similar to those for an existing utility grid.

Tailings storage is one of the most difficult figures to make generalizations about, as so much will depend on local condition of topography and prevailing environmental constraints. The cost estimated should be considered to be a minimum and corresponds to those cases where suitable tailings storage can be developed under favorable conditions of topography, accessibility, and minimal environmental effects within 3 kms of the mill site.

Large quantities of water are utilized in mining and milling operations. Large mines are usually required to reclaim water from their tailings storage area for recycling through the plant. Smaller mines can sometimes operate with fresh water only, providing it is environmentally safe when it re-enters the river system.

An estimator is given for the number of gallons/minute (GPM) of reclaim water required to operate the mill at the site.

Capital costs of general plant services includes provision for such items as the following: a warehouse, maintenance shop for mill, vehicle garages and employees parking lot, change houses and "dry", general purpose vehicles, administration offices, and security and fencing.

The user is asked to provide a figure for the required number of kms of access road during the input prompts for this section. This line in the program estimates the cost of providing a 10 m wide gravelled road with adequate drainage and gradients to allow for concentrate haulage. The topography is assumed to be moderate and local sources of gravel are available.

During the input prompts for this Section, the user is asked whether the mine site is within daily commuting distance of an existing community. If this is the case, provision for housing key staff employees only is made.

In an isolated situation, the total number of employees required is considered. If this is less than 100, MINE-FORECAST estimates the cost for a bunkhouse style camp with single quarters, mess hall, and leisure facilities for 95% of the employees, and family style housing provided for only the remaining 5% of senior staff employees.

If the total number of employees is greater than 100, it is assumed that the operation is large enough to justify a family style townsite. It is assumed that 65% of the employees will be housed in family style detached or apartment units while the remaining 35% will require single accommodation.

The sum of all the above items is incorporated into the determination of a figure for Utilities and General Service Capital Costs.

3.8 Project Overhead Costs

No user input is required for completion of this module. It has been broken into 3 components: Feasibility Studies, Project Supervision and General Administration. The following overhead percentages have been used:

Direct Project Costs for Mine, Mill and Utilities:

	< $100 Million	> $100,000
Feasibility Studies	5%	8%
Project Supervision	8%	10%
General Administration	4%	7%

3.9 Sustaining Capital Costs

An annual figure is obtained here which is required to cover the cost of capital items requiring replacement in the ongoing mining and milling operations.

3.10 Summary

For convenient reference, the previously derived figures are summarized in this module. New figures that appear in the summary list for the first time include the Total Pre-production Capital Costs and the Payback (in years).

The Pre-production Period in years is also defined in this section. The allowances for the length of the pre-production period depend on the annual mine capacity and vary from one year to five years.

3.11 Cash Flow Table

The total Pre-production Capital Costs are divided by the number of pre-production years and distributed accordingly into the table. The sum of the pre-produciton years plus the mine lfietime is used to decide in which year the Working Capital will be recaptured.

The Annual Revenue, Sustaining Capital Costs, and Annual Operating Costs are then placed into the table for the years

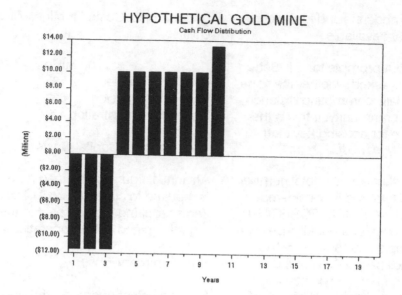

HYPOTHETICAL GOLD MINE
Cash Flow Distribution

of operation. Provision is made for a total of up to 20 years of operation plus pre-production period. The sum of revenues and costs are displayed after year 20 for the lifetime of the operation. The cash flow, either positive or negative, is shown for each year and the whole is summarized in a graph.

3.12 Financial Parameters

Below the table, a summary of financial statistics is provided which gives the net present value of the cash flow discounted at rates of 8, 10, 12, 14, and 15%. The Discounted Cash Flow Rate of Return (DCFROR) or Internal Rate of Return (IRR) for the cash flow is also displayed.

4.0 OVERRIDING MINE-FORECAST ESTIMATES

In some cases hard data may be available on such items as metallurgical recoveries which the user would prefer to incorporate instead using of the estimates made by MINE-FORECAST. This is simply done by editing the appropriate cell in the spreadsheet to contain the desired value.

If the user feels that estimates are either too low or too high to suit their particular circumstances, the MINE-FORECAST estimates can be adjusted. An appropriate adjusting factor is inserted by multiplying the estimating equation in any cell by a factor such as 0.8 (80% of estimate) or 1.2 (120% of estimate).

5.0 SENSITIVITY ANALYSIS

Having constructed a basic spreadsheet model for the deposit, the user is able to rapidly carry out sensitivity analyses using the built-in features packaged with LOTUS 1-2-3.

The simplest approach is to change any

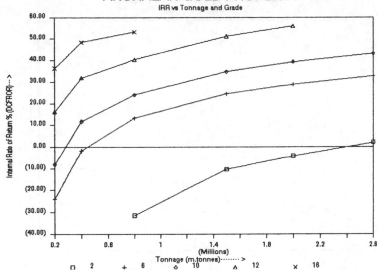

ARCHAEAN GOLD PROPERTY

IRR vs Tonnage and Grade

Internal Rate of Return % (DCFROR)···>

Tonnage (m.tonnes)········>

□ 2   + 6   ◇ 10   △ 12   × 16

of the basic variables initially supplied by the user. For example, the gold price cell could be changed in steps from $300/oz to $600/oz. By pressing "F9" (the recalculate key), the spreadsheet, including any graphs, is updated to reflect this new value. By changing individual items, such as ore grades, tonnages, stoping widths, etc., an idea can be rapidly gained of the sensitivity of the model to these various parameters.

LOTUS provides a very useful feature for enhancing this procedure, known as the Data Table Function. A table is constructed using ore grades varying, for example, from 4 g/tonne Au through 20 g/tonne Au. Column heads are then inserted for variables dependent on this parameter. Examples would be Annual Revenue, Payback, Metallurgical Recovery, Internal Rate of Return, etc. LOTUS will then automatically complete the table and fill in the appropriate values.

It is also possible to look at two independent variables, such as tonnage and grade, construct a table using these as row and column heads, and then have LOTUS complete the table with values of a variable dependent on these two parameters. A useful example would be to examine the variation of the Internal Rate of Return as a function of grades and tonnages.

Using LOTUS graphics the data from such sensitivity studies can be readily incorporated into appropriate illustrations for use in summary reports.

6.0 CURRENT DEVELOPMENT

MINE-FORECAST II is currently under development. This is building on the basic software of the original design with added enhancements to make the program more useful. Menu selections have been emphasized in this version.

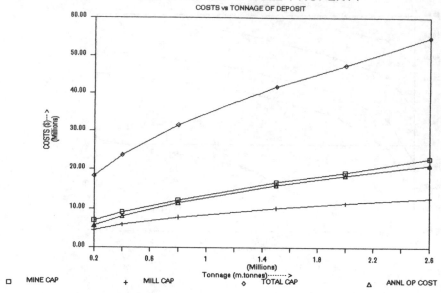

HYPOTHETICAL GOLD PROPERTY
COSTS vs TONNAGE OF DEPOSIT

| □ MINE CAP | + MILL CAP | ◇ TOTAL CAP | △ ANNL OP COST |

7.0 CONCLUSION

The sequence of the program stops at the end of each module and allows the user to cycle back through. In this way, the effects of using, for example, different mining methods can be evaluated before a final choice is incorporated into the model.

Sufficient data for a wider range of metals, ore types, and metallurgical processes (such as heap leach gold recovery) are now available and are being incorporated.

Wage and cost differentials for Canadian Provinces are being incorporated as well as provision for tax and royalty calculations appropriate to the jurisdiction for the project being modelled.

Initial feedback from users, which include major mining companies and government agencies, indicate that the software is fulfilling the role it was designed for, i.e., allowing for a rapid means of evaluating a project's viability at an early stage, at minimum cost and maximum convenience.

REFERENCES

Canmet, 1986, "Estimating Preproduction and Operating Costs of Small Underground Deposits", Project No. 140901.

Mular, A.C., 1987, "Mining and Mineral Processing Equipment Costs and Preliminary Capital Cost Estimations", Canadian Institute of Mining and Metall., Special Volume 25.

O'Hara, T. Alan, 1980, "Quick Guides to the Evaluation of Orebodies", Canadian Mining and Metall., Bull. 73(814):87-99.

O'Hara, T. Alan, 1982, "Mine Evaluation", Chapter 6, in "Mineral Industry Costs", Northwest Mining Association.

O'Hara, T. Alan, 1987, "Quick Guides to Mine Operating Costs and Revenue", Paper No. 186, CIM, 89th Annual General Meeting, Toronto, May 1987.

U.S. Bureau of Mines, 1975, "Capital and Operating Cost Estimating System Manual for Mining and Beneficiation of Metallic and Nonmetallic Minerals Except Fossil Fuels in the United States and Canada".

U.S. Bureau of Mines, 1987, "Bureau of Mines Cost Estimating System Handbook", I.C. 9142.

Computer Applications in the Mineral Industry, Fytas, Collins & Singhal (eds)
© 1988 Balkema, Rotterdam. ISBN 90 6191 760 3

Orebodies and mine planning 3-D modelling, Phase I: Geological contours and ore reserves assessment

A.Boyer, N.Billette & R.Boyle
CANMET Mining Research Laboratories, EMR Canada, Ottawa

ABSTRACT: In 1985, the Mining Research Laboratories carried out preliminary resources and project planning to permit future in-house research directed at developing mine expert systems. With the acquisition of the necessary graphical computer equipment and completion of previous research commitments, a project was initiated in 1986 to develop a 3-D mine modelling and planning system.

The first phase of the project which was completed earlier this year involved the adaptation of the CANMET's well known CADD/GEM package to permit the 2-D and 3-D modelling of orebodies. The adaptation is the subject of the presentation.

RÉSUMÉ: En 1985, les Laboratoires de recherche minière de CANMET ont fait un examen approfondi de leurs ressources et projets, afin de définir l'orientation de leur recherche interne en matière de systèmes experts miniers. L'acquisition des outils informatiques et graphiques requis et l'aboutissement des projets en cours ont permis de démarrer en 1986 un projet de modélisation tridimensionnelle des gisements et leur planification informatisée.

La première phase du projet qui a été complétée plus tôt cette année concernait la conversion en Fortran 77 du progiciel CADD/GEM bien connu, afin de permettre la modélisation graphique des gisements, bidimensionnelle d'abord, puis tridimensionnelle par la suite. La présentation se limite à décrire les résultats de ce travail.

INTRODUCTION

The Mining Research Laboratories' Ore Reserves Assessment Group (ORAG) was established in 1975 to provide independant assessment of Canada's uranium reserves and mine production capabilities to meet Federal uranium policy information requirements. In carrying out the reserve assessment task, the group was faced with handling a large quantity of core log data from occurrences, deposits and orebodies across Canada. To efficiently store and manipulate the data and subsequently carry out ore reserve estimation studies, ORAG developed data acquisition and ore reserve assessment systems for use on Control Data Corporation CYBER computers. Both systems have since been commercialized and are presently used by companies and agencies in Canada and abroad in the assessment of very diverse commodities.

The CADD system (Coding and Analysis of Drillhole Data) is used to code, edit and register basic drillhole data for use in the subsequent evaluation of reserves. The GEM system (Geostatistical Evaluation of Mines) is a set of computer programs allowing generation of statistics, variogram computation and kriging estimations with respect to the reserves. The two systems combined consist of twenty-five FORTRAN and COBOL computer programs.

Recently, the CADD and GEM programs have been converted to FORTRAN 77, to end machine dependency. Debugging is now completed; the new GEM software can also be used on VAX and SUN systems. Presently, effort is being made to modify some of the GEM programs to take advantage of graphics modelling.

BACKGROUND

Despite the persistent need of geologists and mining engineers to visualize complex geological structures in space, ameanable or not to economic exploitation, there has been resistence on their part to use mathematical models as an aid. This attitude stems from the complexities of ore reserve delineation and the oversimplifications that can be provided by mathematical models. Also, when only imprecise data is available, the use of rough methods of estimation and planning are reasonable.

With the drop of commodity prices in the 1980s, geologists developed a need for upgraded raw geological data for decision making purposes. The availability of upgraded data bases has led to greater use of improved analytic methods in ore reserve assessment. It is still necessary, however, for a mine planner using these improved methods to understand the structure and how they should be zoned for modelling.

Visual representation of ore deposits for study purposes, however, remains a major handicap to the development of mathematical models in the mineral industries. Quantum advances in computer miniaturization over the last decade, however, has resulted in new graphical tools that are capable of representing orebodies and other structures in three dimensions. Until now, most software has developed to provide only 2-D representation of underground structures. However, three-dimensional modelling is becoming increasingly popular, but at additional costs to the user. Its use also requires a more powerful computer system – a supermicro or better, if a suitably fast response graphics terminal is to be available for design studies.

It is MRL view that it is important for the mining industry to develop a system which will permit the three-dimensional representation of geological and mine structures with facilities. Adaptation of the GEM packages for use on super-micro computers to meet this objective will make more accessible to geologists and mine planners models to more realistically assess limits, grades and stope sizes of orebodies.

CADD SYSTEM

Over the last decade, ORAG accumulated a very extensive primary geological data base on Canadian uranium deposits using various coding techniques. The analysis of large amounts of drillhole data is required to perform valid ore reserve calculations. The purpose of the CADD system is to code and edit such data, as well as to prepare data for use by GEM packages. The CADD system was designed using a sequential rather than a randomly indexed file system for the following reasons:

1. with a few exceptions, the amount of drillhole data per file is not excessive and can be handled by sequential files;
2. drillhole data sets are complete when received by ORAG; and
3. for most uranium deposits, all drillhole data are used in reserve estimation.

CADD can accomodate the entry of data from any standard logging system used by a mine

geologist. Preferably, the data should include the following drillcore information:
- drillhole collar location;
- drillhole deviations;
- drill core density;
- percent core recovery;
- chemical assay of core, and/or radiometric readings;
- simple geological, lithological and structural information.

With respect to the last point, it is often necessary for the geologist entering the data into the CADD system to sort and select the necessary lithological information from total available drill log information.

The coding, correction and adaptation of drillhole data to permit ore reserve estimation studies, in ORAG's experience, is a major task. A major percentage of a reserve estimation study must be spent in data validation. Error detection requires that specialized staff be used to create the computer data files. Drillhole coordinate or grade errors can have dramatic effects on forthcoming estimations. Most errors can be easily identified, when expert technical staff is used in data assemblage, but there are always some undetected errors in the final files. Some of these will be identified later in the final stage of the estimation study, because values are provided which are inconsistent with other geological information.

To aid user use of the initial card system, an interactive data entry has been developed, called Computer Drilling Data Entry/version 1.0 (CDDE/1.0). The present version is written in VAX 11 Fortran – and presently being used on a VAX 11/750 VMS.

The CDDE/1.0 program was designed to utilize standard drillhole information, and to allow very rapid data entry and validation(Fig. 1 & Fig. 2):
- by utilizing the *distance from the collar* as basic program information;
- by entering data columnwise, rather than rowwise;
- by having a program provision to enter repetitive data extremely rapidly;
- by having a program provision to rapidly correct errors of commission or omission.

After completion of each form, and again at the completion of all forms, the user has an opportunity to view, correct, modify the completed form(s), and/or to proceed. After data storage, however, no more data modification is possible with the CDDE/1.0 program. Further modification of entered data is possible with system editors.

The same data entry procedure may be adapted in the future for use with the Laboratories' graphics terminal, using a mouse or window menu-driven software. However, it is not a priority area of development for the Laboratory. It is presently planned to enter data to the Laboratory's SUN system from the Laboratory's VAX mini-computer.

GEM SYSTEM

The GEM system is comprised of about 20 software programs assembled into four different option groups (Fig. 3). The first software option is oriented towards providing the user means of displaying diamond drillhole intersections, a profile of grades and economic intersections in two dimensions (Fig. 4 & 5). The second software option divides the data into equal length segments and evaluates reserves using moving average or inverse squared distance methods. The third software option produces for further analysis basic statistics, histograms and other graphics (Fig. 6), variograms by either the sector or the blocking methods, and trend and generalized covariance data (Fig. 7). The fourth software option permits access to various estimation procedures: polygons, kriging – point, block, cluster or universal, and generates isocurves or tables of reserve inventory.

685

It is thus clear that the GEM system can access many different estimation procedures. These procedures can be classified as follows (Fig. 3):
— geometrical models: sectional or polygonal;
— inverse distance models: squared or moving average;
— geostatistical models: block, cluster, normal, lognormal or universal.

Evaluations made on tabular or 2-D deposits are distinct from evaluations made on all other types of deposits (3-D).

The GEM software used to construct geometrical models starts the process by computing composite intersections (Fig. 4) or equal length segments along drillholes. The location of the composite segments defined by the geometrical modelling are indicated as 2-D model plans. Graphical representation can also be made as isocurves when the variable to highlight is distributed on a regular grid.

For the purpose of discussion, the inverse distance models are considered a moving average method of estimation. For stratabound or tabular orebodies, two-dimensional variables are defined. The same procedure of composite intersections is used to generate basic data for the model. Programs are then run to assess the average grade of regular blocks using the moving average model. Maps are printed to scale, showing tonnages, amounts of metal and average grade of each block and of the deposit, in relation to cut-off grades as specified by the user. Mineralized block contour maps can also be generated.

Discussion of GEM's geostatistical capabilities will be limited to discussion of three-dimensional geostatistical estimation. In the simplest case, only one variable, the grade, will be assessed using equal length segments. With the GEM system, three-dimensional experimental variograms are first produced (Fig. 7) and with them a study is carried out of the proportional effect, looking at local variances and means.

Theoretical models are then fitted on experimental variograms and used for kriging purposes: normal, lognormal, block, random or universal. Estimated variable within blocks with its kriging variance is then printed on maps generated to scale. Isocurves of estimated variables and variances can be produced as well.

All original software was developed using a Control Data Corporation Fortran compiler with extensions. In the last few years, our Laboratory has converted all GEM software to Fortran 77, a universal language for most computers. GEM and CADD programs can now be run as well on IBM and VAX machines. It is evident that a VAX 11/750 system cannot produce results at the same speed as mainframe computers. Until now, however, this has not been a major handicap.

MRL is presently in the process of modifying the GEM package so that it can be used on the Laboratories SUN computer, a super-micro enhanced with a colour graphics monitor. Initially, only 2-D software will be transferred. In a few months, 3-D graphics will be considered. Mining Research Laboratories' personnel believes that 3-D modelling of orebodies is an essential key to better and safer mine design.

CONCLUSION

The Ore Reserve Assessment Group of CANMET's Mining Research Laboratories has been active in developing tools permitting more precise and elaborate ore reserve estimation studies of orebodies. Information contained within the national Uranium reserve data bank is made available to owners in various ways for their use on request. Recently, the new owners of a prospect asked to be provided with the most updated information about their new deposit.

The first phase of modernization of the CADD/GEM package is underway. On the second phase of the project, it is planned to develop a mine modelling and planning system for one mining method, which will permit simulation of openings and assessment of stability, taking into consideration the uncertainties of orebody shape. In subsequent phases, research will be directed at extending the project so that other mining methods can be treated, ore reserves assessment carried out and expert criteria for mining method selection become available for use.

CANMET is interested in providing the Canadian mining industry with the most advanced tools for making the best possible decisions concerning resource utilization. The introduction of 3-D graphics is seen within CANMET as a major step in improving decision making. MRL is responding to this perceived need by transforming existing proven software to a suitable new computer system. Future steps will involve the development of new software to support 3-D modelling of mining methods, and simulation of long term stability. CANMET hopes to provide industry with a new sophisticated decision tool in the next few years.

REFERENCES

Boyer, A. & R. Boyle 1987. A user-manual of the CDDE/1.0 computer program – optimization of drill log data entry into a computer data bank. Mining Research Laboratories Division report MRL 87-8 (TR), 96 p.

Sabourin, R. 1983. Utilisation de l'ordinateur pour l'évaluation des ressources – Systèmes CANMET pour l'évaluation des réserves minérales. Laboratoires de recherche minière, rapport de division LRM 83-15(OP), 18p.

Sabourin, R. 1983. Made in Canada software brings mining into computer age. GEOS, Vol. 12, No. 3, pp.20-21.

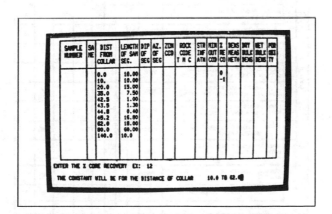

Fig. 1. CADD system — Fast data entry procedure: repetitive data in % core recovery column

687

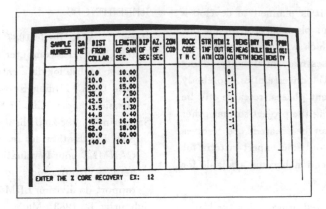

Fig. 2. CADD system —
% core recovery column completed

DESCRIPTION OF COMPUTER PROGRAMS			CANMET PROGRAM NAME	TRADITIONAL RESOURCE EVALUATION METHODOLOGY: GEOMETRIC METHODS – SECTION		GEOMETRIC METHODS – POLYGON		MOVING AVERAGE AND DISTANCE WEIGHTING METHODS		GEOSTATISTICAL METHODS	
TABULAR DEPOSITS 2 DIMENSIONAL				✓		✓		✓		✓	
ALL DEPOSITS 3 DIMENSIONAL					✓		✓		✓		✓
CADD		DATA PREPARATION	1000 1600	✓	✓	✓	✓	✓	✓	✓	✓
GEM SYSTEM	GROUP 1	ECONOMIC INTERSECTIONS	100	✓	✓	✓		✓		✓	
GEM SYSTEM	GROUP 1	GROUPING ECONOMIC INTERSECTIONS	2010			✓		✓		✓	
GEM SYSTEM	GROUP 1	GRAPHIC PROJECTION OF DRILLHOLES ON SECTION OR PLAN	3000	✓	✓	✓	✓				
GEM SYSTEM	GROUP 2	EQUAL LENGTH SEGMENTS ALONG DRILLHOLES	2040				✓		✓		✓
GEM SYSTEM	GROUP 2	GROUPING EQUAL LENGTH SEGMENTS WITHIN UNIT BLOCKS	2050					✓	✓		✓
GEM SYSTEM	GROUP 2	EVALUATING BLOCK GRADE USING MOVING AVERAGE OR INVERSE DISTANCE	2055					✓	✓		
GEM SYSTEM	GROUP 3	BASIC STATISTICS - HISTOGRAMS DISTRIBUTION TESTS - REGRESSIONS	2100							✓	✓
GEM SYSTEM	GROUP 3	VARIOGRAMS AND TRENDS COMPUTED BY THE SECTORS METHOD	2200							✓	✓
GEM SYSTEM	GROUP 3	VARIOGRAMS AND TRENDS COMPUTED BY THE BLOCKING METHOD	2250							✓	✓
GEM SYSTEM	GROUP 3	TREND AND GENERALIZED COVARIANCE ANALYSIS •	2350							✓	
GEM SYSTEM	GROUP 4	NORMAL - LOGNORMAL - CLUSTER - UNIVERSAL POINT - BLOCK - POLYGON - KRIGING	2300							✓	✓
GEM SYSTEM	GROUP 4	NORMAL - UNIVERSAL - POINT - BLOCK KRIGING •	2350							✓	
GEM SYSTEM	GROUP 4	TABLES OF RESERVES INVENTORY	2400							✓	✓
GEM SYSTEM	GROUP 4	ISOCURVES TRACING	2500			✓	✓	✓	✓	✓	✓

• This program is not available

Fig. 3. GEM system — evaluation methodologies
(after R. Sabourin's article in GEOS, 1983/3)

```
COMBINED WITH        ZZX0039        DEPARTURE                7600.00   DDH TYPE                     AXT       U308 CUT-OFF GRADE      2.00
COMPANY DRILL HOLE NUMBER   77032   LATITUDE                 8540.00   HOLE DIAMETER (INCHES)   0.000      MIN. MINING WIDTH      10.00
SPECIAL DRILL-HOLE IDENT.            COLLAR ELEVATION         1210.50   INITIAL CASING TYPE                    MIN. WASTE THICKNESS    0.00
DRILLING STARTED       15/10/77     HOLE LENGTH                60.50   INITIAL CASING LENGTH     0.00      MAX. DIL. WIDTH         0.00
DRILLING COMPLETED     15/10/77     DIP OF HOST FORMATION       -60   FINAL CASING TYPE                     TOP CUT WIDTH           0.00
                                    AZIMUTH OF DIP OF H.F.      170   FINAL CASING LENGTH       0.00      ZONE(S)    ALL
LOCATION  UNDERGROUND HOLE , 200 LEVEL , K ZONE                        NUMBER OF SEGMENTS          20
```

```
                                                    U 3 0 8          C M.     D I L U T E D / M I N E D - O U T   P O R T I O N
                      DIST.  SEGMENT                PERC.            C M.              U308         METAL PER UNIT AREA
         SAMPLE       FROM   THICKNESS  SEGMENT R  R  CORE.A C       G O.DE    AVG. GRADE   G R O S S     N E T      THICKNESS     ID
         NUMBER       COLLAR APPAR. APPAR.DIP AZI. C  T  REC.M M  GRADE I I.II DIL. MINED  DIL.  MINED  DIL.  MINED DIL. MINED  D M

          32-1    0.00   2.80   2.80  0 180  K     100 C E   0.00 -
          32-1   -2.80   3.80   3.80  0 180  K     100 S B   1.56 -
          32-2   -6.60   2.10   2.10  0 180  K     100 S B   4.86 +   D1    5.11       9.11        5.54       21.40     01
          32-3   -8.70   2.80   2.80  0 180  K     100 S B   7.86 +   D1
          32-4  -11.50   2.30   2.30  0 180  K     100 S B   9.24 +   D1
          32-5  -13.80   2.50   2.50  0 180  K     100 S B   4.34 +   D1
          32-6  -16.30   3.70   3.70  0 180  K     100 S B   4.30 +   D1
          32-7  -20.00   1.80   1.80  0 180  K     100 S B   3.66 +   D1
          32-8  -21.80   0.50   0.50  0 180  K     100 S B   3.66 +   D1
          32-9  -22.30   1.20   1.20  0 180  K     100 S B   9.36 +   D1
          32-10 -23.50   1.50   1.50  0 180  K     100 S B   0.16 -   D1
          32-11 -25.00   1.70   1.70  0 180  K     100 S B   3.08 +   D1
          32-12 -26.70   1.30   1.30  0 180  K     100 S B   3.08 +   D1
          32-13 -28.00   0.90   0.90  0 180        100 S B   0.98 -
          32-14 -28.90   1.10   1.10  0 180        100 S B   0.98 -
          32-15 -30.00   1.80   1.80  0 180        100 S B   1.74 -
          32-16 -31.80   1.70   1.70  0 180        100 S B   0.62 -
                -33.50   1.20   1.20  0 180        100 C E   0.00 -
                -34.70   3.60   3.60  0 180        100 C E   0.00 -
                -38.30  22.20  22.20  0 180        100 C E   0.00 -
```

```
              DIST  APPAR.  TONS       U 3 0 8                     THO2
PORTION.      FROM  THICK-  PER  AVG.  METAL PER AREA.  AVG. METAL/. AVG. METAL/. AVG. METAL/. AVG. METAL/. AVG. METAL/. AVG. METAL/.
   ID.     . COLLAR  NESS   AREA GRADE  GROSS     NET .  GRADE AREA .GRADE AREA .GRADE AREA .GRADE AREA .GRADE AREA .GRADE AREA .

D 01TT     -6.6  21.4 1.783  5.11  9.11   5.54   0.33  0.59  0.00  0.00  0.00  0.00  0.00  0.00  0.00  0.00  0.00  0.00
E 01TT     -6.6  21.4 1.783  5.11  9.11   5.54   0.33  0.59  0.00  0.00  0.00  0.00  0.00  0.00  0.00  0.00  0.00  0.00
```

Fig. 4.   GEM system — exemple of economic intersection results (program 100)

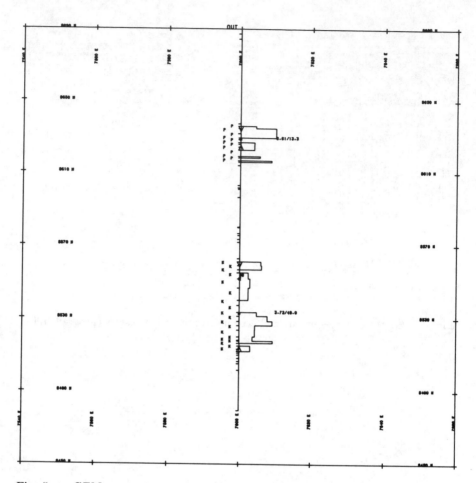

Fig. 5.   GEM system — graphical output of drillhole data with economic intersections (program 3000)

```
           SECTION  4   -   STATISTICS OF VARIABLE  1  (IREJ=0)

     TITLE - STATISTICS OF GRADE (LBS/TON)                                38 ACCEPTED OBSERVATIONS USED
                                                                             TO PRODUCE THE FOLLOWING
                                                                             STATISTICS


           38  NOBS                    3.080000  MEDIAN                 12.580000  RANGE
     3.763843  ARITHMETIC MEAN         2.348163  GEOMETRIC MEAN          1.053832  HARMONIC MEAN
    10.804060  VARIANCE                3.286953  STANDARD DEVIATION      0.873297  COEFFICIENT OF VARIATION
     0.000000  SICHEL MEAN             0.000000  SICHEL STAN. DEV.       0.000000  SICHEL COEF. OF VAR.
     3.543333  QUENOUILLE MEAN         2.865000  QUENOUILLE STAN. DEV.   0.808561  QUENOUILLE COEF. VAR.
    -0.408349  DAVIES COEF. OF SYM.    0.137559  WIJS COEF. VAR.


  FIRST FOUR MOMENTS ABOUT ZERO        0.37638426E+01   0.24686251E+02   0.21221013E+03   0.21112009E+04
  FIRST FOUR MOMENTS ABOUT THE MEAN   -0.75290080E-06   0.10519742E+02   0.40105671E+02   0.41253690E+03
  FIRST FOUR K-STATISTICS             -0.75290080E-06   0.10804059E+02   0.43477917E+02   0.11785933E+03

     1.175434  SKEWNESS (ESTIMATED FROM MOMENTS)        1.224301  SKEWNESS  (ESTIMATED FROM K-STATISTICS)
     0.727800  KURTOSIS (ESTIMATED FROM MOMENTS)        1.009695  KURTOSIS  (ESTIMATED FROM K-STATISTICS)



           SECTION  5   -   HISTOGRAM AND CUMULATIVE FREQUENCY DIAGRAM


     TITLE - STATISTICS OF GRADE (LBS/TON)

         NO. OF OBSERVATIONS =    38

         MINIMUM =   0.100000    MAXIMUM =   12.6800    MEAN =   3.76384

  UPPER   FREQ   REL    CUM  1      RELATIVE FREQUENCY DIAGRAM      1        CUMULATIVE FREQUENCY DIAGRAM
  LIMIT          FREQ   FREQ 1                                     1
   0.477    3    7.89    7.89 1****                                1****
   3.764   19   50.00   57.89 1************************            1*********************************
   7.051   10   26.32   84.21 1*************                      1********************************************
  10.338    4   10.53   94.74 1*****                              1************************************************
> 10.338    2    5.26  100.00 1***                                1**************************************************


           SECTION 5A - PROBABILITY PLOT OF CUMULATIVE FREQUENCY DISTRIBUTION
     TITLE - STATISTICS OF GRADE (LBS/TON)

                                            PROBABILITY PLOT OF CUMULATIVE FREQUENCY DISTRIBUTION.
  UPPER   CUM. PCT. .01    .1        1      5  10 16  25       50     75 84 90  95       99      99.9   99.99
  LIMIT
                     +--------+---------+------+--+-+--+---+-------+---+--+--+---+-------+-------+------+
   0.477    7.895    -                              *
   3.764   57.895    -                                          *
   7.051   84.211    -                                               *
  10.338   94.737    -                                                     *
> 10.338  100.000    -
                     +--------+---------+------+--+-+--+---+-------+---+--+--+---+-------+-------+------+
                    .01    .1        1      5  10 16  25       50     75 84 90  95       99      99.9   99.99
```

Fig. 6.   GEM system — exemple of statistics for the grade variable (program 2100)

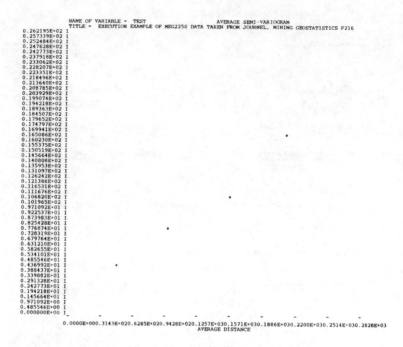

Fig. 7.　　GEM system — average semi-variogram (program 2250)

Computer Applications in the Mineral Industry, Fytas, Collins & Singhal (eds)
© *1988 Balkema, Rotterdam. ISBN 90 6191 760 3*

PC-XPLOR – The application of a microcomputer based data management and graphics system in exploration

Peter J.Franklin
GEMCOM Services Inc., Vancouver, BC, Canada

ABSTRACT: PC-XPLOR is a micro-computer based, integrated, data management and graphics software system for exploration data. PC-XPLOR is designed to store, manage, report and display any type of spatially located data. In exploration projects this could typically include field mapping data such as outcrops, features, rock types and faults; and regional geochemical sampling data such as locations, soil and rock samples and analysis results; trenching data with analysis results; drillhole data such as lithology and down hole assays; and polygonal data such as claims, property boundaries, underground workings and surface features.

The PC-XPLOR system provides a database manager that allows extremely flexible database structures, easy data entry and editing, comprehensive reporting, filtering sorting and data manipulation.

The integrated graphics provide extensive capabilities for plan and section display. Various options are provided for structure maps; X-Y plots; symbol plots; traverse maps; vertical, inclined or horizontal sections through all data; and polygon plots.

Additional analysis tools can be added to PC-XPLOR for drillhole compositing; polygon reserves; single and multi-variate statistics; geo-statistics; and gridding and contouring.

This paper describes the various modules of PC-XPLOR and illustrates the use of the system through a number of case studies from different types of exploration projects.

1. THE PC-XPLOR SYSTEM

PC-XPLOR is a data management and graphics system designed specifically for all facets of exploration data, ranging from regional mapping to geochemical surveys, from drillhole data to claim boundaries and underground workings. The system has been designed specifically for micro-computers, in particular those that utilize the MS-DOS operating system. PC-XPLOR has been designed to be an integrated package, to be relatively easy to learn and operate, and to provide a direct upwards compatible path to a complete range of other software products that allow such activities as ourebody modelling and mine design, all the way through to mine surveying and management of daily operations, to be carried out on a single, common database.

As an integrated package, PC-XPLOR is modular and menu driven. There are three main modules in the system; for Database Management; for Reporting and Manipulation; and for Graphical output.

Additional modules are available, as options, for drillhole analysis, statistics, geostatistics and contouring.

The keystone of all the modules in PC-XPLOR is a relational database designed specifically for geological and engineering data. The database offers a large degree of flexibility for the storage and management of exploration data yet is kept extremely simple in concept. With a completely user defined data structure, the database manager allows the explorationist to build different databases for different types of exploration data. As all exploration data can be readily broken down into series of related tables, users are free to specify the number of tables that constitute a database, the number and type of data values (fields) that make up tables, and series of values to be used for data checking, validation and default values. The database can hold data-types ranging from special co-ordinates through real numbers, to dates and character strings. The database sizes are restricted only by the physical storage capability of the microcomputer being used.

1.1 Module 1 - Database Management

The first module in PC-XPLOR, for Database Management, provides a series of utilities for the definition of the numbers and types of databases, individual database structures and various basic project parameters such as descriptions, locations and unit types. The module also contains an interactive screen editor that provides customized data entry and editing screens with fully automatic navigation through the database. The editor also has integrated digitizing capability so map data can be entered and edited extremely rapidly. Additional functionality is also provided in the Database Management module for direct loading of ASCII or text data from flat files with any tabular format, and for merging of data directly from ASCII or text files in a form typically provided by modern assay laboratories. The merging utility includes full indexing based on contents of specified fields, such as sample numbers, for example.

1.2 Module 2 - Reporting and Manipulation

Once data has been entered into the databases, the Reporting and Manipulation module provides facilities for filtering and sorting of subsets of data from the database and then reporting to the computer screen, to a printer or to ASCII or text files.

Reporting is through an easy to learn procedural reporting language that enables an infinite variety of output formats to be generated from the database. Reports can be made in summary form, or in as little, or as much detail as to user requires and the database allows.

Data extraction is a tool that allows subsets of data (in any form) to be extracted from any database and the location of the data converted to X, Y and Z co-ordinate form. This utility is a primary link from PC-XPLOR to other orebody modelling systems such as PC-MINE.

The ability to manipulate data within a database, for example to transform co-ordinates from one system to another or to convert assays from one unit of measurement to another, is valuable, and PC-XPLOR provides this ability through either straightforward or conditional "expressions" entered interactively by the user.

1.3 Module 3 - Graphics

Geologists tend to represent their data graphically, through various types of maps and cross-sections. The graphics module is designed to produce typical maps showing sample locations and values, trenches, drillhole locations, claim boundaries and other polygonal outlines; and also cross sections through drillhole data and three-dimensional polygon data.

The mapping capabilities include geological plans of rock types, outcrops, contacts, faults and other features digitized into the database; posting maps showing sample locations, numbers, values and descriptions; symbol maps showing sample values by an appropriately scaled symbol sizes; trench and traverse maps showing the location of samples and their values; and any features such as roads, claim boundaries, topography or underground workings that have been digitized into a polygon database.

The display of drillhole data on section is a function that is applicable to nearly all exploration projects. PC-XPLOR has an extensive section plotting capability that allows the geologist to project drillhole information onto any plane in three dimensional space from within a definable projection distance. All information outside of this projection distance is "clipped" from the plot. Any information held in a database can be used to annotate the drillhole - for example, the collar of the hole can be annotated with the hole name, the date drilled or the length of the hole, and the drillhole trace can be annotated on one side with lithological descriptions and on the other with sample numbers and a selection of assay values.

All graphical data produced by PC-XPLOR can be viewed from within the system on high resolution colour monitors, plotted to scale on a variety of pen plotters or printers, or transfer to a CAD package such as AutoCAD for detailed graphical editing and enhancement.

1.4 Additional Modules

Further analysis of exploration data is also an integral part of PC-XPLOR. Separate modules can be added to the system to provide detailed drillhole analysis; single and multi-variate statistics; geostatistics; and gridding with contouring.

Detailed drillhole analysis provides the geologist with the capability to composite assay or numeric data down drillholes by a variety of weighted average methods such

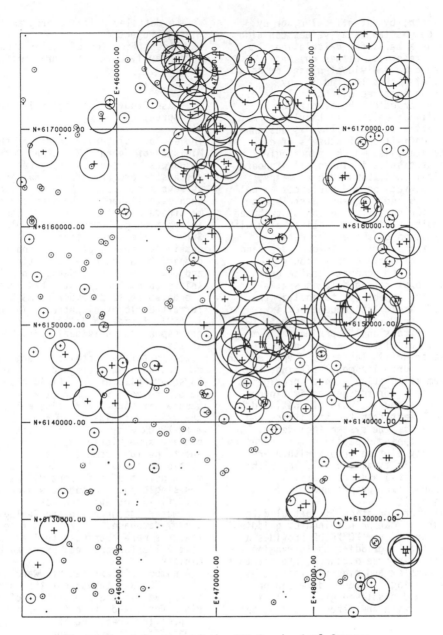

Figure 1 - A Portion of the GSB Geochemical Survey

as by length, by cut-off value, or by bench. Composited assay values can also be used as a basis for calculating polygonal reserves, either on, or projected onto individual planes or sets of parallel planes located in any orientation in three dimensional space. Polygon reserves can be displayed graphically as well as in tabular format.

The statistical capability is designed to allow definition of subsets of data prior to analysis, which, for single variate statistics includes all standard statistical indicators such as means, standard deviations and variances as well as histogram and frequency distribution analysis for normally and log distributed data. Multi-variate statistics, also with the definition of subsets of data, provides the capability to do regression analysis, calculate correlation coefficients and generate scattergrams.

As most exploration data is spatially related, the capability to analyze the degree of sample interdependency is important, so PC-XPLOR provides both two and three-dimensional experimental semi-variogram calculation on any point, traverse or drillhole data. Complete with interactive graphics model fitting, the variogram calculation program provides a valuable tool for further estimation of ore reserves and analysis of spatial data.

To complement both the statistics and geostatistics, the system includes a flexible graphing capability for producing high quality plotted graphs with user definable axis types (normal, logarithms and probability) and graph styles (histograms, line graphs and scattergrams).

Visual representation of spatial data distributions can be obtained by gridding and contouring, and PC-XPLOR provides a variety of data gridding tools ranging from simple inverse distance interpolators to more complex and rigorous techniques such as kriging. Gridded data can be contoured and displayed graphically in conjunction with any other data from the database.

2. CASE STUDIES

To illustrate the applicability of PC-XPLOR in a variety of exploration environments, three different case studies are presented, each of which utilizes a different aspect of the PC-XPLOR system.

The case studies are:
. A regional geochemical program undertaken by the Government of British Columbia

. A detailed drilling program for a mining project
. Identification of exploration targets in a heavily worked historic mining area in Utah.

2.1 Case Study 1 - A regional Geochemical Survey

The Geological Survey Branch of the B.C. Ministry of Energy, Mines and Petroleum resources has been carrying out a systematic program of regional geochemical surveys since 1976. A large database has been established on the Ministry's mainframe computer that holds the information gathered by the Geological Survey. The information is all public domain but most exploration and mining companies do not have facilities to access and make use of this data. To help solve this problem the ministry has downloaded the database to a number of MS-DOS formatted floppy diskettes. The data is in ASCII format in datafiles that correspond to individual 1:250,000 NTS map sheets.

Each datafile contains a large number of samples with an average density of one per 13 square kilometres. Field observations have been made for characteristics of the sample site and for routine multi-element analysis of field- dried sediment samples.

PC-XPLOR was used in this case to prepare some plotted output of the data and to perform some simple statistical analysis on selected elements for a number of different NTS map sheets.

A single point database was defined and data fields established for each of the parameters recorded in the Geological Survey database. PC-XPLOR's file import features were then used to load in the text files (unmodified) provided by the Ministry.

A number of text reports were made from the PC-XPLOR database but the main objective was the preparation of symbol plots for the distribution of mercury over a specified area. PC-XPLOR's indexing capability prepared a subset of data that covered the area of interest and then the Graphics module was used to prepare a symbol plot of this data. The symbol plot used circles of varying diameter plotted at each sample location to represent the magnitude of the mercury content of each sample. A portion of that plot is illustrated in Figure 1.

The statistics capabilities of PC-XPLOR were then used for the same data set to produce the histogram and frequency distributions shown in Figure 2.

2.2 Case Study 2 - A Detailed Drilling Program

An international exploration company, City Resources (Canada) Ltd., is developing a large gold deposit located in British Columbia with the aim of establishing a producing mine. The deposit, located on Graham Island in the Queen Charlotte group is due to begin production 1989. The PC-XPLOR system was used to store and manage some 120,000 feet of exploration drilling. The drilling program consisted of over 300 diamond and reverse circulation drillholes and over 15,000 assays.

PC-XPLOR was first used in the field offices for data entry on a portable computer and was then relocated back to Vancouver, to the main office of City Resources, where the software was run on a larger COMPAQ 386 with a plotter and digitizer attached. During this time the structure of the database evolved as the relevance of certain information changed and the users became more experienced with the operation of the system.

After the database had been checked and finalized, the Graphics module was used to prepare a number of sets of drillhole cross-sections, showing firstly lithology, and secondly assay data. A typical cross-section is shown in Figure 3.

Geologists then used the plotted sections as a basis for their structural interpretation. This process was aided by another software package called GEO-MODEL, designed specifically for geological interpretation on section and level plans.

The combination of PC-XPLOR and GEO-MODEL operating together was an invaluable tool for the determination of geological reserves for this structurally complex deposit.

2.3 Case Study 3 - Identification of Exploration Targets

In this case study PC-XPLOR was used in conjunction with AutoCAD to assist in the identification of exploration targets in a heavily mined area of Utah. The client held mining rights to a large number of claims covering a large underground vein mine that ceased production some twenty to thirty years ago due to poor economics. The only data available about the claims was a set of plans some thirty feet long that detailed the underground development. Records indicated that there was still substantial ore tonnage left in veins within the mine but there was no information to indicate the location of this ore other than the mine development.

Handling of the large plans was impractical and the plans also gave no easy visual indications of the vein structures.

PC-XPLOR was therefore used to build a large three dimensional database of the underground workings. A polygonal database was defined and all the underground workings were digitized in three dimensions into the database. A plan view, produced by PC-XPLOR, of the mine workings is shown in Figure 4. Once the database had been completed and checked, sets of parallel sections were cut though the workings at areas of interest and plotted to scale (Figure 5). These sections crudely showed intersections of haulages, sub-levels and raises and formed the basis for a manual interpretation of the vein structure from the underground workings.

This was done by transferring the crude cross-sections of mines working to AutoCAD and then manually inserting the probable vein locations and structure onto the section. Finished, presentation quality plots were also generated (Figure 6).

After all sections had been completed in this manner, two primary areas of interest for further exploration were defined as Block A and Block B and a drilling program designed to intersect these ares. A combined plot showing a longitudinal projection of the mine workings and a general underground mine layout plan with detail of the two potential areas of interest (Figure 7) was then prepared for future reference.

3. SUMMARY

Exploration data, by its very nature, consists of extremely large volumes of spatial, numeric and descriptive information. Such data, provided it is collected in a consistent manner is ideally suited for computerization. With suitable hardware and software such as PC-XPLOR, a geologist can manage, store, report, graphically display and manipulate his data rapidly, accurately and effectively.

The benefits of computerization can be measured in terms of the increase in conference of the accuracy of data, the speediness of turnaround, the precision of graphical data and the knowledge that statistical analysis can provide. Management can review data soon after it arrives from the field, draw conclusions and re-direct exploration efforts within days. In a business that is competitive and often unpredictable, this efficiency

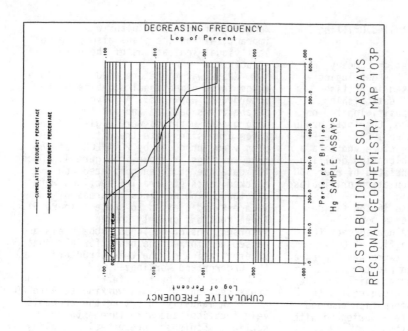

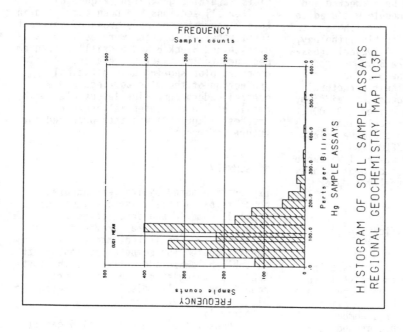

Figure 2 - Output from the Statistics Module

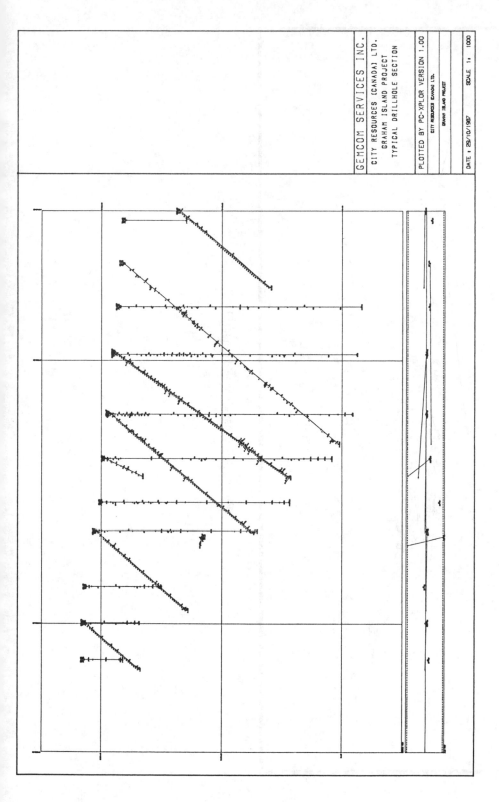

GEMCOM SERVICES INC.

CITY RESOURCES (CANADA) LTD.
GRAHAM ISLAND PROJECT
TYPICAL DRILLHOLE SECTION

PLOTTED BY PC-XPLOR VERSION 1.00

CITY RESOURCES (CANADA) LTD.

GRAHAM ISLAND PROJECT

DATE : 29/10/1987   SCALE 1: 1000

Figure 3 - City Resources - Typical Drillhole Section

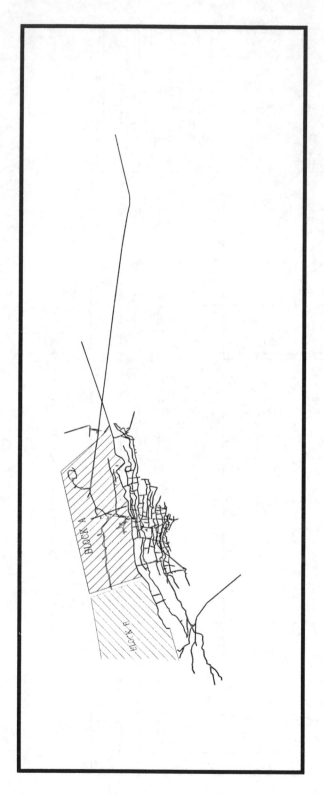

Figure 4 - General Mine Layout Plan

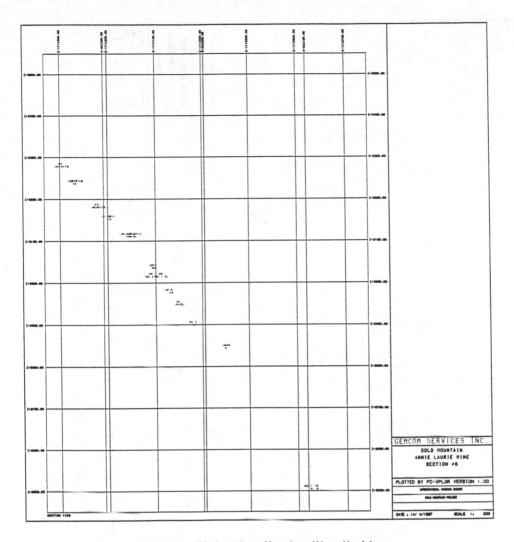

Figure 5 - PC-XPLOR Section Showing Mine Workings

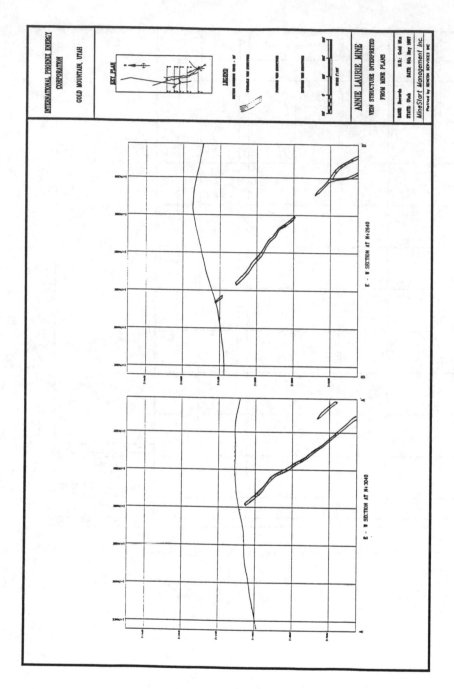

Figure 6 - Cross-Sections Showing Vein Structure

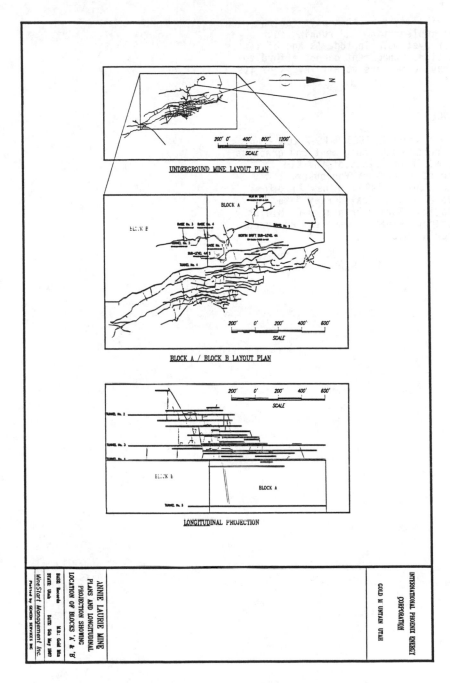

INTERNATIONAL PHOENIX ENERGY
CORPORATION

GOLD N UNTAIN UTAH

ANNIE LAURIE MINE
PLANS AND LONGITUDINAL
PROJECTION SHOWING
LOCATION OF BLOCKS 'A' & 'B'

BASE: Records   M.B.: Gold Mtn
STATE: Utah   DATE: 5th May 1987

MineStart Management Inc.
Plotted by GEMCOM SERVICES INC.

Figure 7 - Plans and Section Through Mine

and flexibility more than pays for the
cost of implementing and running such
computer systems. In today's age of
information, management cannot afford to
be without computers at all levels of its
activity.

REFERENCES

Franklin, Peter J. 1987. PC-XPLOR - An
    integrated data management and graphics
    system. Proceedings of the CIM
    District 6 Meeting, Vancouver. B.C.
Matysek, Paul F. 1986. A new look for
    regional geochemical survey data.
    Geological Survey Branch, B.C. Ministry
    of Energy, Mines & Petroleum Resources,
    Victoria, B.C.

15 Maintenance management

Computer Applications in the Mineral Industry, Fytas, Collins & Singhal (eds)
© 1988 Balkema, Rotterdam. ISBN 90 6191 760 3

Computer-based maintenance management – Australian experiences

Gordon J.Melvin & Peter McKenzie
Mincom Pty Ltd, Australia

ABSTRACT: The Australian mining industry is highly export-oriented, with international competitiveness being the key element in the survival and profitability of many mining operations. The cost of performing equipment maintenance is a significant component in the total operating costs of capital-intensive mining operations. Because of the dependence of equipment maintenance on other functional areas of the minesite, an integrated approach to the control of maintenance costs is necessary.

1 INTRODUCTION

With world markets for coal and minerals depressed, the international competitiveness of a mining company is determined by its ability to contain operating costs. Companies are now directing their attention towards close monitoring and control of maintenance and associated operational areas such as materials supply and inventory management, with a view to containing costs. Because equipment maintenance affects several departments in the operating mine – workshops, maintenance planning, purchasing, warehousing, accounting and mine planning – and since it consumes the major resources of labour, materials and equipment, it is apparent that an effective system for managing maintenance requires integration of information from a wide range of functional areas. Such a task is ideally suited to the power of modern computers.

In Australian mines, the maintenance manpower is typically 30% to 40% of the mine-site workforce and the operating costs for maintenance labour and materials is typically 50% to 60% of the mine operating costs and rising. Thus, maintenance is a significant element in the mine's cost structure and, as such, represents an area where management can focus attention on controlling costs and improving labour productivity. The maintenance function in mines is affected not only by the economic forces of the international market place, but also by the harsh environmental conditions prevalent at many minesites. Remoteness, temperature extremes and dust affect the minesite personnel as well as the equipment. It is generally difficult to recruit and retain skilled maintenance personnel. Against this background, it is apparent that management efforts to control maintenance costs must take into account the diverse skill levels of the maintenance personnel as well as labour turnover.

2 THE INTEGRATED APPROACH

The maintenance function in a mining operation does not exist in isolation. Indeed, the maintenance department has everyday interactions with the warehouse and purchasing departments, the accounting department, the personnel department and the mining department. An integrated system enables these interactions to be addressed at the mine-site by the people who need the information to make the day-to-day decisions.

A systematic and integrated computer-based approach can improve equipment maintenance by:

1. improved utilisation of labour resources (by allocating the right people to the right tasks at the right time; by defining those tasks, etc).

2. reducing the inventories of spare parts and materials whilst maintaining or improving the service efficiency of the warehouse.

3. reducing shop floor problems - removing the panic operation mode and personnel frustrations, for example, for a 3000-hour service on a coal hauler, when the job order is initiated, the stores inventory for all parts would be checked to see if everything needed was available, the parts and materials would be transferred to a "job bin" and delivered to the shop floor prior to job commencement.

4. providing dynamic scheduling for more accurate scheduling and job continuity.

5. providing history files which can be used to: analyse equipment performance; design modifications to equipment; design and specify new equipment; store data for estimating upcoming jobs and for future budget projections; provide continuity of equipment experience, maintenance standards and work procedures despite discontinuities in maintenance personnel as a result of labour turnover; and accurately report the costs of owning and operating a piece of equipment through its life cycle, a key element in the decision of when to replace items.

3 THE MINCOM INTEGRATED MINE MANAGEMENT SYSTEMS

3.1 Systems overview

The integrated mine management systems are a set of modular computer packages which comprise a comprehensive methodology both for control of production, maintenance and materials management activities, and for collecting, processing and displaying day-to-day mine management information. The information is available throughout the workshops, warehouse and office on visual display units, and the system collects data at a high level of detail, allowing useful reporting to all levels of management from the shop floor to the general manager. The major modules of the system are the following:

1. operations and maintenance information
2. maintenance scheduling
3. operating statistics
4. condition monitoring
5. materials management
6. labour management

The modules are integrated, as shown diagramatically in the figure, so that information maintained by any one application area is accessible as required by all other applications. The systems are not only integrated among themselves but also provide for data interface with corporate financial and management information systems.

3.2 Operations and maintenance information

This system is a collection of sub-systems which together provide comprehensive facilities for recording and reporting information on significant operations and maintenance activity at the mine-site. The sub-systems include the following:

1. Equipment and component registers. Many applications of the mine management systems are related to the operation and maintenance of mine equipment. Identification of equipment and components is therefore vital to the total system and is the major role of the equipment and component registers.

The component register provides formats for various types of components, so that static information peculiar to that type can be recorded. Only registered components can be subject to component tracing.

2. Work orders. The main purpose of the work order sub-system is to record the operational circumstances and costs associated with a job. The facility provides a comprehensive tool useful in both operations and maintenance areas. A job is identified for the system as one of several types. For example, work orders can be created for one-off breakdown maintenance, standing work orders to cover many small jobs of the same type, without costing individual jobs, standing work orders for collecting the period costs of scheduled maintenance work, project work orders for jobs that are part of a project in the project costing

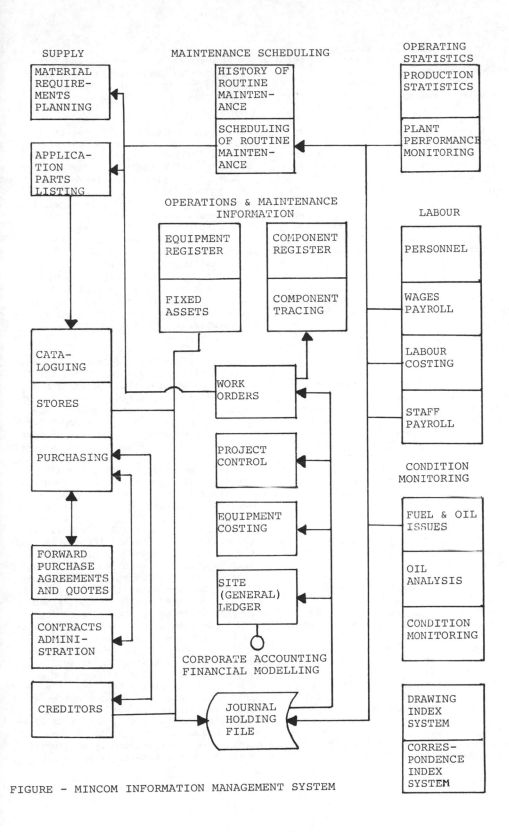

FIGURE - MINCOM INFORMATION MANAGEMENT SYSTEM

sub-system, and training work orders to accumulate individuals' time and cost for on-the-job training.

The system provides powerful recording, reporting and analysis mechanisms, using both coded and narrative information. A series of codes are available to categorise a job comprehensively. In addition, extensive English narrative can be enterd by completing the work order in such a way as to permanently record facts about the job.

3. Component tracing. Component tracing provides an additional facility for tracing individual rotable spares through their life cycle of purchase, fitment, removal and rebuild. The information held in the materials system and work order files, for each of these documents, is used to present a comprehensive picture of the operating and maintenance history of individual components.

4. Equipment costing. An important element of mine-site expenditure is the full cost of owning, operating and maintaining equipment. To provide detailed information on this cost element, the equipment costing facility combines work order costs related to an equipment item with costs which have been charged directly to that item. This enables the presentation of a comprehensive cost picture of each individual piece of equipment.

Costs reported are based on normal accounting periods, and include the following:

a. routine work orders for each equipment number.
b. standing work orders opened to cover multiple occurrences of the same job or a series of small jobs on several equipment numbers (that is, all light bulb replacements on a fleet of trucks). The system segregates costs for the period for each equipment number appearing on these work orders. This feature is convenient for grouping small jobs which are not important enough for individual work orders, while still allowing costs to be assigned to individual equipment numbers for equipment costing purposes.

c. fuel and oil costs from transactions supplied by the fuel and oil module.
d. items charged directly to equipment numbers from individual transactions that are not part of a work order.

5. Project control. This facility is used to control construction and similar projects falling in either capital or operating cost categories. The system supports the engineer in controlling costs against estimates and provides appropriate accumulations for transfer to fixed asset accounts based on taxation units of property. It also supports planning and cost control for large infrequent operating jobs such as major plant shutdowns. Project control provides for reporting committed costs as well as actual costs, when orders for goods and services for projects are placed through the mine management materials system.

6. Cost ledger (site ledger). The cost ledger facility is used primarily for preparation of mine-site operating budgets as well as reporting of actual costs against budgets for use by maintenance and operating personnel. The provision of a cost ledger devoted to on-site mine costs separate from the company's general ledger provides a number of distinct advantages. Generally, operations and maintenance personnel require accounts at a detailed level in order to properly classify and control costs. Such information needs to be integrated with job costing information from the work order system, equipment costing information and operating statistics for variance analysis. At this more detailed level, cost ledger information is an important segment of the mine manager's total information requirements. On the other hand, the general ledger of the company is orientated towards statutory and accounting requirements at a higher summary level. The general ledger also includes additional complexities of foreign exchange and revenue accounting not required by the mine-site. The mine management system therefore provides a cost ledger geared to the budget and costing information requirements of mine-site management, and capable of transmitting any level of summary information to the corporate general ledger.

4 MAINTENANCE SCHEDULING

The maintenance scheduling module is devoted to two functions:

1. Schedules of preventative maintenance; provide lists of recurring maintenance tasks to be performed on specific equipment and components by individual maintenance groups working in rostered shifts. The system can schedule tasks based on either the elapsed time of operating hours and can be extended to incorporate other scheduling criteria.

2. History of preventative maintenance; reports history of scheduled and performed occurrences of selected tasks. This history can include measurements, readings, English comments and references to related work orders opened as a consequence of the routine maintenance.

The objective of the maintenance scheduling facility is to perform those routine clerical tasks necessary to remind maintenance personnel when recurring tasks need to be done. This approach encourages easy acceptance of the system by maintenance staff who receive immediate benefit from system use. Computerised recording of routine maintenance procedures reduces the impact of personnel turnover and provides a basis for implementation of more sophisticated planning procedures. The system is designed to be totally controlled and directly operated by maintenance personnel.

5 OPERATING STATISTICS SYSTEM

5.1 Plant performance monitoring

Performance statistics on individual pieces of plant and other productive units are an important segment of the total information requirements of the mine-site. Not only do such statistics highlight problem areas of low availability and poor utilisation; they can also indicate areas of excessive availability, possibly caused by over maintenance.

5.2 Production statistics

A wide range of production statistics can be maintained and displayed by the system, such as: haul truck trips made

from particular mine blocks or pits to particular dumping points, the average speed, distance travelled, wet and dry tons hauled; loader usage in terms of the number of trucks loaded, wet and dry tons loaded; process plant production by number of wet and dry tons input and produced, water consumption, primary screen hours; and ship loader statistics on wet and dry tons of each product loaded per weightometer for each ship, time taken to load, etc.

5.3 Conditioning monitoring

Monitoring the condition of equipment by measuring and tracking indicator parameters, eg vibration analysis, oil chemistry analysis, etc.

This module will record all readings, calculate predicted life, and react to warning levels and danger levels.

6 MATERIALS SYSTEM

The mine management supply system includes six interrelated functions:

6.1 Cataloguing

Cataloguing is a mechanism within the mine management materials system which provides for disciplined identification and description of material used in the operation of the mine. Each item of material is identified by a unique stock code assigned by the system, but it can also be fully cross-referenced by colloquial names and manufacturers' part numbers, terms of reference more commonly used by maintenance personnel. On-line system access is provided by part number, colloquial name, stock code, and application part listing (APL) number.

6.2 Stores

The stores sub-system of the mine management materials system provides for the accounting and control of all items held as inventory. The main functional areas supported are the receipt of items into inventory, either at a warehouse or off-site receiving point; controlled issue of material to authorised requestors; control of the quantities of stock on hand and their physical location within the warehouse; performance of cyclical or periodic stock audits, in the form of physical inspection and count, and technical assessment; and provision of detailed and summarised information for the management, control and analysis of inventory. The system provides for multiple warehouses, and multiple locations for each stock code.

6.3 Purchasing

The purchasing function of the mine management materials system provides for the ordering of all material required for the operation. An important feature of the purchasing facility allows inventory controllers and purchasing officers to pre-select those items which they wish to review before purchasing, thus allowing the system to automatically provision those items which do not require their attention.

6.4 Forward purchase agreements

The forward purchase agreement (FPA) facilty is designed to support the negotiation of contract for the future purchase of material required for the operation. The negotiation of such contracts has a number of advantages. They are: negotiation based on annual usage rather than single order volumes can effect lower prices; purchasing information which remains stable for a set period, allowing the system to automatically purchase a large portion of the material required, thus freeing purchasing officers to spend more time in negotiation, expediting and follow-up; and, alternative part numbers and alternative sources are often volunteered by suppliers for review by supply management, providing more complete information on purchasing options.

The system also supports the preparation of tender documents, the evaluation of quotes, the notification of successful suppliers and the updating of materials system files with new purchasing information.

6.5 Creditors

The creditors sub-system of the mine management materials system provides for recording and payment of monies owing to creditors. The system caters for the approval and payment of liabilities incurred both for all ordered goods and services and for items for which specific orders are not placed (rent, fees, etc).

6.6 Application parts listing

The application parts listing (APL) sub-system of the mine management materials system provides for identification of material (that is, parts) required for a particular maintenance job or making up a particular piece of equipment, as the basis for subsequent requisitioning. The sub-system serves an important role as a major communication link between the maintenance department as users of material and the supply departments as providers of that material. APL's are also an integral part of the cataloguing function, providing vital fitment (where used) information and the ability to identify obsolete stock items.

7 LABOUR SYSTEM

The mine management labour system provides three basic facilities:

1. Payroll. Detailed calculation and reporting of employees' pay. Pay envelopes can include a report of the employee's earnings at the timesheet-entry level of detail.

2. Labour costing. Extraction of costing data from timesheet entries. Labour costs may be charged to work orders or directly to individual pieces of equipment.

3. Personnel. Maintenance and reporting of employment history, skills inventory, leave entitlements and absences.

8 DOCUMENTS INDEX

This system provides a comprehensive means of indexing drawings and technical information in regard to the plant. Using the same codes as in the other maintenance system, the operator can retrieve the location references to drawings and other documents for areas, of plant, equipment numbers, construction contracts and other criteria. The system allows for multiple locations for documents, and maintains information on the form (print, operation card, microfilm) and version number maintained at each location. Companies can therefore have a number of drawing offices working on the same base set of drawings and documentation without danger of updating out-of-date versions. The documents registered can be cross-referenced to application parts listing in the supply system to link drawing and bill of material information.

9 AUSTRALIAN CASE STUDIES

9.1 Comalco bauxite mine

The first implementation of these systems was at Comalco's bauxite mine, Weipa, in remote northern Australia. Comalco's experiences were documented by Melvin (1983), in a management environment which insisted on "user managers" taking the lead role in tailoring and implementing the systems.

This approach "proved to be very successful in that the users get what they want - a system that they understand and are very happy to use". (Melvin, 1983)

The systems are used in the mining operation, beneficiation plant, ore handling, shiploading, power station and township.

The systems are used throughout the mine by "shop floor personnel" - tradesmen, inventory clerks, supervisors, by planners, and by line management.

Observed benefits include:

1. steadily decreasing costs of maintaining major plant and equipment, despite an increase in general prices of goods and services in Australia.

2. significant reduction in the dollar value of inventory - some 60% in real dollar terms.

3. an increase in inventory service levels - that is, the percentage of requests for parts/materials which can be met from stock.

4. an increase in equipment availabili-

ties, despite the ageing of the fleet.

9.2 BHP Minerals

The systems have been installed at four mines in the BHP Minerals Group as part of the group's overall strategy for reducing costs by identifying sources of high operating costs and acting to alleviate those causes.

The four mines are:

1. Riverside coal mine - a large strip mine utilising draglines, drills, shovels, coal haulers, mobile equipment, materials handling systems and a modern coal preparation plant.

2. Moura coal mine - which comprises a large strip mine and two underground mines.

3. Koolan Island iron ore mine - an open pit operation in the remote north-west of Western Australia.

4. Groote Eylandt manganese mine - an open pit operation in the remote Northern Territory.

The experiences at Riverside coal mine were documented by BHP's house magazine in 1987.

This review highlighted the critical success factors for MIMS being:

a. management belief that the financial success of the mine relied upon the smooth operation of the equipment and plant, with a minimum of downtime.

b. involvement of the maintenance engineers and supervisors throughout the implementation of the system.

At Riverside, the "information base" provided by MIMS forms the basis of a whole range of judgments and decisions about maintenance trends, long-term equipment performance and reliablity, and present and foreseeable problem areas.

The information from MIMS is used to prepare budgets, analyse high equipment repair costs and determine the cost savings available from the common usage of standard parts.

By way of concrete example, the maintenance personnel in the coal preparation plant area have been able to make real cost savings by:

1. using the lifespan prediction data to replace certain components just before predicted failure. This work is planned for a scheduled preventive maintenance shutdown day, minimising the need to shut down the plant at unplanned times to replace broken components.

2. reducing stores inventory by having replacement parts delivered just prior to the expected replacement date.

3. being able to reliably analyse the performance of alternative materials, alternative suppliers, etc and to adopt the most cost-effective option.

4. prolonging the life of major plant by changing maintenance practices.

5. generally increasing the plant uptime.

9.3 Mt Newman Mining

Mt Newman is a very large (around 30mtpa) open pit iron ore mine in the remote north-west of Western Australia, employing large fleets of shovels, drills and rear dump haulers. Because of the size of the operation, the equipment maintenance modules were implemented in a progressive manner across the organisation, starting with the truck workshop, followed by shovels, drills, light vehicles, mobile plant, then the fixed plant in the ore crushers, beneficiation plant, powerhouse, in-pit crusher, electrical workshops and ultimately township maintenance.

This implementation philosophy ensured that the internal project team could train one group to a high level of user competence before moving to the next group of users.

Whilst it is, at the time of writing this paper (mid 1987), too early to analyse the result of the system's implementation, Mt Newman are expecting improvements of up to 10 percent in equipment availability and up to 5 percent reduction in operating costs.

10 CONCLUSION

The rising costs of equipment maintenance can be monitored and controlled by the use of computer-based management systems which address the needs of the users. The rapid availability, upon request, of detailed and accurate information to workshop foremen, stores clerks, tradesmen and purchasing officers is a key element in enabling these people to make informed decisions on what course of action to take in a given set of circumstances. The information is available to the front line decision-makers at the place the company's money is actually spent - on the shop floor and in the warehouse. It is here that control of maintenance costs can be achieved and the company's competitiveness maintained and improved.

REFERENCES

BHP REVIEW (April 1987). MIMS - the word for Riverside Maintenance.

Melvin, G.J. 1983. Computerised maintenance systems at Weipa. In Proc. Computers in Mining - 1983 Symposium. Australasian Institute of Mining and Metallurgy

MT NEWMAN CHRONICLE (May 1987). MIMS - the word for cutting costs.

STEWART, K.W. 1983. Management of computer systems implementation within an operating mine. Keynote address to the Computers in Mining - 1983 Symposium. Australasian Institute of Mining and Metallurgy.

Computer Applications in the Mineral Industry, Fytas, Collins & Singhal (eds)
© 1988 Balkema, Rotterdam. ISBN 90 6191 760 3

Increasing dragline production through an improved operating and maintenance monitoring system at the Syncrude Oilsands Mine

D.Milner & W.N.McKee
Syncrude Canada Limited, Fort McMurray, Alberta

ABSTRACT: Syncrude Canada Ltd. utilizes four large draglines to mine oilsand into windrows for later transfer to the Extraction Plant by bucketwheel reclaimer and conveyor. Averaging a total of 170,000 bank cubic metres per day, the draglines are each equipped with a production monitoring system to assist in maintaining a consistent production level. The existing systems were installed in late 1978/early 1979 and continue to be operational despite increasing problems. To meet current and future production levels it has become necessary to replace the existing production monitoring system to achieve greater reliability, increased data collection, improved operational control, and monitoring of maintenance functions. This paper will describe the history and operation of the existing monitoring system, the requirements and design of a replacement system, and potential future applications of the system at the Syncrude Mine.

1 BACKGROUND.

Syncrude Canada Ltd., located 42 kilometres (km) north of Fort McMurray, Alberta, is an integrated operation that has been producing synthetic crude oil from an oilsand deposit since 1978. The operation, consisting of an Open Pit Mine, an Extraction Plant, a Bitumen Upgrader, and a Utilities Plant, is currently capable of producing 48 million barrels of synthetic crude oil per year. Work is underway on projects that could potentially increase the production level to well over 60 million barrels per year.

The Open Pit Mine produces an average of 150,000 bank cubic metres (BCM) of oilsand per day from an area approximately 5 km long by 7 km wide. The Mine area is divided into four independent production systems, each consisting of a large dragline, a bucketwheel reclaimer, and a series of conveyors. Each dragline utilizes a 60 cubic metre bucket to excavate the oilsand from an insitu state to a windrow on the operating bench. Since the oilsand deposit includes both ore and waste, the draglines must reject an average of 20,000 BCM of waste per day, bringing the total daily production of the draglines to 170,000 BCM.

The four draglines in operation at Syncrude include two Bucyrus-Erie 2570's and two Marion 8750's.

2 INTRODUCTION.

Dragline mining at Syncrude started in 1977, in an operating mode significantly different than anything attempted with draglines in the past. With the equipment located above a 45 to 55 metre highwall consisting of ore and waste, the following differences are noted:

1. With the draglines producing ore instead of removing overburden, they become an integral part of the production system, putting an increased emphasis on availability and scheduling of maintenance outages.

2. With the draglines required to separate ore and waste bands in the highwall, the digging depth and relative dumping location of the two materials are more critical in the Syncrude operation than in conventional dragline operations.

3. With the draglines required to hoist 90% of the highwall material onto the operating bench within reach of a bucketwheel reclaimer, hoist times and swing angles are larger than a conventional application.

Shortly after start-up of the Syncrude draglines it became evident that the above noted differences, along with the high production levels required, would necessitate the acquisition of a monitoring system capable of collecting production data and providing the operator with instantaneous operating information. Within a year an off-the-shelf product, General Electric's "Excavator Production Reporting System", was purchased and installed. This system recorded second by second readings from six sensors onto tapes, collected at the end of each shift for processing. Unfortunately, the system did not provide operating information to the operator.

This initial system proved to be unsatisfactory, and Syncrude and General Electric (GE) formed a joint venture for the development of the present Digmate system. Syncrude provided the operator and office requirements of the new system along with the testing and evaluating facilities. GE designed and implemented the system, using technology available at the time.

As production levels increased over the first ten years of operation, the information needs of Operations, Maintenance, and Engineering staff became more sophisticated. At the same time, the Digmate monitoring equipment became less reliable, with parts difficult or impossible to obtain. It was increasingly evident that a third generation system with advanced features would be necessary.

3 ORIGINAL MONITORING SYSTEM

3.1 SYSTEM DESIGN

The originally installed "Excavator Production Reporting System" (EPRS) from GE was made up of two separate parts, including an on-board system and an office system. The logic diagram for the overall system is shown in Figure 1.

The on-board system mounted on each dragline consisted of a General Automation (GA) computer, thumb-wheel dials, an indicator light, and six sensors. The computer (with 32K memory) was an integrated, self-contained data-recording subsystem, housed in a rugged, shock- mounted enclosure with cooling and air filtering. It recorded input from the sensors and operator's thumb-wheel, and transferred the time-related sensor data to a magnetic tape. The data collected by the sensors

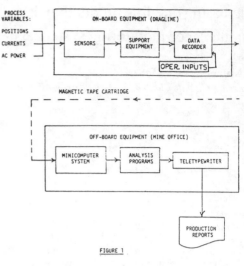

FIGURE 1

GENERAL ELECTRIC EXCAVATOR PRODUCTION REPORTING
SYSTEM DIAGRAM

in the on-board system included total excavator power, hoist and drag armature amps, hoist and drag rope travel, and swing angle.

Operator input for identifying production and maintenance status of the dragline was achieved through a wall-mounted, oil-tight enclosure housing two thumb-wheel dials and an on-off indicator light. The light gave an indication of whether the computer was recording the dial code and time of the reading. Codes from 00 to 99 were proportioned into six categories as shown in the first half of Table 1. These codes were found to be insufficient to properly monitor dragline performance and were later revised to ten categories, also shown in Table 1.

The office system consisted of a GA mini-computer, tape reader, printer, and software to produce shift and daily reports. The magnetic tape cartridges were collected from the on-board systems and taken to the office, where the data was analyzed to produce the required reports.

3.2 SYSTEM INSTALLATION AND OPERATION

The EPRS units were installed on the four Syncrude draglines in September, November, and December 1978, and February 1979. Considerable testing and debugging was required, and it was not

Table 1: OPERATOR INPUT CODES FOR THE EPRS

Initial Input Codes	
Code Number Range	Code Type
00 - 19	Operating (digging) codes
20 - 29	Various delays and downtime
30 - 39	Electrical maintenance (scheduled)
40 - 49	Electrical downtime (unschedule)
50 - 69	Mechanical maintenance (scheduled)
70 - 89	Mechanical maintenance (unscheduled)
90 - 99	Spares

Revised Input Codes	
Code Number Range	Code Type
00 - 11	Operating (Production) codes
12 - 19	Miscellaneous operating delays
20 - 29	Operating Maintenance
30 - 45	Scheduled mechanical maintenance
46 - 61	Unscheduled mechanical maintenance
62 - 73	Scheduled electrical maintenance
74 - 85	Unscheduled electrical maintenance
86 - 88	P.M. Day
89 - 96	External delays
97 - 99	Downstream delays

until late June 1979 that all four units were operating satisfactorily.

Within a year it was obvious that the initial system would not meet the total Syncrude requirements. Major problems existed with the lack of operator feedback and inaccuracies in the recording of bucket position and load. Providing operating information to the operator proved to be a necessary factor in gaining their cooperation. With the operator unable to confirm the accuracy of input information or gain an understanding of what the system was recording, he was inclined to ignore it or intentionally shut it down. The other problem, involving inaccuracies in output, was equally serious, since end-users could not trust the data being produced by the system.

Syncrude and GE utilized the experience gained from the EPRS units to design an upgraded system that became the original Digmate product. Between late 1980 and early 1981 the necessary modifications were made to the on-board system as follows:
- computer memory upgraded to 64K
- software enhanced to improve accuracy and add bucket depth indicator
- CRTs added onboard for operator feedback
- keyboards added for operator input
- instantaneous production information fed to CRTs
- print-outs made available on-board

These modifications proved to be successful in overcoming most of the problems of the initial system. The operators were satisfied with the information available to them, and their cooperation increased significantly. Instant feedback to the operator allowed comparison of production performance against established standards and other shifts, and more accurate separation of ore and waste bands in the highwall.

Unfortunately the problem of data accuracy was not completely resolved and Syncrude found it necessary to continue making enhancements to increase the usefulness of the system. Over the 1981 to 1982 period the following changes were made to the on-board and office systems:
- algorithms improved for determining bucket load and differentiating dig cycle from multiple passes
- software improved to account for total shift time
- remote job entry link installed to allow transferring and storing data on IBM mainframe
- software written for IBM, to add faster and more versatile data reporting capabilities
- reporting capabilities added for rope and tooth monitoring.

The Digmate system was utilized over the ensuing five-year period without the necessity of further modification. Recently though, system reliability has become a serious problem requiring a complete change-out of the system.

3.3 SYSTEM PROBLEMS

Significant progress has been made in the development of a dragline monitoring and reporting system over the period from 1978 to the present, and there continues to be a need for further improvement. Experience gained through use of the Digmate system has allowed a very critical analysis of the capabilities and problems of the existing system. Advances in technology over the same period have also made possible new and desirable features not available when the original Digmate system was designed.

Problems with the existing system can be found in most aspects of the hardware and software, with the major problem being reliability and availability of parts. The original Digmate system, including the GA computer, have become obsolete, and parts are no longer available through the manufacturers. Syncrude is currently using the last of the available parts and beginning to cannibalize systems to keep others running. The situation requires a tight schedule for design and replacement of the entire system.

Other problems with the Digmate system are less severe in terms of operability, but nevertheless have limited the effectiveness of the system. These problems are related to sensors, software, and hardware.

The six sensors used to monitor the status of various operating functions are inadequate in number and accuracy for the complexity of the equipment. Additional sensors are necessary to monitor various maintenance functions of the dragline, and provide an indication of deterioration or imminent failure. The sensors will form part of the annunciation system which will provide ample warning to avoid prolonged outages.

The existing software in the Digmate system does not have a procedure for zeroing bucket weight to eliminate the effect of oilsand carry-back. The weather and type of material encountered at Syncrude make bucket carry-back a common occurrance, and with no way to eliminate the effect of the dead load, the Digmate system records a higher production level than actually achieved. The software for the bucket depth indicator is accurate for a level bench, but loses accuracy under normal conditions where the bench is at a small gradient. The subsequent error in bucket depth is enough to adversely affect the separation of ore and waste in the highwall.

The office system software is capable of producing an assortment of reports meeting the minimum needs of the various mining groups involved. The production data is not stored in a data management system and is therefore not available for ad-hoc queries. A change in report format is a lengthy process involving a trained programmer.

The architecture of the Digmate hardware and software utilizes outdated 8-bit technology, that prevents any significant upgrading of capabilities. In addition, the available ports and memory are fully utilized, eliminating any possibility of handling more monitoring functions.

4 REPLACEMENT MONITORING SYSTEM

4.1 ALTERNATIVES

A decision to replace the existing monitoring system was made based on known problems as well as anticipated future problems related to meeting higher dragline production levels. An accurate and comprehensive monitoring system was felt to be an essential ingredient in the process to increase dragline utilization and productivity. A review of the alternatives available for obtaining an acceptable system has led to the quick elimination of the "do nothing" and "upgrade" options for reasons stated previously.

The available alternatives also include the purchase of an existing product, and the construction and installation of an internally- developed product. The purchase of an existing product had some features to make it attractive, including the potential to save development time and cost, and the potential to obtain a proven system. The negative factors of the purchase option eventually outweighed the positive though, as follows:
- significant modification could be required to integrate an off-the-shelf system into the existing Syncrude data management system
- ongoing trouble-shooting or modification would require external assistance
- an off-the-shelf system would not incorporate the same hardware as other Syncrude installations, increasing maintenance and parts-stocking needs.

The final alternative, that of developing an in-house system, was given serious consideration and found favourable as a way to avoid the problems listed under the purchase option. The use of components similar to existing Syncrude installations provides a significant benefit in terms

718

of parts-stocking and maintenance staff familiarity. The use of common components in a new system also has the benefit of being proven in the Syncrude environment. An internally-developed system could be integrated with other Syncrude systems and therefore meet the same technical standards. Meeting the changing requirements of the various users would also be easier with a system designed for maximum flexibility and expandibility.

The decision on acquiring a replacement monitoring system was given serious consideration before choosing the option of in-house development. A major factor in the choice was the availability of Syncrude staff with the skills necessary to carry out the design and construction of the required system.

4.2 FUNCTIONAL REQUIREMENTS

With the decision made on acquiring a replacement system, efforts were turned toward the establishment of functional requirements for the new system. This phase involved lengthy discussions with Operations, Maintenance, Engineering, and Systems personnel to obtain the specific requirements of each area and prioritize into musts and wants. Subsequent meetings of combined groups and management were necessary to finalize the requirements from the initial wish lists. The requirements were divided into on-board and office system requirements with each of these further divided into general, Operations, and Maintenance areas.

TABLE 2. GENERAL FUNCTIONAL REQUIREMENTS OF THE ON-BOARD SYSTEM.

Production	o Record ore dig, multiple pass, waste dump and waste rehandle cycles
	o Incorporate bucket zeroing procedure
	o Calculate an accurate weight for each cycle
	o Provide an accurate bucket depth indicator
Telemetry	o Transmission of all necessary information to office system within one minute of measurement
	o Make continuous back-up of telemetry data stream on permanent storage media
	o Transmit all production related sensor data
	o Transmit all alarm annunciations and the related sensor data
	o Transmit all operator-entered event and activity data
	o Transmit other data on an as-needed basis, initiated from the office system.

TABLE 3. OPERATIONS FUNCTIONAL REQUIREMENTS OF ON-BOARD SYSTEM

Input	o A means to easily enter activity codes
	o A menu display of activity codes
	o A means to record replacement of bucket, teeth and adaptors
	o A means to calibrate zero bucket load
	o A means to establish the swing angle reference point
Bucket and Boom Positioning	o Continuous display of bucket depth with respect to bench elevation
	o Operator adjustable bucket depth alarm to aid in positioning for ore/waste separation
	o Operator adjustable swing angle alarm to aid in correct positioning of ore and waste windrows
Production Information	o Continuous display of:
	- volume
	- productivity rate
	- fill factor
	- swing angle
	- cycle time for previous cycle, shift average cycle and shift total
	o Update of production display after each cycle
Reports	o A report of shift production information for current and previous four shifts via on-board printer

TABLE 4. MAINTENANCE FUNCTIONAL
REQUIREMENTS OF ON-BOARD SYSTEM

Boom	o Monitor hoist and drag rope motion to prevent damage from the bucket under static and dynamic conditions o Provide operator alarm for problems, and over-ride master motion controls
Propel Synchronization	o Monitor shoe positions and synchronization o Provide operator alarm for problems, and automatically trip propel excitation
Temperature	o Monitor temperature of critical components including bearings, motors, and house o Provide operator alarm for high and low condititions
Lubrication	o Monitor the operation of critical lube components o Provide operator alarm for malfunction
Ventilation	o Monitor the operation of ventilation system o Provide operator alarm for malfunction

TABLE 5. GENERAL FUNCTIONAL
REQUIREMENTS OF OFFICE SYSTEM

Data Reception	o Receive telemetry transmission of data o Detect loss of contact with dragline, and record missing time o Accept data from alternate sources to fill missing time up to four days later
Reports	o Provide display of all dragline junctions and production, on demand o Provide routine and ad-hoc reports o Provide graphing capabilities

TABLE 6. OPERATIONS FUNCTIONAL
REQUIREMENTS OF OFFICE SYSTEM

Production Information	o Provide real-time monitoring of dragline status o Provide running shift production totals o Provide up-to-date query capabilities for production and downtime information o Maintain production information on-line for one week period
Editing	o Provide capability to edit shift production data without loss of original data
Reports	o Provide routine and ad-hoc reports on dragline consumable items o Maintain consumables information on-line for a one year period o Incorporate existing rope monitoring program

TABLE 7. MAINTENANCE FUNCTIONAL
REQUIREMENTS OF OFFICE SYSTEM

Maintenance Information	o Provide real-time monitoring of transmitted data o Provide monitoring on demand for data not normally transmitted o Provide schematic diagrams of sensors' locations, and their values. o Provide real-time indication of alarm annunciation and associated sensor data
Reports	o Provide routine and ad-hoc reports on all maintenance and alarm data o Maintain maintenance information on-line for one week period

The functional requirements for the on-board system are detailed for the general area in Table 2, the Operations area in Table 3, and the Maintenance area in Table 4. The maintenance alarm functions must take priority and pre-empt any other functions in progress.

The functional requirements for the office system are broken out into three areas, with the general requirements in Table 5, Operations in Table 6, and Maintenance in Table 7. The office system features must be available on demand 24 hours a day, 7 days a week to monitor the continuous operation.

4.3 SYSTEM DESIGN

The Syncrude Dragline Monitoring and Data Management System, will consist of three major parts including the on-board system, the office system and the data management system. Figure 2 provides a schematic view of the system design and data flow.

The basic component of the on-board system will be a programmable logic controller (PLC) dedicated to each dragline. The PLC will be capable of monitoring the required 250 production and maintenance sensors, carrying out data calculations and checks, and providing the data for display and transmission. Dedicating a PLC to each dragline will ensure an uninterrupted flow of essential data to the operator for monitoring vital functions. The alternative of having the processing capability as part of the office system, does allow for an efficient processing system, but relies on having a fail-safe telemetry system to send critical information to the office and back to the operator.

The basic component of the office system will be a Micro Vax 3500 computer configured with 16 MB memory, 456 MB direct access storage, and a VMS operating system. The Micro Vax will receive the transmitted data from each dragline and direct it to the appropriate data bases. In the case of interrupted transmission, the Micro Vax will be capable of downloading from other sources. Normally, data will be processed into the required format and made available for real-time display on office terminals to meet the needs of operating and maintenance supervision, and engineering staff. The terminals will also have access to information from databases in user-formatted reports.

The data management system will utilize a relational data base manager (DBMS) on the existing Syncrude IBM mainframe. Production and maintenance summaries will be checked and corrected if necessary before transfer into the data management system. The only source of error will be dragline operator coding of activities, and this is

TABLE 8. DRAGLINE MONITORING SYSTEM DEVELOPMENT SCHEDULE

Office System	o Jan. 1988 - Logic design complete and software selected. System sized and hardware selected.
	o Mar. 1988 - Physical design complete and computer installed.
	o Aug. 1988 - Programming underway. Communications hardware installed. Bench testing of office system underway.
	o Dec. 1988 - Parallel testing initiated with first on-board system.
	o 1st. Quarter 1989 - Office system complete.
	o 3rd. Quarter 1989 - Remaining three draglines brought on-line.
	o 4th. Quarter 1989 - Existing Digmate System turned off.
On-Board	o Apr. 1988 - On-board system bench tested.
	o Sept. 1988 - Dragline house changes started to accomodate system.
	o Dec. 1988 - First on-board system installed
	o 3rd. Quarter 1989 - Remaining three on-board systems installed
	o 4th. Quarter 1989 - Complete turnover of new system

expected to be minimal after an initial period of familiarity.

With all data stored in the DBMS, it is readily accessible for authorized users to carry out data analyses aimed at improving the dragline operation. Existing application programs utilizing data from the existing monitoring system will be modified to accept data from the new system.

Estimates of total development cost have been completed but are not available for publication.

4.4 DEVELOPMENT SCHEDULE

With the short remaining life of the existing system, replacement must occur over the next two years. The schedule shown in Table 8 was developed reflecting an aggressive development, testing, and implementation program.

PLC

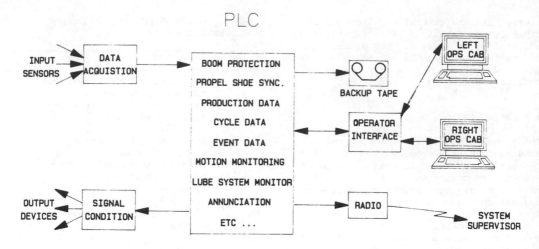

DRAGLINE ONBOARD SYSTEM (TYPICAL FOR ALL D/Ls)

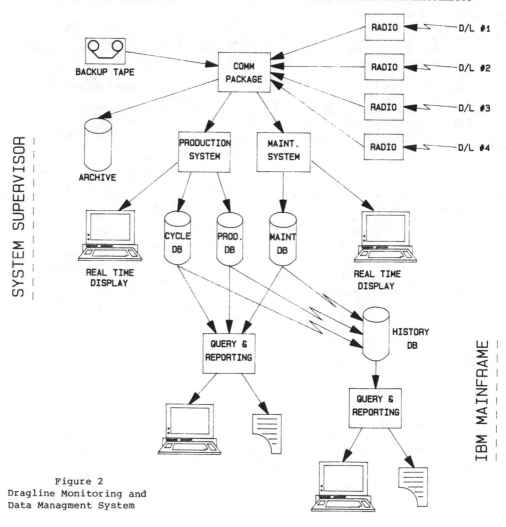

Figure 2
Dragline Monitoring and
Data Managment System

4.5 EXPECTED BENEFITS

The time and expense involved in replacing the existing monitoring system have been justified based on the benefits derived from existing and past systems. A new and enhanced system will also provide additional functions unavailable on the existing system, further justifying the change-out.

The major benefits will be improved dragline reliability, increased productivity, and an improved ore grade in the windrow. Improved reliability will be achieved through close monitoring of mechanical and electrical components, early detection of equipment problems, and problem identification and resolution through historical data analyses.

Increased productivity can be expected from better informed operators, more accurate and reliable production data, and closer monitoring of the operation by supervision. Analysis of historical production data will allow the identification of trends and the establishment of productivity standards.

The windrow ore grade is expected to improve through the improved selectivity of ore and waste bands. Improved ore grade will lead to a significant unit cost reduction for mining and ore processing.

Other benefits will be realized through the operation and maintenance of a Syncrude designed and constructed system. Troubleshooting and modification will be much simpler to achieve through use of maintenance staff familiar with components of the system common to other installations.

Given the aggressive nature of the Syncrude production and cost targets of the future, it is essential that such quality systems be implemented to provide the basic tools and data needed to achieve the objectives.

5.0 FUTURE DEVELOPMENT

Development of the new system is being undertaken in such a way that critical functions are being given priority. The requirements listed in Tables 2 through 6 are included in this category. As the system is brought on line and experience gained with its operation, additional functions will be added to generate further efficiencies in dragline operation. The additional functions listed here, as well as those not thought of yet, exist in previously untried areas that could potentially pay large dividends. The payback for projects with even minor productivity gains or cost reductions is significant, and provides a good incentive and opportunity for future development.

Future development ideas for the Dragline Monitoring and Data Management System include the following:

- Through integration of the geological model and the dragline monitoring system, it may be possible for the system to determine bucket contents in terms of ore and waste. The determination would be signalled to the operator for appropriate dumping. In an ore body where the ore and waste bands are highly irregular and difficult to differentiate visually, this could significantly improve selectivity and the resultant ore grade. The success of this idea will rely very heavily on the accuracy of the geological model, and a dragline positioning system to locate the bucket within the three-dimensional model.

- Automatic coding of ore and waste handling activities may be possible, eliminating the need for operator input. This would be achieved by inputting engineered excavation plans either through the office or on-board systems.

- Installation of tub-levelling and ground pressure sensors would assist the operator in evaluating bench preparation quality, and possibly prevent tub deterioration over the long term.

- Integration of the dragline monitoring system and an existing mechanical/electrical lock-out system would allow closer monitoring of lock-outs. An alarm system would warn of lock-out violations.

- A manual system could be added for ordering dragline supplies from a central dispatch area. This would lead to an improved supply handling system and a reduction in the use of the overloaded two-way radio system.

- The dragline maintenance work order system could be added to the system to allow accessing and modifying work order status from the on-board system. This would lead to improved information available to field mechanics when undertaking general maintenance work on the draglines.

- The possibility of having the dragline crew closely monitor essential electrical and mechanical components for unacceptable trends, could lead to fast preventative measures that significantly increase component life.

6.0 CONCLUSION

The plan to implement a state-of-the-art dragline monitoring system on the Syncrude draglines has been driven by the need to replace the existing obsolete system as quickly as possible to ensure continued equipment productivity. Despite the time constraints on the replacement project, the new system design is being carefully thought out to ensure it is comprehensive, accurate, and flexible. Significant gains are expected with the new system in place, both in terms of equipment reliability and productivity.

Development of the new system in-house is a significant departure from previous experience, but is felt to be feasible due to the skills of existing staff and the experience already gained with common components. The use of in-house expertise is thought to offer significant benefits in terms of system troubleshooting and future enhancement. Future development ideas have been generated, and will no doubt constitute subsequent development phases.

The new system will prove to be a valuable tool in the drive toward higher utilization and lower cost of dragline operations in an oilsand surface mine. The long term success of oilsand mining rests heavily on technological advancement such as described here.

ACKNOWLEDGEMENTS

The authors wish to acknowledge the contributions made by members of the project team, including Derrick Kershaw, Doug Martel, Paul Heising, Dave Hisey, Jack Jansen, Dawn Logan, and Brian Underhill. We also wish to thank Syncrude Canada Ltd. for permission to publish this paper.

Computer Applications in the Mineral Industry, Fytas, Collins & Singhal (eds)
© *1988 Balkema, Rotterdam. ISBN 90 6191 760 3*

Expert systems for failure diagnostic improvement in mining equipment

C.Santa Cruz & J.L.Kohler
Department of Mineral Engineering, Mine Electrical Laboratory, Pennsylvania State University, USA

ABSTRACT: The components of the mine maintenance cost structure are amenable to substantial reduction if the failure diagnostic is qualitatively and quantitatively improved. It is in this area that mine maintenance personnel could be helped with the implementation of expert systems for equipment troubleshooting.

Due to its nature, equipment troubleshooting requires highly specialized personnel in a narrow domain. This is one type of domain that is suitable for the implementation of expert systems.

This paper briefly reviews expert systems, and it includes a literature review of expert systems developed for equipment troubleshooting and diagnosis. Domain suitability and the diagnosis problem are discussed and different approaches to troubleshooting are examined.

1 INTRODUCTION

Effectiveness in maintenance is one of the most important factors in any successful mining operation. Several decades ago when mechanization in the mining industry was just beginning, a very small part of the total mining cost went into maintenance. Today, things are changing dramatically. During the last 40 years there has been an increasing trend for mechanization of mining operations without much consideration for maintenance. Historically, maintenance needs have been subservient to those of production.

Now mine management is recognizing that good maintenance is essential in achieving production goals. Maintenance costs can now be five to seven times more than was the case 30 years ago (del Gallego, 1984).

Equipment availability must also be considered in any discussion of maintenance. The larger the equipment in operation, the fewer the number of the machines to do the job. The fewer the number of machines to do the job, the greater the share of each machine in the production output. Consequently, when one of these machines becomes idle, for whatever reason, the negative impact on the operation is very significant.

Therefore, higher equipment availability through improved maintenance is imperative.

Mine managers have made several efforts to improve availability. For example, they have tried to improve personnel performance through training and more selective hiring policies. Efforts to provide adequate facilities to personnel have also been made. The implementation of management information systems, better control and coordination are being considered as well (Hunter, 1974; Kohler, Ramani and Frantz, 1986).

This paper examines the role of knowledge based expert systems for improving equipment maintainability, and hence equipment availability, through the improvement of troubleshooting and failure diagnostic techniques.

2 EXPERT SYSTEMS

Expert systems are a product of artificial intelligence, a field of long standing, dedicated to the study of problem-solving using heuristics, construction of symbolic representations of knowledge about the world, communication processes in a natural language, and machine learning. An expert system is a program which uses AI techniques to

do the same type of task as an expert
does, i.e., complex inferential
reasoning based on a wide knowledge of a
limited domain (Welbank, 1983).

An expert system must exhibit "expertise" in a certain domain. "Expertise"
is often defined as the body of
knowledge that is acquired along many
years of experience with some specific
problem. One of the distinguishing
characteristics of an expert system is
that it is built from the interaction of
a domain expert and a knowledge
engineer.

2.1 Expert system features

Accuracy, speed, and cost-effectiveness
of information-gathering techniques
should be characteristics of an expert
system. In addition, an expert system
should embody the following:
  - The capability to explain and
    justify answers.
  - The ability for using reasoning
    procedures similar to those
    used by human experts.
  - The capability of reasoning under
    uncertainty or incomplete infor-
    mation about the current problem.
  - The ability to summarize and point
    out features of the problem
    situation that were most important
    in leading to an answer.
  - The use of verbal or symbolic
    encodings for knowledge, most
    of which are readily communicated in
    natural language.
  - The capability to grow gradually by
    adding new pieces of knowledge,
    usually in the context of solving an
    unfamiliar problem.

2.2 Expert systems structure

The main components of an expert system
are shown in Figure 2.1. The User
Interface in the figure is shown as a
collection of capabilities: knowledge
acquisition, debugging and experimenting
with the knowledge base, running test
cases, generating summaries of
conclusions, explaining the reasoning
that led to a conclusion, and evaluating
system performance.

The main computation engine contains
search guidance and inference components.

It searches the knowledge base for
applicable knowledge, and makes inferences on the basis of current problem

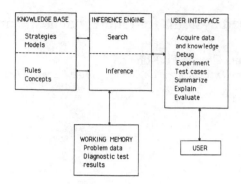

Figure 2.1 Expert systems components

data. The search guidance component
selects which portion of the knowledge
base is most important to try to apply
at any point in the problem-solving
session. It may use general knowledge
considerations, or it may use
user-specified strategic rules (called
meta-knowledge). The inference
component evaluates individual rules
and interconnections among concepts in
the knowledge base, in order to add to
the working memory. The Working Memory
is a store of the current problem data.

The knowledge base is the main
repository for domain-specific heuristics. It is considered to be in four
levels, each one built out of elements
of the next lower level:

1. Concepts: Declarative representation of domain objects, with both
abstract classes and concrete
instances. Complex interrelationships
can usually be represented and used in
making inferences and in constructing
similarities. Usually this knowledge
can be obtained from textbooks, and
includes the basic terms of the problem
domain.

2. Rules: Empirical associations
linking: causes and effects; evidence
and likely to be concluded situations
and desirable actions to perform; etc.
This level of knowledge is obtained
from an expert and is based on
experience. The knowledge is empirical
and may have associated with it
"certainty factors" indicating degrees
of belief in its applicability.

3. Models: Collections of
interrelated rules, usually associated
with a particular problem hypothesis or

726

overall diagnostic conclusion.
Sometimes this represents a subsystem
within a complex mechanical or natural
structure. Rules within models interact
much more strongly with each other than
with rules in other models. This level
of organization is often achieved using
contexts as an organizational device.

4. Strategies: Rules and procedures
to aid in the use of the rest of the
knowledge base, e.g., guiding search and
resolving conflicts when several equally
plausible rules apply to a given
situation.

2.3 Expert system prototype development

Two main stages have been identified in
the development of an expert system
prototype: the knowledge definition
stage and the implementation stage (see
Figure 2.2).

1. Knowledge definition stage
   This is a phase of analysis and definition of the knowledge structure that
precedes actual acquisition of the knowledge and the implementation of the
system. The emphasis at this stage is
to make progress on decomposing a large
and complex problem, while not getting
bogged down in the specifics of the
problem.
Step 1: Familiarization and analysis of
the problem.
   The objective of this step is to
determine the scope and complexity of
the task. The process is initiated by a
combination of relatively unstructured
interviews and observation sessions. It
is convenient to pick a sample problem
to work on that is more or less
representative of the task for which one
wishes to build an expert system. It is
advisable to either watch the expert
solve the problem or talk to the expert
about how the problem is more easily
solved.
   The information collected from these
sessions should be examined systematically to produce a list of sentences
that are representative expressions of
the facts and rules the expert has given
to the knowledge engineer.
Step 2: Organizing knowledge
   As the information collected in Step 1
gets larger, it becomes difficult to
manage. At the same time, it should
begin to exhibit some regularity in the
sense that expressions of similar form
reappear frequently. This step is to
capture these regularities by building a

KNOWLEDGE DEFINITION STAGE   PROTOTYPE IMPLEMENTATION STAGE

1.- Familiarization	4.- Acquiring knowledge
2.- Organizing knowledge	5.- Inference strategy design
3.- Representing knowledge	6.- Interface design
	7.- Testing

Figure 2.2 Expert systems development stages

knowledge acquisition grammar to
express the facts and rules in the
paper knowledge base.
Step 3: Representing knowledge
   Once the English-knowledge
acquisition grammar has been specified,
it is used to the process of deciding
how the knowledge is to be represented
in a prototype expert system. The most
simple way to do this is to begin with
the categories of the semantic grammar.
Nearly all these categories will be
meaningful from the stand point of a
representation. It is necessary to
determine a specific form for storing
instances of the category.
   2. Implementation stage
   Once the external and internal
knowledge base formats have been
defined, they can be used to guide the
implementation of a prototype expert
system. The implementation process
consists in acquiring the knowledge
base, building an inference engine,
building an appropriate interface and
testing the prototype.
Step 4: Acquiring knowledge
   Once a semantic grammar and a mapping
to internal rule formats have been
defined, it is possible to make an
effort to acquire knowledge relevant to
a particular task. One way to
accomplish this step is to use a tool
that allows the expert to generate
English expressions conforming to the
knowledge acquisition grammar and
translates these automatically into the
target formats.
Step 5: Inference strategy design
   Once a partial knowledge base has
been acquired, it is time to build or
select an inference engine to process
the knowledge base. Because their
prior exercises in acquiring and
building the knowledge base, the
project team has a much better
familiarity with the requirements of
the problem and is able to make a more
educated choice of inference
strategies.

Step 6: Interface design
     The design of an effective inter-
face is very important in delivering
acceptable expert systems. Generally
this involves trying to discover what
parts of the task are routine and can be
handled in an effective interface.
Step 7: Prototype test
   The prototype must be incrementally
tested. In this step the system
designer challenges the knowledge system
with several sample problems that it is
supposed to be able to solve. If the
execution of any of the test cases
produces erroneous results, it is
necessary to go back to the expert (see
Figure 2.3). The expert can generally
tell why the program produced the
erroneous result and recommends what the
knowledge system should know or do so as
not to repeat the same mistake. This
new knowledge is encoded in the form of
new rules and added to the system, which
is tested again with more sample test
cases until the designer in coordination
with the user believes it convenient.
   If a development tool is used for
expert system development the process is
simpler; step 5 and step 6 are not
necessary because the inference engine
and the interface facility are usually
incorporated in the tool. The use of a
development tool speeds up the expert
system development process but the
knowledge engineer has less flexibility
because a tool normally is not tailored
precisely to the current task.
   Development tools for building expert
systems have not had the success that
might have been expected. Shells have
been found too restrictive for both
knowledge representation and control,
whereas high level languages are too
demanding on the knowledge engineer
(Cowan, 1985).

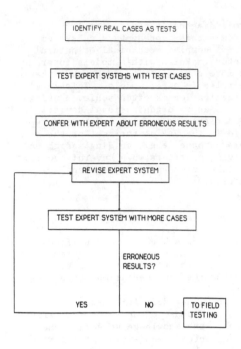

Figure 2.3 Test stage

3 ARTIFICIAL INTELLIGENCE APPLICATIONS
   TO MAINTENANCE

Mine maintenance systems present deficiencies such as: long maintenance times,
resources wasted in unnecessary and
inefficient maintenance actions, and
systems out of action which need not be.
Correction these problems is important
in order to improve equipment
availability.
   To achieve a significant improvement
in maintenance will require the
application of drastic changes to the
technology. One possibility is the
application of Artificial Intelligence
(AI) techniques to maintenance. AI is
beginning to see application to
practical problems in many disciplines,
and hence is potentially capable of
being implemented in the near future in
the mining industry.

3.1 Literature review

There are several applications of
expert systems in the area of diagnosis
and troubleshooting of equipment and
machinery, the following is a summary
of the most important found in the
literature.
   AUTO-MECH is an expert system which
diagnoses automobile fuel systems. The
problems that this system is able to
diagnose are represented as nodes within a hierarchy. Each node has knowledge
about how to confirm or reject the
problem hypothesis, as well as
knowledge about what nodes to consider
next. AUTO-MECH is implemented in
CSRL, a language which is specifically
intended for building diagnostic expert
systems (Tanner and Bylander, 1983).

728

DELTA (Diesel Electric Locomotive Troubleshooting Aid) is an expert system for diesel electric locomotive repair. DELTA is a rule-based expert system which guides the troubleshooter in this task, enforcing some disciplined troubleshooting procedures that will minimize the cost and time of corrective maintenance. The prototype system contains approximately 530 rules, partially representing the knowledge of a senior field service engineer. The troubleshooting system uses a mixed-configuration inference engine based on backward and a forward chaining (Bonissone and Johnson, 1983).

ACE (Automated Cable Expert) is a knowledge-based system designed to direct preventive maintenance activities in the local telephone network. ACE was developed at AT&T Bell laboratories to demonstrate and evaluate the potential for using expert systems technology in cable maintenance. ACE selects equipment for preventive maintenance by analyzing historical data on problems in the loop telephone plant. ACE differs from other expert systems in that it gathers information to analyze from a data base system, rather than through interaction with users (Miller, Coop, Vesonder and Zielinski, 1983).

LES (Lockheed Expert System) has been designed to guide less-experienced maintenance personnel in the fault diagnosis of electronic systems. LES uses not only the knowledge of the diagnostician, but also knowledge about the structure, function, and causal relations of the device under study to perform rapid isolation of the module causing the failure. LES can also explain its reasoning and actions to the user, and can provide extensive database retrieval and graphics capabilities (Laffey, Perkins and Firschein, 1983).

CRIB (Computer Retrieval Incidence Bank) helps computer engineers and maintenance personnel to locate computer hardware and software faults. The user gives to the system a description of his observations. CRIB matches this against the data base of known faults. By successively matching larger and larger groups of symptoms with the incoming description, CRIB arrives at a subunit which is either repairable or replaceable. If a subunit is reached and the fault is not restored the system backtracks automatically to the last decision point and tries to find another match. CRIB models the machine under diagnosis as a simple hierarchy of subunits in a semantic net. The system is written in CORAL 66. It was developed in a combined effort by International Computers Limited (ICL), the Research and Advanced Development Centre (RADC), and Brunel University. CRIB reached the stage of a research prototype (Addis, 1980 and Hartley, 1984).

DART (Diagnostic Assistance Reference Tool) assists in diagnosing faults in computer hardware systems using information about the design of the device being diagnosed. DART uses a device-independent inference procedure that is similar to a type of resolution theorem proving, where the system attempts to generate a proof related to the cause of the device's malfunction. The system was developed at Stanford University and reached the stage of a research prototype (Bennet and Hollander, 1981 and Genesereth, 1984).

FOREST isolates and diagnoses faults in electronic equipment. The system supplements the fault detection and isolation capabilities of current Automatic Test Equipment (ATE) diagnostic software. FOREST's knowledge includes experiential rules of thumb from expert engineers, knowledge of the use of circuit diagrams and general electronic troubleshooting principles. This knowledge is encoded using rules with certainty factors and explanation facility. It was developed at the University of Pennsylvania in cooperation with RCA Corporation and reached the stage of a demonstration prototype (Dincbas, 1980).

IN-ATE helps a technician to troubleshoot a Tektronix Model 465 oscilloscope by analyzing symptoms and producing a decision tree of test points to be checked by the technician. Knowledge is represented as rules incorporating probabilistic measures of belief about circuit malfunctions. As the rules are applied, the beliefs are updated through a procedure similar to minimaxing. The system is implemented in FRANZ LISP. It was developed at the Naval Research Laboratory and reached the stage of a research prototype (Cantone, Pipitone, Lander and Marrone, 1983).

NPPC (Nuclear Power Plant Consultant) helps nuclear power plant operators determine the cause of some abnormal event by applying rules in conjunction with a model of plant operation. The model consists of a common sense

729

algorithm network that accesses appropriate diagnostic rules. The system was developed at the Georgia Institute of Technology and reached the stage of a research prototype (Underwood, 1982 and William, 1984).

REACTOR assists reactor operators in the diagnosis and treatment of nuclear reactor accidents by monitoring instrument readings, to identify deviations from normal operating conditions such as feed-water flow and containment radiation level. When an abnormal situation is detected, the system evaluates it and recommends appropriate action, using knowledge about the reactor's configuration, the functional relations of its components and knowledge about the expected behavior of the reactor under known accident conditions. REACTOR is implemented in LISP as a rule-based system which uses both forward and backward chaining. It was developed by EGG Idaho and reached the stage of a research prototype (Nelson, 1982).

3.2 Domain suitability

"Consider expert systems only if expert system development is possible, justified and appropriate" [Waterman, D]. There are some specific requirements that a specific problem domain must fulfill to make possible the application of expert systems (see Figure 2.4).

Once of the most important requirements is that real experts must exist. Without the extensive and powerful knowledge that the experts represent, the development of an expert system will not produce a skillful program. The experts in some domain should agree about the choice and accuracy of the solutions concerning a particular problem. The expert must also be able to articulate and express the methods that they use in solving the domain problems, otherwise the knowledge acquisition would be a very painful stage, if not impossible.

The problem solving process must require cognitive skills. However, this does not mean that a problem with a physical component must be ruled out. Another required characteristic is that the task must not be so difficult that the knowledge required to solve it is extremely difficult to capture. Finally, the task should be very well understood and should not demand a

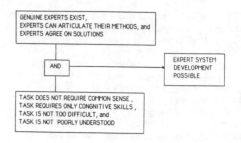

Figure 2.4 Problem domain characteristics required to make expert system development possible.

significant amount of common sense. AI programs face problems with tasks that require common sense reasoning.

The development of an expert system can be justified by the following reasons (see Figure 2.5):
-- Human expertise is scarce
-- Human expertise being lost
-- The task solution has a high payoff
-- Expertise is needed in many locations
-- Expertise needed in hostile environments

The nature, complexity and scope of the problem to be solved determine when it is appropriate to develop an expert system (see Figure 2.6). The nature of the problem should be such that it requires symbol manipulation and heuristic solutions. The problem must not be too easy. It should be a serious problem, in a domain in which it takes a human years to study or practice to achieve the status of an expert. Finally, the problem should have the proper scope to make the problem manageable and sufficiently broad to ensure that the problem has some practical interest.

3.3 The diagnosis problem

What makes a diagnosis difficult is the large amount of knowledge and experience it requires. First, it requires knowledge of the equipment and how it normally operates. Secondly, it requires gathering information about the failed equipment and its fault symptoms. Thirdly, it requires the knowledge to decide what type of equipment information is necessary to

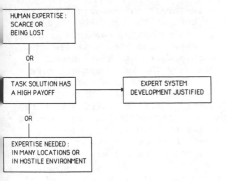

Figure 2.5 Reasons to justify expert system development

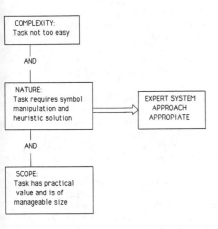

Fig 2.6 Factors that determine when it is appropiate
to develop an expert sytem

gather and which information is relevant
to the fault. Fourthly, it requires the
ability to use the knowledge about the
equipment and the information gathered
to explain how the fault could have
occurred. Finally, it requires the
ability to form a hypothesis and perform
some tests to get back more information
that either confirms or denies the
hypothesis.

The process of gathering information
and formulating and testing hypothesis
may need to be done several times if the
hypothesis formulated turned out
negative (see Figure 2.7). Only at the

end of this process can the expert
repair the fault or replace the
malfunctioning part.

Unfortunately, few people appear to
have the body of knowledge necessary to
make them good troubleshooters. Those
that do tend to be promoted so that
they do not use their expertise
anymore, except to act as a consultant
to a limited number of people. The
remaining troubleshooters may be
competent, but they lack the spark and
instincts that made the expert unique.

3.4 Troubleshooting system goals

The main goals in a troubleshooting
system are: robustness, generality,
efficiency and constructibility.
1. Robustness: It is desirable the
troubleshooting system succedes with a
fault that the designer of the system
did not foresee. One does not want
unexpected inputs or unexpected faults
to blow the troubleshooting system.

2. Generality: One wants the
troubleshooting system to be general
and work on a wide number of cases.

3. Efficiency: Anyone can build
troubleshooting systems that make a
measurement at every node and eventually figure out what is wrong. But
measurements are expensive. Some are
more expensive than others, so there is
a range within which a troubleshooting
system optimizes.

4. Constructibility: The troubleshooting system must be easy to build
and flexible enough to be corrected if
there is something wrong in the design.

4   APPROACHES TO TROUBLESHOOTING

Different approaches can be used in
building a diagnosis and troubleshooting system. Some approaches are:
the use of simple programming languages,
the use of empirical associations,
structural organization, behavioral
organization, organization based on deep
knowledge about fault modes or organization based on a causal model. Other
systems do not follow a rigid scheme
and are based on a combination of some
of the approaches previously mentioned.

4.1 Use of simple programming languages

This approach consists of writing a
troubleshooting program in a common

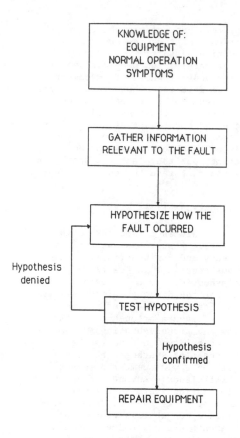

KNOWLEDGE OF:
EQUIPMENT
NORMAL OPERATION
SYMPTOMS

GATHER INFORMATION
RELEVANT TO THE FAULT

HYPOTHESIZE HOW THE
FAULT OCURRED

Hypothesis
denied

TEST HYPOTHESIS

Hypothesis
confirmed

REPAIR EQUIPMENT

Fig. 2.7 Diagnosis of equipment

because the empirical associations are good for both of the troubleshooter's basic tasks; computing the entailments of measurements and proposing new ones.

The main problem of this approach is illustrated by the knowledge engineer trying to fit empirical associations into the knowledge base. Many rules are required and it is difficult to add new rules. When new rules are added it is not easy to know whether they contradict the previous set of rules.

This approach does present some advantages. One is that the knowledge is now explicit in the program. It is also easier to modify and there is some inference, although it is very weak. Its robustness is still not satisfactory due to the large number of rules that have to be written. Efficiency is good because the measurements that it proposes depend on previous measurements. Constructibility is still poor, because extracting a large number of rules is required.

4.3 Structural organization

Considering the pool of rules, it can be broken into subsets associated with circuit modules. That observation is based on the fact that the function of a circuit arises out of this structure. This structure can be used to organize the rules.

If a troubleshooting system incorporates an explicit notion of structure, the number of rules can be reduced. This idea is very simple, but it gives to the knowledge engineer a great amount of leverage.

Similarly, a causal model could be used in the same way to organize the rules. It is not too important what is used to organize the rules. If there is only some truth in what is used to organize the rules, the robustness is going to be improved as it is easier to notice missing rules.

4.4 Behavioral organization

The fact that the device operates as a consequence of the behavior of its parts has been ignored by the three previous approaches. This fourth approach considers the use of deep knowledge about behavior.

Applying this method one should describe the behavior of every component by a constraint that can be

programming language. The knowledge is all implicit in the code. There is no interference occurring at all.

These types of programs have little robustness and they are not general at all. Most of these programs are inefficient in the sense that they make the same set of measurements regardless of the particular situation. These programs do not know when to start or stop. All the tests are performed. Their constructibility is also poor.

4.2 Use of empirical associations

They resemble cause-effect type relationships which are extracted from the expert. This system works better than using simple programming languages,

used in any way -- forward, backward or sideways. There is no notion of forward or backward in constraints. Then, it is possible to use the constraint propagation, the idea of composing constraints with each other, to construct expectations.

The basic idea of deep knowledge about behavior as an approach is that the propagations allow the knowledge engineer to construct expectations of circuit behavior. If an expectation is constructed and measurements are made at a point of expectation, information is obtained. With this information two things can happen: the measurement can corroborate the expectation, or the measurement can disagree with the propagation. Based on that information we can arrive at a conclusion.

Using this approach, the number of rules required is reduced considerably. A fault in a component is now a violation of its own law, not a guess that some designer made in advance. Generality is also much better because the approach can be applied to more cases. Its efficiency is good and its constructibility is better because of the smaller number of rules. With regard to the knowledge, there is little knowledge in the system about the particular case. Instead, there is some deep knowledge about the problem domain in general and that deep knowledge applies to all cases.

4.5 Organization based on deep knowledge about fault modes

Faults have been considered as being a deviation from expected behavior. This approach is based on the fact that the devices also have fault modes.

This alternative introduces a second propagator. The first propagator propagates behavior, the second one propagates errors or faults. The second propagator inverts expectations that have already been constructed by the first propagator.

4.6 Organization based on a causal model

Rules are organized in such a way that the system is able to trace paths of causality. Paths of causal interaction play a central role in diagnosis. An important part of the knowledge about a domain is understanding the mechanisms and pathways by which one component can affect another.

5 COMMENTS

There are several potential benefits associated with the implementation of expert systems for the diagnosis and troubleshooting of mining equipment. Some of them are the following:

-- Knowledge institutionalization: Expert systems offer a way to preserve and protect the trouble shooter's expertise and to make that expert a consultant to many people.

-- Reduced overtime: This cost is directly related to emergency work. A reduction in emergency work will result in a corresponding overtime reduction.

-- Better manpower use: The more efficient the troubleshooting operation is, the more efficient the manpower utilization.

-- Less downtime and more production: Because the efficient trouble-shooting reduces the emergency work, consequently total downtime can be reduced.

-- Reduced maintenance costs: This is achieved due to the efficient failure diagnostic and to the better use of manpower.

-- Less stand by equipment needed: Downtime reduction will increase mechanical availability therefore, less stand by equipment will be necessary.

-- Training of personnel: Although the idea of a fully automated tutor remains an elusive end point in a continuum of computer-based training systems, it is possible to use AI devices in a number of other ways in technical training.

-- Design for maintainability: Often, systems are becoming too complex to be effectively maintained. It is possible that expert systems could help to decrease the psychological complexity of the devices and thereby make their maintenance more feasible.

The robustness, generality, efficiency and constructibility of a trouble-shotting system can be improved by the implementation of behavioral (normal or faulty) and/or causal organization during the system building process.

There are also some concerns about the application of this new technology.

733

Some of them are the following:
- -- Lack of robustness: It is well known that typical, unstructured rule-based systems are not very robust. An expert system carefully designed for one application may be totally useless when applied to even a slightly different one.
- -- Diagnosis is not a primary goal: Repair is often a goal that supercedes diagnosis. Maintenance systems must be able to decide just how precise a diagnosis must be before a repair should be effected. Policy factors also figure heavily in the goals of the maintainer; the attitudes of managers may often determine false replacement rates and other properties of a maintenance operation. Thus, while troubleshooting and diagnosis remain an important problem in maintenance systems, they sometimes are far from primary goals.
- -- Distributed systems and lack of accessibility: Many of the systems that are most difficult to troubleshoot are widely distributed. Components that are accessible in some situations may not be in others. Hence, an expert system that demands access to these components may have to deal with ubiquitous and unpredictable physical constraints.

Even though expert systems are an infant technology at the present time, it shows potential for a successful future application in the mining industry, especially in maintenance. Highly mechanized mining operations in which equipment availability is a key factor for the operations success are suitable for the future implementation of this new technology.

6 REFERENCES

Addis, T. R. 1980. Expert Systems: an evolution in information retrieval. International Computers Limited Technical Journal.

Barr, A. and Feigenbaum, E. A. 1981. The handbook of artificial intelligence. Vol. I and II. Los Altos, CA: William Kaufmann, Inc.

Bennet, J. S. and Hollander, C. R. 1981. DART: an expert system for computer fault diagnosis. Proceedings IJCAI-81, pp. 843-845.

Bonissone, P. P. 1983. DELTA: an expert system to troubleshoot diesel electrical locomotives. Proceedings ACM, pp. 44-45, New York.

Cammarota, A. F. 1986. Artificial Intelligence: training and education. Design News, Vol. 42, n. 5, pp. 122-126.

Cantone, R. R., Pipitone, F. J., Lander, W. B. and Marrone, M. P. 1983. Model-based probabilistic reasoning for electronics troubleshooting. Proceedings IJCAI-83.

Charniak, E. and McDermott, D. 1986. Introduction to artificial intelligence. Addison-Wesley Publishing Co.

Dincbas, M. 1980. A knowledge-based expert system for automatic analysis and synthesis in CAD. Information Processing 80. IFIPS Proceedings, pp. 705-710.

Freiling, M., Alexander, J., Messick, S., Rehfuss, S. and Shulman, S. 1985. Starting a knowledge engineering project: a step-by-step approach. The AI Magazine, Vol., 6, n. 3, pp. 150-164.

Hart, A. 1985. Knowledge elicitation: issues and methods. Computer Aided Design, Vol. 17, n. 9, pp. 455-462.

Hartley, R. T. 1984. CRIB: computer fault-finding through knowledge engineering. Computer, Vol. 17, n. 3.

Kohler, J. L. 1986. The ventilation expert for underground coal mines. Advance Monitoring System Short Course, The Pennsylvania State University.

Kohler, J. L., Ramani, R. V., Koharchik, G. and Bhaskar, R. 1986. Conceptual investigation of a management information system for underground coal mines. BOM, a mining research contract report.

Kohler, J. L., Ramani, R. V., Bieniawski, Z. T. Expert systems: An overview and a mining example. Preprint, Society of Mining Engineers of AIME.

Laffey, T. J. and Firschein, O. 1984. LES: a model based expert system for electronic maintenance. Proceedings of the Joint Services Workshop on AI in Maintenance, pp. 1-7.

Nelson, W. R. 1982. REACTOR: an expert system for diagnosis and treatment of nuclear reactor accidents. Proceedings AAAI-82, pp. 296-301.

Niebel, B. W. 1985. Engineering
  maintenance management. New York:
  Marcel Deeker, Inc.
Oomen, M. Santa Cruz, C. and Kohler,
  J. L. 1987. Maintainability improve-
  ment of mining machines using
  knowledge-based expert systems.
  Proceedings of the IEEE Mining
  Industry Technical Conference.
Oomen, M. 1986. Insight 2+ and the PRL
  environment. The Pennsylvania State
  University Mine Electrical Laboratory,
  Internal Report.
Rayson, P. T. 1985. Review of expert
  systems principles and their role in
  manufacturing systems. Robotica.
  Vol. 3, No. 4, pp. 279-287.
Richardson, J. 1985. Artificial
  intelligence in maintenance. New
  Jersey: Noyes Publications.
Santa Cruz, C. 1986. Miner-Expert: an
  expert system for the trouble-
  shooting of the 120L Heliminer
  hydraulic system. The Pennsylvania
  State University Mine Electrical
  Laboratory, Internal Report.
Waterman, D. 1985. A guide to expert
  systems. Reading: Addison-Wesley
  Publishing Company.
Underwood, W. E. 1982. A CSA model-
  based nuclear power plant consultant.
  Proceedings AAAI-82, pp. 302-305.

Computer Applications in the Mineral Industry, Fytas, Collins & Singhal (eds)
© 1988 Balkema, Rotterdam. ISBN 90 6191 760 3

Implantation du logiciel d'entretien Main Saver aux Mines Selbaie

Denis Roy
Les Mines Selbaie, Joutel, Quebec, Canada

ABSTRACT: Ce texte se veut une présentation des applications et des
implications de l'introduction d'un logiciel de gestion de l'entretien en
milieu minier. Au moment d'écrire ces lignes, l'implantation est à peine
amorcée. Nous insisterons donc sur la phase préparatoire à l'utilisation.
De plus, nous n'aborderons la structure interne du logiciel que dans la
mesure où cela peut faciliter la compréhension de son utilisation.

CONSIDERATIONS GENERALES

Les Mines Selbaie sont situées dans
le Nord-Ouest du Québec et extraient
quotidiennement environ 7 000 tonnes
de minerai. La plus grande part
provient de l'exploitation à ciel
ouvert, les 1 600 t/j restant étant
produites par la mine souterraine.
Le minerai est concentré sur place.

La mise en exploitation récente de
la mine à ciel ouvert, en
quadruplant la production a amené
les dirigeants à se tourner du côté
de l'informatique pour combler les
nouveaux besoins en gestion de
l'entretien qu'exigeait l'arrivée
d'équipement à haute productivité.

Deux années d'étude sur ce projet
ont permis de dégager une première
conclusion intéressante:
l'informatisation est en soi un
problème en tout point semblable à
celui qu'elle doit régler.

Informatiser un système
d'entretien, c'est comme entretenir
de l'équipement. Tout le monde est
d'accord, personne ne sait ce qu'il
faut faire au juste et il ne faut
surtout pas déranger la production.
Dans ces conditions, pourquoi
remettre à demain ce qu'on peut
faire après-demain...

1. LES PREREQUIS

L'informatique ne fait aucune
concession, si votre système
d'entretien ne tient pas debout, il
ne tiendra pas non plus dans un
ordinateur. Le premier pas est donc
de définir la manière dont
l'entretien est géré et où doivent
porter les gains de productivité
apportés par l'informatique.

Dans toute industrie normale, ces
tâches sont rapidement déléguées aux
échelons subalternes ou confiées à
des consultants pendant qu'on
s'attaque au choix du logiciel
idéal.

On manque alors l'occasion de
redécouvrir les bases d'un système
d'entretien qui sont: les
spécifications, les procédures et
les bons de travail. Les
spécifications définissent l'état
dans lequel l'équipement doit être
maintenu. Les procédures donnent la
façon dont on le remet en état. Le
bon de travail permet l'exécution
efficace de ces tâches et la
récupération de l'information.

La gestion de l'entretien
ressemble le plus souvent à la
digestion d'une quantité incroyable
d'informations. Tout équipement
doit:

1. Exister (posséder un numéro d'équipement normalisé et une description)

2. Etre entretenu (maintenir les conditions de détérioration minimales pendant l'usage)

3. Etre réparé (remplacer les pièces endommagées)

4. Etre remplacé (lorsque sa vie économique est terminée)

Le prérequis à tout système d'entretien et donc à l'informatisation sera l'établissement d'un fichier central avec une liste d'équipements numérotés et un dossier pour chaque équipement avec ses spécifications, ses procédures d'entretien et ses procédures de réparation. Quant à la gestion de ce système, c'est l'utilisation des bons de travail qui en est la clé.

2. LES BONS DE TRAVAIL

On l'a dit, gérer c'est avant tout une question d'information. En entretien, le bon de travail est l'outil de communication par excellence. Sur le bon de travail, on doit retrouver toute l'information concernant l'exécution d'une tâche. Consigner cette information sur le bon de travail la rend disponible pour tous et permet une coordination efficace des travaux d'entretien.

Il est bon de souligner ici l'importance de l'exactitude des renseignements. Si dans le cas des fichiers d'équipement, on peut compter sur du personnel clérical, dans le cas des bons de travail, ce sont les personnes sur le chantier qui doivent consigner l'information et souvent, avec un soin très variable d'un individu à l'autre. C'est une contrainte avec laquelle il faut vivre.

L'utilisation des bons de travail a pour effet d'emmener les gens à mieux définir, prévoir et planifier leur besoins en services d'entretien. Par conséquent, le département d'entretien saura mieux définir, prévoir et planifier les ressources humaines et matérielles allouées aux services clients en fonction des priorités établies par la direction.

En planifiant la plus grande part possible des tâches, on évitera les mauvaises surprises causées par la non-disponibilité de pièces, de main-d'oeuvre ou de l'équipement pour l'entretien.

La planification se fait le plus souvent par l'élaboration manuelle de cédules de travail.

Le compagnon souvent méconnu du bon de travail est la procédure. Toute tâche doit être accomplie selon certaines normes de qualité, de compétence et, bien sûr, de sécurité. L'ensemble de ces normes forme une procédure. Une grande portion et parfois toute la procédure elle-même est souvent implicite; c'est-à-dire que le travailleur l'appliquera de manière automatique par compétence ou expérience. Cependant, la procédure doit exister afin de favoriser une standardisation des méthodes et permettre aux apprentis d'apprendre par formation plutôt que par "expérience" (lire échecs répétés).

3. INFORMATISATION

Quand un système d'entretien possède tous ces éléments, on songe aussitôt à des moyens d'en augmenter l'efficacité. La grande quantité d'information à traiter et la diversité des intervenants suppose beaucoup de travail clérical et des échanges constants. On en vient à passer plus de temps à manipuler l'information qu'à l'utiliser.

L'informatisation apporte un support de trois manières principales:

On peut stocker toute l'information sur un seul fichier central accessible à plusieurs utilisateurs dispersés.

On peut échanger de l'information en temps réel.

On peut se faire assister par l'ordinateur dans l'élaboration des cédules d'entretien et la production de rapports administratifs.

Les avantages de l'informatisation se situeront plus au niveau de la diminution du temps improductif, de l'amélioration de l'état de l'équipement et de la qualité de la gestion, que des quelques économies dues à l'accélération du travail clérical. En fait, celui-ci devrait plutôt s'accroître.

4. LE LOGICIEL

Les Mines Selbaie ont opté pour le logiciel MainSaver. L'essai antérieur du logiciel ABC a montré qu'il ne correspondait pas au système souhaité, l'achat ayant précédé l'analyse des besoins. La souplesse dans la gestion des bons de travail fut un élément clé.

MainSaver se divise en quatre modules de gestion: les bons de travail, l'entretien préventif, l'inventaire d'entrepôt et les rapports administratifs. Ces modules gèrent une base de données contenant tous les renseignements sur les équipements (numéros, spécifications et procédures) et sur les bons de travail (historique, travail en cours) auxquels se greffent l'inventaire des pièces et de la main-d'oeuvre disponibles.

Pour des raisons administratives, l'utilisation des modules d'inventaire et de certains rapports administratifs n'est pas effective à ce stade-ci. L'efficacité du système s'en trouvant affectée, nous espérons que cette lacune sera comblée dans un proche avenir. Les deux modules essentiels sont cependant en fonction: l'entretien préventif et les bons de travail.

5. ENTRETIEN PREVENTIF

L'entretien préventif vise à corriger la détérioration inévitable de certaines pièces lors de l'opération d'un équipement. Les exemples les plus communs sont les changements d'huile et de filtres, le nettoyage, la calibration, le changement de revêtement et les inspections.

L'entretien préventif a la caractéristique d'être prévisible par l'observation d'une variable telle: heures d'opération, tonnage manipulé, jours écoulés, etc.. Une fois un équipement inscrit au module, il suffit de fournir régulièrement l'état de sa variable de référence et un bon de travail d'entretien sera automatiquement émis par l'ordinateur à la fréquence prescrite.

6. BONS DE TRAVAIL

Ce module émet et fait le suivi de toute demande de travail d'entretien ou de réparation. Lorsque le travail est complété, l'information devient disponible sous forme d'historique sous la rubrique de l'équipement.

On peut aussi obtenir des cédules des travaux planifiés pour la journée, la semaine ou pour plusieurs semaines à venir.

Enfin, le retard accumulé (backlog) permet d'évaluer la force de travail disponible versus les tâches à accomplir.

7. BASE DE DONNEES

La base de données auquel sera couplé MainSaver est Ingress. Soulignons ici que cette base de données est déjà utilisée dans les différents services de BP et que cette standardisation permettra des échanges fructueux entre MainSaver et d'autres logiciels de gestion.

MainSaver permet de sortir des rapports taillés sur mesure à partir de la base de données elle-même. Avec la capacité de capter les coûts de pièces et de main-d'oeuvre au niveau de chaque bon de travail, on a là un outil de gestion et d'analyse budgétaire très prometteur.

Parmi les exigences du service d'entretien, le traitement des composantes majeures des équipements et des besoins spécifiques en électricité et instrumentation ont nécessité certaines modifications au logiciel.

On peut donc suivre une composante dans son déplacement d'un équipement à l'autre, tout en conservant sa portion de vie passée sur un équipement rattachée à l'historique de cet équipement.

Dans le même ordre d'idée, les moteurs électriques sont suivis comme des équipements tout en demeurant composantes des vrais équipements.

Pour ce qui est de l'instrumentation, les tables de calibration en font des composantes très particulières. L'entretien de plusieurs composantes identiques sur autant d'équipements différents impose pour des raisons d'efficacité, de mettre l'accent sur les groupes de composantes identiques plutôt que sur les équipements eux-mêmes. Encore là, l'historique demeure rattaché à l'équipement.

8. MAINSAVER DANS LES MINES

Du point de vue de l'entretien,
l'application au secteur minier
n'apparaît pas différente de celle
dans l'industrie lourde en général.
On peut cependant souligner quelques
points.

Le secteur minier nécessite une
grande diversité d'équipement et de
corps de métiers pour fonctionner.
Cela rend plus complexe
l'implantation d'un système
d'entretien structuré préalable à
l'informatisation. Cela peut
cependant laisser espérer une
amélioration encore plus marquée de
l'entretien lorsque l'implantation
se fait avec succès.

Historiquement, les mines n'ont
pas formé l'avant-garde en terme de
technologie. Il est donc normal de
voir une réticence plus marquée à
l'arrivée d'outils modernes de
gestion.

Enfin, une mine est une mine et
quand l'action se passe sous-terre
loin des ordinateurs, on est tenté
de s'en remettre à l'"expérience"
des contremaîtres plutôt qu'aux
méthodes de gestion. Encore là, la
technologie devra s'imposer et il
n'est pas loin le jour où ce même
contremaître aura son écran
cathodique à 300m en-dessous...et
s'en servira!

Computer Applications in the Mineral Industry, Fytas, Collins & Singhal (eds)
© 1988 Balkema, Rotterdam. ISBN 90 6191 760 3

Système de planification des entretiens et contrôles

Pierre Savard
La Compagnie Minière Quebec Cartier, Canada

OBJET: Le Système de Planification des Entretiens et Contrôles communément appelé SPEC est un système dont l'objectif principal est d'assister le planificateur de l'entretien dans son travail quotidien. Il permet un suivi en temps réel (online system) de tous les équipements et génère en temps différé (batch processing) des analyses statistiques et de contrôles. Il opère conjointement avec les systèmes de PAIE et MMS pour les données de main-d'oeuvre et de matériel.

1. INTRODUCTION

La Compagnie Minière Québec Cartier a son siège social à Port-Cartier sur la rive nord du St-Laurent. Elle exploite le gisement de minerai de fer de Mont-Wright situé dans la fosse Québec-Labrador, un chemin de fer d'une longueur de 300 Km qui assure le transport du concentré de fer et de marchandises, une usine de bouletage et des installations portuaires pouvant accueillir des bateaux jaugeant jusqu'à 160,000 tonnes. La capacité annuelle des installations minières et de bouletage est respectivement de 19 et 8 millions de tonnes métriques.

Ses opérations variées et des conditions climatiques rigoureuses l'ont amené à développer un système de Planification des Entretiens et de Contrôles. Avec un suivi quotidien de tous les équipements, ce système a inévitablement permis une réduction des coûts d'entretien.

2. STRUCTURE DE SPEC

Toutes les transactions accumulées dans SPEC se rattachent à l'une des entités suivantes soit "équipement" ou "composant". Le composant est défini comme étant une partie constituante d'un équipement. Certains qualificatifs viennent s'ajouter à ces entités pour former des ensembles d'équipements ou de composants. Ainsi, on peut avoir la représentation suivante:

. Équipement maître
 . Équipement secondaire
  . Composants maîtres
   . Composants secondaires
    . Composants sec. multiples
  . Composants ordinaires
   . Composants ord. multiples
 . Composants maîtres
  . Composants secondaires
   . Composants sec. multiples
 . Composants ordinaires
  . Composants ord. multiples

où,

. l'équipement maître désigne un équipement auquel l'on a associé (ou pas) des équipements;
. l'équipement secondaire désigne un équipement associé à un équipement maître;
. le composant maître désigne un composant sur lequel l'on a installé d'autres composants;
. le composant ordinaire désigne un composant directement installé sur un équipement;
. le composant ordinaire multiple désigne un composant multiple directement installé sur un équipement;
. le composant secondaire désigne un composant installé sur un composant maître;
. le composant secondaire multiple désigne un composant multiple installé sur un composant maître.

Voici maintenant les modules qui complè-
tent la structure de SPEC.

Flottes. Module d'identification des
flottes avec les unités d'opération
respectives. Ces unités sont des heures
d'opérations, des mètres cubes, des
kilomètres, etc.

Fiche signalétique. Ce module est le
cardex du système. Tous les équipements
et composants suivis par SPEC doivent
posséder une fiche signalétique où
l'on retrouve une description et la
fiche technique. Les équipements/com-
posants sont regroupés par modèle et
par flotte.

Association des modèles. Module où l'on
identifie les modèles de composants qui
peuvent être installés sur un modèle
d'équipement.

Entretiens. Ce module permet la mise
à jour des entretiens à effectuer sur
un équipement. Un entretien comprend
une description, une fréquence, les
spécifications du travail à effectuer
et un estimé de main-d'oeuvre.

Identification de pièces. Module de
mise à jour des pièces d'un équipement.
La liste est de type "générale" si
elle se réfère à un modèle et de type
"particulière" si elle se réfère
équipement ou un composant.

Moyenne de vie estimée. Module permettant d'identifier la durée de vie moyenne d'un composant par rapport à un
modèle d'équipement.

3. FICHE OPÉRATIONNELLE

La fiche opérationnelle est la feuille
de route des équipements/composants.
Elle contient les données d'opération
et toutes les transactions effectuées
sur chaque équipement/composant. Ces
transactions sont la mise en marche,
l'installation, l'enlèvement, la réparation sur place en atelier ou chez un
fournisseur.

La saisie des données se fait de deux
manières: manuelle et mécanique. La
saisie manuelle se fait en temps réel.
Elle est grandement simplifiée étant
donné qu'elle se fait au niveau supérieur de la structure organisationelle.
Une donnée saisie au niveau de l'équipement maître est cumulée à tous les

équipements et composants qui lui
sont rattachés par un traitement en
temps différé. La saisie mécanique
provient du transfert des données
d'opération de systèmes informatiques
indépendants, tel le contrôle des
camions de production, le traitement
du minerai et le contrôle des trains.

Les transactions décrites précédemment qui sont incluses dans la fiche
opérationnelle proviennent du bon
de travail.

4. GÉNÉRATEUR DE TRAVAUX

Ce module relie la structure de
SPEC avec la fiche opérationnelle.
Les entretiens et leurs fréquences
sont définis dans la structure et
la fiche opérationnelle contient
les cumulatifs d'opération, il est
donc possible d'émettre mécaniquement
le bon de travail lorsqu'un équipement a atteint son échéance d'entretien. C'est ce que l'on appelle le
"dépistage". A cette étape, les composants qui ont atteints leur moyenne
de durée de vie sont identifiés afin
d'être remplacés lors de l'entretien.

Les informations contenues dans
le bon de travail sont: la date d'émission, la date d'échéance, l'émetteur,
le numéro d'équipement, une description
du travail à effectuer, les heures
cédulées et travaillées par atelier,
les coûts de main-d'oeuvre et de matériel et les coûts liés à des réparations externes.

Il existe trois types de bon de
travail:

- Bon de travail pour entretien préventif. Ce type de bon de travail
  est émis mécaniquement lors du
  dépistage des entretiens. Il peut
  aussi être déclenché par le planificateur d'entretien si l'équipement
  est arrêté pour une raison quelcon-
  que et que l'on veuille profiter
  de cette occasion pour effectuer
  l'entretien.

· Bon de travail pour entretien
  inopiné ou curatif. Ce type est
  généré manuellement et est généralement utilisé pour les bris
  imprévus qui demandent une réparation immédiate.

- Bon de travail pour entretien planifiable. Ce type couvre tous les
autres entretiens que l'on peut
reporter sur la cédule des travaux.

Il est aussi possible de rattacher
un ensemble de bon de travail pour
former un bon de travail "PARENT".
Ce dernier permet de suivre une
réparation majeure sur un équipement. Le
planificateur peut assigner une priorité
de 1 à 5 au bon de travail, i.e. de
"urgent" à "planifiable".

Suite à l'émission du bon de travail,
il est possible de consulter le module
des pièces.

Module de pièces. Ce module comprend
la liste des pièces qui composent un
équipement. Cette liste est disponible
en temps réel et est subdivisée selon
les parties constituantes de l'équipement
ou des composants. A partir de cette
liste, il est possible de commander
le matériel nécessaire à la réparation.
Avant de passer la commande, ce module
calcule la valeur du matériel commandé
et la compare avec la valeur de remplacement. Cette comparaison permet une
prise de décision quant à la réparation
ou le remplacement de l'équipement/compo-
sant.

Cédule des travaux. Cette cédule permet au planificateur de céduler
les bons de travail par atelier selon
leur priorité et la disponibilité
de la main-d'oeuvre.

5. ANALYSES

Différentes analyses sont disponibles au planificateur afin de lui
permettre de prendre les meilleures
décisions. Ces analyses peuvent
être:

- l'identification des composants qui
  n'atteignent pas la moyenne de
  durée de vie;
- la fréquence des réparations;
- l'utilisation et la disponibilité
  des équipements;
- les coûts d'entretien;
- bon de travail en cours;
- etc..

6. CONCLUSION

En plus d'être un outil de travail
quotidien efficace, ce système a

permis une réduction des coûts d'entretien en main-d'oeuvre et matériel.
La planification des entretiens pré-
ventifs a aussi permis de prolonger
la vie utile des équipements.

Computer Applications in the Mineral Industry, Fytas, Collins & Singhal (eds)
© 1988 Balkema, Rotterdam. ISBN 90 6191 760 3

Computer applications in mine maintenance

John Melnick & Ron Wilson
Suncor Inc., Fort McMurray, Alberta, Canada

ABSTRACT: This is an overview of the application of digital computers to the Maintenance Planning function and to the Maintenance Engineering support role of a large open pit BWE/conveyor mining operation. The benefits to Suncor Mine Maintenance are highlighted so this paper will be of interest to any group contemplating the installation of computers to improve productivity and flexibility.

1 INTRODUCTION

Suncor Inc. Oil Sands Group consists of a 130,000 short ton per day Mine and Bitumen Extraction Plant, 53,000 barrel per day Upgrader and a Utilities Plant providing 53,000,000 pounds per day of steam along with 70 megawatts of electrical generating capability.

Up until the development of the Management Information System began in 1983, information systems were conventional and satisfied OSG's needs under the conditions prevailing at the time. With plant expansion and the emphasis on improving production reliability, it was recognized that a system was required which would provide Maintenance personnel the access to most of the information for their job in one consolidated location.

In order to cover the wide spectrum of needs for maintaining a mobile equipment fleet, Bucketwheel Excavator mining equipment, a concentrator (Extraction Plant), the Utilities plant and a hydrocarbon Upgrader, the system would have to be applicable to both fixed plants and mobile equipment. Repetitive tasks would have to be based on calendar time, running hours, tons throughput or any other reasonable measure. To help to ensure its acceptance and to minimize the disruptions usually experienced with the introduction of a new system, the Management Information System (MIS) would have to be as similar as possible to systems currently in use.

2 MATERIALS

The catalogue for stock materials and their availability was stored on microfiche which was generated in Edmonton from our receiving and consumption figures. Turnaround time for this process was about one week during which time the information in use at the Plant was becoming more and more outdated. This lag vented useful consumption monitoring and the management of inventory levels down to the lowest practical level.

Orders were phone-in which was tedious, error prone and resulted in wasted time as a result of being put on "hold" while Materials personnel dealt with a variety of requests from all over plantsite. In all, the previous system was only useful as a catalogue and an accounting tool.

3 WORK ORDER SYSTEM

Writing, planning, closing and filing of Work Orders was done manually. Copies of active Work Orders were sorted by equipment into hanging files on a mobile workstation which could only be conveniently accessed by one Planner or Clerk at a time. Labour hours were accrued on separate reports based on keypunch entries which were in turn derived from timesheets. Most statistics, such as "backlog" or "Work Orders held for materials", had to be manually calculated from a variety of sources. As a result, the Work Order System was also mainly an accounting tool rather than an information system managing the Maintenance function.

4 PREVENTIVE MAINTENANCE

Mine Preventive Maintenance was totally manual and consisted of about 2000 Callup cards (tickler files) which were supported by printed checksheets. Tracking and recording compliance was made difficult by the large volume of activities and frequent changes to scheduled downtime due to production requirements. Since the system was manual, it was cumbersome to change inspection frequencies or to refine the technical content of the program.

5 EQUIPMENT HISTORY

Equipment History was fragmented among Inspection files, vibration analysis files, oil analysis files and Area history files. The information which was recorded was spotty since even routine facts such as date and Work Order Number had to be manually entered on Kardex or Visifiles whenever time permitted. Engineers were unable to cross-reference to category, equipment and component or to other information systems such as the drawing index. As a result, trends and "the big picture" were difficult to establish thereby hampering professional failure analysis and corrective action.

6 DEVELOPMENT OF THE NEW MIS

Since few Maintenance jobs can begin without the proper materials, this was the first priority and the Suncor Inventory and Materials System or SIMS was developed in five stages:
- Warehousing
- Inventory Control
- Purchasing
- Management Reporting
- Integration with the Computerized
  Maintenance Information System or
  CMIS (Integration of the two systems
was with the objective of SIMS
providing the CMIS with timely, accurate materials information and the CMIS in turn supplying the SIMS with planned materials forecasts).

Once the SIMS was well underway, work began on the CMIS which also proceeded in five stages:
- Work Order System
- Equipment Records
- Preventive Maintenance
- Integration with SIMS
- Management Reporting

This overlap is illustrated on the barchart in the Appendix.

Development was guided by a steering committee consisting of representatives from Planning, Maintenance Engineering, Materials Management and Information Services with an overall Project Manager and program coordinator. Effort to date totals 8 man-years of analysts' time and the Users' steering committee has invested about 8 man-years. The development has been on schedule and on budget for the past four years!

As an aid to debugging and enhancing the system, it was first implemented in Mine Maintenance through the use of prototyping versions followed by improved production versions which were then recycled until Mine needs were met. The improved program was then taken to other Areas and the necessary changes made to adapt it to the specifics of their operations.

The system was developed in house using COGNOS SOFTWARE PRODUCTS' "POWERHOUSE" - a highly successful fourth generation language* developed in Canada.

7 DESCRIPTION OF NEW MIS SYSTEM

Overview
The Management Information System (MIS) consists of four modules which will be fully integrated*:

1 The SIMS is the first module and provides complete information on parts - where they are used and their availability. The next three modules are part of CMIS.

2 The Work Order System facilitates entering, tracking, updating and filing of Work Order data. It supports Planning of Work Orders; and completion of Work Orders resulting in automatic Work Order and equipment history records. Options are provided to add standard jobs to Work Orders.

3 The Preventive Maintenance module generates lists of activities due according to a master schedule which is stored in the computer. Reporting Work Order completions against this schedule tracks compliance while all the time allowing changes to be made to the frequency of checks and content of the the checksheets. The master checksheets are hardcopy* but checksheet facsimilies are also stored in the computer as a backup and to provide field crews access to the

information outside office hours.

4 The Equipment History module provides a method to enter and update equipment and component records including repair and changeout history as noted by the field on complete Work Orders. There is also provision for recording of equipment and component oil sampling and vibration analysis data.

Security of data in the MIS is accomplished in two ways: Passwords for initial entry and restricting the ability to change records to those who must be able to do this in order to perform their jobs, eg. Planners. Entry is blocked if the password is not recognized or the name of the person attempting entry does not match with a list of authorized users of the function. To date there have not been any problems with security and it is important to note that there are no restrictions on entering the system to obtain information.

Access to all screens is through menus* and on-line* help is available as an aid to new or infrequent users. Manuals are also available for a more conventional introduction to the system. For anyone needing a permanent record of their transactions, screens and reports can be printed at several locations on Plantsite which are located to be within easy reach of users' workstations. A copy of the opening menu is included in the Appendix.

Anyone who sees an opportunity for an improvement can initiate changes to the system by Service Request. This provides a method to control and test enhancements before their release thereby ensuring that the system continues to serve the needs of the majority of users.

8 DESCRIPTION OF THE SIMS:

Essentially, the SIMS system will provide all current information on materials and their status. This includes information about stock on hand, location in the Warehouse for stock items, where the material is used, how much of it is on order at any time, economic order point and order quantity and stock movement history. These facts can be accessed by stock code, by identifying the material as to its end use or by keyword* search. The first two menus and a typical output screen of the SIMS are included in the Appendix and these give an overall appreciation of its capabilities.

In addition to supplying information, requests for materials from authorized people are printed directly in the warehouse or the toolcribs as issue slips. This eliminates the phone-in system!

9 DESCRIPTION OF THE WORK ORDER SYSTEM

The Work Order system handles the full spectrum of Work Order activities ranging from Maintenance Work Requests (small jobs with no requirement for planning) through regular Work Orders and Standing Work Orders of routine work such as lubrication.

Emergencies (which comprise <3% of total Work Order volume) are handled by entering and closing them after the fact.

Linkage to our Plant accounting system is through the General Ledger which makes information on costs for labour and materials available to the Work Order System. These accounting functions are resident on another mainframe* and are brought together daily with the Work Order system so that any materials charges and labour are at most 24 hours old.

Administrative data such as date, account number, equipment number, etc. are entered using a primary screen when a Work Order is first opened. This entry is facilitated by defaults* and linkages between fields eg. entry of an equipment number results in a suggested account number appearing in its respective field. Code numbers for Planners and entries in other fields also result in the minimum number of keystrokes.

After initial entry, four optional screens can be used to plan the Work Order by adding information on material estimates, crafts estimates, comments and additional data. A particularly useful feature allows the Planner to bring in standard jobs, such as gearbox overhauls, by copying all specifications, job steps and parts list of the standard job to the Work Order being planned. Since it takes several minutes for large standard jobs (more than 50 lines) to be copied, provision is made for the Planner to start the copying process and then work with other screens during the transfer. This feature continues to build and 4,000 of these standard jobs were in the system as of April, 1987!

Once a Work Order is signed off as complete, this is reported to the system and the information on the Work Order becomes part of the equipment history.

Both entry and completion of Work Orders are accomplished clerically to economize on Planner time. Other time saving measures include Work Order status changes which take place automatically depending on the phrase of the job. For example, completion of the planning screens for labour and materials results in a Work Order status change from "On Request" to "Ready for scheduling" and when labour is first reported to a Work Order number, status changes to "In Progress". Routine arithmetic is performed to calculate labour costs based on estimated hours and rates for the respective Trades and material costs are calculated using Warehouse book value. The screen format forces consideration of all fields to help ensure a thorough and consistent job of planning.

At any given time 12,000 Work Orders will be in the system and 150 + Work Orders per day are entered against an average 5 week backlog.

Some typical Work Order screens* are included in the Appendix.

10 DESCRIPTION OF THE PREVENTIVE MAINTENANCE SYSTEM

Preventive Maintenance activities are initiated from a master schedule of frequencies based on intervals which have been recommended by Maintenance Engineering. The calendar date, equipment running hours, tons handled or other criteria are reported to the system and this information is used to calculate and print the list of activities due.

Checks are not scheduled to be done on a specific date; rather a window equal to the time interval between checks is printed out along with the number of the applicable checksheet. This gives the Field enough flexibility to arrange their schedule to do the "down" checks during outages and the "running" checks when the equipment is operating. The previous manual system had this feature which was necessary to cope with the dynamic production demands on the Mine.

If a check has not been reported as "complete" within the prescribed time frame, it becomes "overdue" and the overage is reported in the appropriate column of the report. Mine policy requires that these overdue checks become top priority which in effect forces their early completion. As can be seen from a copy of the schedule in the Appendix, the answer to the question "Are you doing it when you are supposed to be doing it" is apparent. As a safeguard to ensure that

no automatic assumptions are made which could result in missed checks or false statistics, Preventive Maintenance completions must be specifically reported to the CMIS by the Maintenance Clerk. Checks can be reported as "complete" in blocks however, to reduce the workload which would be required for individual keying.

Reports are available which calculate the costs of labour and materials for doing Preventive Maintenance by activity plus the costs for the total year. This information can be used for optimization.

A typical checksheet is also included in the Appendix.

11 DESCRIPTION OF EQUIPMENT HISTORY

The Equipment History module captures all basic data regarding maintenance history. This allows us to track what was worked on, what was done and why we did it. These activities can be tracked by equipment number and/or serial number down to the component level, eg. electric motors (components) can be tracked around the entire Plant by their serial number. Oil sample results and vibration history are currently entered manually but preset flags highlight any values which exceed the limits set by Maintenance Engineering.

Approximately 90 reports are available from the system based on this data and the reports are organized into five levels:

- Zone
- Plant
- System
- Major Machine
- Equipment

Sub-codes for equipment include component, serial number and part numbers which are linked to the parts list module of the SIMS. Reports are also available by Class of equipment which could be considered another level eg. trucks, pumps etc.

Features such as analysis, trending, Mean Time Before Failure and Mean Time To Repair are not currently available.

More detail on this module in the forms of output can be found in the Appendix.

12 PROBLEMS ENCOUNTERED TO DATE

As with any change supported by a service group outside a maintenance organization, there was difficulty in communicating needs versus system capabilities between users and Systems Analysts. This was quickly overcome by the prototyping/ recycling strategy mentioned earlier. It was also important to foster the principle that ownership is with the user and the system is only as good as the information put into it on a consistent basis. On a less philosophical note, quite often during the early development it was found that there are more users than terminals or that system response was unacceptably slow. This has been relieved to a large extent by the purchase of more powerful hardware although this cannot always be used as a solution since there are economic limits. As a result of this upgrade, average access time is now 5 seconds during 90% of office hours. Hardware problems such as system "crashes" have not been a problem since the start-up phase and the system is now 98% "up" in general and 99% "up" during prime time. These statistics are impressive considering the overall usage:

	Potential Users	Signons*	Accesses*
Maintenance	250	100	Av. 6x/Day
Materials	Total	150	Same

Training, although not difficult in itself, was complicated by the need to conduct it efficiently in larger groups and balancing this against the loss of productive manhours. Four people per group was found to be optimum.

13 FUTURE DEVELOPMENT PLANNED

The MIS is a dynamic system with new additions as required and functions being deleted when they are no longer appropriate to our business. The ability to do this is a significant advantage over "canned" packages.

The printing of Preventive Maintenance checksheets and Work Orders directly by the system is under development.

A direct link with the vibration monitoring program will allow data to be downloaded* directly, thereby saving keypunch time.

For standard jobs the estimates could be refined using Maximost engineered time standards.

Schedules will be calculated using information on outstanding Work Orders.

Job costs to be available on-line will include invoiced items such as contracts and crane rentals thereby allowing better control of expenditures while there is still time to take corrective action.

When our Payroll system goes on-line

(from a timesheet based system) then labour costs will be available in real time* also.

Preventive Maintenance history can be integrated with equipment history to give an overall picture of maintenance. Steps are also being taken to merge the operational history available in our Mine Productivity System to give the total service history of the equipment.

Other development considered but not committed includes making the files containing our Engineering drawings index and the drawings themselves directly accessible through the MIS.

Also planned are a series of higher level "Management Reports" which will give an overview of performance in the areas of Parts, the Work Order System, Preventive Maintenance and Equipment History.

14 SYSTEM BENEFITS TO DATE

Considering that time on the system is used:

60% for accessing parts information and ordering
35% for opening/closing and planning Work Order
5% for studying equipment history

Anything which makes the process quicker and more complete by providing one source for all information such as stock status, materials requirements, standard job information and Work Order history will result in savings which can best be looked at in terms of tangible and intangible (but real) benefits. Here is a sampling of tangible benefits:

Approximately 90% fewer jobs are cancelled or rescheduled due to wrong parts or lack of parts.

In Extraction, the use of the system reduced the time required to produce a basic maintenance schedule from 16 hours to 4.5 hours.

Work Orders are opened and closed in real time so there are fewer unmatched labour and material flags. Previously, timesheets and material slips were sent to the accounting system, processed and reports were mailed to the Areas. This turnaround of information was frequently out of phase with the life cycle of the Work Order. As noted earlier, Work Order History was posted manually onto Kardex and retrieval of information from a closed Work Order would require going to the equipment history file to find the Work Order number, looking up a filed hardcopy Work Order for some of the details and going to separate accounting reports to get labour and materials usage. Now, all of this information is in one place along with Preventive Maintenance History, oil analyses and vibration. As a comparison, tracking pulley rebuild history by serial number took two weeks using the old system and 1/2 day under the new system. Our Mobile Equipment Maintenance has experienced more success in warranty claims due to better documentation of warranty hours and equipment problems. In its last manual phase in 1983, the Preventive Maintenance system in the Mine resulted in the issue of 75 checksheets per week which were returned with a 60% compliance rate. This cycle required 24 manhours weekly to initiate and record as complete. Now with more equipment installed in the Mine (+25%) and with checksheets at a rate of 125/week, compliance is 80% + and consumes 2 manhours weekly to administer. Furthermore, an equivalent workload of electrical checksheets is coming on line which will only add an additional 2 manhours per week! The high compliance is partly due to the system driving the Preventive Maintenance policy decision and committment process plus the fact that it is readily changed (flexible) thereby adjusting to the real world (format, scheduling, frequency, etc.) A material review resulted in a reduction of inventory from 14.8 to 11.25 million dollars. There were problems created by elimination of some spare parts but they were administrative; not related to the MIS. The point is that the system had the capability to supply the information in a useable form.

The Planning function is now carried out by 33% more people now as compared to 1980 levels. However, equipment under maintenance has doubled since then and these people are now handling an estimated 5 times the volume of information; therefore individual planning productivity is much improved. This volume of information comes from the ability to supply more information to more people in more different formats during more of the day (unrestricted access) and tailor it to meet individual needs. Field personnel get the information they need when they need it either through the Planners or directly on their own. Allocation of materials for scheduled jobs was done by such means as tagging or physical segregation of stock. Now the computer places a reservation on material which has been identified for a Work Order thereby eliminating double handling. Previously, follow-up of Purchase Orders was through buyers who had to make phone calls and do file searches – this was distracting to them. Now individual customers can use the system to enquire as to the status of their materials. Also, re-order points and economic order quantities were once done manually and are now calculated automatically based on usage history and delivery lead times. These features result in more buyer time to perform their prime functions such as negotiating with a variety of vendors for more favourable quotes. More materials requirements are planned (from information shared by CMIS and SIMS) which result in less premium freight charges. All of this adds up to savings for Suncor.

These point and others like them accrue to a bottom line estimate of savings for Mine Maintenance of 1/2 million dollars/year. Then there are the Intangible Benefits:

Planning quality is improved because the system brings together more information in a shorter time using historical data and standard jobs wherever possible. Estimates are less arbitrary since they can be compared to historical averages.

Planners can estimate a group of Work Orders quickly and put together a project complete with all estimates. This flexibility allows us to capitalize on opportune maintenance.

The MIS is more flexible than previous manual systems as illustrated by the development of a database (within five weeks start-finish) to track the 500 portable radios on site by serial number. Another example of flexibility is the fact that the new system is not restricted to serially numbered Work Orders and recognizes each Work Order as an individual job. This flexibility eliminates the confusion in the previous system which was caused by duplicate Work Orders in the system or Work Orders out of sequence.

Development of the MIS supplied a framework for maintenance plantwide to focus on a common interest. A spinoff of this was that 7 Work Order forms previously used have now been reduced to one form. Planners exchange information freely and automatically each time they access the database. No formal exchange program is necessary.

Hardware consists of a Hewlett-Packard HP 3000 configured like this:

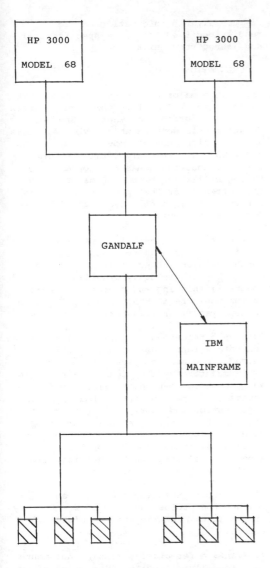

WORKSTATIONS AROUND PLANTSITE

The GANDALF is a communications device to allow data to flow between the mainframes, minicomputer and workstations.

An HP3000 Model 48 not shown is also available for system development and test.

This package was chosen from a field of 11 potential vendors because it represented the most cost-effective solution; was demonstratably easy to use and operate; had already been installed in other Plants and Mines (Syncrude, Luscar); was supported by excellent software with an established track record and the Company had a presence in Fort McMurray with a good service history.

15 COMPUTER APPLICATIONS IN MINE MAINTENANCE SUPPORT

Maintenance Engineering is continually looking for ways in which to improve productivity and professionalism. The group has been using a basic microcomputer* for the past three years and with the improved economics of hardware and the wide range of available software we are broading our applications.

INFORMATION RETRIEVAL

A stand alone* computer can be used to store any information that is useful to have in that format (text/statistics/ numeric data). Although a basic computer is marginally adequate for this purpose, a hard disk* to improve storage capacity and reduce retrieval time is recommended. Two examples of systems which lend themselves to this format are the Work Request database used for gathering statistical data on our service delivery and the Suncor drawing Index used for retrieving techincal drawings.

Networking* is the next stage which will be implemented. This will allow all of the computers in Maintenance Engineering to exchange information freely without having to physically move floppy disks* from one machine to another. This stage involves interconnection with the mainframes on plantsite thereby giving us access to all databases which will aid in the analysis of life cycles, repair costs and individual component reliability.

Interconnection of Suncor's computers to the outside world is a logical extension of networking. With payment of subscription fees, sign-on fees and long distance charges, access to virtually all information on computers around the world is available. In practice this means that such things as databases for technical literature can be searched using Personal Computers.

Some of Suncor's outside contractors maintain databases of "as left" specifications for components that have been rebuilt in their shop. Vendors such as Browning Engineering are cataloguing their drives, pulleys and couplings in databases that also provide such features as automatic selection of the most economical product. All suppliers in the near future will probably develop comparable systems to be competitive. For Suncor this means up-to-date electronic catalogues and historical files from which to choose goods and services.

With one of the Maintenance Engineering Personal Computers left on line at all times, a machine at home would be used to access work-related information stored on it thereby allowing Engineers to answer technical questions from their homes. For anyone expecting enquiries (on call or involved in a major project), Suncor has portables which can be signed out and taken home for this purpose. Others in the Maintenance Engineering Group have their own machines capable of this communication.

SUPPORT FOR PROJECTS

During the life of a Maintenance Engineering Project, there are several opportunities to use Personal Computers which will enhance professionalism and productivity.

All recommendations must have with them the lowest possible ratios of cost: benefit and spreadsheet programs help with these analyses.

The accuracy of repair scopes (cutting and replacing structural members for example) can be more rigorous using basic finite element analysis rather than rules of thumb.

Statistical information such as component history can be readily handled and manipulated using database and spreadsheet programs. Once data has been entered, it is a simple matter to produce graphs as an aid to analysis or making presentations.

Larger projects can be planned, tracked, costed and reported using commercially available project management packages.

For administration, the spreadsheet programs are useful for budgetting - especially when changes and forecasting are required on short notice.

Drawings and text associated with standard repair procedures can be developed and stored on Personal Computers. This information can be readily modified to suit a specific task and issued very quickly.

DOCUMENTATION

One of the major obstacles to overcome in using Personal Computers for documentation is the perception that when an Engineer/ Technician is working at a keyboard, it is an inefficient use of expensive specialist time. This is true only until the time when the individual has developed basic typing skills and becomes familiar with the software. At that point productivity improves significantly over manual methods for these reasons:

1) When a professional is typing, he or she is not just reading someone elses' rough draft and entering it as a typist does. They are actually creating the document; thinking and composing exactly as you would be if they were writing with pen and paper; only faster.

2) When considering the time it takes to generate a report using conventional methods, one must consider the entire cycle relative to the document - drafting, submission for typing, rearranging for desired effect, proofing, submission for distribution and lastly filing/retrieval. Many of these steps are made shorter or eliminated completely when a document is created using a Personal Computer - especially if stenographic services are remote from the work location.

3) Electronic spelling checkers can enhance the professionalism of the work by virtually eliminating typographical errors.

4) Standard formats for routine documents such as failure reports can be assured by using a template in a master file which an Engineer/Technologist can call up and fill in when writing a report. Any change or enhancements can then be accomplished just by changing templates.

5) Only drawings and rough notes need to be filed in conventional ways, text and numerical data are electonically filed. This results in a saving of space and retrieval is easier.

16 CONCLUSION

It is our conclusion that the application of computers to the Maintenance function has resulted in several benefits to

Suncor. In particular, the success of the MIS is due to these factors:

1) The conversion of workable manual systems into computer-based counterparts. For example, only forms and procedures recognized Plantwide were used. This was done rather than adapting to a commercial program (which could not be found anyway in spite of a long search). Wide user acceptance is the result.

2) The system was developed in-house by dedicated, qualified people who shared a common goal.

3) The project was broken down into manageable modules and a realistic, achievable schedule.

4) We had Management support.

In closing, it is our conculsion that Suncor could do anything with this system but that this must be weighed against the effort and expense involved - given procedures now in place. All of this must also be in the context of a rapidly changing business climate. Anyone contemplating such a program would do well to consider these points. It is intended that as the costs of hardware decreases and software becomes more useable, individual access to computers will significantly improve productivity in the Maintenance function.

GLOSSARY

Accesses - while signed on to a system, the user assesses the various modules in order to make use of their unique features. Each signon usually results in several assesses but the user must sign on first.

Database - an accumulation of information in an organized, retrievable form. A telephone book is a database.

Defaults - preprogrammed assumptions which are submitted to the computer as data when certain conditions are met. This reduces operator workload. For example, if an estimate is entered for Millwrighting manhours, the hourly rate will default to the value for the Millwright trade.

Download - transferring data from one computer to another electronically.

Floppy Disks - a data storage device made by coating a flexible vinyl disk with a magnetic medium similar to audio tape. These disks are spun under heads which convert electrical impulses from the computer into digital magnetic fields. These fields result in a record on the disk medium which can then be read back by the computer at a later time.

Fourth Generation Language - a relatively simple set of commands to generate the code necessary for screen design and database management - two functions which can be quite tedious in third level languages such as FORTRAN or COBOL. The language also checks for potential errors or "bugs" automatically thereby helping to ensure that a relatively minor change to one part of the system does not have serious adverse affects on another, seemingly unrelated function.

Hardcopy - output which has been printed or plotted on paper as compared to data stored on disk or within the computer's memory.

Hard Disk - a device which stores and retrieves much larger amounts of information more quickly than the soft or floppy disks in common use. Ten or twenty megabytes for a hard disk as compared to 360 kilobytes for a typical floppy. (A byte is one character such as a letter or number).

Hardware - the electronics which make up a computer system.

Integrated - a system designed with the greatest possible commonality. Modules of a computerized system are designed to perform different and unique functions while sharing the same database and responding to user commands with common syntax. The user is able to move freely from one module to another and use commands which are easy to learn.

Keyword - a computerized search technique whereby the desired information is located through a search for a word which is unique to that portion of the data that is to be isolated.

Main Frame - a large central computing facility. Microcomputers, minicomputers and mainframes all operate on the same principles; the differences in the terminology are arbitrarily based on scale, speed and memory size.

Menus - a list of available features from which a user can choose and access.

753

Microcomputers - small personal units
with basic peripherals (such as printers
and disk drives) with storage capacity of
one megabyte or less.

Networking - interconnecting computers
electronically so information can be
shared.

On-line - to become available to the
users on a regular basis. Also refers to
a system whereby information such as
labour hours are entered directly into
the computer without going through a
conventional timecard/timesheet as an
intermediate step.

Real Time - participating or being
informed of an event as it is happening
rather than after the fact. Real time
information allows you to influence
events and their eventual outcome. For
Management this is a distinct advantage
compared to being informed after the fact
when the only option is to take steps
that are intended to prevent a
recurrence.

Screens - electronic forms which appear
on the users' computer monitor to
facilitate data entry or to present
computed information in a standard
format.

Signons - the action of entering the
overall system to work with its
features.

Software - the programmed instructions
which direct a computer what to do with
each piece of information it receives and
how to make decisions based on that
information.

Stand Alone Computer - no connections to
the outside world or other computers.

Template - an overlay which gives form
and organization to information. A T1 tax
form is a template.

ACKNOWLEDGEMENTS

We would like to thank these people for
their assistance in the preparation of
this paper:

Dan Boilieau - Senior Systems Analyst,
Information Services

Jane Millions - Maintenance Clerk/
Scheduler, Mine Maintenance

Yvonne Campbell - Planning Clerk, Mine
Maintenance

Jim Miller - Mechanical Specialist, Mine
Maintenance

Izim Okeren - Coordinator, Client
Computing, Information Services

Doug Shorr - Superintendent, Support
Services, Information Services

Richard Townell - Staff Engineer,
Maintenance Support

Dick White - Senior Engineer, Maintenance
Support

APPENDIX

- Original Project Schedule

- Typical Work Order Screens

- Preventive Maintenance Reports

- Preventive Maintenance Checksheet

- Equipment History Reports

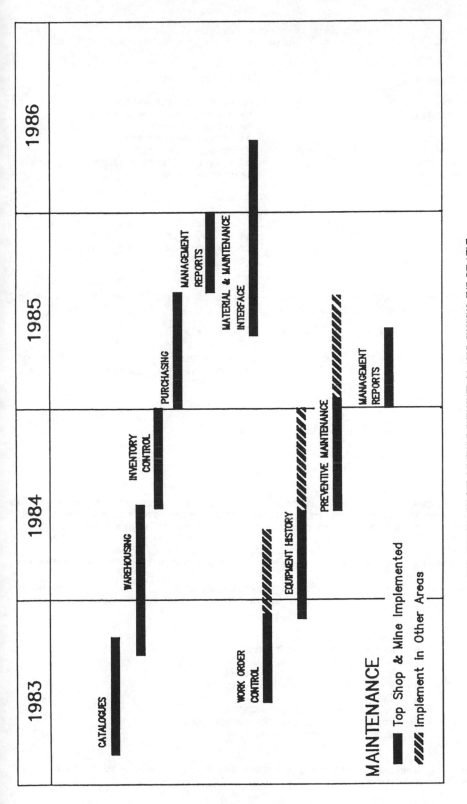

THE ORIGINAL TARGET SCHEDULE PUBLISHED IN 1983 SHOWING THE RELATIVE
TIMING OF MAINTENANCE AND MATERIALS SYSTEMS IMPLEMENTATION

WORK ORDER ENTRY/UPDATE   ADDITIONAL MENUS

01 WORK ORDER ENTRY   08 MATERIALS INFORMATION
02 WORK ORDER PLANNING   09 WORK ORDER INFORMATION
03 SCHEDULE A SINGLE WORK ORDER   10 EQIUPMENT DATA
04 SCHEDULE WORK ORDERS   11 EQUIPMENT HISTORY
05 NEW SCHEDULE WORK ORDERS   12 STANDARD JOB DATA
06 COMPLETE WORK ORDERS   13 PREVENTIVE MAINTENANCE
07 DATE OF LAST LABOUR UPDATE   14 SYSTEM CODES

REPORTS

15 WORK ORDER REPORTS   18 PERFORMANCE REPORTS MENU
16 EQUIPMENT REPORTS   19 PERFORMANCE INDICES MENU
17 EQUIPMENT CLEANUP REPORTS

THE MAIN MENU WHICH IS FIRST ENCOUNTERED AFTER SIGNING ON TO THE SYSTEM. THIS ALLOWS
ACCESS TO THE FOUR MAIN MODULES OF THE MIS.

* *

01   MATERIAL IDENTIFICATION AND STATUS
02   STOCK CODE INQUIRY MENU
03   HISTORICAL INFORMATION MENU
04   NON-STOCK MATERIAL INQUIRY MENU

THIS SIMS MENU ALLOWS THE USER TO IDENTIFY THE STOCK CODE FOR MATERIAL WHEN THIS FACT IS
NOT KNOWN.   IT ALSO ALLOWS THE USER TO DETERMINE THE STATUS OF STOCK ITEMS, SEE THE
RECORD OF USAGE OF STOCK ITEMS AND TO CHECK ON THE STATUS OF NON-STOCK (SPECIAL ORDER)
MATERIALS.

* *

01   STOCK CODE STATUS
02   STOCK CODES FOR PART NUMBER
03   EQUIPMENT PARTS
04   MANUFACTURER INQUIRY BY NAME
05   STOCK CODES FOR MANUFACTURER NUMBER
06   CATEGORY INDEX
07   SUBCATEGORIES FOR CATEGORY
08   STOCK CODES FOR CATEGORY/SUBCATEGORY
09   STOCK CODES FOR KEYWORD

THIS SIMS MENU FOR MATERIAL IDENTIFICATION ALLOWS THE USER THE OPTION TO DETERMINE THE
STOCK CODE IN SEVERAL DIFFERENT WAYS.

STOCK CODE   52058125   REMARKS:
DESCRIPTION   BARSTOCK   OSR 1771

                        1-3/4 IN.
U.O.I.   FT.
                                                        ADDITIONAL DESCRIPTION
QTY ON HAND   36 LOCATION: 120991   GRADE B7, ASTM A193
QTY COMMITTED   R.O.P.   2   CHROME MOLY STEEL,
QTY ON ORDER   E.O.Q.   13   CONTINUOUS THREAD LESS NUTS
SUGG ORDER QTY   LAST   8 TPI, 12 FT. LENGTHS
FORCED QTY   ACTIVE   85/07/23
INCEPTION DATE   72/06/01   ICR NO.
PACKAGE MULTI   12
AVE UNIT PRICE $12.1406
LEAD TIME   13
BUYER NAME   MONICA BIBEAU   743-6925
MANUFACTURER
CATEGORY   405   FASTENERS IMPERIAL STUDS
SUBCATEGORY   002   BARSTOCK
01 SELECT FOR STOCK CODE INQUIRY MENU

ONCE THE STOCK IS IDENTIFIED, THE STATUS OF THE MATERIAL IN THE WAREHOUSE CAN BE
DETERMINED. TECHNICAL INFORMATION IS INCLUDED TO AID IN THE SELECTION OF THE CORRECT
MATERIAL FOR AN APPLICATION.

* *

EQUIPMENT =   CATEGORY =   WORK ORDER
PRIORITY =   CAUSE =   NO.
W.O. STATUS =   ORIGINAL AREA =   01
AREA =

DATE WRITTEN   DATE REQUIRED   EQUIPMENT   ACCOUNT   PRIORITY   TAX

02   03   04   05   06   07

WCAT   JOB CODE   COMP. CODE   WO TYPE   WO STAT   CAUSE   PLAN AREA   ORIG AREA   TRB SHEET

08   09   10   11   12   13   14   15   16

REQUESTOR   DESCRIPTION   INSPECTION REPORT

17   18   19

20 ENTER MORE DESCRIPTION   21 ADD OR MODIFY A CRAFT
22 COMPLETE THE WORK ORDER   23 ADD OR MODIFY A MATERIAL

THE WORK ORDER ENTRY SCREEN WHICH ALLOWS THE MAINTENANCE CLERK TO FIRST ENTER THE
INFORMATION IN THE SYSTEM.

```
MODE: F ACTION: L            MAINTENANCE INFORMATION SYSTEM              87-06-19
USER: MPLAN                       WORK ORDER PLANNING

WORK ORDER NUMBER   560270     EQUIPMENT: 02G98XX
                               ACCOUNT  : 450200


01 DESCRIPTION        BLD STOCK SCAFF EQUIPMENT_____GEN_____HOURS_____
04 REQUESTOR          N. TRAFFORD                  DURATION    EST.     ACT.
05 DATE WRITTEN       87/06/15
06 DATE REQUIRED      87/06/15               07       80      160
08 PLANNER            3399  G. SMITH _____ WELD        11  TAX CODE:    0
   CONTRACTOR                                               PLAN AREA:   5
12 APPROVED BY        N. TRAFFORD                          ORIG. PLAN:  5
   STANDARD JOB                                            INSP. REP.
16 WORK ORDER STATUS: 3 READY FOR SCHED.        SELECT THE FOLLOWING TO:
   PRIORITY           2
   CAUSE              X                         19 PLAN THE WORK ORDER
   CATEGORY           3                         20 SCHEDULE THE WORK ORDER
   WORK ORDER TYPE    4                         21 WORK ORDER DETAILED SCHEDULING
                                                22 MORE WORK ORDER DATA
                                                23 PRINT PLANNED WORK ORDER REPORT


_____ESTIMATED COSTS_____
MATERIAL      NON-STOCK        LABOUR

26                          3,680.00       3,680.00 (TOTAL)

   SHOULD YOU NOW CHANGE THE STATUS OF THE WORK ORDER?

ONE OF THE PLANNING SCREENS USED BY THE PLANNERS TO ADD INFORMATION TO THE WORK ORDERS.

* * * * * * * * * * * * * * * * * * * * * * * * * * * * * * * * * * * * * * * * *

REPORT NO. MTZ061               MAINTENANCE INFORMATION SYSTEM       PAGE NO. 1
MINE                            LABOUR MANHOUR DISTRIBUTION          DATE 18 JUN 87
MINE MECHANICAL                                                     TIME 20:12
LAST LABOUR UPDATED FOR CATCO = 87/06/12
                     SUNCOR = 87/06/12


   CATEGORY              _____LAST WEEK_____        _____YEAR TO DATE_____

                        HOURS WORKED   PERCENT OF TOTAL    HOURS WORKED   PERCENT OF TOTAL

   PREVENTIVE MTCE.        214.00           4.59               9,628.00          8.25
   WORK FROM PM         1,128.00           24.18              23,458.50         20.09
   PERIODIC MTCE.       1,124.75           24.11               4,981.00          4.27
   ROUTINE MTCE.          128.00            2.74               1,793.50          1.54
   CORRECTIVE MTCE.     1,155.50           24.77              26,519.25         22.71
   EMERGENCY MTCE.        249.50            5.35              13,451.75         11.52
   ASSIST OPERATION        68.00            1.46                 367.00           .13
   ALTERATIN/CAPITL        81.00            1.74              12,826.00         10.98
   NON-PRO TIME           515.50           11.05              23,735.00         20.33

   TOTALS               4,664.25          100.00             116,760.25        100.00
```

A TYPICAL MANAGEMENT LEVEL REPORT SHOWING THE DISTRIBUTION OF LABOUR HOURS CHARGED TO
WORK ORDERS FOR BOTH SUNCOR AND CONTRACTOR EMPLOYEES.

MAINTENANCE INFORMATION SYSTEM
MINE PM LUBE DOWN_____2T15   PAGE 1
P.M. WEEKLY SCHEDULE (MAJOR CRAFT - - LUBRICATION)

DOWN CHECKS

CHECK SHEET	DESCRIPTION		DUE DATE	FREQUENCY	WEEKS TO GO OVER
5LD1TA 2T-15	LUBE LOAD BOOM	(R65) 0.7 HR	JUN 24/87	2 WEEKS	0
5LD1TB 2T-15	LUBE SUPERSTRUCTURE	(R66) 0.3 HR	JUN 17/87	1 WEEK	0
5LD1TC 2T-15	LUBE FILTERS	(R67) 1.0 HR	JUL 08/87	4 WEEKS	2
5LD1TD 2T-15	LUBE SAMPLE & CHGE	(R68) 0.8 HR	AUG 19/87	13 WEEKS	8
5LD1TE 2T-15	LUBE SAMPLE/CHG	(R69) 2.0 HR	JUN 01/88	52 WEEKS	49

THE SCHEDULE FOR "DOWN" LUBRICATION CHECKS ON A PIECE OF EQUIPMENT. NOTE HOW THE
OVERDUE CHECK STANDS OUT. THE "0" WEEKS OVER INDICATED THAT IT IS DUE IN THE CURRENT
WEEK. NEXT TIME THE SCHEDULE IS ISSUED THIS WOULD BECOME A "1" IF THE CHECK HAS NOT BEEN
REPORTED COMPLETE.

EQUIPMENT		BWE
02E00		

TYPE	CLASS	FREQUENCY
MECHANICAL	RUNNING	1

CRAFT	WORK ORDER	MANHOURS
MILLWRIGHT	599301	0.7

ITEM	T A S K D E S C R I P T I O N
	L/B:
01	-#2 DRIVE MOTOR
02	- BRAKE ASSEMBLY
03	- GEARBOX
04	-#2 HINGE PIN
05	-DRIVE PULLEY
06	-UPPER M.W.D. MOTOR
07	- CARDAN SHAFT
08	- BRAKE ASSEMBLY
09	- GEARBOX
10	- CLUTCH
11	-LOWER M.W.D MOTOR
12	- CARDAN SHAFT
13	- BRAKE ASSEMBLY
14	- GEARBOX
15	- CLUTCH
16	-#1 DRIVE MOTOR
17	- BRAKE ASSEMBLY
18	- GEARBOX
19	-#1 HINGE PIN
20	-RETURN PULLEY
21	-PADDLE WHEEL MOTOR
22	- TORQUE ARM
23	
24	

THE FACE OF A TYPICAL CHECKSHEET. THIS SHOWS THE ADMINISTRATIVE AND TECHNICAL
INFORMATION ASSOCIATED WITH A RUNNING CHECK OF A BUCKETWHEEL BOOM. A GUIDE AS TO WHAT
HAS TO BE COMPLETED TO PERFORM AN ADEQUATE CHECK IS AVAILABLE TO THE CHECKERS.

EQUIPMENT	TYPE	CLASS	FREQ	ROUTE
1300 BWE	MECHANICAL	RUNNING	1	A 02

THE REVERSE PAGE OF A TYPICAL CHECKSHEET SHOWING THE LOGICAL ROUTE TO BE TAKEN TO PERFORM

THE CHECK MOST EFFICIENTLY. THIS IS ALSO USEFUL INFORMATION FOR TRAINING!

EQUIPMENT NUMBER   02E005D   1300 L/B DRIVE ASS'Y

	WORK ORDER	DESCRIPTION	DATE COMPLETED	EST.	ACT.
01	501263	C/O OIL LOAD BOOM #2 DRIVE --------1300	87/02/15	4	6.00
02	501166	REPLACE RUBBER ELEMENT #1 DR L/B --1300	87/02/12	4	4.00
03	550168	RBLD BRAKE SHOES ------------------1300	87/12/30	4	
04	501075	ADJUST/BRAKES #2 L/B DR. ----------1300	87/01/06	2	2.00
05	501062	FAB PADDLE WHEEL PADDLES	87/02/05	12	20.00
06	501054	C/O OIL L/B SPILL DRIVE G/BOX -----1300	87/04/19	4	
07	650225	MACHINE BUSHINGS AS PER SAMPLE ----1300	86/12/05	4	13.00
08	550152	RBLD BRAKE SHOES ------------------1300	86/11/25	4	
09	500925	INST. HINGE ON PADDLE WHEEL COVER -1300	86/12/02	2	6.00
10	500892	BUILD GUARDS FOR B/B DRIVES -------1300L/B	86/11/05	24	24.00
11	500862	ADJUST BRAKES #2 L/B DRIVE --------1300	86/11/01	1	
12	500815	FAB/INST GUARDS L/B DRIVE ---------1300	86/12/16	32	

AN EQUIPMENT HISTORY SCREEN SHOWING THE NUMBERS, DATES, DESCRIPTION AND LABOUR OF WORK
ORDERS COMPLETED AGAINST THE EQUIPMENT NUMBER.

* *

EQUIPMENT NO.   02E005D   DESCRIPTION: 1300 L/B DRIVE ASS'Y
CONTAINER NO.   #1

NO.	DATE	IRON	COP.	LEAD	SIL.	CHR.	ALM.	TIN	WATER	VIS.	
		0	0	0	0	0	0	0	.00	12.0-LO	STD
		350	125	60	150	0	0	0	.50	20.0-HI	STD
#1	87/05/29	25	2	1	2	0	0	0	.07	16.0	
#1	87/04/23	20	1	0	5	0	0	0	.08	14.7	
#1	87/03/26	14	2	1	3	0	0	0	.12	14.4	
#1	87/02/25	9	1	1	4	0	0	0	.06	15.2	
#1	87/01/30	10	2	2	2	0	0	0	.08	13.8	
#1	87/01/02	6	2	0	7	0	0	0	.04	14.4	
#1	86/12/07	9	2	2	0	0	0	0	.04	15.6	
#1	86/03/19	20	3	3	11	0	0	0	.06	15.3	
#1	86/03/19	22	2	2	8	0	0	0	.04	16.0	
#1	86/01/15	30	7	3	12	0	0	0	.04	13.5	

THIS IS A RECORD OF THE LUBE OIL ANALYSIS RESULTS FOR SAME PIECE OF EQUIPMENT AS ON THE
PREVIOUS SCREEN.

```
EQUIPMENT NO.   02E005D      DESCRIPTION:   1300 L/B DRIVE ASS'Y
01 CONTAINER NO. #1    FOR: IRON
        0               70             140            210            280            350
VAL
   25   *****
   20   ****
   14   **
    9   *
   10   **
    6   *
    9   *
   20   ****
   22   ****
   30   ******
```

GRAPHICAL REPRESENTATION OF IRON CONTENT OF THE LUBE OIL ANALYSIS.

Computer Applications in the Mineral Industry, Fytas, Collins & Singhal (eds)
© *1988 Balkema, Rotterdam. ISBN 90 6191 760 3*

Index of authors